최신 대기과학용어집
Atmospheric Sciences Terminology
大氣科學用語集

(사)한국기상학회 · 기상청 편찬

Σ시그마프레스

최신 대기과학용어집

발행일 | 2015년 2월 5일 1쇄 발행

편찬 | (사)한국기상학회 · 기상청
편집 | 대기과학용어집 편찬위원회
발행인 | 강학경
발행처 | (주)시그마프레스
등록번호 | 제10-2642호
주소 | 서울특별시 영등포구 양평로 22길 21 선유도코오롱디지털타워
　　　　 A401~403호
전자우편 | sigma@spress.co.kr
홈페이지 | http://www.sigmapress.co.kr
전화 | (02)323-4845, (02)2062-5184~8
팩스 | (02)323-4197
ISBN | 978-89-6866-236-2

발간사

기상학은 인간의 일상생활과 가장 밀접한 학문으로서 자연과학의 여러 학문 중에서도 가장 오래된 학문입니다. 날씨와 기후변화를 예측함으로써 학문 중에 유일하게 미래지향적인 학문이기도 합니다. 18세기 말에 시작된 산업혁명 이후에 세계 각국의 산업화는 결과적으로 지난 수십 만 년 동안 인류가 경험하지 못한 대기의 상태를 만들어 내고 있습니다. 자동차와 산업체로부터 대기로 배출되는 이산화탄소를 비롯한 메탄가스와 오존 등의 온실기체와 검댕이(블랙카본)를 비롯한 각종 에어로졸의 급속한 증가는 인류가 과거에 경험하지 못한 기후변화를 유발할 것으로 속속 밝혀지고 있습니다. 대기 중에 미량의 온실기체와 에어로졸이 지구의 기후를 변화시키고, 구름의 생성을 변화시키고 따라서 강수의 변화도 초래하고 있습니다.

지난 20여 년간, 대기 중의 유기탄소화합물에 대한 관측과 모델링, 즉 대기화학 연구가 활발히 이루어졌고, 이는 과거의 전통적인 기상학에서는 다루지 않았던 새로운 분야입니다. 대기과학 학문의 범주가 전통적인 기상학에서 대기화학을 비롯한 수치예보학, 대기환경학, 기후과학을 포함하는 등 확대일로에 있으므로, 이들을 통합하는 대기과학 영역의 새로운 용어가 날로 늘어가고 있습니다.

현재 널리 사용되고 있는 "대기과학용어집"은 1996년에 발간된 것으로 급속하게 발전하는 대기화학, 수치예보학, 기후과학 등의 많은 용어들을 담지 못하고 있습니다. 새로이 대두되는 전문용어에 대한 통합된 표준화 작업이 시급해져 한국기상학회는 기상청의 후원으로 "대기과학용어집"을 다시 편찬하게 되었습니다. 이 용어집의 발간이 우리나라 대기과학의 발전과 국가 기상업무의 발전 및 기후변화 대응에 크게 기여할 것으로 기대합니다.

끝으로, 오랜 기간 대기과학용어집 편찬을 위하여 수고해주신 전종갑 교수님과 용어집 편집에 열정을 다하여 노력하신 오성남 편집위원장을 비롯한 많은 편집인 여러분의 노고에 존경과 깊은 감사를 드리는 바입니다.

이 용어집의 편찬을 위하여 연구비를 지원해주신 기상청에 감사드립니다.

2013년 3월

한국기상학회장 윤순창

추천사

기후변화 시대에 대기과학에 대한 이해와 활용은 이제 국민적 차원에서 국가의 과학과 경제 분야 등에 필수적 전략적 요소로 되었다. 기후와 기상현상에 민감한 현대에 학술적·상업적·정치적·전략적 개발과 소통에 대기과학의 올바른 전달은 국가적 의무이다.

정보화시대에 대기과학과 관련한 국내의 각종 사업과 연구 활동이 다양해지고 확대됨에 따라 대기과학 분야 전문가뿐만 아니라 학생, 일반인 등에게도 대기과학 용어의 사용이 크게 확장되고 있다. 2,000년대 이후 급속히 첨단화 되고 성장해온 대기과학 분야에는 많은 신용어가 발생하였고 이에 대한 활용성이 크게 높아지고 있다. 이와 함께 다양한 분야에서 통일되지 않고 표준화되지 않은 대기과학용어를 무분별하고 부정확하게 사용하는 경우가 빈번하다.

대기과학용어 중에는 타 학문분야에서 사용하는 용어와 상이하게 사용되는 것이 상당수 포함되어 있어 표현과 이해에 혼란을 야기시키고 때로는 정확하지 못한 전달로 국가의 경제적 손실도 초래할 수 있어 이에 대한 용어의 표준화가 시급한 실정이다. 대기과학 용어의 표준화와 함께 주요 용어의 해설은 국민적 기초과학으로써 대기과학의 이해와 활용에 중요한 좌표가 된다.

따라서 새로 발간되는 "대기과학용어집"을 바탕으로 이러한 문제를 해결하고 대기과학 기반조성을 위한 기초 자료로써 국민적 바른 소통을 위하여 그 준비가 시급히 요구되는 실정이다.

본 용어집의 발간과 관련하여 전 분야의 대기과학 관련 용어를 취합하고 편집하는 쉽지 않은 사업을 수행하신 편찬위 관계자 여러분의 노고에 감사드리며, 본 용어집이 많은 학자, 전문가, 관리자 및 학생 등 관련되는 국민들에게 크게 도움이 될 것으로 확신하는 바이다.

2013년 3월
기상청장

머리말

2000년대 이후 급속히 첨단화되고 성장해 온 대기과학 분야에는 많은 현대 신용어가 생성되었고 이에 대한 활용성도 크게 높아지고 있다.

그 동안 우리나라에서는 각 분야에 통용되고 있는 기상용어가 심의 제정되어 왔고 이에 따라 수차례 대기과학용어집을 발간한 바 있다. 대표적으로 1996년에 편찬된 "대기과학용어집"은 1985년 한국기상학회 용어심의위원회를 발족시킨 이후 10년 동안 준비하여 온 용어집으로서 대기과학용어를 바르게 사용하는 데 크게 기여하였다. 그러나 이 "대기과학용어집"은 15년이 지난 현대의 첨단화된 대기과학 분야의 지식과 융합된 과학정보를 위한 용어집으로서는 크게 미치지 못하고 있다. 더구나 많은 전문분야에서 기상용어가 혼용되고 있고 사용하기에 부적절한 용어가 다수 있는가 하면 새로 제정하여야 할 용어도 많아 대기과학용어집을 크게 보완할 필요가 제기되었다. 아울러 현재의 대기과학용어 중에는 타 학문분야에서 상이하게 사용되는 것이 상당수 포함되어 있어 표현과 이해에 혼란을 야기하고, 때로는 정확하지 못한 전달로 국가의 경제적 손실도 초래할 수 있어 이에 대한 용어의 표준화가 시급한 실정이다.

또한 국내의 각종 포털사이트에서 검색되고 있는 대기과학용어에는 서로 다른 표현이 있는가 하면, 때로는 시대적 흐름을 반영하지 못한 의미가 발견되기도 한다. 최근 기후변화와 기상재해 예측 그리고 각종 첨단 관측기기 개발 등 다양한 분야에서 새로운 대기과학용어가 생산됨에 따라 정보화 시대에 도서편찬만으로 사용자들의 수요를 만족시키는 것은 매우 어려운 상황이 되고 있다. 따라서 누구나 쉽게 접하여 웹상에서 용어를 검색할 수 있도록 제공함으로써 모든 분야에 대한 대기과학의 역할을 크게 높일 수 있다.

이에 따라 한국기상학회는 대기과학용어가 꾸준히 심의 제정되어 보급되어야 한다는 회원들의 의견을 반영하여 대기과학용어심의위원회를 바탕으로 기상청과 협의하여 2009년 대기과학용어집의 개정을 위한 편찬위원회를 구성하게 되었다. 우리나라 대기과학의 발전과 국제화에 기여하고 새로운 용어를 수집하여 학제 간 대기과학용어를 표준화할 수 있는 "대기과학용어집"을 발간하며 용어의 정의와 바른 해설을 위하여 "대기과학용어사전"을 편찬하여 보급함이 편찬위원회의 목적이다. 더불어 정보화 시대에 다양한 대기과학용어의 신속한 이용과 사용자의 올바른 활용을 위하여 웹기반 자동용어검색시스템(XML)의 개발도 포함하였다. 이를 위하여 기존의 "대기과학용어집" 내용을 먼저 갱신하고 현재 사용되고 있는 새로운 용어를 첨가하여 전문가의 의견을 수렴함으로써 각 용어를 표준화하였다.

대기과학용어의 재 제정은 대학 2학년 이상의 수준에서 국내 대기과학 관련 전문기관인 기상청을 포함하여 대학, 연구기관, 기상 사업자 및 일반인에게까지 활용될 수 있는 대기과학 관련 모든 용어를 총망라하였다. 용어의 범위는 기상학, 기후학, 물리학, 화학, 생물학, 환경학, 수문학, 지리학 등 기초과학 분야와 농업, 토목학, 건축학, 산림학, 토양학, 원격탐사, 지형정보학, 위성, 레이더 등 공학 및 산업응용 분야를 모두 포함한다. 따라서 본문에는 21,000여 개의 대기과학용어를, 부록에는 국제적으로 사용되고 있는 970여 개의 대기과학 분야 약어를 수록하였다.

용어의 출처는 미국기상학회(AMS)와 미국기상청(NOAA)을 비롯하여 유럽, 일본, WMO 등에서 활용되고 있는 용어와 국내 대기과학 분야 대학교 등에서 사용하는 전문용어를 중심으로 1996년도 대기과학용어집에 수록된 용어를 확대 개편하였다. 용어의 수집과 표준화 작업에서 본 편집위원회가 의도한 바를 충분히 나타내지 못한 아쉬움과 시대의 바뀜에 따라 용어의 변화가 너무나 다양하여 예상 밖의 어려움도 있었다. 이제 대기과학 소통에 크게 기여할 수 있는 대다수의 용어를 수록하였다고 자부하면서 이 용어집을 이용하는 전문가부터 학생에 이르기까지 효율적으로 활용하기를 기대한다.

그 동안 본 용어집이 성공적으로 출판될 수 있도록 처음부터 열정적으로 참여하여 수많은 검토와 수정에 기꺼이 협조해 주신 편집 전문가들에게 존경과 감사를 표하는 바이다. 끝으로 본 용어집이 나오기까지 뒷받침하고 성원하여 준 기상청에 깊이 감사드리며 제작에 수고를 아끼지 않았던 (주)시그마프레스 관계자 여러분들에게 감사드린다.

2013년 3월
대기과학용어심의위원회
위원장 및 편집위원장 오성남

일러두기

1. 용어의 범위와 수준 :

 본 용어집에 수록되어 있는 용어에는 대기과학과 관련된 수학, 물리학, 화학, 생물학, 환경학, 지리학, 지질학, 해양학 등 기초과학분야와 농학, 산림학, 토목학, 수문학, 건축학, 토양학, 원격탐사, 지형정보학, 위성, 레이더 등 공학 및 응용분야에서 활용되는 대기과학 관련 용어가 포함되었다. 용어 수준은 국내 대기과학 관련 전문기관인 기상청을 비롯하여 대학, 연구기관 등에서 대학 2학년 이상의 학생 수준에서 활용될 수 있는 모든 대기과학관련 상용학술용어를 총망라하였다.

2. 표제어와 용어의 배열 :

 모든 용어는 영문과 한글, 한자로 표기하고 그 뜻, 발음 및 용법에 관계없이 "표제어"의 철자 순으로 배열하였고 영한편에서는 ABC 순으로, 한영편에서는 가나다 순으로 배열하였으며 각 낱말을 표준화하여 영문과 한글을 표제어로 하였다.

 각 "표제어"의 언어가 다양하게 활용되고 의미가 각 전문분야마다 상이하다고 인정되는 용어는 해설어를 모두 기입하였다.

3. 표준어가 아닌 용어(예 : 북한어, 방언)는 표준어로 정비하고 한글 맞춤법 규정에 어긋난 용어는 우리말 어문 규정에 따라 정비하였다.

 용어의 띄어쓰기는 '전문 용어는 단어별로 띄어 씀을 원칙으로 하되, 붙여 쓸 수 있다'는 띄어쓰기 원칙에 따랐다. 다만, 붙여 써야 하는 용어를 띄어쓰기로 하였을 경우에는 이를 붙여 사용하였다. 그러나 구 단위가 명백한 경우(용언의 관형형＋명사)에는 이를 띄어 쓰고, '명사＋명사＋명사' 등과 같이 3개 이상의 명사가 나열될 때는 필요에 따라 적절하게 띄어 씀을 원칙으로 하였다.

4. 용어의 정비와 활용 :

 용어의 정비는 〈표준국어대사전〉에 등재된 형태로 정비하되, 등재되지 않은 용어의 경우에는 어문 규정에 제시된 사이시옷 규정에 맞추어 정비하였다. 적절치 못한 용언의 활용형이 쓰인 용어는 용례를 참고하여 해당 전문 분야에서의 관용성을 인정해 적절한 용어를 받아 들였다.

5. 외래어 표기 용어 :
1차적으로 〈표준국어대사전〉을 따라 정비하였고, 등재되지 않은 용어는 국립국어원의 외래어표기용례집을 따랐다. 그 밖의 용어는 영어사전과 용례 검색을 토대로 하여 정비함을 기본으로 하였다.

6. 용어 또는 원어 정보에 오타가 있는 용어는 영어 원어 정보, 한자 정보 및 한글 정보를 바탕으로 오타를 수정하였다.

7. 용어의 한자 보충어는 우리말에 합당한 정사체를 사용하였으며, 우리말 표제어나 해설어가 외래어로서 한자화할 수 없는 용어는 '-' 표기로 나타내었고 1996년도에 출판된 대기과학용어집에서 빠진 한자를 가능한 찾아 보충하였다(예 : 임계깊이 臨界-, 임계고도 臨界高度).

8. 줄표의 일관성 :
표제어가 여러 개의 명사로 나열되어 있는 경우에는 단어 사이에 줄표를 삽입하였고, 형용사와 명사의 나열로 이루어지는 용어에는 띄어쓰기를 하였다(예 : Q-band, Q-띠; 이 경우에 알파벳은 큐-띠와 같이 음차하지 않고 영문 알파벳을 그대로 사용하였다.).

9. 합성어 중 정비 원칙에는 띄어 씀이 옳으나, 관용성을 인정하여 한국기상학회 용어심의위원회에서 결정한 붙여 쓰기를 원칙으로 하였다(예 : active permafrost 활동성영구언땅).

2013년 3월
편집위원회

대기과학용어집 편찬위원회

편찬위원장	전종갑(서울대학교)
편집위원장	오성남(연세대학교)
용어선정 및 편집위원	김경익(경북대학교)　　　　오재호(부경대학교) 김성중(극지연구소)　　　　윤일희(경북대학교) 김영준(광주과학기술원)　　이규태(강릉원주대학교) 김영준(한국수치예보개발사업단)　이동규(서울대학교) 김영철(한서대학교)　　　　이승재(서울대학교) 김준(서울대학교)　　　　　이우진(기상청) 김준(연세대학교)　　　　　이재원(기상청) 김해동(계명대학교)　　　　이태영(연세대학교) 박록진(서울대학교)　　　　정영근(전남대학교) 반기성(K-Weather)　　　　차은정(기상청) 소선섭(공주대학교)　　　　하경자(부산대학교) 양영규(가천대학교)　　　　허복행(기상청) 염성수(연세대학교)　　　　홍성길(기상전문인협회)
편집 보조	김현경(기상산업진흥원) 신우주(연세대학교) 김태민(가천대학교)
언어감수	강현화(연세대학교 국문과)
1996년 대기과학용어 심의위원	강인식　곽종흠　김광식　김문일　김보희　김성삼 김정우　노재식　문승의　민경덕　박경윤　박순웅 박용대　변희룡　봉종헌　성락도　심경섭　윤순창 이광호　이동규　이승만　이종범　이태영　전종갑 정용승　정을영　정창희　조봉구　조희구　최　효 최희승　한영호　홍성길

참고문헌

월간전자기술 편집부(2005), 전자용어사전. 성안당.

임정남(1987), 농업기상용어해설. 향문사.

한국과학기술단체총연합회(1988), 과학기술용어집. 아카데미서적.

한국기상학회(1996), 대기과학용어집(영한중일). (주)교학사.

한국물학술단체연합회(2011), 물용어집. (주)시그마프레스.

한국지구과학회(2003), 지구과학 학술용어집. (주)시그마프레스.

한국해양학회(2005), 해양과학용어사전. 아카데미서적.

Adrien, N. C.(2004), *Computational Hydraulics and Hydrology: An Illustrated Dictionary*. CRC Press.

Air Pollution Training Institute(1975), *Air Pollution Meteorology*.

Allaby, M.(2002), *Encyclopedia of Weather and Climate, Volume I(A-L)*. Facts On File, Inc.

Allaby, M.(2002), *Encyclopedia of Weather and Climate, Volume II(M-Z)*. Facts On File, Inc.

American Meteorological Society(2000), *Glossary of Meteorology, Second Edition*.

Arya, S. P.(1999), *Air Pollution Meteorology and Dispersion*. Oxford University Press.

Arya, S. P.(2001). *Introduction to Micrometeorology, Second Edition*. Academic Press.

Blackhart, K., D. G. Stanton, and A. M. Shmada(2006), *NOAA Fisheries Glossary*.

BMTC(Bureau of Meteorology Training Center, Australia)(1991), *Radar Principles*.

Doviak, R. J. and Zrnic, D. S.(1993), *Doppler Radar and Weather Observations*. Academic Press.

Edde, B.(1993), *Radar: Principles, Technology, Application*. Prentice Hall.

Glickman, T. S. (Editor)(2000), *Glossary of Meteorology, Second Edition*. American Meteorological Society.

Mclntosh, D. H.(1972), *Meteorological Glossary, 5th Edition*. Chemical Publishing Company, Inc.

Raghavan, S.(2003), *Radar Meteorology*. Kluwer Academic Publishers.

Rinehart, R. E.(2004), *Radar for Meteorologists*. Rinehart Publishing.

Sauvageot, A.(1991), *Radar Meteorology,* Artech House.

Schneider, S. H. (Editor)(1996). *Encyclopedia of Climate and Weather, Volume 1*. Oxford

University Press.

Schneider, S. H. (Editor)(1996). *Encyclopedia of Climate and Weather. Volume 2.* Oxford University Press, pp. 461-929.

Scott, A. W.(1993), *Understanding Microwaves.* Wiley-Interscience.

Skolnik, M.(2001), *Introduction to Radar System, 3rd Edition.* McGraw Hill.

Stephens, G. L.(1994), *Remote Sensing of the Lower Atmosphere: An Introduction.* Oxford University Press.

W.M.O.(2006), *Instruments and Observing Methods.* Report No. 88, Training Material on Weather Radar Systems. WMO/TD-No. 1308.

Wallace, J. M. and P. V. Hobbs(2006), *Atmospheric Science, An Introductory Survey, Second Edition.* Elsevier.

Encyclopedia of Meteorology(1995). 東京堂出版, 大橋信夫(in Japanese).

地球科學技術 Handbook(2001), Frontier Research System 發行(in Japanese).

Met Office, cited 2012: Climate Change Glossary.

[Available online at http://www.metoffice.gov.uk/climate-change/guide/glossary.html]

Met Office, cited 2012: Marine Forecasts Glossary.

[Available online at http://metoffice.gov.uk/weather/marine/guide/glossary.html]

National Aeronautics and Space Administration, cited 2012: Glossary.

[Available online at http://science.nasa.gov/glossary/]

National Oceanic and Atmospheric Administration, cited 2012: Glossary of NHC Terms.

[Available online at http://www.nhc.noaa.gov/aboutgloss.shtml]

National Oceanic and Atmospheric Administration, cited 2012: National Weather Service Glossary.

[Available online at http://w1.weather.gov/glossary/]

University Corporation for Atmospheric Research, cited 2012: Glossary.

[Available online at http://www.dpc.ucar.edu/VoyagerJr/glossary.html]

University Corporation for Atmospheric Research, cited 2012: Thunderstorm Glossary.

[Available online at http://www2.ucar.edu/news/backgrounders/thunderstorm-glossary]

University Corporation for Atmospheric Research, cited 2012: Weather Glossary.

[Available online at http://eo.ucar.edu/webweather/glossary.html]

차례

영한편

A, a

영문	한글	한자
abatement of wind	풍력감쇠	風力減衰
abbreviated ship code	약식선박부호	略式船舶符號
ABC bucket	ABC 두레박(표면수온측정용)	(表面水溫測定用)
abduction	외전	外轉
aberration	수차, 광행차	收差, 光行差
aberwind	해동바람〈알프스〉	解凍-
abnormal	이상	異常
abnormal climate	이상기후	異常氣候
abnormal cold wave	이상한파	異常寒波
abnormal dryness	이상건조	異常乾燥
abnormality	이상, 이상도	異常, 異常度
abnormal lapse rate	이상감률	異常減率
abnormal refraction	이상굴절	異常屈折
abnormal weather	이상기상	異常氣象
above ground level	지상고도이상	地上高度以上
above mean sea level	평균해발고도이상	平均海拔高度以上
above normal	정상이상	正常以上
above sea level	해발고도이상	海拔高度以上
ab-polar current	이탈극류	離脫極流
Abrolhos squall(= abroholos)	아브롤로스스콜	
abrupt climate change	급격기후변화	急激氣候變化
abscissa	가로좌표	橫座標
absolute acceleration	절대가속도	絶對加速度
absolute altitude	절대고도	絶對高度
absolute angular momentum	절대각운동량	絶對角運動量
absolute annual range of temperature	온도 절대연교차	溫度絶對年較差
absolute cavity radiometer	절대공동복사계	絶對空洞輻射計
absolute ceiling	절대실링	絶對-
absolute circulation	절대순환	絶對循環

A

absolute convergence	절대수렴	絕對收斂
absolute coordinate system	절대좌표계	絕對座標系
absolute dating	절대연대측정	絕對年代測定
absolute drought	절대가뭄	絕對-
absolute error	절대오차	絕對誤差
absolute extreme	절대극값	絕對極-
absolute frequency	절대빈도	絕對頻度
absolute gradient current	절대경도류	絕對傾度流
absolute humidity	절대습도	絕對濕度
absolute hygrometer	절대습도계	絕對濕度計
absolute index of refraction	절대굴절률	絕對屈折率
absolute instability	절대불안정	絕對不安定
absolute instrument	절대계기	絕對計器
absolute inequality	절대부등식	絕對不等式
absolute isohypse	절대등고선	絕對等高線
absolute linear momentum	절대선형운동량	絕對線形運動量
absolute moisture of the soil	토양절대수분함량	土壤絕對水分含量
absolute momentum	절대운동량	絕對運動量
absolute monthly maximum temperature	월절대최고온도	月絕對最高溫度
absolute monthly minimum temperature	월절대최저온도	月絕對最低溫度
absolute parallax	절대시차	絕對視差
absolute pollen frequency	절대꽃가루빈도	絕對-頻度
absolute potential vorticity	절대위치소용돌이도	絕對位置-度
absolute pressure	절대압력, 절대기압	絕對壓力, 絕對氣壓
absolute pyrheliometer	절대직달일사계	絕對直達日射計
absolute radiation scale	절대복사규모	絕對輻射規模
absolute reference frame	절대기준틀	絕對基準-
absolute refractive index	절대굴절률	絕對屈折率
absolute scale	절대눈금	絕對-
absolute stability	절대안정도	絕對安定度
absolute standard barometer	절대표준기압계	絕對標準氣壓計
absolute temperature	절대온도	絕對溫度
absolute temperature extreme	절대온도극값	絕對溫度極-
absolute temperature scale	절대온도눈금	絕對溫度-
absolute topography	절대지형, 절대높낮이	絕對地形
absolute unit	절대단위	絕對單位
absolute vacuum	절대진공	絕對眞空
absolute value	절댓값	絕對-
absolute variability	절대변동(률)	絕對變動(率)
absolute velocity	절대속도	絕對速度
absolute vorticity	절대소용돌이도	絕對-度
absolute vorticity maximum	절대소용돌이도 최대	絕對-度最大
absolute zero	절대영도	絕對零度
absorbance	흡광도	吸光度

absorbate	흡수물	吸收物
absorbed dose	흡수선량	吸收線量
absorbed solar radiation	흡수태양복사	吸收太陽輻射
absorbent	흡착제	吸着劑
absorber	흡수기, 흡수체, 흡수제	吸收器, 吸收體, 吸收劑
absorbing layer	흡수층	吸收層
absorbing medium	흡수매질	吸收媒質
absorbing surface	흡수면	吸收面
absorptance	흡수도, 흡수율	吸收度, 吸收率
absorption	흡수	吸收
absorption band	흡수대, 흡수띠	吸收帶
absorption coefficient	흡수계수	吸收係數
absorption cross-section	흡수단면(적)	吸收斷面(積)
absorption efficiency	흡수효율	吸收效率
absorption factor	흡수인자	吸收因子
absorption features analysis	흡수특징분석	吸收特徵分析
absorption hygrometer	흡수습도계	吸收濕度計
absorption line	흡수선	吸收線
absorption loss	흡수손실	吸收損失
absorption number	흡수수	吸收數
absorption of radiation	복사흡수	輻射吸收
absorption of solar radiation	태양복사흡수	太陽輻射吸收
absorption optical depth	흡수광학깊이	吸收光學-
absorption optical thickness	흡수광학두께	吸收光學-
absorption schemes	흡수방식	吸收方式
absorption spectroscopy	흡수분광법	吸收分光法
absorption spectrum	흡수스펙트럼	吸收-
absorption tower	흡수타워	吸收塔
absorptive index	흡수지수	吸收指數
absorptive power	흡수능	吸收能
absorptivity	흡수율	吸收率
abstraction	분리, 추출	分離, 抽出
abyssal plain	심해저평원	深海底平原
accelerated airflow	가속기류	加速氣流
accelerated erosion	가속침식	加速浸蝕
accelerated motion	가속운동	加速運動
acceleration	가속도	加速度
acceleration length scale	가속도 길이규모	加速度-規模
acceleration of air parcel	공기덩이가속도	空氣-加速度
acceleration of gravity	중력가속도	重力加速度
acceleration of melting snow	융설촉진	融雪促進
acceleration of wind	바람가속도	-加速度
accelerator	가속기, 생장촉진기	加速器, 生長促進器
accelerometer	가속도계	加速度計

A

A

acceptance capacity	채택용량	採擇容量
acceptance region	채택역	採擇域
accessory cloud	부속구름	附屬-
accidental drought	우발가뭄	偶發-
accidental error	우연오차	偶然誤差
acclimation	순화, 순응	馴化, 順應
acclimatization	기후적응	氣候適應
accommodation box	수용상자	收容箱子
accommodation coefficient	조절계수	調節係數
accretion	결착	結着
accretion efficiency	결착률	結着率
accretion equation	결착방정식	結着方程式
accumulated cooling	누적냉각	累積冷却
accumulated potential water loss	누적가능물손실	累積可能-損失
accumulated precipitation	누적강수	累積降水
accumulated snow	적설	積雪
accumulated temperature	적산온도	積算溫度
accumulation	적산, 누적	積算, 累積
accumulation mode	누적모드	累積-
accumulation zone	적산대	積算帶
accumulative area	적산면적, 집수면적	積算面積, 集水面積
accumulative rain gauge	적산우량계	積算雨量計
accuracy	정확도	正確度
accuracy assessment	정확도평가	正確度評價
accuracy scatter graph	정확도분산그래프	正確度分散-
acetaldehyde	아세트알데히드	
acetic acid	아세트산, 초산	硝酸
aceton	아세톤	
acetonitrile	아세토니트릴	
acetylene	아세틸렌	
acicular ice	바늘얼음, 침상빙	針狀氷
acid	산	酸
acid cloud	산성구름	酸性-
acid deformation	산성변형	酸性變形
acid deposition	산성침적	酸性沈積
acid dew	산성이슬	酸性-
acid fog	산성안개	酸性-
acid frost	산성서리	酸性-
acid hail	산성우박	酸性雨雹
acid haze	산성박무	酸性薄霧
acidic	산성	酸性
acidic deposition	산성침적	酸性沈積
acidification	산성화	酸性化
acid ion	산성이온	酸性-

acidity	산도	酸度
acidity of water	물의 산도	-酸度
acidity profile	산도연직분포	酸度鉛直分布
acid-neutralizing capacity	산중성화능력	酸中性化能力
acid pollution	산성오염	酸性汚染
acid precipitation	산성강수	酸性降水
acid rain	산성비	酸性-
acid rain distribution	산성비분포	酸性-分布
acid rain effect	산성비효과	酸性-效果
acid smut	산덩어리(스머트)	酸-
acid snow	산성눈	酸性-
acid soot	산성검댕	酸性-
aclinic line	(지)자기적도	(地)磁氣赤道
acoustical scintillation	소리변동	-變動
acoustic array	소리배열	-配列
acoustic backscattering	음향후방산란	音響後方散亂
acoustic cloud	소리구름, 음향반사운	音響反射雲
acoustic detection and ranging	음파레이더	音波-
acoustic dispersion	음향분산	音響分散
acoustic Doppler current profile	초음파 도플러 유속측정기	超音波-流速測定器
acoustic echo sounding	음파에코측심기	音波-測深器
acoustic frequency generator	음파발생기	音波發生機
acoustic gravity wave	음향중력파	音響重力波
acoustic imaging	음파영상	音波映像
acoustic impedance	음향임피던스	音響-
acoustic intensity	음파강도	音波强度
acoustic lens	음파렌즈	音波-
acoustic ocean current meter	음파해류계	音波海流計
acoustic pressure	음파압력	音波壓力
acoustic radar	음파레이더	音波-
acoustic rain gauge	음파우량계	音波雨量計
acoustic reflecting profiling	음파반사탐사	音波反射探査
acoustic reflectivity	음파반사율	音波反射率
acoustic refraction	음파굴절	音波屈折
acoustic resonance	음파공명	音波共鳴
acoustic reverberation	음파반향	音波反響
acoustics	음향학	音響學
acoustic scattering	음파산란	音波散亂
acoustic scintillation	음향섬광	音響閃光
acoustic signature	음파신호	音波信號
acoustic sounder	음파탐측기	音波探測機
acoustic sounding	음파탐측	音波探測
acoustic thermometer	음파온도계	音波溫度計
acoustic topography	음파높낮이	音波-

acoustic transducer	음향변환기	音響變換機
acoustic transponder	음파송수신기	音波送受信機
acoustic trauma	청신경성외상	聽神經性外傷
acoustic velocimeter	음파속도계	音波速度計
acoustic wave	음파	音波
acquisition	수집	收集
acquisition condition	획득조건	獲得條件
acre-foot	에이커풋(1피트 깊이로 1에이커를 채우는 수량)	
acrocyanosis	동상	凍傷
across-jet flight	제트횡단비행	-橫斷飛行
Ac standing lenticular	정체렌즈고적운	停滯-高積雲
actinic	광화학반응(의)	光化學反應
actinic absorption	광화학흡수	光化學吸收
actinic balance	광화학저울	光化學-
actinic flux	광화학플럭스	光化學-
actinic radiation	광화학복사	光化學輻射
actinic rays	광화학선	光化學線
actiniform	방사형(의)	放射形
actinogram	일사기록	日射記錄
actinograph	자기일사계	自記日射計
actinometer	일사계	日射計
actinometry	일사측정학	日射測定學
actinothermic index	복사능지수	輻射能指數
action	작용	作用
action center	작용중심	作用中心
action integral	작용적분	作用積分
action variable	작용변수	作用變數
activated carbon	활성탄소	活性炭素
activated carbon process	활성탄소과정	活性炭素過程
activated charcoal	활성탄	活性炭
activated complex	활성착물, 활성복합체	活性錯物, 活性複合體
activated complex theory	활성복합체 이론	活性複合體理論
activation	활성화	活性化
activation energy	활성화에너지	活性化-
active accumulated temperature	활동적산온도	活動積算溫度
active anafrontal surface	능동적 상승활주면	能動的上昇滑走面
active basin area	활동성분지지역	活動性盆地地域
active center	활동중심	活動中心
active cloud	활동성구름	活動性-
active front	활동성전선	活動性前線
active glacier	활동성빙하	活動性氷河
active instrument	능동형계기, 능동측정기	能動型計器, 能動測定器
active layer	활동층	活動層
active microwave instrument	능동형극초단파장치	能動型極超短波裝置

A

active network	(채층의)활동성망상조직	活動性網狀組織
active nitrogen	활성질소	活性窒素
active permafrost	활동성영구 언땅	活動性永久-
active radar	능동형레이더	能動形-
active region	활동영역	活動領域
active remote sensing	능동형원격탐사	能動形遠隔探査
active remote sensor	능동원격감지부	能動遠隔感知部
active sensor	능동감지	能動感知
active site	활성부위	活性部位
active solar collector	능동태양열집열기	能動太陽熱集熱器
active surface	능동면, 활동표면	能動面, 活動表面
active system	능동시스템, 능동계	能動系
activity	활성도	活性度
activity coefficient	활성도계수	活性度係數
activity of a foyer of atmospherics	대기공전강도	大氣空電强度
actual density of the soil	토양실제밀도	土壤實際密度
actual elevation	실제고도	實際高度
actual evaporation	실제증발량	實際蒸發量
actual evapotranspiration	실(유효)증발산(량)	實(有效)蒸發散(量)
actual pressure	실제압력	實際壓力
actual time of observation	실제관측시간	實際觀測時間
adabatic chart	단열선도	斷熱線圖
adaptability	적응성, 적응도	適應性, 適應度
adaptation	적응	適應
adaptation brightness	적응밝기	適應-
adaptation field	적응장	適應場
adaptation illuminance	적응조도	適應照度
adaptation level	적응수준	適應水準
adaptation luminance	적응휘도	適應輝度
adaptation strategy	적응전략	適應戰略
adaptation type	적응형	適應型
adaptive control	적응조절	適應調節
adaptive convergence	적응수렴	適應收斂
adaptive disease	적응병	適應病
adaptive grid	적응격자, 가변격자	適應格子, 可變格子
adaptive observation	맞춤관측, 최적관측	最適觀測
adaptive observational network	최적관측망	最適觀測網
adaptive radiation	적응복사	適應輻射
additive color	가색법	加色法
adfreezing	빙착	氷着
adhesion	부착력	附着力
adhesion efficiency	부착효율	附着效率
adhesion tension	부착장력	附着張力
adhesive water	접착수	接着水

adiabat	단열선	斷熱線
adiabatic	단열(의)	斷熱
adiabatically enclosed system	단열폐쇄계, 단열닫힌계	斷熱閉鎖系
adiabatic approximation	단열근사	斷熱近似
adiabatic atmosphere	단열대기	斷熱大氣
adiabatic change	단열변화	斷熱變化
adiabatic chart	단열도	斷熱圖
adiabatic compression	단열압축	斷熱壓縮
adiabatic condensation	단열응결	斷熱凝結
adiabatic condensation point	단열응결점	斷熱凝結點
adiabatic condensation pressure	단열응결압력	斷熱凝結壓力
adiabatic condensation temperature	단열응결온도	斷熱凝結溫度
adiabatic condition	단열조건	斷熱條件
adiabatic cooling	단열냉각	斷熱冷却
adiabatic curve	단열곡선	斷熱曲線
adiabatic diagram	단열선도	斷熱線圖
adiabatic differential equation	단열미분방정식	斷熱微分方程式
adiabatic effect	단열효과	斷熱效果
adiabatic equation	단열방정식	斷熱方程式
adiabatic equilibrium	단열평형	斷熱平衡
adiabatic equivalent temperature	단열상당온도	斷熱相當溫度
adiabatic expansion	단열팽창	斷熱膨脹
adiabatic heating	단열가열	斷熱加熱
adiabatic invariant	단열불변량	斷熱不變量
adiabatic lapse rate	단열감률	斷熱減率
adiabatic layer	단열층	斷熱層
adiabatic liquid water content	단열액체수	斷熱液體水
adiabatic method	단열법	斷熱法
adiabatic motion	단열운동	斷熱運動
adiabatic process	단열과정	斷熱過程
adiabatic reference state	단열기준상태	斷熱基準狀態
adiabatic region	단열구역, 단열영역	斷熱區域, 斷熱領域
adiabatics	단열선	斷熱線
adiabatic saturation point	단열포화점	斷熱飽和點
adiabatic saturation pressure	단열포화압력	斷熱飽和壓力
adiabatic saturation temperature	단열포화온도	斷熱飽和溫度
adiabatic solid water content	단열고체수 함량	斷熱固體水含量
adiabatic temperature change	단열온도변화	斷熱溫度變化
adiabatic temperature gradient	단열온도경도	斷熱溫度傾度
adiabatic trail	단열흔적	斷熱痕跡
adiabatic trial	단열시행	斷熱施行
adiabatic warming	단열승온	斷熱昇溫
adiabatic wet-bulb temperature	단열습구온도	斷熱濕球溫度
A-display	에이 디스플레이	

adjacent spot test	인접점시험	隣接點試驗
adjoint assimilation	수반자료동화	隨伴資料同化
adjoint equation	수반방정식, 딸림방정식	隨伴方程式
adjoint model	수반모형	隨伴模型
adjoint radiative transfer equation	수반복사전달방정식	隨伴輻射傳達方程式
adjoint sensitivity	수반민감도	隨伴敏感度
adjustable cistern barometer	수은조조절식기압계	水銀槽調節式氣壓計
adjustment	조절	調節
adjustment process	조절과정	調節過程
admissible concentration limit	허용농도한계	許容濃度限界
admittance function	허용함수	許容函數
adsorbate	피흡착질	被吸着質
adsorbent	흡착제	吸着劑
adsorption	흡착(작용)	吸着(作用)
adsorption isotherm	흡착등온선	吸着等溫線
advanced earth observing satellite	선진지구관측위성	先進地球觀測衛星
advanced earth observing system	선진지구관측시스템	先進地球觀測-
advanced microwave sounding unit	선진초단파탐측장치	先進超短波探測裝置
advanced TIROS-N	선진 TIROS-N	先進-
Advanced Very High Resolution Radiometer (AVHRR)	선진고분해능복사계	先進高分解能輻射計
advanced vidicon camera system	선진비디콘카메라시스템	先進-
advection	이류	移流
advectional inversion	이류역전	移流逆轉
advection-diffusion equation	이류-확산 방정식	移流擴散方程式
advection equation	이류방정식	移流方程式
advection fog	이류안개	移流-
advection frost	이류서리	移流-
advection inversion layer	이류역전층	移流逆轉層
advection layer(= advective region)	이류층	移流層
advection model	이류모형	移流模型
advection region	이류역	移流域
advection scale	이류규모	移流規模
advection solenoid	이류솔레노이드	移流-
advection term	이류항	移流項
advection time scale	이류시간규모	移流時間規模
advection upslope cloud	이류활승구름	移流活昇-
advective boundary layer	이류경계층	移流境界層
advective change	이류변화	移流變化
advective change of temperature	온도이류변화	溫度移流變化
advective effect	이류효과	移流效果
advective flux	이류플럭스	移流-
advective-gravity flow	이류-중력 흐름	移流重力-
advective hypothesis	이류가설	移流假說

advective layer	이류층	移流層
advective model	이류모형	移流模型
advective pressure tendency	이류압력경향	移流壓力傾向
advective region	이류역	移流域
advective term	이류항	移流項
advective thunderstorm	이류성뇌우	移流性雷雨
adverse effect	역효과, 유해효과	逆效果, 有害效果
advisory	주의보	注意報
advisory area	주의보영역	注意報領域
advisory committee on climate applications and data	기후응용 및 자료에 관한 자문위원회	氣候應用資料諮問委員會
advisory forecast	주의보	注意報
aeolian	풍성(질)	風成(質)
aeolian anemometer	바람소리풍속계	-風速計
aeolian sound	바람소리	
aeolian tone	바람가락	
aeration	환기, 통풍	換氣, 通風
aerial	공중	空中
aerial cloud seeding	항공구름씨뿌리기	航空-
aerial experiment	항공실험	航空實驗
aerial photos	항공사진	航空寫眞
aerial plankton	공중플랑크톤	空中-
aeroallergen	에어로알레르겐	
aerobiology	공중생물학	空中生物學
aeroclimatology	고층기후학	高層氣候學
AERO code	항공기상전문형식	航空氣象電文型式
aerodrome advisory	비행장주의보	飛行場注意報
aerodrome control tower	비행장관제탑	飛行場管制塔
aerodrome control weather information system	비행장관제기상정보시스템	飛行場管制氣象情報-
aerodrome elevation	비행장표고	飛行場標高
aerodrome forecast	비행장예보	飛行場豫報
aerodrome meteorological minimum	비행장기상최저값	飛行場氣象最低-
aerodrome meteorological observation system	비행장기상관측시스템	飛行場氣象觀測-
aerodrome security service	비행장안전업무	飛行場安全業務
aerodrome special report	비행장특보	飛行場特報
aerodrome warning	비행장경보	飛行場警報
aerodynamic	공기역학(의)	空氣力學
aerodynamically rough surface	공기역학적 거친 면	空氣力學的-面
aerodynamically smooth surface	공기역학적 매끄러운 면	空氣力學的-面
aerodynamic approach	공기역학접근법	空氣力學接近法
aerodynamic balance	공기역학균형	空氣力學均衡
aerodynamic contrail	공기역학비행운	空氣力學飛行雲
aerodynamic damping	공기역학감쇠	空氣力學減衰
aerodynamic drag	공기역학항력	空氣力學抗力

aerodynamic equivalent diameter	공기역학상당직경	空氣力學相當直徑
aerodynamic force	공기역학힘	空氣力學-
aerodynamic forcing	공기역학강제	空氣力學强制
aerodynamic instability	공기역학불안정	空氣力學不安定
aerodynamic method	공기역학방법	空氣力學方法
aerodynamic property	공기역학성질	空氣力學性質
aerodynamic resistance	공기역학저항	空氣力學抵抗
aerodynamic rough flow	공기역학 거친 흐름	空氣力學-
aerodynamic roughness	공기역학거칠기	空氣力學-
aerodynamic roughness length	공기역학거칠기길이	空氣力學-
aerodynamics	공기역학	空氣力學
aerodynamic scale	공기역학규모	空氣力學規模
aerodynamic smooth flow	공기역학 매끈한 흐름	空氣力學-
aerodynamic smoothness	공기역학 매끄러움	空氣力學-
aerodynamic trail	공기역학비행운	空氣力學飛行雲
aeroembolism	고공병	高空病
aerogenerator	풍차	風車
aerogram	에어로그램	
aerograph	자기고층계	自記高層計
aerographical chart	고층기상도	高層氣象圖
aerography	대기지, 기술기상학	大氣誌, 記述氣象學
aerolite	운석	隕石
aerologation	최소항로	最小航路
aerological analysis	고층분석	高層分析
aerological ascent	고층탐측	高層探測
aerological days	고층관측의 날	高層觀測-
aerological diagram	고층기상선도	高層氣象線圖
aerological instrument	고층기상관측기	高層氣象觀測器
aerological observation	고층기상관측	高層氣象觀測
aerological observation system	고층기상관측시스템	高層氣象觀測-
aerological sounding	고층기상탐측	高層氣象探測
aerological station	고층기상관측소	高層氣象觀測所
aerological table	고층기상표	高層氣象表
aerological theodolite	측풍용경위(의)	測風用經緯
aerology	고층기상학	高層氣象學
aerometeorograph	고층기상기록계	高層氣象記錄計
aeronautical climatology	항공기후학	航空氣候學
aeronautical fixed telecommunication network	항공고정통신망	航空固定通信網
aeronautical information publication	항공정보간행물	航空情報刊行物
aeronautical information service	항공정보업무	航空情報業務
aeronautical information service unit	항공정보업무기관	航空情報業務機關
aeronautical meteorological observation	항공기상관측	航空氣象觀測
aeronautical meteorological service	항공기상업무	航空氣象業務
aeronautical meteorological station	항공기상관측소	航空氣象觀測所

A

aeronautical meteorology	항공기상학	航空氣象學
aeronautical mobile service	항공이동업무	航空移動業務
aeronautical safe aids	항공보안시설	航空保安施設
aeronautical safe radio aids	항공보안무선시설	航空保安無線施設
aeronautical telecommunication station	항공통신소	航空通信所
aeronautical traffic service	항공교통업무	航空交通業務
aeronomy	고층대기물리학	高層大氣物理學
aeropause	고층권계면	高層圈界面
aerophotography	항공사진술	航空寫眞術
aeroplane antenna	항공기안테나	航空機-
aeroscopy	대기관측	大氣觀測
aerosol	에어로졸, 미세입자	微細粒子
aerosol asymmetry factor	에어로졸비대칭인자	-非對稱因子
aerosol direct effect	에어로졸직접효과	-直接效果
aerosol direct radiative effect	에어로졸직접복사효과	-直接輻射效果
aerosol electricity	에어로졸전기	-電氣
aerosol index	에어로졸지수	-指數
aerosol indirect effect	에어로졸간접효과	-間接效果
aerosol loading	에어로졸하중	-荷重
aerosol mass extinction efficiency	에어로졸질량소멸효율	-質量消滅效率
aerosol optical depth	에어로졸광학깊이, 에어로졸광학두께	-光學-
aerosol phase function	에어로졸위상함수	-位相函數
aerosol radiative forcing	에어로졸복사강제력	-輻射强制力
aerosol radiative forcing efficiency	에어로졸복사강제력효율	-輻射强制力效率
aerosol semi-direct effect	에어로졸반직접효과	-半直接效果
aerosol single-scattering albedo	에어로졸일차산란알베도	--次散亂-
aerosol size distribution	에어로졸크기분포	-分布
aerospace	우주공간	宇宙空間
aerosphere	대기권	大氣圈
aerostat	기구, 비행선	氣球, 飛行船
aerostatic balance	공기저울	空氣-
aerostatics	공기정역학	空氣靜力學
aerostat meteorograph	기구용기상기록계	氣球用氣象記錄計
aerovane	(프로펠러식)풍향풍속계	風向風速計
aestivation	여름잠, 하면	夏眠
afforestation	조림, 식림	造林, 植林
Afghanets	아프가네츠(아프가니스탄으로부터 불어오는 아무다리아 강 상류의 강풍)	
Africa Center of Meteorological Applications for Development	아프리카기상개발응용센터	-氣象開發應用-
Africa drought	아프리카가뭄, 아프리카한발	-旱魃
African jet	아프리카제트	
African Monsoon Multidisciplinary Analysis	아프리카몬순다분야분석	-多分野分析

African wave	아프리카파	-波
afterglow	잔광, 저녁놀	殘光
afterheat	늦더위	
afterimage	잔상	殘像
afternoon effect	오후효과	午後效果
afternoon mixing depth	오후혼합두께	午後混合-
after summer	가을더위	
Aftonian interglacial	아프토니아간빙기	-間氷期
agenda	비망록	備忘錄
agent of precipitation enhancement	강수증가촉매제	降水增加觸媒劑
age-of-air	공기연령	空氣年齡
age of tide	조령	潮齡
ageostrophic	비지균	非地均
ageostrophic advection	비지균이류	非地均移流
ageostrophic advection term	비지균이류항	非地均移流項
ageostrophic component	비지균성분	非地均成分
ageostrophic effect	비지균효과	非地均效果
ageostrophic method	비지균방법	非地均方法
ageostrophic model	비지균모형	非地均模型
ageostrophic motion	비지균운동	非地均運動
ageostrophic wind	비지균풍	非地均風
ageostrophic wind component	비지균풍성분	非地均風性分
agglomeration	결집	結集
agglutination	교착, 점착	膠着, 粘着
aggregate of ice particle	얼음입자 부착	-粒子附着
aggregation	부착	附着
aggressive water	센물	
agnostic chart	(아무도 믿지 않는)예상도	豫想圖
agonic line	무편각선	無偏角線
agradient airflow	아경도공기흐름	亞傾度空氣-
agricultural climate	농업기후	農業氣候
agricultural climatology(= agroclimatology)	농업기후학	農業氣候學
agricultural drought	농업가뭄	農業-
agricultural meteorological station	농업기상관측소	農業氣象觀測所
agricultural meteorological station for specific purpose	특수목적농업기상관측소	特殊目的農業氣象觀測所
agricultural meteorology(= agrometeorology)	농업기상학	農業氣象學
agricultural revolution	농업혁명	農業革命
agricultural season	농사계절	農事季節
agricultural source	농업오염원	農業汚染源
agricultural weather forecast	농업일기예보	農業日氣豫報
agriculture	농업	農業
agroclimatic classification	농업기후구분	農業氣候區分
agroclimatic index	농업기후지수	農業氣候指數

agroclimatic region	농업기후구	農業氣候區
agroclimatology	농업기후학	農業氣候學
agroforestry	산림농업	山林農業
agrometeorological forecast	농업기상예보	農業氣象豫報
agrometeorological station	농업기상관측소	農業氣象觀測所
agrometeorological station for specific purposes	특수목적농업기상관측소	特殊目的農業氣象觀測所
agrometeorology	농업기상학	農業氣象學
agrotopoclimatology	농업지형기후학	農業地形氣候學
Agulhas current	아굴라스해류	-海流
Agulhas stream	아굴라스류	-流
Agung eruption	아궁화산분화	-火山噴火
air	공기, 대기	空氣, 大氣
air acidity	공기산도	空氣酸度
air acoustic ranging sensor	음파탐사센서	音波探査-
air base microburst	비행장미규모폭풍	飛行場微規模暴風
airborne	공기(의), 공중부유(의), 항공기탑재(의)	空氣, 空中浮遊, 航空機搭載
airborne antarctic ozone experiment	항공기탑재남극오존실험	航空機搭載南極-實驗
airborne arctic stratospheric expedition	항공기탑재북극성층권탐사	航空機搭載北極成層圈探査
airborne dust analysis	공기먼지분석	空氣-分析
airborne expendable bathythermograph	항공기투하식 수심수온계	航空機投下式水深水溫計
airborne fraction	공기비율	空氣比率
airborne observation	항공기탑재관측	航空機搭載觀測
airborne particle size	공기입자크기	空氣粒子-
airborne particles	공중부유입자	空中浮遊粒子
airborne radar	항공레이더	航空-
airborne radiation thermometer	항공복사온도계	航空輻射溫度計
airborne search radar	항공기탑재탐사레이더	航空機搭載探査-
airborne sensing system	항공기탑재감지시스템	航空機搭載感知-
airborne southern hemisphere ozone experiment	항공기탑재남반구오존실험	航空機搭載南半球-實驗
airborne weather radar	항공기탑재기상레이더	航空機搭載氣象-
air bump	공기범프, 요동	空氣-, 搖動
air burst	공기파열	空氣破裂
air capacity	공기용량	空氣容量
air cascade	공기폭포	空氣瀑布
air cataract	기폭	氣瀑
air circulation	공기순환	空氣循環
air cleaning	공기청정	空氣淸淨
air column	공기기둥, 기주	空氣-, 氣柱
air compressibility	공기압축성	空氣壓縮性
air conditioning	공기조절	空氣調節
air conductivity	대기전도도	大氣傳導度
aircraft	항공기	航空機
aircraft-borne sensor	항공기탑재센서	航空機搭載-

aircraft ceiling	항공기실링	航空機-
aircraft communications addressing and reporting system	항공기운항정보교신시스템	航空機運航情報交信-
aircraft control	항공기통제	航空機統制
aircraft data collection system	항공운항정보시스템	航空運航情報-
aircraft electrification	항공기대전	航空機帶電
aircraft ice accretion(= aircraft icing)	항공기착빙	航空機着氷
aircraft impactor	항공기탑재형	航空機搭載型
aircraft measurement	항공기측정	航空機測定
aircraft meteorological station	항공기상관측소	航空氣象觀測所
aircraft observation	항공기관측	航空機觀測
aircraft platform	항공기플랫폼	航空機-
aircraft power distribution system	항공기전력공급시스템	航空機電力供給-
aircraft reconnaissance	항공기정찰	航空機偵察
aircraft report	항공기보고	航空機報告
aircraft sounding	항공기탐측	航空機探測
aircraft thermometry	항공기온도측정법	航空機溫度測定法
aircraft-to-satellite data relay	항공기-위성자료중계	航空機衛星資料中繼
aircraft trail	항공기궤적	航空機軌跡
aircraft trajectory	항공기궤적	航空機軌跡
aircraft turbulence	항공기난류	航空機亂流
aircraft weather reconnaissance	항공기기상정찰	航空機氣象偵察
air current	기류	氣流
air density	공기밀도	空氣密度
air discharge	공중방전	空中放電
air drainage	배기	排氣
air dried soil	풍건토	風乾土
air dust analysis	공기면지분석	空氣-分析
air-earth conduction current	공지전도전류	空地傳導電流
air-earth current	공지전류	空地電流
airfield color code	비행장색깔전문형식	飛行場-電文型式
air filter	공기여과기	空氣濾過機
air flow	공기흐름, 기류	空氣-, 氣流
airflow over mountain	산월기류	山越氣流
airfoil section	날개단면	-斷面
Air Force Weather Wing	공군기상단	空軍氣象團
air fountain	공기샘	空氣-
airframe icing	항공기착빙	航空機着氷
air freezing-index	지상결빙지수	地上結氷指數
air frost	지상서리	地上-
air-fuel ratio	공기연료비	空氣燃料比
air gauge	기압계	氣壓計
airglow	대기광	大氣光
air hoar	나무서리	

air humidity	습도	濕度
air issue	공기이슈, 대기과제	空氣-, 大氣課題
airlight	공기광	空氣光
airlight formula	공기광공식	空氣光公式
air line correction	항로보정	航路補正
air line sounding	항로탐측	航路探測
airman's meteorological information	저층악기상정보	底層惡氣象情報
air-mass	기단	氣團
air-mass analysis	기단분석	氣團分析
air-mass calendar	기단력	氣團歷
air-mass characteristic	기단의 특성	氣團特性
air-mass chart	기단도	氣團圖
air-mass classification	기단분류	氣團分類
air-mass classification system	기단분류시스템	氣團分類-
air-mass climatology	기단기후학	氣團氣候學
air-mass diagram	기단선도	氣團線圖
air-mass discontinuity	기단불연속	氣團不連續
air-mass fog	기단안개	氣團-
air-mass frequency	기단빈도	氣團頻度
air-mass identification	기단식별	氣團識別
air-mass meteorology	기단기상학	氣團氣象學
air-mass modification	기단변질	氣團變質
air-mass precipitation	기단강수	氣團降水
air-mass property	기단성질	氣團性質
air-mass shower	기단소나기	氣團-
air-mass source	기단발원지	氣團發源地
air-mass source region	기단발원지역	氣團發源地域
air-mass temperature	기단온도	氣團溫度
air-mass thunderstorm	기단뇌우	氣團雷雨
air-mass transformation	기단변질	氣團變質
air-mass transformation experiment	기단변질실험	氣團變質實驗
air-mass transport	기단이동	氣團移動
air-mass-type diagram	기단유형도표	氣團類型圖表
air-mass weather	기단날씨	氣團日記
air-mass wind shear	기단윈드시어	氣團-
airmeter	미풍계	微風計
air molecule	공기분자	空氣分子
air motion	공기운동	空氣運動
air parcel	공기덩이	空氣-
airparcel dynamic method	공기덩이역학방법	空氣塊力學方法
airparcel trajectory	공기덩이궤적	空氣-軌跡
air particle	공기입자	空氣粒子
air particle index	공기입자지수	空氣粒子指數
airpath	항로	航路

air permeability	통기성	通氣性
airplane meteorograph	비행기기상기록계	飛行機氣象記錄計
airplane observation	비행기관측	飛行機觀測
air plankton	공중플랑크톤	空中-
air pocket	에어포켓	
air poise	공기저울	空氣-
air pollutant	공기오염물질, 대기오염물질	空氣汚染物質, 大氣汚染物質
air pollution	대기오염	大氣汚染
air pollution alert	대기오염경보	大氣汚染警報
air pollution control	대기오염통제	大氣汚染統制
air pollution control act	대기오염방지법	大氣汚染防止法
air pollution episode	대기오염사건	大氣汚染事件
air pollution meteorology	대기오염기상학	大氣汚染氣象學
air pollution observation station	대기오염관측소	大氣汚染觀測所
air pollution potential	대기오염퍼텐셜	大氣汚染-
air pollution prevention	대기오염방지	大氣汚染防止
airport	공항	空港
airport elevation	공항표고	空港標高
airport forecast	공항예보	空港豫報
airport height	공항고도	空港高度
airport model	공항모형	空港模型
air-position indicator	공중위치지시기	空中位置指示器
air pressure	기압	氣壓
air pressure tendency	기압경향	氣壓傾向
air quality	대기질	大氣質
air quality act	대기질관련조례	大氣質關聯條例
air quality control region	대기질제어지역	大氣質制御地域
air quality criteria	대기오염기준, 대기질범주	大氣汚染基準, 大氣質範疇
air quality cycle	대기질순환	大氣質循環
air quality display model	대기질디스플레이모형	大氣質-模型
air quality index	대기질지수	大氣質指數
air quality legislation	대기질법률제정	大氣質法律制定
air quality management	대기질관리	大氣質管理
air quality modeling	대기질모형화	大氣質模型化
air quality standard	대기질표준	大氣質標準
air report	항공기보고	航空機報告
air resistance	공기저항	空氣抵抗
Air Resources Regional Pollution Assessment (ARRPA) Model	대기자원지역 오염평가모형	大氣資源地域汚染評價模型
air route	항로	航路
air sampler	공기포집기	空氣捕集器
air-sea interaction	대기-해양 상호작용	大氣海洋相互作用
air shed	대기분수계	大氣分水系
air shower	공기소나기, (외연)공기샤워	空氣-, (外延)空氣-

air sink	공기흡원	空氣吸源
air space	공역	空域
airspeed	대기속력	大氣速力
air-sphere	기권	氣圈
air stagnation model	기단정체모형	氣團停滯模型
air stream	기류	氣流
air stream concept	기류개념	氣流槪念
air temperature	기온	氣溫
air temperature field	기온장	氣溫場
air temperature inversion	기온역전	氣溫逆轉
air temperature profile	기온연직분포	氣溫鉛直分布
air temperature sensor	기온감부	氣溫感部
air thawing-index	지상해빙지수	地上解氷指數
air thermoscope	공기온도경, 공기측온계	空氣溫度鏡, 空氣測溫計
air torrent	공기격류	空氣激流
air toxic	대기독성, 유해대기오염물질	大氣毒性, 有害大氣汚染物質
air toxin	공기독소	空氣毒素
air traffic control	항공교통관제, 항공관제	航空交通管制, 航空管制
air traffic control radar	항공관제레이더	航空管制-
air trajectory	공기궤적	空氣軌跡
air transport	공기수송	空氣輸送
air trap	공기트랩	空氣-
air wave	공기파	空氣波
airway	항공로	航空路
airway forecasting	항공로예보	航空路豫報
airways code	항공로부호, 항공로코드	航空路符號
airways observation	항공로관측	航空路觀測
airways shelter	항공로백엽상	航空路百葉箱
airway weather	항공로일기	航空路日氣
Airy disk	에어리원판	-圓板
Airy function	에어리함수	-函數
Aitken counter	에이킨계수기	-計數器
Aitken dust counter	에이킨먼지계수기	-計數器
Aitken nucleus	에이킨핵	-核
Aitken nucleus counter	에이킨핵계수기	-核計數器
Aitken particle	에이킨입자	-粒子
alarm level	경보수위	警報水位
Alaska current	알래스카해류	-海流
Alaskan Stream	알래스카스트림	
albedo	알베도, 반사율	反射率
albedograph	알베도기록계, 반사율기록계	反射率記錄計
albedometer	반사율계, 알베도계	反射率計
albedo of the earth	지구알베도	地球-
albedo-temperature feedback	알베도-온도 되먹임	-溫度-

albedo value	알베도값	
Alberta clipper	앨버타클리퍼	
Alberta cyclone	앨버타저기압	-低氣壓
Alberta low	앨버타저기압	-低氣壓
alcohol	알코올	
alcohol biomass fuel	알코올바이오매스연료	-燃料
alcohol-in-glass thermometer	알코올온도계	-溫度計
alcohol thermometer	알코올온도계	-溫度計
alcyone days(= hylcyon days)	정온기	靜穩期
aldehyde	알데히드	
Alecto unit	요드화은 에어로졸 발생장치	-銀-發生裝置
alert concentration	경계농도	警戒濃度
Aleutian current	알류샨해류	-海流
Aleutian low	알류샨저기압	-低氣壓
Alexander's dark band	알렉산더의 검은 띠	
Alfvén wave	알프벤파	-波
algae	조류	藻類
algal bloom	적조	赤潮
algal reef	조초	藻礁
Algerian current	알제리해류	-海流
algorithm	알고리즘	
algorithmic language	알고리즘언어	-言語
aliasing	위신호, 에일리어싱	僞信號
aliasing error	위신호오차	僞信號誤差
alidade	시준의	視準儀
alignment	정렬	整列
alignment chart	계산도표	計算圖表
alimentation	증량(빙하의)	增量
aliphatic compound	지방족화합물	脂肪族化合物
Alisov's classification of climate	알리솝기후구분	-氣候區分
alkalinity	알칼리도, 염기도	鹽基度
alkane	알케인, 알켄	
alkenone	알케논	
alkylperoxy radical	알킬퍼록시라디칼	
alkyne	알킨	
Allard's law	알라드법칙	-法則
allergy	알레르기	
Allerød period	알레뢰드기(마지막 빙하시대 말기에 나타난 온난습윤기)	-期
All-hallow summer(= All Saints' summer)	만성절여름	萬聖節-
allobar	변압구역	變壓區域
allobaric	변압(의)	變壓
allobaric wind	변압풍	變壓風
allocation	배치, 배분	配置, 配分

allohypsic wind	변고풍	變高風
allohypsography	고도변화도법	高度變化圖法
allometric growth	상대생장	相對生長
allowable error	허용오차	許容誤差
allowed transition	허용된 전이	許容轉移
all-sky camera	전천카메라	全天-
all-sky photometer	전천광도계	全天光度計
alluvial	충적(의)	沖積
alluvial aquifer	충적대수층	沖積帶水層
all-weather airport	전천후공항	全天候空港
all-weather landing	전천후착륙	全天候着陸
almanac	역서, 연감	曆書, 年監
Almeria-Oran effect	알메리아-오랑 효과	-效果
Almeria-Oran front	(서부지중해) 알메리아-오랑 전선	(西部地中海)-前線
almucantar	고도권	高度圈
almwind	암윈드(폴란드 크라코우의 남쪽 타트라 산맥을 가로질러 불어오는 푄바람)	
along-jet flight	제트평행비행	-平行飛行
along-slope wind system	경사면을 따라서 부는 바람시스템	傾斜面-
along track direction	궤도방향	軌道方向
along track resolution	궤도방향해상도	軌道方向解像度
along track scanners	궤도를 따른 주사	軌道走査
along track scanning radiometer	궤도를 따라서 주사하는 복사계	軌道走査輻射計
along-valley wind	계곡을 따라서 부는 바람	溪谷-
along-valley wind system	계곡을 따라서 부는 바람시스템	溪谷-
alpach(= aberwind, aperwind)	해동풍	解凍風
alpenglow(= Alpenglühen)	고산광	高山光
alpha decay	알파붕괴	-崩壞
alphanumeric data	문자숫자데이터	文字數字-
alpha-particle	알파입자	-粒子
alpha-ray(= α-ray)	알파선	-線
alpine climate	고산기후	高山氣候
alpine experiment	고산실험	高山實驗
alpine glacier	고산빙하	高山氷河
alpine glow	고산광	高山光
alpine tundra(= mountain tundra)	고산툰드라	高山-
alpine vegetation	고산식생	高山植生
alternate airport	대체공항	代替空港
alternate forecast	예비비행장일기예보	豫備飛行場日氣豫報
alternate transmission	교차송신	交叉送信

alternating current	교류	交流
alternating tensor	교대텐서	交代-
alternating unit tensor	교대단위텐서	交代單位-
alternative energy	대체에너지	代替-
Alter shield	우량계용바람막이	雨量計用-
altichamber	기압검정상자, 기압실험실	氣壓檢定箱子, 氣壓實驗室
alti-electrograph	고공전위기록계	高空電位記錄計
altigraph	자기고도계	自記高度計
altimeter	고도계	高度計
altimeter corrections	고도계보정	高度計補正
altimeter equation	고도계방정식	高度計方程式
altimeter setting	고도설정(값)	高度設定
altimeter setting indicator	고도설정지시기	高度設定指示器
altimetry	측고법	測高法
altithermal period	건조기	乾燥期
altitude	(해발)고도	(海拔)高度
altitude calculation	고도계산	高度計算
altitude correction	고도보정	高度補正
altitude information	고도정보	高度情報
altitude of the sun	태양고도	太陽高度
altocumulus	고적운	高積雲
altocumulus castellanus	탑고적운	塔高積雲
altocumulus cloud	고적운	高積雲
altocumulus duplicatus	겹고적운	-高積雲
altocumulus floccus	송이고적운	-高積雲
altocumulus glomeratus	밀집고적운	密集高積雲
altocumulus informis	무정형고적운	無定形高積雲
altocumulus lacunosus	벌집고적운	-高積雲
altocumulus lenticularis	렌즈고적운	-高積雲
altocumulus mamma	유방고적운	乳房高積雲
altocumulus nebulosus	면사포고적운	面紗布高積雲
altocumulus opacus	불투명고적운	不透明高積雲
altocumulus perlucidus	틈새고적운	-高積雲
altocumulus radiatus	방사고적운	放射高積雲
altocumulus stratiformis	층고적운	層高積雲
altocumulus translucidus	반투명고적운	半透明高積雲
altocumulus undulatus	파도고적운	波濤高積雲
altocumulus virga	꼬리고적운	-高積雲
altostratocumulus	고층적운	高層積雲
altostratus	고층운	高層雲
altostratus cloud	고층운	高層雲
altostratus densus	농밀고층운	濃密高層雲
altostratus duplicatus	겹고층운	-高層雲
altostratus fractus	조각고층운	-高層雲

A

altostratus lenticularis	렌즈고층운	-高層雲
altostratus maculosus	무늬고층운	-高層雲
altostratus mamma	유방고층운	乳房高層雲
altostratus opacus	불투명고층운	不透明高層雲
altostratus opacus undulatus	불투명파도고층운	不透明波濤高層雲
altostratus pannus	토막고층운	-高層雲
altostratus precipitation	강수고층운	降水高層雲
altostratus precipitus	강수고층운	降水高層雲
altostratus radiatus	방사고층운	放射高層雲
altostratus translucidus	반투명고층운	半透明高層雲
altostratus undulatus	파도고층운	波濤高層雲
altostratus virga	꼬리고층운	-高層雲
aluminum oxide humidity element	산화알루미늄제감습소자	酸化-感濕素子
amateur forecast	아마추어예보	-豫報
amateur weather station	아마추어기상관측소	-氣象觀測所
Amazon basin	아마존분지	-盆地
Amazon forest	아마존삼림	-森林
Amazonia area	아마존강유역	-江流域
Amazon River	아마존강	-江
ambient	주변	周邊
ambient air	주변공기, 주위공기	周邊空氣, 周圍空氣
ambient air quality guideline	주변공기질가이드라인	周邊空氣質-
ambient air standard	주변공기질표준	周邊空氣質標準
ambient air temperature	주변기온	周邊氣溫
ambient liquid water content	주변물함량	周邊-含量
ambient pressure	주변압력	周邊壓力
ambient temperature	주변온도	周邊溫度
ambiguity function	모호성함수	模糊性函數
Amble diagram	앰블선도	-線圖
amended aerodrome forecast	수정비행장예보	修正飛行場豫報
Amended Area Forecast	수정공역예보	修正空域豫報
Amended Route Forecast	수정항공로예보	修正航空路豫報
American Geophysical Union	미국지구물리연합	美國地球物理聯合
American Meteorological Society(AMS)	미국기상학회	美國氣象學會
Ameri Flux	아메리플럭스	
ammeter	전류계	電流計
ammonia	암모니아	
ammonification	암모니아생성	-生成
ammonium	암모늄	
ammonium chloride	염화암모늄	鹽化-
ammonium hydroxide	수산화암모늄	水酸化-
ammonium ion	암모니아이온	
ammonium nitrate	질산암모늄	窒酸-
ammonium sulfate	황산암모늄	黃酸-

ammonium sulfate haze	황산암모늄박무	黃酸-薄霧
ammonium sulphate	암모늄황산염	-黃酸鹽
Amontons's law	아몽통법칙	-法則
amorphous	무정형, 비정질(의)	無定形, 非晶質
amorphous cloud	무정형구름	無定形-
amorphous cloud cluster	무정형구름무리	無定形-
amorphous frost	무정형서리	無定形-
amorphous sky	무정형구름하늘	無定形-
amorphous snow	무정형눈	無定形-
amount of clouds	구름양, 운량	雲量
amount of evaporation	증발량	蒸發量
amount of precipitation	강수량	降水量
amount of snow cover	총강설지역	總降雪地域
ampere	암페어	
amphibia	양서류	兩棲類
amphidrome	무조점	無潮點
amphidromic point	무조점	無潮點
amphidromic region	무조역	無潮域
amplification	증폭	增幅
amplification matrix	증폭행렬	增幅行列
amplitude	진폭	振幅
amplitude distribution	진폭분포	振幅分布
amplitude modulated indicator	진폭변조지시기	振幅變調指示器
amplitude modulation	진폭변조	振幅變調
amplitude spectrum	진폭스펙트럼	振幅-
amplitude vacillation	진폭동요	振幅動搖
anabatic	활승(의)	滑昇
anabaric(= ananllobaric)	기압증가(의)	氣壓增加
anabaric center	활승중심	滑昇中心
anabatic cloud	활승구름	滑昇-
anabatic flow	활승흐름	滑昇-
anabaric front	활승전선	滑昇前線
anabatic wind	활승바람	滑昇-
anabolism	동화작용	同化作用
Anadyr Current	아나디리해류	-海流
anaerobic	혐기성(의)	嫌氣性
anaerobic bacteria	혐기성박테리아	嫌氣性-
anaerobic condition	혐기성조건	嫌氣性條件
anaerobic digestors	혐기성처리장치	嫌氣性處理裝置
anaerobic respiration	혐기호흡	媒氣呼吸
anaflow	활승류	滑昇流
anafront	활승전선	滑昇前線
anallobar	등기압상승선	等氣壓上昇線
anallobaric	기압상승(의)	氣壓上昇

A

anallobaric center	기압상승중심	氣壓上昇中心
analog climate model	아날로그기후모형	-氣候模型
analog computer	아날로그전산기	-電算機
analog digital converter	아날로그디지털변환기	-變換器
analog forecasting	유사예보	類似豫報
analog method	아날로그방법	-方法
analog model	아날로그모형	-模型
analogous map	유사일기도	類似日氣圖
analogous organ	상사기관	相似器管
analog prefiltering	아날로그 프리필터링	
analog-to-digital conversion	아날로그-디지털 변환	-變換
analog(ue)	유사(물), 아날로그	類似(物)
analogue data transmission	아날로그데이터 전송	-電送
analogue method	유사법	類似法
analogue model(= analogus model)	유사모형	類似模型
analysed chart (or map)	분석도	分析圖
analysis center	분석센터	分析-
analysis increment	분석증분	分析增分
analysis method	분석방법	分析方法
analysis of variance	분산분석	分散分析
anaphalanx	온난전선면	溫暖前線面
Anasazi people	아나사지인	-人
anchor balloon	초고압기구	超高壓氣球
anchor(ed) trough	정체기압골	停滯氣壓-
anchor ice	바닥얼음	
Andes glow	안데스광	-光
Andes light	안데스빛	
Andes lightning	안데스번개	
Andes mountains	안데스산맥	-山脈
andhis	앤드히스(인도 북서지방에 부는 모래폭풍)	
anelastic	비탄성(의)	非彈性
anelastic approximation	비탄성근사	非彈性近似
anelastic assumption	비탄성가정	非彈性假定
anemo-biagraph	압력관풍속기록계	壓力管風速記錄計
anemo-cinemograph	아네모시네모그래프	
anemoclinometer	경풍계	傾風計
anemogram	경풍계	傾風計
anemograph	자기풍속계	自記風速計
anemology	측풍학	測風學
anemometer	풍속계	風速計
anemometer level	풍속계고도	風速計高度
anemometer mast (or tower)	바람관측탑	-觀測塔
anemometry	풍속측정법	風速測定法
anemophily	풍매	風媒

anemoscope	풍향계	風向計
anemovane	풍향풍속계	風向風速計
aneroid	아네로이드	
aneroid barograph	아네로이드자기기압계	-自記氣壓計
aneroid barometer	아네로이드기압계	-氣壓計
aneroid capsule	아네로이드공합	-空盒
aneroid element	아네로이드소자	-素子
aneroidogram	아네로이드기압기록	-氣壓記錄
aneroidograph	아네로이드자기기압계	-自記氣壓計
angel	이상회파	異常回波
angel echo	에인절에코	
angel of elevation	고도각, 올려본 각	高度角
angina pectoris	협심증	狹心症
angle of aperture	구경각	口徑角
angle of arrival	도달각	到達角
angle of circumference	원주각	圓周角
angle of declination	편각	偏角
angle of deflection	편전각	偏轉角
angle of deviation	편향각	偏向角
angle of friction	마찰각	摩擦角
angle of incidence	입사각	入射角
angle of inclination	경사각	傾斜角
angle of indraft	유입각	流入角
angle of minimum deviation	최소편향각	最小偏向角
angle of reflection	반사각	反射角
angle of refraction	굴절각	屈折角
angle of rotation	회전각	回轉角
angle of scattering	산란각	散亂角
angle of view	시각	視角
Anglian glacial	앵글리아빙하	-氷河
Anglo-Brazilian climate observation study (ABRACOS)	영국-브라질 아마존 기후관측연구	英國-氣候觀測研究
Angola-Benguela front	앙골라-벵겔라 전선	-前線
Angola current	앙골라해류	-海流
Ångström	옹스트룀(거리단위)	
Ångström compensation	옹스트룀보상	-補償
Ångström compensation pyrheliometer	옹스트룀보상직달일사계	-補償直達日射計
Ångström equation	옹스트룀방정식	-方程式
Ångström exponent	옹스트룀지수	-指數
Ångström pyrgeometer	옹스트룀지구복사계	-地球輻射計
angular acceleration	각가속도	角加速度
angular area	각면적	角面積
angular displacement	각변위	角變位
angular drift	각표류	角漂流

A

angular frequency	각진동수	角振動數
angularity correction	향사도수정	向斜度修正
angular momentum	각운동량	角運動量
angular momentum balance	각운동량균형	角運動量均衡
angular resolution	각분해능	角分解能
angular spreading	각퍼짐	角-
angular spreading factor	각퍼짐인자	角-因子
angular velocity	각속도	角速度
angular velocity of rotation	자전각속도	自轉角速度
angular velocity of the earth	지구각속도	地球角速度
angular width	각도너비, 각폭	角度-, 角暴
anhyetism	결우성	缺雨性
animal climate	동물기후	動物氣候
animal fog	동물안개	動物-
animal production	동물생산	動物生産
animation of imagery	영상애니메이션, 동영상	映像-, 動映像
anion	음이온	陰-
anisobaric	비등압(의)	非等壓
anisotropic	비등방성(의)	非等方性
anisotropic hydrometeor	비등방수상체	非等方水狀體
anisotropization	비등방성화	非等方性化
anisotropy	비등방성	非等方性
anniversary wind	연주기바람	年週期-
annual anomaly	연편차	年偏差
annual-basis pentad	연중반순	年中半旬
annual climatological report	기상연보	氣象年報
annual crop	일년생 작물	一年生作物
annual cycle	연주기	年週期
annual eclipse	금환식	金環蝕
annual exceedance series	연초과치계열	年超過値系列
annual flood	연홍수위	年洪水位
annual flood series	연홍수계열	年洪水系列
annual layer	연층	年層
annual maximum series	연최대치계열	年最大値系列
annual mean	연평균	年平均
annual mean temperature range	연평균온도교차	年平均溫度交差
annual minimum series	연최저치계열	年最低値系列
annual range	연교차	年較差
annual report	연보	年報
annual runoff	연유출	年流出
annual series	연계열	年系列
annual snow line	연설선	年雪線
annual storage	연저장	年貯藏
annual temperature range	연온도교차	年溫度交差

annual temperature wave	연온도변화파	年溫度變化波
annual variation	연변동	年變動
annual wave	연변화파	年變化波
annular mode	환상(고리모양)모드	環狀-
annulus experiment	애뉼러스(환대)실험	-(環帶)實驗
anomalous	이상(의)	異常
anomalous circulation	이상순환	異常循環
anomalous (cloud) line	이상(구름)열	異常-列
anomalous dispersion	이상분산	異常分散
anomalous gradient wind	이상경도풍	異常傾度風
anomalous propagation	이상전파	異常傳播
anomalous radio propagation	이상전자기파전파	異常電磁氣波傳播
anomalous refraction	이상굴절	傳播屈折
anomalous sound propagation	이상음파전파	異常音波傳播
anomaly	편차, 이상	偏差, 異常
anomaly chart	평년편차도	平年偏差圖
anomaly correction	이상보정	異常補正
anomaly correlation	이상상관관계	異常相關關係
anomaly of dynamic height	역학고도편차	力學高度偏差
anomaly of geopotential difference	지오퍼텐셜차 편차	-差偏差
anomaly of specific volume	비부피편차	比-偏差
anoxemia	산소결핍(증)	酸素缺乏(症)
anoxia	산소결핍증	酸素缺乏症
anoxic water	무산소물	無酸素水
anricycloniic shear	고기압성시어	高氣壓性-
Antarctic	남극(의)	南極
Antarctica	남극대륙	南極大陸
Antarctic air (mass)	남극기단	南極氣團
Antarctic anticyclone	남극고기압	南極高氣壓
Antarctic bottom water	남극저층수	南極底層水
Antarctic Circle	남극권	南極圈
Antarctic circumpolar current	남극순환해류	南極循環海流
Antarctic circumpolar water	남극순환수	南極循環水
Antarctic circumpolar wave	남극순환파동	南極循環波動
Antarctic climate	남극기후	南極氣候
Antarctic cold reversal	남극추위반전, 남극한랭반전, 남극한랭역전	南極-反轉, 南極寒冷反轉, 南極寒冷逆轉
Antarctic convergence	남극수렴	南極收斂
Antarctic deep water	남극심층수	南極深層水
Antarctic divergence	남극발산	南極發散
Antarctic front	남극전선	南極前線
Antarctic glaciation history	남극빙하역사	南極氷河歷史
Antarctic high	남극고기압	南極高氣壓
Antarctic ice sheets	남극빙상	南極氷床

Antarctic intermediate water	남극중층수	南極中層水
Antarctic Ocean	남빙해	南氷海
Antarctic oscillation	남극진동	南極振動
Antarctic ozone hole	남극오존구멍	南極-
Antarctic ozone layer	남극오존층	南極-層
Antarctic polar current	남극한대해류	南極寒帶海流
Antarctic polar vortex	남극한대소용돌이	南極寒帶-
Antarctic Pole	남극	南極
Antarctic sea ice	남극바다얼음, 남극해빙	南極海氷
Antarctic sea ice history	남극해빙역사	南極海氷歷史
Antarctic sea smoke	남극바다김안개	南極-
Antarctic stratospheric vortex	남극성층권소용돌이	南極成層圈-
Antarctic surface water	남극표층수	南極表層水
Antarctic system	남극시스템, 남극계	南極系
Antarctic zone	남극대	南極帶
antecedent	선행(의)	先行
antecedent precipitation index	선행강수지수	先行降水指數
antecedent soil moisture	선행토양수분, 이전토양수분	先行土壤水分, 以前土壤水分
antenna	안테나	
antenna feed	안테나피드	
antenna gain	안테나이득	-利得
antenna limit	안테나한계	-限界
antenna pattern	안테나패턴	
antenna reflector	안테나반사체, 안테나반사경	-反射體, -反射鏡
antenna temperature	안테나온도	-溫度
anthelic arcs	맞무리호	-弧
anthropocene	인류세	人類世
anthropoclimatology	인류기후학	人類氣候學
anthropogenic aerosol	인위적 에어로졸	人爲的-
anthropogenic climatic change	인위적 기후변화	人爲的氣候變化
anthropogenic climatic effect	인위적 기후효과	人爲的氣候效果
anthropogenic CO_2 increase	인위적 이산화탄소 증가	人爲的二酸化炭素增加
anthropogenic effluent	인위적 방출	人爲的放出
anthropogenic emission	인위적 배출	人爲的排出
anthropogenic forcing	인위적 강제	人爲的强制
anthropogenic global warming	인위적 지구온난화	人爲的地球溫暖化
anthropogenic heat	인위적 열	人爲的熱
anthropogenic heat input	인위적 열입력	人爲的熱入力
anthropogenic process	인위적 과정	人爲的過程
anthropogenic release	인위적 방출	人爲的放出
antibaric flow	반압류	反壓流
anticlockwise	반시계(방향)	反時計(方向)
anticorona	반대광환	反對光環
anticrepuscular arch	반대박명호	反對薄明弧

anticrepuscular ray	반대박명광	反對薄明光
anticyclogenesis	고기압발생	高氣壓發生
anticyclolysis	고기압소멸	高氣壓消滅
anticyclone	고기압	高氣壓
anticyclone movement	고기압이동	高氣壓移動
anticyclone subsidence	고기압성침강	高氣壓性沈降
anticyclonic	고기압성	高氣壓性
anticyclonic blocking	고기압저지, 고기압블로킹	高氣壓沮止
anticyclonic bora	고기압성보라	高氣壓性-
anticyclonic circulation	고기압성순환	高氣壓性循環
anticyclonic curvature	고기압성곡률	高氣壓性曲率
anticyclonic divergence	고기압성발산	高氣壓性發散
anticyclonic eddy	고기압성맴돌이	高氣壓性-
anticyclonic flow	고기압성흐름, 고기압성기류	高氣壓性氣流
anticyclonic gloom	고기압성흐림	高氣壓性-
anticyclonicity	고기압성	高氣壓性
anticyclonic phase	고기압성위상	高氣壓性位相
anticyclonic region	고기압지역	高氣壓地域
anticyclonic rotation	고기압성회전	高氣壓性回轉
anticyclonic vortex	고기압성소용돌이	高氣壓性-
anticyclonic vorticity	고기압성소용돌이도	高氣壓性-度
anticyclonic weather	고기압성날씨	高氣壓性日氣
anticyclonic weather type	고기압성날씨형	高氣壓性日氣型
antidynamo	반다이나모	反-
anti-ENSO event	반ENSO사건	反-事件
anti-hail gun	우박억제포	雨雹抑制砲
anti-hail rocket	우박억제로켓	雨雹抑制-
antihelion(= anthelion)	맞무리해	
anti-icing	착빙억제	着氷抑制
Antilles current	안틸해류	-海流
antilog	안티로그	
antimonsoon	반대계절풍	反對季節風
anti-mountain wind	반대산풍	反對山風
antipersistence	반지속성	反持續性
antipleion	반편차중심	反偏差中心
antiproton	반양성자	反陽性子
antiradiance	복사방지	輻射防止
antisea breeze	반대해풍	反對海風
antiselena	맞무리달	
anti-solar point	대일점	對日點
anti-symmetric tensor	비대칭텐서	非對稱-
anti-symmetrical	비대칭(의)	非對稱
antitrades	반대무역풍	反對貿易風
antitrade wind	반대무역풍	反對貿易風

antitriptic wind	마찰풍	摩擦風
anti-twilight	반대박명	反對薄明
anti-twilight arch	반대박명호	反對薄明弧
antivalley wind	반대계곡풍	反對溪谷風
anvil	모루	
anvil cloud	모루구름	
anvil dome	모루돔	
aperiodic	비주기적	非週期的
aperiodic motion	비주기운동	非週期運動
aperiodic oscillation	비주기진동	非週期振動
aperture	구경	口徑
aperwind	해동풍	解凍風
aphelion	원일점	遠日點
Apheliotes	아펠리오테스, 동풍〈그리스어〉	東風
apical cell	정단세포	頂端細胞
apob	항공관측	航空觀測
apocenter	원궤점	遠軌點
apogean range	원지점조차	遠地點潮差
apogean tide	원지점조석	遠地點潮汐
apogean wind	원지점바람	遠地點-
apogee	원지점	遠地點
apostilb	아포스틸브	
Appalachian mountains	애팔래치아산맥	-山脈
apparatus	계기	計器
apparent brightness	겉보기밝기	
apparent diameter	겉보기지름	
apparent expansion	겉보기팽창	-膨脹
apparent force	겉보기힘	
apparent form of the sky	하늘겉보기모양	-模樣
apparent freezing point	겉보기어는점	-點
apparent gravity	겉보기중력	-重力
apparent groundwater velocity	겉보기지하수속도	-地下水速度
apparent horizon	겉보기수평선	-水平線
apparent luminance	겉보기휘도	-輝度
apparent noon	겉보기정오	-正午
apparent solar day	겉보기태양일	-太陽日
apparent solar time	겉보기태양시	-太陽時
apparent specific gravity	겉보기비중	-比重
apparent stress	겉보기응력	-應力
apparent temperature	겉보기온도	-溫度
apparent velocity	겉보기속도	-速度
apparent wind	겉보기바람	
Appleton-Hartree equation	애플턴-하트리 방정식	-方程式
Appleton layer	애플턴층	-層

application technology satellite	응용기술위성	應用技術衛星
applied climatologist	응용기후학자	應用氣候學者
applied climatology	응용기후학	應用氣候學
applied hydrology	응용수문학	應用水文學
applied meteorology	응용기상학	應用氣象學
applied synoptic climatological study	응용종관기후학연구	應用綜觀氣候學研究
applied weather	응용기상	應用氣象
approach channel	접근수로	接近水路
approach-light contact height	진입등 교신고도	進入燈交信高度
approach velocity	접근속도	接近速度
approach visibility	접근시정	接近視程
approximate absolute temperature scale	근사절대온도눈금	近似絕對溫度-
approximate coordinate system	근사좌표계	近似座標系
approximate expression	근사식	近似式
approximate thermal wind equation	근사온도풍방정식	近似溫度風方程式
approximate value	근삿값	近似-
approximation	근사	近似
approximation spread	근사확대	近似擴大
April shower	4월 소나기	
a priori probability	사전확률, 선험적 확률	事前確率, 先驗的確率
a priori reason	사전이유	事前理由
aqueous solution	수용액	水溶液
aqueous vapour	수증기	水蒸氣
aquiclude	난투수층	難透水層
aquifer	투수층	透水層
aquifer system	대수층시스템	帶水層-
aquifer test	대수성시험	帶水性試驗
aquifuge	비투수층	非透水層
aquitard	준대수층	準帶水層
Arabian desert	아라비아사막	-砂漠
Arabian sea	아라비아바다	
Arago distance	아라고거리	-距離
aragonite	아라고나이트, 선석	霰石
Arago point	아라고점	-點
Arago's neutral point	아라고중립점	-中立點
Arakawa Jacobian	아라카와자코비안	
Aral Sea	아랄해	-海
arc cloud	활모양구름	
arc cloud line	활모양구름선	-線
arc discharge	활모양방전	-放電
arch cloud	아치구름	
arch twilight	박명호	薄明弧
Archean	시생대	始生代
Archean environment	시생대환경	始生代環境

arched squall	아치형스콜	-型-
archeological data	고고학자료	考古學資料
Archimedean buoyant force	아르키메데스부력	-浮力
Archimedes's principle	아르키메데스원리	-原理
architectural climatology	건축기후학	建築氣候學
architectural meteorology	건축기상학	建築氣象學
architecture	건축	建築
archived data	저장자료	貯藏資料
archives	고문서, 기록보관소	古文書, 記錄保管所
archiving data	기록보관소자료	記錄保管所資料
arc lines	활모양선	弧線
arcsecond	각초	角秒
arcs of Lowitz	로비츠호	-弧
arctic	북극(의)	北極
arctic air (mass)	북극기단	北極氣團
arctic-alpine	북극고산	北極高山
arctic anticyclone	북극고기압	北極高氣壓
arctic basin	북극분지	北極盆地
arctic blackout	북극권무전감쇠현상	北極圈無電減衰現象
arctic bottom water	북극저층수	北極底層水
Arctic Canada	북극캐나다	北極-
arctic Circle	북극권	北極圈
arctic climate	북극기후	北極氣候
arctic climate impact assessment	북극기후영향평가	北極氣候影響評價
arctic climate system study	북극기후시스템연구	北極氣候-研究
arctic continental air (mass)	북극대륙기단	北極大陸氣團
arctic desert	북극사막	北極砂漠
arctic front	북극전선	北極前線
arctic haze	북극연무	北極煙霧
arctic high	북극고기압	北極高氣壓
arctic hurricane	북극허리케인	北極-
arctic intermediate water	북극중층수	北極中層水
arcticization	극지화	極地化
arctic jet stream	북극제트류, 북극제트(기류)	北極-流, 北極-(氣流)
arctic lower stratosphere	북극하층성층권	北極下層成層圈
arctic maritime air mass	북극해양성기단	北極海洋性氣團
arctic maritime airstream	북극해양성기류	北極海洋性氣流
arctic mist	북극엷은안개	北極-
arctic ocean	북빙양	北氷洋
arctic oscillation	북극진동	北極振動
arctic pack	북극부빙	北極浮氷
arctic polar front	북극한대전선	北極寒帶前線
Arctic Pole	북극	北極
arctic sea ice	북극바다얼음, 북극해빙	北極海氷

arctic sea smoke	북극바다김안개	北極-
arctic smoke	북극김안개	北極-
arctic stratospheric vortex	북극성층권소용돌이	北極成層圈-
arctic surface water	북극저층수	北極底層水
arctic system	북극시스템, 북극계	北極系
arctic tree line	북극수목선	北極樹木線
arctic tropopause	북극대류권계면	北極對流圈界面
arctic vortex observation	북극소용돌이관측	北極-觀測
arctic weather station	북극기상관측소	北極氣象觀測所
arctic whiteout	북극화이트아웃	北極-
arctic wind	북극바람	北極-
arctic zone(= north frigid zone)	북극대	北極帶
arcus	아치구름	
ARDC Model Atmosphere	ARDC 모형대기	-模型大氣
ardometer	광복사고온측정계	光輻射高溫測定計
area average(= area averaging)	면적평균, 영역평균	面積平均, 領域平均
area averaging problem	면적평균문제, 영역평균문제	面積平均問題, 領域平均問題
area conversion factor	면적변환인자	面積變換因子
area-elevation curve	면적-고도 곡선	面積高度曲線
area forecast	지역예보	地域豫報
area forecast center	지역예보센터	地域豫報-
area forecast code	지역예보전문형식	地域豫報電文型式
areal precipitation	면적강수	面積降水
areal rainfall	면적(강)우량	面積(降)雨量
areal reduction factor	면적감소인자	面積減少因子
areal velocity	면적속도	面積速度
area of coverage	피복역, 관측역, 영향역	被覆域, 觀測域, 影響域
area of high pressure	고기압역	高氣壓域
area of influence	영향지역	影響地域
area source	면오염원	面汚染源
area source solution	면오염원해법	面汚染源解法
area subject to disaster	재해위험지역	災害危險地域
Argo float	아르고플로트	
argon	아르곤	
Argo Program	아르고프로그램	
argument of latitude	위도편각	緯度偏角
argument of perigee	근지점의 편각	近地點偏角
arheic region	무하지역	無河地域
arid	건조한	乾燥-
arid boundary	건조한계	乾燥限界
arid climate	건조기후	乾燥氣候
arid cycle	건조주기	乾燥週期
aridity	건조도	乾燥度
aridity coefficient	건조계수	乾燥係數

aridity factor	건조인자	乾燥因子
aridity index	건조지수	乾燥指數
arid region	건조구역	乾燥區域
arid zone	건조지대	乾燥地帶
arid zone hydrology	건조대 수문학	乾燥帶水文學
arithmetic mean	산술평균	算術平均
arithmatic mean method	산술평균법	算術平均法
arithmetic mean temperature	산술평균기온	算術平均氣溫
armoured thermometer	표면온도계(해면 온도측정용)	表面溫度計
aromatic compound	방향족 화합물	芳香族化合物
aromatic hydrocarbon	방향족 탄화수소	芳香族炭化水素
aromatics	방향족 화합물	芳香族化合物
Arrhenius equation	아레니우스식	-式
Arrhenius expression	아레니우스방정식	-方程式
arrowhead chart	화살표머리 도표	-圖表
arroyo	아로요(물이 마른 수로)	
artesian aquifer	피압대수층	被壓帶水層
artesian basin	찬정분지	鑽井盆地
artesian ground water	자분지하수	自噴地下水
artesian well	자분우물, 자분정	自噴井
artificial assistance	인공보조	人工補助
artificial chemiluminescence cloud	인공발광운	人工發光雲
artificial climate	인공기후	人工氣候
artificial cloud	인공구름	人工-
artificial control	인공조절	人工調節
artificial heat	인공열	人工熱
artificial horizon	인공수평선	人工水平線
artificial ice nucleus	인공빙정핵	人工氷晶核
artificial increasing of rain	인공증우	人工增雨
(= artificial rain increasing)		
artificial intelligence	인공지능	人工知能
artificial intelligence system	인공지능시스템, 인공지능계	人工知能系
artificially induced rainfall	인공강우	人工降雨
artificial neural network	인공신경망	人工神經網
artificial nucleation	인공핵생성	人工核生成
artificial nucleus	인공핵	人工核
artificial precipitation	인공강수	人工降水
artificial radioactivity	인공방사능	人工放射能
artificial rain enhancement	인공증우	人工增雨
artificial rainfall	인공강우	人工降雨
artificial recharge	인공재충전	人工再充電
artificial satellite	인공위성	人工衛星
artificial snow enhancement	인공증설	人工增設
artificial snowfall	인공강설	人工降雪

asbestos	석면	石綿
asbestosis	석면침착증	石綿沈着症
ascendent	상승도	上昇度
ascending air	상승공기	上昇空氣
ascending air current	상승기류	上昇氣流
ascending condensation level	상승응결고도	上昇凝結高度
ascending motion	상승운동	上昇運動
ascending node	승교점	昇交點
ascending pass	상승궤도	上昇軌道
ascension (hydrothermal) theory	상승(열수)설	上昇(熱水)說
ascent	상승	上昇
ascent curve	상승곡선	上昇曲線
ascent of air	공기상승	空氣上昇
ash	재	灰
Ashchurch experiment	(잉글랜드)애시처치실험	-實驗
ash devils	(화산)재 회오리	(火山)灰-
ash fall	낙회	落灰
ash-grey light	회색광	灰色光
Asia	아시아	亞細亞
Asia flood	아시아홍수	亞細亞洪水
Asian Brown Cloud	아시아갈색연무	亞細亞褐色煙霧
Asian cold wave	아시아한파	亞細亞寒波
Asian dust phenomenon	황사현상	黃砂現象
Asian monsoon	아시아몬순	亞細亞-
Askesian Society	아스케시안학회	-學會
aspect	경사방위	傾斜方位
aspect ratio	종횡비	縱橫比
aspirated hygrometer	통풍습도계	通風濕度計
aspirated thermometer	통풍온도계	通風溫度計
aspiration condenser	통풍콘덴서	通風-
aspiration meteorograph	통풍기상기록계	通風氣象記錄計
aspiration psychrometer	통풍건습계	通風乾濕計
aspiration thermograph	통풍자기온도계	通風自記溫度計
aspiration thermometer	통풍온도계	通風溫度計
aspirator	통풍기	通風器
assessment	평가	評價
assimilation	동화작용, 흡수	同化作用, 吸收
assimiliation of carbon	탄소동화	炭素同化
assisted migration	보조이주	補助移住
Assmann psychrometer	아스만건습계	-乾濕計
Assmann's aspiration psychrometer	아스만통풍건습계	-通風乾濕計
Assmann ventilated psychrometer	아스만통풍건습계	-通風乾濕計
associative law	결합법칙	結合法則
astatic	불안정한, 무정위(의)	不安定-, 無定位

asteroid	소행성	小行星
asteroid extinction hypothesis	소행성멸종가설	小行星滅種假設
asteroid impact	소행성충돌	小行星衝突
asthenosphere	연약권	軟弱圈
asthma	천식	喘息
astraphobia	천둥번개공포	-恐怖
astrapophobia	천둥번개공포증	-恐怖症
astrometerorology	천체기상학	天體氣象學
astronautics	우주항행학	宇宙航行學
astronomical day	천문일	天文日
astronomical forcing	천문강제	天文强制
astronomical horizon	천문지평선	天文地平線
astronomical latitude	천문위도	天文緯度
astronomical periodicities	천문주기성	天文週期性
astronomical refraction	천문굴절	天文屈折
astronomical scintillation	천문섬광	天文閃光
astronomical seeing	천문시상	天文視像
astronomical theory	천문이론	天文理論
astronomical tide	천문조석	天文潮汐
astronomical twilight	천문박명	天文薄明
astronomical unit	천문단위	天文單位
astronomy	천문학	天文學
Aswan Dam	아스완댐	
asymmetrical center	비대칭중심	非對稱中心
asymmetric circulation pattern	비대칭순환패턴	非對稱循環-
asymmetric factor	비대칭인자	非對稱因子
asymmetry	비대칭(성)	非對稱(性)
asymptote	점근선	漸近線
asymptote of convergence	수렴점근선	收斂漸近線
asymptote of divergence	발산점근선	發散漸近線
asymptotic expansion	점근전개	漸近展開
asymptotic limit	점근극한	漸近極限
asymptotic matching	점근조화	漸近調和
asymptotic mixing length	점근혼합길이	漸近混合-
asymptotic observation	점근관측	漸近觀測
asymptotic point	점근점	漸近點
asymptotic similarity theory	점근상사이론	漸近相似理論
asynchronous communication	비동기통신	非同期通信
asynchronous coupling	비동기결합	非同期結合
asynoptic data	비종관(기상)자료	非綜觀(氣象)資料
asynoptic meteorological observation	비종관기상관측	非綜觀氣象觀測
asynoptic observation	비종관관측	非綜觀觀測
Atacama Desert	아타카마사막	-砂漠
athermancy	불투열성	不透熱性

athermous	불투열성(의)	不透熱性
Athos fall wind	아토스 치내리바람(그리스 아토스산맥 부근에서 부는 바람)	
Atlantic basin	대서양분지	大西洋盆地
Atlantic blocking episode	대서양블로킹에피소드	大西洋-
Atlantic conveyor	대서양컨베이어	大西洋-
Atlantic depression	대서양저기압	大西洋低氣壓
Atlantic high	대서양고기압	大西洋高氣壓
Atlantic hurricane	대서양허리케인	大西洋-
Atlantic multidecadal oscillation	대서양십년단위진동	大西洋十年單位振動
Atlantic time	대서양시간	大西洋時間
atlas	지도책	地圖冊
atmidometer	증발계	蒸發計
atmidometry	증발측정법	蒸發測定法
atmidoscope	습도지시기	濕度指示器
atmology	수증기학	水蒸氣學
atmometer	증발계	蒸發計
atmometry	증발측정법	蒸發測定法
atmophile	친기	親氣
atmoradiograph	공전빈도기록계	空電頻度記錄計
atmosphere	대기, 대기권, 대기압	大氣, 大氣圈, 大氣壓
atmosphere absorption	대기흡수	大氣吸收
atmosphere attenuation	대기감쇠	大氣減衰
atmosphere-mixed layer ocean model	대기-혼합층 해양모형	大氣混合層海洋模型
atmosphere-ocean general circulation model	대기-해양 대순환모형	大氣海洋大循環模型
atmosphere-ocean interaction	대기-해양 상호작용	大氣海洋相互作用
atmospheric	대기(의)	大氣
atmospheric absorption	대기흡수	大氣吸收
atmospheric acoustics	대기음향학	大氣音響學
atmospheric aerosol	대기에어로졸	大氣-
atmospheric attenuation	대기감쇠	大氣減衰
atmospheric barometer	기압계	氣壓計
atmospheric billow	대기물결	大氣-
atmospheric blocking	대기블로킹	大氣-
atmospheric boil	아지랑이	
atmospheric boundary layer	대기경계층	大氣境界層
atmospheric cell	대기세포	大氣細胞
atmospheric chemistry	대기화학	大氣化學
atmospheric chemistry of combustion emission	연소배출 대기화학	燃燒排出大氣化學
atmospheric circulation	대기순환	大氣循環
atmospheric circulation model	대기순환모형	大氣循環模型
atmospheric cloud	대기구름	大氣-
atmospheric complexity	대기복잡성	大氣複雜性
atmospheric composition	대기조성	大氣組成

atmospheric conductivity	대기전도도	大氣傳導度
atmospheric correction	대기보정	大氣補正
atmospheric counter radiation	하향대기복사	下向大氣輻射
atmospheric data	대기자료	大氣資料
atmospheric demand	대기수요량	大氣需要量
atmospheric density	대기밀도	大氣密度
atmospheric deposition	대기침적	大氣沈積
atmospheric deposition networks	대기침적망	大氣沈積網
atmospheric diffusion	대기확산	大氣擴散
atmospheric diffusion equation	대기확산방정식	大氣擴散方程式
atmospheric disaster	대기질병	大氣疾病
atmospheric dispersion	대기분산	大氣分散
atmospheric disturbance	대기요란	大氣擾亂
atmospheric drought	대기건조	大氣乾燥
atmospheric dust	대기먼지	大氣-
atmospheric dynamics	대기역학	大氣力學
atmospheric eddy	대기맴돌이	大氣-
atmospheric effect	대기효과	大氣效果
atmospheric electric field	대기전기장	大氣電氣場
atmospheric electricity	기상전기학, 대기전기	氣象電氣學, 大氣電氣
atmospheric emission	대기방사	大氣放射
atmospheric emissivity	대기방출률	大氣放出率
atmospheric energetics	대기에너지론	大氣-論
atmospheric energy balance	대기에너지균형	大氣-均衡
atmospheric engine	대기열기관	大氣熱機關
atmospheric equator	대기적도	大氣赤道
atmospheric evolution	대기진화	大氣進化
atmospheric extinction	대기소산, 대기감쇄	大氣消散, 大氣減殺
atmospheric flow	대기흐름	大氣-
atmospheric general circulation model	대기대순환모형	大氣大循環模型
atmospheric geology	대기지질학	大氣地質學
atmospheric gravity wave	대기중력파	大氣重力波
atmospheric gravity wind	대기중력풍	大氣重力風
atmospheric hazard	대기위험	大氣危險
atmospheric heat budget	대기열수지	大氣熱收支
atmospheric impurities	대기불순물	大氣不純物
atmospheric influences	대기영향	大氣影響
atmospheric infrared sounder	대기연직구조적외탐측기	大氣鉛直構造赤外探測機
atmospheric instability	대기불안정(도)	大氣不安定(度)
atmospheric interference	대기간섭	大氣干涉
atmospheric ion	대기이온	大氣-
atmospheric ionization	대기이온화	大氣-化
atmospheric lapse rate	대기감률	大氣減率
atmospheric layer	대기층	大氣層

atmospheric lifetime	대기수명	大氣壽命
atmospheric mass	대기질량	大氣質量
atmospheric mesoscale wet season campaign	대기중규모 우기 캠페인	大氣中規模雨期-
atmospheric metamorphism	대기변질	大氣變質
atmospheric mixing	대기혼합	大氣混合
atmospheric model	대기모형	大氣模型
atmospheric model intercomparison project	대기모형상호비교프로젝트	大氣模型相互比較-
atmospheric moisture	대기수분	大氣水分
atmospheric motion	대기운동	大氣運動
atmospheric noise	대기잡음	大氣雜音
atmospheric opacity	대기불투명도	大氣不透明度
atmospheric optics	대기광학	大氣光學
atmospheric oscillation	대기진동	大氣振動
atmospheric oxidation	대기산화	大氣酸化
atmospheric ozone	대기오존	大氣-
atmospheric particle	대기입자	大氣粒子
atmospheric particulate matter	부유입자상물질	浮遊粒子狀物質
atmospheric path length	대기경로길이	大氣經路-
atmospheric phenomenon	대기현상	大氣現象
atmospheric physics	대기물리학	大氣物理學
atmospheric polarization	대기분극(편광)	大氣分極
atmospheric pollutants	대기오염물질	大氣汚染物質
atmospheric pollution	대기오염	大氣汚染
atmospheric pressure	대기압	大氣壓
atmospheric pressure of sea level	해면기압	海面氣壓
atmospheric properties	대기성질	大氣性質
atmospheric radiance module	대기복사모듈	大氣輻射-
atmospheric radiation	대기복사	大氣輻射
atmospheric radiation budget	대기복사수지	大氣輻射收支
atmospheric refraction	대기굴절	大氣屈折
atmospheric region	대기지역	大氣地域
Atmospheric Research and Environment Program	대기 및 환경연구프로그램	大氣環境研究計劃
atmospheric resource management	대기자원관리	大氣資源管理
atmospheric retrieval problem	대기복원문제	大氣復元問題
atmospheric river	대기천, 하늘강	大氣川
atmospheric scales	대기규모	大氣規模
atmospheric scattering	대기산란	大氣散亂
atmospheric scattering body	대기산란체	大氣散亂體
atmospheric scavening effect	대기세정효과	大氣洗淨效果
atmospheric sciences	대기과학	大氣科學
atmospheric sciences in complex terrain program	복잡지형대기과학프로그램	複雜地形大氣科學-
atmospheric scientific(meteorological) temperature	대기과학(기상학)온도	大氣科學(氣象學)溫度
atmospheric shell	대기각, 대기층	大氣殼, 大氣層
atmospheric shimmer	아지랑이	

A

atmospheric sounding projectile	대기탐측로켓	大氣探測-
atmospherics	공전, 공전학	空電, 空電學
atmospherics rocket observation	대기로켓관측	大氣-觀測
atmospheric stability	대기안정도	大氣安定度
atmospheric statics	대기정역학	大氣靜力學
atmospheric stratification	대기성층	大氣成層
atmospheric structure	대기구조	大氣構造
atmospheric studies in complex terrain	복잡지형대기연구	複雜地形大氣研究
atmospheric surface layer	대기지표층	大氣地表層
atmospheric system	대기시스템	大氣-
atmospheric teleconnection	대기원격상관	大氣遠隔相關
atmospheric temperature	기온, 대기온도	氣溫, 大氣溫度
atmospheric temperature profile	기온연직분포	氣溫鉛直分布
atmospheric thermodynamics	대기열역학	大氣熱力學
atmospheric tide	대기조석	大氣潮汐
atmospheric trace molecular spectroscopy	대기추적분자형광분광술	大氣追跡分子螢光分光術
atmospheric tracer	대기미량원소	大氣微量元素
atmospheric transmission	대기투과	大氣透過
atmospheric transport	대기수송	大氣輸送
atmospheric turbidity	대기혼탁도	大氣混濁度
atmospheric turbulence	대기난류	大氣亂流
atmospheric-turbulence wind tunnel	대기난류풍동	大氣亂流風洞
atmospheric urban heat island	대기도시열섬	大氣都市熱島
atmospheric variables	대기변수	大氣變數
atmospheric vertical structure	대기연직구조	大氣鉛直構造
atmospheric vortex	대기소용돌이	大氣-
atmospheric water budget	대기물수지	大氣-收支
atmospheric wave	대기파	大氣波
atmospheric wave motion	대기파동	大氣波動
atmospheric window	대기창	大氣窓
atom	원자	原子
atomic bomb tests	핵폭탄실험	核爆彈實驗
atomic crystal	원자결정	原子結晶
atomic mass	원자질량	原子質量
atomic model	원자모형	原子模型
atomic number	원자번호	原子番號
atomic oxygen	산소원자	酸素原子
atomic structure	원자구조	原子構造
atomic symbol	원자기호	原子記號
atomic weight	원자량	原子量
attached thermometer	부착온도계	附着溫度計
attachment	부착	附着
attended station	유인관측소	有人觀測所
attenuation	감쇠	減衰

attenuation coefficient	감쇠계수	減衰係數
attenuation constant	감쇠상수	減衰常數
attenuation correction	감쇠보정	減衰補正
attenuation error	감쇠오류	減衰誤謬
attenuation factor	감쇠인자	減衰因子
attenuation length	감쇠길이	減衰-
attenuation of solar radiation	태양복사감쇠	太陽輻射減衰
attenuator	감쇠기	減衰器
attitude	방향, 자세	方向, 姿勢
attitude of satellite	위성자세	衛星姿勢
attraction	흡인, 인력	吸引, 引力
attractive potential	인력퍼텐셜	引力-
attractor	끌개, 흡인자	吸引子
audibility	가청도	可聽度
audibility meter	청도계	聽度計
audibility zone	가청역	可聽域
audio frequency band	가청진동수역	可聽振動數域
audio-modulated radiosonde	가청주파수변조라디오존데	可聽周波數變調-
Auger shower	오거소나기	
Aulacoseira granulata	규조류	硅藻類
aureole	주변광	周邊光
aurora	극광	極光
aurora arc	극광호	極光弧
aurora australis	남극광	南極光
aurora bands	극광대	極光帶
aurora borealis	북극광	北極光
aurora corona	극광코로나	極光-
aurora draperies (or curtains)	극광장막(커튼)	極光帳幕
aurora image	오로라영상	-映像
auroral bands	극광대	極光帶
auroral corona	극광코로나	極光-
auroral curtains	극광장막	極光帳幕
auroral draperies	극광장막	極光帳幕
auroral electrojet	극광전자류	極光電子流
auroral green line	극광초록선	極光草錄線
auroral hiss	극광음	極光音
auroral oval	극광타원	極光楕圓
auroral rays	극광선	極光線
auroral storm	극광스톰	極光-
auroral zone	극광대	極光帶
aurora polaris	북극광	北極光
aurora ray	극광선	極光線
aurora spectrum	극광스펙트럼	極光-
aurora streamer	극광스트리머	極光-

A

aurora zone	극광지대	極光地帶
Austalasian mediterranean water	호아지중해수	濠亞地中海水
Austausch	오스타쉬, 교환	交換
Austausch coefficient	오스타쉬 계수, 교환계수	交換係數
Australian Bureau of Meteorology	오스트레일리아 기상청	-氣象廳
Australian monsoon	오스트레일리아 몬순	
Australian outback	오스트레일리아 미개척지	-未開拓地
Australian summer monsoon	오스트레일리아 여름몬순	
austru	오스트루〈뛴의 루마니아어〉	
authorized station	정규관측소	正規觀測所
auto-alarm	자동경보장치	自動警報裝置
autobarotropic	자동순압	自動順壓
autobarotropic atmosphere	자동순압대기	自動順壓大氣
autobarotropy	자동순압(의)	自動順壓
auto-convariance	자기공분산	自己共分散
autoconvection	자동대류	自動對流
autoconvection gradient	자동대류경도	自動對流傾度
autoconvective instability	자동대류불안정	自動對流不安定
autoconvective lapse rate	자동대류감률	自動對流減率
autoconversion	자동변환	自動變換
autocorrection	자동교정	自動校正
autocorrelation	자기상관	自己相關
autocorrelation coefficient	자기상관계수	自己相關係數
autocorrelation function	자기상관함수	自己相關函數
autocovariance function	자기공분산함수	自己共分散函數
autogenous electrification	자생대전	自生帶電
autographic record	자기기록	自己記錄
autoionization	자동이온화반응	自動-化反應
automated balloon launcher	자동기구비양장치	自動氣球飛揚裝置
automated search	자동검색	自動檢索
automated shipboard aerological program	선박용 자동고층관측 프로그램	船舶用自動高層觀測-
automated surface observing system	자동지상관측시스템	自動地上觀測-
automatic climatological station	자동기후관측소	自動氣候觀測所
automatic data editing and switching system	자동자료편집시스템	自動資料編輯-
automatic data exchange system	자동자료교환시스템	自動資料交換-
automatic data processing	자동자료처리	自動資料處理
automatic error request equipment	자동오차교정장치	自動誤差校正裝置
automatic evaporation station	자동증발관측소	自動蒸發觀測所
automatic frequency control	자동진동수제어, 자동주파수제어	自動振動數制御, 自動周波數制御
automatic gain control	자동이득제어	自動利得制御
automatic meteorological observing station	자동기상관측소	自動氣象觀測所
automatic meteorological oceanographic buoy	자동해양기상관측부이	自動海洋氣象觀測-

automatic multifrequency ionospheric recorder	자동다중주파수이온권기록계	自動多重周波數-圈記錄計
automatic picture transmission	자동화상운송	自動畵像運送
automatic radiometeorograph	자동무선기상계기	自動無線氣象計器
automatic radio rain gauge	자동무선우량계	自動無線雨量計
automatic rain recording gauge	자기우량계	自己雨量計
automatic recorder	자동기록계	自動記錄計
automatic sensibility control	자동감도조절	自動感度調節
automatic snow meter	자동적설계	自動積雪計
automatic standard magnetic observatory	자동표준자기관측소	自動標準磁氣觀測所
automatic standard magnetic observatory-remote	자동식표준원격자기관측소	自動式標準遠隔磁氣觀測所
automatic station keeping	자동관측소관리	自動觀測所管理
automatic tide gauge	자동검조기	自動檢潮器
automatic tracking	자동추적	自動追跡
automatic weather observation system	자동기상관측시스템	自動氣象觀測-
automatic weather station	자동기상관측소	自動氣象觀測所
Automatic Weather System	자동기상관측장비	自動氣象觀測裝備
automobile exhaust	자동차배기가스	自動車排氣-
automobile number	자동차대수	自動車臺數
automotive weather observation system	차량형기상관측시스템	車輛型氣象觀測-
autoregression method	자기회귀법	自己回歸法
autoregressive model	자기회귀모형	自己回歸模型
autoregressive process	자기회귀과정	自己回歸過程
autoregressive series	자기회귀수열	自己回歸數列
autosonde	오토존데, 자동존데	自動-
autumnal equinox	추분	秋分
autumn changma	가을장마	
autumn equinoctial period	추분기	秋分期
autumn equinox tide	추분조석	秋分潮汐
autumn ice	가을얼음	
auvergnasse	오베르나제(프랑스 중부 산간지대의 찬 북서풍)	
auxiliary agricultural meteorological station	보조농업기상관측소	補助農業氣象觀測所
auxiliary chart	보조일기도	補助日氣圖
auxiliary data	보조자료	補助資料
auxiliary(adjoint) line	보조선	補助線
auxiliary ship	보조선박	補助船舶
auxiliary ship station	보조선박관측소	補助船舶觀測所
auxiliary station	보조관측소	補助觀測所
auxiliary thermometer	보조온도계	補助溫度計
available energy	가용에너지	可用-
available head	가용수두	可用水頭
available moisture	유효수분	有效水分
available moisture of the soil	가용토양수분	可用土壤水分
available potential energy	가용위치에너지	可用位置-
available soil moisture	가용토양수분	可用土壤水分

available solar radiation	가용·태양복사	可用太陽輻射
available storage capacity	가용·저장용량	可用貯藏容量
available water	가용수분	可用水分
available water capacity	유효수분함량	有效水分含量
avalanche	눈사태, 사태	-沙汰
avalanche blast	사태돌풍	沙汰突風
avalanche path	설붕도	雪崩道
avalanche wind	사태바람	沙汰-
avationroutine meteorological report	항공기상실황전문	航空氣象實況電文
average	평균, 평균값	平均
average acceleration	평균가속도	平均加速度
average annual precipitation	연평균강수량	年平均降水量
average departure	평균편차	平均偏差
average deviation	평균편차	平均偏差
average dissipation rate	평균소산율	平均消散率
average drag coefficient	평균항력계수	平均抗力係數
average error	평균오차	平均誤差
average interstitial velocity	평균틈새속도	平均-速度
average kinetic energy	평균운동에너지	平均運動-
average net balance	평균순균형	平均純均衡
average power	평균전력	平均電力
average rainfall	평균우량	平均雨量
average(mean) rate of change	평균변화율	平均變化率
average sea level pressure	평균해수면기압	平均海水面氣壓
average speed	평균속력	平均速力
average value	평균값	平均-
average variability	평균변동률	平均變動率
average wind speed	평균풍속	平均風速
averaging	평균화	平均化
averaging kernel	평균커널, 평균핵	平均核
averaging time	평균화시간	平均化時間
AVHHR imagery	AVHHR 영상	-映像
avian malaria	닭 말라리아	
aviation	항공	航空
aviation climatology	항공기후학	航空氣候學
aviation forecast	항공예보	航空豫報
aviation forecast zone	항공예보구	航空豫報區
aviation medicine	항공의학	航空醫學
Aviation Meteorological Agency	항공기상청	航空氣象廳
aviation meteorological element	항공기상요소	航空氣象要素
aviation meteorological message	항공기상전문	航空氣象電文
aviation meteorological service	항공기상업무	航空氣象業務
aviation meteorology	항공기상학	航空氣象學
aviation observation	항공관측	航空觀測

A

aviation turbulence	항공난류	航空亂流
aviation weather	항공기상	航空氣象
aviation weather forecast	항공기상예보	航空氣象豫報
aviation weather information	항공기상정보	航空氣象情報
aviation weather observation	항공기상관측	航空氣象觀測
Avogadro constant	아보가드로상수	-常數
Avogadro's hypothesis	아보가드로가설	-假說
Avogadro's law	아보가드로법칙	-法則
Avogadro's number	아보가드로수	-數
avre	아브레(겨울철의 따뜻한 바람과 여름철의 서늘한 바람이 교대로 부는 바람의 프랑스 방언)	
avulsion	적출	摘出
awakening of insects	경칩	驚蟄
aweather	바람맞이의	
axially symmetric flow	축대칭흐름	軸對稱-
axial precession	축세차	軸歲差
axial symmetry	축대칭(의)	軸對稱
axial tilt	축경사	軸傾斜
axis	축	軸
axis of anticyclone	고기압의 축	高氣壓軸
axis of contraction	수축축	收縮軸
axis of coordinate	좌표축	座標軸
axis of depression	저기압축	低氣壓軸
axis of dilatation	팽창축	膨脹軸
axis of earth	지축	地軸
axis of inflow	유입축	流入軸
axis of low	저기압축	低氣壓軸
axis of outflow	유출축	流出軸
axis of ridge	기압마루축	氣壓-軸
axis of trough	기압골축	氣壓-軸
axisymmetric	축대칭	軸對稱
axisymmetric flow	축대칭류, 축대칭흐름	軸對稱流
axisymmetric hill	축대칭언덕	軸對稱丘
axisymmetric turbulence	축대칭난류	軸對稱亂流
ayalas	아얄라스(프랑스 중부 산간지대에서 남동쪽으로부터 불어오는 대륙성 열풍)	
azel-scope	방위-고도지시기	方位高度指示器
azimuth	방위각	方位角
azimuthal perturbation	방위섭동	方位攝動
azimuthal resolution	방위분해능	方位分解能
azimuth angle	방위각	方位角
azimuth distortion	방위왜곡	方位歪曲
azimuth elevation	방위고도각	方位高度角
azimuth error	방위오차	方位誤差

azimuth marker	방위표지	方位標識
azimuth resolution	방위해상도, 방위분해능	方位解像度, 方位分解能
azimuth scale	방위눈금	方位-
azimuth wavenumber	방위각파수	方位角波數
Azores anticyclone	아조레스고기압	-高氣壓
Azores current	아조레스해류	-海流
Azores high	아조레스고기압	-高氣壓
azran	방위거리	方位距離

B, b

영문	한글	한자
Babinet's point	바비네중립점	-中立點
Babinet's principle	바비네원리	-原理
back-bent occlusion	후굴폐색	後屈閉塞
back-bent warm front	후굴온난전선	後屈溫暖前線
backdoor cold front	뒷문한랭전선	後門寒冷前線
back flow	역류	逆流
background air pollution monitoring network	배경대기오염관측망	背景大氣汚染觀測網
background contrast	배경대비	背景對比
background error covariance	배경오차공분산	背景誤差共分散
background field	배경장	背景場
background level	배경수준, 바탕수준	背景水準
background luminance	배경휘도	背景輝度
background pollution	배경오염, 바탕오염	背景汚染
background radiation	배경복사, 바탕복사	背景輻射
background signal	배경신호	背景信號
backing	반전	反轉
backing wind	반전바람	反轉-
backlash	헛돌음	
backlobe	(안테나)후엽	後葉
backpropagation	후방전파	後方傳播
back radiation	역복사	逆輻射
backscatter	후방산란체	後方散亂體
backscattered depolarization technique	후방산란편극 소거기법	後方散亂偏極消去技法
backscattered signal	후방산란신호	後方散亂信號
back-scattering	후방산란	後方散亂
back-scattering coefficient	후방산란계수	後方散亂係數
back-scattering cross-section	후방산란단면	後方散亂斷面
back-scattering differential phase	후방산란차등위상	後方散亂差等位相
back-scattering parameter	후방산란인자	後方散亂因子

back-scattering phase shift	후방산란위상변이	後方散亂位相變移
back-scattering ultraviolet radiometer	후방산란자외선복사계	後方散亂紫外線輻射計
back-scattering ultraviolet spectrometer	후방산란자외선분광계	後方散亂紫外線分光計
backscatter to extinction ratio	소광대후방산란비	消光對後方散亂比
back-sheared anvil	후방층밀린모루	後方層-
backstay of the sun	부챗살 빛	
backtracking	되추적	-追跡
backward chaining	역방향추론	逆方向推論
backward difference	후방차분	後方差分
backward differencing	후방차분법	後方差分法
backward equation	후방방정식	後方方程式
backward radiation	후방복사	後方輻射
backward scatter	후방산란자	後方散亂子
backward scattering	후방산란	後方散亂
backward scattering coefficient	후방산란계수	後方散亂係數
backwash	역류세정, 역세정	逆流洗淨, 逆洗淨
backwater	역수	逆水
backwater curve	역수곡선	逆水曲線
Baffin current	배핀해류	-海流
Baffin Island	배핀아일랜드	
baffle	정류벽	整流壁
bag filter	백필터	
baghouse	백하우스, 푸대집	
baguio(= bagio, bagyo, vaguio, vario)	바기오(필리핀에 내습하는 열대 저기압, 태풍의 지방적 호칭)	
bai	매(황사)	霾
Baiu	매우, 장마, 바이우	梅雨
Baiu front	바이우전선	-前線
balance	균형	均衡
balanced approximation model	평형근사모형	平衡近似-
balanced force	균형힘, 균형력	均衡力
balanced model	균형모형	均衡模型
balanced wind	균형바람	均衡-
balance equation	균형방정식	均衡方程式
balance height(of aerostat)	(기구의)균형고도	均衡高度
balance meter	복사수지계	輻射收支計
balance of solar radiation	태양복사수지	太陽輻射收支
balancer	균형기	均衡器
balance system	균형계	均衡系
balance year	균형년	均衡年
bali	발리(이탈리아 가르다 호의 호수바람)	
Bali wind	발리바람(자바 동부에서 부는 강한 동풍의 일종)	
ball ice	둥근 얼음	
balling	구형화	球形化

B

ballistic density	탄도밀도	彈道密度
ballistics	탄도학	彈道學
ballistics temperature	탄도온도	彈道溫度
ballistics wind	탄도바람	彈道-
ballistic wind	탄도풍	彈導風
ball lightning	구상번개	球狀-
ballonet	보조기낭	補助氣囊
	(기구·비행선의 부력조정용)	
ballonet ceiling	기구상승한계	氣球上昇限界
balloon	기구	氣球
balloon basket	기구바구니	氣球-
balloon ceiling	기구실링	氣球-
balloon cover	기구커버	氣球-
balloon drag	견인기구	牽引氣球
balloon measurement	기구측정	氣球測定
balloon observation	기구관측	氣球觀測
balloon shroud	기구덮개	氣球-
balloonsonde	탐측기구	探測氣球
balloon sounding	기구탐측	氣球探測
balloon theodolite	측풍경위(의)	測風經緯
balloon trajectory	기구추적	氣球追跡
balloon transit	측풍트랜짓	測風-
ball pyranometer	구형일사계	球形日射計
ballute(= ball + parachute)	발류트	
Baltic region	발트지역	-地域
Baltic Sea	발트해	-海
band	띠	
band absorption	띠흡수	-吸收
band analysis	대역분석	帶域分析
Banda Sea Water	(인도네시아)반다해수	-海水
band cirrus	띠권운	-卷雲
band designation	밴드지정	-指定
banded precipitation	띠강수	-降水
banded structure	띠구조	-構造
band gap	띠간격	-間隔
banding effect	띠효과	-效果
banding eye	태풍의 눈(직경이 40km	颱風-
	이내 크기의 눈)	
band graph	띠그래프	
band lightning	띠번개	
band method	밴드방법	-方法
band model	밴드모형	-模型
band of cloud	구름띠	
band of precipitation	강수띠	降水-

B

band pass filter	대역 통과 필터, 대역 통과 여과기	帶域通過濾過器
band pattern	밴드패턴, 띠모양	-模樣
band spectrum	띠스펙트럼	
band width	띠너비, 띠폭	帶幅
ban-gull	반걸(스코틀랜드의 여름 해풍)	
bank	둑, 해안, 하안	海岸, 河岸
bank fog	뚝모양안개	
bankful stage	만조수위	滿槽水位
banking of the current	기류퇴적(현상)	氣流堆積
banking process(= pile-up process)	퇴적과정	堆積過程
bank storage	하안저류	河岸貯流
banner cloud	깃발구름	
Baquios	바퀴오스(남반구의 태풍)	
bar	바(기압의 단위)	
barb	깃가지(풍속기호)	
Barbados oceanographic and meteorological experiment	바베이도스해양기상실험	-海洋氣象實驗
barbed arrow	바람표시화살	-表示-
barber	악폭풍	惡暴風
barchan	바르한(초생달 모양)	
barchan dune	바르한사구	-砂丘
bare ice	맨얼음	
Barents ice sheet	바렌츠빙상	-氷床
bare soil	맨땅, 나지	裸地
baric analysis	기압분석	氣壓分析
baric area	기압구역	氣壓區域
baric tendency	기압경향	氣壓傾向
baric topography	기압높낮이	氣壓-
baric wind law	기압바람법칙	氣壓-法則
barih	바리	
barium fluoride film hygrometer	불화바륨박막습도계	弗化-薄膜濕度計
barker beetle	껍질딱정벌레	
barkhan(= barchane, barchan)	바르칸(바람에 의해 생긴 초승달 모양의 눈쌓임 혹은 모래언덕)	
barocline	경압	傾壓
barocline state	경압상태	傾壓狀態
baroclinic	경압(의)	傾壓
baroclinic atmosphere	경압대기	傾壓大氣
baroclinic condition	경압조건	傾壓條件
baroclinic convection feedback loop	경압성대류 되먹임고리	傾壓性對流-
baroclinic development model	경압발달모형	傾壓發達模型
baroclinic disturbance	경압요란	傾壓搖亂
baroclinic effect	경압효과	傾壓效果

baroclinic field	경압장	傾壓場
baroclinic flow	경압흐름	傾壓-
baroclinic fluid	경압유체	傾壓流體
baroclinic forecast	경압예보	傾壓豫報
baroclinic instability	경압불안정	傾壓不安定
baroclinicity	경압성	傾壓性
baroclinicity condition	경압성조건	傾壓性條件
baroclinic leaf cloud	경압성구름, 경압잎새구름	傾壓性-
baroclinic medium	경압매질	傾壓媒質
baroclinic model	경압모형	傾壓模型
baroclinic model atmosphere	경압모형대기	傾壓模型大氣
baroclinic motion	경압운동	傾壓運動
baroclinic planetary boundary layer	경압행성경계층	傾壓行星境界層
baroclinic term	경압항	傾壓項
baroclinic torque vector	경압토크벡터	傾壓-
baroclinic trough cloud pattern	경압골구름패턴	傾壓-
baroclinic wave	경압파	傾壓波
baroclinic zone	경압구역, 경압대	傾壓區域, 傾壓帶
baroclinity	경압성	傾壓性
baroclinity parameter	경압매개변수	傾壓媒介變數
baroclinity vector	경압벡터	傾壓-
barocliny	경압성	傾壓性
baro-diffusion	압력확산	壓力擴散
baro-diffusion coefficient	압력확산계수	壓力擴散係數
baro-diffusion ratio	압력확산비	壓力擴散比
barogram	기압기록	氣壓記錄
barograph	기압기록계	氣壓記錄計
barograph trace	자기기압기록	自記氣壓記錄
barometer	기압계	氣壓計
barometer box	기압계상자	氣壓計箱子
barometer case	기압계상자	氣壓計箱子
barometer cistern	기압계수은조	氣壓計水銀槽
barometer column	기압계수은주	氣壓計水銀柱
barometer corrections	기압계보정	氣壓計補正
barometer elevation	기압계표고	氣壓計標高
barometer maximum	기압최곳값	氣壓最高-
barometer reading	기압계읽기	氣壓計-
barometer reduction	기압경정	氣壓更正
barometric	기압(의)	氣壓
barometric altimeter	기압고도계	氣壓高度計
barometric altimetry	기압측고법	氣壓測高法
barometric characteristics	기압특성	氣壓特性
barometric column	기압기둥	氣壓柱
barometric constant	기압상수	氣壓常數

B

barometric correction	기압보정	氣壓補正
barometric correction table	기압계기차보정표	氣壓計器差補正表
barometric depression	저기압	低氣壓
barometric effect	기압효과	氣壓效果
barometric equation	기압방정식	氣壓方程式
barometric error	기압오차	氣壓誤差
barometric fluctuation	기압변동	氣壓變動
barometric formula	측고공식	測高公式
barometric gradient	기압경도	氣壓傾度
barometric height	기압고도	氣壓高度
barometric height formula	기압고도공식	氣壓高度公式
barometric high	고기압	高氣壓
barometric hypsometry(= barometric altimetry)	기압측고법	氣壓測高法
barometric law	기압법칙	氣壓法則
barometric low	저기압	低氣壓
barometric maximum	고(기)압, 기압최고값	高(氣)壓, 氣壓最高-
barometric mean temperature	측고평균온도	測高平均溫度
barometric minimum	저(기)압, 기압최저값	低(氣)壓, 氣壓最低-
barometric pressure	기압	氣壓
barometric rate	기압변화율	氣壓變化率
barometric reduction table	기압경정표	氣壓更正表
barometric ripple	기압미파	氣壓微波
barometric switch	기압스위치	氣壓-
barometric tendency	기압경향	氣壓傾向
barometric wave	기압파	氣壓波
barometry	기압측정법	氣壓測定法
baromil	바로밀	
baroswitch	기압스위치	氣壓-
barothermograph	자기기압온도계	自記氣壓溫度計
barothermohygrograph	자기기압온도습도계	自記氣壓溫度濕度計
barothermohygrometer	기압온습계	氣壓溫濕計
barotropic	순압(의)	順壓
barotropic atmosphere	순압대기	順壓大氣
barotropic boundary layer	순압경계층	順壓境界層
barotropic condition	순압조건	順壓條件
barotropic disturbance	순압요란	順壓搖亂
barotropic field	순압장	順壓場
barotropic fluid	순압유체	順壓流體
barotropic forecast	순압예보	順壓豫報
barotropic instability	순압불안정(성, 도)	順壓不安定(性, 度)
barotropic medium	순압매질	順壓媒質
barotropic model	순압모형	順壓模型
barotropic planetary boundary layer	순압행성경계층	順壓行星境界層
barotropic pressure function	순압압력함수	順壓壓力函數

B

barotropic stability	순압안정도	順壓安定度
barotropic state	순압상태	順壓狀態
barotropic steering	순압스티어링	順壓-
barotropic torque vector	순압토크벡터	順壓-
barotropic vorticity equation	순압소용돌이도 방정식	順壓-度方程式
barotropic wave	순압파	順壓波
barotropy	순압성	順壓性
barrage cloud	뚝구름	
barrens	불모지	不毛地
barrier	장벽	障壁
barrier berg	평판빙산	平板氷山
barrier iceberg	평판빙산	平板氷山
barrier jet	장벽제트	障壁-
barrier layer	장벽층	障壁層
barrier theory	장벽설	障壁說
barrier theory of cyclone	저기압장벽설	低氣壓障壁說
barycentric velocity	질량중심속도	質量中心速度
barye	바라이(기압의 단위, 1dyne/cm^2, 0.001mb)	
basal heating	밑면가열	-面加熱
basal ice	바다얼음, 빙하, 빙하기저	氷河, 氷河基底
basal metabolism	기초대사	基礎代謝
basal sliding	밑면사태	-面沙汰
basal slip	밑면미끄럼	-面-
base data	기초자료	基礎資料
base flow	기저유출	基底流出
baseflow recession	기저유출후퇴	基底流出後退
baseflow storage	기저유출저장	基底流出貯藏
base length(= base period)	기저시간	基底時間
base level of erosion	침식기선	浸蝕基線
baseline	기준선	基準線
baseline check	기준선점검	基準線點檢
baseline monitoring	기준선감시	基準線監視
base map	기초도, 백지도	基礎圖, 白地圖
base runoff	기저유출	基底流出
base station	기준관측소	基準觀測所
base width	저폭	底幅
basic computer model	기본컴퓨터모형	基本-模型
basic equation	기본방정식	基本方程式
basic equational system	기초방정식계	基礎方程式系
basic flow	기본흐름	基本-
basic intake rate	기본흡입률, 기본침투율	基本吸入率, 基本浸透率
basic(base) quantity	기준량	基準量
basic runway weather	기본활주로기상	簡易滑走路氣象
basic state	기본상태	基本狀態

basic surface radiation network	기본지표면복사네트워크 (세계기후연구프로그램)	基本地表面輻射觀測網
basic system	기본시스템	基本-
basin	유역	流域
basin accounting	유역계산	流域計算
basin(= catchment, catchment area, drainage area, watershed)	유역	流域
basin fog	분지안개, 분지무	盆地霧
basin lag	유역지체	流域遲滯
basin outlet	유역배출구	流域排出口
basin recharge	유역함양	流域涵養
basin response	유역반응	流域反應
basis function	기본함수	基本函數
basis vector	기본벡터	基本-
batch simulation	회분식모의	回分式模擬
bath plug vortex	욕탕플러그모양소용돌이	浴湯-模樣-
bathymetric chart	해저지형도	海底地形圖
bathymetry	측심법	測深法
bathythermograph	심해수온기록계	深海水溫記錄計
bathythermograph grid	심해수온기록계격자	深海水溫記錄計格子
bathythermograph print	심해수온기록계프린트	深海水溫記錄計-
bathythermograph slide	심해수온기록계슬라이드	深海水溫記錄計-
battle field billet weather	야전숙영기상	野戰宿營氣象
battlefield climate	전장기후	戰場氣候
baud	보드(전신신호 전송속도의 단위)	
baud rate	보드율	-率
bay ice	항만얼음	港灣-
Bay of Bengal Water	벵골만 해수	-灣海水
Baysian theorem	베이지안 정리	-定理
B-curve	B 곡선	-曲線
B-display	B 디스플레이	
beach ice	바닷가얼음	
beacon	비콘	
beaded lightning	구슬번개	
beam	빔	
beam angle	빔각	-角
beam blocking	빔블록킹(차폐)	-(遮蔽)
beam broadening	빔퍼짐	
beam divergence	빔발산	-發散
beam filling	빔채움, 빔충만	-充滿
beam filling meteorological target	빔충만 기상학적 표적	-充滿氣象學的標的
beam number	빔수	-數
beam pattern	빔패턴	
beamsplitter	빔(광)쪼개기	-(光)-

beam spreading	빔퍼짐, 빔분산	-分散
beam swinging	빔진동	-振動
beam wave	빔파	-波
beam width	빔폭	-幅
beam width distortion	빔폭 왜곡	-幅歪曲
beam wind	옆바람	
bearing	방위	方位
beat frequency	맥놀이횟수	-回數
beat frequency oscillator	맥놀이주파수발진기	-周波數發振器
Beaufort force	보퍼트풍력	-風力
Beaufort notation	보퍼트일기기호	-日氣記號
Beaufort number	보퍼트수	-數
Beaufort (wind) scale	보퍼트(풍력)계급	-(風力)階級
Beaufort weather notation	보퍼트기상표기	-氣象表記
Beaumont period	보먼트기간	-期間
beavertail antenna	비버꼬리안테나	
Beckmann thermometer	베크만온도	-溫度
becquerel	베크렐(방사능 피폭의 단위)	
Beekley gauge	비클리우량계	-雨量計
Beer-Bouguer law	베르-부게 법칙	-法則
Beer's law	베르법칙	-法則
beetles	초시류	鞘翅類
beginning of Changma	장마시작	-始作
beginning of flowering	개화시기	開花始期
behavioral model	행동모형	行動模型
bel	벨(음향강도측정단위, 10데시벨)	
Bellamy method	벨라미 방법	-方法
bellow	공합, 주름통	空盒, -桶
bell taper	벨 테이퍼	
below minimum flight weather condition	비행불가기상조건	飛行不可氣象條件
below minimums	최저조건미만	最低條件未滿
below normal	정상미만	正常未滿
below radar beam	레이더 빔 아래	
beltane-ree	벨탄-리	
belt of convergence	수렴대	收斂帶
belt of fluctuation	파동대	波動帶
Beltrami equation	벨트라미방정식	-方程式
Beltrami flow	벨트라미흐름	
Bemporad's formula	벰포라드공식	-公式
Bénard cell	베나르세포	-細胞
Bénard problem	베나르문제	-問題
Bénard-Rayliegh convection	베나르-레일리 대류	-對流
bench mark	수준점	水準點
bench mark station	수준점관측소	水準點觀測所

bending mode	굽힘방식	-方式
bending type lodging	만곡도복	彎曲倒伏
Benguela current	벵구엘라해류	-海流
bent-back occlusion	후굴폐색	後屈閉塞
benthic foraminifera	저서유공충	底棲有孔蟲
benthic habitats	저서서식지	底棲棲息地
Bentley Subglacial Trench	벤틀리 해구	-海溝
bent stem earth thermometer	곡관지중온도계	曲管地中溫度計
bent stem thermometer	곡관온도계	曲管溫度計
benzene	벤젠	
benzopyrene	벤조피렌	
Bergen Geophysical Institute	베르겐지구물리연구소	-地球物理研究所
Bergen School	베르겐학교(학파)	-學校(學派)
Bergeron	베르게론	
Bergeron Climatic Classification System	베르게론기후구분시스템	-氣候區分-
Bergeron confluence	베르게론합류	-合流
Bergeron effect	베르게론효과	-效果
Bergeron-Findeisen mechanism	베르게론-핀다이젠 메커니즘	
Bergeron-Findeisen process	베르게론-핀다이젠 과정	-過程
Bergeron-Findeisen theory	베르게론-핀다이젠 이론	-理論
Bergeron process	베르게론과정	-過程
berg wind	베르그바람, 산바람	山-
	(남아프리카 남해안 열풍의 일종)	
bergy bit	소빙산	小氷山
Bering current	베링해류	-海流
Bering Sea	베링해	-海
Bering slope current	베링경사류	-傾斜流
Bering Strait	베링해협	-海峽
Bermuda-Azores high	버뮤다-아조레스 고기압	-高氣壓
Bermuda high	버뮤다고기압	-高氣壓
Bernal-Fowler rule	버널-파울러 규칙	-規則
Bernoulli distribution	베르누이분포	-分布
Bernoulli effect	베르누이효과	-效果
Bernoulli mechanism	베르누이메커니즘	
Bernoulli principle	베르누이원리	-原理
Bernoulli's equation	베르누이방정식	-方程式
Bernoulli's theorem	베르누이정리	-定理
Berson winds (or westerlies)	베르손 서풍대	-西風帶
berylliosis	베릴륨중독증	-中毒症
beryllium(Be)	베릴륨(원자번호 4인 금속원소)	
beryllium copper	베릴륨구리	
Bessel equation	베셀방정식	-方程式
Bessel function	베셀함수	-函數
Besson comb nephoscope	베송 빗모양측운기	-模樣測雲器

Best Linear Unbiased Estimation	최량선형 비편향추정	最良線形非偏向推定
best number	최적수	最適數
best track	최적경로	最適經路
beta decay	베타붕괴, 베타감쇠	-崩壞, -減衰
beta drift	베타표류	-漂流
beta effect	베타효과	-效果
beta effectiveness	베타효과성, 베타유효	-效果性, -有效
beta gyre	베타환류, 베타자이어	-還流
beta particle	베타입자	-粒子
beta plane	베타평면	-平面
beta plane approximation	베타면근사, 베타평면근사	-面近似, -平面近似
beta ray(β-ray)	베타선	-線
beta spiral	베타나선	-螺線
Betts Miller scheme	BM 방안	-方案
between layers	층과 층 사이, 층간	層間
Bhaskara satellite	바스카라 위성	-衛星
bhoot	브후트	
bias	치우침, 편중도, 편의	偏重度, 偏倚
biased variance	편의분산	偏倚分散
bi-directional reflectance	양방향반사율	兩方向反射率
bi-directional reflectance distribution function	양방향반사도분포함수	兩方向反射度分布函數
bi-directional reflectance factor	양방향반사인자	兩方向反射因子
bi-directional reflection function	양방향반사함수	兩方向反射函數
bi-directional vane	양방향풍향계	兩方向風向計
bi-directional wind vane	양방향풍향계	兩方向風向計
biennial oscillation	이년진동	二年振動
biennial wind oscillation	이년바람진동	二年-振動
bifilar electrometer	두줄전위계	-電位計
bifurcation	분기, 분기점	分岐, 分岐點
bifurcation point	두갈래치기지점	-地點
Bigelow's evaporation formula	비겔로증발공식	-蒸發公式
big leaf model	큰잎모형	-模形
bijang	비장	比張
Bikini atoll	비키니환초	-環礁
bilateral symmetry	좌우대칭	左右對稱
Bilham screen	빌엄스크린(백엽상)	(百葉箱)
bilinear interpolation method	양선형보간법	兩線形補間法
billow cloud	물결구름	
billow wave	물결구름파동	-波動
bimetallic actinograph	바이메탈자기일사계	-自記日射計
bimetallic actinometer	바이메탈일사계	-日射計
bimetallic thermograph	바이메탈자기온도계	-自記溫度計
bimetallic thermometer	쌍금속온도계	雙金屬溫度計
bimetal strip	바이메탈판	-板

B

bimetal sunshine sensor	바이메탈일조계	-日照計
bimetal thermograph	쌍금속판자기온도계	雙金屬板自記溫度計
bimetal thermometer	바이메탈온도계	-溫度計
bimodal amplitude distribution	쌍봉우리 진폭분포	雙峰-振幅分布
bimodal distribution	쌍봉우리 분포	雙峰-分布
bimodal droplet distribution	쌍봉우리 작은물방울분포	雙峰-分布
bimodality	쌍봉성	雙峰性
bimodal spectrun	쌍봉우리 스펙트럼	雙峰-
bin	빈, 레이더 표본체적	-標本體積
binary class validation	이진분류검증	二進分類檢證
binary cyclone	쌍저기압	雙低氣壓
binary encoding classification	이진부호분류	二進符號分類
binary operation	이진연산	二進演算
binary-to-decimal converter	이진수-십진수 변환기	二進數十進數變換器
binary universal format representation	2진법범용형식표현,	二進法汎用型式表現,
	이진보편양식표기	二進普遍樣式表記
binary variable	이진변수	二進變數
binding energy	결합에너지	結合-
binomial coefficient	이항계수	二項係數
binomial distribution	이항분포	二項分布
binomial equation	이항방정식	二項方程式
binomial expression	이항식	二項式
binomial random variable	이항확률변수	二項確率變數
binomial theorem	이항정리	二項定理
bioaccumulation	생물누적	生物累積
bio-assay	생물검정	生物檢定
biochore	생태역	生態域
biocide	살생물제	殺生物劑
biocimatology	생(물)기후학	生(物)氣候學
bioclim	최적서식지역	最適棲息地域
bioclimate	생기후	生氣候
bioclimatic law	생기후법칙	生氣候法則
bioclimatics	생(물)기후학	生(物)氣候學
bioclimatic zone	생물기후대	生物氣候帶
bioclimatograph	생물기후도	生物氣候圖
bioclimatological data	생기후학 자료	生氣候學資料
biodiversity	생물다양성	生物多樣性
biodiversity hot spot	생물다양성열점	生物多樣性熱點
bioefflurnt	생체배기	生體排氣
biofilter	생물학적 필터	生物學的-
biofog	생물안개	生物-
biofuel	바이오연료	-燃料
biogas	바이오가스	
biogenesis	생물발생설	生物發生說

biogenic	자연적 발생(의)	自然的發生-
biogenic ice neucleus	생물성빙정핵	生物性氷晶核
biogenic pollutant	생물기원오염물질	生物起源汚染物質
biogenic sediment	생물기원퇴적물	生物起源堆積物
biogenic sedimentary rock	생물기원퇴적암	生物起源堆積巖
biogenic trace gas	생물기원미량기체	生物起源微量氣體
biogeochemical	생지화학	生地化學
biogeochemical cycle	생지화학순환	生地化學循環
biogeochemical flux	생지화학플럭스	生地化學-
biogeochemistry	생지화학	生地化學
biological aerosol	생물학적 에어로졸	生物學的-
biological climate	생물기후	生物氣候
biological contaminant	생물학적 오염물질	生物學的汚染物質
biological evolution	생물학적 진화	生物學的進化
biological minimum temperature	생물학적 최저온도	生物學的最低溫度
biological oxygen demand(BOD)	생물학적 산소요구량	生物學的酸素要求量
biological period	생물기간	生物期間
biological production	생물생산	生物生産
biological scatterer	생물산란체	生物散亂體
biological species	생물학적 종	生物學的種
biological warfare weather	생물학전쟁 기상	生物學戰爭氣象
biological yield	생물생산량	生物生産量
biological zero point	생물학적 영점	生物學的零點
biology	생물학	生物學
bioluminescence	생물발광	生物發光
biomagnification	생물농축	生物濃縮
biomass	생체량, 생물자원, 바이오매스	生體量, 生物資源
biomass burning	생물체량(바이오매스) 소각, 바이오매스 연소	生物體量燒却, -燃燒
biomass cookers	바이오매스 조리기구	-調理器具
biomass enhancement ratio	바이오매스 향상비	-向上比
biomass fuel	생물자원연료	生物資源燃料
biome	생물군계	生物群系
biome model	생물군계모형	生物群系模型
biometeorology	생물기상학	生物氣象學
bionics	생체공학	生體工學
biophenology	생물계절학	生物季節學
bioscrubber	생물학적 스크러버	生物學的-
biosphere	생(물)권	生(物)圈
biosphere-atmosphere interaction	생물-대기상호작용	生物大氣相互作用
biosphere reserve	생물권저장소	生物圈貯藏所
biostratigraphic method	생층서학	生層序學
biota	생물군	生物群

biotemperature	바이오온도	-溫度
biotic proxy	생물대리요소	生物代理要素
bipolar circulation pattern	쌍극순환형	雙極循環型
bipolar pattern	쌍극유형	雙極類型
bipolar see-saw	쌍극시소	雙極-
bipolar video	쌍극비디오	雙極-
birainy	쌍우기	雙雨期
birainy climate	쌍우기기후	雙雨期氣候
bird burst	버드버스트(항공기 엔진 속으로 새가 빨려 들어가도록 하는 내리바람), 새내리 바람	
bird echo	조류에코	鳥類-
bird strike	새공습, 버드스트라이크	-空襲
birefringence	복굴절	復屈折
Birkeland current	버크랜드해류	-海流
bise	찬 북동풍	-北東風
Bishop's corona	비숍광환	-光環
Bishop's ring	비숍고리	
Bishop wave	비숍파	-波
bispectral technique	쌍스펙트럼기술	雙-技術
bistatic	바이스태틱, 쌍상태	雙狀態
bistatic radar	쌍안테나 레이더, 이중안테나 레이더	雙-, 二重-
bit	비트	
biting wind	강추위바람	强-
bit rate	비트율	-率
bitter cold	강추위	强-
bittern	간수	-水
bivane	두방향풍향계	-方向風向計
Bjerknes cyclone model	비야크네스 저기압모형	-低氣壓模型
Bjerknes linearization procedure	비야크네스 선형화절차	-線形化節次
Bjerknes model	비야크네스 모형	-模型
Bjerknes's circulation theorem	비야크네스 순환정리	-循環定理
black blizzard	검은 먼지보라	
black body	흑체	黑體
black body radiation	흑체복사	黑體輻射
black body radiation curve	흑체복사곡선	黑體輻射曲線
black body radiation theory	흑체복사이론	黑體輻射理論
black body spectrum	흑체스펙트럼	黑體-
black body temperature	흑체온도	黑體溫度
black-bulb thermometer	흑구온도계	黑球溫度計
black carbon aerosol	검정탄소에어로졸	-炭素-
Blackdar mixing length	블랙카다르 혼합길이	-混合-
Black Death	흑사병	黑死病
blackening of snow surface	설면흑화	雪面黑化

black fog	검은 안개	
black frost	검은 서리	
black ice	검은 얼음	
black lightning(= dark lightning)	검은 번개	
black northeaster	검은 북동풍	
blackout	정전	停電
black rain	검은 비	
Black Sea flood hypothesis	흑해홍수가설	黑海洪水假說
black smoke	검은 연기	-煙氣
black squall	검은 스콜	
black storm	검은 폭풍(중앙아시아의 karaburan과 같은 먼지보라)	-暴風
black stratus	검은 층운	黑層雲
BLAG hypothesis	BLAG 가설	-假說
Blaney-Criddle model	블래니-크리들 모형	-模型
Blasius profile	블라시우스 연직분포	-鉛直分布
blast	돌풍, 폭발풍	突風, 爆發風
blast disease	도열병	稻熱病
blast effect	폭발효과	爆發效果
blast wave	폭발파	爆發波
Blaton's formula	블라통공식	-公式
bleaching	백화현상	白化現象
Bleeker humidity diagram	블리커습도선도	-濕度線圖
blended orography	혼합지형	混合地形
blending height	블렌딩높이	
blight	(식물의)마름병	-病
blind drainage	막힌 배수	-排水
blind roller	큰 노을	
blini	블리니	
blink	반짝임	
blip	에코	
blirty	변덕날씨	變德-
blizzard	눈보라	
blizzard warning	눈보라경보	-警報
blob	(레이더 에코의)얼룩	
block	저지	沮止
blockade weather	봉쇄기상	封鎖氣象
block averaging	구역평균	區域平均
block code	구역코드	區域-
block diagram	블록다이어그램	
blocked flow	저지류	沮止流
blocking	저지	沮止
blocking action	저지작용	沮止作用
blocking area	저지구역	沮止區域

blocking effect	저지효과	沮止效果
blocking high	저지고기압	沮止高氣壓
blocking low	저지저기압	沮止低氣壓
blocking oscillator	저지발진기	沮止發振器
blocking phenomenon	저지현상	沮止現象
blocking situation	저지상황	沮止狀況
blocking system	저지계, 저지시스템	沮止系
blocking wave	저지파	沮止波
block number	구역번호	區域番號
block off	차단효과	遮斷效果
blood flow	혈류	血流
blood rain	붉은 비	
blood snow	붉은 눈	
blossom shower	꽃잎(벚꽃)소나기	
blout(= blouter, clowther, blowthir)	블라우트(스코틀랜드의 돌발적 폭풍우 혹은 우박)	
blowby	(바람에)날린	
blowby gas	블로바이가스	
blowdown	(바람으로)넘어진	
blowholes	통풍구	通風口
blowing dust	높날림먼지	
blowing sand	높날림모래	
blowing snow	높날림눈	
blowing snow advisory	높날림눈주의보	-注意報
blowing spray	날린 물보라	
blowout	파열	破裂
bloxam	블록잼순평균법	-旬平均法
blue band	푸른 띠	
blue flash	푸른 섬광	-閃光
blue-green flame	청록불꽃	青綠-
blue-green propagation	청록색전파	青綠色傳播
blue-green pulse	청록색펄스	青綠色-
blue ice	푸른 얼음	
blue ice area	푸른 얼음 지역	-地域
blue jets	푸른 제트	
blue moon	푸른 달	
blue noise	블루노이즈	
blue of the sky	하늘푸름	
blue shift	청색편이	青色偏移
blue sky	푸른 하늘	
blue sky scale	푸른 하늘 척도	-尺度
blue sun	푸른 태양	-太陽
blue tilt	블루틸트	
bluff body	둔체, 흐름박리현상	鈍體, -剝離現象
bluff body effect	흐름박리현상효과	-剝離現象效果

blunk	블런크(잉글랜드의 돌발 스콜)	
body chemistry	몸화학, 신체화학	身體化學
body wave	실체파	實體波
bog	보그	
bogus	모조태풍	模造颱風
bogus data	모조자료	模造資料
bogusing	태풍보거싱	
bogus observation	모조관측	模造觀測
bogus vortex	모조소용돌이	模造-
Bohr's hypothesis	보어가설	-假說
Bohr's model	보어모형	-模型
boil	끓임, 비등	沸騰
boiling	끓음, 비등	沸騰
boiling point	끓는점	-點
boiling point elevation	끓는점 오름	-點-
boiling point thermometer	끓는점 온도계	-點溫度計
boiling spring	비등천	沸騰泉
Bolgiano scaling	볼기아노스케일링(규모조정)	
bolide	불덩이유성	-流星
Bølling-Allerød interstadial	뵐링-알레뢰드 아간빙기,	-亞間氷期, -溫暖期
	뵐링-알레뢰드 온난기	
bologram	복사기록	輻射記錄
bolograph	복사기록계	輻射記錄計
bolometer	열복사계	熱輻射計
bolster eddy	볼스터 맴돌이	
bolt	번개	
Boltzmann constant	볼츠만상수	-常數
Boltzmann distribution	볼츠만분포	-分布
Boltzmann equation	볼츠만방정식	-方程式
Boltzmann factor	볼츠만인자	-因子
Boltzmann's constant	볼츠만상수	-常數
bomb	화산탄	火山彈
bomb cyclone	폭탄저기압	爆彈低氣壓
bond	결합	結合
bond cycle	결합주기	結合週期
bond energy	결합에너지	結合-
bond length	결합길이	結合-
bond polarity	결합극성	結合極性
Bonneville Salt Flats	보네빌 소금 평원	-平原
bootstrap	부트스트랩	
bora	보라	
bora fog	보라안개	
borasco(= borasca, bourrasque)	보라스코(지중해상의 강한 뇌우를 동반하는 폭풍)	
bora scura	저기압형보라	低氣壓型-

B

bora wind	보라바람	
border ice	가장자리얼음	
border spring	경계샘	境界-
border zone	경계역	境界域
bore	보어	
boreal	북녘(의), 북부(의)	北部
boreal climate	한대기후	寒帶氣候
boreal forest	한대수림	寒帶樹林
boreal period	한대기	寒帶期
boreal region	한대지역	寒帶地域
boreal woodland	한대침엽수림지대	寒帶針葉樹林地帶
boreal zone	한대지역	寒帶地域
Boreas(= borras)	보레아스〈그리스어〉	
borehole climatology	시추공기후학	試錐孔氣候學
Borgas solution	보가스해법	-解法
borino	약한 보라	弱-
borras	보라스	
Bosporus Strait	보스포루스해협	-海峽
botanical zone	식물대	植物帶
bottleneck	병목	瓶-
bottle post	표류병	漂流瓶
bottle thermometer	병온도계	瓶溫度計
Bottlinger's ring	보틀링거고리	
bottom boundary condition	바닥경계조건	底層境界條件
bottom current	저층해류	底層海流
bottom friction	바닥마찰, 저층마찰	底層摩擦
bottom ice	바닥얼음	
bottom layer	바닥층	-層
bottom temperature	바닥온도	-溫度
bottom topography	바닥높낮이	
bottom-up approach	상향식 접근법	上向式接近法
bottom water	저층수	底層水
Bouguer's halo	부게의 무리	
Bouguer's law	부게법칙	-法則
boundary	경계	境界
boundary condition	경계조건	境界條件
boundary current	경계류	境界流
boundary layer	경계층	境界層
boundary layer climatologist	경계층기후학자	境界層氣候學者
boundary layer climatology	경계층기후학	境界層氣候學
boundary layer convergence line	경계층수렴선	境界層收斂線
boundary layer effect	경계층효과	境界層效果
boundary layer equation	경계층방정식	境界層方程式
boundary layer height	경계층높이	境界層-

boundary layer hypothesis	경계층가설	境界層假設
boundary layer parameterization	경계층매개변수화	境界層媒介變數化
boundary layer pumping	경계층펌핑	境界層-
boundary layer radar	경계층레이더	境界層-
boundary layer roll	경계층롤	境界層-
boundary layer scaling	경계층스케일링	境界層-
boundary layer separation	경계층분리	境界層分離
boundary layer theory	경계층이론	境界層理論
boundary mixing	경계혼합	境界混合
boundary of saturation	포화경계	飽和境界
boundary surface	경계면	境界面
boundary value problem	경곗값문제	境界-問題
boundary wave	경계파	境界波
bound charge	구속전하	拘束電荷
boundedness	(해의)유계성	有界性
bounded weak echo region	제한 약한 에코영역, 경계부-약에코역	制限弱-領域, 境界部弱-域
bound water	결합수	結合水
Bourdon thermometer	부르동온도계	-溫度計
Bourdon tube	부르동관	-管
Bourdon tube thermometer	부르동관 온도계	-管溫度計
bourrasque	부라스크	
Boussinesq approximation	부시네스크근사	-近似
Boussinesq assumption	부시네스크가정	-假定
Boussinesq equation	부시네스크방정식	-方程式
Boussinesq number	부시네스크수	-數
bow	무지개	
bow echo	활에코, 활모양 강수띠	-模樣降水-
Bowen-Ludlam process	보엔-루들럼 과정	-過程
Bowen ratio	보엔비	-比
Bowen ratio similarity	보엔비상사	-比相似
boxcar function	박스카함수	-函數
Box-Jenkins method	박스-젠킨스 방법	-方法
box kite	상자연	箱子鳶
box model	상자모형	箱子模型
Boyden index	보이든지수	-指數
Boyle-Charles' law	보일-샤를 법칙	-法則
Boyle-Mariotte law	보일-마리오트 법칙	-法則
Boyle's law	보일법칙	-法則
Boys camera	보이스카메라	
BPI pan	BPI증발계	-蒸發計
Bq	베크렐	
brackish water	기수, 반염수	汽水, 半鹽水
Bragg condition	브래그조건	-條件

Bragg scattering	브래그산란	-散亂
braincase size	두개골 크기	
brain size	뇌 크기	
brash	얼음조각뭉치	
brash ice	조각유빙	-流氷
brave west wind	사나운 서풍	-西風
Brazil current	브라질해류	-海流
breadth-first search	폭우 선탐색	幅優先探索
break	단절	中斷
breakaway depression	유리저기압	遊離低氣壓
breakaway low	분할저기압	分割低氣壓
breakdown	붕괴	崩壞
breakdown field	붕괴장	崩壞場
breakdown potential	붕괴퍼텐셜	崩壞-
breaker	쇄파	碎波
breaker depth	쇄파수심	碎波水深
breaking-drop theory	물방울 분열이론	-分裂理論
breaking-off process(= cutting-off process)	분리과정	分離過程
break line	단절선	斷絕線
break of Changma	장마휴식	-休息
break of the monsoon	몬순휴식	-休息
breaks in overcast	틈흐림	
breakthrough curve	누출곡선	漏出曲線
breakup	해빙	解氷
breakup period	해빙주기	解氷周期
breakup season	해빙철	解氷-
breakwater	방파제	防波堤
breather	열대회오리바람	熱帶-
breathing space condition	소강상태	小康狀態
bred mode	브레드모드	
bred vector	브레드벡터	
breeder reactor	증식형 원자로	增殖型原子爐
breeding method	육종법	育種法
breeding vector(BV)	브리딩벡터, 자람벡터	
breeze	미풍	微風
breeze front	미풍전선	微風前線
Bremstrahlung effect	제동복사효과	制動輻射效果
Brewer-Dobson circulation	브루어-도브슨 순환	-循環
Brewster's point	브루스터점	-點
bridled anemometer	제동풍속계	制動風速計
bridled-cup anemometer	제동컵풍속계	制動-風速計
bridled pressure plate	제동압력판	制動壓力板
briefing	브리핑	
Brier score(BS)	브라이어 점수	-點數

bright band	밝은 띠	
bright band echo	밝은 띠 에코	
brightness	밝기	
brightness criteria	밝기기준	-基準
brightness level	밝기수준	-水準
brightness temperature	밝기온도	-溫度
brightness value	밝기값, 휘도	輝度
bright network	(채층의) 밝은 망상조직	-網狀組織
bright segment	밝은 부분	-部分
bright sunshine	일사	日射
bright sunshine duration	일사시간	日射時間
brine	소금물	
brine slush	소금물찌꺼기	
briny wind	조풍	潮風
bristlecone pine	브리슬콘소나무	
British ice sheet	브리티시빙상	-氷床
British thermal unit	영국열단위	英國熱單位
broadband	광대역	廣帶域
broadband emissivity	광대역방출	廣帶域放出
broadband radiation	광대역복사	廣帶域輻射
broadband sensor	광대역수감	廣帶域隨感
broadcast	방송	放送
broadcast meteorology	방송기상학	放送氣象學
broad-crested weir	광정웨어	廣頂-
Brocken bow	반대광환	反對光環
Brocken bowl	브로켄사발	
Brocken spectre	브로켄괴물	-怪物
Brocken spectrum	브로켄스펙트럼	
broken	튕흐림	
broken cloud	튕흐림	
broken level	튕고도	-高度
broken sky	튕흐림하늘	
bromine	브롬	
bromine compound	브롬화합물	-化合物
bromine monoxide	브롬일산화물	--酸化物
bronchitis	기관지염	氣管支炎
brontides	땅울림	
brontograph	뇌우기록계	雷雨記錄計
brontometer	뇌우계	雷雨計
Bronze Age	청동기시대	靑銅器時代
Brookhaven dispersion diagram	브룩헤븐 분산다이어그램	-分散-
Brørup interstadial	브뢰럽간빙기	-間氷期
brown carbon aerosol	갈색탄소에어로졸	褐色炭素-
brown cloud	갈색구름, 갈색연무	褐色-, 褐色煙霧

Brownian diffusion	브라운확산	-擴散
Brownian motion	브라운운동	-運動
Brownian rotation	브라운회전	-回轉
browning	갈변	褐變
brown smog	갈색스모그	褐色-
brown snow	갈색눈	褐色-
brubu	브루부(인도 동부의 회오리바람의 일종)	
Bruckner cycle	브뤼크너주기	-週期
bruma	브루마(칠레 연안에서 오후의 해풍으로 생기는 연무)	
brume	브룸〈스페인어〉	
Brunt-Douglas isallobaric wind	브런트-더글러스 변압풍	-變壓風
Brunt-Douglas wind	브런트-더글러스 바람	
Brunt-Väisälä frequency	브런트-바이살라 진동수	-振動數
brush discharge	브러시방전	-放電
bryochore	서식대	棲息帶
B scope	B스코프	
bubble	거품	
bubble bursting	버블터짐, 기포터짐	氣泡-
bubble convection	거품대류	-對流
bubble gauge	거품수위계, 기포수위계	氣泡水位計
bubble high	거품고기압	-高氣壓
bubble ice	거품얼음	
bubble nucleus	거품핵, 기포핵	氣泡核
bubble policy	거품정책	-政策
bubbler	분수식 빨대	分數式-
bubbly ice	거품얼음	
bubonic plague	가래톳 흑사병	-黑死病
Buchan spells	버캔냉난기	-冷暖期
bucket method	물통법, 버킷법	-桶法
bucket temperature	물통온도	-桶溫度
bucket thermometer	물통온도계	-桶溫度計
Buckingham pi theorem	버킹엄 파이 정리	-定理
bud burst	발아	發芽
budget	수지	收支
budget operator	수지연산자	收支演算子
budgets of atmospheric species	대기종의 수지	大氣種收支
budget year	수지년	收支年
Budyko classification	부디코분류	-分類
Budyko Climatic Classification System	부디코기후분류시스템	-氣候分類-
Budyko number	부디코수	-數
Budyko ratio	부디코비	-比
buffer factor	완충인자	緩衝因子
buffering capacity	완충용량	緩衝容量
buffering Mach number	버퍼링마하수	-數

buffer zone	완충지대	緩衝地帶
Buger's vortex	버거의 소용돌이	
building climatology	건축기후학	建築氣候學
building design	건축설계	建築設計
building geometry	건축기하학	建築幾何學
building material effect	건축자재효과	建築資材效果
build-up index	증강지수	增强指數
bulb	구부	球部
bulk	부피, 용적	容積
bulk aerodynamic drag formulation	총체공기역학항력공식화	總體空氣力學抗力公式化
bulk aerodynamic method	총체공기역학방법	總體空氣力學方法
bulk approach	총체접근방법	總體接近方法
bulk average	총체평균	總體平均
bulk density of soil	총체토양밀도	總體土壤密度
bulk heat flux	총체열플럭스	總體熱-
bulk method	총체법, 벌크법	總體法
bulk mixed-layer model	부피혼합층모형	-混合層模型
bulk modulus	부피탄성률	-彈性率
bulk-Richardson number	총체-리처드슨 수	總體-數
bulk stable boundary growth	총체적 안정경계성장	總體的安定境界成長
bulk stomatal resistance	총체기공저항	總體氣孔抵抗
bulk transfer	총체전달	總體傳達
bulk transfer coefficient	총체전달계수	總體傳達係數
bulk transfer law	총체전달법칙	總體傳達法則
bulk turbulence scale	총체난류규모	總體亂流規模
bulk variable	총체변수	總體變數
bulk viscosity	총체점성	總體粘性
bulletin	회보	會報
bull's eye squall	불스아이 스콜	
Bultink-Malet stability	벌틴크-말레 안정도	-安定度
bummock	해먹	
bump	충격	衝激
bumpiness	충격도	衝激度
bumpy air	난류성공기	亂流性空氣
B-unit	B단위	-單位
buoy	부표	浮標
buoyancy	부력	浮力
buoyancy acceleration analogy	부력가속도유사성	浮力加速度類似性
buoyancy curvature analogy	부력곡률유사성	浮力曲率類似性
buoyancy destruction rate	부력파괴율	浮力破壞率
buoyancy effect	부력효과	浮力效果
buoyancy energy	부력에너지	浮力-
buoyancy factor	부력인자	浮力因子
buoyancy flux	부력플럭스	浮力-

buoyancy flux profile	부력플럭스연직분포	浮力-鉛直分布
buoyancy force	부력	浮力
buoyancy frequency	부력진동수	浮力振動數
buoyancy length scale	부력길이규모	浮力-規模
buoyancy lift	부력상승	浮力上昇
buoyancy oscillation	부력진동	浮力振動
buoyancy parameter	부력매개변수	浮力媒介變數
buoyancy scale	부력규모	浮力規模
buoyancy subrange	부력아영역	浮力亞領域
buoyancy velocity	부력속도	浮力速度
buoyancy wave	부력파	浮力波
buoyancy wavenumber	부력파수	浮力波數
buoyant convection	부력대류	浮力對流
buoyant energy	부력에너지	浮力-
buoyant force	부력	浮力
buoyant instability	부력불안정도	浮力不安定度
buoyant production	부력생산	浮力生産
buoyant production rate	부력생산율	浮力生産率
buoy weather station	부이 기상관측소	-氣象觀測所
Bureau of Meteorology	(오스트레일리아)기상청	氣象廳
Burger number	버거수	-數
Burgers equation	버거스방정식	-方程式
Burgers material	버거스물질	-物質
Burgers vector	버거스벡터	
burial flux	매몰플럭스	埋沒-
burning index	연소지수	燃燒指數
burn-off	소산	消散
burst	버스트, 파열, 돌발, 위상신호	破裂, 突發, 位相信號
bursting height	파열고도	破裂高度
bursting point	파열점	破裂點
burst of the monsoon	몬순돌발	-突發
burst phenomena	파열현상	破裂現象
burst point of convection	대류파열점, 대류발생점	對流破裂點, 對流發生點
Busch lemniscate	부쉬선	-線
bush	먼지덤불	
bushfire	(오스트레일리아)들불	
business as usual scenario	축소 없는 시나리오	
Businger-Dyer relationship	부싱어-다이어 관계	-關係
Businger-Dyer wind profile	부싱어-다이어 바람프로파일	
buster	파열	破裂
butadiene	부타디엔(합성 고무 제조에 쓰이는 무색의 탄화수소 가스)	
butane	부탄	

butterfly effect	나비효과	-效果
Buys-Ballot rule	바이스발롯 규칙	-規則
Buys-Ballot's law	바이스발롯 법칙	-法則
Byram anemometer	바이램풍속계	-風速計

C, c

영문	한글	한자
C3 pathway	C3 경로	-經路
C4 pathway	C4 경로	-經路
cabelling	등밀도혼합	等密度混合
cacimbo	카심보(남아프리카의 앙골라 지방에서 겨울철에 밤에는 바다로부터 안개가 밀려들고, 낮에는 걷히는 현상)	
cadmium	카드뮴(원자번호 48, 원자량 112.41)	
Cainozoic climate	신생대기후	新生代氣候
cajú rains	카주의 소나기	
cake ice	덩어리 얼음	
calculation of altitude	고도계산	高度計算
calendaricity	특이상태	特異狀態
calendar year	역년	曆年
calf	부빙조각	浮氷-
caliber	직경, 구경	直徑, 口徑
calibration	교정	較正
calibration equation	교정방정식	較正方程式
calibration error	교정오차	較正誤差
calibration interval	교정주기	較正週期
calibration of an instrument	기기보정	器機補正
calibration of meteorological instruments	기상기기 검증	氣象器機檢證
calibration problem	보정문제	補正問題
calibration tank	보정수조	補正水曹
California current	캘리포니아해류	-海流
California fog	캘리포니아안개	
California method	캘리포니아방법	-方法
California norther	캘리포니아북풍	-北風
California plotting position	캘리포니아확률도시법	-確率圖視法
calina	캘리나(스페인에서 여름에 많은 연무)	

Callao painter	칼라오페인터(페루연안에 종종 나타나는 안개)	
callendar effect	캘런더효과	-效果
calm	고요(풍력계급 0), 무풍	無風
calm belt	무풍띠	無風-
calm central eye	무풍중심눈	無風中心-
calm layer	무풍층	無風層
calms of Cancer	북반구아열대무풍대	北半球亞熱帶無風帶
calms of Capricorn	남반구아열대무풍대	南半球亞熱帶無風帶
calm zone	무풍대	無風帶
caloradiance	열복사강도	熱輻射强度
caloric equation	칼로리양방정식	-量方程式
calorific value	열량	熱量
calorimeter	열량계	熱量計
calorimetry	열량측정법	熱量測定法
calved ice	부빙조각	浮氷-
calving	빙하분리	氷河分離
calvus	대머리구름	
camanchaca(= garúa)	카만차카(페루 연안에서 겨울철에 바다로부터 밀려오는 짙은 안개와 이슬비)	
Campbell-Stokes sunshine recorder	캠벨-스토크스 일조계	-日照計
Canadian Climate Model	캐나다기후모형	-氣候模型
Canadian hardness gauge	캐나다굳기측정계	-測定計
canalization	수로화	水路化
canal theory	수로이론	水路理論
Canaries current	카나리해류	-海流
cancellation ratio	소거비, 소거율	消去比, 消去率
candela	칸델라(촉광단위)	(燭光單位)
candle	촉광	燭光
candle ice	양초모양 얼음	洋-模樣-
Candlemas Eve wind	성촉절 이브 바람	聖燭節-
candle power	촉광, 광도	燭光, 光度
canonical equation	정준방정식	正準方程式
canonical pattern	정준패턴	正準-
canopy	캐노피, 군락, 천공	群落, 天空
canopy architecture	임관구조, 초관구조	林冠構造, 草冠構造
canopy density	캐노피밀도	-密度
canopy drag	캐노피항력	-抗力
canopy height	캐노피높이	
canopy parameters	캐노피매개변수	-媒介變數
canopy photosynthesis	군락광합성	群落光合成
canopy photosynthesis model	군락광합성모형	群落光合成模型
canopy resistance	캐노피저항	-抵抗
canopy structure	군락구조, 캐노피구조	群落構造
canopy temperature	캐노피온도	-溫度

Canterbury northwester	(뉴질랜드)캔터베리북서풍	-北西風
canting angle	경사각	傾斜角
Cantor set	칸토르집합	-集合
canyon	협곡	峽谷
canyon wind	협곡바람	峽谷-
capacitance rain gauge	용량형우량계	容量型雨量計
capacitive element	용량성소자	容量性素子
capacity	용량	容量
capacity correction	용량보정	容量補正
capacity dimension	용량차원	容量次元
capacity of the wind	(바람의)취주량	吹走量
cap cloud	모자구름	
Cape Hatteras low	해터라스곶저기압	-低氣壓
cape scrub	케이프스크럽(남아프리카 관목)	
cape Verde type	베르데곶형	-型
	(한의 기후구분의 하나)	
capillarity	모세관현상	毛細管現象
capillarity correction	모세관보정	毛細管補正
capillary action	모세관작용	毛細管作用
capillary collector	모세관채수기	毛細管採水器
capillary conductivity	모세관전도도	毛細管傳導度
capillary depression	모세관강하	毛細管降下
capillary diffusion	모세관확산	毛細管擴散
capillary electrometer	모세관전위계	毛細管電位計
capillary force	모세관힘	毛細管-
capillary fringe	모세관물띠	毛細管-
capillary head	모관수두	毛管水頭
capillary hysteresis	모세관이력현상	毛細管履歷現象
capillary interstice	모세관공극	毛細管空隙
capillary layer	모세관층	毛細管層
capillary moisture capacity	모세관용수량	毛細管用水量
capillary phenomenon	모세관현상	毛細管現象
capillary potential	모세관퍼텐셜	毛細管-
capillary pressure	모세관압력	毛細管壓力
capillary ripple	모세관파	毛細管波
capillary rise	모세관상승	毛細管上昇
capillary suction	모세관흡입	毛細管吸入
capillary water	모세관수	毛細管水
capillary wave	모세관파	毛細管波
capillatus	털보구름	
capped column	모자 쓴 기둥	
capping inversion	착모역전층, 마개역전	着帽逆轉層, -逆轉
capping stable layer	모자안정층	帽子安定層
captive balloon	계류기구	繫留氣球

captive balloon sounding	계류기구탐측	繫留氣球探測
captive breeding	포획사육	捕獲飼育
capture	포착	捕捉
capture cross-section	포착단면(적)	捕捉斷面(積)
capture process	포착과정	捕捉過程
capture rate	포착률	捕捉率
carabine	카라비네	
caracenet	카라세넷	
carbonaceous aerosol	탄소에어로졸, 에어로졸	炭素-
carbon assimilation	탄소동화작용	炭素同化作用
carbonate compensation depth	탄산염보상심도	炭山鹽補償深度
carbonate conservation depth	탄산염보존깊이	炭酸鹽保存-
carbonate rock	탄산염바위	炭酸鹽-
carbon balance	탄소균형	炭素均衡
carbon black	카본블랙	
carbon black seeding	탄소가루 씨뿌리기	炭素-
carbon bond mechanism	탄소결합메커니즘	炭素結合-
carbon compound	탄소화합물	炭素化合物
carbon cycle	탄소순환	炭素循環
carbon-dating	탄소연대측정법	炭素年代測定法
carbon dioxide	이산화탄소	二酸化炭素
carbon dioxide atmospheric concentration	이산화탄소대기농도	二酸化炭素大氣濃度
carbon dioxide band	이산화탄소띠	二酸化炭素-
carbon dioxide compensation point	이산화탄소보상점	二酸化炭素補償點
carbon dioxide emission	이산화탄소배출	二酸化炭素排出
carbon dioxide equivalence	이산화탄소환산	二酸化炭素換算
carbon dioxide fertilization	이산화탄소시비	二酸化炭素施肥
carbon dioxide fertilizing effect	이산화탄소시비효과	二酸化炭素施肥效果
carbon dioxide information analysis center	이산화탄소정보분석센터	二酸化炭素情報分析-
carbon dioxide molecule	이산화탄소분자	二酸化炭素分子
carbon disulfide	이황화탄소	二黃化炭素
carbon emission	탄소방출	炭素放出
carbon exchange	탄소교환	炭素交換
carbon-film hygrometer element	탄소막습도계소자	炭素膜濕度計素子
carbon footprint	탄소발자국	炭素-
carbonic acid	탄산	炭酸
Carboniferous	석탄기(고생대)	石炭期
carbon intensity	탄소강도	炭素强度
carbon isotope	탄소동위원소	炭素同位元素
carbon monoxide	일산화탄소	一酸化炭素
carbon pool	탄소풀	炭素-
carbon sequestration	탄소격리	炭素隔離
carbon sink	탄소흡(수)원	炭素吸(收)源
carbon source	탄소발생원	炭素發生源

carbon storage	탄소저장	炭素貯藏
carbon tax	탄소세	炭素稅
carbon tetrachloride	사염화탄소	四鹽化炭素
carbon trading	탄소무역	炭素貿易
carbonyl compound	카르보닐화합물	-化合物
carbonyl sulfide	황화카르보닐	黃化-
carboxyhemoglobin	당화혈색소	糖化血色素
carboxylation	카르복실화	-化
carburetor icing	기화기착빙	氣化器着氷
carcinogen	발암물질	發癌物質
carcinogenic compound	발암성물질	發癌性物質
carcinogenicity	발암성	發癌性
cardinal direction	방위	方位
cardinal point	4방위점(동, 서, 남, 북)	四方位點
cardinal temperature	임계온도	臨界溫度
cardinal winds	4방위바람(동, 서, 남, 북풍)	四方位-
cardiovascular disease	심혈관질환	心血管疾患
Caribbean current	카리브해류	-海流
caribou	순록, 카리부	馴鹿
Carnot cycle	카르노순환	-循環
Carnot efficiency	카르노효율	-效率
Carnot engine	카르노기관	-機關
Carnot's cycle process	카르노순환과정	-循環過程
Carnot's theorem	카르노정리	-定理
carrier-balloon system	수송기구시스템	輸送氣球-
carrier frequency	반송주파수	搬送周波數
carrier wave	반송파	搬送波
carrying capacity	탑재량	搭載量
carry-over	이월	移越
carry-over effect	이월효과	移越效果
Cartesian coordinates	카테시안좌표	-座標
Cartesian tensor	카테시안텐서	
cartographical climatology	도해기후학	圖解氣候學
cartwheel satellite	수레바퀴형 위성	-型衛星
cascade	다단폭포, 계단	多段瀑布, 階段
cascade impactor	다단충돌채집기	多段衝突採集器
cascade of energy	에너지다단분산	-多段分散
cascade process	다단과정, 캐스케이드과정	多段過程, -過程
cascade shower	다단샤워	多段-
cascading water	유수	流水
case-based reasoning	사례기반추론	事例基盤推論
Casella's siphon rainfall recorder	카셀라의 사이폰식 우량기록계	-式雨量記錄計
case study	사례연구	事例研究
case weather	사례날씨	事例-

Caspian Sea	카스피해	-海
Cassegrainian mirror	카세그레인거울	
Cassegrainian reflector	카세그레인반사망원경	-反射望遠鏡
Cassegrainian telescope	카세그레인망원경	-望遠鏡
castellanus	탑모양(구름)	塔模樣-
catafront(= katafront)	활강전선	滑降前線
Catalina eddy	카타리나 맴돌이	
catalyst	촉매	觸媒
catalytic agent	촉매	觸媒
catalytic converter	촉매컨버터, 촉매변환기	觸媒變換機
catalytic cycle	촉매순환	觸媒循環
catalytic incineration	촉매소각장치	觸媒燒却裝置
catalytic incinerator	촉매소각로	觸媒燒却爐
cataphalanx	한랭전선면	寒冷前線面
catastrophe	큰 재앙	-災殃
catastrophic windthrow	바람재앙	-災殃
catastrophism	격변설	激變說
catathermometer	카타온도계	-溫度計
catch	강수량, 승수	降水量, 承水
catchment area	집수역	集水域
catchment basin(= drainage area, drainage basin, watershed)	유역면적	流域面積
catchment boundary	유역경계	流域境界
catchment yield	유역총유출양	流域總流出量
categorical forecast	범주예보	範疇豫報
category forecast	범주예보	範疇豫報
cathetometer	고도차측정기	高度差測定器
cathode-ray direction finding	음극선방향결정	陰極線方向決定
cathode-ray oscillograph	음극선오실로그래프	陰極線-
cathode-ray oscilloscope	음극선오실로스코프	陰極線-
cathode-ray radiogoniometer	음극선무선방위계	陰極線無線方位計
cathode-ray tube	음극선관	陰極線管
cat ice	살얼음(얼음과 물 사이에 공간이 있음)	
cation	양이온	陽-
cation exchange capacity	양이온치환용량	陽-置換容量
cat's paw	캣포우(고요한 수면에 잔물결을 일으키는 약한 바람)	
Cauchy distribution	코시분포	-分布
Cauchy number	코시수	-數
Cauchy-Riemann equation	코시-리만 방정식	-方程式
cauliflower cloud	꽃양배추구름	-洋-
causality	인과관계	因果關係
cause-effect relationship	원인-결과 관계	原因結果關係
caustic scrubbing	부식제거법	腐蝕除去法
cavaburd(= kavaburd)	캐바버드(스코틀랜드 북동 쉐틀랜드섬의 큰 눈)	

cavaliers	카발리에(프랑스 몽펠리에 부근에서 대기화학월말부터 대기환경월초에 걸쳐 mistraral이 가장 강한 날)	
cavitation	공동화작용	空洞化作用
cavity	구멍, 공동	空洞
cavity pyrheliometer	공동직달일사계	空洞直達日射計
cavity radiometer	공동복사계	空洞輻射計
CAVT table	씨에이브이티(CAVT)표, 일정절대소용돌이도궤적표	一定絕對-度軌跡表
C-band	C밴드	
Cb cluster	적란운 무리	積亂雲-
ceiling	실링, 조건부운고, 상승한계 (항공기, 기구, 로켓 등의)	條件附雲高, 上昇限界
ceiling alarm	실링경보기	-警報器
ceiling and visibility unlimited	청명	淸明
ceiling balloon	운고기구	雲高氣球
ceiling classification	운고분류	雲高分類
ceiling detector	운고탐지기	雲高探知器
ceiling height	운고	雲高
ceiling height indicator	운고지시기	雲高指示器
ceiling light	운고관측등	雲高觀測燈
ceiling of convection	대류고도	對流高度
ceiling projector	운고투광기	雲高投光器
ceiling zero	실링영(운고 15m 이하)	-零
ceilometer	운고계	雲高計
celerity	민첩속도	敏捷速度
celestial body	천체	天體
celestial dome	천구	天球
celestial equator	천구적도	天球赤道
celestial horizon	천구지평선	天球地平線
celestial pole	천구극	天球極
celestial sphere	천구	天球
cell	단위, 세포, 소자	單位, 細胞, 素子
Cellini's halo(= heiligenschein)	셀리니의 무리	
cell of convection	대류세포	對流細胞
cell outline	세포윤곽	細胞輪廓
cell theory	세포설	細胞說
cellular automation model	세포자동모형	細胞自動模型
cellular circulation	세포형순환	細胞型循環
cellular cloud pattern	세포형구름패턴	細胞型-
cellular convection	세포형대류	細胞型對流
cellular hypothesis	세포가설	細胞假說
cellular movement	세포형운동	細胞型運動
cellular pattern	세포형태	細胞形態
cellular structure	세포구조	細胞構造

cellular vortex	세포형소용돌이	細胞型-
Celsius	섭씨	攝氏
Celsius-Fahrenheit conversion	섭씨-화씨온도환산	攝氏華氏溫度換算
Celsius-Kelvin conversion	섭씨-절대온도환산	攝氏絕對溫度換算
Celsius scale	섭씨눈금	攝氏-
Celsius temperature	섭씨온도	攝氏溫度
Celsius temperature scale	섭씨온도눈금	攝氏溫度-
Celsius thermometer	섭씨온도계	攝氏溫度計
Celsius thermometric scale	섭씨온도눈금	攝氏溫度-
Cenozoic climate	신생대기후	新生代氣候
centered difference	중앙차분	中央差分
centered-in-time scheme	시간중앙차분방식	時間中央差分方式
center jump	중심도약	中心跳躍
center of action	작용중심	作用中心
center of buoyancy	부력중심	浮力中心
center of curvature	곡률중심	曲率中心
center of cyclone	저기압중심	低氣壓中心
center of falls	하강중심	下降中心
center of gravity	중력중심	重力中心
center of pressure	압력중심	壓力中心
center of pressure system	기압계중심	氣壓系中心
center of rise	상승중심	上昇中心
center of rotation	회전중심	回轉中心
center of symmetry	대칭중심	對稱中心
center of variation	변동중심	變動中心
centibar	센티바	
centigrade	백분도	百分度
centigrade scale	백분도눈금	百分度-
centigrade temperature scale	백분도온도눈금	百分度溫度-
centigrade thermometer	백분도온도계	百分度溫度計
centimeter-gram-second system	cgs계	-系
centipoise	센티포이즈(점도의 단위)	
central angle	중심각	中心角
central calm	중심무풍구역	中心無風區域
central core	중심핵	中心核
central dense overcast	원형의 두터운 구름	圓形-
central depression	중심저기압	中心低氣壓
central dogma	중심원리	中心原理
Central England climate record	중앙잉글랜드 기후기록	中央-氣候記錄
Central England temperature	중앙잉글랜드 온도	中央-溫度
central estimate	중앙어림	中央-
central finite-difference approximation	중앙유한차분근사	中央有限差分近似
central force	중심력	中心力
central force field	중심력장	中心力場

central forecasting office	중앙예보소	中央豫報所
central limit theorem	중심극한정리	中心極限定理
central line	중심선	中心線
central moment	중심모멘트	中心-
central pressure	중심압력	中心壓力
Central Standard Time	중부표준시간(미국)	中部標準時間
central tendency	중심경향	中心傾向
central wavenumber	중앙파수	中央波數
centrifugal acceleration	원심가속도	遠心加速度
centrifugal effect	원심효과	遠心效果
centrifugal force	원심력	遠心力
centrifugal instability	원심불안정	遠心不安定
centrifugal potential	원심퍼텐셜	遠心-
centrifuge	원심분리기	遠心分離機
centrifuge moisture equivalent	원심함수당량	遠心含水當量
centripetal	구심	求心
centripetal acceleration	구심가속도	求心加速度
centripetal force	구심력	求心力
centroid	중심	重心
centrosome	중심체	中心體
ceraunograph	천둥번개기록계	-記錄計
ceraunometer	천둥번개측정계	-測定計
cers	세르(미스트랄의 일종)	
certain energy	특정에너지	特定-
certainty factor	확신율	確信率
Cesàro sum	체자로 합	-合
C figure	C수	-數
chaff	금속조각	金屬-
chaff seeding	금속조각살포	金屬-撒布
chain length	연쇄반응길이	連鎖反應-
chain lightning	연쇄번개	連鎖-
chain reaction	연쇄반응	連鎖反應
chain rule	연쇄규칙	連鎖規則
chance expectation	확률기댓값	確率期待-
Chandler wobble	챈들러 흔들림	
change	변화	變化
change chart	변화도	變化圖
change of phase	상변화	相變化
change of state	상태변화	狀態變化
change of type	형태변화	形態變化
changes with altitude	고도변화	高度變化
change threshold	변화문턱	變化門-
change vector analysis	변화벡터분석	變化-分析
Changma	장마	

Changma front	장마전선	-前線
Changma index	장마지수	-指數
Changma jet stream	장마제트류	-流
channel	통로, 경로, 유로, 채널	通路, 經路, 流路
channel control	물길조절	-調節
channeled high wind	수로강풍	水路强風
channeled scablands	화산용암지대	火山鎔岩地帶
channel flow	유로흐름, 채널흐름	流路-
channel inflow	수로유입	水路流入
channelling	채널링	
channel precipitation	수로상강수(량)	水路上降水(量)
channel storage	하도저류	河道貯溜
chaos	혼돈	混沌
chaos theory	혼돈이론	混沌理論
chaotic behavior	혼돈거동	混沌擧動
chaparral	샤파렐(관목의 일종)	
Chapman cycle	채프먼순환	-循環
Chapman layer	채프먼층	-層
Chapman mechanism	채프먼메커니즘	
Chapman ozone chemistry	채프먼오존화학	-化學
Chapman profile	채프먼프로파일	
Chapman region	채프먼역	-域
Chappicus band	찹퓌스띠	
Chappuis	찹퓌스	
characteristic curve	특성곡선	特性曲線
characteristic equation	특성방정식	特性方程式
characteristic form	특성형	特性型
characteristic frequency	특성진동수	特性振動數
characteristic function	특성함수	特性函數
characteristic length	특성길이	特性-
characteristic line	특성선	特性線
characteristic number	특성수	特性數
characteristic of the pressure tendency	압력경향특성	壓力傾向特性
characteristic point	특성점	特性點
characteristic reduced frequency	특성환산주파수	特性換算周波數
characteristics	특징, 특성, 특성곡선	特徵, 特性, 特性曲線
characteristic scale	특성규모	特性規模
characteristic specific humidity	특성비습	特性比濕
characteristic surface energetics	특성지상에너지학	特性地上-學
characteristic temperature	특성온도	特性溫度
characteristic value	특성값	特性-
characteristic value problem	특성값문제	特性-問題
characteristic velocity	특성속도	特性速度
characteristic virtual temperature	특성가온도	特性假溫度

charduy	차르듀이	
charge	전하, 대전, 하전	電荷, 帶電, 荷電
charge coupled device	전하결합소자(CCD)	電荷結合素子
charged cloud	대전구름	帶電-
charge density	전하밀도	電荷密度
charged particle	하전입자	荷電粒子
charged plate	대전판	帶電板
charge number	전하수	電荷數
charge separation	전하분리	電荷分離
Charles-Gay-Lussac law	샤를-게이-뤼삭 법칙	-法則
Charles's law	샤를법칙	-法則
Charney-Drazin criterion	차니-드라진 범주	-範疇
Charnock approach	차녹접근법	-接近法
Charnock's constant	차녹상수	-常數
Charnock's formula	차녹공식	-公式
Charnock's relation	차녹관계	-關係
chart	도, 도표, 해도, 일기도	圖, 圖表, 海圖, 日氣圖
chart datum	해도기준	海圖基準
chart plotter	기입자	記入者
chart (or map) plotting	일기도기입	日氣圖記入
check	점검	點檢
check observation	점검관측	點檢觀測
Chelford interstadial	첼퍼드간빙기	-間氷期
chemical affinity	화학친화력	化學親和力
chemical bond	화학결합	化學結合
chemical cell(battery)	화학전지	化學電池
chemical composition	화학성분	化學成分
chemical composition of air	대기의 화학적 조성	大氣化學的組成
chemical composition of precipitation	강수화학조성	降水化學組成
chemical energy	화학에너지	化學-
chemical equation	화학반응식	化學反應式
chemical equilibrium	화학평형	化學平衡
chemical erosion	화학적침식	化學的浸蝕
chemical hygrometer	화학습도계	化學濕度計
chemical kinetics	화학운동학	化學運動學
chemical lifetime	화학수명	化學壽命
chemical mechanism	화학기작	化學機作
chemical potential	화학퍼텐셜	化學-
chemical propulsion system	화학연료추진체계	化學燃料推進體系
chemical prospecting	화학탐사	化學探查
chemical reaction	화학반응	化學反應
chemical reaction rate	화학반응률	化學反應率
chemical remanent magnetism	화학적 잔류자성	化學的殘留磁性
chemical sediment	화학침전물	化學沈澱物

chemical sedimentary rock	화학침전암	化學沈澱巖
chemical tracers	화학미량원소	化學微量元素
chemical transmitter	화학전달물질	化學傳達物質
chemical transport model	화학수송모형	化學輸送模型
chemical variable	화학변수	化學變數
chemical weathering	화학적풍화(작용)	化學的風化(作用)
chemiluminescence	화학발광	化學發光
chemisorbent	흡착제	吸着劑
chemopause	화학권계면	化學圈界面
chemoreceptor	화학수용기	化學受容器
chemosphere	화학권	化學圈
chergui	체르구이	
cherry-blossom front	벚꽃전선	-前線
Chézy equation	셰지방정식	-方程式
Chézy formula	셰지공식	-公式
Chicago School	시카고학교(학파)	-學校(學派)
Chicxulub impact	칙쇼루브충돌	-衝突
chilblains	동상	凍傷
Chile current	칠레해류	-海流
chili	칠리(튀니지아의 온난하고 건조한 하강풍으로서 시로코와 유사)	
chill	한랭	寒冷
chilled mirror dew-point thermometer	냉각거울이슬점온도계	冷却-點濕度計
chilled mirror hygrometer	냉각거울습도계	冷却-濕度計
chill factor	한랭인자	寒冷因子
chill hour	냉각시간	冷却時間
chilling	냉각	冷却
chilling injury	냉해	冷害
chilling requirement	저온요구성	低溫要求性
chill unit	냉각단위	冷却單位
chill wind factor	냉각바람인자	冷却-因子
chilly	쌀쌀한	
chimney cloud	굴뚝구름	
chimney current	굴뚝기류	-氣流
chimney height	굴뚝높이	
Chimu flood	치무홍수	-洪水
China coastal current	중국연안류	中國沿岸流
Chinook	치누크	
Chinook arch	치누크아치(럭키산맥 위에 구름둑 같이 나타나는 푄구름)	
Chinook wall cloud	치누크벽구름	-壁-
Chinook wind	치누크바람	
chinophile	눈을 좋아하는	
chip log	칩로그(배의 속력을 잴 수 있는 도구)	

chi-square distribution	카이제곱분포	-分布
chi-square test	카이제곱검정	-檢定
chlorine	염소	鹽素
	(원자번호 17, 원자량 35.453)	
chlorine atom	염소원자	鹽素原子
chlorine compound	염소화합물	鹽素化合物
chlorine dioxide	이산화염소	二酸化鹽素
chlorine monoxide	일산화염소	一酸化鹽素
chlorine monoxide dimer	일산화염소이합체	一酸化鹽素二合體
chlorine monoxide radical	일산화염소라디칼	一酸化鹽素-
chlorine nitrate	염소질산염	鹽素窒酸鹽
chlorine oxide	(ClOx)염소산화물	鹽素酸化物
chlorinity	염소량	鹽素量
chlorofluorocarbon	염화불화탄소	鹽化弗化炭素
chloroform	클로로포름	
chlorophyll	엽록소	葉綠素
chloroplast	엽록체	葉綠體
chlorosis	황백화(현상), 황화병, 백화현상(식물의)	黃白化(現象), 黃化病, 白化現象
chlorosity	염소도	鹽素度
chlorotic dwarf disease	아염소산위축병	亞鹽素酸萎縮病
chocolate gale	초콜릿 큰바람	
chop	촙(짧은 주기를 가진 표면중력파)	
chopper	촙퍼, 단절기	斷截機
choppy sea	놀치는 바다	
choroisotherm	공간등온선	空間等溫線
chota bursat	초타버르사트(인도에서 몬순 우기전 1~2일 동안 비오는 날의 힌두어)	
chou lao hu	추라오후	
Christoffel symbol	크리스토펠기호	-記號
chromatic aberration	색수차	色收差
chromatic scintillation	색섬광	色閃光
chromosphere	채층	彩層
chronic acidification	만성산성화	慢性酸性化
chronic allergic rhinitis	만성알레르기비염	慢性-鼻炎
chronic bronchitis	만성기관지염	慢性氣管支炎
chronic disease	만성질환	慢性疾患
chronic exposure	만성피폭, 만성노출	慢性被爆, 慢性露出
chronic obstructive lung disease	만성폐쇄성폐질환	慢性閉鎖性肺疾患
chronoanemoisothermal diagram	시간풍향등온선도	時間風向等溫線圖
chronograph	시간기록기, 크로노그래프	時間記錄器
chronoisotherm	시간등온선	時間等溫線
chronology	연대학	年代學
chronometer	크로노미터	

chronometric radiosonde	시계식라디오존데	時計式-
chronothermometer	시계형온도계	時計形溫度計
Chukwoodae(= a pedestal of Korean raniguage)	측우대	測雨臺
Chukwookee(= Korean raingauge)	측우기	測雨器
chun fung	봄바람(중국)	
chunyuh	봄비(중국)	
churada	추라다(마리아나제도에서 북동몬순기에 내리는 심한 비)	
chytrid fungus	항아리 곰팡이	
cierzo	시에르조(스페인 에브로 계곡에서 부는 강한 돌풍)	
cigarette smoking	담배흡연	-吸煙
cilia	솜털, 섬모	纖毛
C index	C지수	-指數
cipher(= cypher)	아라비아 숫자, 0의 기호, 暗號	
	암호	
circle of illumination	조명원	照明圓
circle of inertia	관성원	慣性圓
circle-type seeding	원형씨뿌리기	圓形-
circuit	회로	回路
circular cylindrical coordinates	원기둥좌표	圓-座標
circular depolarization ratio	원편광회복률	圓偏光回復率
circular frequency	원진동수	圓振動數
circular horizontal motion	원형수평운동	圓形水平運動
circular motion	원운동	圓運動
circular oscillation	원진동	圓振動
circular polarization	원편광	圓偏光
circular polarization ratio	원형편파비	圓形偏波比
circular polarized wave	원편파	圓偏波
circular symmetry	원대칭	圓對稱
circular variable	원형변수	圓形變數
circular vortex	원형소용돌이	圓形-
circular waveguide	원형도파관	圓形導波管
circulation	순환	循環
circulation anomaly	순환이상	循環異常
circulation cell	순환세포	循環細胞
circulation flux	순환플럭스	循環-
circulation index	순환지수	循環指數
circulation integral	순환적분	循環積分
circulation model	순환모형	循環模型
circulation of air	공기순환	空氣循環
circulation of atmosphere	대기순환	大氣循環
circulation pattern	순환모양	循環模樣
circulation theorem	순환정리	循環定理
circulation type	순환형	循環型
circulator	서큘레이터	

circulatory system	순환계	循環系
circumference	원둘레	圓-
circumferential velocity	원둘레속도	圓-速度
circumhorizontal arc	환수평호	環水平弧
circumpolar cyclone	극둘레저기압	極-低氣壓
circumpolar flow	극둘레흐름	極-
circumpolar map	극둘레지도	極-地圖
circumpolar trough	극둘레기압골	極-氣壓-
circumpolar vortex	극둘레소용돌이	極-
circumpolar westerlies	극둘레편서풍	極-偏西風
circumpolar whirl	극둘레회오리	極-
circumscribed halo	외접무리	外接-
circumscribed circle	외접원	外接圓
circumscribed polygon	외접다각형	外接多角形
circumscription	외접	外接
circumsolar radiation	환태양복사	環太陽輻射
circumzenithal arc	천정호	天頂弧
cirque glacier	권곡빙하	圈谷氷河
cirques	권곡	圈谷
cirriform	권운형	卷雲形
cirriform cloud	권운형구름	卷雲形-
cirrocumulus	권적운	卷積雲
cirrocumulus castellanus	탑권적운	塔卷積雲
cirrocumulus cloud	권적운	卷積雲
cirrocumulus floccus	송이권적운	-卷積雲
cirrocumulus lacunosus	벌집권적운	-卷積雲
cirrocumulus lenticularis	렌즈권적운	-卷積雲
cirrocumulus mackerel	비늘권적운	-卷積雲
cirrocumulus mamma	유방권적운	乳房卷積雲
cirrocumulus nebulosus	면사포권적운	面紗布卷積雲
cirrocumulus stratiformis	층권적운	層卷積雲
cirrocumulus undulatus	파도권적운	波濤卷積雲
cirrocumulus virga	꼬리권적운	-卷積雲
cirro-film (or thread)	실권운	-卷雲
cirro-macula	무늬권운	-卷雲
cirro-nebula	면사포권운	面紗布卷雲
cirro-ripple	잔물결권운	-卷雲
cirrostratus	권층운	卷層雲
cirrostratus cloud	권층운	卷層雲
cirrostratus communis	보통권층운	普通卷層雲
cirrostratus cumulosus	적운형권층운	積雲形卷層雲
cirrostratus duplicatus	(두)겹권층운	-卷層雲
cirrostratus fibratus	섬유권층운	纖維卷層雲
cirrostratus filosus	섬유권층운	纖維卷層雲

C

cirrostratus nebulosus	면사포권층운	面紗布卷層雲
cirrostratus undulatus	파도권층운	波濤卷層雲
cirrostratus vittatus	띠권층운	-卷層雲
cirro-velum	면사포권운	面紗布卷雲
cirrus	권운	卷雲
cirrus aviaticus	비행권운	飛行卷雲
cirrus canopy	권운덮개	卷雲-
cirrus castellanus	탑권운	塔卷雲
cirrus caudatus	꼬리권운	-卷雲
cirrus cloud	권운	卷雲
cirrus deck	납작지붕권운	-卷雲
cirrus densus	짙은 권운	-卷雲
cirrus duplicatus	(두)겹권운	-卷雲
cirrus excelsus	높은 권운	-卷雲
cirrus fibratus(= cirrus filosus)	섬유권운	纖維卷雲
cirrus floccus	송이권운	-卷雲
cirrus intortus	엉킨 권운	-卷雲
cirrus mamma	유방권운	乳房卷雲
cirrus nebula	면사포권운	面紗布卷雲
cirrus nothus	짙은 권운	-卷雲
cirrus plume	깃털권운	-卷雲
cirrus radiatus	방사권운	放射卷雲
cirrus sheet	권운시트	卷雲-
cirrus shield	방패권운	防牌卷雲
cirrus spissatus	짙은 권운	-卷雲
cirrus-type cloud	권운형구름	卷雲形-
cirrus uncinus	갈퀴권운	-卷雲
cirrus uncinus cloud	갈퀴권운	-卷雲
cirrus ventosus(= windy cirrus)	바람권운	-卷雲
cirrus vertebratus	갈비권운	-卷雲
cistern	수은조	水銀槽
cistern barometer	물통기압계	-桶氣壓計
citizen suit	시민소송	市民訴訟
city climate	도시기후	都市氣候
city fog	도시안개	都市-
Civil Aeronautical Administration	민간항공관리국	民間航空管理局
civil day	상용일	常用日
civil time	상용시	常用時
civil twilight	상용박명	常用薄明
Clapeyron equation	클라페이론방정식	-方程式
Clapeyron's diagram	클라페이론선도	-線圖
clapotis	클라포티, 정상파〈프랑스어〉	定常波
clarity(of gas)	명료성, 청정도(기체의)	明瞭性, 淸淨度
Clarke orbit	클라크궤도	-軌道

Clark's nutcrackers	클락의 호두까기	
class	계급	階級
class-A pan	A급증발계	-級蒸發計
classical condensation theory	고전응결이론	古典凝結理論
classical hydrodynamics	고전유체역학	古典流體力學
classic supercell	고전적 초대형세포	古典的超大型細胞
classification	분류	分類
classification of air mass	기단분류	氣團分類
classification of climate	기후구분	氣候區分
classification of clouds	구름분류	-分類
classification of front	전선분류	前線分類
classification of wave forms of atmospherics	공전파형분류	空電波形分類
class interval	계급구간	階級區間
class mark	계급값	階級-
clathrate hydrate	가스하이드레이트	
Clausius-Clapeyron equation	클라우시우스-클라페이론 방정식	-方程式
Clausius equation of state	클라우시우스상태방정식	-狀態方程式
Clausius inequality	클라우시우스부등식	-不等式
Clausius-Kirchhoff relation	클라우시우스-키르히호프 관계	-關係
Clausius theorem	클라우시우스원리	-原理
clay atmometer	진흙증발계	-蒸發計
Clayden effect	클레이든효과	-效果
clean air	맑은 공기	-空氣
clean air act	대기청정법	大氣淸淨法
clean development mechanism	청정개발체제	淸淨開發體制
clear	맑은, 맑음	
clear air	청천	晴天
clear air echo	청천대기에코	晴天大氣-
clear air radar	청천대기레이더	晴天大氣-
clear-air return echo	청천공기에코, 청천되돌이에코	晴天空氣-, 晴天-
clear air turbulence	청천난류	晴天亂流
clearance	운량감소(시각), 비행허가	雲量減少, 飛行許可
clear and pure atmosphere	청결대기	淸潔大氣
clear column radiance	맑은 구역복사휘도	-區域輻射輝度
clear day	맑은 날	
clear ice	맑은 얼음	
clear icing	맑은 착빙	-着氷
clearing	운량감소	雲量減少
clearing house	정보교환센터	情報交換-
clear layer	투명층	透明層
clear line of sight	시계명확선	視界明確線
clear night	맑은 밤	
Clear Skies Act	청정대기법개정안	淸淨大氣法改正案

clear sky	맑음(전운량 1/8 이하)	
clear sky radiation	청천복사	晴天輻射
cliff eddy	낭떠러지맴돌이	
climagram	기후도표	氣候圖表
climagraph	기후도표	氣候圖表
climamatology	기후학	氣候學
CLIMAT broadcast	CLIMAT 방송	-放送
climate	기후	氣候
climate accident	기후사고	氣候事故
climate analogs	기후유사	氣候類似
climate anomaly	기후이상, 기후편차	氣候異常, 氣候偏差
climate atlas	기후도	氣候圖
climate change	기후변화	氣候變化
climate change convention	기후변화협약	氣候變化協約
climate change detection	기후변화탐지	氣候變化探知
climate change feedback	기후변화되먹임	氣候變化-
climate change impacts review group	기후변화영향분석그룹	氣候變化影響分析-
climate change projection	기후변화투영	氣候變化投影
climate change recorded in ice	얼음속기록기후변화	-記錄氣候變化
climate change scenario	기후변화시나리오	氣候變化-
climate change science program	기후변화과학프로그램	氣候變化科學-
climate classification	기후구분, 기후분류(법)	氣候區分, 氣候分類(法)
climate clonological table	기후연표	氣候年表
climate comfort	기후쾌적	氣候快適
climate cycle	기후주기	氣候週期
climate data	기후자료	氣候資料
climate database	기후데이터베이스	氣候-
climate data output	기후자료출력	氣候資料出力
climate diagram	기후도	氣候圖
climate divide	기후분할	氣候分割
climate division	기후구분	氣候區分
climate drift	기후표류	氣候漂流
climate dynamics	기후역학	氣候力學
climate element	기후요소	氣候要素
climate-environment response	기후-환경반응	氣候-環境反應
climate feedback	기후되먹임	氣候-
climate feedback mechanism	기후되먹임기구	氣候-機構
climate fluctuation	기후요동	氣候搖動
climate forcing	기후강제력	氣候强制力
climate forecast	기후예보	氣候豫報
climate forecasting	기후예보	氣候豫報
climate-human feedback	기후-인간 되먹임	氣候-人間-
climate impact assessment	기후영향평가	氣候影響評價
climate impact models	기후영향모형	氣候影響模型

climate impact study	기후영향연구	氣候影響研究
climate index	기후지수	氣候指數
climate information and prediction service	기후정보예측서비스	氣候情報豫測-
climate issue	기후쟁점	氣候爭點
climate-leaf analysis multivariate program	기후-엽다변량분석계획	氣候葉多變量分析計劃
climate-leisure interactions	기후-레저상호작용	氣候-相互作用
climate model	기후모형	氣候模型
climate modelling	기후모델링	氣候-
climate modification	기후조절	氣候調節
climate monitoring	기후감시	氣候監視
climate monitoring and diagnostics laboratory	기후감시진단연구소	氣候監視診斷研究所
climate near the ground	접지기후	接地氣候
climate noise	기후잡음	氣候雜音
climate of eternal frost	영구동토기후	永久凍土氣候
climate outlook	기후전망	氣候展望
climate periodicity	기후주기성	氣候週期性
climate point	기후점	氣候點
climate prediction	기후예측	氣候豫測
climate prediction center	기후예측센터	氣候豫測-
climate projection	기후투영	氣候投影
climate proxy	기후대리요소	氣候代理要素
climate proxy data	기후프록시자료	氣候-資料
climate refugee	기후난민	氣候亂民
climate region	기후구	氣候區
climate research	기후연구	氣候研究
climate research unit	기후연구소	氣候研究所
climate scenario	기후시나리오, 기후각본	氣候脚本
climate science	기후과학	氣候科學
climate sensitivity	기후민감도	氣候敏感度
climate services	기후서비스	氣候-
climate signal	기후신호	氣候信號
climate simulation	기후모의, 기후시뮬레이션	氣候模擬
climate snow line	기후설선	氣候雪線
climate-society study	기후-사회 연구	氣候社會研究
climate statistics	기후통계학	氣候統計學
climate system	기후계	氣候系
climate type	기후형	氣候型
climate variability	기후변동	氣候變動
climate watch	기후감시	氣候監視
climate zone	기후대	氣候帶
climatic alteration	기후변형	氣候變形
climatic amelioration	기후개량	氣候改良
climatic analogue	기후유사	氣候類似
climatic anomaly	기후편차, 기후이상	氣候偏差, 氣候異常

climatic barrier	기후장애	氣候障碍
climatic belt	기후대	氣候帶
climatic boundary	기후경계	氣候境界
climatic chamber	인공기후실	人工氣候室
climatic change	기후변화	氣候變化
climatic change convention	기후변화협약	氣候變化協約
climatic chart	기후도	氣候圖
climatic classification	기후구분, 기후분류(법)	氣候區分, 氣候分類(法)
climatic contrast	기후대조	氣候對照
climatic control	기후조절(인자)	氣候調節(因子)
climatic cycle	기후순환	氣候循環
climatic data	기후자료	氣候資料
climatic diagram(= climatograph)	기후도표	氣候圖表
climatic discontinuity	기후불연속	氣候不連續
climatic divide	기후분할	氣候分割
climatic division	기후구분	氣候區分
climatic element	기후요소	氣候要素
climatic environment	기후환경	氣候環境
climatic extremes	기후극값	氣候極-
climatic factor	기후인자	氣候因子
climatic fluctuation	기후변동	氣候變動
climatic forcing	기후강제	氣候强制
climatic forecast	기후예보	氣候豫報
climatic formula	기후공식	氣候公式
climatic hypothesis of human evolution	인간진화 기후가설	人間進化氣候假說
climatic index	기후지수	氣候指數
climatic instability	기후불안정(성)	氣候不安定(性)
climatic jump	기후점프, 기후비약	氣候飛躍
climatic landscape	기후경관	氣候景觀
climatic limit	기후한계	氣候限界
climatic limit of crop	작물기후한계	作物氣候限界
climatic map	기후도	氣候圖
climatic mapping and prediction project	기후지도작성 및 예측계획, 기후지도작성예측계획	氣候地圖作成豫測計劃
climatic model	기후모형	氣候模型
climatic noise	기후노이즈, 기후잡음	氣候雜音
climatic normal	기후평균값, 평년값	氣候平均-, 平年-
climatic optimum	기후최적기	氣候最適期
climatic oscillation	기후진동	氣候振動
climatic pathology	기후병리학	氣候病理學
climatic periodicity	기후주기성	氣候週期性
climatic physiology	기후생리학	氣候生理學
climatic prediction	기후예측	氣候豫測
climatic productivity index	기후생산력지수	氣候生産力指數

climatic projection	기후전망	氣候展望
climatic province	기후구	氣候區
climatic psychology	기후심리학	氣候心理學
climatic record	기후기록	氣候記錄
climatic regime	기후체제	氣候體制
climatic region	기후구	氣候區
climatic resources	기후자원	氣候資源
climatic retardation	기후지연	氣候遲延
climatic revolution	기후장기변동	氣候長期變動
climatic rhythm	기후리듬	氣候-
climatic risk	기후위기	氣候危機
climatic season	기후계절	氣候季節
climatic seasonal observation	기후계절관측	氣候季節觀測
climatic sensitivity	기후민감도	氣候敏感度
climatic shift	기후편이	氣候偏異
climatic signal	기후신호	氣候信號
climatic snowline	기후설선	氣候雪線
climatic stress	기후스트레스	氣候-
climatic subdivision	아기후구	亞氣候區
climatic symbol	기후기호	氣候記號
climatic system	기후시스템	氣候-
climatic test	기후시험	氣候試驗
climatic therapy	기후요법	氣候療法
climatic tolerances	기후내성	氣候耐性
climatic trend	기후경향	氣候傾向
climatic type	기후형	氣候型
climatic unit	기후단위	氣候單位
climatic vacillation	기후요동	氣候搖動
climatic variability	기후변동성	氣候變動性
climatic variation	기후변동	氣候變動
climatic year	기후년	氣候年
climatic zone	기후대	氣候帶
climatic zone by dynamic climatology	동기후학기후대	動氣候學氣候帶
climatic zone by statical climatology	정기후학기후대	靜氣候學氣候帶
climatization	기후적응	氣候適應
climatizer	기후실험실	氣候實驗室
CLIMAT message	CLIMAT 메시지	
climatogenesis	기후생성	氣候生成
climatogram	기후도	氣候圖
climatograph	기후도표	氣候圖表
climatography	기후지	氣候誌
climatological	기후학적	氣候學的
climatological approach	기후학적 접근	氣候學的接近
climatological archives	기후저장소	氣候貯藏所

climatological atlas	기후도감	氣候圖鑑
climatological chart	기후도	氣候圖
climatological conductance	기후전도도	氣候傳導度
climatological data	기후자료	氣候資料
climatological data annual report	기후자료연보	氣候資料年報
climatological diagram	기후선도	氣候線圖
climatological dispersion model	기후분산모형	氣候分散模型
climatological division	기후학적 분할	氣候學的分割
climatological forecast	기후학적 예보	氣候學的豫報
climatological front	기후전선	氣候前線
climatological intra-seasonal oscillations	기후학적 계절간진동	氣候學的季節間振動
climatological mean profile	기후학적 평균연직분포	氣候學的平均鉛直分布
climatological method	기후학적 방법	氣候學的方法
climatological network	기후관측망	氣候觀測網
climatological observation	기후관측	氣候觀測
climatological outlook	기후학적 전망	氣候學的展望
climatological resistance	기후학적 저항	氣候學的抵抗
climatological scale	기후학적 규모	氣候學的規模
climatological standard normals	기후표준평년값(1981~2010)	氣候標準平年-
climatological station	기후관측소	氣候觀測所
climatological station elevation	기후관측소표고	氣候觀測所標高
climatological station for specific purposes	특수목적기후관측소	特殊目的氣候觀測所
climatological station network	기후관측소망	氣候觀測所網
climatological station pressure	기후관측소기압	氣候觀測所氣壓
climatological statistics	기후통계(학)	氣候統計(學)
climatological substation	기후관측분소	氣候觀測分所
climatological summary	기후개황	氣候概況
climatological year	기후년	氣候年
climatologist	기후학자	氣候學者
climatology	기후학	氣候學
climatostratigraphy	기후층서학	氣候層序學
climatotherapy	기후요법	氣候療法
climax	극상	極相
climax vegetation	극상식생	極相植生
climogram	클라이모그램	
climograph	클라이모그래프	
clinometer	앙각계	仰角計
clo	클로, 인체보온지수	人體保溫指數
clock-star	시계별	時計-
clockwise	시계방향	時計方向
clockwise wind	시계방향바람	時計方向-
clockwork anemometer	시계장치풍속계	時計裝置風速計
Cloncurry low	클론커리저기압	-低氣壓
close air support operation weather	근접공중지원작전기상	近接支援作戰氣象

closed basin	닫힌 만	
closed cells	닫힌 (구름)세포	-細胞
closed-cell stratocumulus	닫힌 세포층적운	閉細胞層積雲
closed circuit wind tunnel	순환형풍동, 회유형	循環型風洞, 回流型
closed circulation	닫힌 순환	-循環
closed curve	닫힌 곡선	-曲線
closed drainage	닫힌 유역	-流域
closed high	닫힌 고기압	-高氣壓
closed interval	닫힌 구간	-區間
closed isobar	닫힌 등압선	-等壓線
closed lake	닫힌 호수	-湖水
closed low	닫힌 저기압	-低氣壓
closed system	닫힌 계	-系
closed weather	닫힌 날씨	-氣象
close-in fallout	근접낙진	近接落塵
close pack ice	밀유빙	密遊氷
closure	종결	終結
closure assumption	종결가정	終結假定
closure problem	종결문제	終結問題
closure scheme	종결방안	終結方案
closure technique	종결기법	終結技法
closure turbulence	종결난류	終結亂流
clothing climate	의복기후	衣服氣候
cloud	구름	
cloud absorption	구름흡수	-吸收
cloudage	구름양	-量
cloud albedo	구름알베도, 구름반사율	-反射率
cloud algorithm	구름알고리즘	
cloud amount	운량, 구름양	雲量
cloud and vapo(u)r tracking	구름과 수증기 추적	-水蒸氣追跡
cloud atlas	구름도감	-圖鑑
cloud attenuation	구름감쇠	-減衰
cloud band	구름띠	
cloud bank	구름둑	
cloud banner	구름깃발	
cloud bar	구름줄	
cloud base	구름밑면	-面
cloud base recorder	구름밑면기록계	-面記錄計
cloud base temperature	구름밑면온도	-面溫度
cloud boundary	구름경계	-境界
cloudbow	구름무지개	
cloud-breeze	구름미풍	-微風
cloudburst	폭우	暴雨
cloudburster	구름소산기	-消散器

cloud camera	구름용카메라	-用-
cloud canopy	구름캐노피	
cloud cap(= cap cloud)	구름모자	-帽子
cloud ceiling	운고	雲高
cloud cells	구름세포	-細胞
cloud chamber	구름상자	-箱子
cloud chart	구름도	-圖
cloud classification	구름분류	-分類
cloud classification scheme	구름분류방법	-分類方法
cloud climatology	구름기후, 구름기후학	-氣候, -氣候學
cloud cluster	구름무리	
Cloud Condensation Level(CCL)	구름응결고도	-凝結高度
cloud condensation nuclei counter	구름응결핵계수기	-凝結核計數機
cloud condensation nucleus	구름응결핵	-凝結核
cloud cover	운량	雲量
cloud cover radiometer	구름양 복사계	-量輻射計
cloud crest	삿갓구름	
cloud current	구름(속)기류	-氣流
cloud deck	구름상부면	-上部面
cloud depth	구름깊이, 구름두께	
cloud-detection radar	구름탐지레이더	-探知-
cloud detrainment	구름배기, 구름유출	-排氣, -流出
cloud development	구름발달	-發達
cloud diary	구름일지	-日誌
cloud direction	구름방향	-方向
cloud discharge	구름방전	-放電
cloud dispersal (or dissipation)	구름소산	-消散
cloud drift winds	구름표류바람	-漂流-
cloud drop	구름방울, 운적	雲滴
cloud droplet	작은 구름방울	
cloud droplet collector	구름방울포착기	-捕捉器
cloud droplet growth	(작은)구름방울성장	-成長
cloud droplet size	구름방울 크기	
cloud drop sampler	구름방울채집기	-採集器
cloud due to volcanic eruption	화산구름	火山-
cloud dynamics	구름역학	-力學
cloud echo	구름에코	
cloud electrification	구름대전	-帶電
cloud element	구름요소	-要素
cloud emittance	구름방출도	-放出度
cloud entrainment	구름흡입	-吸入
cloud-environment system	구름-환경계	-環境系
cloud estimation	구름추정	-推定
cloud étage	구름고도역	-高度域

cloud feedback	구름되먹임	
cloud filtering	구름필터링	
cloud flash(= cloud discharge)	구름 속 방전	-放電
cloud flying	구름 속 비행	-飛行
cloud forest	구름숲	
cloud form	구름모양, 운형	-模樣, 雲形
cloud formation	구름생성, 구름형성	-生成, -形成
cloud-free area	맑은 구역	-區域
cloud-free line of sight	구름 없는 시선	-視線
cloud frequency	구름빈도	-頻度
cloud from explosions	폭발구름	爆發-
cloud from fire	화재구름	火災-
cloud from imager	영상계구름	映像計-
cloud from industry	산업구름	産業-
cloud from volcanic eruptions	화산분출구름	火山噴出-
cloud from waterfall	폭포구름	瀑布-
cloud front	한랭전선	寒冷前線
cloud genera	구름종류	-種類
cloud genus	구름종류	-種類
cloud growth	구름성장	-成長
cloud gun	구름총	-銃
cloud head	구름상부, 구름머리	-上部
cloud height	구름높이, 운고	雲高
cloud height indicator	구름높이지시기	-指示器
cloud height measurement method	운고측정법	雲高測定法
cloud height meter	운고계	雲高計
cloud icing nucleus	구름빙정핵	-氷晶核
cloud identification	구름식별	-識別
cloudier	인조구름	人造-
cloud image animation	구름동영상	-動映像
cloud indexing	구름지수화	-指數化
cloudiness	구름양	-量
cloud layer	구름층	-層
cloud leaf	구름조각	
cloudless	맑음(전운량 1/8 이하)	
cloudlet	구름조각	
cloud level	구름고도, 구름층	-高度, -層
cloud line	구름선	-線
cloud liquid water	구름수분량	-水分量
cloud liquid water content	구름수분함량	-水分含量
cloud luminance	구름휘도	-輝度
cloud mass	구름덩이	
cloud microphysical model	구름미세물리학모형	-微細物理學模型
cloud microphysics	구름미세물리	-微細物理學

C

cloud microstructure	구름미세구조	-微細構造
cloud mirror	구름거울	
cloud model	구름모형	-模型
cloud model technique	구름모형기법	-模型技法
cloud modification	구름변조	-變造
cloud motion vector	구름이동벡터	-移動-
cloud movement	구름이동	-移動
cloud nucleus	구름핵	-核
cloud observation	구름관측	-觀測
cloud of spiral pattern	나선형 구름	螺旋形-
cloud optical depth	구름광학두께, 구름광학깊이	-光學-
cloud parameterization	구름매개변수화	-媒介變數化
cloud particle	구름입자, 구름알갱이	-粒子
cloud particle camera	구름입자카메라, 운립자카메라	雲粒子-
cloud particle counter	운립자계수기	雲粒子計數器
cloud particle imager	구름입자영상기, 운립자영상기	雲粒子映像機
cloud particle sampler	운립자채집기	雲粒子採集器
cloud particle sonde	운립자존데	雲粒子-
cloud pattern	구름패턴	
cloud phase	구름위상	-位相
cloud phase chart	구름상도	-相圖
cloud photogrammetry	구름사진측량기법	-寫眞測量技法
cloud photography	구름촬영술	-撮影術
cloud photopolarimeter	구름(망원사진)편광계	-(望遠寫眞)偏光計
cloud physical implication	구름물리학적함축	-物理學的含蓄
cloud physics	구름물리학	-物理學
cloud physics observation instrument	구름물리관측장비	-物理觀測裝備
cloud picture	구름사진	-寫眞
cloud radar	구름레이더	
cloud radiative forcing	구름복사강제력	-輻射强制力
cloud reflectance	구름반사도	-反射度
cloud reflector	구름반사거울	-反射-
cloud resolving model	구름분해능모형	-分解能模型
cloud sea	구름바다, 운해	雲海
cloud searchlight	구름탐조등	-探照燈
cloud sector	구름구역	-區域
cloud seeding	구름씨뿌리기	
cloud seeding agent	구름씨뿌리기물질	-物質
clouds from sounders	탐사구름	探査-
clouds from water falls	폭포운	瀑布雲
cloud shadow	구름그림자	
cloud sheet	구름시트	
cloud shield	구름방패	-防牌
cloud simulator	구름시뮬레이터	

cloud species	구름의 종	-種
cloud speed	구름속도	-速度
cloud spiral	구름나선	-螺線
cloud stage	구름단계	-段階
cloud streaks	구름띠	
cloud street	클라우드스트리트	
cloud structure	구름구조	-構造
cloud symbol	구름기호	-記號
cloud system	구름계	-系
cloud system of an depression	저기압구름계	低氣壓-系
cloud tags	구름꼬리표	
cloud thickness	구름두께	
cloud-to-cloud discharge	구름사이방전	-放電
cloud-to-ground discharge	벼락, 구름과 땅 사이의 방전	-放電
cloud-to-ground flash	구름-지면 사이 섬광	-地面-閃光
cloud-to-ground lightning	구름-지면 사이 번개	-地面-
cloud top	운정, 구름꼭대기	雲頂
cloud-top entrainment instability	운정유입불안정도	雲頂流入不安定度
cloud top height	운정고도	雲頂高度
cloud-top indicator	운정지시기	雲頂指示器
cloud-topped boundary layer	구름낀 경계층	-境界層
cloud top pressure	운정기압	雲頂氣壓
cloud top temperature	운정온도	雲頂溫度
cloud tracker	구름탐색기	-探索機
cloud-track winds	구름경로바람	-經路-
cloud transmissivity	구름투과도	-透過度
cloud turbulence	구름난류	-亂流
cloud type	구름형	-型
cloud variety	구름변종	-變種
cloud veil	구름막	-幕
cloud velocity	구름속도	-速度
cloud water content	구름수분함량, 구름함양수	-水分含量, -涵養水
cloud winds	구름바람	
cloud with great vertical development	강한 연직운	-鉛直雲
cloud with vertical development	연직발달구름	鉛直發達-
cloudy	흐림	
cloudy convection	흐린 날 대류, 구름대류	-對流
cloudy day	흐린 날	
cloudy sky	흐린 하늘	
cloudy weather	흐린 날씨	
cloudy zone	구름대	-帶
clumping	부착	附着
cluster	무리	
cluster analysis	클러스터분석	-分析

cluster diffusion	클러스터확산	-擴散
clustering	무리짓기	
cluster ion	클러스터이온	
cluster map	클러스터일기도	-日氣圖
cluster of thunderstorms	뇌우클러스터, 뇌우복합	雷雨複合
clutter	클러터	
clutter filter	클러터필터	
clutter map	클러터지도	-地圖
clutter rejection	클러터제거	-除去
cnoidal wave	크노이드파	-波
CO_2 compensation point	이산화탄소보상점	二酸化炭素補償點
CO_2 fertilization effect	이산화탄소시비효과	二酸化炭素施肥效果
coagulation	결집, 부착	結集, 附着
coalescence	병합(구름알갱이)	倂合
coalescence coefficient	병합계수	倂合係數
coalescence effect	병합효과	倂合效果
coalescence efficiency	병합효율	倂合效率
coalescence process	병합과정	倂合過程
coalescence theory	병합설	倂合說
coarse-grained snow	거친 알갱이눈	
coarse mesh	성긴 격자	-格子
coarse mesh grid	성긴 그물격자	-格子
coarse particle	거대입자, 조대입자	巨大粒子, 粗大粒子
coarse particle fraction	거대입자분율	巨大粒子分率
coastal boundary layer	해안경계층	海岸境界層
coastal climate	해안기후	海岸氣候
coastal community	해안집단	海岸集團
coastal desert	해안사막	海岸砂漠
coastal desert climate	해안사막기후	海岸沙漠氣候
coastal diffusion	해안확산	海岸擴散
coastal effect	해안효과	海岸效果
coastal environments	해안환경	海岸環境
coastal erosion	해안침식	海岸浸蝕
coastal flooding	연안홍수	沿岸洪水
coastal flood warning	연안홍수경보	沿岸洪水警報
coastal flood watch	연안홍수감시보	沿岸洪水監視報
coastal fog	연안안개	沿岸-
coastal front	해안전선	海岸前線
coastal impact	해안영향	海岸影響
coastal infiltration weather	해안침투기상	海岸浸透氣象
coastal jet	해안제트	海岸-
coastal/lakeshore flood advisory	연안/호안 홍수주의보	沿岸/湖岸洪水注意報
coastal/lakeshore flooding	연안/호안 홍수	沿岸/湖岸洪水
coastal/lakeshore flood warning	연안/호안 홍수경보	沿岸/湖岸洪水警報

coastal/lakeshore flood watch	연안/호안 홍수감시보	沿岸/湖岸洪水監視報
coastally trapped wave	연안간힘파	沿岸-波
coastal night fog	해안밤안개	海岸-
coastal plains	해안평야	海岸平野
coastal region	연안지역	沿岸地域
coastal route	연안항로	沿岸航路
coastal route forecast	연안항로예보	沿岸航路豫報
coastal sea	연안바다	沿岸-
coastal storm	해안폭풍	海岸暴風
coastal upwelling	연안용승	沿岸湧昇
coastal water	연안물	沿岸-
coastal water forecast	연안물예보	沿岸-豫報
coastal zone	연안대	沿岸帶
coastal zone color scanner	연안대색주사기	沿岸帶色走査機
coastline	해안선	海岸線
coastline change	해안선변화	海岸線變化
coastline length	해안선길이	海岸線-
coaxial cable	동축케이블	同軸-
coaxial collinear dipole	동축쌍극	同軸雙極
coaxial correlation method	동축상관법, 공축상관법	同軸相關法, 共軸相關法
cobblestone turbulence	약한 난류	弱-亂流
coccolith	인편모충, 유공충	鱗鞭毛蟲, 有孔蟲
cockeyed bob	콕아이밥(오스트레리아 북서안에서 여름철에 일어나는 뇌우를 동반하는 스콜)	
co-cumulative spectrum	공누적스펙트럼	共累積-
code figure	부호수	符號數
code form	부호형식	符號型式
code group	부호무리	符號-
code letter	부호문자	符號文字
code-sending radiosonde(= code-type radiosonde)	부호송신라디오존데	符號送信-
code sonde	부호존데	符號-
code specification	부호설명	符號說明
code symbol	부호기호	符號記號
code table	부호표	符號表
code-type radiosonde	부호형라디오존데	符號型-
coding	부호화	符號化
codomain	공역	共域
coefficient	계수	係數
coefficient of absorption	흡수계수	吸收係數
coefficient of air permeability	통기계수	通氣係數
coefficient of aridity	건조계수	乾燥係數
coefficient of barotropy	순압계수	順壓係數
coefficient of compressibility	압축계수	壓縮係數
coefficient of consolidation	경화계수	硬化係數

coefficient of continentality	대륙도계수	大陸度係數
coefficient of correlation	상관계수	相關係數
coefficient of cubical expansion	체적팽창계수	體積膨脹係數
coefficient of diffusion	확산계수	擴散係數
coefficient of dynamic viscosity	역학점성계수	力學粘性係數
coefficient of eddy conduction	맴돌이전도계수	-傳導係數
coefficient of eddy diffusion	맴돌이확산계수	-擴散係數
coefficient of eddy viscosity	맴돌이점성계수	-粘性係數
coefficient of excess	과잉계수	過剩係數
coefficient of exchange	교환계수	交換係數
coefficient of expansion	팽창계수	膨脹係數
coefficient of extinction	소산계수	消散係數
coefficient of friction	마찰계수	摩擦係數
coefficient of haze	안개계수	-係數
coefficient of heat conduction	열전도계수	熱傳導係數
coefficient of humidity	습도계수	濕度係數
coefficient of kinematic viscosity	운동점성계수	運動粘性係數
coefficient of kinetic friction	운동마찰계수	運動摩擦係數
coefficient of molecular viscosity	분자점성계수	分子粘性係數
coefficient of multiple correlation	다중상관계수	多重相關係數
coefficient of mutual diffusion	상호확산계수	相互擴散係數
coefficient of piezotropy	압성계수	壓性係數
coefficient of polytropy	다방계수	多方係數
coefficient of runoff, flood coefficient	유출계수	流出係數
coefficient of skewness	왜도계수	歪度係數
coefficient of skin friction	표피마찰계수	表皮摩擦係數
coefficient of static friction	정지마찰계수	靜止摩擦係數
coefficient of storage	저류계수	貯留係數
coefficient of tension	장력계수	張力係數
coefficient of thermal conduction	열전도계수	熱傳導係數
coefficient of thermal expansion	열팽창계수	熱膨脹係數
coefficient of transmission	투과계수	透過係數
coefficient of transparency	투명계수	透明係數
coefficient of turbidity	혼탁계수	混濁係數
coefficient of viscosity	점성계수	粘性係數
cognitive task analysis	인지과제분석	認知課題分析
coherence	동조, 간섭성, 상관성, 점착, 부착	同調, 干涉性, 相關性, 粘着, 付着
coherence element	동조요소	同調要素
coherence time	동조시간	同調時間
coherent backscatter radar	간섭후방산란 레이더	干涉後方散亂-
coherent detection	동기검파	同期檢波
coherent Doppler lidar	위상정합 도플러 라이다	位相整合-
coherent echo	동조에코	同調-

coherent error integrator	간섭성오차 적분회로	干涉性誤差積分回路
coherent integration	정합적분, 간섭적분, 동조적분	整合積分, 干涉積分, 同調積分
coherent light detection and ranging	간섭광거리탐지기	干涉光距離探知器
coherent memory filter	간섭기억필터	干涉記憶-
coherent optical radar	간섭광학레이더	干涉光學-
coherent oscillator	간섭발진기	干涉發振器
coherent radar	간섭성레이더	干涉性-
coherent radiation	간섭복사	干涉輻射
coherent receiver	코히어런트 수신기, 간섭수신기	干涉受信機
coherent scattering	간섭성산란	干涉性散亂
coherent signal	코히어런트 신호, 정합신호, 간섭신호	-整合信號, 干涉信號
coherent structure	동조구조, 응집구조	同調構造, 凝集構造
coherent target	가간섭성목표물	可干涉性目標物
coherent transmitter	간섭송신기	干涉送信機
coherent type	위상정합형	位相整合型
cohesion	응집	凝集
coincidence error	동시성오류	同時性誤謬
coincidental correlation	동시상관	同時相關
cokriging	공동크리깅	共同-
col	안장부	鞍裝部
co-latitude	여위도	餘緯度
cold	추위, 차가움	
cold advection	한랭이류	寒冷移流
cold air	찬 공기	-空氣
cold air advection	찬 공기이류	-空氣移流
cold air avalanche	찬 공기사태	-空氣沙汰
cold air dam	찬 공기댐, 냉기댐	-空氣-, 冷氣-
cold air damming	찬 공기 가두기	-空氣-
cold air dome	찬 공기돔	-空氣-
cold air drainage	냉기류, 냉기배기	冷氣流, 冷氣排氣
cold air drop(= cold pool)	찬 공기강하	-空氣降下
cold air flow	찬 공기흐름, 냉기류	冷氣流
cold air funnel	찬 공기깔때기, 한랭공기깔때기	寒冷空氣-
cold air lake	찬 공기호수, 냉기호	-空氣湖水, 冷氣湖
cold air-mass	한랭기단	寒冷氣團
cold air pool	찬 공기풀	-空氣-
cold air sink	찬 공기싱크	-空氣-
cold air injection	찬 공기유입	-空氣流入
cold air outbreak	찬 공기터져나감	-空氣-
cold anticyclone	한랭고기압	寒冷高氣壓
cold belt	한랭대	寒冷帶

cold bias	한랭편향	寒冷偏向
cold box	냉각상자	冷却箱子
cold cap	극한대	極寒帶
cold climate	한랭기후	寒冷氣候
cold cloud	한랭구름	寒冷-
cold conveyor belt	한랭컨베이어벨트	寒冷-
cold core	한랭핵	寒冷核
cold core anticyclone	한랭고기압	寒冷高氣壓
cold core cyclone	한랭저기압	寒冷低氣壓
cold-cored system	한랭핵시스템	寒冷核-
cold core high	한랭고기압	寒冷高氣壓
cold core low	한랭저기압	寒冷低氣壓
cold core ring	한랭핵고리	寒冷核-
cold core thunderstorms	한랭핵뇌우	寒冷核雷雨
cold core vortex	한랭소용돌이	寒冷-
cold current	한류	寒流
cold cyclone	한랭저기압	寒冷低氣壓
cold damage	냉해	冷害
cold desert	한랭사막	寒冷砂漠
cold dew	찬 이슬, 한로	寒露
cold district	한랭지	寒冷地
cold dome	한랭돔	寒冷-
cold downslope wind	한랭하강풍	寒冷下降風
coldest month	최한월	最寒月
cold front	한랭전선	寒冷前線
cold frontal surface	한랭전선면	寒冷前線面
cold front occlusion	한랭전선폐색	寒冷前線閉塞
cold front precipitation	한랭전선형 강수	寒冷前線型降水
cold front rain	한랭전선비	寒冷前線-
cold front thunderstorm	한랭전선뇌우	寒冷前線雷雨
cold-front-type occlusion	한랭전선형폐색	寒冷前線型閉塞
cold front wave	한랭전선파	寒冷前線波
cold half year	한후기	寒候氣
cold hardiness	내한성	耐寒性
cold high	한랭고기압	寒冷高氣壓
cold injury	한해	寒旱
cold lightning	한랭번개	寒冷-
cold low	한랭저기압	寒冷低氣壓
coldness	한랭	寒冷
coldness index	한랭지수	寒冷指數
cold occlusion	한랭폐색	寒冷閉塞
cold occlusion depression	한랭폐색저기압	寒冷閉塞低氣壓
cold period	한랭기	寒冷期
cold polar climate	한랭한대기후, 한대기후	寒冷寒帶氣候, 寒帶氣候

cold pole	한극	寒極
cold pool	찬 공기풀	-空氣-
cold protection	추위막이, 방한	防寒
cold rain	찬비	
cold resistance	내한성	耐寒性
cold season	한랭철	寒冷-
cold sector	추운 계절	-季節
cold soak	한랭호우	寒冷豪雨
cold source	한랭원	寒冷源
cold spell	한랭기간	寒冷期間
cold stroke(= frostbite)	동상	凍傷
cold summer damage	냉하해, 냉해	冷夏害, 冷害
cold surge	한기쇄도	寒氣殺到
cold thunderstorm	한랭뇌우	寒冷雷雨
cold tongue	한랭혀	寒冷-
cold trough	한랭골	寒冷-
cold-type occluded front	한랭형 폐색전선	寒冷形閉塞前線
cold-type occlusion	한랭형 폐색	寒冷型閉塞
cold vortex	한랭소용돌이	寒冷-
cold wall	한랭벽	寒冷壁
cold water coral	찬 물산호	-珊瑚
cold water resistance	내냉수성	耐冷水性
cold wave	한파	寒波
cold wave advisory	한파주의보	寒波注意報
cold wave warning	한파경보	寒波警報
cold weather in blooming season	꽃샘추위	
cold weather operation	극한지작전	極寒地作戰
cold weather resistance	냉기내성	冷氣耐性
cold wedge	한랭쐐기	寒冷-
cold wind injury, damage	한풍해	寒風害
cold zone	한랭대	寒冷帶
collar cloud	깃구름	
collecting method	채집법	採集法
collection	채집	捕捉
collection efficiency	채집효율	捕捉效率
collective	집단, 공동체	集團, 共同體
collector	채집기, 집진기	採集器, 集塵機
colligative property	용액특성	溶液特性
collimator	시준의	視準儀
collision	충돌	衝突
collisional deactivation	충돌붕괴	衝突崩壞
collisional excitation	충돌들뜸	衝突-
collision broadening	충돌확장	衝突擴張
collision coalescence process	충돌병합과정	衝突倂合過程

collision efficiency	충돌효율	衝突效率
collision frequency	충돌빈도	衝突頻度
collision integral	충돌적분	衝突積分
collision theory	충돌이론	衝突理論
colloid	콜로이드	
colloidal dispersion	콜로이드분산	-分散
colloidal instability	콜로이드불안정(도)	-不安定(度)
colloidally stable	콜로이드적안정	-的安定
colloidal metastable	콜로이드준안정	-準安定
colloidal solution	콜로이드용액	-溶液
colloidal stability	콜로이드안정(도)	-安定(度)
colloidal stable	콜로이드적안정	-的安定
colloidal suspension	콜로이드부유	-浮游
colloidal system	콜로이드계	-系
colloid meteorology	콜로이드기상학	-氣象學
colonization	식민지건설, 식민지화	植民地建設, 植民地化
Colorado cyclone	콜로라도저기압	-低氣壓
Colorado low	콜로라도저기압	-低氣壓
Colorado sunken pan	콜로라도증발계	-蒸發計
Colorado wave cyclone	콜로라도파동저기압	-波動低氣壓
colorant	색소	色素
color bar	색눈금, 컬러바	色-
color charge	색전하	色電荷
color composite	색채합성	色彩合成
color criteria	색채기준	色彩基準
colored rain	색깔비	色-
color formation process	색채형성과정	色彩形成課程
color perception	색인식, 색지각	色認識, 色知覺
color sense	색각	色覺
color temperature	색온도	色溫度
colpus	발아구	發芽口
column	기둥(빙정의)	
column abundance	기둥함량비	-含量比
columnar crystal	기둥모양결정	-模樣結晶
columnar recombination	기둥재결합	-再結合
columnar resistance	기둥저항	-抵抗
columnar snow crystals	기둥눈결정	-結晶
columnar vortex	기둥모양소용돌이	-模樣-
column concentration	칼럼농도	-濃度
column max	칼럼최대치	-最大値
column model	기둥모형	-模型
column ozone	오존전량	-全量
comanchaca	코만차카(남아메리카의 낮은 층구름에서 유래된 짙은 안개)	

COMBAR code	콤바부호	-符號
combat aircraft code	전투기용전문형식	戰鬪機用電文型式
comber	부서지는 파도	-波濤
combination	조합	組合
combination band	결합밴드	結合-
combination coefficient	결합계수	結合係數
combination equation	결합방정식	結合方程式
combination model	결합모형	結合模型
combination principle	결합원리	結合原理
combined recording wind vane and fan-anemograph	풍차형 풍향풍속계	風車型風向風速計
combined satellite and radar imagery	위성-레이더 조합영상	衛星-組合映像
combining observation	연합관측	聯合觀測
comb nephoscope	빗모양측운기	-模樣測雲器
combustible element	가연원소	可燃元素
combustion	연소	燃燒
combustion dust	연소먼지	燃燒-
combustion exhaust gas	연소배기가스	燃燒排氣-
combustion nucleus	연소핵	燃燒核
combustion reaction	연소반응	燃燒反應
comet	혜성	彗星
comfortable temperature	쾌적온도	快適溫度
comfort chart	쾌적도표	快適圖表
comfort curve	쾌적곡선	快適曲線
comfort index	쾌적지수	快適指數
comfort standard	쾌적기준	快適基準
comfort zone	쾌적대	快適帶
comma cloud	쉼표모양구름	-標模樣-
comma cloud system	쉼표모양구름시스템	-標模樣-
comma head	콤마머리, 쉼표머리	-標-
command and data acquisition station	통제자료수신소	統制資料受信所
comma-shaped cloud	쉼표모양구름	-標模樣-
comma tail	쉼표구름꼬리	-標-
commercial satellite	상업용위성	商業用衛星
commercial weather	상업날씨	商業-
comminution	분쇄	粉碎
commission error	포함오차	包含誤差
common difference	공차	公差
common factor analysis	공통요인분석	共通要因分析
common law	상용법칙	常用法則
common logarithm	상용대수	常用對數
common water	상용수	常用水
commonwealth scientific and industrial research organization	(오스트레일리아) 연방과학산업연구원	聯邦科學産業研究院

communicating climate change	기후변화 커뮤니케이션	氣候變化-
communication center	통신센터	通信-
communication oceanic and meteorological satellite	통신해양기상위성	通信海洋氣象衛星
communication satellite	통신위성	通信衛星
communication station	통신소	通信所
community	군집	群集
community climate model	군집기후모형	群集氣候模型
community noise standard	지역소음표준, 군집잡음표준	地域騷音標準, 群集雜音標準
commutative law	교환법칙	交換法則
commutator	정류자, 정류기	整流子, 整流器
commutator radiosonde	정류식라디오존데	整流式-
compact differencing	집적차분화	集積差分化
compaction	압축, 치밀화작용	壓縮, 緻密化作用
compactness	수축, 축소	收縮, 縮小
comparability	비교성	比較性
comparative meteorology	비교기상학	比較氣象學
comparative rabal	비교레이벌	比較-
compass	나침반	羅針盤
compatibility condition	적합성조건	適合性條件
compensated air thermometer	보정온도계	補正溫度計
compensated pyrheliometer	보정직달일사계	補正直達日射計
compensated scale barometer	보정눈금기압계	補正-氣壓計
compensating pyrheliometer	보정직달일사계	補正直達日射計
compensation	보상	補償
compensation current	보상기류, 보상해류	補償氣流, 補償海流
compensation of instrument	측기보정	測器補正
compensation point	보상점	補償點
competence of the wind	바람의 고형물수송력	-固形物輸送力
complementarity principle	상보성원리	相補性原理
complementary angle	여각	餘角
complementary color	보색	補色
complementary light	보색광	補色光
complete(perfect) combustion	완전연소	完全燃燒
complete(perfect) elastic collision	완전탄성충돌	完全彈性衝突
complete freeze-up	완전동결	完全凍結
complex climatology	복합기후학	複合氣候學
complex cross spectrum	복잡교차스펙트럼	複雜交叉-
complex hydrograph	복잡수문곡선	複雜水文曲線
complex index of refraction (= complex refractive index)	복합굴절률	複合屈折率
complexity	복잡성	複雜性
complex low	복합저기압	複合低氣壓
complex multiplication	복소수승법	複素數乘法

complex number	복소수	複素數
complex number field	복소수장	複素數場
complex plane	복소평면	複素平面
complex quality control	복잡품질관리	複雜品質管理
complex refractive index	복합굴절률지수	複合屈折率指數
complex signal	복합신호	複合信號
complex singularity	복합특이성	複合特異性
complex system	복합시스템, 복합계	複合系
complex terrain	복잡지형	複雜地形
complex variable	복소수변수	複素數變數
component	성분	成分
component of force	힘의 성분	-成分
component vector	성분벡터	成分-
component velocity	성분속도	成分速度
composite analysis	합성분석	合成分析
composite chart(= composite map)	합성도	合成圖
composite flash	복합방전	複合放電
composite forecast chart	합성예보도	合成豫報圖
composite glacier	합성빙하	合成氷河
composite hygrograph	합성수문곡선	合成水文曲線
composite moisture stability	복합습윤안정도	複合濕潤安定度
composite prognostic chart	합성예상도	合成豫想圖
composite reflectivity	합성반사율	合成反射率
composite stroke	합성벼락	合成-
composite variogram	합성배리오그램	合成-
composite vertical cross-section	합성연직단면(도)	合成鉛直斷面(圖)
composite water sample	합성수샘플	合成水-
composite wave	합성파	合成波
composition	조성	組成
composition of atmosphere	대기조성	大氣組成
composition of force	힘의 합성	-合成
composition of vector	벡터합성	-合成
composition sounding	조성탐측	組成探測
compound	화합물	化合物
compound centrifugal force	복합원심력	複合遠心力
compound hydrograph	복합수문곡선	複合水文曲線
compound probability	복합확률	複合確率
compound proportion	복비례	複比例
compound-X	미지화합물	未知化合物
compressibility	압축성	壓縮性
compressibility coefficient	압축계수	壓縮係數
compressibility wave	압축파	壓縮波
compressible	압축성(의)	壓縮性
compressible fluid	압축성유체	壓縮性流體

compression	압축, 압력	壓縮, 壓力
compressional creep	압축변형	壓縮變形
compressional warming	압축온난화	壓縮溫暖化
compression ignition engine	압축점화엔진	壓縮點火-
compression ratio	압축비	壓縮比
compression wave	압축파	壓縮波
Compton effect	콤프턴효과	-效果
Compton electron	콤프턴전자	-電子
computational accuracy	계산정확도	計算正確度
computational constraint	전산구속조항	電算拘束事項
computational diffusion	계산확산	計算擴散
computational dispersion	계산적 분산, 계산분산	計算的分散, 計算分散
computational instability	계산불안정(성)	計算不安定(性)
computational mechanics	전산역학	電算力學
computational mode	계산모드, 계산방식	計算方式
computational resource	전산자원	電算資源
computer simulation	컴퓨터 모의	-模擬
concave	오목면	-面
concentration	함량, 농도	含量, 濃度
concentration basin	농도구덩이	濃度-
concentration convection	농도대류	濃度對流
concentration distribution	농도분포	濃度分布
concentration gradient	농도경도	濃度傾度
concentration of particle	수농도	數濃度
concentration time	집결시간	集結時間
concentration variance	농도분산	濃度分散
concentric eyewall	동심눈벽	同心眼壁
concentric eyewall cycle	동심눈벽순환	同心眼壁循環
conceptual model	개념모형	槪念模型
conceptual tool	개념도구	槪念道具
concrete minimum temperature	콘크리트최저온도	-最低溫度
condensability	응결성	凝結性
condensation	응결	凝結
condensation adiabat	응결단열선	凝結斷熱線
condensation efficiency	응결효율	凝結效率
condensation growth process	응결성장과정	凝結成長過程
condensation layer	응결층	凝結層
condensation level	응결고도	凝結高度
condensation limit	응결한계	凝結限界
condensation nucleus	응결핵	凝結核
condensation nucleus counter	응결핵계수기	凝結核計數器
condensation pressure	응결압력	凝結壓力
condensation process	응결과정	凝結過程
condensation stage	응결단계	凝結段階

condensation temperature	응결온도	凝結溫度
condensation trail	응결자국	凝結-
condensation wave	응결파	凝結波
condenser discharge anemometer	방전식풍속계	放電式風速計
conditional conservatism	조건부보존성	條件附保存性
conditional distribution	조건분포	條件分布
conditional equation of perturbation	섭동조건식	攝動條件式
conditional equilibrium	조건부평형	條件附平衡
conditional inequality	조건부등식	條件不等式
conditional instability	조건부불안정(성)	條件附不安定(性)
conditional instability of the second kind	제2종조건부불안정(성)	第二種條件附不安定(性)
conditionally unstable air	조건부불안정공기	條件附不安定空氣
conditionally unstable atmosphere	조건부불안정대기	條件附不安定大氣
conditional mean	조건평균	條件平均
conditional probability	조건부확률	條件附確率
conditional probability density function	조건부확률밀도함수	條件附確率密度函數
conditional sampling	조건부샘플링	條件附-
conditional stability	조건부안정(성)	條件附安定(性)
conditional symmetric instability	조건부대칭불안정	條件附對稱不安定
conditioned climate	조절기후	調節氣候
conditioned particle technique	조건부입자기법	條件附粒子技法
condition of autobarotropy	자동순압조건	自動順壓條件
condition of particle invariance	입자불변성조건	粒子不變性條件
conditions of readiness	대비단계	對備段階
conductance	전도도	傳導度
conductance for moisture	수분전도도	水分傳導度
conduction	전도	傳導
conduction current	전도전류	傳導電流
conduction electron	전도전자	傳導電子
conductive capacity	전도용량	傳導容量
conductive equilibrium	전도평형	傳導平衡
conductivity	전도성	傳導性
conductivity current	전도성전류	傳導性電流
conductivity meter	전기전도도계	電氣傳導度計
conductivity-temperature-depth profile	전도도-온도-깊이 프로파일	傳導度溫度-
conductor	도체	導體
cone	원뿔, 풍향자루, 폭풍신호	圓-, 風向-, 暴風信號
cone angle	원뿔각	圓-角
cone of depression	원뿔형수위강하	圓-型水位降下
cone of escape	탈출원뿔	脫出圓-
cone of impression	수위저하곡면	水位低下曲面
cone of influence	영향권원뿔	影響圈圓-
cone of visibility	시정원뿔	視程圓-
cone of vision	시계원뿔	視界圓-

cone radiometer	원뿔복사계	圓-輻射計
confidence band	신뢰밴드, 신뢰대역	信賴帶域
confidence coefficient	신뢰계수	信賴係數
confidence degree	신뢰도	信賴度
confidence figure	신뢰도	信賴圖
confidence interval	신뢰구간	信賴區間
confidence level	신뢰수준	信賴水準
confidence limit	신뢰한계	信賴限界
confined aquifer	압력대수층, 피압대수층	壓力帶水層, 被壓帶水層
confined groundwater	압력지하수, 피압지하수	壓力地下水, 被壓地下水
confining bed	가압층, 불투수층, 제한층	加壓層, 不透水層, 制限層
confining unit	난투수층	亂透水層
confining zone	난투수대	亂透水帶
conflict	모순, 충돌	矛盾, 衝突
confluence	합류	合流
confluence zone	합류구역	合流區域
confluent deformation	합류변형	合流變形
confluent flow	합류	合流
confluent jet	합류제트	合流-
confluent theory	합류설	合流說
confluent thermal ridge	합류형온도마루	合流型溫度-
confluent thermal trough	합류형온도골	合流型溫度-
conformal conic projection	등각원뿔투영(법)	等角圓-投影(法)
conformal map	등각지도	等角地圖
conformal projection	등각도법, 등각투영	等角圖法, 等角投影
congelation	응결	凝結
congelifraction	동결풍화작용	凍結風化作用
congeliturbation	물질요동현상	物質搖動現象
congestus	웅대구름	雄大-
congruence condition	합동조건	合同條件
conical beam	원뿔빔	圓-
conical scan	코니컬 스캔, 원추형 주사	圓錐形走査
conical scanning	원뿔주사	圓-走査
conic section	원뿔곡선	圓-曲線
coniferous	침엽(의)	針葉
conimeter	먼지측정계	-測定計
coning	원뿔형형성	圓-型形成
coning plume	원추형 플룸	圓錐形-
coniology(= koniology)	먼지학	-學
coniscope(= koniscope)	먼지계	-計
conjugate coordinate	켤레좌표, 공액좌표	共軛座標
conjugate image	켤레상	-像
conjugate of complex number	켤레복소수	-複素數
conjugate power	켤레멱	-冪

conjugate power law	켤레멱법칙	-冪法則
conjunction	합	合
connate water	동시생성수	同時生成水
connecting discharge	연결방전	連結放電
conjunctive use	결합용도	結合用度
connectivity	연결도	連結度
consecutive mean	이동평균	移動平均
consensus average	일치평균	一致平均
consensus average check	일치평균검사	一致平均檢查
consensus averaging	일치평균화	一致平均化
consentric circle group	동심원군	同心圓群
consequent	당연한 결과	當然結果
conservation equation of angular momentum	각운동량보존식	角運動量保存式
conservation exponent	보존멱지수	保存冪指數
conservation law	보존법칙	保存法則
conservation law of absolute angular momentum	절대각운동량보존법칙	絕對角運動量保存法則
conservation of absolute angular momentum	절대각운동량보존	絕對角運動量保存
conservation of absolute momentum	절대운동량보존	絕對運動量保存
conservation of absolute vorticity	절대소용돌이도보존	絕對-度保存
conservation of angular momentum	각운동량보존	角運動量保存
conservation of charge	전하보존	電荷保存
conservation of energy	에너지보존	-保存
conservation of kinetic energy	운동에너지보존	運動-保存
conservation of mass	질량보존	質量保存
conservation of moisture	수증기보존	水蒸氣保存
conservation of momentum	운동량보존	運動量保存
conservation of potential vorticity	위치소용돌이도보존, 와위보존	位置-度保存, 渦位保存
conservation of vorticity	소용돌이도보존	-度保存
conservation variable	보존변수	保存變數
conservatism	보존성	保存性
conservative	보존성	保存性
conservative element	보존요소	保存要素
conservative field	보존장	保存場
conservative force	보존력	保存力
conservative pollutant	보존오염물질	保存汚染物質
conservative property	보존성질	保存性質
conservative quantity	보존량	保存量
conservative scattering	비흡수형산란	非吸收形散亂
conservative scheme	보존방식	保存方式
conserved parameter diagram	보존매개변수선도	保存媒介變數線圖
conserved quantity	보존량	保存量
conserved variable	보존변수	保存變數
conserved variable diagram	보존변수선도	保存變數線圖

C

consistency	견지성	堅持性
consistent numerical scheme	일치수치방식	一致數值方式
consolidated ice	합병얼음	合併-
consolidation	압밀	壓密
constancy	일정성(바람의)	一定性
constancy of angular momentum	각운동량불변성	角運動量不變性
constancy of climate	기후불변성	氣候不變性
constancy of wind	바람의 일정성	-一定性
constant	상수, 일정한, 불변(의)	常數, 一定-, 不變
constant absolute vorticity	일정절대소용돌이도	一定絶對-度
constant absolute vorticity trajectory	일정절대소용돌이도궤적	一定絶對-度軌跡
constant absolute vorticity trajectory method	일정절대소용돌이도궤적법	一定絶對-度軌跡法
constant altitude	일정고도	一定高度
constant altitude balloon	정고도기구	定高度氣球
constant altitude plan position	일정고도평면위치	一定高度平面位置
constant altitude PPI	일정고도 PPI	一定高度-
constant earth temperature layer	지온불역층	地溫不易層
constant flux layer	일정플럭스층	一定-層
constant function	상수함수	常數函數
constant gas	영구기체	永久氣體
constant-height chart	정고도면도	定高度面圖
constant-height surface	정고도면	定高度面
constant humidity	정습도	定濕度
constant humidity line	정습도선	定濕度線
constant-level balloon	정고도기구	定高度氣球
constant-level chart	정고도면도, 정고도면일기도	定高度面圖, 定高度面日氣圖
constant-level surface	정고도면	定高度面
constant pressure	정압	定壓
constant pressure balloon	정압기구	定壓氣球
constant pressure chart	정압면도	定壓面圖
constant pressure gas thermometer	정압기체온도계	定壓氣體溫度計
constant pressure level chart	정압면일기도	定壓面日氣圖
constant pressure map	정압면도	定壓面圖
constant pressure pattern flight	정압패턴비행	定壓-飛行
constant pressure surface	정압면	定壓面
constant-rate dilution gauging	정율희석검량	定率稀釋檢量
constant stress layer	응력일정층	應力一定層
constant term	상수항	常數項
constant volume balloon	정적기구	定積氣球
constant volume gas thermometer	정적기체온도계	定積氣體溫度計
constant wind	일정바람	一定-
constituent of tide	조석성분	潮汐成分
constructive respiration	구성호흡, 형성호흡	構成呼吸, 形成呼吸
contact anemometer	접촉식풍속계	接觸式風速計

contact angle	접촉각	接觸角
contact cooling	접촉냉각	接觸冷却
contact-cup anemometer	접점컵풍속계	接點-風速計
contact flight	접촉비행	接觸飛行
contact freezing process	접촉결빙과정	接觸結氷過程
contact nucleus	접촉핵	接觸核
contact resistance	접촉저항	接觸抵抗
contact surface	접촉면	接觸面
contact weather	시계비행날씨	視界飛行-
contaminant	오염물	汚染物
contamination	오염	汚染
content	함량	含量
content analysis	함유량분석	含有量分析
contessa di vento	콘테사 디 벤토(고립된 봉우리의 풍하쪽에 생기는 원판모양의 구름. 시칠리아섬에서잘 나타남)	
continental aerosol	대륙에어로졸	大陸-
continental air (mass)	대륙기단	大陸氣團
continental anticyclone	대륙고기압	大陸高氣壓
continental borderland	대륙연변땅	大陸緣邊-
continental broadcast	대륙방송	大陸放送
continental climate	대륙(성)기후	大陸(性)氣候
continental cloud	대륙구름	大陸-
continental collision	대륙충돌	大陸衝突
continental crust	대륙지각	大陸地殻
continental cumulus	대륙적운	大陸積雲
continental cyclone	대륙(성)저기압	大陸(性)低氣壓
continental drift	대륙이동	大陸移動
continental glacier	대륙빙하	大陸氷河
continental hemisphere	대륙반구	大陸半球
continental high	대륙고기압	大陸高氣壓
continental ice	대륙얼음	大陸-
continental ice sheet	대륙빙상	大陸氷床
continental interior regime	대륙내륙체제	大陸內陸體制
continentality	대륙도	大陸度
continentality effect	대륙도효과	大陸度效果
continentality factor	대륙도인자	大陸度因子
continentality index	대륙도지수	大陸度指數
continental platform	대륙대지	大陸臺地
continental polar air (mass)	대륙성한대기단	大陸性寒帶氣團
continental sediment	육성층	陸成層
continental shelf	대륙붕	大陸棚
continental shelf wave	대륙붕파	大陸棚波
continental slope	대륙사면	大陸斜面
continental sphere	대륙권	大陸圈

C

continental subarctic climate	대륙아북극기후	大陸亞北極氣候
continental thermal low	대륙열저기압	大陸熱低氣壓
continental tropical air (mass)	대륙성열대기단	大陸性熱帶氣團
continental weather system	대륙날씨계	大陸日氣系
continental wind	대륙(성)바람	大陸(性)-
contingency angle	분할각	分割角
contingency table	분할표	分割表
continuing current	연속전류	連續電流
continuity	연속성	連續性
continuity chart	연속도	連續圖
continuity equation	연속방정식	連續方程式
continuous absorption	연속흡수	連續吸收
continuous curve	연속곡선	連續曲線
continuous distribution	연속분포	連續分布
continuous drought days	한발계속일수	旱魃繼續日數
continuous dry spell days	무강수계속일수	無降水繼續日數
continuous emission monitor	연속배출감시	連續放出監視
continuous function	연속함수	連續函數
continuous leader	연속선도(방전)	連續先導(放電)
continuously emitting point source	연속배출점오염원	連續排出汚染源
continuous plankton recorder	연속플랑크톤기록계	連續-記錄計
continuous precipitation	연속(성)강수	連續(性)降水
continuous probability distribution	연속확률분포	連續確率分布
continuous rain	연속(성)비	連續(性)-
continuous record	연속기록	連續記錄
continuous snow cover	근설	根雪
continuous spectrum	연속스펙트럼	連續-
continuous thunder and lightning	연속성천둥번개	連續性-
continuous variable	연속변수	連續變數
continuous wave	연속파	連續波
continuous wave radar	연속파레이더	連續波-
continuum	연속체	連續體
continuum absorption	연속체흡수	連續體吸收
continuum hypothesis	연속체가설	連續體假說
continuum theory	연속체이론	連續體理論
contour	등고선, 등치선	等高線, 等値線
contour analysis	등고선분석	等高線分析
contour change line	고도등변화선	高度等變化線
contour chart	등고선도	登高線圖
contour code	등고선코드	等高線-
contour curvature	등고선곡률	等高線曲率
contour gradient	등고선경도	等高線傾度
contour height	등고선고도	等高線高度
contour height error	등고선오차	等高線誤差

contour interval	등고선간격	等高線間隔
contour line	등고선	等高線
contour map	등고선도	等高線圖
contour microclimate	지형미기후	地形微氣候
contour tillage	등고선재배	等高線栽培
contraction	수축	收縮
contraction axis	수축축	收縮軸
contraction field	수축장	收縮場
contrail	비행운	飛行雲
contrail analysis curve	비행운분석곡선	飛行雲分析曲線
contrail formation curve	비행운형성곡선	飛行雲形成曲線
contrail formation graph	비행운형성그래프	飛行雲形成-
contra solem	저기압성	低氣壓性
contrast	차이, 대조, 대비	差異, 對照, 對備
contrast area	대조구역	對照區域
contraste	콘트라스테(지브롤터 해협의 근접한 대기역학접에서 역풍이 부는 현상〈항해용어〉)	
contrast of luminance	휘도대조	輝度對照
contrast stretching	명암대비스트레칭	明暗對比-
contrast threshold	대조문턱값	對照門-
contravariant differentiation	반변미분	反變微分
contravariant vector	반변벡터	反變-
contributing region	기여지역	寄與地域
contribution function	기여함수	寄與函數
control	제어	制御
control action	제어작용	制御作用
control area	대조구역(기상조절)	對照區域
control case	조절사례	調節事例
control cloud	대조구름(기상조절)	對照-
control day	대조일(기상조절)	對照日
control grid	제어격자	制御格子
controlled airspace	관제공역	管制空域
control tower visibility	관제탑시정	管制塔視程
control unit module	제어단위모듈	制御單位-
control well	대조우물	對照-
convection	대류	對流
convectional circulation	대류순환	對流循環
convectional precipitation	대류성강수	對流性降水
convectional rain	대류성비	對流性-
convectional theory	대류이론	對流理論
convection cell	대류세포	對流細胞
convection cloud	대류구름	對流-
convection condensation level	대류응결고도	對流凝結高度
convection current	대류성기류, 대류전류	對流性氣流, 對流轉流

convection equation	대류방정식	對流方程式
convection street	대류줄	對流-
convection tank	대류탱크	對流-
convection theory of cyclone	저기압대류설	低氣壓對流說
convection velocity	대류속도	對流速度
convective acceleration	대류가속	對流加速
convective activity	대류활동	對流活動
convective adjustment	대류조절	對流調節
convective adjustment scheme	대류조절방안	對流調節方案
convective atmosphere	대류성대기	對流性大氣
convective available potential energy	대류가용잠재에너지	對流可用潛在-
convective boundary layer	대류경계층	對流境界層
convective boundary layer depth	대류경계층두께	對流境界層-
convective bubble	대류거품	對流-
convective cell	대류세포	對流細胞
convective circulation	대류순환	對流循環
convective cloud	대류구름	對流-
convective cloud height diagram	대류구름높이선도	對流-線圖
convective cluster	대류무리	對流-
convective complex	대류구름무리	對流-
convective component	대류성분	對流成分
convective condensation level	대류응결고도	對流凝結高度
convective current	대류기류	對流氣流
convective echo	대류에코	對流-
convective element	대류요소	對流要素
convective equilibrium	대류평형	對流平衡
convective flux	대류플럭스	對流-
convective heating	대류가열	對流加熱
convective impulse	대류임펄스	對流-
convective index	대류지수	對流指數
convective inhibition	대류억제	對流抑制
convective instability	대류불안정(도)	對流不安定(度)
convective instability of the second kind	제2종대류불안정도	第二種對流不安定度
convective interval	대류간격	對流間隔
convective inversion	대류역전	對流逆轉
convective layer	대류층	對流層
convective lifting	대류치올림	對流-
convectively stable	대류안정	對流安定
convectively unstable	대류불안정	對流不安定
convectively unstable line	대류불안정선	對流不安定線
convective mass flux	대류질량플럭스	對流質量-
convective matching layer	대류적합층	對流適合層
convective mixed layer	대류혼합층	對流混合層
convective overturn	대류전복	對流顚覆

convective phenomenon	대류현상	對流現象
convective plume	대류플룸	對流-
convective precipitation	대류성강수	對流性降水
convective rain	대류성비	對流性-
convective region	대류구역	對流區域
convective Richardson number	대류리처드슨수	對流-數
convective roller	대류성두루마리구름	對流性-
convective scale	대류규모	對流規模
convective scale interaction	대류규모상호작용	對流規模相互作用
convective scaling	대류스케일링	對流-
convective showers	대류성소나기	對流性-
convective SIGMET	대류시그멧	對流-
convective stability	대류안정(도)	對流安定(度)
convective storm initiation mechanism	대류성폭풍기폭메커니즘	對流性暴風起爆-
convective storms	대류폭풍, 대류스톰	對流暴風
convective stratiform technique	대류층운형기법	對流層雲形技法
convective system	대류계	對流系
convective temperature	대류온도	對流溫度
convective term	대류항	對流項
convective theory of cyclogenesis	저기압생성 대류이론	低氣壓生成對流理論
convective thunderstorm	대류뇌우	對流雷雨
convective time scale	대류시간규모	對流時間規模
convective tornado	대류성토네이도	對流性-
convective transport	대류수송	對流輸送
convective transport theory	대류수송이론	對流輸送理論
convective turbulence	대류난류	對流亂流
conventional radar	재래식레이더	在來式-
convergence	수렴	收斂
convergence field	수렴장	收斂場
convergence lens	수렴렌즈	收斂-
convergence line	수렴선	收斂線
convergence model	수렴모형	收斂模型
convergence region	수렴구역	收斂區域
convergence zone	수렴대	收斂帶
convergent air current	수렴기류	收斂氣流
convergent effect	수렴효과	收斂效果
convergent line	수렴선	收斂線
convergent margin	수렴주변부	收斂周邊部
convergent numerical scheme	수렴수치방안	收斂數值方案
convergent point	수렴점	收斂點
convergent rain	수렴성비	收斂性-
convergent wind	수렴바람	收斂-
converse	역	逆
conversion	전환	轉換

C

conversion factor	전환인자	轉換因子
converter	변환기	變換器
conveyance	운반, 전달	運搬, 傳達
conveyor belt	컨베이어벨트	
conveyor belt model	컨베이어벨트모형	-模型
convolution	소용돌이, 회선	回旋
cooking snow	젖은 눈	
Cook method	쿡(채표) 방법	-(採表)方法
cooling	냉각	冷却
cooling curve	냉각곡선	冷却曲線
cooling degree day	냉방도일	冷房度日
cooling power	냉각률	冷却率
cooling power anemometer	냉각능풍속계	冷却能風速計
cooling process	냉각과정	冷却過程
cooling rate	냉각률	冷却率
cooling temperature	냉각온도	冷却溫度
cooling tower plume	냉각타워플룸	冷却-
coolometer	냉각계	冷却計
cool summer	냉하	冷夏
cool summer damage	냉해	冷害
cool summer damage by delayed growth	지연형냉해	遲延型冷害
cool wave	냉파	冷波
Cooperative Holocene Mapping Project	홀로세지도작성프로젝트	-地圖作成-
cooperative observer	협력관측자	協力觀測者
coordinate	좌표	座標
coordinate axis	좌표축	座標軸
coordinate line	좌표선	座標線
coordinate plane	좌표평면	座標平面
coordinate rotation	좌표회전	座標回轉
coordinate simplification	좌표단순화	座標單純化
coordinate space	좌표공간	座標空間
coordinate system	좌표계	座標系
coordinate transformation	좌표변환	座標變換
Copenhagen water	코펜하겐해수(IAPSO 표준해수)	-海水
copepod	요각류	境脚類
coplane scanning	동일면주사	同一面走査
coplane technique	동일면기법	同一面技法
copolar component	동일편파성분	同一偏波成分
copolar correlation coefficeint	동일편파송수신상관계수	同一偏波送受信相關係數
copolarized signal	동일편광신호	同一偏光信號
coral	산호	珊瑚
coral atoll	환초	環礁
coral band	산호띠	珊瑚-
coral bleaching	산호백화작용-	珊瑚白化作用

coral reef	산호초	珊瑚礁
corange line	등조차선	等潮差線
Cordillera	코르디예라	
Cordillera ice sheet	코르디예라빙상	-氷床
Cordonazo	코르도나조(중앙아메리카와 멕시코의 서쪽 앞바다에 발생하는 수명이 짧은 허리케인)	
core analysis	코어분석	-分析
core sample	코어시료	-試料
core sensor	핵심감지	核心感知
Coriolis acceleration	코리올리가속도	-加速度
Coriolis deflection	코리올리편향	-偏向
Coriolis effect	코리올리효과	-效果
Coriolis factor	코리올리인자	-因子
Coriolis force	코리올리힘	
Coriolis parameter	코리올리매개변수	-媒介變數
corner effect	모서리효과	-效果
corner reflector	모서리반사경	-反射鏡
corner stream	모서리흐름	
corn heat unit	옥수수열지수	-熱指數
cornice	눈처마	
corn snow	봄눈(녹았다 얼었다 한 눈)	
corona	광환, 코로나	光環
corona current	코로나전류	-電流
corona discharge	코로나방전	-放電
coronagraph	코로나그래프	
coronal holes	코로나구멍	
coronal mass injection	코로나질량방출	-質量放出
corona method	코로나방법	-方法
corposant	코로나방전	-放電
corpuscular radiation	미립자복사	微粒子輻射
corpuscular ray	미립자선	微粒子線
corpuscular theory	미립자설	微粒子說
corpuscular theory of light	빛입자설	-粒子說
corrasion	바람침식	-浸蝕
corrected airspeed	보정공기속도	補正空氣速度
corrected altitude	보정고도	補正高度
corrected reflectivity	보정반사도	補正反射度
correction for instrument error	기기오차보정, 기차보정	器機誤差補正, 器差補正
correction line	수정고도	修正高度
correction of (solar) insolation	일사보정	日射補正
correlation analysis	상관분석	相關分析
correlation coefficient	상관계수	相關係數
correlation diagram	상관도	相關圖
correlation factor	상관인자	相關因子

correlation forecasting	상관예보	相關豫報
correlation function	상관함수	相關函數
correlation method	상관법	相關法
correlation product	상관곱	相關-
correlation ratio	상관비	相關比
correlation table	상관표	相關表
correlation tensor	상관텐서	相關-
correlation triangle	상관삼각형	相關三角形
correlative meteorology	상관기상학	相關氣象學
correlogram	상관도	相關圖
correspondence	대응	對應
corresponding point	대응점	對應點
corrosion	부식	腐蝕
cosine law of illumination	조명도 코사인법칙	照明度-法則
cosine response	코사인반응	-反應
cosine tapering	코사인테이퍼링	
cosmical meteorology	우주기상학	宇宙氣象學
cosmic dust	우주먼지	宇宙-
cosmic meteorology	우주기상학	宇宙氣象學
cosmic radiation	우주복사	宇宙輻射
cosmic ray	우주선	宇宙線
cosmic ray ionization	우주선이온화	宇宙線-化
cosmic ray shower	우주선소나기	宇宙線-
cosmogenic radioisotope	우주성방사성동위원소	宇宙性放射性同位元素
cosmogenic radionuclide	우주기원방사성핵종	宇宙起源放射性核種
cosmos	우주	宇宙
cosmos satellite series	코스모스위성시리즈	-衛星-
cospectral similarity	공스펙트럼유사성	共-類似性
co-spectrum	공스펙트럼	共-
co-spectrum peak	공스펙트럼 피크	共-
cost benefit analysis	비용편익분석	費用便益分析
cost function	비용함수	費用函數
cotidal hour	등조시	等潮時
cotidal line	등조선	等潮線
cotton ball cloud	솜덩이구름	
cotton belt climate	목화띠기후	木花-氣候
cotton region shelter	목화지역백엽상	木花地域百葉箱
Couette flow	쿠에테흐름	
coulomb	쿨롱(전기양의 SI 단위)	
counter	계수기, 계산기, 반대, 역	計數器, 計算器, 反對, 逆
countercirculation	반순환	反循環
counterclockwise	반시계방향	反時計方向
counterclockwise wind	반시계방향바람	反時計方向-
counterclockwise wind shift	반시계방향풍향전환	反時計方向風向轉換

countercurrent	반류	反流
counter current heat exchange	반류열교환	反流熱交換
counter flow	반류	反流
counterglow	대일조	對日照
countergradient diffusion	반경도확산	反傾度擴散
countergradient flow	반경도흐름	反傾度-
countergradient flux	반경도플럭스	反傾度-
countergradient transfer	반경도전달	反傾度傳達
countergradient wind	반경도풍	反傾度風
counter parhelion	반무리해	反-
counterradiation	역복사	逆輻射
counter sun(= anthelion)	맞무리해	
countertrades	반대무역풍	反對貿易風
countertwilight	반대박명	反對薄明
counting anemometer	계수풍속계	計數風速計
country breeze	시골바람, 전원풍	田園風
coupled atmospheric and oceanic general circulation model	대기-해양 대순환결합모형	大氣海洋大循環結合模型
coupled climate system	결합기후시스템, 결합기후계	結合氣候系
coupled equation	결합방정식	結合方程式
coupled model	결합모형	結合模型
coupled model system	결합모형시스템, 결합모형계	結合模型系
coupled ocean-atmosphere model	해양-대기결합모형	海洋大氣結合模型
couplet	이행연귀	二行連句
coupling	결합, 접합	結合, 接合
coupling model	결합모형	結合模型
Courant-Friedrichs-Lewy condition	쿠랑-프리데리흐스-레위 조건	-條件
Courant-Friedrichs-Lewy stability criterion	쿠랑-프리데리흐스-레위 안정도기준	-安定度基準
Courant number	쿠랑수	-數
covalent bond	공유결합	共有結合
covalent bond length	공유결합길이	共有結合-
covalent crystal	공유결정	共有結晶
covariable	공변수	共變數
covariance	공분산	共分散
covariance matrix	공분산행렬	共分散行列
covariance tensor	공분산텐서	共分散-
covariant curvature tensor	공분산곡률텐서	共分散曲率-
covariant tensor	공분산텐서	共分散-
covariant time operator	공분산시간연산자	共分散時間演算子
coverage	범위, 관할구역	範圍, 管轄區域
coverage diagram	범위도	範圍圖
cover of cloud	구름양	-量
cow-quaker	카우-퀘이커(영국에서 대기전산월에 일어나는 폭풍우)	

crab angle	게각도	-角度
crachin	크라친(낮은 층운이나 안개 등에 동반되는 보슬비가 내리는 날씨)	
crack	갈라진 틈새, 빙하틈새	氷河-
crackcase ventilation	(자동차의)통기장치	通氣裝置
Crank-Nicholson method	크랭크-니콜슨 방법	-方法
Crank-Nicholson scheme	크랭크-니콜슨 방안	-方案
crater	크레이터, 분화구	噴火口
creep	크리프, 하강점동	下降漸動
creeping flow	크리핑흐름	
creeping of aneroid barometer	아네로이드기압계 뒤짐현상	-氣壓計-現象
crepuscular arch	박명호	薄明弧
crepuscular ray	부챗살빛	
crepuscule(= twilight)	땅거미	
crest cloud	삿갓구름	
crest elevation	마루표고	-標高
crest gauge	최고수위계	最高水位計
crest profile	마루종단면	-縱斷面
crest stage	마루단계	-段階
Cretaceous climate	백악기기후	白堊期氣候
Cretaceous period	(중생대)백악기	白堊期
Cretaceous-Tertiary(K-T) boundary	K-T 경계층, 백악기-제3기 경계층	-境界層, 白堊期第三期境界層
Cretaceous-Tertiary(K-T) extinction	백악기-제3기 멸종	白堊期第三期滅種
Cretaceous warm climate	백악기온난기후	白堊期溫暖氣候
Creutzburg's climatic zone	크루츠버그기후대	-氣候代
crevasse	크레바스	
crevasse hoar	크레바스서리	
crevasse traces	크레바스흔적	-痕迹
crevasse wall	크레바스벽	-壁
Criegee biradicals	크리게바이라디칼	
Criegee intermediate	크리게중간단계	-中間段階
crinoids	바다나리류	-流
crisp variable	보통변수	普通變數
criteria document	기준지침서, 판정지침서	基準指針書, 判定指針書
criteria pollutant	기준오염물질	基準汚染物質
criterion	판별기준, 표준	判別基準, 標準
critical angle	임계각	臨界角
critical area	임계역	臨界域
critical bursting point	임계폭발점	臨界爆發點
critical condition	임계조건	臨界條件
critical constant	임계상수	臨界常數
critical depth	임계깊이	臨界-
critical depth control	임계깊이조절	臨界-調節

critical discharge	임계방전	臨界放電
critical downwind distance	임계배풍거리	臨界背風距離
critical drop radius	임계물방울반지름	臨界-
critical flow	임계흐름	臨界-
critical flux Richardson number	임계플럭스리처드슨수	臨界-數
critical frequency	임계주파수	臨界周波數
critical gradient	임계경도	臨界傾度
critical height	임계높이	臨界-
critical lapse rate	임계감률	臨界減率
critical level	임계고도	臨界高度
critical level instability	임계고도불안정	臨界高度不安定
critical level of escape	임계탈출고도	臨界脫出高度
critical liquid water content	임계함수량	臨界含水量
critical mass	임계질량	臨界質量
critical period of growth	임계생장기간	臨界生長期間
critical phenomenon	임계현상	臨界現象
critical point	임계점	臨界點
critical pressure	임계압	臨界壓
critical radius	임계반경	臨界半徑
critical Rayleigh number	임계레일리수	臨界-數
critical region	임계영역	臨界領域
critical Reynolds number	임계레이놀즈수	臨界-數
critical Richardson number	임계리처드슨수	臨界-數
critical rotation number	임계회전수	臨界回轉數
critical sampling	임계샘플링	臨界-
critical state	임계상태	臨界狀態
critical success index	임계성공지수	臨界成功指數
critical temperature	임계온도	臨界溫度
critical tidal level	임계조위기준면	臨界潮位基準面
critical time	임계시간	臨界時間
critical value	임계값	臨界-
critical velocity	임계속도	臨界速度
critical volume	임계부피	臨界-
critical wave length	임계파장	臨界波長
crochet d'orage(= crochet degrain)	돌풍, 뇌우코	突風
Cromerian interglacial	크로메리아 간빙기	-間氷期
Cromwell current	크롬웰해류	-海流
crop calendar	작물력	作物曆
crop climate	작물기후	作物氣候
crop coefficient	작물계수	作物係數
crop forecast	수확량예측	收穫量豫測
crop forecasting	작황예보	作況豫報
crop growth	작물성장	作物成長
crop heat unit	농작물열단위	農作物熱單位

crop moisture index	농작물습윤지수	農作物濕潤指數
crop period	작물기간	作物期間
cropping season	작계	作季
crop situation	작황	作況
crop type	작물종류	作物種類
crop weather	작물기상	作物氣象
crop yield	농작물산출량	農作物産出量
crop yield model	농작물산출량모형	農作物産出量模型
cross	십자무리	十字-
cross contamination	상호오염	相互汚染
cross contour flow	등고선횡단흐름	等高線橫斷-
cross correlation	교차상관	交叉相關
cross correlation coefficient	교차상관계수	交差相觀係數
cross correlation method	교차상관방법	交叉相關方法
cross covariance	교차공분산	交叉共分散
cross cutting relationships	관입법칙	貫入法則
cross differentiation	교차미분	交叉微分
crossed vortex	교차소용돌이	交叉-
cross isobar angle	등압선교각	等壓線交角
cross isobar flow angle	등압선교차흐름각	等壓線交叉-角
cross-over experiment	교차실험	交叉實驗
cross-over phenomenon	교차현상	交叉現象
cross polar component	교차편파성분	交叉偏波成分
cross polar correlation coefficient	교차편파송수신상관계수	交叉偏波送受信相關係數
cross polarization	교차편광	交叉偏光
cross polarization signal	교차편파신호	交叉偏波信號
cross product	외적	外積
cross sea	역풍파	逆風波
cross-section	단면	斷面
cross-sectional analysis	단면분석	斷面分析
cross-section diagram	단면도	斷面圖
cross-section test	단면검사	斷面檢査
cross spectrum	교차스펙트럼	交叉-
cross totals index	교차총량지수	交叉總量指數
cross-track direction	트랙교차방향	-交叉方向
cross-track resolution	트랙교차해상도	-交叉解像度
cross-valley circulation	계곡횡단순환	溪谷橫斷循環
cross-valley wind	계곡횡단바람	溪谷橫斷-
crosswind	옆바람, 측풍	測風
crosswind component	옆바람성분	-成分
crosswind gustiness	옆바람돌풍성	-突風性
crosswise vorticity	비스듬한 소용돌이도	-度
crown	수관	樹冠
crown flash	관상방전	冠狀放電

CRSTER model	CRSTER 모형	-模型
crude data	원시자료	原始資料
cruising	순항	巡航
cruising efficiency	순항성능, 항속성능	巡航性能, 航續性能
cruising level	순항고도	巡航高度
cruising range	순항거리, 항속거리	巡航距離, 航續距離
cruising speed	순항속도	巡航速度
crust	크러스트	
cryochore	빙설대	氷雪帶
cryoconite hole	얼음구멍	
cryogenian period	크라이오제니아기	-期
cryogenic hygrometer	저온습도계	低溫濕度計
cryogenic limb array etalon spectrometer	저온가지배열에타중간자분광기	低溫-配列-中間子分光器
cryogenic period	저온기	低溫期
cryology	빙설학	氷雪學
cryopedology	동결학	凍結學
cryopedometer	동결측정계	凍結測定計
cryoplanation	얼음침식	-浸蝕
cryoplankton	저온플랑크톤	低溫-
cryosphere	빙설권	氷雪圈
cryoturbation	얼음교란	-攪亂
cryptoclimate	실내기후	室內氣候
cryptoclimatology	실내기후학	室內氣候學
crystal	결정	結晶
crystal form	결정형	結晶形
crystal habit	결정습성	結晶習性
crystal lattice	결정격자	結晶格子
crystalline frost	결정서리	結晶-
crystallinity	결정성(도)	結晶性(度)
crystallization	결정화	結晶化
crystallization nucleus	결정핵	結晶核
crystallographic	결정학적	結晶學的
crystal structure	결정구조	結晶構造
cube	정육면체	正六面體
cubic equation	삼차방정식	三次方程式
cubic inequality	삼차부등식	三次不等式
cubic system	입방결정계	立方結晶系
cultivation	경작	耕作
cultivation period	작기	作期
cultivation under rain shelter	비가림재배	-栽培
cultivation with supplementary light	보강재배	補強栽培
cum sole	고기압성	高氣壓性
cumulant	누적률	累積率
cumulation	누적	累積

cumulative curve(→ mass curve)	누가곡선	累加曲線
cumulative distribution function	누적분포함수	累積分布函數
cumulative error	누적오차	累積誤差
cumulative frequency	누적도수	累積度數
cumulative frequency distribution	누적빈도분포	累積頻度分布
cumulative frequency graph	누적도수그래프	累積度數-
cumulative precipitation	누적강수량	累積降水量
cumulative relative frequency	누적상대도수	累積相對度數
cumulative temperature	누적온도	累積溫度
cumuliform	적운형	積雲形
cumuliform cloud	적운형구름	積雲形-
cumuliformis floccus	포기적운	-積雲
cumulogenitus	적운성	積雲性
cumulonimbus	적란운	積亂雲
cumulonimbus arcus	아치적란운	-積亂雲
cumulonimbus calvus	대머리적란운	-積亂雲
cumulonimbus capillatus	복슬적란운	-積亂雲
cumulonimbus incus	모루적란운	-積亂雲
cumulonimbus mammatus	유방적란운	乳房積亂雲
cumulonimbus pannus	토막적란운	-積亂雲
cumulonimbus pileus	두건적란운	頭巾積亂雲
cumulonimbus praecipitatio	강수적란운	降水積亂雲
cumulonimbus tuba	깔때기적란운	-積亂雲
cumulonimbus velum	면사포적란운	面紗布積亂雲
cumulonimbus virga	꼬리적란운	-積亂雲
cumulus	적운	積雲
cumulus arcus	아치적운	-積雲
cumulus base	적운밑면	積雲-
cumulus calvus	대머리적운	-積雲
cumulus capillatus	복슬적운	-積雲
cumulus cloud	적운	積雲
cumulus congestus	웅대적운	雄大積雲
cumulus congestus cloud	웅대적운	雄大積雲
cumulus convection	적운대류	積雲對流
cumulus field model	적운장모형	積雲場模型
cumulus fractus	조각적운	-積雲
cumulus friction	적운마찰	積雲摩擦
cumulus humilis	넙적적운	-積雲
cumulus lenticularis	렌즈적운	-積雲
cumulus mediocris	중간적운	中間積雲
cumulus model	적운모형	積雲模型
cumulus pannus	토막적운	-積雲
cumulus pileus	두건적운	頭巾積雲
cumulus praecipitatio	강수적운	降水積雲

cumulus radiatus	방사적운	放射積雲
cumulus stage	적운단계	積雲段階
cumulus tuba	깔때기적운	-積雲
cumulus undulatus	파도적운	波濤積雲
cumulus velum	면사포적운	面紗布積雲
cumulus virga	꼬리적운	-積雲
Cunnane plotting position	커네인도시위치	-都市位置
Cunningham correction factor	커닝엄보정계수	-補正係數
Cunningham slip correction	커닝엄미끄러짐보정	-補正
cup anemometer	컵풍속계	-風速計
cup anemometer overspeeding	컵풍속계과속	風杯風速計過速
cup contact anemometer	전접컵풍속계	電接-風速計
cup counter anemometer	계수컵풍속계	計數-風速計
cup crystal	컵결정	-結晶
cup crystal frost	컵결정서리	-結晶-
cup generator anemometer	컵발전풍속계	-發電風速計
curie	퀴리(방사능단위)	
Curie's principle	퀴리원리	-原理
curie temperature	퀴리온도	-溫度
curl	컬	
current	흐름, 전류, 해류	電流, 海流
current chart	해류도	海流圖
current cross section	해류단면	海流斷面
current curve	조류곡선	潮流曲線
current ellipse	조류타원	潮流楕圓
current function	흐름함수	-函數
current meter	유속계	流速計
current pole	유속측정막대기	流速測定-
current profile	유속연직분포	流速鉛直分布
current rose	해류장미	海流薔薇
current table	조석표	潮汐表
current weather	현재일기	現在日氣
curtain	커튼	
curtain aurora	커튼극광	-極光
Curtis-Godson approximation	커티스-고드슨 근사	-近似
curvature effect	곡률효과	曲率效果
curvature tensor	곡률텐서	曲律-
curvature vorticity	곡률소용돌이도	曲率-
curved airflow	굴곡기류	屈曲氣流
curved flow	굴곡흐름	屈曲
curved motion	곡선운동	曲線運動
curve fitting	곡선맞춤	曲線-
curve of growth	성장곡선	成長曲線
curvilinear coordinates	곡선좌표	曲線座標

cusp cloud pattern	뽀족구름패턴	
custard wind	커스터드바람(잉글랜드 북동해안에서 부는 차가운 동풍)	
customized forecast	주문제작예보, 맞춤형 예보	注文製作豫報, -形豫報
cuticular transpiration	큐티클증산	-蒸散
cut-off center	분리중심	分離中心
cut-off effect	차단효과	遮斷效果
cut-off frequency	차단주파수	遮斷周波數
cut-off high	분리고기압	分離高氣壓
cut-off low	분리저기압	分離低氣壓
cut-off vortex	분리소용돌이	分離-
cutting-off process	분리과정	分離過程
C weather	C날씨	
CW radar	지속파레이더	持續波-
cyanometer	사이아노미터	
cyanometry	푸름도측정법	-測定法
cycle	주기	週期
cycle of three cold and four warm days	삼한사온	三寒四溫
cycle process	주기과정	週期過程
cyclic forcing	주기강제	週期强制
cyclic frequency	주기주파수	週期周波數
cyclic sedimentation	윤회성퇴적작용(윤회층)	輪廻性堆積作用(輪廻層)
cyclogenesis	저기압발생	低氣壓發生
cyclo-geostrophic current	저기압성지형류	低氣壓性地衡流
cyclohexane	시클로헥산	
cyclolysis	저기압소멸	低氣壓消滅
cyclone	(온대성)저기압	(溫帶性)低氣壓
cyclone cellar	지하대피실	地下待避室
cyclone collector	사이클론집진기	-集塵器
cyclone dust collector	사이클론집진장치	-集塵裝置
cyclone family	저기압가족	低氣壓家族
cyclone model	저기압모형	低氣壓模型
cyclone movement	저기압이동	低氣壓移動
cyclone path	저기압경로	低氣壓經路
cyclone precipitation	저기압성강수	低氣壓性降水
cyclone-prone area	저기압다발구역	低氣壓多發區域
cyclone rain	저기압강우	低氣壓降雨
cyclone scale	사이클론규모, 저기압규모	-規模, 低氣壓規模
cyclone scrubber	사이클론스크러버	
cyclone separator	사이클론분리기	-分離機
cyclone track	저기압경로	低氣壓經路
cyclonette	꼬마저기압	-低氣壓
cyclone warning	저기압경보	低氣壓警報
cyclone wave	저기압파	低氣壓波
cyclonic	저기압성	低氣壓性

cyclonic air motion	저기압성공기운동	低氣壓性空氣運動
cyclonic bora	저기압성보라	低氣壓性-
cyclonic circulation	저기압순환	低氣壓循環
cyclonic curvature	저기압성곡률	低氣壓性曲率
cyclonic extra-tropical storms project	저기압성중위도폭풍연구계획	低氣壓性中緯度暴風研究計劃
cyclonic flow	저기압성흐름	低氣壓性-
cyclonicity	저기압성	低氣壓性
cyclonic lifting	저기압성치올림, 저기압성상승	低氣壓性上昇
cyclonic motion	저기압성운동	低氣壓性運動
cyclonic phase	저기압성단계	低氣壓性段階
cyclonic precipitation	저기압성강수	低氣壓性降水
cyclonic rain	저기압성비	低氣壓性-
cyclonic rotation	저기압성회전	低氣壓性回轉
cyclonic scale	저기압규모	低氣壓規模
cyclonic sense	저기압성방향	低氣壓性方向
cyclonic shear	저기압성시어	低氣壓性-
cyclonic storm	저기압성폭풍우	低氣壓性暴風雨
cyclonic thermal involution	저기압성온도퇴화	低氣壓性溫度退化
cyclonic thunderstorm	저기압성뇌우	低氣壓性雷雨
cyclonic tornado	저기압성토네이도	低氣壓性-
cyclonic vorticity	저기압성소용돌이도	低氣壓性-度
cyclonic wave	저기압성파	低氣壓性波
cyclonic wind	저기압성바람	低氣壓性-
cyclostrophic balance	선형평형	旋衡平衡
cyclostrophic convergence	선형풍수렴	旋衡風收斂
cyclostrophic divergence	선형풍발산	旋衡風發散
cyclostrophic flow	선형흐름	旋衡-
cyclostrophic function	선형함수	旋衡函數
cyclostrophic motion	선형운동	旋衡運動
cyclostrophic transport	선형수송	旋衡輸送
cyclostrophic wind	선형풍	旋衡風
cylindrical coordinates	원기둥좌표	圓-座標
cylindrical coordinate system	원통좌표계	圓筒座標系
cylindrical function	원기둥함수	圓-函數
cylindrical polar coordinates	원기둥극좌표	圓-極座標
cylindrical projection	원통도법	圓筒圖法
cylindrical radiometer	원기둥복사계	圓-輻射計
cylindrical rain gauge	원통형우량계	圓筒形雨量計
cylindrical snow gauge	원통형설량계	圓筒形雪量計

D, d

영문	한글	한자
dabacle	해빙	解氷
DAD analysis	DAD분석, DAD해석	-分析, -解析
DAD value	DAD값	
Daehan current	대한해류	大韓海流
daily compensation point	일보상점	日補償點
daily extreme	일극값	日極-
daily forecast	일예보	日豫報
daily maximum temperature	일최고온도	日最高溫度
daily mean	일평균(값)	日平均
daily mean amount of cloud	일평균운량	日平均雲量
daily mean humidity	일평균습도	日平均濕度
daily mean temperature	일평균온도	日平均溫度
daily mean wind speed	일평균풍속	日平均風速
daily minimum temperature	일최저온도	日最低溫度
daily observation	일관측	日觀測
daily periodism	일주기성	日週期性
daily precipitation	일강수량	日降水量
daily range	일교차	日較差
daily range of temperature	온도일교차	溫度日較差
daily rate	일변화율	日變化率
daily storage	일저류량	日貯流量
daily surplus production	일잉여생산량	日剩餘生産量
daily synoptic series	일별종관연속	日別綜觀連續
daily temperature range	일온도교차, 일교차	日溫度較差, 日較差
daily variation	일변동	日變動
d'Alembert's paradox	달랑베르역설	-逆說
d'Alembert's solution	달랑베르해	-解
Dalton approach	돌턴접근법	-接近法
Dalton number	돌턴수	-數

Dalton's law	돌턴법칙	-法則
damage	재해	災害
Damköhler number	담쾰러수	-數
damming(= blocking)	저지현상	沮止現象
damp air	습한 공기	濕-空氣
damped mode	감쇠모드	減衰-
damped oscillation	감쇠진동	減衰振動
damped wave	감쇠파	減衰波
damp haze	습한 연무	濕-煙霧
damping	감쇠, 억제	減衰, 抑制
damping depth	감쇠깊이, 제동깊이	減衰-, 制動-
damping factor	감쇠인자	減衰因子
damping ratio	감쇠비	減衰比
D-analysis	D-분석	-分析
dancing dervish	먼지회오리	
dancing devil	먼지회오리	
danger line	위험선	危險線
dangerous half	위험반원	危險半圓
dangerous interference	위험간섭	危險干涉
dangerous semicircle	위험반원	危險半圓
Danielsen's model of upper tropospheric folding	상부대류권 접힘의 다니엘센모형	上部對流圈-模型
Dansgaard-Oeschger(D-O) cycle	단스고-외슈거 주기	-週期
Dansgaard-Oeschger event	단스고-외슈거 사건	-事件
Darcian velocity	다시속도	-速度
Darcy's law	다시법칙	-法則
dark adaptation	어둠적응	-適應
dark band	어둠띠	
darkening	암화	暗化
dark light	어두운 빛	
dark lightning	어두운 번개	
dark line	어둠선	-線
dark radiation	어두운 복사	-輻射
dark ray	어두운 복사선	-輻射線
dark respiration	암호흡	暗呼吸
dark segment	어둠띠	
darling shower	달링소나기	
dart leader	화살선도방전	-先導放電
data	자료	資料
data acquisition	자료수집	資料蒐集
data acquisition and transmission subsystem	(위성)자료수신/송신부시스템	(衛星)資料受信送信副-
data acquisition facility	자료수집시설	資料蒐集施設
data acquisition system	자료수집체계, 자료수집장치	資料蒐集體系, 資料蒐集裝置
data analysis center	자료분석센터	資料分析-

data archive	자료저장	資料貯藏
data assimilation	자료동화	資料同化
data assimilation problem	자료동화문제	資料同化問題
data assimilation system	자료동화시스템, 자료동화계	資料同化系
data bank	자료은행	資料銀行
data block indentifier	자료유형구분자	資料類型區分子
data buoy	부이자료	-資料
data collection	자료수집	資料蒐集
data collection platform	자료수집대	資料蒐集臺
data communication system	데이터통신시스템	-通信-
data correction	자료보정	資料補正
data directory	자료명부	資料名簿
data function	자료함수	資料函數
data logger	자료집록장치	資料集錄裝置
data processing	자료처리	資料處理
data requirement	자료요구	資料要求
data sample	자료표본	資料標本
data series	자료군	資料群
dataset	데이터세트	
dataset catalog	데이터세트목록	-目錄
dataset directory	데이터세트명부	-名簿
data system test	데이터시스템시험	-試驗
data transmission	자료전송	資料傳送
data tropical number	자료열대수	資料熱帶數
data utilization station	자료이용소	資料利用所
data volume	자료크기	資料-
data window	자료윈도	資料-
date line	날짜선	-線
date of red coloring of leaves	홍엽일	紅葉日
date of yellow coloring of leaves	황엽일	黃葉日
datum level	기준면, 기본수준면	基準面, 基本水準面
daughter isotope	딸동위원소, 자원소	-同位元素, 子元素
Davidson-Bryant equation	데이비드슨-브라이언트방정식	-方程式
Davidson current	데이비드슨해류(해)	-海流
Davies number	데이비스수	-數
dawn chorus	새벽공전음	-空電音
day breeze	낮미풍	-微風
day degree(= degree day)	일도	日度
dayglow	낮대기광	-大氣光
daylight	일광	日光
daylight saving time	일광절약시간	日光節約時間
day of rain	강우일	降雨日
day of precipitation	강수일	降水日

day of snow lying	적설일(수)	積雪日(數)
daytime	낮	
daytime upslope	낮활승류	-滑昇流
daytime visual range	낮시정	-視程
dBZ	dBZ(decibels relative to Z, 레이더 에코 강도의 단위)	
DC circuit	직류회로	直流回路
DDA value	DDA값(시간, 길이, 넓이)	
Deacon wind profile parameter	디콘바람프로파일인자	-因子
deactivation	불활성화	不活性化
dead air	정체공기	停滯空氣
dead glacier	죽은 빙하	-氷河
dead reckoning	추측항법	推測航法
dead water	사수, 정체수	死水, 停滯水
dealiasing	위신호보정, 디에일리어싱	僞信號補整
Deardorff velocity	디어도프속도	-速度
deasil	태양진행방향, 시계방향	太陽進行方向, 時計方向
death assemblage	사체군집	死體群集
de-briefing	사후보고	事後報告
debris	파편	破片
debris flow	쇄설류	碎屑流
decadal oscillation	십년주기변동	十年週期變動
decade	십년	十年
decay	감쇠	減衰
decay area	감쇠지역	減衰地域
decay constant	붕괴상수	崩壞常數
decay distance	감쇠거리	減衰距離
decaying mode	감쇠모드	減衰-
decay of wave	파감쇠	波減衰
decelerate	감속	減速
deceleration of wind	바람감속	-減速
decibar	10바	
decibel(db)	데시벨	
decile	십분위수	十分位數
decimal	소수	小數
decimal coefficient of extinction	상용소산계수	常用消散係數
decimation	부분제거	部分除去
decision framework	결정프레임워크	決定-
decision height	결정고도	決定高度
decision making	의사결정	意思決定
decision tree	결정트리, 결정나무	決定-
deck(= lid)	갑판	甲板
declination	편각, 적위	偏角, 赤緯
decoder	부호해독기	符號解讀器
decoding	부호해독	符號解讀

decomposer	분해자	分解者
decomposition	분해	分解
deconvolution	디콘볼루션(초점이 빗나간 사진과 흔들린 사진 보정 팁)	
decorrelation distance	비상관거리	非相關距離
decorrelation time	감상관시간	減相關時間
decoupling	분리, 해리	分離, 解離
decrease	감소	減少
decreasing availability water balance model	감소 가용성 물균형 모형	減少可用性-均衡模型
deelectric constant	유전상수	誘電常數
deelectric material	유전물질	誘電物質
deep convection	심층대류, 강한 대류현상	深層對流, -對流現象
deep convective index	심층대류지수	深層對流指數
deep easterlies	깊은 편동풍	-偏東風
deepening	(저기압의)심화	深化
deepening cyclone	심화저기압	深化低氣壓
deepening of a depression	저기압심화	低氣壓深化
deepening stage	심화단계	深化段階
deep flood irrigation	심수관개	深水灌漑
deep-sea drilling project	심해굴착계획	深海掘鑿計畵
deep-sea wave	심해파	深海波
deep-space probe	원거리우주탐사체	遠距離宇宙探査體
deep time	장구시간	長久時間
deep trades	두꺼운 무역풍	-貿易風
deep-water wave	심수파	深水波
Defant method	데판트법	-法
Defant's model	데판트모형	-模型
defense meteorological satellite program	방위기상위성계획	防衛氣象衛星計劃
deficiency	결핍증	缺乏症
definite integral	정적분	定積分
deflation hollow	취식와지, 사막포도	吹蝕窪地, 砂漠鋪道
deflecting acceleration	전향가속도	轉向加速度
deflecting factor	전향인자	轉向因子
deflecting force	전향력	轉向力
deflection	전향	轉向
deflection anemometer	편향풍속계	偏向風速計
deflection force	전향력	轉向力
deflection-modulated indicator	편향변조지시기	偏向變調指示器
defoliation time	낙엽기	落葉期
deforestation	산림벌채	山林伐採
deformation	변형	變形
deformational flow	변형흐름	變形-
deformation axis	변형축	變形軸
deformation field	변형장	變形場
deformation plane	변형면	變形面

D

deformation radius	변형반지름, 변형반경	變形半徑
deformation term	변형항	變形項
deformation thermograph	변형온도기록계	變形溫度記錄計
deformation thermometer	변형온도계	變形溫度計
deformation vector	변형벡터	變形-
deformation zone	변형구역	變形區域
defuzzification	퍼지해석	-解釋
degeneration	퇴화	退化
deglacial two-step	퇴빙 2단계	退氷二段階
deglaciation	퇴빙, 해빙, 탈빙하	退氷, 解氷, 脫氷河
degradation of energy	에너지퇴화	-退化
degree	도	度
degree day	도일	度日
degree hour	도시	度時
degree of confidence	신뢰도	信賴度
degree of freedom	자유도	自由度
degree of polarization	편광도	偏光度
degree of saturation	포화도	飽和度
degree of stability	안정도	安定度
degree of superheat	과열도	過熱度
degree of frost	서리도	-度
dehumidification	제습	除濕
dehumidifier	제습기	除濕器
dehumidifying ventilation	제습환기	除濕換氣
deice	제빙	除氷
deicer	제빙기	除氷器
deicing device	제빙장치	除氷裝置
dekad	십일	十日
delay	지연	遲延
delayed automatic gain control	지연자동이득제어	遲延自動利得制御
delayed fallout	지연낙진	遲延落塵
delayed oscillator theory	지연발진기이론	遲延發振器理論
delayed report	지연보고	遲延報告
delay line	지연선	遲延線
delinescope	환등	幻燈
Delinger effect	델린저효과	-效果
deliquescence	조해성	潮解性
delivering time of message	통보시각	通報時刻
del-operator	델연산자	-演算子
delta	삼각주	三角洲
delta region	삼각주, 출구역	三角洲, 出口域
deluge	대홍수	大洪水
demarcation	경계구분	境界區分
demarcation line	분계선	分界線

demodulation	복조, 검파	復調, 檢波
dendrite	나뭇가지모양	-模樣
dendritic crystal	나뭇가지모양결정	-模樣結晶
dendritic snow crystal	나뭇가지모양눈결정	-模樣-結晶
dendrochronological record	연륜연대기록	年輪年代記錄
dendrochronology	연륜학	年輪學
dendroclimatology	연륜기후학	年輪氣候學
Dendy method	덴디방법	-方法
Denekamp interstrade	데네캄프간빙기	-間氷期
denoxification	탈질	脫窒
dense fog	짙은 안개	
dense fog season	농무기, 안개계절	濃霧期, -季節
dense upper cloud	짙은 상층운	-上層雲
densitometer	비중계	比重計
density	밀도	密度
density advection	밀도이류	密度移流
density altitude	밀도고도	密度高度
density channel	밀도채널	密度-
density constant height	밀도불변고도	密度不變高度
density correction	밀도보정	密度補正
density current	밀도류	密度流
density function	밀도함수	密度函數
density matrix	밀도행렬	密度行列
density of dry air	건조공기밀도	乾燥空氣密度
density of flue gas pollutants	매연농도	煤煙濃度
density of moist air	습윤공기밀도	濕潤空氣密度
density of snow	눈밀도	-密度
density of snow gauge	눈밀도계	-密度計
density of soil	토양밀도	土壤密度
density ratio	밀도비	密度比
density stratification	밀도성층	密度成層
density (thickness) advection	밀도(두께)이류	密度移流
densus	짙음(구름)	
denudation	삭박작용	削剝作用
departing wave	이탈파	離脫波
departure	편차	偏差
depegram	데피그램	
dependence	종속	從屬
dependent event	종속사건	從屬事件
dependent meteorological office	보조기상대(항공)	補助氣象臺(航空)
dependent variable	종속변수	從屬變數
depergelation	영구동토풀림	永久凍土-
depletion curve	감수곡선	減水曲線
depolarization	감극	減極

depolarization ratio	편극소거비	偏極消去比
deposit gauge	침착량측정기	沈着量測定器
deposition	침적, 침착, 침강	沈積, 沈着, 沈降
deposition nucleus	침착핵	沈着核
deposition velocity	침착속도	沈着速度
depression	저압부	低壓部
depression angle	내림각	-角
depression belt	저압대	低壓帶
depression family	저기압가족	低氣壓家族
depression in East China Sea	동중국해 저기압	東中國海低氣壓
depression of the dew point	기온이슬점차, 기온노점차	氣溫露點差
depression of the wet bulb	건습구온도차	乾濕球溫度差
depression storage	요지저류량	凹地貯溜量
depth-area curve	우량-면적 곡선	雨量面積曲線
depth-area-duration(DAD) analysis	우량-면적-지속기간 분석	雨量面積持續期間分析
depth-area formula	우량-면적 공식	雨量面積公式
depth-duration-area(DDA) analysis	DDA 분석	-分析
depth-duration curve	우량-지속 곡선	雨量持續曲線
depth-duration-frequency curve	우량-지속-빈도 곡선	雨量持續頻度曲線
depth hoar	속서리	
depth marker	깊이표시기	-表示器
depth of autumn	한가을	
depth of compensation	보상깊이	補償-
depth of run off	유출고	流出高
depth of snow	적설깊이	積雪-
depth of snow cover	적설깊이	積雪-
depth of snow fall	신적설깊이, 강설깊이	新積雪-, 降雪-
depth of spring	한봄	
depth of summer	한여름	
depth of winter	한겨울	
derivative	도함수, 유도체	導函數, 誘導體
derived gust velocity	유도돌풍속도	誘導突風速度
derived unit	유도단위	誘導單位
de Saint-Venant equation	드셍베낭방정식	-方程式
desalination of seawater	해수담수화	海水淡水化
Descartes ray	데카르트광선	-光線
descendent	하강(의)	下降
descending air current	하강기류	下降氣流
descending current	하강류	下降流
descending flow	하강기류	下降氣流
descending node	강교점	降交點
descending pass	하강궤도	下降軌道
descent	하강	下降
descent of air	공기하강	空氣下降

descriptive climatology	서술기후학	敍述氣候學
descriptive meteorology	서술기상학	敍述氣象學
desert albedo	사막알베도	砂漠-
desert belt	사막대	砂漠帶
desert climate	사막기후	砂漠氣候
desert devil	먼지회오리	
desertification	사막화	砂漠化
desert mirage	사막신기루	砂漠蜃氣樓
desert steppe	사막스텝	砂漠-
desert wind	사막바람	砂漠-
desert wind squall	사막돌풍	砂漠突風
desiccation	건조	乾燥
design criteria	설계기준	設計基準
design discharge	설계유량	設計流量
design flood	설계홍수	設計洪水
design storm	설계폭풍	設計暴風
desorption	탈착	脫着
destruction rate	파괴율	破壞率
destructive interference	상쇄간섭	相殺干涉
desulfurization	탈황	脫黃
detached stratus	분리층운	分離層雲
detail weather information	상세기상정보	詳細氣象情報
detection	탐지, 검출, 검파	探知, 檢出, 檢波
detection technique	탐지기법	探知技法
detector	탐지기	探知機
deterioration index	부패지수	腐敗指數
deterioration report	일기악화보고	日氣惡化報告
determinant	행렬식	行列式
determinate forecast	확정예보	確定豫報
determination condition	결정조건	決定條件
deterministic chaos	결정론적 카오스	決定論的-
deterministic term	결정론적 항	決定論的項
detrainment(= discharge, effluent, efflux, outflow, runoff)	유출	流出
detrending	추세제거	趨勢除去
detritus	암설	岩屑
Detroit-Windsor area	디트로이트-윈저 지역	-地域
deuterium	중수소	重水素
deuterium content	중수소함량	重水素含量
deuterium excess	중수소과잉	重水素過剩
devastatiing drought	극심가뭄	極甚旱魃
developing stage	발달단계	發達段階
development index	발달지수	發達指數
development wave	발달파	發達波

Devensian glacial	디벤시아빙기	-氷期
deviating force	편향력	偏向力
deviation	편차, 편각	偏差, 偏角
deviation chart	편차도	偏差圖
devil	먼지회오리	
dew	이슬	
dewbow	이슬무지개	
dew cap	이슬막이	
dew cell	이슬점계	-點計
dew condensation	결로	結露
dewfall	결로	結露
dew gauge	이슬량계	-量計
dewing	결로	結露
dew point	이슬점, 노점	露點
dew point apparatus	이슬점측정계, 노점측정계	露點測定計
dew point deficit	기온이슬점차, 기온노점차	氣溫露點差
dew point depression	이슬점차, 노점차	露點差
dew point formula	이슬점공식, 노점공식	露點公式
dew point front	이슬점전선, 노점전선	露點前線
dew point hygrometer	이슬점습도계, 노점습도계	露點濕度計
dew point instrument	이슬점계, 노점계	露點計
dew point lapse rate	이슬점감률, 노점감률	露點減率
dew point line	이슬점불연속선, 노점불연속선	露點不連續線
dew point radiosonde	이슬점라디오존데	露點-
dew point recorder	이슬점기록계, 노점기록계	露點記錄計
dew point spread	기온이슬점차, 기온노점차	氣溫露點差
dew point temperature	이슬점온도, 노점온도	露點溫度
dew point temperature lapse rate	노점온도감률	露點溫度減率
dew trap	이슬트랩	
dextral	오른쪽(의)	
diabatic	비단열(의)	非斷熱
diabatic cooling	비단열냉각	非斷熱冷却
diabatic effect	비단열효과	非斷熱效果
diabatic heating	비단열가열	非斷熱加熱
diabatic influence function	비단열영향함수	非斷熱影響函數
diabatic process	비단열과정	非斷熱過程
diabatic surface layer	비단열지표층	非斷熱地表層
diabatic temperature change	비단열온도변화	非斷熱溫度變化
diabatic wind profile	비단열바람연직분포	非斷熱-鉛直分布
diagnosis(= diagnose)	진단, 분석	診斷, 分析
diagnostic coupling	진단결합	診斷結合
diagnostic equation	진단방정식	診斷方程式
diagnostic model	진단모형	診斷模型

diagnostic variable	진단변수	診斷變數
diagram	선도	線圖
dialysis	투석	透析
dialysis membrane	투석막	透析膜
diamond dust	다이아몬드먼지	
diapause	휴면기	休眠期
diaphragm	진동판	振動板
diaphragm manometer	진동판압력계	振動板壓力計
diastrophism	지각변동	地殼變動
diathermancy	투열성	透熱性
diatoms	규조류	硅藻類
dicarboxylic acids	디카르복실산	-酸
dichotomous impactor	양분충격장치	兩分衝擊裝置
dichotomous sampler	차상샘플러	叉狀-
die back	줄기마름병	-病
dielectric	유전체	誘電體
dielectric constant	유전율	誘電率
dielectric factor	유전율	誘電率
dielectric material	유전체	誘電體
dielectronic recombination	쌍전자재결합	雙電子再結合
difference	계차	階差
difference analogue	차분아날로그	差分-
difference equation	차분방정식	差分方程式
differentiable	미분가능(한)	微分可能
differential absorption	차분흡수법	差分吸收法
differential absorption lidar	차분흡수법라이다	差分吸收法-
differential advection	차등이류	差等移流
differential analyser	미분분석기	微分分析機
differential analysis	층차분석	層差分析
differential chart	차이도	差異圖
differential coefficient	미분계수	微分係數
differential equation	미분방정식	微分方程式
differential fall speed	차등낙하속력	差等落下速力
differential geometry	미분기하(학)	微分幾何(學)
differential heating	차등가열	差等加熱
differential kinematics	미분운동학	微分運動學
differential operator	미분연산자	微分演算子
differential phase	차등위상, 차등위상차	差等位相, 差等位相差
differential reflectivity	차등반사도	差等反射度
differential thermal advection	차등온도이류	差等溫度移流
diffluence	분류	分流
diffluence zone	분류구역	分流區域
diffluent flow	분류흐름	分流-
diffluent ridge	분류형(기압)마루	分流型(氣壓)-

diffluent thermal jet	분류형온도제트	分流型溫度-
diffluent thermal ridge	분류형온도마루	分流型溫度-
diffluent thermal trough	분류형온도골	分流型溫度-
diffluent trough	분류형(기압)골	分流型(氣壓)-
diffracted ray	회절광선	回折光線
diffraction	회절	回折
diffraction fringe	회절무늬	回折-
diffraction lattice	회절격자	回折格子
diffraction pattern	회절패턴	回折-
diffraction phenomenon	회절현상	回折現象
diffraction spectrum	회절스펙트럼	回折-
diffraction zone	회절역	回折域
diffuse approximation	확산근사	擴散近似
diffuse boundary	확산경계면	擴散境界面
diffused light	확산광	擴散光
diffuse front	희미한 전선	-前線
diffuse illumination	확산조명	擴散照明
diffuse light	확산광	擴散光
diffuse radiation	확산복사	擴散輻射
diffuse reflection	난반사	亂反射
diffuse reflector	난반사체	亂反射體
diffuse scattering	난산란	亂散亂
diffuse short-wave radiation	확산단파복사	擴散短波輻射
diffuse skylight	확산하늘빛	擴散-
diffuse sky radiation	확산하늘복사	擴散-輻射
diffuse solar radiation	확산태양복사	擴散太陽輻射
diffusion	확산	擴散
diffusion analogue	확산유사물	擴散類似物
diffusion coefficient	확산계수	擴散係數
diffusion computation	확산계산	擴散計算
diffusion diagram	확산선도	擴散線圖
diffusion equation	확산방정식	擴散方程式
diffusion equation limit	확산방정식한계	擴散方程式限界
diffusion experiment	확산실험	擴散實驗
diffusion flux	확산플럭스	擴散-
diffusion formula	확산공식	擴散公式
diffusion hygrometer	확산습도계	擴散濕度計
diffusion model	확산모형	擴散模型
diffusion problem	확산문제	擴散問題
diffusion process	확산과정	擴散過程
diffusion theory	확산이론	擴散理論
diffusion velocity	확산속도	擴散速度
diffusive equilibrium	확산평형	擴散平衡
diffusive force	확산력	擴散力

diffusive separation	확산분리	擴散分離
diffusivity	확산율	擴散率
diffusivity equation	확산율방정식	擴散率方程式
diffusograph	자기산란복사계	自記散亂輻射計
diffusometer	산란복사계, 산란일사계	散亂輻射計, 散亂日射計
digital computer	디지털컴퓨터	
digital count	디지털수	-數
digital data processor	디지털자료처리기	-資料處理器
digital data transmission	디지털자료전송	-資料電送
digital filter	디지털필터	
digital forecast	디지털예보	-豫報
digital image	디지털이미지	
digital image processing	디지털영상처리	-映像處理
digital prefiltering	디지털프리필터링	
digital sampling	디지털샘플링	
digital-to-analog conversion	디지털-아날로그 변환	-變換
digital-to-analog converter	디지털-아날로그 변환기	-變換器
digital video integrator processor	디지털비디오통합처리기	-統合處理器
digitiser(= digitizer)	숫자화 장치	數字化裝置
digitized cloud map	숫자구름분포도	數字-分布圖
digitized radar experiment	숫자레이더실험	數字-實驗
dihedral angle	이면각	二面角
dihedral target	이면표적	二面標的
dilatation	신장, 팽창	伸張, 膨脹
dilatation axis	신장축	伸張軸
dilatation field	신장장	伸張場
dilatometer	신장계	伸張計
dilution	희석	稀釋
diluvial hypothesis	홍수가설	洪水假說
dim cloud	회색권층운	灰色卷層雲
dimensional analysis	차원분석	次元分析
dimensional equation	차원방정식	次元方程式
dimensional matrix	차원행렬	次元行列
dimension of reaction volume	반응체적차원	反應體積次元
dimensionless frequency	무차원주파수	無次元周波數
dimensionless height	무차원높이	無次元高度
dimensionless hydrograph	무차원수문곡선	無次元水文曲線
dimensionless number	무차원수	無次元數
dimensionless parameter	무차원매개변수	無次元媒介變數
dimensionless quantity	무차원량	無次元量
dimensionless temperature	무차원온도	無次元溫度
dimensionless temperature gradient	무차원온도경도	無次元溫度傾度
dimensionless wind shear	무차원바람시어	無次元-
dimethyl acetamide	다이메틸아세트아마이드	

D

dimethyl sulfide	다이메틸설파이드, 이유기유황염	二有機硫黃鹽
Dines anemometer	다인스풍속계	-風速計
Dines compensation	다인스보상	-補償
Dines compensation theorem	다인스보상원리	-補償原理
Dines pressure anemograph	다인스풍압풍속기록계	-風壓風速記錄計
Dines radiometer	다인스복사계	-輻射計
dinitrogen pentoxide	오산화이질소	五酸化二窒素
diode	다이오드, 이극관	二極管
diode laser	이극관레이저	二極管-
diopter	디옵터	
dioxane	다이옥산	
dioxins	다이옥신	
dip(= inclination)	복각, 경각	伏角, 傾角
dip circle	지자기경각계	地磁氣傾角計
dip equator	복각적도	伏角赤道
dip of the horizon	수평면경각	水平面傾角
dipole	쌍극자	雙極子
dipole antenna	쌍극안테나	雙極-
dipole moment	쌍극자모멘트	雙極子-
dip pole(= magnetic pole)	자극	磁極
Dirac delta function	디랙델타함수	-函數
Dirac's delta	디랙델타	
direct beam radiation	직달빔복사	直達-輻射
direct capture	직접포착	直接捕捉
direct cell	직접세포	直接細胞
direct circulation	직접순환	直接循環
direct current	직류	直流
direct flow	직접흐름	直接-
direct heating method	직접가열법	直接加熱法
direct interaction approximation	직접상호작용근사	直接相互作用近似
directional antenna	방향안테나	方向-
directional coupler	방향성결합기	方向性結合器
directional distribution	방향분포	方向分布
directional divergence	방향발산	方向發散
directional hemispherical reflectance	방향반구반사율	方向半球反射率
directional hydraulic conductivity	방향수리전도도	方向水理傳導度
directional radiance	방향복사휘도	方向輻射輝度
directional reflectance	방향반사율	方向反射率
directional shear	방향시어	方向-
direction angle	방향각	方向角
direction finder	방향탐지기	方向探知器
direction finding	방향탐지	方向探知
direction number	방향수	方向數

direction of curvature	곡률방향	曲率方向
directivity	지향성	指向性
direct mode	직접방식	直接方式
direct model output	모형자료원본출력	模型資料原本出力
direct numerical simulation	직접수치모의	直接數值模擬
direct product	직접곱	直接-
direct radiation	직달복사	直達輻射
direct reading instrument	직독측기	直讀測器
direct route	직항로	直航路
direct runoff	직접유출	直接流出
direct runoff hydrograph	직접유출수문곡선	直接流出水文曲線
direct solar radiation	직달일사, 직달태양복사	直達日射, 直達太陽輻射
direct tide	직행조	直行潮
direct transmission system	직접전송시스템, 직접전송계	直接傳送系
direct vision nephoscope	직시측운기	直視測雲器
direct vision prism	직시프리즘	直視-
Dirichlet condition	디리클레조건	-條件
Dirichlet program	디리클레프로그램	
Dirichlet's stability theorem	디리클레안정도정리	-安定度定理
dirigible balloon ascent	비행가능기구탐측	飛行可能氣球探側
disaster	재난	災難
disaster aid	재난구조	災難救助
disaster prevention period	방재기간	防災期間
discharge	유출(량), 방전, 방수	流出(量), 放電, 防水
discharge area	배출지역, 유출역	排出地域, 流出域
discharge coefficient	유출계수	流出係數
discharge curve	유출곡선	流出曲線
discharge mass curve	유출량곡선	流出量曲線
discharge measurement	유량측정	流量測定
discharge per unit drainage area	비유출량	比流出量
(= specific discharge)		
discharge section line	유출량구분선	流出量區分線
discharge tube	방전관	放電管
discomfort index	불쾌지수	不快指數
discomfort zone	불쾌지대	不快地帶
discontinuity	불연속(성)	不連續(性)
discontinuity of first order	일차불연속	一次不連續
discontinuity of zero order	영차불연속	零次不連續
discontinuity surface	불연속면	不連續面
discontinuous	불연속(인)	不連續
discontinuous tropopause	불연속대류권계면	不連續對流圈界面
discontinuous turbulence	불연속난류	不連續亂流
discrete Fourier transform	불연속푸리에변환	不連續-變換
discrete-ordinates method	불연속-세로좌표방법	不連續-座標方法

discrete probability (random) distribution	이산확률분포	離散確率分布
discrete spectrum	불연속스펙트럼	不連續-
discretization	분할	分割
discriminant analysis	판별분석	判別分析
disdrometer	우적계	雨滴計
disease resistance	병해저항성	病害抵抗性
disease tolerance	내병성	耐病性
dishpan experiment	회전원판실험	回轉圓板實驗
disintegration	분해, 분열	分解, 分裂
disk hardness-gauge	원판경도계	圓板硬度計
dispersed irrigation	분산관개	分散灌漑
dispersed phase	분산위상	分散位相
disperse system	분산계	分散系
dispersing medium	분산매질	分散媒質
dispersion	분산	分散
dispersion coefficient	분산계수	分散係數
dispersion curve	분산곡선	分散曲線
dispersion-deposition model	분산-침적모형	分散沈積模型
dispersion diagram	분산선도	分散線圖
dispersion medium	분산매질	分散媒質
dispersion model	분산모형	分散模型
dispersion of light	빛분산	-分散
dispersion parameterization scheme	분산매개변수화방안	分散媒介變數化方案
dispersion relation	분산관계	分散關係
dispersion relationship	분산관계	分散關係
dispersion variable	분산변수	分散變數
dispersive flux	분산플럭스	分散-
dispersive medium	분산성매질	分散性媒質
dispersive wave	분산파	分散波
dispersivity	분산도	分散度
displacement	변위	變位
displacement current	변위전류	變位電流
displacement distance	변위거리	變位距離
displacement height	변위고도	變位高度
displacement thickness	변위두께	變位-
displacement vector	변위벡터	變位-
display	전시, 표출	展示, 表出
display resolution	표출분해능	表出分解能
dissemination	분배, 전파	分配, 傳播
dissipating range	소멸범위	消滅範圍
dissipating stage	소산단계, 소멸기	消散段階, 消滅期
dissipation	소산, 소멸	消散, 消滅
dissipation coefficient	소산계수	消散係數
dissipation constant	소산상수	消散常數

dissipation contrail	소산비행운	消散飛行雲
dissipation length scale	소산길이규모	-規模
dissipation of energy	에너지소산	-消散
dissipation of wave	파의 소산	波消散
dissipation range	소산영역	消散領域
dissipation rate	소산율	消散率
dissipation rate of energy	에너지소산율	-消散率
dissipation trail	비행운소산	飛行雲消散
dissociation	분리	分離
dissociation constant	해리상수	解離常數
dissociation curve	해리곡선	解離曲線
dissociation energy	해리에너지	解離-
dissociation rate	해리율	解離率
dissociation wavelength	해리파장	解離波長
dissolution	분해, 용해	分解, 溶解
dissolutions of gas	기체의 용해도	氣體-溶解度
dissolved oxygen amount	용존산소량(DO)	溶存酸素量
distance-altitude cross section	거리-고도단면도	距離高度斷面圖
distance between centers	중심거리	中心距離
distance-neighbor graph	거리-이웃그래프	距離-
distant flash	먼거리방전	-距離放電
distillation	증류	蒸溜
distinct CDO	뚜렷한 CDO	
distinct large eye	큰 태풍눈(직경이 40km를 넘는 크기의 눈)	-颱風-
distinct small eye	작은 태풍눈(직경이 40km 이내 크기의 눈)	-颱風-
distorted raindrop	변형된 빗방울	變形-
distorted wake	왜곡후류기류	歪曲後流氣流
distorted water	왜곡수	歪曲水
distortion	비틀림	
distortional wave	비틀림파	-波
distrail	비행운소산	飛行雲消散
distributed target	분포표적(물)	分布標的(物)
distribution	분배, 분포	分配, 分布
distribution coefficient	분포계수	分布係數
distribution curve	분포곡선	分布曲線
distribution function	분포함수	分布函數
distribution graph	분포그래프	分布-
distribution of air pressure	기압분포	氣壓分布
distribution polygon	분포다각형	分布多角形
distributive law	분배법칙	分配法則
district forecast	구역예보	區域豫報
District Meteorological Office	지방기상대	地方氣象臺

disturbance	요란	擾亂
disturbance line	요란선	擾亂線
disturbance management	요란관리	擾亂管理
disturbed soil sample	교란토양시료	攪亂土壤試料
diurnal	하루의	
diurnal amplitude	일진폭	日振幅
diurnal change	일변화	日變化
diurnal cooling	일냉각	日冷却
diurnal cycle	일주기	日周期
diurnal heating	일가열	日加熱
diurnal inequality	일부등성	日不等性
diurnal periodicity	일주기성	日週期性
diurnal range	일교차	日較差
diurnal surface energy balance	일지표면에너지균형	日地表面-均衡
diurnal temperature range	일(온도)교차	日(溫度)較差
diurnal tide	일일조	一日潮
diurnal trend	일변화경향	日變化傾向
diurnal variation	일변동	日變動
diurnal vertical migration	일연직이동	日鉛直移動
diurnal wave	일변화파	日變化波
diurnal wind	일변화바람	日變化-
divergence	발산	發散
divergence due to directional change	방향변화발산	方向變化發散
divergence effect	발산효과	發散效果
divergence equation	발산방정식	發散方程式
divergence field	발산장	發散場
divergence line	발산선	發散線
divergence region	발산구역	發散區域
divergence signature	발산징후	發散徵候
divergence theorem	발산정리	發散定理
divergence theory of cyclogenesis	저기압발생발산설	低氣壓發生發散說
divergence theory of cyclone	저기압발산이론	低氣壓發散理論
divergent barotropic atmosphere	발산순압대기	發散順壓大氣
divergent margin	발산주변부	發散周邊部
divergent point	발산점	發散點
diversification	다양성, 변형	多樣性, 變形
diversion of water	분수	分水
divide	분수령	分水嶺
divide line	분수선	分水線
dividing crest	분수령	分水嶺
dividing ridge	분수마루	分水-
dividing streamline	분할유선	分割流線
Dixie Alley	딕시길목	
D-layer	D층	-層

Dobson instrument	도브슨기기	-器機
Dobson ozone spectrophotometer	도브슨오존분광광도계	-分光光度計
Dobson spectrophotometer	도브슨분광광도계	-分光光度計
Dobson unit	도브슨단위	-單位
dog day	복중	伏中
doldrum equatorial trough	적도무풍대	赤道無風帶
doldrums	적도무풍대	赤道無風帶
Dole effect	돌효과	-效果
domain	영역	領域
domestic aerodrome forecast	국내비행장예보	國內飛行場豫報
domestic climate	생활기후	生活氣候
domestic climatology	생활기후학	生活氣候學
domestic daily forecast	육상일일예보	陸上日日豫報
domestic local forecast	육상국지예보구역	陸上局地豫報區域
domestic regional forecast	육상광역예보구역	陸上廣域豫報區域
domestic short range forecast	육상단기예보	陸上短期豫報
dominant wind	탁월풍	卓越風
Donau Glacial	도나우빙기	-氷期
Donau-Günz interglacial	도나우-귄츠 간빙기	間氷期
Dong-Nae Forecast	동네예보	-豫報
Donora episode	도노라사건	-事件
Doppler	도플러	
Doppler broadening	도플러넓어짐	
Doppler dilemma	도플러딜레마	
Doppler effect	도플러효과	-效果
Doppler equation	도플러방정식	-方程式
Doppler error	도플러오차	-誤差
Doppler frequency shift	도플러진동수 변이	-振動數變異
Doppler laser radar	도플러레이저 레이더	
Doppler lidar	도플러라이다	
Doppler measurement	도플러측정	-測定
Doppler principle	도플러원리	-原理
Doppler radar	도플러레이더	
Doppler shape	도플러형상	-形象
Doppler shift	도플러변이	-變移
Doppler sodar	도플러소다	
Doppler spectrum	도플러스펙트럼	
Doppler velocity	도플러속도	-速度
Doppler wind measurements	도플러바람관측	-觀測
dormant volcano	휴화산	休火山
dosage	조사적량	照射適量
dose-response	용량-반응	用量反應
dosimeter	자외선측정기, 도시미터	紫外線測定器
dot product(= scalar product)	내적, 스칼라곱	內積

double diffusion	이중확산	二重擴散
double-dot product	이중도트곱	二重-
double eye pattern	이중눈패턴	二重-
double integral	이중적분	二重積分
double mass analysis	이중적산분석	二重積算分析
double register	이중기록계	二重記錄計
double scalar product	이중스칼라곱	二重-
double theodolite observation	쌍경위의관측	雙經緯儀觀測
double theodolite technique	쌍경위의기술	雙經緯儀技術
double tropopause	이중권계면	二重圈界面
double wave	이중파	二重波
doubling method	배증법	倍增法
dough ripe stage	호숙기	糊熟期
Dove's law	도베법칙	-法則
downburst	다운버스트, 하강돌풍	下降突風
downconverter	다운컨버터	
down current	하강기류	下降氣流
downdraft	하강기류	下降氣流
downdraft velocity	하강류속도	下降流速度
downdraught(= downdraft)	하강기류	下降氣流
downglide motion	활강운동	滑降運動
downpour	폭우	暴雨
downrush	막내리바람, 하강급류	下降急流
downscaled data	규모축소자료	規模縮小資料
downscaling	다운스케일링, 규모축소	規模縮小
downshear	순시어	順-
downslide	활강	滑降
downslide surface	활강면	滑降面
downslope airflow	활강흐름	滑降-
downslope flow	활강흐름	滑降-
downslope motion	활강운동	滑降運動
downslope wind	활강바람, 활강풍	滑降風
downstream	풍하측	風下側
downstream development	풍하측발달	風下側發達
downward force	하향력	下向力
downward longwave radiation	하향장파복사	下向長波輻射
downward motion	하향운동	下向運動
downward (total) radiation	하향(전)복사	下向(全)輻射
downward radiation	하향방사	下向放射
downward shortwave radiation	하향단파복사	下向短波輻射
downward terrestrial radiation	하향지구복사	下向地球輻射
downwash	씻어내림	
downwelling	침강, 침강류	沈降, 沈降流
down wind	풍하측, 풍하(쪽)	風下測

draft(= draught)	(소규모의)기류	氣流
drag	항력	抗力
drag acceleration	항력가속(도)	抗力加速
drag coefficient	항력계수	抗力係數
drag crisis	항력위기	抗力危機
drag force	항력	抗力
dragon	용오름	龍-
drag plate	항력판	抗力板
drainage	배수	排水
drainage area	유역, 배출역	流域, 排出域
drainage basin	집수역	集水域
drainage density	배수밀도	排水密度
drainage divide	유역분수계	流域分水界
drainage flow	배출류	排出流
drainage force	배수력	排水力
drainage gauge	배수량계	排水量計
drainage wind	배출풍	排出風
dramundan	장막극광	帳幕極光
drecho	드레이쇼	
D-region	D영역	-領域
drift	표류	漂流
drift bottle	표류병	漂流瓶
drift coefficient	편류계수	偏流係數
drift correction	편류보정	偏流補正
drift correction angle	편류조정각	偏流調整角
drift current	편류, 표류전류	偏流, 漂流電流
drift epoch	표류기	漂流期
drift ice	유빙	遊氷
drifting-blowing dust	땅날림먼지, 날림먼지	
drifting buoy	표류부이, 표류부표	漂流浮漂
drifting dust	땅날림먼지, 낮은 풍진	-風塵
drifting sand	땅날림모래	
drifting snow	땅날림눈, 낮은 땅눈보라	
drift meter	편류계	偏流計
drip irrigation	물방울 관개	-灌漑
driven snow	날려쌓인 눈, 풍적설	風積雪
driving force	추진력	推進力
driving rain index	풍성우량지수	風成雨量指數
drizzle	이슬비	
drizzle drop	이슬비방울	
drizzling fog	이슬비안개	
drogue	부표	浮標
drone	무인비행기	無人飛行機
drop	방울	

D

drop coalescence theory	구름방울병합설	-倂合說
drop collector	구름방울수집기	-收集器
droplet	작은 구름방울	
droplet collector	작은 구름방울수집기	-收集器
droplet phase	작은 구름방울상	-相
droplet-size distribution	작은 구름방울크기분포	-分布
droplet spectrum	작은 구름방울스펙트럼	
drop of water	수적	水滴
drop orientation	수적정렬방향	水滴整列方向
droppable pyrotechnic flare system	투하연소시스템	投下燃燒-
dropping flare	구름씨 투하기구	-投下器具
drop-size distribution	구름방울크기분포	-分布
drop-size meter	구름방울크기측정계	-測定計
drop-size spectrometer	구름방울크기분광계	-分光計
dropsonde	낙하존데	落下-
dropsonde data	낙하존데자료	落下-資料
dropsonde dispenser	낙하존데투하장치	落下-投下裝置
dropsonde observation	낙하존데관측	落下-觀測
drop sounding	낙하탐측	落下探測
drop spectrum	구름방울스펙트럼	
drop theory	저기압장벽설	低氣壓障壁說
dropwindsonde	낙하바람존데	落下-
drosograph	자기이슬량계	自記-計
drosometer	이슬량계	-計
drought	가뭄, 한발	旱魃
drought damage	가뭄피해	-被害
drought disaster	한해	旱害
drought duration	가뭄지속기간	-持續期間
drought frequency	가뭄빈도	-頻度
drought index	가뭄지수	-指數
drought injury	한해	旱害
drought monitoring	가뭄감시	-監視
drought resistance	가뭄저항성	-抵抗性
drought stress	가뭄스트레스	
droughty season	갈수기	渴水期
droxtal(= drop + crystal)	얼음알갱이	
drum (of self-recording instrument)	자기기록계원통	自己記錄計圓筒
drumlin	빙퇴구	氷堆丘
dry adiabat	건조단열	乾燥斷熱
dry adiabatic	건조단열(의)	乾燥斷熱
dry adiabatic atmosphere	건조단열대기	乾燥斷熱大氣
dry adiabatic curve	건조단열곡선	乾燥斷熱曲線
dry adiabatic change	건조단열변화	乾燥斷熱變化
dry adiabatic cooling	건조단열냉각	乾燥斷熱冷却

dry adiabatic lapse rate	건조단열감률	乾燥斷熱減率
dry adiabatic process	건조단열과정	乾燥斷熱過程
dry adiabatic rate	건조단열률	乾燥斷熱率
dry air	건조공기	乾燥空氣
dry airstream	건조기류	乾燥氣流
dry-and-wet-bulb hygrometer	건습구습도계	乾濕球濕度計
dry-and-wet-bulb thermometer	건습구온도계	乾濕球溫度計
dry Baiu	마른 장마	
dry bulb	건구	乾球
dry bulb temperature	건구온도	乾球溫度
dry bulb thermometer	건구온도계	乾球溫度計
dry Changma	마른 장마	
dry climate	건조기후	乾燥氣候
dry convection	건조대류	乾燥對流
dry convection layer	건조대류층	乾燥對流層
dry convective adjustment	건조대류조정	乾燥對流調定
dry damage	건조해	乾燥害
dry deposition	건성침착	乾性沈着
dry downburst	건조하강돌연풍	乾燥下降突然風
dry fog	마른 안개	
dry freeze	마른 결빙	-結氷
dry growth	건조성장	乾燥成長
dry haze	마른 연무	-煙霧
dry ice	드라이아이스	
dry ice seeding	드라이아이스씨뿌리기	
dry impingement	건조분자입사	乾燥分子入射
drying mechanism	건조메커니즘	乾燥-
drying power	건조능	乾燥能
dry instability	건조불안정	乾燥不安定
dry intrusion	건조관입	乾燥貫入
	(차고 건조한 공기 침투)	
dry inversion	건조역전	乾燥逆轉
dryland degradation	건조지대붕괴	乾燥地帶崩壞
drylands	건조지대	乾燥地帶
dry line	건조선	乾燥線
dry line front	건조선전선	乾燥線前線
dry line storm	건조선폭풍	乾燥線暴風
dry matter	건물량	乾物量
dry matter production	건물생산	乾物生産
dry method of exhaust gas desulfurization	건식배연탈황법	乾式排煙脫黃法
dry microburst	건조마이크로버스트,	乾燥瞬間突風
	건조순간돌풍	
dryness	건조도	乾燥度
dryness advisory	건조주의보	乾燥注意報

dryness index	건조도지수	乾燥度指數
dryness warning	건조경보	乾燥警報
dry period	건조기간, 건기	乾燥期間, 乾期
dry permafrost	건조영구 언땅	乾燥永久-
dry region	건조지역	乾燥地域
dry salt lake	건조염호	乾燥鹽湖
dry scrubbers	드라이스크러버, 건식세정기	乾式洗淨器
dry season	건조계절	乾燥季節
dry sepal	건식꽃받침	
dry slot	건조슬롯	乾燥-
dry snow	마른 눈	
dry spell	건조지속기, 건기	乾燥持續期, 乾期
dry stage	건조단계	乾燥段階
dry subhumid climate	건조아습윤기후	乾燥亞濕潤氣候
dry summer subtropical climate	건조여름아열대기후	乾燥-亞熱帶氣候
dry thunderstorm	마른 뇌우	-雷雨
dry tongue	건조혀	乾燥-
dry wind	건조바람	乾燥-
Dst index	자기폭풍지수	磁氣暴風指數
dual channel difference method	이중채널차방법	二重-差方法
dual Doppler	이중도플러	二重-
dual Doppler processing	이중도플러처리	二重-處理
dual frequency	이중진동수	二重振動數
duality	이중성	二重性
dual polarization radar	이중편파레이더	二重偏波-
dual receiver system	이중수신시스템, 이중수신계	二重受信系
dual wavelength radar	이중파장레이더	二重波長-
DuBois area	두보이스지역	-地域
duct	덕트	
ducting	(레이더파의)빔갇힘	
duff	낙엽층	落葉層
dumb-bell depression	아령저기압	啞鈴低氣壓
dummy variable	가변수	假變數
dune	모래언덕, 사구	砂丘
duplexer	송수신전환스위치	送受信轉換-
duplicate ratio	제곱비	-比
duplication	중복	重複
duplicatus	겹구름	
duration curve	지속기간곡선	持續期間曲線
duration of frost-free period	무상기간	無霜期間
duration of growing period	생육기간	生育期間
duration of ice cover	적빙기간	積氷期間
duration of sunshine	일조시간	日照時間
duration statistics	기간통계학	期間統計學

dusk	땅거미	
dust	먼지	
dust aerosol	먼지 에어로졸	
dust avalanche	마른 눈사태	-沙汰
dust bowl	더스트보울, 먼지덩이	
dust burden	먼지짐, 먼지나르기	
dust cloud	먼지구름	
dust collector	집진장치, 먼지채집기	集塵裝置, 塵採集器
dust content	먼지함량	-含量
dust counter	먼지계수기	-計數器
dust devil	먼지회오리	
dust dome	더스트돔, 먼지돔	
duster	먼지보라	
dust fall	낙진	落塵
dust fall jar	낙진채집조	落塵採集槽
dust fog	먼지안개	
dust haze	먼지연무	-烟霧
dust horizon	먼지층선	-層線
dustiness	먼지투성이	
dust loading	먼지농도	-濃度
dust mite	먼지진드기	
dust particle	먼지알갱이	
dust plume	먼지플룸	
dust rain	먼지비	
dust scrubber	세정집진장치	洗淨集塵裝置
dust shower	먼지소나기	
dust storm	먼지폭풍, 먼지보라	-暴風
dust transport	먼지수송, 먼지이동	-輸送, -移動
dust turbidity	먼지혼탁도	-混濁度
dust veil	먼지베일	
dust veil index	먼지베일지수	-指數
dust wall	먼지벽	-壁
dust warnig	먼지경보	-警報
dust whirl	먼지회오리	
duty cycle	듀티주기, 작동주기	作動週期
duty factor	충격계수	衝擊係數
duty of water	용수량	用水量
D-value	D값	
Dvorak analysis	드보락분석법	-分析法
dwelling pollution	주거오염	住居汚染
dwell time	휴지시간	休止時間
dwigh(= dwey, dwoy)	드와이(캐나다 뉴유펀들랜드에서 갑자기 일어나는 소낙비 또는 소낙눈)	
dyad	이가원소	二價元素

dyadic	이가원소(의)	二價元素
dyadic product	다이애드곱	
dynamical concept	역학개념	力學槪念
dynamical convection	역학대류	力學對流
dynamical forecast	역학예보	力學豫報
dynamical structure	역학구조	力學構造
dynamical system	역학시스템	力學-
dynamic anticyclone	역학고기압	力學高氣壓
dynamic boundary condition	역학경계조건	力學境界條件
dynamic boundary surface condition	역학경계지면조건	力學境界地面條件
dynamic characteristics	역학특성	力學特性
dynamic climatology	동기후학	動氣候學
dynamic cloud seeding	동적구름씨뿌리기	動的-
dynamic coefficient of viscosity	역학점성계수	力學粘性係數
dynamic component	역학성분	力學成分
dynamic condition	역학조건	力學條件
dynamic cooling	역학냉각	力學冷却
dynamic cumulus	역학적운	力學積雲
dynamic depth	역학깊이	力學-
dynamic entrainment	역학적 흡입	力學的吸入
dynamic error	역학오차	力學誤差
dynamic global vegetation model	지구식생역학모형	地球植生力學模型
dynamic heating	역학가열	力學加熱
dynamic height	역학고도	力學高度
dynamic ice	역학얼음, 여름극잔존얼음	力學-, -極殘存-
dynamic indifference	역학중립	力學中立
dynamic initialization	역학초기화	力學初期化
dynamic instability	역학불안정(도)	力學不安定(度)
dynamic kilometer	다이나믹킬로미터	
dynamic lift	역학상승	力學上昇
dynamic lifting	역학상승	力學上昇
dynamic linear model	동적선형모형	動的線型模型
dynamic meteorology	기상역학	氣象力學
dynamic meter	다이나믹미터	
dynamic performance	동적성능	動的性能
dynamic pressure	동압	動壓
dynamic range	동적범위	動的範圍
dynamic response	역학반응	力學反應
dynamics	역학	力學
dynamic similarity	역학상사	力學相似
dynamic soaring	동적비상	動的飛上
dynamics of the atmosphere	대기역학	大氣力學
dynamic stability	역학안정(도)	力學安定(度)
dynamic temperature change	역학온도변화	力學溫度變化

dynamic term	역학항	力學項
dynamic trough	역학골	力學-
dynamic unit	역학단위	力學單位
dynamic vertical motion	역학연직운동	力學鉛直運動
dynamic viscosity	역학점성	力學粘性
dynamic warming	역학승온	力學昇溫
dynamic wave routing model	동역학파라우팅모형, 동역학파전달모형	動力學波傳達模型
dynamometer cycles	검력계주기	檢力計週期
dynamo theory	다이나모이론	-理論
dyne	다인	

D

E, e

영문	한글	한자
E$_2$-layer	E$_2$층	-層
Eady problem	이디문제	-問題
eagre(= bore)	보어	
early Eocene climatic optimum	초기에오세기후최적	初期-氣候最適
early fallout	조기낙하	早期落下
early frost	이른 서리	
early frost damage	조상해	早霜害
early Paleozoic climates (Cambrian-Devonian)	고생대전기기후 (캄브리아기-데본기)	古生代前期氣候
early rainfall process	초기강수과정	初期降水過程
early stage Dvoral analysis	초기단계 드보락분석법	初期段階-分析法
early warning system	조기경보체계	早期警報體系
earth-air current	지-공전류	地-空電流
earth atmosphere	지구대기	地球大氣
earth atmosphere radiation budget	지구대기복사수지	地球大氣輻收支
earth atmosphere system	지구대기시스템	地球大氣-
earth axis	지축	地軸
earth climatology	지구기후학	地球氣候學
earth current	땅전류	-電流
earth current storm	땅전류폭풍	-電流暴風
earth curvature correction	지구곡률보정	地球曲率補正
earth ellipsoid	지구타원체	地球楕圓體
earthflow	토류, 토석류	土流, 土石流
earth grounding	접지	接地
earth heat budget	지구열수지	地球熱收支
earthing	접지	接地
earthlight	지구되쬐임, 지구반사광	地球-, 地球反射光
earth magnetism	지자기	地磁氣
earth mound	토제	土堤

earth observing system	지구관측시스템, 지구관측계	地球觀測系
earth observing system program	지구관측시스템프로그램, 지구관측계프로그램	地球觀測系-
earth-ocean-atmosphere system	지구-해양-대기시스템, 지구-해양-대기계	地球海洋大氣系
earth orbit	지구궤도	地球軌道
earth radiation	지구복사	地球輻射
earth radiation budget	지구복사수지	地球輻射收支
Earth Resource Technology Satellite	지구자원탐사위성	地球資源探査衛星
earth rotation	지구자전	地球自轉
earth rotational speed	지구자전속력	地球自轉速力
earth scale	지구규모	地球規模
earth sciences	지구과학	地球科學
earth shadow	지구그림자	地球-
earthshine(= earthlight)	지구광	地球光
earth surface	지표면	地表面
earth stabilization	지구안정화	地球安定化
earth summit	지구정상회담	地球頂上會談
earth-sun relationship	지구-태양 관계	地球-太陽關係
earth surface	지표면	地表面
earth synchronous orbit	지구동시궤도	地球同時軌道
earth system	지구시스템, 지구계	地球系
Earth System Research Center	지구시스템연구센터	地球-研究-
earth system science	지구시스템과학	地球-科學
earth temperature	지중온도	地中溫度
earth thermograph	지중온도기록계	地中溫度記錄計
earth thermometer	지중온도계	地中溫度計
earthwatch program	지구감시프로그램	地球監視-
East African jet	동아프리카제트	東-
East African low-level jet	동아프리카하층제트	東-下層-
East Arabian current	아라비안동안류	-東岸流
East Asian summer monsoon	동아시아여름몬순	東-
East Auckland current	오클랜드동안류	-東岸流
East Australia current	동오스트레일리아해류	東-海流
East Cape current	동케이프해류	東-海流
East China Sea	동중국해	東中國海
east coast boundary current	동안경계류	東岸境界流
east coast climate	동해안기후	東海岸氣候
east coast cyclone	동해안저기압	東海岸低氣壓
east-coast occlusion	동해안폐색	東海岸閉塞
easterlies	편동풍	偏東風
easterly belt	편동풍대	偏東風帶
easterly dip	편동편차	偏東偏差
easterly jet	편동풍제트	偏東風-

easterly trade wind	편동무역풍	偏東貿易風
easterly trough	편동풍골	偏東風-
easterly wave	편동풍파	偏東風波
easterly wind	편동풍	偏東風
Eastern Atlantic pattern	동대서양형	東大西洋型
eastern coastal climate	동안기후	東岸氣候
eastern North Pacific basin	북태평양동부해분	北太平洋東部海盆
East Indian current	인도 동안류	印度東岸流
East Sea	동해	東海
East Sea cyclone	동해저기압	東海低氣壓
ebb	썰물	
ebb current	썰물	
ebb tide	썰물 조류	-潮流
Ebert-Fastie monochromator	에버트-파스티 단색화장치	-單色化裝置
Ebert ion-counter	에버트이온계수기	-計數器
Ebert scanning spectrometer	에버트주사분광기	-走査分光器
ebullioscopy	비등점상승법	沸騰點上昇法
ebullition	비등	沸騰
eccentric angle	이심각	離心角
eccentric anomaly	이심이상	離心異常
eccentricity	이심률	離心率
eccentricity of the earth's orbit	지구공전궤도이심률	地球公轉軌道離心率
eccentric-orbiting geophysical observatory	이심궤도 지구물리관측위성	離心軌道地球物理觀測衛星
echelon cloud	사다리모양구름	-模樣-
echo	에코	
echo amplitude	에코진폭	-振幅
echo area	에코지역	-地域
echo box	에코상자	-箱子
echo free vault	무에코원형덮개	無-圓形-
echo frequency	에코주파수	-周波數
echo power	에코세기	
echo power gradient	에코전력기울기	-電力-
echo power ratio	에코전력비	-電力比
echo pulse	에코펄스	
echo radiosonde	에코라디오존데	
echo signal	에코신호	-信號
echosonde	에코존데	
echo sounding apparatus	음파측심장치	音波測深裝置
echo top	에코정상	-頂上
echo top height	에코정상고도	-頂上高度
Eckart coefficient	에카르트계수	-係數
eclipse	식(일, 월)	蝕
eclipse cyclone	일식저기압	日食低氣壓
eclipse wind	일식바람	日食-

eclipse year	식년	蝕年
ecliptic	황도	黃道
ecnephias	에크네피아스(지중해에서 발생하는 스콜 또는 뇌우)	
ecoclimate	생태기후	生態氣候
ecoclimatology	생태기후학	生態氣候學
eco-efficiency	생태효율	生態效率
ecological climatology	생태기후학	生態氣候學
ecology	생태학	生態學
economical net radiometer	경제적 정미방사계	經濟的正味放射計
economy of radiation	복사수지	輻射收支
ecophysiological model	생리생태학모형	生理生態學模型
ecosystem	생태계	生態系
ecosystem change	생태계 변화	生態系變化
ecosystem diversity	생태계 다양성	生態系多樣性
ecosystem equilibrium	생태계 평형	生態系平衡
ecosystem management	생태계 관리	生態系管理
ecotax	생태세	生態稅
ecotone	추이대	推移帶
ecotype	생태형	生態型
edaphic effect	토지효과	土地效果
eddy	맴돌이, 에디	
eddy accumulation method	맴돌이축적모형	-蓄積模型
eddy advection	맴돌이이류	-移流
eddy available potential energy	맴돌이유효퍼텐셜에너지	-有效-
eddy cascade	에디캐스케이드	
eddy coefficients	맴돌이계수	-係數
eddy conduction	맴돌이전도	-傳導
eddy conductivity	맴돌이전도도	-傳導度
eddy correlation method	맴돌이상관법, 와상관법	渦相關法
eddy covariance	맴돌이공분산	-共分散
eddy covariance method	맴돌이공분산법	-共分散法
eddy covariance software	맴돌이공분산소프트웨어	-共分散-
eddy current	맴돌이흐름	
eddy diameter	맴돌이지름	-直徑
eddy diffusion	맴돌이확산	-擴散
eddy diffusivity	맴돌이확산도	-擴散度
eddy energy	맴돌이에너지	
eddy fluctuation method	맴돌이섭동법	-攝動法
eddy flux	맴돌이플럭스	
eddy friction	맴돌이마찰	-摩擦
eddy heat conduction	맴돌이열전도	-熱傳導
eddy heat flux	맴돌이열속	-熱束
eddy kinetic energy	맴돌이운동에너지	-運動-
eddy momentum	맴돌이운동량	-運動量

eddy motion	맴돌이운동	-運動
eddy resistance	맴돌이저항	-抵抗
eddy resolving model	맴돌이분해모형	-分解模型
eddy shearing stress	맴돌이시어응력	-應力
eddy slope	맴돌이경사	-傾斜
eddy spectrum	맴돌이스펙트럼	
eddy stress	맴돌이응력	-應力
eddy stress tensor	맴돌이응력텐서	-應力-
eddy transport	맴돌이수송	-輸送
eddy turbulence	맴돌이난류	-亂流
eddy turnover time scale	맴돌이반전시간규모	-反轉時間規模
eddy velocity	맴돌이속도	-速度
eddy viscosity	맴돌이점성	-粘性
eddy viscosity coefficient	맴돌이점성계수	-粘性係數
eddy wave	맴돌이파	-波
edge	연변	緣邊
edge enhancement	연변강화	緣邊强化
edge wave	연변파	緣邊波
Eemian interglacial	엠간빙기	-間氷期
effective accumulated temperature	유효적산온도	有效積算溫度
effective aperture	유효구경	有效口徑
effective area	유효면적	有效面積
effective atmosphere	유효대기	有效大氣
effective cloud amount	유효운량	有效雲量
effective collision	유효충돌	有效衝突
effective cross section	유효산란단면적	有效散亂斷面積
effective downwind direction	유효풍향	有效風向
effective drag coefficient	유효항력계수	有效抗力係數
effective earth radius	유효지구반지름, 유효지구반경	有效地球半徑
effective evapotranspiration	유효증발산	有效蒸發散
effective gravity	유효중력	有效重力
effective growing season	유효생육계절	有效生育季節
effective gust velocity	유효돌풍속도	有效突風速度
effective height	유효높이	有效-
effective height of anemometer	풍속계유효높이	風速計有效高度
effective humidity	실효습도, 유효습도	實效濕度, 有效濕度
effective isotropic radiation power	유효등방복사능	有效等方輻射能
effective liquid water content	유효물함량	有效-含量
effective longwave radiation	유효장파복사	有效長波輻射
effectiveness of precipitation	강수효율	降水效率
effective nocturnal radiation	유효야간복사	有效夜間輻射
effective nuclear charge	유효핵전하	有效核電荷
effective optical thickness	유효광학두께	有效光學-
effective penumbra function	실효반영함수	實效半影函數

E

E

effective porosity	유효공극률	有效孔隙率
effective precipitable water	유효가강수량	有效可降水量
effective precipitation	유효강수량	有效降水量
effective pyranometer	유효전천일사계	有效全天日射計
effective radar reflectivity	유효레이더반사도	有效-反射度
effective radar reflectivity factor	유효레이더반사도인자	有效-反射度因子
effective radiant temperature	실효방사온도	實效放射溫度
effective radiating temperature	유효방사온도	有效放射溫度
effective radiation	유효복사	有效輻射
effective radius	유효반경	有效半徑
effective radius of the earth	유효지구반지름, 유효지구반경	有效地球半徑
effective rainfall amount	유효우량, 유효강우(량)	有效雨量, 有效降雨(量)
effective rainfall hyetograph	유효강우도표	有效降雨圖表
effective range	유효구간	有效區間
effective release height	유효방출높이	有效放出-
effective snow melt	유효눈녹음	有效-
effective source height	유효오염원높이	有效汚染源-
effective stack height	유효굴뚝높이	有效-
effective temperature	유효온도	有效溫度
effective terrestrial radiation	유효지구복사	有效地球輻射
effective topography	유효지형도	有效地形圖
effective visibility	유효시정	有效視程
effective wind	유효바람	有效-
effective wind speed	유효풍속	有效風速
effect on ecosystem	생태계영향	生態系影響
effluent	유출물, 배출물, 분사물	流出物, 排出物, 噴射物
effluent buoyancy	배출물부력	排出物浮力
effluent momentum	배출물운동량	排出物運動量
effluent seepage	유출함양삼출, 침출침윤	流出涵養滲出, 浸出浸潤
effluent stream	유출천, 방류하천, 침출천	流出川, 放流河川, 浸出川
effluent velocity	배출물속도	排出物速度
efflux velocity	유출속도	流出速度
e-folding time	e-감쇠시간	-減衰時間
Egnell's law	에그넬법칙	-法則
eguilibrium level	평형고도	平衡高度
Eiffel Tower temperature wave problem	에펠탑온도파문제	-塔溫度波問題
eigenanalysis	고유분석	固有分析
eigencurve	고유곡선	固有曲線
eigendirection	고유방향	固有方向
eigenfrequency	고유진동수	固有振動數
eigenfunction(= characteristic function)	고유함수	固有函數
eigenperiod	고유주기	固有週期
eigenvalue(= charcteristic value, latent value)	고윳값	固有値
eigenvalue equation	고윳값방정식	固有値方程式

eigenvalue problem	고윳값문제	固有値問題
eigenvector(= charcteristic vector, latent vector)	고유벡터	固有-
eigenvector analysis	고유벡터분석	固有-分析
eikonal	아이코날	
Einstein notation	아인슈타인표기법	-表記法
Einstein's theory of relativity	아인슈타인 상대성이론	-相對性理論
Einstein summation convention	아인슈타인합계약정	-合計約定
ejection	사출	射出
ejection chamber	사출함	射出函
Ekman balance	에크만평형	-平衡
Ekman boundary layer	에크만경계층	-境界層
Ekman depth	에크만깊이	
Ekman equation	에크만방정식	-方程式
Ekman force	에크만힘	
Ekman layer	에크만층	-層
Ekman layer depth	에크만층깊이	-層-
Ekman layer instability	에크만층불안정도	-層不安定度
Ekman model	에크만모형	-模型
Ekman number	에크만수	-數
Ekman pumping	에크만펌핑	
Ekman pumping velocity	에크만펌핑속도	-速度
Ekman spiral	에크만나선	-螺線
Ekman transport	에크만수송	-輸送
Ekman turning	에크만전향	-轉向
Ekman volume transport	에크만부피수송	-輸送
elastic	탄성(의)	彈性
elastic barometer	변형기압계, 탄성기압계	變形氣壓計, 彈性氣壓計
elastic body	탄성체	彈性體
elastic collision	탄성충돌	彈性衝突
elastic deformation	탄성변형	彈性變形
elastic force	탄성력	彈性力
elasticity	탄성	彈性
elastic limit	탄성한계	彈性限界
elastic modulus	탄성률	彈性率
elastic rebound	탄성반발	彈性反撥
elastic wave	탄성파	彈性波
E-layer	E층	-層
El Chichon eruption	엘치촌화산분화	-火山噴火
El Chichon volcano	엘치촌화산	-火山
electrical anemometer	전기풍속계	電氣風速計
electrical charge	전하	電荷
electrical conduction	전기전도	電氣傳導
electrical conductivity	전기전도율	電氣傳導率
electrical despun antenna	전기데스펀안테나(인공위성의)	電氣-

E

electrical energy	전기에너지	電氣-
electrical ground support equipment	전자지상지원장비(지상성능 실험용 전기보조장치)	電子地上支援裝備(地上性能 實驗用 電氣補助裝置)
electrical hygrometer	전기습도계	電氣濕度計
electrically scanning microwave radiometer	전동주사마이크로파복사계	電動走査-波輻射計
electrical neutrality	전기적 중성	電氣的中性
electrical phenomenon	전기현상	電氣現象
electrical theromometer	전기온도계	電氣溫度計
electric charge	전하	電荷
electric charge conservation law	전하(량)보존법칙	電荷(量)保存法則
electric circuit	전기회로	電氣回路
electric conductivity	전기전도도, 전기전도율	電氣傳導度, 電氣傳導率
electric conductor	전기도체	電氣導體
electric cup anemometer	전기컵풍속계	電氣-風速計
electric current	전류	電流
electric currents in the atmosphere	대기전류	大氣電流
electric dipole moment	전기쌍극자모멘트	電氣雙極子-
electric discharge	방전	放電
electric disturbance	전기요란	電氣擾亂
electric field	전기장	電氣場
electric field intensity	전기장세기	電氣場-
electric field strength	전기장세기	電氣場-
electric force	전기력	電氣力
electric hygrometer	전기습도계	電氣濕度計
electric intensity	전기장세기	電氣場-
electricity	전기	電氣
electricity of precipitation	강수전기	降水電氣
electric line of force	전기력선	電氣力線
electric oscillation	전기진동	電氣振動
electric pendulum	전기진자	電氣振子
electric potential	전위	電位
electric potential difference	전위차	電位差
electric potential gradient	전위경도	電位傾度
electric power	전력	電力
electric power transmission	전력수송	電力輸送
electric storm	뇌우, 전장요란	雷雨, 電場擾亂
electrification	대전	帶電
electrification ice nucleus	대전빙정핵	帶電氷晶核
electrochemical equivalent	전기화학당량	電氣化學當量
electrochemical sonde	전기화학존데	電氣化學-
electrochemistry	전기화학	電氣化學
electrode effect	전극효과	電極效果
electrodialysis	전기투석	電氣透析
electro explosive device	전기폭발장치	電氣爆發裝置

electrogram	전기장기록	電氣場記錄
electrojet	고층전류	高層電流
electrokinetic potential	전기운동퍼텐셜	電氣運動-
electrolyte	전해물, 전해질	電解物, 電解質
electrolytic strip	전해박판	電解薄板
electromagnet	전자석	電磁石
electromagnetic absorption	전자기흡수	電磁氣吸收
electromagnetic energy	전자기에너지	電磁氣-
electromagnetic field	전자기장	電磁氣場
electromagnetic gas weather	전자기가스기상	電磁氣-氣象
electromagnetic induction	전자기유도	電磁氣誘導
electromagnetic interaction	전자기적 상호작용	電磁氣的相互作用
electromagnetic interference	전자기간섭	電磁氣干涉
electromagnetic oscillation	전자기적 진동	電磁氣的振動
electromagnetic radiation	전자기복사	電磁氣輻射
electromagnetic radiation law	전자기복사법칙,	電磁氣輻射法則,
	전자기방사법칙	電磁氣放射法則
electromagnetic radiation propagation	전자기복사전파,	電磁氣輻射傳播,
	전자기방사전파	電磁氣放射傳播
electromagnetics	전자기학	電磁氣學
electromagnetic spectrum	전자기스펙트럼	電磁氣-
electromagnetic theory	전자기이론	電磁氣理論
electromagnetic unit(emu)	전자기단위	電磁氣單位
electromagnetic wave	전자(기)파	電磁(氣)波
electromagnetic wave spectrum	전자(기)파스펙트럼	電磁(氣)波-
electromagnetism	전자기학	電磁氣學
electrometeor	대기전기현상	大氣電氣現象
electrometer	전위계	電位計
electromotive force	기전력	起電力
electron	전자	電子
electron affinity	전자친화성	電子親和性
electron avalanche	전자사태	電子沙汰
electron beam	전자빔	電子-
electron capture	전자포획	電子捕獲
electron cloud	전자구름	電子-
electron cloud model	전자구름모형	電子-模型
electron donor	전자공여체	電子供與體
electronic balance	전자저울	電子-
electronic charge	전자전하	電子電荷
Electronic Numerical Integrator And Calculator (ENIAC)	전자수치적분계산기	電子數值積分計算機
electronic readout equipment	전자읽음장치	電子-裝置
electronics	전자공학	電子工學
electronic theodolite	전자경위의	電子經緯儀

electronic thermometer	전자온도계	電子溫度計
electronic transition	전자전이	電子轉移
electronic warfare weather	전자전쟁기상	電子戰爭氣象
electron orbit	전자궤도	電子軌道
electron precipitation	전자강수	電子降水
electron shell	전자각, 전자껍질	電子殼
electron specific charge	전자비전하	電子比電荷
electron transfer	전자이동	電子移動
electron volt	전자볼트	電子-
electroscope	검전기	檢電器
electrostatic capacity	정전용량	靜電容量
electrostatic charge	정전하	靜電荷
electrostatic coalescence	정전기병합	靜電氣倂合
electrostatic electrometer	정전전위계	靜電電位計
electrostatic filter	정전필터, 정전집진기	靜電-, 靜電集塵機
electrostatic force	정전기력	靜電氣力
electrostatic generator	정전기 발생장치	靜電氣發生裝置
electrostatic induction	정전기 유도	靜電氣誘導
electrostatic phenomenon	정전기 현상	靜電氣現象
electrostatic precipitator	정전집진장치	靜電集塵裝置
electrostatic screening	정전기 가리기	靜電氣-
electrostatic shielding	정전기 가려막기	靜電氣-
electrostatic unit(esu)	정전기 단위	靜電氣單位
element	요소, 원소	要素, 元素
elemental carbon	원소탄소	元素炭素
elemental carbon aerosol	원소탄소에어로졸	元素炭素-
elemental vortex	기본소용돌이	基本-
elementary electric charge	기본전하량	基本電荷量
elementary particle	기본입자	基本粒子
element data	요소자료	要素資料
elements of weather and climate	기상 및 기후 요소	氣象氣候要素
eletrosonde	전위존데	電位-
Eleuthera	엘류세라섬	
elevated body temperature	상층물체온도	上層物體溫度
elevated convection	상층대류	上層對流
elevated echo	높은 에코	
elevated heat sink	상층열흡원	上層熱吸源
elevated heat source	상층열원	上層熱源
elevated mixed layer	높은 혼합층	-混合層
elevated source	상층오염원	上層汚染源
elevated stable layer	상층안정층	上層安定層
elevation	표고, 해발고도	標高, 海拔高度
elevation angle	앙각	仰角
elevation angle error	앙각오차	仰角誤差

elevation angle of the sun	태양고도	太陽高度
elevation error	고도오차	高度誤差
elevation of boiling point	비등점상승	沸騰點上昇
elevation of ivory point	상아침고도	象牙針高度
elevation of zero point of a barometer	기압계영점고도	氣壓計零點高度
elevation position indicator	앙각위치지시기, 사각지시기	仰角位置指示器, 射角指示器
elevation scan	고도각주사	高度角走査
ELF emission	ELF방출	-放出
ellipse	타원	橢圓
ellipse geometry	타원기하학	楕圓幾何學
ellipsoid	타원면	楕圓面
ellipsoid of revolution	회전타원체	回轉楕圓體
elliptical polarization	타원형편광	楕圓形偏光
elliptic ascent	타원상승	楕圓上昇
Elmo's fire(= St. Elmo's fire)	엘모의 불	
El Niño	엘니뇨	
El Niño event	엘니뇨현상	-現象
El Niño Southern Oscillation	엘니뇨남방진동	-南方振動
Elsasser chart	엘사서차트	
Elsasser's radiation chart	엘사서복사도	-輻射圖
Emagram	에마그램	
emanation	사출	射出
emanometer	라돈측정기	-測定器
embankment	둑, 제방	堤防
embedded circulation	묻힌 순환	-循環
embryo	씨눈(물방울 또는 빙정의)	
emergence	출아	出芽
emergency concentration	위급농도	危急濃度
emergency depot	비상대피소	非常待避所
emergency landing strip	비상활주로	非常滑走路
emergency services	긴급구조대	緊急救助隊
emerging flux region	태양자기방출역	太陽磁氣放出域
Emian (Sangamonian) interglacial	에미안간빙기	-間氷期
emissary sky	전조하늘(저기압 접근의)	前兆-
emission	방출, 배출	放出, 排出
emission charges(= effluent charges)	공해배출부과금	公害排出賦課金
emission control	배출규제	排出規制
emission factor	방출인자	放出因子
emission intensity	방출강도	放出强度
emission layer	방출층	放出層
emission line	방출선	放出線
emission of radiation	복사방출	輻射放出
emission radiation	방출복사	放出輻射
emission rate	방출률	放出率

emission reduction	배출감축	排出減縮
emission reflection	배출반영	排出反影
emissions assessment	배출량평가	排出評價
emission scenario	배출시나리오	排出-
emission spectrum	방출스펙트럼	放出-
emission standard	배출기준	排出基準
emission tax	배출부과금	排出賦課金
emission test cycle	배출검사주기	排出檢查週期
emission theory	배출이론	放出理論
emission trading	배출권거래제	排出權去來制
emission trading scheme	배출량거래체계	排出量去來體系
emission trading system	배출권거래제도	排出權去來制度
emissive power	방출능	放出能
emissivity	방출률	放出率
emissivity correction	방출률수정	放出率修整
emittance	방출률	放出率
emittance of the atmosphere	대기방출률	大氣放出率
emittance of the earth's surface	지표면방출률	地表面放出率
emitter	방출체	放出體
empirical approach	경험적 접근	經驗的接近
empirical climatic classification system	경험기후분류시스템, 경험기후분류계	經驗氣候分類系
empirical formula	경험식	經驗式
empirical orthogonal function	경험직교함수	經驗直交函數
empirical probability	경험적 확률	經驗的確率
enaporative heat regulation	증발열조절	蒸發熱調節
encroachment	침식지	浸蝕池
endangered	멸종위기(의)	滅種危機
endangered species	멸종위기종	滅種危機種
Endangered Species Act	(미국)멸종위기종보호법	滅種危機種保護法
end-Devonian extinction	데본기말멸종	-期末滅種
endeavour	노력	努力
endemics-area relationship	고유종-면적 관계	固有種-面積關係
end-Eocene cooling	에오세말한랭화	-末寒冷化
ending of Changma	장마종료	-終了
endogenetic processes	내인성과정	內因性過程
endothermic reaction	흡열반응	吸熱反應
end-Permian extinction	페름기말멸종	-期末滅種
end-Pleistocene extinction	플라이스토세말 멸종	-末滅種
end point	종말점	終末點
end-Triassic event	트라이아스기말사건	-期末事件
energetics	에너지론	-論
energetics method	에너지학방법	-學方法
energy balance	에너지균형, 에너지수지	-均衡, -收支

energy balance method	에너지균형방법	-均衡方法
energy balance model	에너지균형모형	-均衡模型
energy budget	에너지수지	-收支
energy cascade	에너지캐스케이드	
energy conservation	에너지보존	-保存
energy-containing range	에너지함유범위	-含有範圍
energy conversion	에너지전환	-轉換
energy demand	에너지수요	-需要
energy density	에너지밀도	-密度
energy diagram	에너지선도	-線圖
energy dispersion	에너지분산	-分散
energy dispersive spectrometer	에너지분광기	-分光器
energy dissipation	에너지소산	-消散
energy dissipator	에너지소산자	-消散子
energy efficiency	에너지효율성	-效率性
energy equation	에너지방정식	-方程式
energy exchange	에너지교환	-交換
energy flow	에너지흐름	
energy flux	에너지속, 에너지플럭스	-束
energy form	에너지형태	-形態
energy helicity index	거대뇌우발달지수	巨大雷雨發達指數
energy level	에너지준위	-準位
energy level diagram	에너지준위도	-準位圖
energy model of transport	에너지수송모형	-輸送模型
energy of convection	대류에너지	對流-
energy pathway	에너지경로	-經路
energy release	에너지방출	-放出
energy resource	에너지자원	-資源
energy source	에너지원	-源
energy spectrum	에너지스펙트럼	
energy state	에너지상태	-狀態
energy storage	에너지저장	-貯藏
energy supply	에너지공급	-供給
energy technology	에너지기술	-技術
energy transfer	에너지전달	-傳達
energy transfer function	에너지전달함수	-傳達函數
energy transfer rate	에너지전달률	-傳達率
energy transformation	에너지변환	-變換
energy transport	에너지수송	-輸送
engineer meteorology	공학자기상학	工學子氣象學
engine-exhaust trail(= exhaust trail)	배기구름(비행운)	排氣-(飛行雲)
engine icing	엔진착빙	-着氷
enhanced Fujita intensity scale (EF-scale)	증보후지타강도척도	增補-强度尺度
enhanced greenhouse effect	온실효과강화	溫室效果强化

enhanced image	강조이미지	強調-
enhanced IR technique	적외선강조기법	赤外線强調技法
enhanced wording	강조사항	强調事項
enhancement curves	강조곡선	强調曲線
enhancement of precipitation	강수증가	降水增加
enhancement of rainfall	증우, 강우강조	增雨, 降雨强調
enhancement tables	강조테이블	强調-
enroute advisory service	항로주의보서비스	航路注意報-
ensemble	앙상블	
ensemble data assimilation	앙상블자료동화	-資料同化
ensemble forecast	앙상블예보	-豫報
ensemble forecasting	총합조화예보, 앙상블예보	總合調和豫報
ensemble hydrologic forecasting	앙상블수문예측	-水文豫測
ensemble Kalman filter	앙상블칼만필터	
ensemble mean	앙상블평균	-平均
ensemble prediction	앙상블예보	-豫報
ensemble prediction model	앙상블예측모형	-豫測模型
ensemble prediction system	앙상블예측시스템, 앙상블예측계	-豫測系
ENSO diagnostic discussion	엘니뇨-남방진동 진단토의	-南方振動診斷討議
enstrophy	엔스트로피	
enstrophy flux	엔스트로피플럭스	
enthalpy	엔탈피	
enthalpy flux	엔탈피플럭스	
entrainment	유입, 흡입	流入, 吸入
entrainment coefficient	유입계수, 흡입계수	流入係數, 吸入係數
entrainment constant	유입상수	流入常數
entrainment interfacial layer	유입계면층	流入界面層
entrainment parameter	유입매개변수	流入媒介變數
entrainment rate	흡입률	吸入率
entrainment ratio	유입비	流入比
entrainment zone	유입역	流入域
entrance region	입구역	入口域
entrance to jet streak	제트스트리크입구	-入口
entropy	엔트로피	
entropy constant	엔트로피상수	-常數
entropy death	엔트로피죽음	
entropy potential temperature	엔트로피온위	-溫位
entropy production	엔트로피생산	-生産
entropy pump	엔트로피펌프	
enumeration coordinate	계산좌표	計算座標
envelope	포락선, 기구껍질	包絡線, 氣球-
envelope orography	포락선지형	包絡線地形
environment	환경	環境

environmental action	환경운동	環境運動
environmental change	환경변화	環境變化
environmental crime	환경범죄	環境犯罪
environmental data	환경자료	環境資料
environmental degradation	환경붕괴	環境崩壞
environmental effect	환경효과	環境效果
environmental factor	환경요인	環境要因
environmental flow	주변흐름, 주변장, 환경장	周邊-, 周邊場, 環境場
environmental forecasting services	환경예보서비스	環境豫報-
environmental impact	환경영향	環境影響
environmental impact assessment	환경영향평가	環境影響評價
environmental integrity group	환경보전그룹	環境保全-
environmental issue	환경문제	環境問題
environmentalist	환경론자	環境論者
environmental lapse rate	환경감률	環境減率
environmental meteorology	환경기상학	環境氣象學
Environmental Modification Convention	환경조절금지조약	環境調節禁止條約
environmental noise	환경소음	環境騷音
environmental pollution	환경오염	環境汚染
environmental problems	환경문제	環境問題
environmental protection	환경보호	環境保護
Environmental Protection Agency	(미국)환경청	環境廳
environmental quality standard	환경질표준	環境質標準
environmental resistance	환경저항	環境抵抗
environmental satellite	환경관측위성	環境觀測衛星
environmental sciences	환경과학	環境科學
environmental survey satellite	환경탐사위성	環境探査衛星
environmental temperature	환경온도	環境溫度
environmental temperature sounding	환경온도연직탐측	環境溫度鉛直探測
environmental threat	환경위협	環境威脅
environmental wind	환경바람	環境-
environment of meteorological observation	기상관측환경	氣象觀測環境
environments	환경	環境
Eocene climatic optimum	에오세(신생대)기후최적	-氣候最適
eolation	바람침식	-浸蝕
EOLE	에올(바람의 신, 프랑스 기상위성의 이름)	
eolian	풍성	風成
eolian process	풍성과정	風成過程
eolian sediment	풍성퇴적물	風成堆積物
ephemeral stream	하루살이 내	
ephemeris	천체력	天體歷
epicenter	진앙	震央
epipelagic zone	표해수대	表海水帶
episode	사건	事件

epoch	세	世
Eppley pyrheliometer	에플리 직달일사계	-直達日射計
equal-angle projection	등각투영	等角投影
equal-area chart	등면적도	等面積圖
equal-area map	등면적지도	等面積地圖
equal-area transformation	등면적변환	等面積變換
equal cleavage	등할	等割
equal(equality) sign	등호	等號
equal-surface projection	등면적투영	等面積投影
equation of barotropy	순압방정식	順壓方程式
equation of continuity	연속방정식	連續方程式
equation of harmonic oscillator	조화진동자방정식	調和振動子方程式
equation of hydrostatic equilibrium	정역학평형방정식	靜力學平衡方程式
equation of mass conservation	질량보존방정식	質量保存方程式
equation of motion	운동방정식	運動方程式
equation of normal	법선방정식	法線方程式
equation of piezotropy	압성방정식	壓性方程式
equation of radiative transfer(ERT)	복사전달방정식	輻射傳達方程式
equation of state	상태방정식	狀態方程式
equation of time	균시차	均時差
equation of trajectory	유적선방정식	流跡線方程式
equation of wave motion	파동방정식	波動方程式
equator	적도	赤道
equatorial acceleration	적도가속도	赤道加速度
equatorial air (mass)	적도기단	赤道氣團
equatorial belt of convergence	적도수렴대	赤道收斂帶
equatorial beta-plane	적도베타면	赤道-面
equatorial buffer zone	적도완충대	赤道緩衝帶
equatorial bulge	적도부풀	赤道-
equatorial calm	적도무풍	赤道無風
equatorial calm belt	적도무풍대	赤道無風帶
equatorial climate	적도기후	赤道氣候
equatorial continental air (mass)	적도대륙기단	赤道大陸氣團
equatorial convergence zone	적도수렴대	赤道收斂帶
equatorial countercurrent	적도반류	赤道反流
equatorial dry zone	적도건조대	赤道乾燥帶
equatorial easterlies	적도편동풍	赤道偏東風
equatorial forest	적도림	赤道林
equatorial front	적도전선	赤道前線
equatorial frontal zone	적도전선대	赤道前線帶
equatorial low	적도저기압	赤道低氣壓
equatorial low pressure belt	적도저압대	赤道低壓帶
equatorial low pressure zone	적도저압대	赤道低壓帶
equatorial maritime air (mass)	적도해양기단	赤道海洋氣團

equatorial orbiting satellite	적도궤도위성	赤道軌道衛星
equatorial plane	적도면	赤道面
equatorial plate	적도판	赤道板
equatorial radius of earth	적도지구반경	赤道地球半徑
equatorial rainforest	적도우림	赤道雨林
equatorial Rossby wave	적도로스비파	赤道-波
equatorial stratosphere	적도성층권	赤道成層圈
equatorial tide	적도조석	赤道潮汐
equatorial trough	적도골, 적도기압골	赤道-, 赤道氣壓-
equatorial vortex	적도소용돌이	赤道-
equatorial wave	적도파	赤道波
equatorial westerlies	적도편서풍	赤道偏西風
equatorial zone	적도대	赤道帶
equiangular spiral	등각나선	等角螺線
equiareal transform	등적변형	等積變形
equideparture	등편차	等偏差
equigepotential surface	등지오퍼텐셜면	等-面
equilateral triangle	정삼각형	正三角形
equilibrant	평형력	平衡力
equilibrium concentration	평형농도	平衡濃度
equilibrate convection	평형대류	平衡對流
equilibrium	평형	平衡
equilibrium climate	평형기후	平衡氣候
equilibrium drawdown	평형수위강하	平衡水位降下
equilibrium evaporation	평형증발량	平衡蒸發量
equilibrium flow	평형류, 평형흐름	平衡流
equilibrium level	평형고도	平衡高度
equilibrium lines	평형선	平衡線
equilibrium mixed layer	평형혼합층	平衡混合層
equilibrium of dissolution	용해평형	溶解平衡
equilibrium of force	힘평형	-平衡
equilibrium paraboloid	평형포물체	平衡抛物體
equilibrium phase transformation	평형상전환	平衡狀轉換
equilibrium point	평형점	平衡點
equilibrium shift	평형이동	平衡移動
equilibrium solution	평형해	平衡解
equilibrium spheriod	평형타원체	平衡橢圓體
equilibrium state	평형상태	平衡狀態
equilibrium supersaturation	평형과포화	平衡過飽和
equilibrium surface discharge	평형지표유출량	平衡地表流出量
equilibrium temperature	평형온도	平衡溫度
equilibrium temperature level	평형온도면	平衡溫度面
equilibrium tide	평형조석	平衡潮汐
equilibrium time	평형시각	平衡時刻

E

E

equilibrium vapour	평형증기	平衡蒸氣
equilibrium vapour pressure	평형증기압	平衡蒸氣壓
equilibrium velocity	평형속도	平衡速度
equilibrium wind	평형풍, 평형바람	平衡風
equinoctial colure	춘추분경선	春秋分經線
equinoctial gale	춘추분 큰바람	春秋分-
equinoctial rains	춘추분우기	春秋分雨期
equinoctial storm	춘추분폭풍우	春秋分暴風雨
equinoctial storm effect	춘추분폭풍우효과	春秋分暴風雨效果
equinoctial tide	춘추분조석	春秋分潮汐
equinox	춘추분, 분점	春秋分, 分點
equipartition	등분배	等分配
equipluve	등강수량선	等降水量線
equipotential line	등퍼텐셜선	等-線
equipotential surface	등퍼텐셜면	等-面
equipotential temperature	등온위	等溫位
equiscalar line	등치선	等値線
equiscalar surface	등치면	等値面
equisubstantial surface	등물질면	等物質面
equivalence theorem	등가정리	等價定理
equivalent	등가	等價
equivalent altitude of aerodrome	비행장상당고도	飛行場相當高度
equivalent barotropic atmosphere	상당순압대기	相當順壓大氣
equivalent barotropic height	상당순압고도	相當順壓高度
equivalent barotropic level	상당순압고도	相當順壓高度
equivalent barotropic model	상당순압모형	相當順壓模型
equivalent blackbody temperature	상당흑체온도	相當黑體溫度
equivalent constant wind	탄도바람	彈道-
equivalent depth	상당깊이	相當-
equivalent head wind	상당맞바람	相當-
equivalent height	상당고도	相當高度
equivalent longitudinal wind	상당남북바람	相當南北-
equivalent path	상당경로	相當徑路
equivalent potential temperature	상당온위	相當溫位
equivalent potential temperature diagram	상당온위도	相當溫位圖
equivalent radar reflectivity	상당레이더반사도	相當-反射度
equivalent radar reflectivity factor	상당레이더반사도인자	相當-反射度因子
equivalent reflectance	상당반사율	相當反射率
equivalent reflectivity factor	상당반사율인자	相當反射率因子
equivalent tail wind	상당뒷바람	相當-
equivalent temperature	상당온도	相當溫度
equivalent wavelength	상당파장	相當波長
equivalent width	상당폭	相當幅
era	대, 기	代, 紀

E-region	E영역	-領域
erg	에르그	
Ergodic hypothesis	에르고드 가설	-假說
ergodicity	동질성	同質性
Ergodic process	에르고드 과정	-過程
erosion	침식	浸蝕
erosional susceptibility	침식취약성	侵蝕脆弱性
erosion of thermal	열기포침식	熱氣泡侵蝕
erosion ridge	침식마루	浸蝕-
erratic track	미주진로	迷走進路
error	오차	誤差
error analysis	오차분석	誤差分析
error correction	오차보정	誤差補正
error-detecting code	오차검출부호	誤差檢出符號
error distribution	오차분포	誤差分布
error function	오차함수	誤差函數
error of estimate	추정오차	推定誤差
error of measurement	측정오차	測定誤差
error of representativeness	대표성오차	代表性誤差
Ertel potential vorticity	에르텔위치소용돌이도	-位置-度
Ertel-Rossby invariant	에르텔-로스비 불변량	-不變量
Ertel's conservation theorem	에르텔 보존정리	-保存定理
Ertel's vortex invariant	에르텔 소용돌이불변량	-不變量
Ertel's vortex theorem	에르텔 소용돌이정리	-定理
Ertel vorticity theorem	에르텔 소용돌이도정리	-度定理
ertor	오존층유효온도	-層有效溫度
escape speed	이탈속력	離脫速力
escape velocity	이탈속도	離脫速度
Esperanza (Argentine station)	에스페란자 기지	-基地
Espy-Köppen theory	에스피-쾨펜 이론	-理論
estegram	에스테그램	
estimate	추정	推定
estimated ceiling	목측(어림)실링	目測-
estimated vertical visibility	목측(어림)연직시정	目測鉛直視程
estimated intensity	추정강도	推定强度
estivation	여름잠, 하면	夏眠
Estoque advection line	에스토크 이류선	-移流線
Estoque equation	에스토크 방정식	-方程式
estuary	하구	河口
esturine water	기수, 반염수, 염수	汽水, 半鹽水, 鹽水
esturine zone	기수역, 하구역	汽水域, 河口域
Eta model	에타모형	-模型
eternal frost climate	영구빙설기후	永久氷雪氣候
Etesian climate	지중해기후	地中海氣候

Etesian winds	지중해계절풍	地中海季節風
ethane	에탄	
ethanol	에타놀	
ethylene	에틸렌	
Etna	에트나화산	-火山
Euclid	유클리드(BC 330~275)	
Euclidian norm	유클리드표준	-標準
Euclidian space	유클리드공간	-空間
Euler development	오일러전개	-展開
Euler equation	오일러방정식	-方程式
Eulerian acceleration	오일러가속도	-加速度
Eulerian approach	오일러접근	-接近
Eulerian autocorrelation coefficient	오일러자동상관계수	-自動相關係數
Eulerian change	오일러변화	-變化
Eulerian coordinates	오일러좌표	-座標
Eulerian correlation	오일러상관	-相關
Eulerian correlation function	오일러상관함수	-相關函數
Eulerian derivative	오일러도함수	-導函數
Eulerian diffusion	오일러확산	-擴散
Eulerian equation	오일러방정식	-方程式
Eulerian flow	오일러흐름	
Eulerian integral length scale	오일러적분 길이규모	-積分-規模
Eulerian integral scale	오일러적분규모	-積分規模
Eulerian length scale	오일러길이규모	-規模
Eulerian method	오일러방법	-方法
Eulerian reference system	오일러기준계	-基準系
Eulerian scale	오일러규모	-規模
Eulerian spectrum	오일러스펙트럼	
Eulerian statistics	오일러통계학	-統計學
Eulerian system	오일러계	-系
Eulerian time scale	오일러시간규모	-時間規模
Eulerian time scale transformation	오일러시간규모변환	-時間規模變換
Eulerian turbulence statistics	오일러난류통계	-亂流統計
Eulerian velocity	오일러속도	-速度
Eulerian wind	오일러바람	
Eulerian wind data	오일러바람자료	-資料
Euler number	오일러수	-數
Euler's function	오일러함수	-函數
Euler wind	오일러바람	
euphotic zone	수광대	受光帶
Eurasian high	유라시아고기압	-高氣壓
Eurasian humid continental climate	유라시아 습윤대륙기후	-濕潤大陸氣候
European Centre for Medium-range Weather Forecast(ECMWF)	유럽중기예보센터	-中期豫報-

eustacy	(전지구적)해수면변화	海水面變化
eustatic sea level	승강해수면	昇降海水面
eutrophication	부영양화	富營養化
evanescent wave	감쇠파	減衰波
Evans Signal Laboratory	에반스통신연구소	-通信硏究所
evaporating dish	증발접시	蒸發-
evaporating heat	증발열	蒸發熱
evaporation	증발	蒸發
evaporation capacity	증발능	蒸發能
evaporation curve	증발곡선	蒸發曲線
evaporation fog	증발안개	蒸發-
evaporation frost	증발서리	蒸發-
evaporation gauge	증발계	蒸發計
evaporation hook gauge	증발측정용 후크게이지	蒸發測定用-
evaporation-mixing fog	증발혼합안개	蒸發混合-
evaporation opportunity	증발가능률	蒸發可能率
evaporation pan	증발접시	蒸發-
evaporation pond	증발연못	蒸發-
evaporation power	증발력	蒸發力
evaporation rate	증발률	蒸發率
evaporation tank	증발탱크	蒸發-
evaporation trail	증발흔적	蒸發痕跡
evaporative cooling	증발냉각	蒸發冷却
evaporative emission	증발배출	蒸發排出
evaporative power	증발력	蒸發力
evaporativity	증발률	蒸發率
evaporimeter	증발계	蒸發計
evaporimeter coefficient	증발계계수	蒸發計係數
evaporogram	증발기록	蒸發記錄
evaporograph	증발기록계, 자기증발계	蒸發記錄計, 自記蒸發計
evapotranspiration	증발산	蒸發散
evapotranspirometer	증발산계	蒸發散計
evapotron	증발조절기	蒸發調節器
evening calm	저녁무풍, 저녁뜸	-無風
evening glow	저녁놀	
even number	짝수	-數
even point	짝수점	-數點
Evershed effect	에버셰드 효과	-效果
evogram	에보그램	
evolution	진화	進化
evolutionary response	진화반응	進化反應
evolution equation	진전방정식	進展方程式
evolution theory	진화설	進化說
Ewing-Donn theory	유잉-돈이론(전지구기후변화)	-理論

exact solution	완전해	完全解
exceedance interval	초과간격	超過間隔
excellent visibility	우수시정	優秀視程
exceptional visibility	최우수시정	最優秀視程
excess	과잉	過剩
excess air ratio	과잉공기율	過剩空氣率
excessive cloud seeding	과다구름씨뿌리기	過多-
excessive heat	폭염	暴炎
excessive heat outlook	폭염전망	暴炎展望
excessive heat warning	폭염경보	暴炎警報
excessive heat watch	폭염주의보	暴炎注意報
excessive precipitation	이상강수	異常降水
excessive rain	이상강우	異常降雨
excess rain	이상강우	異常降雨
exchange	교환	交換
exchange coefficient	교환계수	交換係數
exchange coefficient hypothesis	교환계수가설	交換係數假說
exchange integral	교환적분	交換積分
exchange layer	교환층	交換層
excitation	들뜸	
exclusive flood control storage capacity	배타적 홍수조절저류량	排他的洪水調節貯流量
executive committee	집행위원회	執行委員會
exhalation	방출	放出
exhaust contrail	배기비행운	排氣飛行雲
exhaust gas	배기가스	排氣-
exhaust gas recirculation	배기가스 재순환	排氣-再循環
exhaust trail	배기흔적	排氣痕跡
exit region	출구역	出口域
Exner function	엑스너함수	-函數
exosphere	외기권	外氣圈
exothermic reaction	발열반응	發熱反應
expanding perturbation	전개섭동	展開攝動
expansibility	팽창률	膨脹率
expansion	전개(식), 팽창	展開(式), 膨脹
expansional cooling	팽창냉각	膨脹冷却
expansion coefficient	팽창계수	膨脹係數
expansion cooling	팽창냉각	膨脹冷却
expansion wave	팽창파	膨脹波
expansion work	팽창일	膨脹-
expectation	기댓값	期待-
expected value	기댓값	期待-
experiment	실험	實驗
experimental basin	실험유역	實驗流域
experimental meteorology	실험기상학	實驗氣象學

E

experimental satellite	실험위성	實驗衛星
explained variance	설명분산	說明分散
explicit scheme	명시방안	明示方案
explicit smoothing	명시평활	明示平滑
explosion cloud	폭발구름	爆發-
explosion wave	폭발파	爆發波
explosive cyclogenesis	폭발저기압발생	爆發低氣壓發生
explosive cyclone	폭발저기압	爆發低氣壓
explosive deepening	폭발적 발달	爆發的 發達
explosive warming	폭발승온	爆發昇溫
exponent	지수	指數
exponential atmosphere	지수대기	指數大氣
exponential distribution	지수분포	指數分布
exponential equation	지수방정식	指數方程式
exponential function	지수함수	指數函數
exponential inequality	지수부등식	指數不等式
exponential kernel approximation	지수커넬근사	指數-近似
exponential law	지수법칙	指數法則
exposure	노출, 노출광, 노출도	露出, 露出光, 露出度
expression of relation	관계식	關係式
exsiccation	제습건조	除濕乾燥
extended boundary condition method	연장경계조건방법	延長境界條件方法
extended forecast	연장예보	延長豫報
extended Kalman filtering	확장 칼만필터링	擴張-
external boundary layer	외부경계층	外部境界層
external calibration target	외부검정목표	外部檢定目標
external force	외력	外力
external gravity wave	외부중력파	外部重力波
exterior angle	외각	外角
external pressure	외압	外壓
external water circulation	외부물순환 (바다와 육지 사이의)	外部-循環
external wave	외부파	外部波
external work	외부일	外部-
extinction	소산	消散
extinction coefficient	소산계수	消散係數
extinction cross-section	소산단면	消散斷面
extinction efficiency	소산효율	消散效率
extinction efficiency factor	소산효율인자	消散效率因子
extinction episode	멸종에피소드	滅種-
extinction event	멸종사건	滅種事件
extinction of species	멸종	滅種
extinction risk	멸종위기	滅種危機
extinction thickness	소산두께	消散-

E

extra long-range forecast	초장기예보	超長期豫報
extraordinary ray	이상(광)선	異常(光)線
extraordinary wave	이상파	異常派
extrapolation	외삽	外揷
extrapolation method	외삽법	外揷法
extra-terrestrial gravitational force	외계중력	外界重力
extra-terrestrial radiation	외기복사	外氣輻射
extra-terrestrial short-wave radiation	외기단파복사	外氣短波輻射
extra-terrestrial solar spectrum	외기태양스펙트럼	外氣太陽-
extratropical	온대(의)	溫帶
extratropical belt (or zone)	온대	溫帶
extratropical cyclone	온대저기압	溫帶低氣壓
extratropical cyclone model	온대저기압모형	溫帶低氣壓模型
extratropical hurricane	온대허리케인	溫帶-
extratropical low	온대저기압	溫帶低氣壓
extratropical northern hemisphere climate	온대북반구기후	溫帶北半球氣候
extratropical pump	온대성펌프	溫帶性-
extratropical storm	온대폭풍	溫帶暴風
extratropical transition	온대전이	溫帶轉移
extratropical tropopause	온대대류권계면	溫帶對流圈界面
extratropics	온대	溫帶
extreme climate	극단기후	極端氣候
extremely high frequency	극고주파	極高周波
extremely low frequency	극저주파	極低周波
extreme rainfall	강우량극값	降雨量極-
extremes(= extreme value)	극값	極-
extreme temperature	기온극값	氣溫極-
extreme ultraviolet	극자외선	極紫外線
extreme ultraviolet radiation	극자외복사	極紫外輻射
extreme value	극값	極-
extreme weather	극단날씨	極端-
extreme weather event	극단날씨사건	極限-事件
extreme wind warning	강풍경보	強風警報
eye observation	목측	目測
eye of storm	폭풍눈	暴風-
eye of tropical cyclone	열대저기압눈	熱帶低氣壓-
eye of typhoon	태풍눈	颱風-
eye pattern	눈패턴	
eye wall	눈벽	
eye-wall chimney	눈벽굴뚝	

F, f

영문	한글	한자
F_1-layer	F_1 층	-層
F_2-layer	F_2 층	-層
Fabry-Pérot interferometer	파브리-페로 간섭계	-干涉計
facsimile	팩시밀리(모사전송)	(模寫電送)
facsimile chart	팩시밀리도, 모사도	-圖, 模寫圖
facsimile copier equipment	팩시밀리복사장비	-複寫裝備
facsimile equipment	팩시밀리장비	-裝備
facsimile recorder	팩시밀리기록계	-記錄計
facsimile transmission	팩시밀리전송	-電送
factional inequality	분수부등식	分數不等式
factor	인자, 인수	因子, 因數
factorization	인수분해	因數分解
factor theorem	인수정리	因數定理
faculae	백반	白斑
fade out	사라짐	
fading	페이딩, 쇠퇴	衰退
Fahrenheit	화씨	華氏
Fahrenheit scale	화씨눈금	華氏-
Fahrenheit temperature	화씨온도	華氏溫度
Fahrenheit temperature scale	화씨온도눈금	華氏溫度-
Fahrenheit thermometer	화씨온도계	華氏溫度計
fair	갬	
fairness	갬	
fair-weather	갠날(씨)	
fair-weather cirrus	갠날 권운	-卷雲
fair-weather cumulus	갠날 적운	-積雲
fair-weather current	갠날 전류	-電流
fair-weather electric field	갠날 전기장	-電氣場
fair-weather system	갠날씨 계	-系

Falkland Current	포클랜드해류	-海流
fall	낙하, 가을	落下
fall equinox(= autumnal equinox)	추분	秋分
falling limb	하강부	下降部
falling particle	낙하입자	落下粒子
falling speed	낙하속력	落下速力
falling sphere	낙하구	落下球
falling sphere method	낙구법	落球法
falling tide	썰물	
fallout	낙진	落塵
fallout area weather	낙진영역기상	落塵領域氣象
fallout front	낙진전선	落塵前線
fallout pattern	낙진유형	落塵類型
fallout plot	낙진도	落塵圖
fallout wind	낙진바람	落塵-
fallout wind vector plot	낙진바람벡터표시	落塵-標示
fall streaks	꼬리구름	
Fallstreifen	꼬리구름〈독일어〉	
fall velocity	낙하속도	落下速度
fall wind	내리바람	
false cirrus	거짓권운	-卷雲
false color	거짓색깔	-色-
false color image	거짓색깔 영상	-色-映像
false front	거짓전선	-前線
false reflection	거짓반사	-反射
false sun	무리해	
false warm sector	거짓난역	-暖域
false weather forecast	날씨오보	-誤報
false white rainbow	거짓흰무지개	
family of chemical species	화학종가족	化學種家族
family of clouds	구름가족	-家族
family of cyclones	저기압가족(미국)	低氣壓家族
family of depressions	저기압가족(영국)	低氣壓家族
family of tornadoes	토네이도가족	-家族
fan anemometer	팬풍속계	-風速計
fan-beam antenna	부채꼴빔안테나	
fanning	부채꼴연기퍼짐	
fanning plume	부채형 플룸	-型-
farad	패러드(정전용량단위)	(靜電容量單位)
Faraday notation	패러데이기호법	-記號法
Faraday rotation	패러데이회전	-回轉
far field	원거리장	遠距離場
far-infrared	원적외선	遠赤外線
far-infrared radiation	원적외복사	遠赤外輻射

fast Fourier transform	고속푸리에변환	高速-變換
fast ice	고속얼음	高速-
fast ion	고속이온	高速-
fast-response sensor	고속반응-센서	高速反應-
fata morgana	파타 모르가나(신기루)	(蜃氣樓)
fathom	패돔(1.83m, 6feet)	
fathom curve	등심선	等深線
fauna	동물군	動物群
favogn	푄〈스위스〉	
fax	팩스	
fax chart	팩스도	-圖
fax map	팩스도	-圖
feasibility study	타당성조사	妥當性調査
feather	(바람기호의)깃	
feathery crystal	깃털결정	-結晶
feathery stratus	새털층운	-層雲
feedback	되먹임	
feedback factor	되먹임인자	-因子
feedback loop	되먹임루프	
feedback mechanism	되먹임기구	-機構
feedback process	되먹임과정	-過程
feeder band	(태풍의)나선팔	螺線-
feeder cloud	공급자구름	供給者-
Fengshen (typhoon)	펑센	
Fermat's principle	페르마원리	-原理
fern frost	고사리서리	
Ferrel cell	페렐세포	-細胞
Ferrel circulation	페렐순환	-循環
Ferrel's law	페렐법칙	-法則
festoon cloud	꽃줄구름	
fetch	취주거리	吹走距離
fetch effect	취주거리효과	吹走距離效果
fetch length	취주거리	吹走距離
fibratus	명주실구름	明紬-
fibrous ice	섬유얼음	纖維-
Fickian diffusion	픽확산	-擴散
Fickian diffusion equation	픽확산방정식	-擴散方程式
Fickian equation	픽방정식	-方程式
Fick's diffusion law	픽확산법칙	-擴散法則
Fick's equation	픽방정식	-方程式
Fick's law	픽법칙	-法則
fictitious front	가짜전선	-前線
fictitious sun	가짜태양	-太陽
fiducial interval	신뢰구간	信賴區間

fiducial limit	신뢰한계	信賴限界
fiducial mark	기준표지	基準標識
fiducial point	기준점	基準點
fiducial temperature	기준온도	基準溫度
field	장	場
field book	관측야장, 야장	觀測野帳, 野帳
field brightness	장밝기	場-
field capacity	포화토양용수량	飽和土壤容水量
field elevation	장표고	場標高
field-elevation pressure	비행장고도	飛行場高度
field experiment	야외실험	野外實驗
field ice	빙원얼음	氷原-
field intensity	장세기	場-
field luminance	장휘도	場輝度
field maximum moisture capacity	포장최대용수량	圃場最大容水量
field mill	필드밀(전기장측정기)	
field moisture deficiency	포장수분부족량	圃場水分不足量
field of convergence	수렴장	收斂場
field of deformation	변형장	變形場
field of divergence	발산장	發散場
field of pressure	압력장, 기압장	壓力場, 氣壓場
field of solenoid	솔레노이드장	-場
field of view	시계, 시야	視界, 視野
field of vision	시계	視界
field of vorticity	소용돌이도장	-度場
field operation weather	야외작전기상	野外作戰氣象
field strength	장세기	場-
field variable	장변수	場變數
fiery cloud	노을구름	
figure	그림, 숫자	數字
figure code	숫자부호, 숫자코드	數字符號
filament	필라멘트	
filamental flow	필라멘트전류	-電流
filament channel	필라멘트채널	
filament current	필라멘트전류	-電流
Filchner ice shelf	(남극대륙)필히너빙붕	-氷棚
fillet lightning	띠번개	
filling cyclone	매몰저기압	埋沒低氣壓
filling of a depression	저기압매몰	低氣壓埋沒
filling up	매몰, 채움	埋沒
film	얇은 층, 얇은 막, 필름	-層, -膜
film crust	필름크러스트	
film water	엷은막 물	-膜-
filosus	명주실구름	明紬-

filter	여과기, 여파기, 필터, 거르개	濾過器, 濾波器
filter-capture	필터포착	-捕捉
filter condition	여파조건	濾波條件
filtered barotropic model	여파순압모형	濾波順壓模型
filtered equation	여파방정식	濾波方程式
filtered model	여파모형	濾波模型
filter equation	여파방정식	濾波方程式
filter function	여파함수	濾波函數
filtering	필터링, 여파	濾波
filtering approximation	여파근사	濾波近似
filter paper	여파지	濾波紙
filter radiometer	여과복사계	濾過輻射計
filter wedge spectrometer	필터쐐기형분광계	-形分光計
filtration	여과	濾過
final velocity	끝속도	-速度
final warming	최종승온	最終昇溫
Findeisen-Bergeron nucleation process	핀다이센-베르게론 핵형성과정	-核形成過程
Findlater jet	핀들레터 제트(인도양 상공에서 나타나는 하층 제트)	
fine	맑음, 미세	微細
Fineman's nephoscope	피네만측운기	-測雲器
fine-mesh	미세망	微細網
fine-mesh grid	미세망격자	微細網格子
fine-mesh model	미세망모형	微細網模型
fine particle	미세입자	微細粒子
fine structure	미세구조	微細構造
fingerprint method	지문법	指紋法
finite difference	유한차분	有限差分
finite-difference approximation	유한차분근사	有限差分近似
finite-difference equation	차분방정식	差分方程式
finite-difference method	차분법	差分法
finite-difference ratio	차분비	差分比
finite differencing	유한차분	有限差分
finite element method	유한요소법	有限要素法
finite field	유한체	有限體
finite sequence	유한수열	有限數列
finite set	유한집합	有限集合
fiord(= fjord)	협만	峽灣
fire danger index(FDI)	화재위험지수	火災危險指數
fire-danger meter	화재위험계	火災危險計
fire hazard	화재재해	火災災害
fire meteorology	화재기상학	火災氣象學
fire storm	화재폭풍	火災暴風
fire weather	화재날씨	火災-

firing	점화	點火
firmly bound water	강결합수	强結合水
firn	해묵은 눈, 만년설	萬年雪
firn field	해묵은 눈밭	
firn ice	해묵은 눈얼음	
firnification	만년설화	萬年雪化
firn layer	만년설층	萬年雪層
firn limit	해묵은 눈한계	-限界
firn line	만년설선	萬年雪線
firn snow	해묵은 눈	
firn wind(= glacier wind)	빙하바람	氷河-
first freezing	첫얼음	
first frost	첫서리	
First GARP Global Experiment	FGGE 실험,	
	1차 GARP 전지구실험	一次-全地球實驗
first guess	초기추정	初期推定
first guess field	초기추정장	初期推定場
first gust	첫돌풍	-突風
first hoar frost	첫서리	
first hoar frost day	첫서리날	
first law of motion	운동제1법칙	運動第一法則
first law of thermodynamics	열역학제1법칙	熱力學第一法則
first metric simplification	1차계량단순화	一次計量單純化
first moment	1차모멘트	一次-
first observation date	첫관측일	-觀測日
first-order closure	1차종결	一次終結
first-order closure model	1차종결모형	一次終結模型
first-order closure scheme	1차종결방식	一次終結方式
first-order discontinuity	일차불연속	一次不連續
first path retrieval	초기경로복구	初期徑路復舊
first path sounding	초기경로탐측	初期徑路探測
first principle of thermodynamics	열역학제1원리	熱力學第一原理
first purple light	제1자색광	第一紫色光
first quarter (of the moon)	상현	上弦
first snow	첫눈	
first snowfall	초설	初雪
first snowfall day	초설일	初雪日
first time step	첫시간단계(초기단계)	-時間段階(初期段階)
Fischer-Porter rain gauge	피셔-포터 우량계	-雨量計
fission	분열	分裂
fission product	핵분열생성물	核分裂生成物
fissure	틈새	
fitness figure	적합지수	適合指數
fitting	맞추기	

fitting mode	맞추기모드	
FitzRoy barometer	피츠로이 기압계	-氣壓計
five-and-ten system	5-10 시스템	
five-channel scanning radiometer	5채널주사복사계	五-走査輻射計
five-day wave	5일파	五日波
five-lens aerial camera	5렌즈공중카메라	五-空中-
fixed balloon	고정기구	固定氣球
fixed-beam ceilometer	고정빔운고계	固定-雲高計
fixed-level chart	등고도면도	等高度面圖
fixed point	고정점	固定點
fixed sea platform	고정바다플랫폼	固定-
fixed ship station	정점관측선	定點觀測船
fixed time broadcast	정시방송	定時放送
fixed wave	고정파	固定波
fixing of atmospheric gas	기체고정화	氣體固定化
Fjørtoft condition	피외르토프조건	-條件
Fjørtoft equation	피외르토프방정식	-方程式
Fjørtoft steering method	피외르토프지향법	-指向法
flakes	눈송이	
flame collector	불꽃집전기	-集電器
Flanders storm	플랑드르 폭풍(영국에서 남풍에 의해 일어나는 큰눈)	-暴風
Flandrian interglacial	플란드리아 간빙기	-間氷期
flare	플레어	
flare echo	플레어 에코	
flare echo radar signature	플레어 에코 레이더 징후	-徵候
flare for weather modification	기상조절연소탄	氣象調節燃燒炭
flare oxidation	플레어산화	-酸化
flash	섬광	閃光
flash flood	돌발홍수	突發洪水
flash flood warning	돌발홍수경보	突發洪水警報
flash flood watch	돌발홍수주의보	突發洪水注意報
flash frost	돌발서리	突發-
flashing	섬광	閃光
flash rate	섬광률	閃光率
flash-to-ground	벼락, 대지방전	對地放電
flat low	평탄저기압	平坦低氣壓
flat maximum	평탄형최고	平坦形最高
flat-plate (cone) radiometer	평탄(원뿔)복사계	平坦(-)輻射計
flat-plate reflector	평탄형반사기	平坦型反射器
flat-plate solar collector	평탄형태양집광기	平坦型太陽集光器
flattening	편평도	扁平度
flaw	진풍	陣風
F-layer	F층	-層

F

fleecy cloud	양털구름	
fleecy sky	양털구름하늘	
Fleet Broadcast	선단방송	船團放送
Fleet Weather Central	함대중앙기상부(미국해군)	艦隊中央氣象部
flight altitude	비행고도	飛行高度
flight area	공역	空域
flight briefing	비행브리핑	飛行-
flight control	비행관제	飛行管制
flight cross section	항로기상단면도	航路氣象斷面圖
flight document	비행예보철, 비행문서	飛行豫報綴, 飛行文書
flight documentation	비행서류	飛行書類
Flight Document Service(FDS)	비행예보철지원시스템	飛行豫報綴支援-
flight dossier(= flight forecast-folder)	비행로일기예보철	飛行路日氣豫報綴
flight folder	비행로기상자료철	飛行路氣象資料綴
flight forecast	비행로예보	飛行路豫報
flight information	비행로정보	飛行路情報
flight information center	비행로기상정보센터	飛行路氣象情報-
flight information database	비행로정보데이터베이스	飛行路情報-
flight information region	비행로기상정보구역	飛行路氣象情報區域
flight level	비행고도	飛行高度
flight meteorological watch	비행기상감시	飛行氣象監視
flight plan	비행계획	飛行計劃
flight visibility	비행시정	飛行視程
flight-weather briefing(= pilot briefing)	비행일기브리핑(조종사브리핑)	飛行日氣-
flight weather information system	비행기상정보시스템	飛行氣象情報-
flist	플리스트(스코틀랜드에서 돌풍을 동반하는 소나기)	
float barograph	부표형자기기압계	浮標型自記氣壓計
floating automatic weather station	부유자동기상관측소	浮遊自動氣象觀測所
floating balloon	부유기구	浮遊氣球
floating lysimeter	부유라이시미터	浮遊-
floating pan	부력증발계	浮力蒸發計
floating point	부동점	浮動點
float recording precipitation gauge	부표자기우량계	浮標自記雨量計
float type rain gauge	부표형우량계	浮標型雨量計
flocculi	양털무늬(태양)	
floccus	송이(구름)	
floe	부빙덩이	浮氷-
floeberg	표류빙산	漂流氷山
floe ice	부빙	浮氷
flood	홍수, 만조	洪水, 滿潮
flood atlas	홍수지도	洪水地圖
flood category	홍수분류(계급)	洪水分類(階級)
flood control	홍수조절	洪水調節
flood control reservoir	홍수조절저수지	洪水調節貯水池

flood control storage	홍수조절용량	洪水調節容量
flood crest	홍수정점	洪水頂點
flood current	밀물	
flood damage	홍수재해	洪水災害
flood discharge	홍수방류	洪水放流
flood discharge level	홍수방류수위	洪水放流水位
flooded ice	얼음홍수	-洪水
flood forecast	홍수예보	洪水豫報
flood frequency	홍수빈도	洪水頻度
flood frequency curve	홍수빈도곡선	洪水頻度曲線
flood gate	홍수조절문	洪水調節門
flood hydrograph	홍수수문곡선	洪水水門曲線
flood icing	넘친 물 빙결	-氷結
flooding ice	넘친 물 얼음	
flood loss reduction measure	홍수피해저감대책	洪水被害低減對策
flood of record	기록적 홍수	記錄的洪水
flood plain	범람평원	氾濫平原
flood potential outlook	홍수가능성전망	洪水可能性展望
flood prevention	홍수예방	洪水豫防
flood profile	홍수프로파일	洪水-
flood-prone area	침수우발지역	浸水偶發地域
flood proofing	홍수방지법	洪水防止法
flood recurrence interval	홍수재발간격	洪水再發間隔
flood routing	홍수추산	洪水推算
floods-above-base series	홍수부분시계열	洪水部分時系列
flood stage	홍수위	洪水位
flood statement	홍수공시	洪水公示
flood tide	만조, 밀물	滿潮
flood tide in Baekjung	백중사리	百中-
flood wall	홍수방어벽	洪水防禦壁
flood warning	홍수경보	洪水警報
flood watch	홍수주의보	洪水注意報
flood wave	홍수파	洪水波
flood way	홍수로	洪水路
flora	식물군	植物群
Florida Current	플로리다해류	-海流
flow	흐름	
flow chart	순서도	順序圖
flow curvature	흐름곡률	-曲率
flow depth	흐름깊이	
flow duration curve	유량지속곡선	流量持續曲線
flowering	개화	開花
flowering date	개화일	開花日
flowering time	개화기	開花期

F

flow folding	흐름겹침	
flow function	흐름함수	-函數
flowing well	자분정	自噴井
flow integral	흐름적분	-積分
flow meter	유량계	流量計
flow-off	지표유출	地表流出
flow pattern	흐름패턴	
flow regime	흐름영역	-領域
flow separation	흐름분리	-分離
flow splitting	흐름분지	-分枝
flow vector	흐름벡터	
flow visualization	흐름가시화	-可視化
fluctuating plume model	변동플룸모형	變動-模型
fluctuation	요동, 파동	搖動, 波動
fluctuation-dissipation theorem	변동-소산정리	變動-消散定理
fluctuation velocity	요동속도	搖動速度
flue gas	연도기체	煙道氣體
flue gas denitrogenization	배연탈질	排煙脫窒
flue gas desulfurization	배연탈황	排煙脫黃
flue gas spillage	연소기체유출	燃燒氣體流出
fluid	유체	流體
fluid dynamics	유체(동)역학	流體(動)力學
fluid earth	유체지구	流體地球
fluid element	유체요소	流體要素
fluid flow	유체흐름	流體-
fluid flow type	유체흐름형	流體-型
fluid flux	유체선속	流體線束
fluid inclusion	유체포유물	流體包有物
fluid integral	유체적분	流體積分
fluid mechanics	유체역학	流體力學
fluid parcel	유체덩이	流體-
fluid resistance	유체저항	流體抵抗
fluid surface	유체면	流體面
fluid volume	유체부피	流體-
fluorescence radiation	형광복사	螢光輻射
fluorescent light	형광	螢光
fluoride	불화물	弗化物
fluorocarbon	탄화불소	炭火弗素
fluorosis	불소침착증	弗素沈着症
flurry	소낙눈, 돌풍, 눈사태바람	突風, -沙汰-
flushing time	플러싱시간	-時間
flux	플럭스	束
flux adjustment	플럭스조정	-調整
flux convergence	플럭스수렴	-收斂

flux density	플럭스밀도	-密度
flux density threshold	플럭스밀도문턱값	-密度門-
flux determination	플럭스결정	-決定
flux divergence	플럭스발산	-發散
flux-gradient relationship	플럭스경도관계	-傾度關係
flux-gradient similarity	플럭스경도상사	-傾度相似
flux measurement	플럭스측정	-測定
flux of radiation	복사플럭스	輻射-
flux of solar radiation	태양복사플럭스	太陽輻射-
fluxplate	플럭스판	-板
flux-Richardson number	플럭스-리처드슨수	-數
flux temperature	플럭스온도	-溫度
flux-variance relation	플럭스분산관계	-分散關係
flux-variance similarity	플럭스분산상사	-分散相似
fly ash	비산재	飛散-
flying laboratory	비행실험실	飛行實驗室
flying visibility	비행시정	飛行視程
flyoff	증발산	蒸發散
flyways	철새비행통로	-飛行通路
FM system	기상전보식시스템, 메시지형식시스템	氣象電報式-, -形式-
focal depth	초점깊이	焦點-
focal distance	초점거리	焦點距離
focus	초점, 진원	焦點, 震源
foehn(= föhn)	푄	
foehn air	푄공기	-空氣
foehn bank	푄둑	
foehn break	푄구름틈새	
foehn cloud	푄구름	
foehn cyclone	푄저기압	-低氣壓
foehn effect	푄효과	-效果
foehn gap	푄구름틈새	
foehn island	푄섬	
foehn nose	푄코	
foehn pause	푄쉼, 푄정지	-靜止
foehn period	푄기간	-期間
foehn phase	푄단계	-段階
foehn storm	푄폭풍우	-暴風雨
foehn trough	푄골	
foehn wall	푄벽	-壁
foehn wave	푄파	-波
foehn wind	푄바람	
fog	안개	
fog bank	안개둑	

fogbow	안개무지개	
fog clearing	안개걷힘	
fog crystal	안개결정	-結晶
fog day	안개일	-日
fog dispersal	안개인공소산	-人工消散
fog dispersal operation	안개인공소산작업	-人工消散作業
fog dispersion	안개소산	-消散
fog dissipation	안개소산	-消散
fog drip	안개적하	-滴下
fog drop	안개방울	
fog droplet	안개알갱이	
fog forest	안개숲	
fog frequency	안개빈도수	-頻度數
fog horizon	안개수평선	-水平線
fog microphysics	안개미물리학	-微物理學
fog model	안개모형	-模型
fog observation system	안개관측시스템	-觀測-
fog patch	안개조각	
fog point	안개점	-點
fog precipitation	안개강수	-降水
fog prevention forest	안개막이숲	
fog rain	안개비	
fog region	안개구역	-區域
fog scale	안개계급	-階級
fog shower	안개소나기	
fog signal	안개신호	-信號
fog streamer	안개스트리머	
fog structure	안개구조	-構造
fog wind	안개바람	
foil	금속박편	金屬薄片
foil impactor	포일채취기	-採取機
Fokker-Planck equation	포커-플랑크 방정식	-方程式
folding	접힘	
folding tropopause	대류권계면접힘현상, 접힌대류권계면	對流圈界面-現象, -對流圈界面
foliage density	엽밀도	葉密度
foliage velocity	엽속도	葉速度
following wind	뒷바람	
foot-candle	풋촉광(조도단위)	-燭光
footlambert	풋램버트(밝기단위)	
foot-pound-second system	f-p-s 단위계	-單位系
footprint	발자국	
footprint model	발자국모형	-模型
foracan	포라칸(허리케인의 별명)	

forbidden zone of cloud formation	구름불형성층(고도 30~75km)	-不形成層
force	힘	力
force balance	힘균형	-均衡
forced airflow	강제기류	强制氣流
forced condensation process	강제응결과정	强制凝結過程
forced convection	강제대류	强制對流
forced diffusion	강제확산	强制擴散
forced dropping	강제강하	强制降下
forced orthogonalization	강제직교화	强制直交化
forced oscillation	강제진동	强制振動
forced ventilation	강제환기	强制換氣
forced vibration	강제진동	强制振動
forced wave	강제파	强制波
force of gravitation	(만유)인력	(萬有)引力
force of gravity	중력	重力
force of kinetic friction	운동마찰력	運動摩擦力
force of resistance	저항력	抵抗力
force of static friction	정지마찰력	靜止摩擦力
forcing	강제(력)	强制(力)
forcing advection	강제력이류	强制力移流
forcing factor	강제력인자	强制力因子
forcing function	강제력함수	强制力函數
forcing term	강제력항	强制力項
forecast	예보	豫報
forecast accuracy	예보정확도	豫報正確度
forecast amendment	예보수정	豫報修正
forecast area	예보구역	豫報區域
forecast bulletin	예보발표문	豫報發表文
forecast chart	예보도	豫報圖
forecast circle	예보원	豫報員
forecast data	예보자료	豫報資料
forecast district	예보구역	豫報區域
forecast division	예보부서	豫報部署
forecaster	예보자	豫報者
forecast error	예보오차	豫報誤差
forecaster-user collaboration	예보관-사용자 협력	豫報官使用者協力
forecast evaluation	예보평가	豫報評價
forecast for bathing beach	해수욕장예보	海水浴場豫報
forecast for the heading time	출수예측	出穗豫測
forecast guidance	예보가이던스	豫報-
forecasting center	예보센터	豫報-
forecasting method	예보법	豫報法
forecasting mode	예보모드	豫報-
forecast office	예보관서	豫報官署

forecast organization	예보기관	豫報機關
forecast parameter	예보인자	豫報因子
forecast period	예보기간	豫報期間
forecast quality	예보품질	豫報品質
forecast reversal test	예보역검정	豫報逆檢定
forecast review	예보사후분석	豫報事後分析
forecast service	예보업무	豫報業務
forecast skill	예보기술	豫報技術
forecast system	예보체계	豫報體系
Forecast Systems Laboratory	예보시스템연구소	豫報-硏究所
forecast target area	예보대상구역	豫報對象區域
forecast target period	예보대상기간	豫報對象期間
forecast terminology	예보용어	豫報用語
forecast time	예보시각, 예보시점	豫報時刻, 豫報時點
forecast type	예보유형	豫報類型
forecast valid time	예보유효시간	豫報有效時間
forecast validation	예보평가	豫報評價
forecast value	예보가치	豫報價値
forecast verification	예보검토	豫報檢討
forecast zone	예보구역	豫報區域
Forel scale	포렐척도(바다색깔측정)	-尺度
forensic meteorology	범죄기상학	犯罪氣象學
fore-runner	전조, 선구파	前兆, 先驅波
foreshadow	예측	豫測
forest canopy	숲캐노피	
forest clearing	산림개간	山林開墾
forest climate	삼림기후	森林氣候
forest conversion	산림개조	山林改造
forest decline	삼림쇠퇴	森林衰退
forest environment	삼림환경	森林環境
forest-fire	산불	山-
forest fire cloud	산불구름	山-
forest-fire meteorology	산불기상학	山-氣象學
forest limit	삼림한계	森林限界
forest line	수풀한계선	-限界線
forest meteorology	삼림기상학	森林氣象學
forest micrometeorology	삼림미기상학	森林微氣象學
forest radiation budget	삼림복사수지	森林輻射收支
forestry	임학	林學
forest wind	수풀바람	
forked lightning	가랑이번개	
format	형식, 자료배열	形式, 資料配列
formation reaction	생성반응	生成反應
form drag	형체항력	形體抗力

form friction	형체마찰	形體摩擦
formula	공식	公式
formula weight	화학식량	化學式量
Fortin barometer	포르틴기압계	-氣壓計
Fortin mercury barometer	포르틴 수은기압계	-水銀氣壓計
Forty Saints' storm	사십성폭풍(그리스에서 춘분	四十聖暴風
	조금 전(3월)에 일어나는 강풍)	
forward difference	전방차분	前方差分
forward-in-space difference approximation	전방공간차분근사	前方空間差分近似
forward model	전방모형	前方模型
forward path probability	전방경로확률	前方經路確率
forward scatter	전방산란체	前方散亂體
forward scatter continuous wave radar	전방산란CW 레이더	前方散亂-
forward scattering	전방산란	前方散亂
forward time difference	전방시간차분	前方時間差分
forward-upstream difference	전방상류차분	前方上流差分
forward visibility	앞시정	-視程
fossil fuel	화석연료	化石燃料
fossil fuel burning	화석연료연소	化石燃料燃燒
fossil fuel combustion	화석연료연소	化石燃料燃燒
fossil fuel energy	화석연료에너지	化石燃料-
fossil ice	화석얼음	化石-
fossil permafrost	화석영구 언땅	化石永久-
fossil pollen	화석꽃가루	化石-
fossil record	화석기록	化石記錄
fossil turbulence	화석난류	化石亂流
four-dimensional analysis	4차원분석	四次元分析
four-dimensional data assimilation	4차원자료동화	四次元資料同化
four-dimensional variational data assimilation framework	4차원변수자료동화구성틀	四次元變數資料同化構成-
Fourier analysis	푸리에분석	-分析
Fourier coefficient	푸리에계수	-係數
Fourier cosine series	푸리에코사인급수	-級數
Fourier integral	푸리에적분	-積分
Fourier number	푸리에수	-數
Fourier series	푸리에급수	-級數
Fourier's law	푸리에법칙	-法則
Fourier's law of heat conduction	푸리에열전도법칙	-熱傳導法則
Fourier spectrum	푸리에스펙트럼	
Fourier transform	푸리에변환	-變換
foyer of atmospherics	대기공전원	大氣空電源
fractal dimension	프랙탈차원	-次元
fractal instability	프랙탈불안정도	-不安定度
fractal variability	프랙탈변동성	-變動性

F

fraction	분수	分數
fractional bias	부분바이어스	部分-
fractional Brownian motion	부분브라운운동	部分-運動
fractional canopy cover	단편캐노피덮개	斷片-
fractional cloud cover	단편구름양	斷片-量
fractional equation	분수방정식	分數方程式
fractional error	비례오차	比例誤差
fractional expression	분수식	分數式
fractional function	분수함수	分數函數
fractional power law	분수멱법칙	分數冪法則
fractional step	분할스텝	分割-
fractional uncertainty	부분불확정성	部分不確定性
fractional volume abundance	부분용적비	部分容積比
fractionation	분류	分溜
fraction of saturation	포화비율	飽和比率
fracto-cumulus	조각적운	-積雲
fracto-nimbus	조각난운	-亂雲
fracto-stratus	조각층운	-層雲
fracturing	균열생성	龜裂生成
fractus	조각구름	
fragmentation	분열화	分裂化
fragmentation nucleus	분열핵	分裂核
frame	틀	
Framework Convention on Climate Change	기후변화협약	氣候變化協約
F-ratio	F-비	-比
Fraunhofer band	프라운호퍼 띠	
Fraunhofer's line	프라운호퍼 선	-線
frazil ice	바늘얼음	
Fréchet derivative	프레셰 도함수	-導函數
Fredholm integral equation	프레드홀름 적분방정식	-積分方程式
free aerostat	자유고공기구	自由高空氣球
free air	자유대기	自由大氣
free air foehn	고공 푄	高空-
free atmosphere	자유대기	自由大氣
free atmosphere mesoscale phenomenon	자유대기중규모현상	自由大氣中規模現象
free balloon	자유기구	自由氣球
free convection	자유대류	自由對流
free convection layer	자유대류층	自由對流層
free convection scaling temperature	자유대류 규모화 온도	自由對流規模化溫度
free electron	자유전자	自由電子
free energy	자유에너지	自由-
free enthalpy	자유엔탈피	自由-
free expansion	자유팽창	自由膨脹
free fall acceleration	자유낙하가속도	自由落下加速度

free falling	자유낙하	自由落下
free foehn	자유 푄	自由-
free lift	자유상승	自由上昇
free lifting force	자유상승력	自由上昇力
free oscillation	자유진동	自由振動
free oxygen atom	자유산소원자	自由酸素原子
free path	자유경로	自由經路
free period	자유주기	自由週期
free radical	자유기	自由基
free space	자유공간	自由空間
free spectral range	자유스펙트럼역	自由-域
free streamline	자유유선	自由流線
free-stream Mach number	자유기류마하수	自由氣流-數
free surface	자유표면	自由表面
free surface condition	자유면조건	自由面條件
free surface wave	자유표면파	自由表面波
free turbulence	자유난류	自由亂流
free wave	자유파	自由波
freeze data	결빙자료	結氷資料
freeze-free period	무결빙기간	無結氷期間
freezer burn	동상	凍傷
freeze-thaw pattern	동결-해빙 패턴	凍結-解氷-
freeze-up	완전결빙	完全結氷
freeze-up date	결빙일	結氷日
freeze warning	결빙경보	結氷警報
freezing	결빙, 얼음	結氷
freezing burst index	동파지수	凍破指數
freezing damage	동해	凍害
freezing degree-day	결빙도일	結氷度日
freezing drizzle	어는 이슬비	
freezing drizzle advisory	어는 이슬비주의보	-注意報
freezing fog	어는 안개	
freezing hardiness	내동성	耐凍性
freezing index	결빙지수	結氷指數
freezing level	결빙고도	結氷高度
freezing level chart	결빙고도도	結氷高度圖
freezing nucleus	결빙핵	結氷核
freezing nucleus spectrum	결빙핵 스펙트럼	結氷核-
freezing point	결빙점, 어는점	結氷點
freezing point depression	어는점 내림	-點-
freezing point line	결빙점선, 어는 점선	結氷點線
freezing precipitation	어는 강수	-降水
freezing rain	어는 비	
freezing rain advisory	어는 비주의보	-注意報

freezing season	어는 계절	-季節
freezing spray	결빙물보라	結氷-
freezing spray advisory	결빙물보라 주의보	結氷-注意報
freezing temperature	어는 온도	-溫度
F-region	F영역	-領域
Frenet-Serret formula	프레네-세레 공식	-公式
freon	프레온	
frequency	빈도, 주파수, 진동수	頻度, 周波數, 振動數
frequency analysis	빈도분석	頻度分析
frequency band	주파수대	周波數帶
frequency carrier	진동수반송, 주파수반송파	振動數搬送, 周波數搬送波
frequency condition	진동수조건	振動數條件
frequency curve	빈도곡선	頻度曲線
frequency distribution	빈도분포	頻度分布
frequency distribution curve	도수분포곡선	度數分布曲線
frequency distribution polygon	도수분포다각형	度數分布多角形
frequency distribution table	도수분포표	度數分布表
frequency domain	진동수영역	振動數領域
frequency equation	진동수방정식	振動數方程式
frequency function	진동수함수	振動數函數
frequency meter	진동수계	振動數計
frequency-meter anemometer	빈도표시풍속계	頻度表示風速計
frequency-modulated continuous-wave radar	주파수변조 연속파 레이더	周波數變調連續波-
frequency-modulated radar	주파수변조 레이더	周波數變調-
frequency modulation	주파수변조	周波數變調
frequency of occurrence	발생빈도	發生頻度
frequency of radiation	복사주파수	輻射周波數
frequency polygon	빈도다각형	頻度多角形
frequency response	진동수반응	振動數反應
frequency shift	주파수편이, 진동수편이	周波數偏移, 振動數偏移
frequency smoothing	주파수평활	周波數平滑
frequency spectrum	빈도스펙트럼	頻度-
fresh air	신선공기	新鮮空氣
fresh air mass	신선기단	新鮮氣團
fresh breeze	흔들바람(보퍼트풍력계급5)	
freshet	눈녹은 홍수	-洪水
fresh gale	큰 바람	
fresh snow avalanche	신설사태	新雪沙汰
fresh snow cover	신적설	新積雪
fresh water	민물	
fresh-water lens	담수렌즈	淡水-
fresh-water system	담수계	淡水系
Fresnel scatter	프레넬산란체	-散亂體
Fresnel scattering model	프레넬산란모형	-散亂模型

Fresnel zone	프레넬대	-帶
friagem(= vriajem)	프리아젬	
friction	마찰	摩擦
frictional coefficient	마찰계수	摩擦係數
frictional convergence	마찰수렴	摩擦收斂
frictional discontinuity	마찰불연속	摩擦不連續
frictional dissipation	마찰소멸	摩擦消滅
frictional divergence	마찰발산	摩擦發散
frictional drag	마찰항력	摩擦抗力
frictional effect	마찰효과	摩擦效果
frictional force	마찰력	摩擦力
frictional forcing	마찰강제력	摩擦强制力
frictional height	마찰고도	摩擦高度
frictional layer	마찰층	摩擦層
frictional secondary flow	마찰2차흐름	摩擦二次-
frictional stress	마찰응력	摩擦應力
frictional torque	마찰토크	摩擦-
frictional velocity	마찰속도	摩擦速度
friction layer	마찰층	摩擦層
friction temperature	마찰온도	摩擦溫度
friction velocity	마찰속도	摩擦速度
frigid zone	한대	寒帶
frigorigraph	냉각력기록계	冷却力記錄計
frigorimeter	냉각력측정계	冷却力測定計
Fritsch-Chappell cumulus parameterization scheme	프리치-차펠 적운매개변수화 방식	-積雲媒介變數化方式
frog storm	개구리폭풍	-暴風
front	전선	前線
frontal action	전선활동	前線活動
frontal analysis	전선분석	前線分析
frontal boundary	전선경계	前線境界
frontal circulation	전선순환	前線循環
frontal classification	전선분류	前線分類
frontal cloud	전선구름	前線-
frontal cloud band	전선구름대	前線-帶
frontal cloud formation	전선구름생성	前線-生成
frontal contour	전선등고선	前線等高線
frontal contour chart	전선등고선도	前線等高線圖
frontal cyclone	전선저기압	前線低氣壓
frontal decay	전선소멸	前線消滅
frontal depression	전선저기압	前線低氣壓
frontal fog	전선안개	前線-
frontal inversion	전선역전	前線逆轉
frontal inversion layer	전선역전층	前線逆轉層

frontal jet	전선제트	前線-
frontal layer	전선층	前線層
frontal lifting	전선치올림	前線-
frontal line	전선	前線
frontal low	전선저기압	前線低氣壓
frontal mass	전선기단	前線氣團
frontal model	전선모형	前線模型
frontal occlusion	전선폐색	前線閉塞
frontal passage	전선통과	前線通過
frontal passage fog	전선통과안개	前線通過-
frontal precipitation	전선(성)강수	前線(性)降水
frontal profile	전선프로파일	前線-
frontal rainfall	전선(성)강우	前線(性)降雨
frontal slope	전선기울기, 전선경사	前線-, 前線傾斜
frontal speed	전선속력	前線速力
frontal strength	전선강도	前線強度
frontal strip	전선대	前線帶
frontal structure	전선구조	前線構造
frontal surface	전선면	前線面
frontal system	전선계	前線系
frontal theory	전선이론	前線理論
frontal thunderstorm	전선뇌우	前線雷雨
frontal topography	전선면형태, 전선면높낮이	前線面形態
frontal trapping	전선차단	前線遮斷
frontal trough	전선골	前線-
frontal turbulence	전선난류	前線亂流
frontal upgliding	전선활승	前線滑昇
frontal wave	전선파	前線波
frontal wave model	전선파모형	前線波模型
frontal wave motion	전선파동	前線波動
frontal weather	전선(성)날씨	前線(性)-
frontal wedging	전선쐐기	前線-
frontal zone	전선대	前線帶
front lobe	전엽	前葉
front modification	전선변질	前線變質
frontogenesis	전선발생, 전선발달	前線發生, 前線發達
frontogenetical area	전선발생역, 전선발달역	前線發生域, 前線發達域
frontogenetical front	발달성전선	發達性前線
frontogenetical function	전선발생함수	前線發生函數
frontogenetical sector	전선발생역	前線發生域
frontogenetic effect	전선발생효과	前線發生效果
frontology	전선론	前線論
frontolysis	전선소멸, 전선쇠약	前線消滅, 前線衰弱
frontolytical area	전선소멸역, 전선쇠약역	前線消滅域, 前線衰弱域

frontolytical front	소멸성전선, 쇠약성전선	消滅性前線, 衰弱性前線
frontolytical sector	전선쇠약역	前線衰弱域
front passage fog	전선통과안개	前線通過-
front shear line	전선시어선	前線-線
front-trough system model	전선-기압골시스템모형	前線-氣壓-模型
frost	서리	
frost action	동결작용	凍結作用
frost advisory	서리주의보	-注意報
frost belt	동결띠	凍結-
frostbite	동상	凍傷
frost blister	언 언덕	
frostburn	동상	凍傷
frost climate	서리기후	-氣候
frost column	서릿발	
frost dam(= frost belt)	동결띠	凍結-
frost damage	서리피해	-被害
frost day	서릿날	
frost descent	상강	霜降
frost fan	서리막이팬	
frost feather(= ice feather)	서리깃	
frost flake(= ice fog)	서리조각	
frost flower	서리꽃	
frost fog(= ice fog)	서리안개	
frost-free season	서리 없는 계절	-季節
frost-freeze warning	서리경보	-警報
frost hardiness	내상성	耐霜性
frost hazard	서리피해	-被害
frost haze	서리연무	-煙霧
frost heaving	동상	凍上
frost heaving damage	동상해	凍上害
frost hole	상혈	霜穴
frost hollow	상혈	霜穴
frost hygrometer	서리습도계	-濕度計
frost icing	서리착빙	-着氷
frost injury	동상해	凍霜害
frost line	동결깊이선	凍結-線
frost mist	서리박무	-薄霧
frost mound	서리언덕	
frost pillar	서릿발	
frost plant	서리식물	-植物
frost pocket	서리주머니	
frost point	서릿점	-點
frost-point hygrometer	서릿점습도계	-點濕度計
frost-point technique	서릿점기법	-點技法

frost-point temperature	서릿점온도	-點溫度
frost prevent forest	서리막이숲	
frost prevention	서리방지	-防止
frost protection	서리보호	-保護
frost resistance	서리저항	-抵抗
frost ring	서리나이테	
frost season	서리계절	-季節
frost smoke	서리연기(언 안개)	-煙氣
frost snow	서리눈	
frost splitting	서리갈라짐	
frost table	언면	-面
frost tolerance	서리내성	-耐性
frost way	상도	霜道
frost wedging	동결쐐기	凍結-
frost zone	언층	-層
Froude number	프루드수	-數
frozen dew	언 이슬	
frozen fog	언 안개	
frozen magnetic field	동결자기장	凍結磁氣場
frozen precipitation	언 강수	-降水
frozen rain	언비	
frozen sea test	언 바다(빙해) 시험	-(氷海)試驗
frozen snow crust	언 눈표피	-表皮
frozen turbulence	동결난류	凍結亂流
frozen turbulence hypothesis	동결난류가설	凍結亂流假說
frozen wave approximation	동결파동근사	凍結波動近似
frozen wave hypothesis	동결파동가설	凍結波動假說
F scale	후지타규모	-規模
F-test	F검정	-檢定
fuel air ratio	연료공기비	燃料空氣比
fuel cell	연료전지	燃料電池
fuel combustion source	연료연소원	燃料燃燒源
fuel moisture	연료수분	燃料水分
Fujita intensity scale	후지타강도등급	-强度等級
Fujita-Pearson scale	후지타-피어슨 등급	-等級
Fujita scale	후지타등급(토네이도)	-等級
Fujita tornado intensity scale	후지타토네이도강도등급	-强度等級
Fujita tornado scale	후지타토네이도등급	-等級
Fujiwhara effect	후지와라효과	-效果
fulchronograph	뇌격전류계	雷擊電流計
fulgurite	풀구라이트	
full moon	보름달	
full radiator	전파장방출복사체	全波長放出輻射體
fumigation	그을리기, 연기가라앉음	煙氣-

fumigation event	연기가라앉음사건, 훈증사건	煙氣-事件, 熏蒸事件
fumulus	연기구름	煙氣-
functional determinant	함수행렬식	函數行列式
functional validation bench	(위성시스템)기능검사실험장치	(衛星-)機能檢査實驗裝置
fundamental equation of hydrodynamics	기본유체역학방정식	基本流體力學方程式
fundamental interaction	기본상호작용	基本相互作用
fundamental tensor	기본텐서	基本-
fundamental tissue system	기본조직계	基本組織系
fundamental unit	기본단위	基本單位
fundamental vibration	기본진동	基本振動
funding meteorological service	자금기상서비스	資金氣象-
funnel(= funnel column)	깔때기구름	
funnel cloud	깔때기구름	
funnel effect	깔때기효과	-效果
funneling	깔때기효과	-效果
funneling effect	깔때기효과	-效果
funnel stand	깔때기대	-臺
furacana(= furacan(e), furicane, furicano)	허리케인	
furrow irrigation	이랑관개	-灌漑
fusion	녹음, 융해, 핵융합	融解, 核融合
fuzzification	퍼지화	-化
fuzzy logic	퍼지논리	-論理

F

G, g

영문	한글	한자
Gaia hypothesis	가이아가설	-假說
gain	증폭, 게인	增幅
gain factor	증폭인자	增幅因子
gain matrix	증폭행렬	增幅行列
gale	큰 바람(풍력계급 8)	
gale cone	큰 바람표지	-標識
Galerkin method	갤러킨방법	-方法
gale signal	큰 바람신호	-信號
gale warning	큰 바람경보	-警報
gale watch	큰 바람감시	-監視
galvanometer	검류계	檢流計
gamma radiation	감마복사	-輻射
gamma ray	감마선	-線
gamma ray snow gauge	감마선 적설계	-線積雪計
Ganges type	간디스형(기온변화)	-型
gap filling	틈새채우기	
gap model	틈새모형	-模型
Garnier's climate classification	가르니에기후분류	-氣候分類
GARP Atlantic Tropical Experiment	GATE 실험, GARP 대서양열대실험	-實驗, -大西洋熱帶實驗
Garrett-Munk model	가렛-뭉크 모형	-模型
garua(= camanchaca)	가루아	
gas	기체	氣體
gas chromatograph	가스크로마토그래프(기체농도측정장치)	
gas constant	기체상수	氣體常數
gas correction spectroscopy	기체보정분광법	氣體補正分光法
gas correlation spectroscopy	기체상관분광법	氣體相關分光法
gas electric discharge	기체방전	氣體放電
gaseous constituent	기체성분	氣體成分

gaseous diffusion	기체확산	氣體擴散
gaseous discharge	기체유출	氣體流出
gaseous effluent material	기체배출물질	氣體排出物質
gaseous electric discharge	기체방전	氣體放電
gaseous state	기체상태	氣體狀態
gas equation	기체방정식	氣體方程式
gas exchange	기체교환	氣體交換
gasification	기화, 기체화	氣化, 氣體化
gas law	기체법칙	氣體法則
gasometer	기체계량기	氣體計量器
gas phase	기체상	氣體相
gas phase reaction	기체상반응	氣體相反應
gas-sphere(= air-sphere)	기(체)권	氣(體)圈
gas theory	기체론	氣體論
gas thermometer	기체온도계	氣體溫度計
gas uptake	기체흡입	氣體吸入
gateway hypothesis	통로가설	通路假說
gauge	계기	計器
gauge pressure	계기압력	計器壓力
gauss	가우스(자장의 단위)	
Gauss curve	가우스곡선	-曲線
Gaussian curve	가우스곡선, 정규분포곡선	-曲線, 正規分布曲線
Gaussian diffusion model	가우스확산모형	-擴散模型
Gaussian dispersion model	가우스분산모형	-分散模型
Gaussian distribution	가우스분포	-分布
Gaussian equation	가우스방정식	-方程式
Gaussian grid	가우스격자	-格子
Gaussian latitude	가우스위도	-緯度
Gaussian model	가우스모형	-模型
Gaussian noise	가우스잡음	-雜音
Gaussian plume dispersion	가우스플룸분산	-分散
Gaussian plume equation	가우스플룸방정식	-方程式
Gaussian probability function	가우스확률함수	-確率函數
Gaussian process	가우스과정	-過程
Gaussian puff diffusion	가우스퍼프확산	-擴散
Gaussian puff model	가우스퍼프모형	-模型
Gaussian shape	가우스형태	-形態
Gaussian turbulence	가우스난류	-亂流
Gaussian year	가우스해	
Gauss's divergence theorem	가우스발산정리	-發散定理
Gauss's theorem	가우스정리	-定理
Gay-Lussac experiment	게이뤼삭실험	-實驗
Gay-Lussac's law	게이뤼삭법칙	-法則
GCA minimum	GCA 최저기상조건	-最低氣象條件

geg	게그(중국과 티벳사막의 먼지회오리)	
gegenschein	대일조	對日照
Geiger counter	가이거계수기	-計數器
general baroclinic vortex theorem	일반경압소용돌이정리	一般傾壓-定理
general circulation	(대기)대순환	(大氣)大循環
general circulation model	(대기)대순환모형	(大氣)大循環模型
general circulation of atmosphere	대기대순환	大氣大循環
general current	일반류	一般流
general direction	일반방향	一般方向
general forecast	일반예보	一般豫報
generalized absorption coefficient	일반화흡수계수	一般化吸收係數
generalized coordinates	일반화좌표	一般化座標
generalized hydrostatic equation	일반화정역학방정식	一般化靜力學方程式
generalized linear wave equation	일반화선형파동방정식	一般化線形波動方程式
generalized scale invariance	일반화규모불변성	一般化規模不變性
generalized transmission function	일반화투과함수	一般化透過函數
generalized vertical coordinate	일반화연직좌표	一般化鉛直座標
generalized vertical velocity	일반화연직속도	一般化鉛直速度
general meteorology	일반기상학	一般氣象學
genera of cloud	구름의 종류	-種類
genesis	발생, 기원	發生, 起源
genetic classification of climate	기후형성분류(법)	氣候形成分流(法)
Geneva Convention on Long-Range Transboundary Air Pollution	장거리 월경대기오염 제네바 협약	長距離越境大氣汚染-協約
genitus	파생구름	派生-
Genoa cyclone	제노아 저기압	-低氣壓
Genoa-type depression	제노아형 저기압	-型低氣壓
Gentilli's climate classification	젠틸리 기후분류	-氣候分類
gentle breeze	산들바람(풍력계급3)	
geocentric horizon	지심지평	地心地平
geocentric latitude	지심위도	地心緯度
geocentric longitude	지심경도	地心經度
geocentric reference system	지심좌표계	地心座標計
geocentric zenith	지심천정	地心天頂
geochemical climate model	지구화학 기후모형	地球化學氣候模型
geochemical model	지구화학 모형	地球化學模型
geochemical proxy	지구화학 대용물	地球化學代用物
geochemical proxy data	지구화학 대용자료	地球化學代用資料
geochemical tracer	지구화학 미량기체	地球化學微量氣體
geochemist	지구화학자	地球化學者
geocorona	지구 코로나	地球-
geodesic line	측지선	測地線
geodesic visibility	최단시정	最短視程
geodesy(= geodetics)	측지학	測地學

G

geodetic curve	측지곡선	測地曲線
geodynamic height	지구역학고도	地球力學高度
geodynamic meter	지오다이나믹미터	
geoengineering	지구공학	地球工學
geographical classification	지리분류	地理分類
geographic balance	지리평형	地理平衡
geographic(al) coordinates	지리좌표	地理座標
geographic effect	지리효과	地理效果
geographic grid	지리격자	地理格子
geographic horizon	지리지평선(면)	地理地平線(面)
geographic information system	지리정보시스템, 지리정보계	地理情報系
geographic(al) latitude	지리위도	地理緯度
geographic location	지리위치	地理位置
geographic(al) longitude	지리경도	地理經度
geographic map	지리도, 지도	地理圖, 地圖
geographic meridian	지리자오선	地理子午線
geographic position	지리위치	地理位置
geographic scope	지리범위	地理範圍
geography	지리학	地理學
geoid	지오이드	
geoidal surface	지오이드면	-面
geoisotherm	등지온선	等地溫線
geologic age	지질시대	地質時代
geological age	지질시대	地質時代
geological climate	지질시대기후	地質時代氣候
geological era	지질대	地質代
geological-geochemical proxy	지질학-지구화학 대용물	地質學-地球化學代用物
geological time scale	지질연대규모	地質年代規模
geologic epoch	지질세	地質世
geologic era	지질대	地質代
geologic period	지질기	地質期
geologic time	지질연대	地質年代
geology	지질학	地質學
geomagnetic coordinates	지자기좌표	地磁氣座標
geomagnetic equator	지자기적도	地磁氣赤道
geomagnetic field	지자기장	地磁氣場
geomagnetic index	지자기지수	地磁氣指數
geomagnetic latitude	지자기위도	地磁氣緯度
geomagnetic map	지자기도	地磁氣圖
geomagnetic meridian	지자기자오선	地磁氣子午線
geomagnetic pole	지자기극	地磁氣極
geomagnetic storm	지자기폭풍	地磁氣暴風
geomagnetism	지자기	地磁氣
geomedicine	의학지리	醫學地理

geometrical horizon	기하수평선	幾何水平線
geometrical mean	기하평균	幾何平均
geometrical optics	기하광학	幾何光學
geometrical ray tracing	기하광선추적	幾何光線追跡
geometric differential	기하미분	幾何微分
geometric distribution	기하분포	幾何分布
geometric mean	기하평균	幾何平均
geometric optics	기하광학	幾何光學
geometric(geometrical) probability	기하(학)적 확률	幾何(學)的確率
geometric scale	기하규모	幾何規模
geometric sequence (progression)	등비수열	等比數列
geometric similarity	기하상사	幾何相似
geometry	기하학	幾何學
geomorphology	지형학	地形學
geophysical day	지구물리일	地球物理日
geophysical fluid dynamics	지구물리유체역학	地球物理流體力學
Geophysical Fluid Dynamics Laboratory	지구물리유체역학연구소	地球物理流体力學研究所
Geophysical Year	지구물리년	地球物理年
geophysicist	지구물리학자	地球物理學者
geophysics	지구물리학	地球物理學
geopotential	지오퍼텐셜, 지위	地位
geopotential altitude	지오퍼텐셜고도, 지위고도	-高度, 地位高度
geopotential coordinates	지오퍼텐셜좌표, 지위좌표	-座標, 地位座標
geopotential field	지오퍼텐셜장, 지위장	-場, 地位場
geopotential height	지오퍼텐셜고도, 지위고도	-高度, 地位高度
geopotential height gradient	지오퍼텐셜고도경도	-高度傾度
geopotential height surface	지오퍼텐셜고도면	-高度面
geopotential meter	지오퍼텐셜미터	
geopotential surface	지오퍼텐셜면, 지위면	地位面
geopotential thickness	지오퍼텐셜두께, 지위두께	地位-
geopotential topography	지오퍼텐셜높낮이, 지위높낮이	地位-
geosphere	지권, 지구권	地圈, 地球圈
Geostationary Earth Radiation Budget Satellite	정지지구복사수지위성	靜止地球輻射收支衛星
geostationary meteorological satellite	정지기상위성	靜止氣象衛星
geostationary meteorological satellite system	정지기상위성계	靜止氣象衛星系
geostationary ocean color imager	정지궤도해색센서	靜止軌道海色-
geostationary operational environmental satellite	정지실용환경위성	靜止實用環境衛星
geostationary orbit	정지궤도	靜止軌道
geostationary orbit satellite	정지궤도위성	靜止軌道衛星
geostationary satellite	정지위성	靜止衛星
geostation microwave imager	지구정거장 마이크로파 영상기	地球停車場-波映像器
geostation microwave sounder	지구정거장 마이크로파 탐측기	地球停車場-波探測器
geostrophic	지균(의)	地均
geostrophic adjustment	지균조절	地均調節

G

geostrophic advection	지균이류	地均移流
geostrophic advection scale	지균이류규모	地均移流規模
geostrophical wind scale	지균풍척도	地均風尺度
geostrophic approximation	지균근사	地均近似
geostrophic balance	지균균형	地均均衡
geostrophic coefficient	지균계수	地均係數
geostrophic coordinates	지균좌표	地均座標
geostrophic current	지균류	地均流
geostrophic deformation	지균변형	地均變形
geostrophic departure	지균편차	地均偏差
geostrophic deviation	지균편차	地均偏差
geostrophic distance	지균거리	地均距離
geostrophic divergence	지균발산	地均發散
geostrophic divider	지균풍 디바이더	地均風-
geostrophic drag	지균항력	地均抗力
geostrophic drag coefficient	지균항력계수	地均抗力係數
geostrophic equation	지균방정식	地均方程式
geostrophic equilibrium	지균평형	地均平衡
geostrophic flow	지균흐름	地均-
geostrophic flux	지균플럭스	地均-
geostrophic forcing	지균강제력	地均强制力
geostrophic frontogenesis	지균전선발생	地均前線發生
geostrophic height	지균고도	地均高度
geostrophic motion	지균운동	地均運動
geostrophic Richardson number	지균리처드슨수	地均-數
geostrophic shear	지균시어	地均-
geostrophic streamfunction	지균유선함수	地均流線函數
geostrophic trajectory	지균궤적	地均軌跡
geostrophic transport	지균수송	地均輸送
geostrophic turbulence	지균난류	地均亂流
geostrophic value	지균값	地均-
geostrophic velocity	지균속도	地均速度
geostrophic vorticity	지균소용돌이도	地均-度
geostrophic wind	지균풍	地均風
geostrophic wind departure	지균풍편차	地均風偏差
geostrophic wind level	지균풍고도	地均風高度
geostrophic wind relation	지균풍관계	地均風關係
geostrophic wind scale	지균풍척도	地均風尺度
geostrophic wind shear	지균풍시어	地均風-
geostrophic wind vector	지균풍벡터	地均風-
geostrophic wind velocity	지균풍속	地均風速
geostrophy	지균성	地均性
geosynchronous orbit	정지궤도, 지구동기궤도	靜止軌道, 地球同期軌道
geosynchronous satellite	정지위성	靜止衛星

geothermal energy	지열에너지	地熱-
geothermal flux	지열플럭스	地熱-
geothermal gradient	지열경도	地熱傾度
geothermal heat flow	지열류	地熱流
geothermal heating	지열가열	地熱加熱
geothermal resource	지열자원	地熱資源
geothermic depth	지온상승깊이	地溫上昇-
geothermic step	지온경도	地溫傾度
geothermometer	지중온도계	地中溫度計
Gerdien aspriator	게르디엔통풍기	-通風器
germination date	발아일	發芽日
Gerstner wave	거스트너파	-波
GHOST Balloon Project	GHOST 기구계획	-氣球計劃-
giant nucleus	거대핵	巨大核
Gibbs distribution	기브스분포	-分布
Gibbs-Duhem equation	기브스-뒤엠 방정식	-方程式
Gibbs-Duhem relation	기브스-뒤엠 관계	-關係
Gibbs energy	기브스에너지	
Gibbs free energy	기브스자유에너지	-自由-
Gibbs function	기브스함수	-函數
Gibbs' fundamental equation	기브스기본방정식	-基本方程式
Gibbs-Helmholtz equation	기브스-헬름홀츠 방정식	-方程式
Gibbs' phase rule	기브스상규칙	-相規則
Gibbs phenomenon	기브스현상	-現象
Ginzburg-Landau theory	긴츠부르그-란다우 이론	-理論
glacial	빙하	氷河
glacial advance	빙하전진	氷河前進
glacial anticyclone	빙하고기압	氷河高氣壓
glacial anticyclone theory	빙하고기압이론	氷河高氣壓理論
glacial basin	빙하유역	氷河流域
glacial climate	빙하기후	氷河氣候
glacial coastline	빙하해안선	氷河海岸線
glacial drift	빙하표류	氷河漂流
glacial earthquake	빙하지진	氷河地震
glacial epoch	빙하기	氷河期
glacial erosion	빙하침식	氷河侵蝕
glacial erratic	빙하표석	氷河漂石
glacial evidence	빙하증거	氷河證據
glacial geology	빙하지질학	氷河地質學
glacial geomorphology	빙하지형학	氷河地形學
glacial high	빙하고기압	氷河高氣壓
glacial interstadial	아간빙기	亞間氷期
glacial isostasy	빙하평형	氷河平衡
glacial limit	빙하한계	氷河限界

G

glacial mass balance observation	빙하질량균형관측	氷河質量均衡觀測
glacial maximum	최대빙하기, 최대빙하장소	最大氷河期, 最大氷河場所
glacial outwash	빙하성유수	氷河性流水
glacial period	빙하기	氷河期
glacial phase	빙하기	氷河期
glacial retreat	빙하후퇴	氷河後退
glacial sediment	빙하퇴적물	氷河堆積物
glacial theory	빙하이론	氷河理論
glacial varves	빙하침전층	氷河沈澱層
glacial wind	빙하바람	氷河-
glaciated cloud	빙하구름	氷河-
glaciation	빙하화, 빙하작용	氷河化, 氷河作用
glaciation limit	빙하한계고도	氷河限界高度
glaciation period	빙하작용기간	氷河作用期間
glacier	빙하	氷河
glacier breeze	빙하바람	氷河-
glacier fall	빙하폭포	氷河瀑布
glacier front	빙하전선	氷河前線
glacier ice	빙하얼음	氷河-
glacier period	빙하기	氷河期
glacier plain	빙하평원	氷河平原
glacier wind	빙하바람	氷河-
glacioclimatology	빙하기후학	氷河氣候學
glaciofluvial sediment	융빙수퇴적물,	融氷水堆積物,
	빙하성유수퇴적물	氷河性流水堆積物
glaciology	빙하학	氷河學
glaciomarine sediment	빙하해양퇴적물	氷河海洋堆積物
glacon	작은 부빙	-浮氷
glare	반짝빛, 섬광	閃光
glare ice	거울얼음	
glass-house (or greenhouse) climate	온실기후	溫室氣候
glaze	비얼음(미국)	
glazed frost	비얼음(영국)	
glaze ice	비얼음	
glaze storm	비얼음보라	
Gleissberg cycle	글라이스베르크 주기	-週期
gliding flight	활공비행	滑空飛行
gliding mean	활공평균	滑空平均
glime	반투명얼음	半透明-
	(비얼음과 상고대의 중간)	
global air pollution	지구대기오염	地球大氣汚染
global albedo	지구알베도, 지구반사율	地球反射率
Global Area Coverage Oceans Pathfinder Project	전지구해양탐험자계획	全地球海洋探險者計劃
Global Atmosphere Watch(GAW)	지구대기감시	地球大氣監視

Global Atmospheric Research Program(GARP)	지구대기연구계획	地球大氣研究計劃
Global Atmospheric Research Project	지구대기연구프로젝트	地球大氣研究-
global change scenario	지구변화시나리오	地球變化-
Global Change System for Analysis, Research and Training(GCSART)	분석·연구·훈련용 지구변화시스템	分析·研究·訓練用 地球變化-
global circulation	전지구순환	全地球循環
global circulation model	전지구순환모형	全地球循環模型
global circulation system	전지구순환계	全地球循環系
global climate	전지구기후	全地球氣候
global climate change	전지구기후변화	全地球氣候變化
Global Climate Observing System(GCOS)	지구기후관측시스템	地球氣候觀測-
global climate warming	전지구기후온난화	全地球氣候溫暖化
global cooling	전지구냉각화	全地球冷却化
global coverage	전지구범위	全地球範圍
global data assimilation and prediction system	전지구자료동화예측시스템	全地球資料同化豫測-
global data processing observation system	전지구자료처리관측시스템	全地球資料處理觀測-
global data processing system	전지구자료처리시스템	全地球資料處理-
global dimming	지면일조량 감소	地面日照量減少
global distribution	전구분포	全球分布
global electric circuit	지구전기회로	地球電氣回路
Global Energy and Water Cycle Experiment	지구에너지물순환실험	地球-循環實驗
global energy budget	전구에너지수지	全球-收支
global environment	지구환경	地球環境
Global Environmental Monitoring System	지구환경감시시스템	地球環境監視-
global environment facility	지구환경시설	地球環境施設
global environment monitoring index	지구환경감시지수	地球環境監視指數
global forecast system	전지구예보체계	全地球豫報體系
Global Historical Climatology Network	지구역사기후학네트워크	地球歷史氣候學-
global horizontal sounding technique	지구수평탐측기술	地球水平探測技術
global hydrologic cycle	전지구수문순환	全地球水文循環
Global Hydrology Center	지구수문센터	地球水文-
Global Inventory Monitoring and Modeling Systems	전지구배출목록감시 모델링시스템	全地球排出目錄監視-
global-mean energy balance	전지구평균에너지수지	全地球平均-收支
global-mean temperature variation	전구평균온도변동	全球平均溫度變動
global meteorological observing system	지구기상관측시스템	地球氣象觀測-
global meteorological satellite system	지구기상위성시스템	地球氣象衛星-
global model	전지구모형	全地球模型
global near-surface temperature	전지구지상온도	全地球地上溫度
global observation system	전지구관측시스템	全地球觀測-
global observing system	전지구관측시스템	全地球觀測-
Global Ozone Observing System(GOOS)	지구오존관측시스템	地球-觀測-
Global Positioning System(GPS)	지구위치시스템, 위치정보시스템	地球位置-, 位置情報-

Global Positioning System(GPS) sonde	전지구위치측정 존데	全地球位置測定-
global radiation	전천복사	全天輻射
global radiation budget	지구복사수지	地球輻射收支
global radiation spectrum	전천복사스펙트럼	全天輻射-
global radiative equilibrium	전지구복사평형	全地球輻射平衡
Global Resource Information Database	지구자원정보데이터베이스	地球資源情報-
global-scale diffusion	지구규모확산	地球規模擴散
global-scale diffusion model	지구규모확산모형	地球規模擴散模型
global-scale dispersion	지구규모확산	地球規模擴散
global shortwave radiation	전천단파복사	全天短波輻射
global solar radiation	전천일사	全天日射
global telecommunication system	세계통신시스템	世界通信-
global warming	지구온난화	地球溫暖化
global warming controversy	지구온난화논쟁	地球溫暖化論爭
global warming period	지구온난화기간	地球溫暖化期間
global warming potential	지구온난화잠재력	地球溫暖化潛在力
global water balance equation	지구물수지방정식	地球-收支方程式
global water budget	지구물수지	地球-收支
global water cycle	전지구물순환	全地球-循環
Global Weather Experiment(GWE)	지구기상실험	地球氣象實驗
global weather reconnaissance	전지구기상정찰	全地球氣象偵察
global wind energy council	세계풍력협의회	世界風力協議會
globe lightning(= ball lightning)	구상번개	球狀-
g-load	중력부하	重力負荷
globular projection	구상투영	球狀投影
glomeratus	덩어리(구름)	
glory	그림자광륜	-光輪
glove thermometer	글로브온도계	-溫度計
glow	노을	
glow discharge	글로방전	-放電
glucose	포도당	葡萄糖
gnomon	(고대)해시계	-時計
gnomonic projection	심사도법	心射圖法
Gobi desert	고비사막	-砂漠
Gold slide	골드계산척	-計算尺
goldbeater's skin hygrometer	금박막습도계	金薄膜濕度計
Gondwana Ice Age	곤드와나빙하시대	-氷河時代
goniometer	측각계	測角計
goniometry direction finding	측각방향탐지	測角方向探知
Goddard Institute of Space Sciences	고더드우주과학연구소(미국)	-宇宙科學硏究所(美國)
GOES Satellite	GOES 위성	-衛星
goodness of fit	적합도	適合度
goodness-of-fit test	적합도검정	適合度檢定
gorge wind	협곡바람	峽谷-

gosling blast	새끼거위돌풍	-突風
gosling storm	새끼거위폭풍	-暴風
governing equation	지배방정식	支配方程式
Gowganda glaciation	(북아메리카의) 고건다 빙하작용	-氷河作用
gowk storm	뻐꾸기폭풍	-暴風
Goyder Line	(사우스오스트레일리아의) 고이더(강우)경계선	-(降雨)境界線
Görtler number	괴틀러수	-數
Görtler vortex	괴틀러소용돌이	
gradient	경도, 기울기, 물매	傾度
gradient current	경도류	傾度流
gradient diffusion approximation	경도확산근사	傾度擴散近似
gradient equation	경도방정식	傾度方程式
gradient flow	경도흐름	傾度-
gradient level	경도풍고도	傾度風高度
gradient Richardson number	경도리처드슨수	傾度-數
gradient transport theory	경도수송이론	傾度輸送理論
gradient velocity	경도풍속도	傾度風速度
gradient wind	경도풍	傾度風
gradient wind equation	경도풍방정식	傾度風方程式
gradient wind level	경도풍고도	傾度風高度
gradient wind speed	경도풍속	傾度風速
graduated glass cylinder	눈금실린더(우량측정용)	(雨量測定用)
graduation	눈금매김	
grain full	소만	小滿
grain in ear	망종	芒種
grain of ice	얼음싸라기	
grain rain	곡우	穀雨
gram atomic volume	그램원자부피	-原子-
gram atomic weight	그램원자량	-原子量
gram equivalent	그램당량	-當量
gram molecular weight	그램분자량	-分子量
granular snow	알갱이눈	
graphical addition	도식가법	圖式加法
graphical analysis	도식분석, 그래프분석	圖式分析
graphical calculation	도식계산	圖式計算
graphical method	도해법	圖解法
graphical solution	도해	圖解
graphical statics	도해역학	圖解力學
graphical subtraction	도식감법	圖式減法
graphic method	도해법	圖解法
graphic representation	그래프표시	-表示
graphing board	그래프판	-板

G

Grashof number	그라쇼프 수	-數
grass	잡음	雜音
grassland	초원	草原
grassland climate	초원기후	草原氣候
grassland experiment	초원실험	草原實驗
Grassmann rule	그라스만규칙	-規則
grass minimum	초상최저	草上最低
grass minimum temperature	초상최저온도	草上最低溫度
grass minimum theromometer	초상최저온도계	草上最低溫度計
grass temperature	초상온도	草上溫度
grassy soil	초지	草地
grating	격자	格子
grating spectrometer	(회절)격자분광기	(回折)格子分光器
grating spectroscope	회절분광기	回折分光器
graupel(= snow pellet)	싸락눈	
gravitation	중력, 만유인력	重力, 萬有引力
gravitational acceleration	중력가속도	重力加速度
gravitational attraction	지구인력	地球引力
gravitational constant	중력상수	重力常數
gravitational contraction	중력수축	重力收縮
gravitational convection	중력대류	重力對流
gravitational equilibrium	중력평형	重力平衡
gravitational field	중력장	重力場
gravitational force	만유인력, 중력	萬有引力, 重力
gravitational instability	중력불안정	重力不安定
gravitational interaction	중력상호작용	重力相互作用
gravitational mass	중력질량	重力質量
gravitational moist-convective instability of the second kind	제2종중력습윤대류불안정	第二種重力濕潤對流不安定
gravitational potential	중력퍼텐셜	重力-
gravitational potential energy	중력위치에너지	重力位置-
gravitational pressure	중력압	重力壓
gravitational separation	중력분리	重力分離
gravitational settling	중력침강	重力沈降
gravitational stability	중력안정, 중력안정도	重力安定, 重力安定度
gravitational tide	중력조석	重力潮汐
gravitational unit of force	힘의 중력단위	-重力單位
gravitational unit system	중력단위계	重力單位系
gravitational water	중력수	重力水
gravitational wave	중력파	重力波
gravity	중력	重力
gravity correction	중력보정	重力補正
gravity current	중력류	重力流
gravity force	중력	重力

gravity-free	무중력	無重力
gravity gradient stabilization	중력경도안정화	重力傾度安定化
gravity vector	중력벡터	重力-
gravity water	중력수	重力水
gravity wave	중력파	重力波
gravity wind	중력바람	重力-
gray atmosphere	회색대기	灰色大氣
graybody	회색체	灰色體
graybody absorption	회색체흡수	灰色體吸收
graybody method	회색체방법	灰色體方法
Gray code	그레이코드	
gray smog	회색스모그	灰色-
grease ice	연한 얼음	軟-
Great Basin high	그레이트 베이슨 고기압	-高氣壓
great circle	대원	大圓
great-circle course	대원경로	大圓經路
great-circle navigation	대원항해, 대원운항	大圓航海, 大圓運航
greatest lower bound	하한	下限
Great Ocean Conveyor Belt	심층순환, 그레이트해양 컨베이어벨트	深層盾環, -海洋-
Great Plain Experiment(GPE)	대평원실험	大平原實驗
great radiation belt	대방사능대	大放射能帶
Great Red Spot	큰 붉은 점	-點
green belt(= frostless zone)	녹지대	綠地帶
green flash	녹색섬광	綠色閃光
green growth	녹색성장	綠色成長
greenhouse climate	온실기후	溫室氣候
greenhouse effect	온실효과	溫室效果
greenhouse gas	온실기체	溫室氣體
greenhouse gas inventory	온실기체방출목록	溫室氣體放出目錄
greenhouse weather	온실날씨	溫室-
Greenland anticyclone	그린랜드고기압	-高氣壓
Greenland high	그린랜드고기압	-高氣壓
green moon	초록달	草綠-
green oxygen line	산소녹색선	酸素綠色線
green ray	녹색선	綠色線
Green's function	그린함수	-函數
green sky	초록하늘	草綠-
green snow	초록눈	草綠-
Green's theorem	그린정리	-定理
green sun	초록태양	草綠太陽
Greenwich apparent time	그리니치시(태양시)	-時
Greenwich civil time	그리니치상용시	-常用時
Greenwich mean time	그리니치평균태양시	-平均太陽時

G

Greenwich meridian	그리니치자오선	-子午線
Greenwich sidereal time	그리니치항성시	-恒星時
Greenwich time	그리니치시	-時
G-region	G-구역	-區域
grenade	측풍유탄	側風榴彈
grey absorber	회색흡수체	灰色吸收體
greybody	회색체	灰色體
greybody radiation	회색체복사	灰色體輻射
grid	격자	格子
grid constant	격자상수	格子常數
gridding	격자화	格子化
grid heading	격자방향	格子方向
grid meridian	격자자오선	格子子午線
grid method	격자법	格子法
grid navigation	격자항법	格子航法
grid nephoscope	격자모양측운기	格子模樣測雲器
grid north	격자북	格子北
grid-point	격자점	格子點
grid-point representation	격자점표현	格子點表現
grid-point value	격자점값	格子點-
grid system	격자계	格子系
grid telescoping	격자압축	格子壓縮
gross-austausch	대규모소용돌이교환	大規模-交換
gross photosynthesis	총광합성	總光合成
gross production	총생산량	總生産量
Grosswetterlage	대형일후형세〈독일어〉	大型日候形勢
ground acquisition and command station	지상수신 및 통제소	地上受信-統制所
ground air meteorological forecast	저층기상예보	底層氣象豫報
ground avalanche	지상사태	地上沙汰
ground burner	지상연소기	地上燃燒器
ground check	지상점검	地上點檢
ground check chamber	지상점검상자	地上點檢箱子
ground clutter	지면반사에코	地面反射-
ground-controlled approach	지상관제접근	地上管制接近
ground discharge	벼락, 땅방전	-放電
ground experiment	지상실험	地上實驗
ground fallout plot	지면낙진도	地面落塵圖
ground fog	땅안개	
ground frost	지면동결	地面凍結
ground generator	지상발생기(인공강우용)	地上發生器
ground ice	땅얼음	
ground ice mound	땅얼음언덕	
ground icing	지상착빙	地上着氷
ground inversion	접지역전	接地逆轉

ground layer	지표층	地表層
ground level	지표면	地表面
ground phenomenon	지면현상	地面現象
ground run	지상활주	地上滑走
ground sensor	지면센서, 지면감부	地面感部
ground snowfall	지상강설	地上降雪
ground speed	대지속도	對地速度
ground station	지상관측소	地上觀測所
ground streamer	땅스트리머	
ground swell	땅너울	
ground temperature	지온	地溫
ground thermometer	지온계	地溫計
ground-to-cloud discharge	땅-구름방전	-放電
ground track	지적선	地跡線
ground truth	지상참값	地上-
ground visibility	지상시정	地上視程
ground water	지하수	地下水
ground water depletion curve	지하수감소곡선	地下水減少曲線
ground water level(= undergorund water level)	지하수위	地下水位
ground water recession	지하수감퇴	地下水減退
ground water table	지하수면	地下水面
ground wave	지상파	地上波
grouping	그룹화	-化
group velocity	군속도	群速度
growing degree-day	생육도일	生育度日
growing season	성장계절	成長季節
growler	작은 빙산	-氷山
growth curve	생장곡선	生長曲線
growth movement	생장운동	生長運動
growth respiration	생장호흡	生長呼吸
guess field	추정장	推定場
Guilbert's rule	길베르규칙	-法則
Guinea current	기니아해류	-海流
Guldberg-Mohn wind	굴드버그-몬 바람	
Gulf Stream	멕시코만류	-灣流
Gulf Stream System	멕시코만류계	-灣流系
gully erosion	협곡침식	峽谷浸蝕
gully washer	집중호우	集中豪雨
gunprobe	발포탐사	發砲探査
gush(= cloudburst)	폭우	暴雨
gust	돌풍	突風
gust amplitude	돌풍진폭	突風振幅
gust decay time	돌풍감쇠시간	突風減衰時間
gust duration	돌풍기간	突風期間

G

gust formation time	돌풍형성시간	突風形成時間
gust frequency	돌풍빈도	突風頻度
gust frequency interval	돌풍빈도간격	突風頻度間隔
gust front	돌풍전선	突風前線
gust-gradient distance	돌풍경도거리	突風傾度距離
gustiness	돌풍도(성), 바람의 숨	突風度(性)
gustiness component	돌풍성분	突風成分
gustiness factor	돌풍률	突風率
gust load	돌풍부하	突風負荷
gustnado	돌풍토네이도	突風-
gust peak speed	돌풍최대풍속	突風最大風速
gustsonde	돌풍관측존데	突風觀測-
gust vector	돌풍벡터	突風-
Guti weather	구티날씨	
guttation	일액	溢液
guttra	구트라	
gyro-frequency	자이로주파수	-周波數
gyromagenetic frequency	회전자기빈도	回轉磁氣頻度
gyro reference system	자이로참조시스템	-參照-

H, h

대기과학용어집

영문	한글	한자
haar	하르	
habitat	서식지	棲息地
haboob(= habbub, haboub, hubbob, hubbub)	하부브	
Hadley cell	해들리세포	-細胞
Hadley circulation	해들리순환	-循環
Hadley regime	해들리영역	-領域
Hadley's principle	해들리원리	-原理
Hagen-Poiseuille flow	하겐-푸아죄유 흐름	
hail	우박	雨雹
hail cannon	우박대포	雨雹大砲
hail contamination	우박오염	雨雹汚染
hail damage	우박피해	雨雹被害
hail embryo	우박핵	雨雹核
hail fallout zone	우박지대	雨雹地帶
hail generation zone	우박생성역	雨雹生成域
hail growth	우박성장	雨雹成長
hail growth zone	우박성장역	雨雹成長域
hail impactor	충돌식우박측정기	衝突式雨雹測定器
hail index	우박지수	雨雹指數
hail lobe	우박돌기	雨雹突起
hail mitigation	우박소산	雨雹消散
hailpad	우박측정판	雨雹測定板
hail path	우박경로	雨雹經路
hail prevention	우박방재	雨雹防災
hail research experiment	우박연구실험	雨雹研究實驗
hail shower	우박소나기	雨雹-
hail signal	우박신호	雨雹信號
hail spike	우박못	雨雹-
hail stage	우박단계	雨雹段階

hailstone	우박덩이	雨雹-
hail storm	우박스톰, 우박보라	雨雹-
hailstreak	우박선	雨雹線
hail suppression	우박억제	雨雹抑制
hailswath	우박구역	雨雹區域
Haines index	하인스지수	-指數
hair hygrograph	자기모발습도계	自記毛髮濕度計
hair hygrometer	모발습도계	毛髮濕度計
halcyon day(= alcyone days)	핼시온기간	-期間
half-arc angle	반호각	半弧角
half-life	반감기	半減期
half moon	반달	半-
half-period zone	반주기역	半週期域
half-power point	반전력점	半電力點
half reaction	반쪽반응	-反應
half-tide level	평균조위(면)	平均潮位(面)
	(만조와 간조의 중간)	
half-value period	반감기	半減期
half-width	반폭	半幅
Hall effect	홀효과	-效果
halny wiatr	할니비아트르(높새바람)〈폴란드어〉	
halo	무리	
halocline	염분약층	鹽分躍層
halo of Hevelius	헤벨리우스무리	
halo of 22°	22도 무리	
halo of 46°	46도 무리	
halting voltage	정지전압	靜止電壓
hand anemometer	휴대용풍속계	携帶用風速計
hanging glacier	걸린 빙하	-氷河
haracana	하라카나	
hardening	경화	硬化
hard freeze	굳은 결빙	-結氷
hard frost	된서리	
hard radiation	경질방사선	硬質放射線
hard rain	폭우	暴雨
hard rime	굳은 상고대	-霜高臺
harmattan	하마탄	
harmonic	조파, 조화량	調波, 調和量
harmonic analyser	조화분석기	調和分析器
harmonic analysis	조화분석	調和分析
harmonic dial	조화다이얼	調和-
harmonic function	조화함수	調和函數
harmonic motion	조화운동	調和運動
harmonic oscillation	조화진동	調和振動

harmonic radiation	조화복사	調和輻射
harmonic series	조화급수	調和級數
haroucana	하로우카나	
harrua	하루아	
Hartley band	하틀리띠	
Hartley-Huggins band	하틀리-허긴스 띠	
haster	헤이스터(잉글랜드의 거센 폭풍우)	
haud	하우드(스코틀랜드의 스콜)	
Haurwitz-Helmholtz wave	하우르비츠-헬름홀츠 파	-波
hay fever	고초열, 꽃가루병	枯草熱, -病
hazard assessment	위험도평가	危險度評價
hazardous air pollutant	유해대기오염물질	有害大氣汚染物質
hazardous dust	유해분진	有害粉塵
hazardous in-flight weather advisory service	위험비행기상주의보서비스	危險飛行氣象注意報-
hazardous seas warning	위험바다경보	危險-警報
hazardous seas watch	위험바다감시보	危險-監視報
hazardous weather	위험날씨	危險-
hazardous weather management consultation	위험기상관리상담	危險氣象管理相談
hazardous weather outlook	위험기상전망	危險氣象展望
haze	연무	煙霧
haze aloft	고공연무	高空煙霧
haze droplet	연무방울	煙霧-
haze horizon	연무선	煙霧線
haze layer	연무층	煙霧層
haze level	연무고도	煙霧高度
haze line	연무경계선	煙霧境界線
hazemeter	헤이즈미터	
Hazen method	헤이즌법	-法
HCMM Satellite	HCMM 위성	-衛星
heading	(항행 또는 비행)방향	方向
head wind	맞바람	
heap cloud	퇴적구름	堆積-
heat	열	熱
heat accumulation	열축적	熱蓄積
heat advection	열이류	熱移流
heat advisory	열주의보	熱注意報
heat balance	열평형	熱平衡
heat budget	열수지	熱收支
heat capacity	열용량	熱容量
Heat Capacity Mapping Mission Satellite	열용량지도제작임무위성	熱容量地圖製作任務衛星
heat conduction	열전도	熱傳導
heat conductivity	열전도율	熱傳導率
heat content	열함유량	熱含有量
heat convection	열대류	熱對流

heat cumulus	열적운	熱積雲
heat cycle	열순환	熱循環
heat energy	열에너지	熱-
heat engine	열기관	熱機關
heat equator	열적도	熱赤道
heat exchange	열교환	熱交換
heat exchanger	열교환기	熱交換器
heat exhaustion	더위탈진, 열사병	-脱盡, 熱射病
heat flow	열류	熱流
heat flow plate	열류판	熱流板
heat flow transducer	열류변환기	熱流變換機
heat flux	열플럭스, 열속	熱束
heat function	열함수	熱函數
heat generation	열생성	熱生成
heat index	열지수	熱指數
heating climate	난방기후	暖房氣候
heating degree-day	난방도일	暖房度日
heating index	난방지수	暖房指數
heating value	발열량	發熱量
heat island	열섬	熱-
heat island effect	열섬효과	熱-效果
heat lightning	열번개	熱-
heat low(= thermal low)	열저기압	熱低氣壓
heat of combination	화합열	化合熱
heat of condensation	응결열, 서림열	凝結熱
heat of congelation	동결열, 응결열	凍結熱, 凝結熱
heat of crystallization	결정열	結晶熱
heat of decomposition	분해열	分解熱
heat of dissolution	용해열	溶解熱
heat of evaporation	증발열	蒸發熱
heat of freezing	결빙열, 얼림열	結氷熱
heat of fusion	융해열	融解熱
heat of solidification	응고열	凝固熱
heat of solution	용해열	溶解熱
heat of sublimation	승화열	昇華熱
heat of vaporization	기화열, 증발열	氣化熱, 蒸發熱
heat partition	열분배	熱分配
heat radiation	열복사	熱輻射
heat ray	열선	熱線
heat reservoir	열저장체	熱貯藏體
heat shimmer	아지랑이	
heat sink	열흡수원	熱吸收源
heat source	열원	熱源
heat storm	열폭풍	熱暴風

heat stress index	열스트레스지수	熱-指數
heat stroke	열사병	熱射病
heat thunderstorm	열뇌우	熱雷雨
heat tolerance	내열성, 내서성	耐熱性, 耐暑性
heat transfer	열전달	熱傳達
heat transfer coefficient	열전달계수	熱傳達係數
heat transport	열수송	熱輸送
heat unit	열단위	熱單位
heat wave	열파	熱波
heat wave advisory	폭염주의보	暴炎注意報
heat wave warning	폭염경보	暴炎警報
heave	부침	浮沈
heaven	하늘	天
heavenly body	천체	天體
heaving(= frost heaving)	동상	凍傷
Heaviside layer	헤비사이드층	-層
heavy fog	짙은 안개	
heavy hail	큰 우박	
heavy ice	두꺼운 얼음	
heavy ion	중이온	重-
heavy overcast	짙은 흐림	
heavy rain	큰비, 호우	豪雨
heavy rain advisory	호우주의보	豪雨注意報
heavy rain associated with Changma front	장마전선호우	-前線豪雨
heavy rainfall	폭우	暴雨
heavy rain warning	호우경보	豪雨警報
heavy shower	센 소나기	
heavy snow	큰눈, 대설	大雪
heavy snow advisory	대설주의보	大雪注意報
heavy snowfall	폭설	暴雪
heavy snow warning	대설경보	大雪警報
heavy snow with large flakes	함박눈	
heavy squall	센 스콜	
heavy surf	거친 파도	-波濤
heavy surf advisory	거친 파도주의보	-波濤注意報
heavy winter clothing wear weather	방한복착용기상	防寒服着用氣象
hecto-	헥토(10^2을 표시하는 SI접두어)	
hectopascal(hPa)	헥토파스칼	
height(= altitude)	높이, 고도	高度
height analysis method	고도분석법	高度分析法
height-azimuth-range position indicator	고도-방위-거리 위치지시기	高度方位距離位置指示器
height calculation	고도계산	高度計算
height-change chart	고도변화도	高度變化圖
height-change line	고도변화선	高度變化線

height coordinate	고도좌표	高度座標
height-finding radar	고도탐지레이더	高度探知-
height gain	높이이득	高度利得
height gradient	고도경도, 높이경도	高度傾度
height nomogram	고도계산척	高度計算尺
height of cloud	구름높이	
height of cloud base	구름밑면고도, 운저고도	雲底高度
height of land surface	토지표면고도	土地表面高度
height pattern	고도패턴	高度-
Heiligenschein(= Cellini's halo)	하일리겐샤인〈독일어〉	
Heinrich event	하인리히사건	-事件
helical antenna	나선안테나	螺旋-
helical scanning	나사선주사	螺絲線走査
helicity	나선도, 헬리시티	螺線度
helicoidal anemometer	나선풍속계	螺旋風速計
heliograph	일광계	日光計
heliometric index	전천복사평균강도지수	全天輻射平均强度指數
heliophil(e)	호광식물	好光植物
heliophobe	혐광식물	嫌光植物
heliosphere	헬리오층	-層
heliostat	헬리오스태트, 정일경	定日鏡
heliotropic wind	일사풍	日射風
helm cloud	투구구름	鬪狗-
Helmholtz equation	헬름홀츠방정식	-方程式
Helmholtz free energy	헬름홀츠자유에너지	-自由-
Helmholtz function	헬름홀츠함수	-函數
Helmholtz gravitational wave	헬름홀츠중력파	-重力波
Helmholtz instability	헬름홀츠불안정도	-不安定度
Helmholtz reciprocity principle	헬름홀츠상호성원리	-相互性原理
Helmholtz wave	헬름홀츠파	-波
hemisphere	반구	半球
hemisphere broadcast	반구방송	半球放送
hemisphere exchange center	반구교환센터	半球交換-
hemispheric jet	반구제트	半球-
hemispheric model	반구모형	半球模型
hemispheric wave number	반구파수	半球波數
Herlofson diagram	헐로프슨선도	-線圖
hertz	헤르츠(주파수의 단위)	
Hertz's diagram	헤르츠선도	-線圖
Herzberg band	헤르츠베르크띠	
heterodyne	헤테로다인	
heterogeneity	비균질성	非均質性
heterogeneous	비균질(의)	非均質
heterogeneous fluid	비균질유체	非均質流體

heterogeneous nucleation	비균질핵화	非均質核化
heterogeneous substance	비균질체	非均質體
heterogeneous system	비균질계	非均質系
heterosphere	비균질권	非均質圈
Hevelian halo	헤빌리우스무리	
Hevelius's parhelia	헤빌리우스무리해	
hexagonal column	육각기둥	六角-
hexagonal platelet	육각판	六角板
hexagonal system	육방계	六方系
HHLL method	HHLL법	-法
hibernal	겨울(의), 한랭한	寒冷-
hibernating period	동면기	冬眠期
hibernation	겨울잠, 동면	冬眠
hig(= ig)	히그(잉글랜드의 지속시간이 짧은 폭풍)	
high	높은, 고기압	高氣壓
high air burst weather	고공폭발기상	高空爆發氣象
high aloft	상층고기압	上層高氣壓
high altitude	고공	高空
high altitude bombing weather	고공폭격날씨	高空爆擊-
high altitude low opening weather	고공강하날씨	高空降下-
high-altitude observatory	고산관측소	高山觀測所
high-altitude station	산악관측소	山岳觀測所
high artic climate	북극기후	北極氣候
high atmospheric pressure	고기압	高氣壓
high belt	고압대	高壓帶
high cloud	상층구름	上層-
highest minimum sector	최저구역고도	最低區域高度
highest minimum sector altitude	구역최고제한고도	區域最高制限高度
highest temperature	최고온도	最高溫度
high foehn	상층 푄	上層-
high fog	높은 안개	
high frequency	고주파	高周波
high frequency amplifier	고주파증폭기	高周波增幅器
high frequency oscillator	고주파발진기	高周波發振器
high-impact weather	고영향날씨	高影響-
high index	높은 지수	-指數
high-inversion fog	높은 역전안개	-逆轉-
highland	고랭지	高冷地
highland climate	고지기후	高地氣候
high land fog	고원안개	高原-
highland glacier	고지빙하	高地氷河
highland ice	고지얼음	高地-
high latitudes	고위도	高緯度
high-level analysis	고층분석	高層分析

H

high-level anticyclone	고층고기압	高層高氣壓
high-level cloud	고층구름	高層-
high-level cyclone	고층저기압	高層低氣壓
high-level forecast(= HLFOR)	고층예보	高層豫報
high-level ridge	고층(기압)마루	高層(氣壓)-
high-level significant weather prognostic chart	고층중요기상예상도	高層重要氣象豫想圖
high-level thunderstorm	고층뇌우	高層雷雨
high-level trough	고층(기압)골	高層(氣壓)-
highlight on visible imagery	가시영상의 강조	可視影像強調
high overcast	높은 흐림	
high polar glacier	두꺼운 극빙하	-極氷河
high precipitation supercell	고강수초대형세포	高降水超大型細胞
high pressure	고압, 고기압	高壓, 高氣壓
high pressure area	고기압지역	高氣壓地域
high pressure belt of middle latitudes	중위도고압대	中緯度高壓帶
high pressure center	고기압중심	高氣壓中心
high pressure ridge	고기압능	高氣壓-
high pressure system	고기압계	高氣壓系
High Rate Information Transmission	고속정보전송	高速情報電送
high reference signal	고기준신호	高基準信號
high reflectance surface	고반사면	高反射面
high resolution infrared radiometer	고분해적외복사계	高分解赤外輻射計
high resolution infrared sounder	고분해적외탐측계	高分解赤外探測計
high resolution X-ray diffractometer	고분해엑스선회절기	高分解-線回折器
high sea	거친 바다	
high seas forecast	외해예보	外海豫報
high spatial resolution	공간고분해능	空間高分解能
high spatial resolution sensor	공간고분해능감부(센서)	空間高分解能感部
high speed facsimile	고속팩시밀리	高速-
high speed stream	고속류	高速流
high surf	고쇄파	高碎波
high surf advisory	고쇄파주의보	高碎波注意報
high surf warning	고쇄파경보	高碎波警報
high temperature and humidity	고온다습	高溫多濕
high temperature injury	고온장해	高溫障害
high tide	고조	高潮
high tide advisory	고조주의보	高潮注意報
high tide warning	고조경보	高潮警報
high vacuum	고진공	高眞空
high velocity drop weather	고속투하기상	高速投下氣象
high volume precipitation spectrometer probe	항공거대강수입자스펙트로미터	航空巨大降水粒子-
high volume sampler	고용량에어샘플러	高容量-
high water	최고조위, 최고수위	最高潮位, 最高水位
high water discharge	고수위방수	高水位放水

high water interval	고조간격	高潮間隔
high water level	고수위, 고조위	高水位, 高潮位
high water mark	고수위표지, 고조위표지	高水位標識, 高潮位標識
high wind	강풍	强風
high wind advisory	강풍주의보	强風注意報
high wind warning	강풍경보	强風警報
high wind watch	강풍감시	强風監視
high wind wave advisory	풍랑주의보	風浪注意報
high wind wave warning	풍랑경보	風浪警報
high zonal circulation	센 동서순환	-東西循環
hill fog	산안개	山-
histogram	히스토그램	
historical climate	역사기후	歷史氣候
historical sequence	역사적 순서	歷史的順序
history of paleoclimatology	고기후학의 역사	古氣候學歷史
HLLH method	HLLH법	-法
hoar	서리	
hoar crystal	서리결정	-結晶
hoarfrost	하얀 서리	
Hobb's theory	홉스이론	-理論
hodograph	호도그래프	
hohlraum	(복사)공동	(輻射)空洞
hollerith system	홀러리드장치	-裝置
holocene climate	홀로세기후	沖積世氣候
holocene treeline fluctuations	홀로세수목한계선변동	-樹木限界線變動
hologram	홀로그램	
holography	홀로그래피	
holography method	호로그래피법	-法
holosteric barometer	공합기압계	空盒氣壓計
homobront	등뇌우시선	等雷雨時線
homoclime	동일기후	同一氣候
homodyne	호모다인	
homogeneity	균질성	均質性
homogeneous area	균질역	均質域
homogeneous atmosphere	균질대기	均質大氣
homogeneous condensation	균질응결	均質凝結
homogeneous fluid	균질유체	均質流體
homogeneous freezing	균질결빙	均質結氷
homogeneous medium	균질매체	均質媒體
homogeneous nucleating material	균질핵화물질	均質核化物質
homogeneous nucleation	균질핵화	均質核化
homogeneous series	균질계열	均質系列
homogeneous sublimation	균질승화	均質昇華
homogeneous system	균질계	均質系

H

homogeneous turbulence	균질난류	均質亂流
homogenization	균질화	均質化
homologous element	동족원소	同族元素
homologous turbulence	균질난류	均質亂流
homopause	균질권계면	均質圈界面
homosphere	균질권	均質圈
hook cloud pattern	갈고리구름유형	-類型
hook echo	갈고리에코	
hook echo radar signature	갈고리에코 레이더 신호	-信號
hook gauge	갈고리계기	-計器
Hooke's Law	훅법칙	-法則
Hopfield bands	홉필드띠	
horizon	지평, 수평	地平, 水平
horizon distance	수평거리	水平距離
horizon sensor	수평감지기, 수평센서	水平感知器
horizontal	수평	水平
horizontal advection	수평이류	水平移流
horizontal angle	방위각, 수평각	方位角, 水平角
horizontal circle	수평분도원	水平分度圓
horizontal component	수평성분	水平成分
horizontal convergence	수평수렴	水平收斂
horizontal distribution	수평분포	水平分布
horizontal divergence	수평발산	水平發散
horizontal flow	수평흐름	水平-
horizontal inertia motion	수평관성운동	水平慣性運動
horizontal insolation intensity	수평일사강도	水平日射強度
horizontal instability	수평불안정(도)	水平不安定(度)
horizontal leaved canopy	수평엽형군락	水平葉型群落
horizontal mixing	수평혼합	水平混合
horizontal motion	수평운동	水平運動
horizontal motion equation	수평운동방정식	水平運動方程式
horizontal plane	수평면	水平面
horizontal polarization	수평편광	水平偏光
horizontal pressure force	수평압력	水平壓力
horizontal pressure gradient	수평압력경도	水平壓力傾度
horizontal pressure gradient force	수평기압경도력	水平氣壓傾度力
horizontal rainbow	수평무지개	水平-
horizontal resolution	수평해상도	水平解像度
horizontal seismograph	수평지진계	水平地震計
horizontal sounding technique	수평탐측기술	水平探測技術
horizontal stability	수평안정도	水平安定度
horizontal tangent arc	수평접호	水平接弧
horizontal temperature gradient	수평온도경도	水平溫度傾度
horizontal time section	수평시단면	水平時斷面

horizontal variation	수평변동	水平變動
horizontal velocity	수평속도	水平速度
horizontal visibility	수평시정	水平視程
horizontal wave	수평파	水平波
horizontal wind shear	수평윈드시어	水平-
horizontal wind vector	수평바람벡터	水平-
horn antenna	뿔모양안테나	-模樣-
horn card	뿔풍향판	-風向板
horn radiator	뿔모양복사체	-模樣輻射體
horse latitude	말위도	-緯度
horse latitude high	아열대고기압	亞熱帶高氣壓
hot-air balloon	열기구	熱氣球
hot belt	고온대	高溫帶
hot climate	고온기후	高溫氣候
hot-film anemometer	열막풍속계	熱膜風速計
hotplate precipitation gauge	열판강우계측기	熱板降雨計測器
hot spot	열점	熱點
hot thunderstorm	열뇌우	熱雷雨
hot tower	열탑	熱塔
hot wave	열파	熱波
hot wind	열바람	熱-
hot-wire anemometer	열선풍속계	熱線風速計
hour angle	시간각	時間角
hourly distance scale	시간거리표	時間距離表
hourly observation	매시관측	每時觀測
hourly precipitation	시간강수량	時間降水量
house climate	실내기후	室內氣候
house microclimate	실내미기후	室內微氣候
Höiland's circulation theorem	회이란순환정리	-循環定理
Hudson bay low	허드슨만저기압	-低氣壓
Huggins bands	허긴스띠	
human bioclimatology	인간생(물)기후학	人間生(物)氣候學
human biometeorology	인간생기상학	人間生氣像學
human climate	인간기후	人間氣候
human phenology	생활계절학	生活季節學
Humboldt current	훔볼트해류	-海流
humicap	감습콘덴서	感濕-
humid climate	습윤기후	濕潤氣候
humid continental climate	습윤대륙기후	濕潤大陸氣候
humidifier fever	가습기발열	加濕器發熱
humid injury	습윤해	濕潤害
humidity	습도	濕度
humidity coefficient	습도계수	濕度係數
humidity element	습도요소	濕度要素

humidity index	습윤지수	濕潤指數
humidity lapse rate	습윤감률	濕潤減率
humidity province	습윤구역	濕潤區域
humidity sensor	습도센서, 습도감부	濕度感部
humidity slide-rule	습도계산자	濕度計算-
humidity strip	습도소자	濕度素子
humid layer	습윤층	濕潤層
humid mesothermal climate	습윤중열기후	濕潤中熱氣候
humid microthermal climate	습윤미열기후	濕潤微熱氣候
humidometer	습도계	濕度計
humid temperate climate	습윤온대기후	濕潤溫帶氣候
humid zone	습윤대	濕潤帶
humilis	편평구름	扁平-
humistor	감습저항	感濕抵抗
hummock	얼음언덕, 빙구	氷丘
hummocked ice	언덕얼음	
hurly-burly	헐리-벌리(영국과 스코틀랜드의 뇌우)	
hurricane	허리케인, 싹쓸바람(풍력계급 12)	
hurricane analog technique	허리케인진로예상법	-進路類似法
hurricane balloon	허리케인기구	-氣球
hurricane band	허리케인띠	
hurricane bar	허리케인둑	
hurricane beacon	허리케인비컨	
hurricane cloud	허리케인구름	
hurricane eye	허리케인눈, 태풍눈	颱風-
hurricane force wind	허리케인급 바람	
hurricane force wind warning	허리케인급 바람경보	-警報
hurricane microseisms	허리케인미진	-微震
hurricane seeding	허리케인씨뿌리기	
hurricane tracking	허리케인추적	-追跡
hurricane warning	허리케인경보	-警報
hurricane warning system	허리케인경보체계	-警報體系
hurricane watch	허리케인주의보, 태풍주의보	颱風注意報
Huygens's principle	호이헨스원리	-原理
Huygens's wavelets	호이헨스 작은 파	-波
hybrid	잡종, 혼성	雜種, 混成
hybrid retrieval	하이브리드검색, 이종검색	異種檢索
hybrid storm	혼합형스톰	混合形-
hybrid variable	하이브리드변광성, 이종변광성	異種變光星
hydrate	수화물, 함수화물	水化物, 含水化物
hydraulic analogy	수력유사	水力類似
hydraulic conductivity	수리전도도	水理傳導度
hydraulic gradient	수력경도	水力傾度

hydraulic head	수두	水頭
hydraulic jump	물뜀	
hydraulic press	수압기	水壓機
hydraulic radius	수력반지름, 수력반경	水力半徑
hydraulics	수리학, 수력학	水理學, 水力學
hydrocarbons	탄화수소	炭化水素
hydroclimatic unit	수문기후단위	水文氣候單位
hydrodynamically rough surface	유체역학적 거친면	流體力學的-面
hydrodynamic instability	유체역학 불안정	流體力學不安定
hydrodynamic pressure	유체동압력	流體動壓力
hydrodynamics	유체역학	流體力學
hydrodynamic stability	유체역학 안정	流體力學安定
hydrofluorocarbons	하이드로플루오로카본	
hydrogen(-filled) balloon	수소기구	水素氣球
hydrogen bond	수소결합	水素結合
hydrogen-ion concentration	수소이온농도	水素-濃度
hydrogen line	수소선	水素線
hydrogen scale of temperature	수소온도 눈금	水素溫度-
hydrogen thermometer	수소온도계	水素溫度計
hydrograph	수문곡선	水文曲線
hydrographic station	수문관측소	水文觀測所
hydrographic survey	수로측량	水路測量
hydrograph separation	수문곡선분리	水文曲線分離
hydrography	수로학	水路學
hydrokinetics	유체동역학	流體動力學
hydrolapse	수증기감률	水蒸氣減率
hydrologic accounting	수문계산	水文計算
hydrological basin	수역	水域
hydrological cycle	수문순환	水文循環
hydrological drought	수문학적 가뭄	水文學的-
hydrological phenomena	수상	水象
hydrological resource	수문자원	水文資源
hydrological statistics	수문통계학	水文統計學
hydrologic balance (budget)	물평형	-平衡
hydrologic cycle	물순환	-循環
hydrologic equation	수문학방정식	水文學方程式
hydrologic flood routing	수문학적 홍수추적	水文學的洪水追跡
hydrologic forecasting	수문학적 예측	水文學的豫測
hydrologic model	수문학적 모형	水文學的模型
hydrologic year	수문년	水文年
hydrology	수문학	水文學
hydrolysis	가수분해	加水分解
hydromagnetic emission	유체자기방출	流體磁氣放出
hydromagnetics	유체자기역학	流體磁氣力學

H

hydromagnetic wave	자기유체파	磁氣流體波
hydromechanics	수력학, 유체역학	水力學, 流體力學
hydrometeor	대기물현상	大氣-現象
hydrometeorologist	수문기상학자	水文氣象學者
hydrometeorology	수문기상학	水文氣象學
hydrometeor videosonde	대기수상 비디오존데	大氣水象-
hydrometer	액체비중계	液體比重計
hydrometric network, hydrological network	수문관측망	水文觀測網
hydrophilicity	친수성	親水性
hydrophobicity	소수성	疏水性
hydrophotometer	유체광도계	流體光度計
hydroplaning	수막현상	水膜現象
hydrosphere	수권	水圈
hydrostatic	정역학	靜力學
hydrostatic adjustment	정역학조절	靜力學調節
hydrostatic approximation	정역학근사	靜力學近似
hydrostatic assumption	정역학가정	靜力學假定
hydrostatic balance	정역학평형	靜力學平衡
hydrostatic energy	정역학에너지	靜力學-
hydrostatic equation	정역학방정식	靜力學方程式
hydrostatic equilibrium	정역학평형	靜力學平衡
hydrostatic instability	정역학불안정(도)	靜力學不安定(度)
hydrostatic model	정역학모형	靜力學模型
hydrostatic pressure	유체정압	流體靜壓
hydrostatics	유체정역학	流體靜力學
hydrostatic stability	정역학안정(도)	靜力學安定(度)
hydrothermal vent	열수분출공	熱水噴出孔
hydro unit	수문단위	水文單位
hydroxyl emission	수산기방출	水酸基放出
hydroxyl radical	하이드록실라디칼	
hyetal	비(의), 강우(의), 강우지대(의)	降雨, 降雨地帶
hyetal coefficient	우량계수	雨量係數
hyetal equator	우량적도	雨量赤道
hyetal region	우량구역	雨量區域
hyetograph	강수도표	降水圖表
hyetographic curve	강수분포곡선	降水分布曲線
hyetography	강수(분포)학	降水(分布)學
hyetology	강수학	降水學
hyetometer	우량계	雨量計
hygiene meteorology(= hygieno meteorology)	위생기상학	衛生氣象學
hygristor	습도소자	濕度素子
hygrodeik	건습구습도계	乾濕球濕度計
hygrogram	습도기록곡선	濕度記錄曲線
hygrograph	습도기록계	濕度記錄計

hygrokinematics	수분운동학	水分運動學
hygrology	습도학	濕度學
hygrometer	습도계	濕度計
hygrometer of dew point	이슬점습도계	露點濕度計
hygrometric continentality	강수대륙도	降水大陸度
hygrometric equation	습도방정식	濕度方程式
hygrometric formula	습도공식	濕度公式
hygrometric state	습윤상태	濕潤狀態
hygrometric table	습윤표	濕潤表
hygrometry	습도측정법	濕度測定法
hygrophyte	습생식물	濕生植物
hygroscope	검습기, 간이습도계	檢濕器, 簡易濕度計
hygroscopic	흡습성, 흡수성	吸濕性, 吸收性
hygroscopic movement	흡습운동	吸濕運動
hygroscopicity	흡습도	吸濕度
hygroscopic moisture	흡습수분	吸濕水分
hygroscopic nucleus	흡습핵, 흡습성핵물질	吸濕性核, 吸濕性核物質
hygroscopic nucleus cloud seeding	흡습성핵물질 구름씨뿌리기	吸濕性核物質-
hygroscopic particle	흡습성입자	吸濕性粒子
hygroscopic water	흡습수, 흡착수	吸濕水, 吸着水
hygroscopic wettable nucleus	흡습성핵	吸濕性核
hygrothermogram	온습도기록곡선	溫濕度記錄曲線
hygrothermograph	온습도기록계	溫濕度記錄計
hygrothermoscope	온습도계	溫濕度計
hyperbaroclinic	초경압(의)	超傾壓
hyperbola	쌍곡선	雙曲線
hyperbolic point	쌍곡점	雙曲點
hyperon	하이퍼론	
hypersensitivity pneumonitis	과민성폐렴	過敏性肺炎
hypersonic	극초음속	極超音速
hyperspectral classification	초분광분류	超分光分類
hyperspectral sensor	초분광센서, 초분광감부	超分光感部
hyper-velocity	초고속도	超高速度
hypolimnion	심수층	深水層
hypotenuse	빗변	-邊
hypothermia	체온저하	體溫低下
hypothesis	가설	假說
hypothetical global climate	가상전구기후	假想全球氣候
Hypsithermal	힙시서멀(기후최적기)	
hypsography	측고법	測高法
hypsometer	측고계	測高計
hypsometric equation	측고방정식	測高方程式
hypsometric formula	측고공식	測高公式
hypsometry	측고법	測高法

H

hysteresis	이력(현상), 히스테리시스	履歷(現想)
hysteresis characteristic	히스테리시스 특성	-特性
hythergraph	하이서그래프, 온습도	溫濕圖

I, i

영문	한글	한자
ICAO standard atmosphere	ICAO표준대기	-標準大氣
ice	얼음	
ice accretion	착빙	着氷
ice accretion indicator	착빙지시기, 결빙지시기	着氷指示器, 結氷指示器
ice age	빙하시대	氷河時代
ice atlas	얼음지도	-地圖
ice band	얼음띠	
ice bay	얼음만	-灣
ice belt	얼음띠	
iceberg	빙산	氷山
ice bight	얼음만	-灣
ice blink	얼음빛	
ice blister	얼음언덕, 빙구	氷丘
icebreaker	쇄빙선, 쇄빙기	碎氷船, 碎氷機
ice-bulb temperature	빙구온도	氷球溫度
ice calorimeter	얼음열량계	-熱量計
ice cap	얼음모자, 빙모	氷帽
ice cap climate	빙설기후	氷雪氣候
ice cascade	얼음폭포	-瀑布
ice cave	얼음굴	
ice clearing	얼음제거	-除去
ice climate	빙설기후	-雪氣候
ice cloud	얼음구름	
ice column	얼음기둥, 빙주	氷柱
ice core	빙하코어, 빙하핵	氷河核
ice cover	적빙, 얼음덮개, 얼음피복	積氷, -被覆
ice crust	얼음표층	-表層
ice crystal	얼음결정, 빙정	-結晶, 氷晶
ice crystal cloud	얼음결정구름, 빙정구름	-結晶-, 氷晶-

ice crystal effect	얼음결정효과, 빙정효과	-結晶效果, 氷晶效果
ice crystal fog	얼음결정안개, 빙정안개	-結晶-, 氷晶-
ice crystal haze	얼음결정연무, 빙정연무	-結晶煙霧, 氷晶煙霧
ice crystal nucleus	얼음결정핵	-結晶核
ice crystal process	빙정과정	氷晶過程
ice crystal theory	얼음결정이론, 빙정설	-結晶理論, 氷晶說
ice day	참겨울날	
ice deposit	얼음쌓임	
ice desert	빙설지역	氷雪地域
ice-edge vortex	빙산가장자리 소용돌이	氷山-
ice fall	얼음폭포	-瀑布
ice fat	얼음비계	
ice feather	얼음깃털	
ice field	빙원, 얼음벌판	氷原
ice firn	해묵은 언눈	
ice floe	부빙	浮氷
ice flower	얼음꽃	
ice fog	얼음안개	
ice foot	얼음기슭	
ice-formation condition	얼음생성조건	-生成條件
ice-forming nucleus	얼음생성핵, 빙정핵	-生成核, 氷晶核
ice friction theory	얼음마찰이론	-摩擦理論
ice fringe	얼음줄무늬	
ice front	얼음전선	-前線
ice gorge	얼음골짜기	
icehouse climate	아이스하우스기후, 제빙실 한랭기후	製氷室寒冷氣候
ice island	얼음섬	
ice jam	얼음쌓임	
Icelandic low	아이슬랜드 저기압	-低氣壓
ice limit	얼음한계	-限界
ice mound	얼음언덕, 빙구	氷丘
ice multiplication	얼음증식	-增殖
ice needle	얼음바늘	
ice nucleus	빙정핵	氷晶核
ice on runway	활주로결빙	滑走路結氷
ice pack(= pack ice)	얼음팩	
ice particle	얼음알갱이, 얼음입자	-粒子
ice particle imager	강설입자영상기	降雪粒子映像機
ice pellet	얼음싸라기	
ice pillar	서릿발	
ice point	얼음점, 빙점	氷點
ice pole	얼음극, 빙극	氷極
ice prism	얼음각기둥	-角-

ice-rafted debris	빙하운반쇄설물	氷河運搬碎屑物
ice reconnaissance	얼음정찰	-偵察
ice ribbon	얼음리본	
ice rind	얼음껍질	
ice run	얼음갈라짐	
Ice Saints	아이스세인트(5월 11∼13일 유럽에 있어서 서리가 많은 시기)	
ice saturation	얼음면포화	-面飽和
ice sheet(= ice field)	얼음평상, 빙상	-平床, 氷床
ice shelf	얼음선반, 빙붕	氷棚
ice sky	얼음빛하늘	
ice splinter	얼음조각	
ice storm	얼음보라	
ice storm warning	착빙성 폭풍우경보	着氷性暴風雨警報
ice stream	얼음흐름	
ice strip	얼음띠	
ice tongue	얼음혀	
ice water content	빙수량	氷水量
icicle	고드름	
icing	착빙	着氷
icing environment	착빙환경	着氷環境
icing index	착빙지수	着氷指數
icing intensity	착빙강도	着氷强度
icing level	착빙고도	着氷高度
icing meter	착빙계	着氷計
icing mound	얼음언덕, 빙구	氷丘
icing on a hull	선체착빙	船體着氷
icing-rate meter	착빙률계	着氷率計
icy wind	얼음바람	
ideal climate	이상기후	理想氣候
ideal fluid	이상유체	理想流體
ideal foliage	이상군락	理想群落
ideal gas	이상기체	理想氣體
ideal gas equation	이상기체방정식	理想氣體方程式
ideal gas law	이상기체법칙	理想氣體法則
ideal horizon	이상수평	理想水平
ideal liquid	이상액체	理想液體
identification	식별	識別
identity	항등식	恒等式
idle time	휴지시간	休止時間
IFR flight	IFR비행	-飛行
IFR terminal minimum	IFR목적지 최저기상조건	-目的地最低氣象條件
IFR weather	계기비행 기상	計器飛行氣象
ig(= hig)	히그	

igneous meteor	불빛현상	-現象
igneous rock	화성암	火成巖
illite	일라이트	
ill-posed problem	부적절 설정문제	不適切設定問題
illuminance	조도	照度
illumination	조명	照明
illumination climate	조도기후	照度氣候
illuminometer	조도계	照度計
image	영상	映像
image correction	영상보정	映像補正
Image Data Aquisition and Control System	영상자료 취득 및 (전처리)관리시스템	映像資料取得-(前處理)管理-
image differencing	영상차분	映像差分
image dissector camera system	영상분석 카메라시스템	映像分析-
image enhancement	영상개선, 영상강조	映像改善, 映像强調
image enhancement technique	영상강조기법	映像强調技法
image filming system	영상촬영시스템	映像撮影-
image inclination	영상경사	映像傾斜
Image Motion Compensation(IMC)	영상위치보정시스템	映像位置補正-
Image Navigation and Registration	영상위치보정	映像位置補正
Image Preprocessing Subsystem	영상전처리시스템	映像前處理-
imagery	영상	映像
image snow cover meter	영상적설계	映像積雪計
image transfer theory	영상전달이론	映像傳達理論
image zoom	영상확대	映像擴大
imaginary number	허수	虛數
imaginary part	허수부분	虛數部分
imaginary unit	허수단위	虛數單位
imbibition	흡수, 흡입	吸收, 吸入
Immediate Air Support weather	긴급항공지원기상	緊急航空支援氣象
immediate transmission	즉시전송, 실시간전송	卽時電送, 實時間電送
impact	충돌	衝突
impactor	충돌채집기	衝突採集器
imperfect forecast	불완전예보	不完全豫報
impetus	충격	衝擊
impingement	충격, 충돌	衝擊, 衝突
implicit scheme	암시적 방안	暗示的方案
implicit smoothing	암시평활	暗示平滑
impound	저류	貯留
improved TIROS operational satellite	개량TIROS운영위성	改良-運營衛星
improvement report	일기회복보고	日氣回復報告
impulse	충격, 역적	衝擊, 力積
impulsive force	충격력	衝擊力
impulsive warming	폭발승온	暴發昇溫

impurity	불순도	不純度
inactive front	비활동전선	非活動前線
inactive region	비활성지역	非活性地域
inactivation	비활성화	非活性化
inadvertent climate modification	비의도적 기후개조	非意圖的氣候改造
inadvertent cloud modification	비의도적 구름조절	非意圖的-調節
inadvertent modification	비의도적 조절	非意圖的調節
inadvertent weather modification	비의도적 기상조절	非意圖的氣象調節
incendiaries agent operation weather	소이작용제작전기상	燒夷作用劑作戰氣象
incidence	입사	入射
incident light	입사광	入射光
incident point	입사점	入射點
incident pulse	입사펄스	入射-
incident radiation	입사복사	入射輻射
incident ray	입사광선	入射光線
incident wave	입사파	入射波
incipient low	발생기저기압	發生期低氣壓
incipient stage	발생단계, 발생기	發生段階, 發生期
inclination	경사	傾斜
inclination angle	경사각	傾斜角
inclination of the axis of a cyclone	저기압축경사	低氣壓軸傾斜
inclination of the axis of an anticyclone	고기압축경사	高氣壓軸傾斜
inclination of the wind	바람경각	-傾角
in-cloud lightning	구름 속 번개	
incoherent integration	비간섭적분	非干涉積分
incoherent receiver	비정합수신기	非整合受信機
incoherent scatter radar	비간섭산란레이더	非干涉散亂-
incoherent scatter technique	비간섭산란기법	非干涉散亂技法
incoherent signal	비간섭신호	非干涉信號
incoming radiation	입사복사	入射輻射
incoming radiation type	입사복사형, 수열형	入射輻射型, 受熱型
incoming solar radiation	입사태양복사	入射太陽輻射
incomplete combustion	불완전연소	不完全燃燒
incompressibility	비압축성	非壓縮性
incompressible fluid	비압축성유체	非壓縮性流體
inconstant cirrus	변덕권운	變德卷雲
increase variable	증가변수	增加變數
increasing	증가	增加
increasing function	증가함수	增加函數
increasing of snow	증설	增雪
increasing state	증가상태	增加狀態
increment	증분	增分
incremental analysis update	증분분석갱신	增分分析更新
incursion	침입	侵入

incus	모루구름	
independent event	독립사건	獨立事件
independent sample	독립표본	獨立標本
independent variable	독립변수	獨立變數
indestructibility	비파괴성	非破壞性
index	지수	指數
index arm of sextant	육분의 지표막대	六分儀指標-
index correction	지표보정	指標補正
index cycle	지수순환	指數循環
index error	지표오차	指標誤差
index number	지수, 찾아보기 번호	指數, -番號
index number of station	지점번호	地點番號
index of absorption	흡수지수	吸收指數
index of aridity	건조지수	乾燥指數
index of continentality	대륙도지수	大陸度指數
index of instability	불안정지수	不安定指數
index of refraction	굴절률	屈折率
Indian monsoon	인도계절풍	印度季節風
Indian Ocean dipole mode	인도양쌍극자모드	印度洋双極子-
Indian summer	인디언서머	
India summer monsoon	인도여름몬순	印度-
indicated airspeed	지시공기속도	指示空氣速度
indicated air temperature	지시기온	指示氣溫
indicated altitude	지시고도	指示高度
indicator	지시기	指示器
indicator diagram	지시선도	指示線圖
indicator lamp	지시등	指示燈
indicator organism	지표생물	指標生物
indicator plant	지표식물	指票植物
indifferent air mass	중성기단	中性氣團
indifferent equilibrium	중립평형	中立平衡
indifferent stability	중립안정도	中立安定度
indirect aerological analysis	간접고층분석	間接高層分析
indirect aerology	간접고층기상학	間接高層氣象學
indirect cell	간접세포	間接細胞
indirect circulation	간접순환	間接循環
indirect hit (or effect)	간접영향	間接影響
indirect source	간접오염원	間接汚染源
individual change	실질적 변화	實質的變化
individual derivative	개별도함수	個別導函數
individual droplet	개별방울, 개별수적	個別水滴
individual nucleus	개별핵	個別核
indoor air	실내공기	室內空氣
indoor air distribution	실내공기분포	室內空氣分布

indoor air flow	실내기류	室內氣流
indoor air pollution	실내대기오염	室內大氣汚染
indoor climate	실내기후	室內氣候
indoor temperature	실내온도	室內溫度
Indo-Pakistan low	인도-파키스탄 저기압	印度-低氣壓
induced charge	유도전하	誘導電荷
induced current	유도전류	誘導電流
induced electromotive force	유도기전력	誘導起電力
induction	귀납법	歸納法
induction experiment	유도시험	誘導試驗
induction icing	흡입착빙	吸入着氷
induction method	유도법	誘導法
inductive lightning	유도뢰	誘導雷
industrial climatology	산업기후학	産業氣候學
industrial meteorology	산업기상학	産業氣象學
inelastic collision	비탄성충돌	非彈性衝突
inequality	부등식	不等式
inert element	비활성원소	非活性原素
inert gas	비활성기체	非活性氣體
inertia	관성	慣性
inertia circle	관성원	慣性圓
inertial coordinate system	관성좌표계	慣性座標系
inertial flow	관성흐름	慣性-
inertial force	관성력	慣性力
inertial forecast	관성예보	慣性豫報
inertial frequency	관성진동수	慣性振動數
inertial gravity wave	관성중력파	慣性重力波
inertial instability	관성불안정(도)	慣性不安定(度)
inertial lag	관성지연	慣性遲延
inertial mass	관성질량	慣性質量
inertial oscillation	관성진동	慣性振動
inertial period	관성주기	慣性週期
inertial radius	관성반지름, 관성반경	慣性半徑
inertial range	관성영역	慣性領域
inertial resistance	관성저항	慣性抵抗
inertial stability	관성안정(도)	慣性安定(度)
inertial sublayer	관성아층	慣性亞層
inertial subrange	관성아구간	慣性亞區間
inertial system	관성계	慣性(座標)系
inertial wave	관성파	慣性波
inertia term	관성항	慣性項
inertia velocity	관성풍속	慣性風速
inertia wave	관성파	慣性波
inertio-gravitational wave	관성중력파	慣性重力波

Infantry Tank Combined Arms Operation weather	보전협동작전기상	步戰協同作戰氣象
inferior mirage	아래신기루	-蜃氣樓
infiltration	침투	浸透
infiltration capacity	침투능	浸透能
infiltration capacity curve	침투능곡선	浸透能曲線
infiltration coefficient	침투계수	浸透係數
infiltrometer	침투계	浸透計
infinite	무한	無限
infinite(nonterminating) decimal	무한소수	無限小數
infinite geometric(geometrical) sequence	무한등비수열	無限等比數列
infinite geometric(geometrical) series	무한등비급수	無限等比級數
infinite series	무한급수	無限級數
infinity	무한	無限
inflexion point	변곡점	變曲點
inflight evaluation	비행 중 평가	飛行中評價
inflow	유입(량)	流入(量)
inflow-storage-discharge curve	유입-저수-유출곡선	流入貯水流出曲線
influence function	영향함수	影響函數
informis	부정형(구름)	不定形-
infralateral tangent arcs	하부측면접호	下部側面接弧
infrared	적외(선)	赤外(線)
infrared absorption hygrometer	적외흡수습도계	赤外吸收濕度計
infrared atmospheric window	적외대기창	赤外大氣窓
infrared band	적외선대역	赤外線帶域
Infra-Red Earth Sensor	적외지구센서	赤外地球-
infrared gas analyzer	적외선가스분석계	赤外線-分析計
infrared image	적외영상	赤外映像
infrared imager	적외화상기	赤外畫像器
infrared imagery	적외영상	赤外映像
infrared imaging	적외영상	赤外映像
infrared interferometer spectrometer	적외간섭분광계	赤外干涉分光計
infrared picture	적외선사진	赤外線寫眞
infrared radiation	적외복사	赤外輻射
infrared radiometry	적외복사측정	赤外輻射測定
infrared ray	적외(광)선	赤外(光)線
infrared region	적외역	赤外域
infrared satellite imagery	적외위성영상	赤外衛星映像
infrared spectrum	적외스펙트럼	赤外-
infrared telescope	적외선망원경	赤外線望遠鏡
infrared temperature profile radiometer	연직적외온도복사계	鉛直赤外溫度輻射計
infrared thermal radiation	적외열방사	赤外熱放射
infrared wavelength	적외선파장	赤外線波長
infrared window	적외선창	赤外線窓

infrasonic	초저음(의)	超低音
infrasound	초저음	超低音
Inhalable Particulate Matter	흡입성입자상 물질	吸入性粒子狀物質
inhibitor	저해물질	沮害物質
inhomogeneous cloud nucleating material	비균질구름핵화물질	非均質-核化物質
inhomogeneous medium	불균질매체, 불균질매질	不均質媒體, 不均質媒質
initial	초기	初期
initial abstraction, initial loss	초기손실	初期損失
initial condensation phase	초기응결단계	初期凝結段階
initial condition	초기조건	初期條件
initial detention	초기역류	初期逆留
initial infiltration rate	초기침투율	初期浸透率
initialization	초기값설정, 초기화	初期値設定, 初期化
initial lift	초기상승	初期上乘
initial phase	초기위상	初期位相
initial reading	첫읽기	
initial state	초기상태	初期狀態
initial step(= first time step)	초기단계	初期段階
initial value problem	초깃값문제	初期値問題
initial velocity	초기속도, 첫속도	初期速度
injection temperature	주입온도	注入溫度
injury by warm temperature in winter	난동해	暖冬害
injury due to dry air	건조공기피해	乾燥空氣披害
injury due to dryness	건조해	乾燥害
inland	내륙	內陸
inland climate	내륙기후	內陸氣候
inland desert	내륙사막	內陸沙漠
inland fog	내륙안개	內陸-
inland freshwater wetlands	내륙담수습지	內陸淡水濕地
inland ice	내륙얼음, 내륙빙하	內陸氷河
inlandity	내륙도	內陸度
inland water	내륙수	內陸水
inner audible zone	내청역	內聽域
inner friction	내부마찰	內部摩擦
inner planet	내행성	內行星
inner product	내적	內積
In-Orbit Test	초기궤도시험	初期軌道試驗
	(발사 후 수개월 동안 위성시스템의 정상작동 검사)	
in-phase(I-component)	위상일치(I-성분)	位相一致
input	입력	入力
Insat	Insat위성	-衛星
inscribed angle	원주각	圓周角
insect echo	곤충에코	昆蟲-
insecticide	살충제	殺蟲劑

insect resistance	내충성	耐蟲性
inshore operation	연안작전	沿岸作戰
inshore wind	연안풍	沿岸風
inside of cloud	운중, 구름 속	雲中
in situ	현장채취	現場採取
insolation	일사	日射
insolation duration	일사시간, 일조시간	日射時間, 日照時間
insoluble wettable nucleus	불용습성핵	不溶濕性核
inspection	검사, 검정	檢査, 檢定
inspectional analysis	검사분석	檢査分析
instability	불안정(도)	不安定(度)
instability chart	불안정도도표	不安定度圖表
instability energy	불안정에너지	不安定-
instability line	불안정선	不安定線
instability of circular vortex	원형소용돌이 불안정도	圓形-不安定度
instability pattern	불안정패턴	不安定-
instability shower	불안정소나기	不安定-
instabilizing factor	불안정인자	不安定因子
installation	장치	裝置
instantaneous	순간적	瞬間的
instantaneous acceleration	순간가속도	瞬間加速度
instantaneous dipole moment	순간쌍극자모멘트	瞬間雙極子-
instantaneous field of view	순간시야, 순간관찰영역	瞬間視野, 瞬間觀察領域
instantaneous rate of change	순간변화율	瞬間變化率
instantaneous speed	순간속력	瞬間速力
instantaneous velocity	순간속도	瞬間速度
instantaneous wind speed	순간풍속	瞬間風速
instant occlusion	순간폐색	瞬間閉塞
instrumental error	계기오차(기차)	計器誤差(器差)
instrumentation ship	관측선	觀測船
instrument constant	측기상수	測器常數
instrument correction	계기보정	計器補定
instrument error	계기오차	計器誤差
instrument error table	계기오차표	計器誤差表
instrument exposure	계기노출	計器露出
instrument flight	계기비행	計器飛行
instrument flight rule	계기비행규칙	計器飛行規則
instrument icing	계기착빙	計器着氷
instrument landing system	계기착륙장치	計器着陸裝置
instrument meteorological condition	계기비행 기상조건	計器飛行氣象條件
instrument noise	계기잡음	計器雜音
instrument outage	측기내구년수	測器耐久年數
instrument shelter	백엽상	百葉箱
instrument weather	계기날씨	計器-

insulation	절연	絶緣
insulation wall	단열벽	斷熱壁
insurance damage from storm and flood	풍수해보험	風水害保險
in-tandem model	직렬모형	直列模型
integer	정수	整數
integrable	적분가능	積分可能
integral calculus	적분학	積分學
integral constant	적분상수	積分常數
integral constraint	통합억제	統合抑制
integrated indicators of stability	통합안정도척도	統合安定度尺度
Integrated Ocean Drilling Program	통합해저시추프로그램	統合海底試錐-
integrated pest management	종합적해충관리	綜合的害蟲管理
integrated terminal weather system	통합비행장기상시스템	統合飛行場氣象-
integration	적분	積分
integration by parts	부분적분(법)	部分積分(法)
intensification	강화	強化
intensity	강도	強度
intensity-duration formula	강도-시간 공식	強度時間公式
intensity-modulated indicator	강도-변조 지시기	強度變調指示器
intensity modulation	강도변조	強度變調
intensity of illumination	조도	照度
intensity of radiation	방사강도	放射強度
intensity of turbulence	난류강도	亂流強度
intensive quantity	내연량	內延量
intentional weather modification	고의적 기상조절	故意的氣象調節
interaction	상호작용	相互作用
interannual oscillation	경년변동	經年變動
interannual variability	경년변동성	經年變動性
interannual variation	경년변동	經年變動
intercalary day	윤일	閏日
intercept parameter	절편매개변수	切片媒介變數
interception	차단	遮斷
interception of precipitation	강수차단	降水遮斷
interception of rainfall	강우차단	降雨遮斷
interceptometer	차단우량계	遮斷雨量計
interchange coefficient	교환계수	交換係數
interchannel regression test	내부채널 회귀검사	內部-回歸檢查
intercirrus region	권운사이구역	卷雲-區域
intercloud discharge	구름사이방전	-放電
intercumulus region	적운사이구역	積雲-區域
Interdiction Area weather	차단지역기상	遮斷地域氣象
interdiurnal pressure variation	나날기압변동, 경일압력변동	經日壓力變動
interdiurnal temperature variation	나날온도변동, 경일온도변동	經日溫度變動
interdiurnal variability	나날변동성, 경일변동성	經日變動性

interface	경계면	境界面
interface control document	접속조절문서 (시스템 간 혹은 부시스템 간 접속규약)	接續調節文書
interfacial tension	계면장력	界面張力
interference	간섭	干涉
interference colour	간섭색	干涉色
interference fringe	간섭무늬	干涉-
interference instrument	간섭계	干涉計
interference lobe	간섭잎	干涉-
interference region	간섭역	干涉域
interferometer	간섭계	干涉計
interflow	중간흐름	中間-
interglacial climate	간빙기기후	間氷期氣候
interglacial period	간빙기	間氷期
interglacial phase	간빙기	間氷期
Intergovernmental Panel on Climate Change	기후변화에 관한 정부 간 협의체	氣候變化-政府-協議體
interhourly variability	시시변동성, 경시변동성	時時變動性, 經時變動性
intermediate fallout	중거리낙진	中距離落塵
intermediate form	중간형식	中間形式
intermediate frequency	중간주파수	中間周波數
intermediate frequency amplifier	중간주파수증폭기	中間周波數增幅器
intermediate hybrid	중간잡종	中間雜種
intermediate insertion	중간입력	中間入力
intermediate ion	중간이온	中間-
intermediate layer	중간층	中間層
intermediate ocean-statistical atmosphere coupled model	중간단계 해양-통계대기 접합모형	中間段階海洋統計大氣 接合模型
intermediate plant	중간식물	中間植物
intermediate scale	중간규모	中間規模
intermediate standard time	중간일기도시각	中間日氣圖時刻
intermediate synoptic observation	중간종관관측	中間綜觀觀測
intermittent rain	단속성비	斷續性-
intermittent stream	간헐천	間歇川
intermittent/integrated sampling	간헐적/통합적 시료채취	間歇的/統合的試料採取
intermolecular force	분자 간 힘	分子間-
inter-monthly pressure variation	경월압력변동	經月壓力變動
inter-monthly temperature variation	경월온도변동	經月溫度變動
inter-monthly variability	경월변동성	經月變動性
inter-monthly variation	경월변동	經月變動
internal atmospheric gravity wave	내부중력파	內部重力波
internal boundary	내부경계면	內部境界面
internal boundary layer	내부경계층	內部境界層
internal calibration target	내부보정목표	內部補正目標

internal energy	내부에너지	內部-
internal friction	내부마찰	內部摩擦
internal gravity wave	내부중력파	內部重力波
internal latent heat	내부 숨은 열	內部-熱
internal mixture	내부혼합	內部混合
internal structure of regime	양식 내부구조	樣式內部構造
internal water circulation	내부물순환	內部-循環
internal wave	내부파	內部波
international aerodrome forecast	국제비행장예보	國際飛行場豫報
International Airways Volcano Watch	국제항공로화산감시	國際航空路火山監視
international analysis code	국제분석전문부호	國際分析電文符號
international candle	국제촉광	國際燭光
International Civil Aviation Agreement	국제민간항공조약	國際民間航空條約
International Civil Aviation Agreement Annex	국제민간항공조약부속서	國際民間航空條約附屬書
International Civil Aviation Organization	국제민간항공기구	國際民間航空機構
International Cloud Abbreviations	국제운형약어	國際雲形略語
international cloud atlas	국제구름도감	國際-圖鑑
international cloud year	국제구름관측년	國際-觀測年
international code	국제기상코드, 국제기상전문형식	國際氣象-, 國際氣象電文型式
International Date Line	국제날짜선	國際-線
International Geophysical Cooperation	국제지구물리협력	國際地球物理協力
International Geophysical Year	국제지구물리관측년	國際地球物理觀測年
international index numbers	(기상)국제지점번호	(氣象)國際地點番號
International Maritime Organization	국제해사기구	國際海事機構
International Meteorological Organization(IMO)	국제기상기구(WMO의 전신)	國際氣象機構
International Meteorological Telecommunication Network(IMTN)	국제기상통신망	國際氣象通信網
International Meteorological Teleprinter Network	국제기상텔레프린터망	國際氣象-網
international period	국제관측기간	國際觀測期間
international polar year	국제극관측년	國際極觀測年
international pyrheliometric scale	국제직달일사척도	國際直達日射尺度
International Quiet Sun Year	국제정온태양관측년	國際靜穩太陽觀測年
international reference atmosphere	국제기준대기	國際基準大氣
International Satellite Broadcast	국제위성통신체제(미국워싱턴)	國際衛星通信體制
international signs	국제부호	國際符號
international standard atmosphere	국제표준대기	國際標準大氣
international symbol	국제부호, 국제기호	國際符號, 國際記號
international synoptic code	국제종관전문부호, 국제종관기상전문형식	國際綜觀電文符號, 國際綜觀氣像電文型式
international synoptic surface observation code	국제종관지상관측전문부호	國際綜觀地上觀測電文符號
international table calorie	국제테이블칼로리	國際-
international temperature scale	국제온도척도	國際溫度尺度
International Time Bureau	국제시간국	國際時間局

International Union of Geodesy and Geophysics	국제측지학지구물리학연합	國際測地學地球物理學聯合
International Units	국제단위	國際單位
international visual storm warning signal	국제폭풍경보신호	國際暴風警報信號
international weather code	국제기상전문부호,	國際氣象電文符號,
	국제기상전문형식	國際氣象電文型式
International Years of the Quiet Sun	국제정온태양관측년	國際靜穩太陽觀測年
interpluvial	간우기(의)	間雨期
interpolation	내삽(법)	內揷(法)
interpolation formula	내삽공식	內揷公式
interpolation scheme	내삽방안	內揷方案
interpretation phase	해석위상	解釋位象
interpreted program	FSW의 실행프로그램	-實行-
interpulse period	펄스간 시간간격	-時間間隔
inter-quartile range	4분위범위	四分位範圍
interrogation recording and location system	조회기록위치확인시스템	照會記錄位置確認-
interrogation sign	검색부호	檢索符號
interseasonal variability	계절사이변동성	季節-變動性
intersect	교차	交差
intersection	교집합	交集合
intersequential variability	서열변동율	序列變動率
interstadial	아온난기, 아간빙기	亞溫暖期, 亞間氷期
interstitial	틈새형	-型
interstroke change	전격사이변화	電擊-變化
intertropical confluence zone	열대합류대	熱帶合流帶
intertropical convergence zone	열대수렴대	熱帶收斂帶
intertropical front	열대전선	熱帶前線
interval	간격, 구간	間隔, 區間
interval estimation	구간추정	區間推定
interval scale image	구간규모영상	區間規模映像
interveinal chlorosis	잎맥간황백화	-脈間黃白化
interveinal necrosis	맥간괴사	脈間壞死
intortus	얽힌 구름	
intracloud discharge	구름속방전	-放電
intraseasonal oscillation	계절안진동	季節內振動
intraseasonal variability	계절안변동성	季節內變動性
intraseasonal variation	계절안변동	季節內變動
intrusion	침입, 관입, 침투	侵入, 貫入, 浸透
inundation	홍수	洪水
invading air	침입공기	侵入空氣
invariable	불변	不變
invariable stratum of soil temperature	지온불변층	地溫不變層
invasion	침입	侵入
invasion of air	공기침입	空氣侵入

inverse function	역함수	逆函數
inverse Lagrangian	역라그랑지	逆-
inverse matrix method	역행열방법	逆行列方法
inverse model	역산모형	逆算模型
inverse problem	역산문제	逆算問題
inverse square law	역제곱법칙	逆-法則
inverse square law of distance	거리의 역자승법칙	距離逆自乘法則
inverse square rule	역제곱법칙	逆-法則
inverse variation	역변동	逆變動
inversion	역전	逆轉
inversion base	역전밑면	逆轉-面
inversion cloud	역전구름	逆轉-
inversion fog	역전안개	逆轉-
inversion haze	역전연무	逆轉煙霧
inversion layer	역전층	逆轉層
inversion nose	역전돌출	逆轉突出
inversion of precipitation	강수역전	降水逆轉
inversion of rainfall	강우역전	降雨逆轉
inverted image	역영상	逆映像
inverted V-pattern	역V형	逆-型
investigative protocol	조사의정서	調査議定書
invierno	열대아메리카 우기	熱帶-雨期
inviscid fluid	비점성유체	非粘性流體
invisible radiation	비가시복사	非可視輻射
ion	이온	
ion accelerator	이온가속기	-加速器
ion-capture theory	이온포착설	-捕捉說
ion chromatograph	이온크로마토그래프	
ion cloud	이온구름	
ion column	이온기둥	
ion concentration(= ion density)	이온농도	-濃度
ion counter	이온계수기	-計數器
ion density	이온밀도	-密度
ion exchange process	이온교환과정	-交換科程
ion exchange resin	이온교환수지	-交換樹脂
ionic bond	이온결합	-結合
ionic conduction	이온전도	-傳導
ionic crystal	이온결정	-結晶
ionic current	이온전류	-電流
ionic equation	이온반응식	-反應式
ionic mobility	이온이동도	-移動度
ionic model	이온모형	-模型
ionization	이온화	-化
ionization chamber	이온화상자	-化箱子

ionization-dissociation interaction	이온화-해리 교환작용	-化解離交換作用
ionization equilibrium	이온화평형	-化平衡
ionization layer	이온층	-層
ionizing radiation	전리방사선	電離放射線
ion mean life	이온평균수명	-平均壽命
ion mobility	이온이동도	-移動度
ionogram	이오노그램	
ionopause	이온권계면	-圈界面
ion pair	이온쌍	-雙
ionosonde	이온존데	
ionosphere	이온권, 전리권	電離圈
ionospheric layer	이온층, 전리층	電離層
ionospheric recorder	이온층기록계	-層記錄計
ionospheric storm	이온층보라, 이온층스톰	-層-
ionospheric tide	이온층조석	-層潮汐
ionospheric tilt	이온층기울기	-層-
ionospheric trough	이온층골	-層-
ionospheric turbulence	이온층난류	-層亂流
ionospheric wind	이온층바람	-層-
iridescence	무지갯빛	
iridescent altocumulus	무지갯빛 고적운	-高積雲
iridescent cloud	무지갯빛 구름, 채운	彩雲
irisation	무지갯빛 현상	-現象
Irminger current	이르밍거해류	-海流
iron wind	쇠바람(중앙아메리카 북동풍)	
irradiance	복사조도	輻射照度
irradiation	쬐임, 복사조사	輻射照射
irrational expression	무리식	無理式
irrational function	무리함수	無理函數
irrational number	무리수	無理數
irregular crystal	불규칙결정	不規則結晶
irregular echo	불규칙에코	不規則-
irregular front	불규칙전선	不規則前線
irregular high-frequency type atmospherics	불규칙고주파형공중전기	不規則高周波型空中電氣
irregular reflection	난반사	亂反射
irregular regime	난류영역, 비정규체제	亂流領域, 非正規體制
irreversible	비가역적	非可逆的
irreversible engine	비가역기관	非可逆機關
irreversible phenomenon	비가역현상	非可逆現象
irreversible process	비가역과정	非可逆過程
irreversible reaction	비가역반응	非加逆反應
irrigation	물대기	
irrigation facilities	관개시설	灌漑施設
irrotational motion	비회전운동	非回轉運動

isabnormal	등이상선	等異常線
isabnormal line	등편차선	等偏差線
isactine	등일사강도선	等日射强度線
isalea	등일사선	等日射線
isallo-	등변화	等變化
isallobar	등기압변화선	等氣壓變化線
isallobaric	등기압변화(의)	等氣壓變化
isallobaric analysis	등기압변화분석	等氣壓變化分析
isallobaric chart	등기압변화도	等氣壓變化圖
isallobaric gradient	등기압변화경도	等氣壓變化傾度
isallobaric high	기압상승중심	氣壓上昇中心
isallobaric low	기압하강중심	氣壓下降中心
isallobaric maximum	등기압변화최대	等氣壓變化最大
isallobaric minimum	등기압변화최소	等氣壓變化最小
isallobaric wind	변압풍	變壓風
isallobaric wind convergence	변압풍수렴	變壓風收斂
isallohypse	등고도변화선	等高度變化線
isallohypsic wind	변고풍	變高風
isallotherm	등온도변화선	等溫度變化線
isametric	등편차선	等偏差線
isanabat	등연직속도선	等鉛直速度線
isanabation	등연직속도선	等鉛直速度線
isanakatabar	등기압교차선	等氣壓較差線
isanakatabaric chart	등기압교차선도	等氣壓較差線圖
isanemone	등풍속선	等風速線
isanomal	등편차선	等偏差線
isanomalous	등편차(의)	等偏差
isanomalous line	등편차선	等偏差線
isanomaly	등편차선	等偏差線
isanthesic line	등개화선	等開花線
isarithm	등치선	等值線
isasteric surface	등적면	等積面
isaurore	극광등빈도선	極光等頻度線
isentrope	등온위선	等溫位線
isentropic analysis	등온위면분석	等溫位面分析
isentropic atmosphere	등온위대기	等溫位大氣
isentropic cap	등온위덮개	等溫位-
isentropic change	등온위변화	等溫位變化
isentropic chart	등온위도	等溫位圖
isentropic condensation level	등온위응결고도	等溫位凝結高度
isentropic convergence	등온위수렴	等溫位收斂
isentropic divergence	등온위발산	等溫位發散
isentropic mixing	등온위혼합	等溫位混合
isentropic motion	등온위운동	等溫位運動

isentropic process	등온위과정, 등엔트로피과정	等溫位過程
isentropic sheet	등온위판	等溫位板
isentropic surface	등온위면분석	等溫位面分析
isentropic thickness chart	등온위두께선도, 등온위선도	等溫位-線圖
isentropic thunderstorm	등온위뇌우	等溫位雷雨
isentropic weight chart	등온위무게선도	等溫位-線圖
iseoric line	등연온도교차선	等年溫度較差線
iso-abnormal	등이상선	等異常線
iso-amplitude	등진폭선	等振幅線
isoanabaric center	등기압상승중심	等氣壓上昇中心
isoanth	등화기선	等花期線
isoatmic	등증발선	等蒸發線
isoaurora	등오로라선	等極光線
isobar	등압선	等壓線
isobar drawing	등압선묘화	等壓線描畵
isobaric	등압(의)	等壓
isobaric analysis	등압면분석	等壓面分析
isobaric channel	등압통로	等壓通路
isobaric chart	등압도	等壓圖
isobaric condition	등압조건	等壓條件
isobaric contour chart	등압고도도	等壓高度圖
isobaric cooling	등압냉각	等壓冷却
isobaric divergence	등압발산	等壓發散
isobaric equivalent potential temperature	등압상당온위	等壓相當溫位
isobaric equivalent temperature	등압상당온도	等壓相當溫度
isobaric isosteric solenoid	등압등적 솔레노이드	等壓等積-
isobaric isosteric tube	등압등적관	等壓等積管
isobaric line	등압선	等壓線
isobaric motion	등압운동	等壓運動
isobaric process	등압과정	等壓過程
isobaric surface	등압면	等壓面
isobaric temperature change	등압온도변화	等壓溫度變化
isobaric thickness chart	등압두께선도	等壓-線圖
isobaric topography	등압높낮이	等壓-
isobaric tube	등압관	等壓管
isobaric vorticity	등압소용돌이도	等壓-度
isobaric wet-bulb potential temperature	등압습구온위	等壓濕球溫位
isobaric wet-bulb temperature	등압습구온도	等壓濕球溫度
isobarometric line	등압선	等壓線
isobars	동중체	同重體
isobar type	기압배치형	氣壓配置型
isobase	등기준선	等基準線
isobath	등심선	等深線
isobathytherm	등심온도선	等深溫度線

isobront	등뇌우선	等雷雨線
isoceraunic line(= isokeraunic line)	등뇌우빈도선	等雷雨頻度線
isochasm	등오로라빈도선	等-頻度線
isocheim	등한선	等寒線
isochion	등설선	等雪線
isochore line	등수심선	等水深線
isochoric change	등적변화	等積變化
isochoric process	등적과정	等積過程
isochrone	등시선	等時線
isochronism	등시성	等時性
isochronous line of the blooming day	개화전선	開花前線
isoclimatic line	등기후선	等氣候線
isoclinal fold	등사습곡	等斜褶曲
isoclinal line(= isoclinic line)	등복각선, 등기울기선	等伏角線
isocline	등기울기선	等-線
iso-coefficient	등계수선	等係數線
isocorrelation	등상관선	等相關線
isocryme	최한기등수온선	最寒期等水溫線
iso-D	등D값선	等-線
isodense	등밀도선	等密度線
iso-dewpoint deficit line	등습수선	等濕數線
isodiaphore	등월변화선	等月變化線
isodop	등도플러속도선	等-速度線
isodrosotherm	등이슬점선, 등노점선	等露點線
isodynamic	등력선	等力線
isodynamical surface	등동력면	等動力面
iso echo	등에코선	等-線
isoecho line	등에코선	等-線
isoelectronic ion	등전자이온	等電子-
isoeral	등춘선	等春線
isogeotherm	등지온선	等地溫線
isogon	등풍향선	等風向線
isogonal line	등각선	等角線
isogonal map	등각지도	等角地圖
isogonic line	등향선	等向線
isogradient	등경도선	等傾度線
isogram	등치선	等值線
isohaline	등염분선	等鹽分線
isoheight	등고선	等高線
isohel(= isohelic)	등일조선	等日照線
isohion	등적설깊이선, 등설일수선, 등설선고도선	等積雪-線, 等積日數線, 等積線高度線
isohume	등습도선	等濕度線
isohyet	등강수량선	等降水量線

isohyetal	등강수량선(의)	等降水量線
isohyetal method	등우선법	等雨線法
isohyets	등강수량선	等降水量線
isohygrometric line	등습도선	等濕度線
isohyle	삼림선	森林線
isohypse	등고선	等高線
isohypsic chart	등고선도	等高線圖
isohypsic surface	등고면	等高面
isokatabaric center	등강압중심	等降壓中心
isokatanabar	등기압교차선	等氣壓較差線
isokeraunic	등뇌전빈도	等雷電頻度
isokeraunic line	등뇌전빈도선	等雷電頻度線
isokinetic	등풍속선	等風速線
isokinetic sampling	등속표본채취	等速標本採取
isokurtic curve	등첨도곡선	等尖度曲線
isolated echo	고립에코	孤立-
isolated showers	가끔(때때로)소나기, 고립소나기	孤立-
isolated system	고립계	孤立系
isolation	절연, 고립, 격리	絶緣, 孤立, 隔離
isoline	등치선	等値線
isoline of dewpoint deficit	등이슬점차선, 등노점차선	等露點差線
isomenal	월평균등치선	月平均等値線
isomer	등비율선	等比率線
isomeric value	강수백분율	降水百分率
isometeoric line	기상요소등치선	氣象要素等値線
isometeorograde	등기상도선	等氣象度線
isometropal	등추선	等秋線
isoneph	등운량선	等雲量線
isoombre	등증발선	等蒸發線
isopag	등동결기선	等凍結期線
isoparallage	등연온도교차선	等年溫度較差線
isopectrics	등결빙일선	等結冰日線
isophane	등발현선	等發現線
isophote	등광도선	等光度線
isophtor	등재해선	等災害線
isopipteses	동시출현선	同時出現線
isopleth	등치선	等値線
isopleth of brightness	밝기등치선	-等値線
isopluvial	등우량선	等雨量線
isopods	등각류	等脚類
isopotential	등퍼텐셜선	等-線
isoprene	이소프렌	
isopycnic	등밀도(선)	等密度(線)

isopycnic line	등밀도선	等密度線
isopycnic surface	등밀도면	等密度面
isoryme	최한월 월평균등온선	最寒月月平均等溫線
isoseismal line	등진도선	等震度線
isoshear	등시어선	等-線
isostasy	지각평형	地殼平衡
isostere	등비부피선	等比-線
isostere surface	등비적면	等比積面
isosteric	등비부피	等比-
isosteric surface	등비부피면	等比-面
isotac	등해동선	等解凍線
isotach	등풍속선	等風速線
isotach analysis	등풍속선분석	等風速線分析
isotach chart	등풍속선도	等風速線圖
isotalant	등연온도교차선	等年溫度較差線
isothene	등기압평형선	等氣壓平衡線
isothere(= isotheral)	등더위선	等-線
isotherm	등온선	等溫線
isothermal	등온	等溫
isothermal atmosphere	등온대기	等溫大氣
isothermal change	등온변화	等溫變化
isothermal compression	등온압축	等溫壓縮
isothermal equilibrium	등온평형	等溫平衡
isothermal expansion	등온팽창	等溫膨脹
isothermal layer	등온층	等溫層
isothermal process	등온과정	等溫過程
isothermal surface	등온면	等溫面
isothermobath	등수온선	等水溫線
isothermobrose	여름평균우량등치선	-平均雨量等值線
isothermohypse	등온선	等溫線
isotherm ribbon	등온선밀집대	等溫線密集帶
isothickness advection line	등두께이류선	等-移流線
isothickness line	등두께선	等-線
isothyme	등증발량선	等蒸發量線
isotimic	정시등치	定時等值
isotimic line	정시등치선	定時等值線
isotimic surface	정시등치면	定時等值面
isotonicity	등장력선	等張力線
isotonic solution	등장액	等張液
isotope	동위원소	同位元素
isotopic analysis	동위원소분석	同位元素分析
isotopic atom	동위원소원자	同位元素原子
isotopic fractionation	동위원소분별	同位元素分別
isotopic tracer	동위원소추적자	同位元素追跡子

isotropic	등방성(의)	等方性
isotropic antenna	등방성안테나	等方性-
isotropic radiation	등방성복사	等方性輻射
isotropic radiator	등방성전파발사기, 등방성복사체	等方性電波發射機, 等方性輻射體
isotropic reflection	등방성반사	等方性反射
isotropic scattering	등방성산란	等方性散亂
isotropic turbulence	등방성난류	等方性亂流
isotropy	등방성	等方性
isovel	등풍속선	等風速線
isovorticity line	등와도선	等渦度線
issuing time of aviation forecast	항공예보발표시각	航空豫報發表時刻
iteration	반복, 반복법	反復, 反復法
ivory point	상아침	象牙針

J, j

영문	한글	한자
Jacobian	야코비안	
Jacob's ladder	야곱사다리	
January thaw	1월 해빙	一月解氷
Japan current	일본해류	日本海流
Jefferson index	제퍼슨지수	-指數
jet	제트(류)	-(流)
jet axis	제트축	-軸
jet core	제트핵	-核
jet effect wind	제트효과바람	-效果-
jet flow	제트류	-流
jetlet	작은 제트	
jet streak	제트스트리크	
jet streak interaction	제트스트리크 상호작용	-相互作用
jet stream	제트류(제트기류)	-流(-氣流)
jet stream axis	제트류축	-流軸
jet stream cirrus	제트류권운	-流卷雲
jet stream cloud	제트류구름	-流-
jet stream core	제트류중심부	-流中心部
jet stream flight	제트류비행	-流飛行
jet stream front	제트류전선	-流前線
Jevons effect	제번스효과	-效果
jimmycane	허리케인	
Jimsphere	짐기구	-氣球
Johnson-Williams liquid water meter	존슨-윌리엄스 수액계	-水液計
Joint Airborne Training Weather	합동공정작전훈련기상	合同空挺作戰訓練氣象
Joint Typhoon Warning Center	합동태풍경보센터	合同颱風警報-
Jordan sunshine recorder	조르단 일조계	-日照計
joule	줄	
Joule's constant	줄상수	-常數

Joule's law	줄법칙	-法則
Jovian planet	목성형행성	木星型行星
J scope	J 스코프	
jump	도약, 계단적 변화, 돌연변화	跳躍, 階段的變化, 突然變化
junction streamer	분기스트리머	分岐-
Junge aerosol layer	융에 에어로졸층	-層
Jupiter	목성	木星
Jurassic climate	쥐라기기후	-期氣候
Jurine's phenomenon	주리나현상 (북아메리카 사막 내의 광학현상)	-現象
Jurisdiction Airspace weather	관할공역기상	管轄空域氣象
Jurisdiction Sea Zone weather	관할수역기상	管轄水域氣象
jury problem	배심문제	陪審問題
juvenile water(= magmatic water)	초생수	初生水
J-W meter	J-W 계	-計

J

K, k

영문	한글	한자
kaavie	카비(스코틀랜드의 줄기찬 눈)	
kal Baisakhi	칼바이사히(방글라데시와 벵갈 지방에서 남서몬순이 시작할 무렵 잠깐 동안 많은 먼지를 몰고 오는 스콜)	
kaléma	칼레마(겨울철에 아프리카의 기니아 연안에 밀어 닥치는 대단히 거센 파도)	
Kálmán-Bucy filter	칼만-부시 필터	
Kalman filter	칼만필터	
Kalman filtering	칼만필터링	
Kames	케임	
kaolinite	캐올리나이트	
karaburan	카라부란	
Kármán constant	카르만상수	-常數
Kármán temperature	카르만온도	-溫度
Kármán vortex street	카르만소용돌이행렬	-行列
katabaric	기압감소	氣壓減少
katabatic	냉기하강류(의)	冷氣下降流
katabatic front	활강전선	滑降前線
katabatic wind	활강바람	滑降-
katafront	활강전선	滑降前線
kata-isallobar	등기압하강선	等氣壓下降線
katallobar	기압하강선	氣壓下降線
katallobaric center	기압하강중심	氣壓下降中心
kataphalanx	한냉전선면	寒冷前線面
katathermometer	카타온도계	-溫度計
kavaburd(= cavaburd)	카바버드(영국 북동쪽 셰틀랜드 제도에서 큰 눈을 가리키는 말)	
K-band	K-밴드	
Keetch-Byrum Drought Index	키치-바이람 가뭄지수	-指數
kelsher	켈셔(영국의 격심한 강우)	

kelvin	켈빈(절대온도단위)	
Kelvin-Celsius conversion	켈빈-섭씨 변환	-變換
Kelvin-Helmholtz billows	켈빈-헬름홀츠 물결구름	
Kelvin-Helmholtz instability	켈빈-헬름홀츠 불안정	-不安定
Kelvin-Helmholtz wave	켈빈-헬름홀츠 파	-波
Kelvin scale	켈빈눈금, 절대온도눈금	絶對溫度-
Kelvin's circulation theorem	켈빈순환정리	-循環定理
Kelvin's equation	켈빈식	-式
Kelvin temperature	켈빈온도, 절대온도	-溫度, 絶對溫度
Kelvin temperature scale	켈빈온도눈금, 절대온도눈금	絶對溫度-
Kelvin wave	켈빈파	-波
Kennelly-Heaviside layer	케널리-헤비사이드 층	-層
Kepler's equation	케플러방정식	-方程式
Kepler's laws	케플러법칙	-法則
keraunograph(= ceraunograph)	뇌전기록계	雷電記錄計
Kern counter(= dust counter)	핵계수기	核計數器
kernel	핵	核
kernel ice	굳은 상고대	-霜高臺
Kern's arc	케른호	-弧
kettle	구혈	甌穴
Kew (pattern) barometer	큐형기압계	-型氣壓計
key day(= control day)	시작일	始作日
khamsin(= khamseen, chamsin)	캄신(북아프리카 및 아라비아 반도에서 늦겨울 또는 초봄에 부는 남풍계열의 뜨겁고 건조한 먼지바람)	
killer storm	살인폭풍	殺人暴風
killing freeze	혹한, 된추위, 엄한	酷寒, 嚴寒
killing frost	된서리	
kilogram calorie	물의 비열(1kg의 물의 온도 1도 올리는 데 필요한 열)	-比熱
kilopascal	킬로 파스칼	
K index	K 지수(뇌우의 발달가능성을 나타내는 지수)	-指數
kinematic analysis	운동학적 분석	運動學的分析
kinematic boundary condition	운동학적 경계조건	運動學的境界條件
kinematic condition	운동학적 조건	運動學的條件
kinematic flux	운동학적 플럭스	運動學的-
kinematic front	운동전선	運動前線
kinematic method	운동학방법	運動學方法
kinematics	운동학, 동역학	運動學, 動力學
kinematic viscosity	동점성	動粘性
kinetheodolite	촬영경위(의)	撮影經緯
kinetic energy	운동에너지	運動-
kinetic friction	동마찰	動摩擦
kinetic instability	위치불안정도	位置不安定度

kinetics	동역학	動力學
kinetic temperature	운동온도	運動溫度
kinetic theory of gas	기체분자운동론	氣體分子運動論
kinetic theory of molecule	분자운동론	分子運動論
kinks on isopleths	등치선꺾임	等值線-
Kirchhoff's law	키르히호프 법칙	-法則
Kirchhoff's radiation law	키르히호프 복사법칙	-輻射法則
kite-ascent	연탐측	鳶探測
kite balloon	연기구	鳶氣球
kite observation	연관측	鳶觀測
kite sounding	연탐측	鳶探測
klydonograph	클리도노그래프	
Klystron	클라이트론	
Knollenberg probe	크놀렌베르크 프로브	
knot	노트	
Knudsen's tables	크누센 표	-表
Kolmogoroff cascade	콜모고로프 캐스케이드	
Kolmogoroff's similarity hypothesis	콜모고로프 유사가설	-類似假說
Kolomogrov spectrum	콜로모그로브 스펙트럼	
kona cyclone	코나저기압(하와이제도에 부는 남서풍을 일으키는 한랭핵 저기압)	-低氣壓
kona storm	코나스톰	
konimeter(= conimeter)	먼지채집기	-採集器
koniology(= coniology)	먼지학	-學
koniscope(= coniscope)	먼지지시기	-指示機
konisphere(= staubosphere)	먼지권	-圈
Köppen classification	쾨펜 구분	
Köppen's classification of climates	쾨펜 기후분류	-氣候分類
Köppen's climatic classification	쾨펜 기후분류	-氣候分類
Köppen-Supan line	쾨펜-수판선	-線
Korea Meteorological Administration	(한국)기상청	氣象廳
Korean rain gauge	측우기	測雨器
Koschmieder's law	코슈미더법칙	-法則
Koschmieder's visibility formula	코슈미더시정공식	-視程公式
Krakatoa wind	크라카토아 바람	
kriging	크리깅	
kryptoclimate	실내기후	室內氣候
kryptoclimatology	실내기후학	室內氣候學
Kurihara grid	구리하라 격자	-格子
Kuroshio	쿠로시오	
Kuroshio countercurrent	쿠로시오반류	-反流
Kuroshio extension	쿠로시오속류	-屬流
Kuroshio system	쿠로시오해류계	-海流系
kurtosis	첨도	尖度

K

Kuso weather	구서훈련기상	驅鼠訓練氣象
K-wind	K-바람	
kytoon(= kite balloon)	계류기구	繫留氣球

L, l

영문	한글	한자
L/HRIT Generation Subsystem	LRIT 및 HRIT 자료생성 부시스템	-資料生成 副-
La Mama	라마마 현상	-現像
La Nina	라니냐	
la serpe	라설페(이탈리아의 에느타산 남단에 발생하는 길고 가느다란 조각모양의 구름)	
labile	불안정(의)	不安定
labile energy	불안정에너지	不安定-
lability	불안정도	不安定度
laboratory	실험실, 연구소	實驗室, 研究所
Labrador current	라브라도해류	-海流
lacunaris	벌집(구름)	
lacunosus	벌집(구름)	
lacustrine sediment	호소퇴적물	湖沼堆積物
Lafond's table	라퐁드표	-表
lag	지연	遲延
lag coefficient	지연계수	遲延係數
lag correlation	지연상관	遲延相關
lag of aneroid barometer	아네로이드기압계 지연	-遲延
lag position	지연위치	遲延位置
lag time	지연시간	遲延時間
Lagrangian coordinates	라그랑지좌표	-座標
Lagrangian correlation	라그랑지상관	-相關
Lagrangian differentiation	라그랑지미분	-微分
Lagrangian equation	라그랑지방정식	-方程式
Lagrangian method	라그랑지방법	-法
Lagrangian similarity	라그랑지유사	-類似
Lagrangian wave	라그랑지파	-波
lake breeze	호수바람	湖水-

lake effect	호수효과	湖水效果
lake effect snowstorm	호수효과 눈스톰	湖水效果-
lake evaporation	호수증발	湖水蒸發
lake level	호수면	湖水面
lake level variation	내수면변화, 호수면변화	內水面變化, 湖水面變化
lake temperature(= lake surface temperature)	호수면온도	湖水面溫度
lamb blast	어린 양의 풍설(잉글랜드에서 봄에 내리는 눈)	-羊-風雪
Lambert	람베르트(밝기 단위)	
Lambert conformal	람베르트등각	-等角
Lambert conformal conic projection	람베르트등각원뿔 투영법	-等角圓-投影法
Lambert conformal projection	람베르트등각도법	-等角圖法
Lambertian reflector	람베르트반사경, 람베르트반사체	-反射鏡, -反射體
Lambert's formula	람베르트공식	-公式
Lambert's (cosine) law	람베르트(코사인)법칙	-法則
Lambert's law of absorption(→Bouguer's law)	람베르트흡수법칙	-吸收法則
Lambrecht's polymeter	람브레흐트 다용도습도계	-多用途濕度計
lamellar vector	기울기벡터	
lamella structure	층상구조	層狀構造
laminar boundary layer	층류경계층	層流境界層
laminar flow	층류	層流
laminar sub-layer	층류하층	層流下層
laminary flow	층흐름	層-
laminated current	층류	層流
land and sea breeze	해륙풍	海陸風
land barometer	육상기압계	陸上氣壓計
land breeze	육풍	陸風
land breeze circulation	육풍순환	陸風循環
land breeze front	육풍전선, 육지바람전선	陸風前線
land cover	지표피복, 토지피복	地表被服, 土地被覆
Land Cover Classification System	지표피복분류체계, 토지피복분류체계	地表被服分類體系, 土地被覆分類體系
land cover mapping	지표피복매핑, 토지피복지도제작	地表被服-, 土地被覆地圖製作
landfall	태풍의 상륙	颱風上陸
landfast ice	접안얼음	接岸-
land fog	육지안개	陸地-
land forecast	육상예보	陸上豫報
land ice	육지얼음	陸地-
Landing Area weather	착륙지역기상	着陸地域氣象
landing force objective weather	상륙군 목표기상	上陸軍目標氣象
landing forecast	착륙예보	着陸豫報
landing roll	착륙활주거리	着陸滑走距離

landing speed	착륙속도	着陸速度
landing weather	착륙기상	着陸氣象
land-lash	랜드래시(잉글랜드 지방의 거센폭풍우)	
landmark	경계표지	境界標識
LANDSAT	랜드새트	
Landsat program	랜드새트프로그램	
landscape climatology	경관기후학	景觀氣候學
landscape complexity errors	경관복잡성 오차	景觀複雜性誤差
landscape ecology	경관생태계	景觀生態系
land sky	땅하늘	
landslide	산사태	山沙汰
land slide erosion	붕괴침식	崩壞浸蝕
land sliding erosion	붕괴침식	崩壞浸蝕
landslip wind	산사태바람	山沙汰-
landspout	육지용오름	陸地龍-
land station	육상관측소, 육지관측소	陸上觀測所, 陸地觀測所
land surface	지표면	地表面
land surface roughness	지표거칠기, 지표면조도	地表面粗度
land surface temperature	지표온도	地表溫度
land use type	토지이용유형	土地利用類型
land weather warning	육상특보	陸上特報
Langevin ion	랑주뱅이온	
Langley	랭글리(cal/cm^2)	
Langmuir chain reaction	랭뮤어연쇄반응	-連鎖反應
Langmuir probe	랭뮤어탐지기	-探知機
Laplace equation	라플라스방정식	-方程式
Laplace operator	라플라스연산자	-演算子
Laplace transform	라플라스변환	-變換
Laplacian operator	라플라스연산자	-演算子
Laplacian speed of sound	라플라스음속	-音速
lapse limit(= tropopause)	대류권계면	對流圈界面
lapse line	감률선	減率線
lapse rate	감률	減率
lapse rate of temperature	온도감률	溫度減率
large amplitude	대진폭	大振幅
large calorie	큰 칼로리	
large evaporimeter	대형증발계	大型蒸發計
large halo	큰 햇무리	
large ion	큰 이온	
large nucleus	큰 핵	-核
large particle	큰 입자	-粒子
large scale	대규모	大規模
large scale atmospheric processes	대규모대기과정	大規模大氣過程
large scale disturbance	대규모요란	大規模搖亂

L

large scale flow	대규모흐름	大規模-
large scale lifting	대규모상승	大規模上昇
large scale mixing	대규모혼합	大規模混合
large scale motion	대규모운동	大規模運動
large scale weather situation	대규모일기상황	大規模日氣狀況
larry	래리(영국의 템스강 하구에 생기는 짙은 안개)	
Larsen ice shelf	라르센 빙붕	-氷棚
laser	레이저	
laser ceilometer	레이저운고계	-雲高計
Laser Detection and Ranging	레이저측정거리	-測定距離
laser-light amplification 　by stimulated emission of radiation	모의복사방출에 의한 레이저광 증폭	模擬輻射放出-光增幅
laser pulse ceilometer	레이저운고계	-雲高計
laser radar	레이저레이더	
last frost	늦서리	
Last Glacial maximum	최대빙하기 말	最大氷河期末
Last Glacial termination	최후빙하기 종료 (최후빙하기에서 급격히 온난화가 시작되는 시기)	最後氷河期終了
late frost	만상	晩霜
late frost damage	만상해	晩霜害
late hoarfrost	늦서리	
Late Paleozoic climate	후기고생대 기후, 고생대말 기후	後期古生代氣候, 古生代末氣候
Late Quaternary climate	후기 제4기 기후	後期第四紀氣候
Late Quaternary-Holocene vegetation modeling	제4기말 홀로세 식생모델링	第四紀末-植生-
Late Quaternary megaflood	제4기말 거대홍수	第四紀末巨大洪水
Late Tertiary climate indicators	제3기 후기 기후지표	第三紀後紀氣候指標
latent energy	숨은 에너지	
latent evaporation	잠재증발	潛在蒸發
latent heat	숨은 열, 잠열	潛熱
latent heat of fusion	융해잠열	融解潛熱
latent heat of vaporization	기화잠열	氣化潛熱
latent heat transfer	잠열수송	潛熱輸送
latent instability	잠재불안정	潛在不安定
lateral boundary condition	측면경계조건	側面境界條件
lateral diffusion	옆확산	-擴散
lateral gustiness	옆돌풍도	-突風度
lateral mirage	옆신기루	-蜃氣樓
lateral mixing	옆혼합	-混合
lateral refraction	옆굴절	-屈折
lateral tangent arcs	옆접호	-接弧
laterite	홍토, 라테라이트	紅土
latest frost	끝서리	
latitude	위도	緯度

L

latitude effect	위도효과	緯度效果
latitude-longitude grid	위도-경도 격자	緯度-經度格子
latitudinal fluctuation	남북변동, 위도변동	南北變動, 緯度變動
lattice point	격자점	格子點
Launch and Early Operation Phase	위성발사 및 초기	衛星發射-初期 軌道進入
	궤도진입 운영단계	運營段階
launching	발사, 비양(기구)	發射, 飛揚
launch profile	위성발사 연직 대기분포	衛星發射鉛直大氣分布
laurence	아지랑이	
Laurentide ice sheet	로렌타이드빙상	-氷床
law of absolute vorticity conservation	절대와도보존법칙	絕對渦度保存法則
law of acceleration	가속도법칙	加速度法則
law of clockwise screw	오른나사법칙	-法則
law of conservation	보존법칙	保存法則
law of constant final yield	최종수량일정의 법칙	最終收量一定法則
law of diffusion	확산법칙	擴散法則
law of dynamical similarity	동역학적 유사법칙	動力學的類似法則
law of energy conservation	에너지보존법칙	-保存法則
law of error	오차법칙	誤差法則
law of exponent	지수법칙	指數法則
law of flow visualization	흐름의 가시화법	-可視化法
law of gaseous reaction	기체반응법칙	氣體反應法則
law of heat conservation	열량보존법칙	熱量保存法則
law of inertia	관성법칙	慣性法則
law of mass conservation	질량보존법칙	質量保存法則
law of momentum conservation	운동량보존법칙	運動量保存法則
law of motion	운동법칙	運動法則
law of partial pressure	분압법칙	分壓法則
law of potential vorticity conservation	와위보존칙	渦位保存則
law of storms	폭풍법칙, 스톰법칙	-法則
law of thermodynamics	열역학법칙	熱力學法則
law of universal gravitation	만유인력법칙	萬有引力法則
Lax-Richtmyer theorem	랙스-리히르마이어 정리	-定理
Lax's equivalence theorem	랙스동등이론	-同等理論
Lax's scheme	랙스방안	-方案
Lax-Wendorff differencing scheme	랙스-웬돌프 차분방안	-差分方案
layer	층	層
layer depth	층깊이	層-
layering	성층	成層
layer of compensation	보상층	補償層
layer of discontinuity	불연속층	不連續層
layover effect	설계효과	設計效果
L-band	L-밴드	
leaching	침출	浸出

L

lead	납	鉛
lead alkyl	알킬납	
leader	선도	先導
leader storke	선도낙뢰	先導落雷
leader streamer	선도방전	先導放電
lead time	선행시간	先行時間
leaf age	엽령	葉齡
leaf angle distribution	잎각도분류	-角度分類
leaf area density	엽면적밀도	葉面積密度
leaf area index	엽면적지수	葉面積指數
leaf canopy	임관	林冠
leaf cloud pattern	잎구름 패턴	
leaf component	잎성분	-成分
leaf-emergence	출엽	出葉
leaf moisture content	잎수분량	-水分量
leaf optical property	잎광학특성	-光學特性
leaf spectral signature	잎파장별 특성	-波長別特性
leaf stress	잎응력	-應力
leaf structural property	잎구조 특성	-構造特性
leaf weight ratio	엽중비	葉重比
leapfrog differencing	등넘기차분	-差分
leapfrog method	개구리뜀방법, 개구리도약법	-方法, -跳躍法
leap month	윤달	閏-
least common multiple(LCM)	최소공배수	最小公倍數
least square method	최소제곱근법	最小-根法
least squares	최소제곱	最小-
least squares interpolation	최소제곱내삽법	最小-內揷法
least-time track	최단시간항로	最短時間航路
least upper bound	상한	上限
lee cyclogenesis	풍하저기압발달	風下低氣壓發達
lee depression	풍하저기압	風下低氣壓
lee eddies	풍하맴돌이	風下-
lee side(= leeward)	내리바람쪽, 풍하측	風下側
lee tide	순풍조	順風潮
lee trough	내리바람기압골, 풍하기압골	-氣壓-, 風下氣壓-
leeward islands	풍하열도	風下列島
leeward side	풍하측	風下側
lee wave	내리바람파, 풍하파	風下波
lee wave region	풍하파지역	風下波地域
lee wave system	풍하파시스템	風下波-
leeway	풍하편류	風下偏流
left circular polarization	좌측 원형편파	左側圓形偏波
left-moving thunderstorm	좌이동뇌우	左移動雷雨
Legendre function	르장드르함수	-函數

legionnaires' disease	재향군인병	在鄉軍人病
Lenard effect	레나르트효과	-效果
length of tangent segment	접선길이	接線-
lenticular altocumulus cloud	렌즈고적운	-高積雲
lenticular cloud	렌즈구름	
lenticular cloud band	렌즈구름띠	
lenticularis	렌즈구름	
leptokurtic	급첨(의)	急尖
leptokurtosis	급첨도	急尖度
levanter	리밴터(남프랑스에서 지브롤타 해협까지 해안에 부는 동풍)	
level	수평면, 수준, 수준기, 층, 水平面, 水準, 水準器, 層 높이	
level of cloud base	운저고도	雲底高度
level of cloud top	운정고도	雲頂高度
level of escape	이탈고도	離脫高度
level of free convection	자유대류고도	自由對流高度
level of non-divergence	무발산고도	無發散高度
level surface(= geopotential surface)	고도면	高度面
lichenometry	지의계측(법)	地衣計測(法)
lidar	라이다	
lidar-light detection and ranging	라이다-광 탐지분류	-光 探知分類
lidar meteorology	라이다기상학	-氣象學
life cycle	일생주기	一生週期
lifetime	잔류시간	殘留時間
lifetime of satellite	위성수명	衛星壽命
life zone	생물대	生物帶
lift	치올림	
lifted index	치올림지수, 상승지수	上昇指數
lift index	치올림지수	-指數
lifting condensation level	치올림응결고도, 상승응결고도	上昇凝結高度
lifting detection	치올림탐지	-探知
lifting fog	치올림안개, 상승안개	上昇-
lifting force	상승력	上昇力
lifting temperature	치올림온도	-溫度
light absorbing aerosol	광흡수 에어로졸	光吸收-
light air	실바람(풍력계급 1)	
light breeze	남실바람(풍력계급 2)	
light climate(= illumination climate)	광기후	光氣候
light compensation point	광보상점	光補償點
light diffraction	빛회절	-廻折
light dispersion	광분산	光分散
light energy	빛에너지	
light exposure	빛노출	-露出

L

light filter	여광기	濾光器
light fog	엷은 안개	
light freeze	살짝 결빙	-結氷
light frost	무서리, 엷은 서리	無-
light intensity	빛강도	-强度
light interference	빛간섭	-干涉
light ion	가벼운 이온	
lightning	번개	
lightning arrester	피뢰침	避雷針
lightning channel	번갯길	
lightning conductor	피뢰침	避雷針
lightning counter	번개계수기	-計數器
lightning current	번개전류	-電流
lightning damage	번개피해	-被害
lightning detection	번개탐지	-探知
lightning diffusion	번개확산	-擴散
lightning discharge	번개방전	-放電
lightning echo	번개에코	
lightning flash	번개섬광	-閃光
lightning imaging sensor	번개영상센서	-映像-
lightning mapper	번개지도작성자	-地圖作成者
lightning mapping system	번개지도화시스템	-地圖化-
lightning protector plate	피뢰판	避雷板
lightning recorder	번개기록계	-記錄計
lightning rod	피뢰침	避雷針
lightning spectrum	번개스펙트럼	
lightning storm	번개보라, 번개스톰	
lightning strike	벼락, 낙뢰, 벽력	落雷, 霹靂
lightning stroke	전격	電擊
lightning suppression	번개억제	-抑制
light-of-the-night-sky(= airglow)	대기야광	大氣夜光
light pillar(= sun pillar)	빛기둥	
light pressure	광압	光壓
light rain	약한 비	
light rays	광선	光線
light saturation	광포화	光飽和
light saturation point	광포화점	光飽和點
light scattering diagram	빛산란선도	-散亂線圖
light scattering table	빛산란표	-散亂表
lightship	등대선	燈臺船
lightship code	등대선전보식	燈臺船電報式
lightship station	등대선관측소	燈臺船觀測所
light shower	벼락동반소나기	-同伴-
light snow	세설	細雪

light squall	약한 스콜	弱-
light use efficiency	광사용효율	光使用效率
light wave	광파	光波
light year	광년	光年
Liman Current	리만해류	-海流
limb radiance	주변복사휘도	周邊輻射輝度
limb radiance inversion radiometer	주변복사휘도 역산복사계	周邊輻射輝度逆算輻射計
limen	최저감광한계	最低感光限界
limestone scrubbing	석회석세정	石灰石洗淨
liminal contrast	문턱대조	門-對照
limit	극한, 한계	極限, 限界
limited-area forecast model	제한구역예보모형	制限區域豫報模型
limited fine-mesh model	제한상세격자망모형	制限詳細格子網模型
Limited Southern Hemisphere	제한남반구영역	制限南半球領域
limit of audibility	가청한계	可聽限界
limit of detection	검출한계	檢出限界
limit of error	오차한계	誤差限界
limit of heat	처서	處暑
limit of integration	적분한계	積分限界
limit of visibility	시정한계	視程限界
limiting angle	한계각	限界角
limiting point	극한점	極限點
limiting streamline	한계유선	限界流線
limnological meteorology	호소(생물)기상학	湖沼(生物)氣象學
limnology	호소학	湖沼學
line absorption	선흡수	線吸收
linear	선형	線形
linear acceleration	선형가속도	線形加速度
linear accelerator	선형가속기	線形加速器
linear computational instability	선형계산불안정	線形計算不安定
linear correlation	선형상관	線形相關
linear density	선밀도	線密度
linear dependence	일차종속	一次從屬
linear depolarization ratio	선편파감소비	線編波減小比
linear echo	선형에코	線形-
linear entropy scale	선형엔트로피 척도	線形-尺度
linear equation	선형방정식	線形方程式
linear expansion	선팽창	線膨脹
linear expression	일차식	一次式
linear field of motion	선형운동장	線形運動場
linear field of temperature	선형온도장	線形溫度場
linear function	선형함수	線形函數
linear inequality	일차부등식	一次不等式
linear instability	선형불안정도	線形不安定度

L

linearity	선형성	線形性
linearization	선형화	線形化
linearized differential equation	선형미분방정식	線形微分方程式
linear momentum	선형운동량	線形運動量
linear operator	선형연산자	線形演算子
linear polarization	선형편파	線形偏波
linear stability	선형안정성	線形安定性
linear term	일차항	一次項
linear theory	선형이론	線形理論
linear transformation	일차변환	一次變換
linear transition	선형전이	線形轉移
linear velocity	선속도	線速度
line broadening	선퍼짐	線-
line-by-line calculation	선대선계산	線對線計算
line cloud	선모양구름	線模樣-
line convection	선형대류	線形對流
line echo	선에코	線-
line echo wave pattern	선에코파동형태	線-波動形態
line gale(= equinoctial gale)	춘추분폭풍	春秋分暴風
line of action	작용선	作用線
line of action of the force	힘의 작용선	-作用線
line of apsis	장축선(타원의)	長軸線
line of cloud	구름선	-線
line of departure weather	공격개시선기상(군사용어)	攻擊開始線氣象
line of discontinuity	불연속선	不連續線
line of echo	에코선	-線
line of frontogenesis	전선발생선	前線發生線
line of showers	소나기선	-線
line of sight	광학센서의 시선경로	光學-視線經路
line-of-sight propagation	시선전파	視線傳播
line-of-sight range	시선범위	視線範圍
line of thunderstorm	뇌우선	雷雨線
line source	선모양발생원	線模樣發生源
line spectrum	선스펙트럼	線-
line squall	선스콜	線-
line storm(= equinoctial gale)	춘추분스톰, 춘추분폭풍	春秋分-
line symmetry	선대칭	線對稱
line-type seeding	선형씨뿌리기	線型-
line width	선너비	線-
lingering summer heat	잔서	殘暑
Linke-scale(= blue sky scale)	링케척도	-尺度
Linke turbidity factor	링케혼탁인자	-混濁因子
liquefaction	액화	液化
liquid	액체	液體

liquid air	액체공기	液體空氣
liquid-bubble tracer	액체거품추적자	液體-追跡子
liquid-column barometer	액체기둥기압계	液體-氣壓計
liquid-glass thermograph	유리관액체 자기온도계	-管液體自記溫度計
liquid nitrogen	액체질소	液體窒素
liquid phase	액체상	液體相
liquid state	액상	液相
liquid(-in-metal) thermograph	금속관액체 자기온도계	金屬管液體自記溫度計
liquid thermometer	액체온도계	液體溫度計
liquid water	액체수	液體水
liquid water content	액체수함량	液體水含量
liquid water mixing ratio	액체수혼합비	液體水混合比
liquid water path	액체수경로, 액수경로	液體水徑路, 液水經路
liquid water potential temperature	액체수온위	液體水溫位
lithium chloride dew-point hygrometer	염화리튬이슬점습도계, 염화리튬노점습도계	鹽化-露點濕度計
lithium chloride dew-point thermometer	염화리튬이슬점온도계, 염화리튬노점온도계	鹽化-露點溫度計
lithometeor	대기먼지현상	大氣-現象
lithosphere	암석권	岩石圈
little brother	리틀브라더(가끔 격렬한 요란을 수반하는 부차적인 열대저기압)	
Little Ice Age(LIA)	소빙하기	小氷河期
little sister	리틀시스터(북반구에서 격렬한 요란을 수반하는 여성이름의 열대저기압)	
littoral climate	해안기후	海岸氣候
littoral current	연안류	沿岸流
livestock meteorology	축산기상학	畜産氣象學
living glacier	활동빙하	活動氷河
Livingstone sphere	리빙스턴구(증발계)	-球
lizard balloon	도마뱀기구	-氣球
llanos(= savanna)	라노(중남미 또는 아메리카 남서부의 사바나)	
loading	부하, 적재(량), 농도, 함량	負荷, 積載(量), 濃度, 含量
lobe	열편	裂片
lobe structure	열편구조	裂片構造
local	국지(의), 지방(의)	局地, 地方
local acceleration	국지가속도	局地加速度
local action	국지작용	局地作用
local angular momentum	국지각운동량	局地角運動量
local anticyclone	국지고기압	局地高氣壓
local area	국지면적	局地面積
local area forecast over sea area	해상국지예보	海上局地豫報
local areas for sea area forecast	해상국지예보구역	海上局地豫報區域
local axis	국지축	局地軸

L

local change	국지변화	局地變化
local circulation	국지순환	局地循環
local civil time	지방상용시	地方常用時
local climate	국지기후	局地氣候
local climatology	국지기후학	局地氣候學
local cyclone	국지저기압	局地低氣壓
local derivative	국지도함수	局地導函數
local discontinuity line	국지적 불연속선	局地的不連續線
local effect	국지효과	局地效果
local extra observation	국지임시관측	局地臨時觀測
local fallout	국지낙진	局地落塵
local flight area weather	국지비행공역기상	局地飛行空域氣象
local forecast	국지예보	局地豫報
local frequency oscillator	저주파진동자	低周波振動子
local haul weather	국지수송기상	局地輸送氣象
local heating	국지가열	局地加熱
local horizon	국지수평선	局地水平線
local inflow	국지유입	局地流入
local isotropy	국소등방성	局所等方性
localized torrential rainfall	집중호우	集中豪雨
local lightning counter	국지번개계수기	局地-計數器
local lightning-flash counter	국지번개섬광계수기	局地-閃光計數器
local Mach number	국지마흐수	局地-數
local mean time	지방평균시	地方平均時
local oscillator	국부진동자	局部振動子
local precipitation	국지성강수	局地性降水
local sidereal time	지방항성시	地方恒星時
local similarity hypothesis	국지유사가설	局地類似假說
local solar time	지방태양시	地方太陽時
local standard time	지방표준시	地方標準時
local storm	국지성스톰, 국지폭풍(우)	局地性-, 局地暴風(雨)
local thermodynamic equilibrium	국지열역학적 평형	局地熱力學的平衡
local thunderstorm	국지성뇌우	局地性雷雨
local time	지방시	地方時
local true time	지방진시	地方眞時
local variation	국지변동, 국지변화	局地變動, 局地變化
local vorticity	국지소용돌이도	局地-度
local weather	국지기상	局地氣象
local weather report	지방날씨(일기)통보	地方-(日氣)通報
local wind	국지바람, 국지풍	局地風
locomotive organ	운동기관	運動器官
locus	궤적	軌跡
locus control	궤적제어	軌跡制御
Lodgement Area weather	거점지역기상	據點地域氣象

lodging	도복	倒伏
lodging resistance	내도복성	耐倒伏性
loess	황토	黃土
loess deposits	황토침전물	黃土沈澱物
lofting	상승형	上昇形
logarithmic receiver	로그수신기	-受信機
logarithmic velocity profile	대수풍속분포	對數風速分布
log book(= field note)	야장	野帳
logistic curve	로지스틱곡선	-曲線
logistics operation weather	병참작전기상	兵站作戰氣象
log-normal distribution	대수정규분포	對數正規分布
lolly ice	편빙	片氷
London fog	런던안개	
London smog	런던스모그	
long-day plant	장일식물	長日植物
long discharge	긴 방전	-放電
longitude	경도	經度
longitude effect	경도효과	經度效果
longitudinal divergence	종발산	縱發散
longitudinal gustiness	세로성분돌풍도	-成分突風度
longitudinal vibration	세로진동	-振動
longitudinal wave	세로파	-波
longitudinal wind	경도방향바람, 세로방향바람	經度方向-
long period	장주기	長週期
long period rain recorder	장기자기우량계	長期自記雨量計
long pulse	긴 펄스	
long-range forecast	장기예보	長期豫報
long-range pluviograph	장기우량기록계	長期雨量記錄計
long-range transport	장거리이동	長距離移動
longshore current(= littoral current)	연안류	沿岸流
long spark	긴 불꽃	
long spell of rain	장우	長雨
long-train atmospherics	긴 행렬공중전기	-行列空中電氣
long wave	장파	長波
long wave anisotropic factor	장파 비등방성인자	長波非等方性因子
long wave channel	장파채널	長波-
long wave equation	장파방정식	長波方程式
long wave formula	장파공식	長波公式
long wave trough	장파곡	長波谷
long wave radiation	장파복사	長波輻射
long wave trough	장파기압골	長波氣壓-
Loofah	루파	
lookup table	색인표	索引表
looming	떠보임, 루밍	

L

loop antenna	루프안테나	
looping	고리모양	-模樣
looping of the front	전선파복	前線波腹
looping plume	고리모양 플룸	-模樣-
loose avalanche	마른 눈사태	-沙汰
loosely bound water	약결합수	弱結合水
Lorentz factor	로렌츠인자	-因子
Lorentz lineshape	로렌츠선모양	-線模樣
Lorentz transformation	로렌츠변환	-變換
Loschmidt's number	로슈미트수	-數
louver	비늘살	
louvered screen	루버백엽상	-百葉箱
louvered thermometer screen	비늘살백엽상	-百葉箱
love wave	러브파	-波
low	저, 저기압	低氣壓
low aloft	상층저기압	上層低氣壓
low altitude bombing weather	저고도폭격기상(군사용어)	低高度爆擊氣象
low angle loft bombing weather	저각도 상승폭격기상(군사용어)	低角度上昇爆擊氣象
low atmospheric pressure	저기압	低氣壓
low cloud	하층운	下層雲
low depth of water warning	저수심경보	低水深警報
low earth orbit	지구저궤도	地球低軌道
low-emission vehicle	저공해자동차	低公害自動車
low energy particle	저에너지입자	低-粒子
lower arc	하단호	下端弧
lower atmosphere	하층대기	下層大氣
lower circumzenithal arc	하단천정호	下端天頂弧
lower cloud	하부구름	下部-
Lower Explosion Limit	폭발하한계	爆發下限界
lower friction layer	하부마찰층	下部磨擦層
lowering melting point	융점강하	融點降下
lower limit	아래끝	
lower mirage	하부신기루	下部蜃氣樓
lower stratosphere	하부성층권	下部成層圈
lower symmetric regime	하부축대칭영역	下部軸對稱領域
lower turbulent zone	저층난류대	低層亂流帶
lowest temperature	최저온도	最低溫度
low frequency	저주파수	低周波數
low frequency oscillator	저주파진동자	低周波振動子
low index	낮은 지수	-指數
low instrument flight rule	최저계기비행규칙	最低計器飛行規則
Lowitz arcs	로비츠호	-弧
low-level cloud	하층운	下層雲
low-level cloud vortex	저층구름소용돌이	低層-

low-level jet	하층제트	下層-
low-level jet stream	하층제트류	下層-流
low-level prog chart	하층예상도	下層豫想圖
low level terrain flight weather	저고도침투기상	低高度浸透氣象
low-level turbulence	저층난류	低層亂流
low-level wind shear	저층윈드시어	低層-
low-level wind shear alert system	저층윈드시어경보시스템	低層-警報-
low-noise amplifier	저잡음증폭기	低雜音增幅器
low parry arc	대기광학현상	大氣光學現象
low pressure	저(기)압	低(氣)壓
low pressure area	저압부	低壓部
low pressure zone	저압대	低壓帶
low-rate information transmission	저속 위성자료 전송 (10~256kbps)	低速衛星資料電送
low-reference signal	저기준신호	低基準信號
low slope	완만경사	緩慢傾斜
low temperature during Changma	장마저온	-低溫
low temperature hygrometry	저온습도측정(법)	低溫濕度測定(法)
low tide	저조, 간조, 썰물	低潮, 干潮
low vacuum	저진공	低眞空
low visibility	저시정	低視程
low visibility level	저시정등급	低視程等級
low volume air sampler	저용량 공기 샘플러	低容量空氣-
low water	저수(위), 저조	低水(位), 低潮
low water level	저수위	低水位
low water mark	저수위표지, 저조위표지	低水位標識, 低潮位標識
low with warm sector	난역저기압	暖域低氣壓
low zonal circulation	약한 동서순환	弱-東西循環
lucimeter	전천일사평균강도계	全天日射平均强度計
lull	일시적 고요	一時的-
lumen	루멘(광속의 단위)	
luminance	휘도	輝度
luminance contrast	휘도대조	輝度對照
luminance flux	광속	光束
luminescence	발광	發光
luminosity	광도, 시감도	光度, 視感度
luminosity class	광도계급	光度階級
luminous body	발광체	發光體
luminous cloud	발광운	發光雲
luminous density	광밀도	光密度
luminous efficiency	발광효율, 시감도율	發光效率, 視感度率
luminous emittance	발광도	發光度
luminous energy	발광에너지	發光-
luminous exitance	발광률	發光率

L

luminous exposure	발광노출	發光露出
luminous flux	광플럭스	光-
luminous flux density	발광속밀도	發光束密度
luminous intensity	발광강도	發光强度
luminous meteor	빛현상	-現象
luminous night cloud	야광운	夜光雲
luminous pillar	광주	光柱
luminous power	발광능	發光能
luminous vapour-trail	발광비행운	發光飛行雲
lunar atmospheric tide	달대기조석	-大氣潮汐
lunar calendar	음력	陰曆
lunar corona	달광환	-光環
lunar day	태음일	太陰日
lunar distance	달거리	-距離
lunar eclipse	월식	月食
lunar gravitational tide	달중력조	-重力潮
lunar halo	달무리	
lunar probe	달탐측기	-探測器
lunar rainbow	달무지개	
lunar tide	달조석, 태음조	-潮汐, 太陰潮
lunar year	태음년	太陰年
lunation	태음월	太陰月
lunitidal interval	월조간격	月潮間隔
lustrum	5년간	-年間
lux	룩스	
Lyapunov exponent	리아푸노프지수	-指數
Lyapunov vector	리아푸노프벡터	
Lyman alpha emission line	라이먼알파방출선	-放出線
lysimeter	증발산량계	蒸發散量計

M, m

영문	한글	한자
Mach number	마하수	-數
mackerel breeze	고등어바람(고등어 잡이에 적합한 강풍)	
mackerel sky	비늘구름하늘	
Macky effect	매키효과	-效果
macroburst	대돌연풍	大突然風
macroclimate	대기후	大氣候
macroclimatology	대기후학	大氣候學
macrometeorology	대기상학	大氣象學
macro-physics	거시물리학	巨視物理學
macroscale	대규모	大規模
macroscale cluster	대규모클러스터, 대규모무리	大規模-
macroscale weather pattern	대규모일기패턴	大規模日氣-
macro-scopic	광대한 범위(의), 거시적	廣大-範圍, 巨視的
macroscopic observation	육안관측	肉眼觀測
macroturbulence	대규모난류	大規模亂流
macro-viscosity	거시점성	巨視粘性
maculosus	얼룩(구름)	
Madden-Julian oscillation	매든-줄리안 진동	-振動
maeu	매우	梅雨
magmatic water(= juvenile water)	마그마물	
magnetically disturbed day	자기폭풍일	磁氣暴風日
magnetically quiet day	자기고요일	磁氣-日
magnetic character figure	자기특성수	磁氣特性數
magnetic crotchet	자기크로치트 (지구자장의 단기간 요란)	磁氣-
magnetic declination	자기편각	磁氣偏角
magnetic deflection	자기편향	磁氣偏向
magnetic dip	복각	伏角
magnetic disturbance	자기요란	磁氣擾亂

magnetic double refraction	자기이중굴절	磁氣二重屈折
magnetic element	자기요소	磁氣要素
magnetic energy	자기에너지	磁氣-
magnetic equator	자기적도	磁氣赤道
magnetic field	자기장	磁氣場
magnetic field intensity	자기장강도	磁氣場强度
magnetic field strength	자기장세기	磁氣場勢氣
magnetic inclination	복각	伏角
magnetic induction	자기유도	磁氣誘導
magnetic intensity	자기강도	磁氣强度
magnetic latitude	자기위도	磁氣緯度
magnetic line of force	자력선	磁力線
magnetic meridian	자기자오선	磁氣子午線
magnetic moment	자기모멘트	磁氣-
magnetic needle	자침	磁針
magnetic north	자북	磁北
magnetic north pole	자기북극(N)	磁氣北極
magnetic orientation control coil	자기방향제어코일	磁氣方向制御-
magnetic pole	자기극, 자극	磁氣極, 磁極
magnetic south	자남	磁南
magnetic southern pole	자남극	磁南極
magnetic storm(= magstorm)	자기폭풍	磁氣暴風
magnetic variation	자기변동	磁氣變動
magnetic wind direction	자기풍향	磁氣風向
magnetism	자기성, 자기학	磁氣性, 磁氣學
magnetism layer	자기층	磁氣層
magneto anemometer	자력풍속계	磁力風速計
magnetogram	자력기록	磁力記錄
magnetograph	자력기록계	磁力記錄計
magnetohydrodynamics	자기유체역학	磁氣流體力學
magnetohydrodynamic wave	자기유체역학파	磁氣流體力學波
magneto-ionic theory	자기이온이론	磁氣-理論
magnetometer	자력계	磁力計
magnetopause	자기권계면	磁氣圈界面
magnetosphere	자기권	磁氣圈
magnetron	마그네트론	
magnification factor	확대인자	擴大因子
magnifying glass	확대경	擴大鏡
magnitude	크기, 등급	等級
magnitude of force	힘의 크기	
Magnus effect	매그너스효과	-效果
main battle area weather	주전투지역기상	主戰鬪地域氣象
main beam	주빔	主-
main current	주류	主流

main defense area weather	주방어지역기상	主防禦地域氣象
main lobe	주방사부	主放射部
main meteorological office	주요기상대, 중앙기상대	主要氣象臺, 中央氣象臺
main precipitation core	주강수중심	主降水中心
main standard time	기본관측시각 (00,06,12,18UTC)	基本觀測時刻
main stroke	주뇌격	主雷擊
main telecommunication network	주통신망	主通信網
maintenance respiration	유지호흡	維持呼吸
main trunk circuit	주간선회로	主幹線回路
major cold	대한	大寒
major heat	대서	大暑
major lobe	주열편	主裂片
major ridge	주(기압)마루	主(氣壓)-
major snow	대설	大雪
major trough	주(기압)골	主(氣壓)-
major wave	주파	主波
malodor	악취	惡臭
mamma	유방(구름)	乳房-
mammato cloud	유방구름	乳房-
mammato cumulus	유방적운	乳房積雲
mammatus	유방구름	乳房-
mandatory layer	지정층	指定層
mandatory level	기준등압면	基準等壓面
mandatory surface	지정기압면	指定氣壓面
maneuvering	기동조종	機動操縱
mango shower	망고소나기(타이랜드 중부 연안 지역에서 2~3월에 내리는 소나기)	
man-machine mix	인간-기계 혼합	人間機械混合
manned balloon	유인기구	有人氣球
manometer	압력계	壓力計
manual method	수화법	手話法
manual tracking	수동식 추적	手動式追跡
manufacturing readiness review	제작준비 검토	製作準備檢討
maparam	마파람, 남풍	南風
map factor	지도인자	地圖因子
maple front	단풍전선	丹楓前線
mapping	사상	寫像
mapping scale	지도규모	地圖規模
map plotting	일기도기입	日氣圖記入
map projection	도법	圖法
map scale	지도축척	地圖縮尺
map spotting	일기도기입	日氣圖記入
marching problem	행진문제	行進問題

M

march of temperature	온도변화과정	溫度變化過程
mares' tails	말꼬리구름, 강수궤적에코	江水軌跡-
marginal ray	주변광선	周邊光線
marginal visual flight rule	최저시계비행규칙	最低視界飛行規則
Margules formula	마르굴레스공식	-公式
Margules frontal slope	마르굴레스전선기울기	-前線-
Margules's equation	마르굴레스방정식	-方程式
marigraph	검조계	檢潮計
marine aerosol	해양에어로졸	海洋-
marine air (mass)	해양기단	海洋氣團
marine angle	해양도각	海洋道角
marine barometer	선박용 기압계	船舶用氣壓計
marine biogenic sediments	해양생물기원퇴적물	海洋生物起源堆積物
marine carbon geochemistry	해양탄소 지구화학적 성질	海洋炭素地球化學的性質
marine clay minerals	해양점토광물	海洋粘土鑛物
marine climate	해양기후	海洋氣候
marine climatology	해양기후학	海洋氣候學
marine fog	해상안개	海上-
marine forecast	해양(일기)예보	海洋(日氣)豫報
marine mercury barometer	마린형 수은기압계	-水銀氣壓計
marine meteorological district	해양기상구(역)	海洋氣象區(域)
marine meteorology	해양기상학	海洋氣象學
marine non-biogenic sediments	해양비생물기원 퇴적물	海洋非生物起源堆積物
marine observation	해양(기상)관측	海洋(氣象)觀測
marine observation satellite	해양관측위성	海洋觀測衛星
marine observatory	해양관측소	海洋觀測所
marine prognostic chart	해양예상도	海洋豫想圖
marine rainbow	바다무지개	
marine short range forecast	해상단기예보	海上短期豫報
marine special report	해상특보	海上特報
marine thermometer	해수온도계	海水溫度計
marine weather observation	해양기상관측	海洋氣象觀測
Mariotte's law(= Boyle-Mariotte law)	마리오테법칙	-法則
maritime air	해양공기	海洋空氣
maritime air fog	해양기단안개	海洋氣團-
maritime air mass	해양기단	海洋氣團
maritime climate	해양기후	海洋氣候
maritime cloud	해양구름	海洋-
maritime meteorology	해양기상학	海洋氣象學
maritime polar air mass	해양성한대기단	海洋性寒帶氣團
maritime tropical air mass	해양성열대기단	海洋性熱帶氣團
maritime tundra	해양성툰드라	海洋性-
marketable permit	배출권거래제도	排出權去來制度
Markov chain	마르코프사슬	

Mars	화성	火星
Marsden chart	마르스덴일기도	-日氣圖
marsh	늪	
marsh gas	늪가스, 메탄가스	
Marshall-Palmer distribution	마샬-팔머 분포	-分布
Marshall-Palmer radar-rainfall function	마샬-팔머 레이더-강우 함수	-降雨函數
Marvin sunshine recorder	마빈일조계	-日照計
masked front	차폐전선	遮蔽前線
mass	질량	質量
mass absorption coefficient	질량흡수계수	質量吸收係數
mass concentration	질량농도	質量濃度
mass convergence	질량수렴	質量收斂
mass curve	누적곡선	累積曲線
mass defect	질량결손	質量缺損
mass divergence	질량발산	質量發散
mass emission coefficient	질량사출계수	質量射出係數
mass extinction	대량멸종	大量滅種
mass extinction coefficient	질량소산계수	質量消散係數
massflow of air	공기교환	空氣交換
mass mean temperature	질량평균기온	質量平均氣溫
mass number	질량수	質量數
mass of atmosphere	대기질량	大氣質量
mass rainfall curve	적산우량곡선	積算雨量曲線
mass ratio	질량비	質量比
mass spectrometer	질량분석기	質量分析器
mass transport	질량수송	質量輸送
mast	지지대, 돛대	支持臺
matched filter	정합필터	整合-
material	물질	物質
material coordinates	물질좌표	物質座標
material derivative	물질도함수	物質導函數
material point(= particle)	질점	質點
material surface	물질면	物質面
material volume	물질부피	物質-
maternal vortex	어미소용돌이	
mathematical climate(= solar climate)	수리기후(태양기후)	數理氣候
mathematical climate zone	수리기후대	數理氣候帶
mathematical expectation	수학적 기댓값	數學的期待-
mathematical forecasting	수학적 예보	數學的豫報
mathematical induction	수학적 귀납법	數學的歸納法
mathematical meteorology	수리기상학	數理氣象學
mathematical physics	수리물리학	數理物理學
mathematical probability	수학적 확률	數學的確率
mathematical statistics	수리통계학	數理統計學

M

matric potential	매트릭 퍼텐셜	
matrix	행렬	行列
matto grosso	마토그로소(브라질의 열대초원)	
mature stage	성숙단계, 성숙기	成熟段階, 成熟期
maturity(= ripening)	성숙	成熟
maulu	마우루(하와이에서의 강한 소나기)	
Maunder minimum	마운더 극소기	-極小期
maximum	최대, 최고, 극대	最大, 最高, 極大
maximum acceptable concentration	최대허용농도	最大許容濃度
maximum and minimum thermometer	최고최저온도계	最高最低溫度計
maximum depth-area-duration data	최대DAD자료	最大-資料
maximum detectable range	최대탐지거리	最大探知距離
maximum fresh snow depth	최심신적설	最深新積雪
maximum frictional force	최대마찰력	最大摩擦力
maximum gust lapse	최대돌풍감쇠	最大突風減衰
maximum gust lapse interval	최대돌풍감쇠간격	最大突風減衰間隔
maximum gust lapse time	최대돌풍감쇠시간	最大突風減衰時間
maximum infrared radiance	최대적외복사	最大赤外輻射
maximum instantaneous wind speed	최대순간풍속	最大瞬間風速
maximum likelihood approach	최우도접근법	最尤度接近法
maximum likelihood ensemble filter	최우앙상블필터	最尤-
maximum likelihood method	최우도법	最尤度法
maximum mixing depth	최대혼합깊이	最大混合-
maximum possible precipitation	최대가능강수(량)	最大可能降水(量)
maximum possible sunshine duration	최대가능일광기간	最大可能日光期間
maximum precipitation	최대강수량	最大降水量
maximum probable flood(MDF)	최대가능홍수량	最大可能洪水量
maximum probable precipitation(MPP)	최대가능강수량	最大可能降水量
maximum radiation thermometer	최대복사온도계	最大輻射溫度計
maximum snow depth	최심적설	最深積雪
maximum speed	최대속도	最大速度
maximum sustained wind	평균최대풍속	平均最大風速
maximum take-off load	최대이륙하중	最大離陸荷重
maximum temperature	최고온도	最高溫度
maximum thermometer	최고온도계	最高溫度計
maximum unambiguous range	최대유효거리	最大有效距離
maximum unambiguous velocity	최대유효속도,	最大有效速度,
	최대설탐지속도	最大設探知速度
maximum updraft	최대상승기류	最大上昇氣流
maximum value	극댓값, 최댓값	極大-, 最大-
maximum value test	최댓값검증	最大-檢證
maximum vapour pressure	최대증기압	最大蒸氣壓
maximum water holding capacity	최대용수량	最大用水量
maximum water vapour tension	최대수증기장력	最大水蒸氣張力

M

maximum wave height	최대파고	最大波高
maximum wind	최대풍향풍속	最大風向風速
maximum wind chart	최대풍도	最大風圖
maximum wind level	최대풍고도	最大風高度
maximum wind speed	최대풍속	最大風速
maximum wind topography	최대풍높낮이	最大風-
Maxwell's equation	맥스웰공식, 맥스웰방정식	-公式, -方程式
Maxwell's law	맥스웰법칙	-法則
M-curve	M-곡선	-曲線
mean	평균	平均
mean temperature	평균기온	平均氣溫
mean absolute error	평균절대오차	平均絶對誤差
mean annual range of temperature	평균온도연교차	平均溫度年較差
mean annual temperature	연평균온도	年平均溫度
mean anomaly	평균편차, 평균이상	平均偏差, 平均異常
mean bonding distance	평균결합길이	平均結合-
mean chart	평균도	平均圖
mean cross section	평균단면(도)	平均斷面(度)
mean daily maximum temperature	일평균최고온도	日平均最高溫度
mean daily maximum temperature for a month	월별 일평균최고온도	月別日平均最高溫度
mean daily minimum temperature	일평균최저온도	日平均最低溫度
mean daily minimum temperature for a month	월별 일평균최저온도	月別日平均最低溫度
mean daily temperature	일평균온도	日平均溫度
mean day-to-day variation	평균일별변동	平均日別變動
meandering course	사행경로	蛇行經路
mean deviation	평균편차	平均偏差
mean difference	평균차	平均差
mean distance	평균거리	平均距離
mean Doppler velocity	평균도플러속도	平均-速度
mean effective diameter	평균유효지름	平均有效-
mean elevation of basin	유역평균고도	流域平均高度
mean environmental wind	평균환경바람	平均環境-
mean equatorial day	평균적도일	平均赤道日
mean equinox	평균분점	平均分點
mean error	평균오차	平均誤差
mean field	평균장	平均場
mean free path	평균자유행로	平均自由行路
mean height field	평균고도장	平均高度場
mean interdiurnal variability	평균일별변동성, 평균일별변동률	平均日別變動性, 平均日別變動率
mean kinetic energy	평균운동에너지	平均運動-
mean life	평균수명	平均壽命
mean lunar day	평균태음일	平均太陰日
mean map(= mean chart)	평균도	平均圖

mean meridional circulation	평균자오면순환, 평균남북순환	平均子午面循環, 平均南北循環
mean molecular weight	평균분자량	平均分子量
mean monthly maximum temperature	월평균최고온도	月平均最高溫度
mean monthly minimum temperature	월평균최저온도	月平均最低溫度
mean monthly temperature	월평균온도	月平均溫度
mean motion constant	평균운동상수	平均運動常數
mean noon	평균정오	平均正午
mean parallax	평균시차	平均視差
mean radial velocity	평균시선속도	平均視線速度
mean radiant temperature	평균복사온도	平均輻射溫度
mean radius	평균반지름, 평균반경	平均半徑
mean reaction rate	평균반응속도	平均反應速度
mean sea level	평균해면	平均海面
mean sea level pattern	평균해면패턴	平均海面-
mean sea level pressure	평균해면기압	平均海面氣壓
mean sidereal day	평균항성일	平均恒星日
mean skin temperature	평균표피온도	平均表皮溫度
mean solar day	평균태양일	平均太陽日
mean solar hour	평균태양시	平均太陽時
mean solar time	평균태양시간	平均太陽時間
mean solar year	평균태양년	平均太陽年
mean sphere depth	평균지각수심	平均地殼水深
mean-square error	평균제곱오차	平均-誤差
mean sun	평균태양	平均太陽
mean synodic lunar month	평균삭망월	平均朔望月
mean temperature	평균온도	平均溫度
mean temperature lapse rate	평균기온감률	平均氣溫減率
mean temperature of air column	공기기둥의 평균온도	空氣-平均溫度
mean time	평균시	平均時
mean tropopause	평균대류권계면	平均對流圈界面
mean upper-air flow	평균상층기류	平均上層氣流
mean value	평균값	平均-
mean value theorem	평균값정리	平均値定理
mean velocity	평균속도	平均速度
mean vertical temperature	평균연직온도	平均鉛直溫度
mean-volume radius	평균부피반지름, 평균부피반경	平均-半徑
mean width of basin	유역평균폭	流域平均幅
mean wind velocity	평균바람속도	平均-速度
measure	측정, 척도	測定, 尺度
measured ceiling	측정실링	測定-
measured value	측정값	測定-
measurement	측정	測定
measurement of rainfall amount	강우량측정	降雨量測定

M

measuring element	측정요소	測定要素
measuring stick	측정척	測定尺
mechanical collector	회전집진기	回轉集塵器
mechanical condensation level	역학응결고도	力學凝結高度
mechanical energy	역학에너지	力學-
mechanical energy conservation law	역학적 에너지보존법칙	力學的-保存法則
mechanical equivalent of heat	열의 일(해)당량	熱-(該)當量
mechanical instability	역학불안정(도)	力學不安定(度)
mechanical internal boundary layer	기계적 내부경계층	機械的內部境界層
mechanically unstable	역학적 불안정한	力學的不安定-
mechanical snow damage	기계적 설해	機械的雪害
mechanical stability	역학안정(도)	力學安定(度)
mechanical turbulence	역학난류	力學亂流
mechanical ventilation	기계난류, 기계적 환기	機械亂流, 機械的換氣
mechanics	역학	力學
mechanism	기작	機作
media forecast	대중매체예보, 미디어예보	大衆媒體豫報
media forecast source	대중매체예보자료	大衆媒體豫報資料
median	중앙값	中央-
median radius	중앙반지름, 중앙반경	中央半徑
median volume diameter	중앙부피지름	中央-
median volume radius	중앙부피반지름, 중앙부피반경	中央-半徑
medical climatology	의료기후학	醫療氣候學
medical meteorology	의료기상학	醫療氣象學
Medieval warm period	중세 온난기	中世溫暖期
mediocris	중간(구름)	中間-
Mediterranean climate	지중해기후	地中海氣候
Mediterranean front	지중해전선	地中海前線
Mediterranean frontal zone	지중해전선대	地中海前線帶
medium	매질, 매체, 중간	媒質, 媒體, 中間
medium-altitude bombing weather	중고도폭격기상	中高度爆擊氣象
medium angle camera	중각카메라	中角-
medium cloud	중층운	中層雲
medium frequency	중주파수	中周波數
medium-level cloud	중층구름	中層-
medium-range forecast	중기예보	中期豫報
medium resolution infrared radiometer	중분해 적외복사계	中分解赤外輻射計
medium scale	중간규모	中間規模
medium-scale disturbance	중간규모요란	中間規模擾亂
medium-scale turbulence	중간규모난류	中間規模亂流
medium-term hydrological forecast	중기수문예보	中期水文豫報
mega-	100만, 메가	-萬
megabarye	메가바리에(기압단위)	
megacycle(= Hertz)	메가사이클	

M

megatherm	고온식생	高溫植生
megathermal climate	고온기후	高溫氣候
megathermal period(= Climatic Optimum)	고온기간	高溫期間
megathermal type	고온식생형	高溫植生形
megathermen	고온대	高溫帶
Mei-yu	메이유, 장마(중국)	梅雨
Mei-yu front	메이유전선, 장마전선	梅雨前線
melanoma	흑색종	黑色腫
melting	녹음	
melting band	녹는 띠	
melting curve	녹음곡선	-曲線
melting layer	융해층	融解層
melting level	녹는 고도	-高度
melting of snow	융설	融雪
melting point	녹는점	-點
melting point depression	녹는점내림	-點-
melting temperature	녹는 온도	-溫度
membership function	소속함수	所屬函數
memory	기억, 기억장치	記憶, 記憶裝置
meniscus	메니스커스	
Mercator projection	메르카토르투영(도법)	-投影(圖法)
mercurial barograph	수은기압기록계,	水銀氣壓記錄計,
	수은자기기압계	水銀自記氣壓計
mercurial barometer	수은기압계	水銀氣壓計
mercurial thermometer	수은온도계	水銀溫度計
mercury	수은	水銀
mercury barometer	수은기압계	水銀氣壓計
mercury column	수은기둥	水銀-
mercury-in-glass thermometer	유리관 수은온도계	-璃管水銀溫度計
mercury-in-steel thermometer	강관수은온도계	鋼管水銀溫度計
mercury manometer	수은압력계	水銀壓力計
mercury thermometer	수은온도계	水銀溫度計
merging	융합	融合
meridian	자오(의), 자오선, 경선,	子午, 子午線, 經線, 子午圈
	자오권	
meridian circle	자오권	子午圈
meridian plane	자오면	子午面
meridian transit time	남중시	南中時
meridional cell	자오면세포	子午面細胞
meridional circulation	자오면순환, 남북순환	子午面循環, 南北循環
meridional exchange	자오면교환, 남북교환	子午面交換, 南北交換
meridional flow	자오면류, 남북류	子午面流, 南北流
meridional front	자오면전선	子午面前線
meridional index	남북지수	南北指數

meridional trough cyclogensis	남북기압골형 저기압발생	南北氣壓谷型低氣壓發生
meridional wind	자오선바람	子午線-
mesh current	그물전류	-電流
mesh scale	격자간격	格子間隔
mesoanalysis	중규모분석	中規模分析
mesochart	중규모분석도	中規模分析圖
mesoclimate	중기후	中氣候
mesoclimatology	중기후학	中氣候學
mesocyclone	중규모저기압	中規模低氣壓
mesohigh	중규모고기압	中規模高氣壓
mesojet	중규모제트류	中規模-流
mesokurtic	중앙첨도(의)	中央尖度
mesokurtosis	중앙첨도	中央尖度
mesolow	중규모저기압	中規模低氣壓
mesometeorology	중기상학	中氣象學
meson	중간자	中間子
mesopause	중간권계면	中間圈界面
mesopeak	중간권 최고온도점	中間圈最高溫度點
mesoscale	중규모	中規模
mesoscale analysis	중규모분석	中規模分析
mesoscale category	중규모범주	中規模範疇
mesoscale circulation	중규모순환, 중간규모순환	中規模循環, 中間規模循環
mesoscale convective complex	중규모대류복합체	中規模對流複合體
mesoscale convective system	중규모대류계	中規模對流系
mesoscale disturbance	중규모요란	中規模搖亂
mesoscale front	중규모전선	中規模前線
mesoscale model	중규모모형	中規模模型
mesoscale motion	중규모운동	中規模運動
mesoscale phenomenon	중규모현상	中規模現象
mesoscale storm	중규모폭풍, 중규모스톰	中規模暴風
mesoscale structure	중규모구조	中規模構造
mesoscale turbulence	중규모난류	中規模亂流
mesosphere	중간권	中間圈
mesospheric clouds(= Noctilucent cloud)	중간권구름	中間圈-
mesospheric jet	중간권제트류	中間圈-流
mesotherm	중온식생(형)	中溫植生(型)
mesothermal climate	중온기후	中溫氣候
mesothermal type	중온식생형	中溫植生型
mesothermen	중온대	中溫帶
Mesozoic climate	중생대기후	中生代氣候
message	전보, 통보, 메시지	電報, 通報
message of special report	특보문	特報文
message switching facility	전보교환시설	電報交換施設
Messinian salinity crisis	메시니아 염분위기	-鹽分危機

M

metabolism	신진대사	新陳代謝
metal barometer	금속기압계	金屬氣壓計
metal corrosion	금속부식	金屬腐蝕
metallic barometer	금속기압계	金屬氣壓計
metallic thermometer	금속온도계	金屬溫度計
metamorphosis of snow	눈의 변형	-變形
METAR	기상항공보고서	氣象航空報告書
metastability	준안정도	準安定度
metastable	준안정	準安定
meteor	유성	流星
meteoric dust	유성먼지	流星-
meteoric shower	유성소나기	流星-
meteoric water	천수, 대기수	天水, 大氣水
meteorite	운석	隕石
meteorite shower	운석우	隕石雨
meteorogical satellite	기상위성	氣像衛星
meteorogram	기상기록(지)	氣象記錄(紙)
meteorograph	기상기록계	氣象記錄計
meteoroid	유성체	流星體
meteorological acoustics	기상음향학	氣象音響學
Meteorological Agency	(일본)기상청	氣象廳
meteorological analysis	기상분석	氣象分析
Meteorological Authority	(이집트)기상국	氣象局
meteorological balloon	기상관측기구	氣象觀測氣球
meteorological briefing	기상해설	氣象解說
meteorological broadcast	기상방송	氣象放送
meteorological bulletin	기상회보	氣象會報
meteorological buoy	기상관측부이	氣象觀測-
meteorological chart	기상도	氣象圖
meteorological code	기상전문부호	氣象電文符號
meteorological code form	기상전문부호형식, 기상전보식	氣象電文符號形式, 氣象電報式
meteorological communications	기상통신	氣象通信
meteorological control	기상조절	氣象調節
meteorological data	기상자료	氣象資料
meteorological data center	기상자료센터	氣象資料-
meteorological debriefing	기상결과보고	氣象結果報告
meteorological disaster	기상재해	氣象災害
meteorological disease	기상병	氣象病
meteorological display	기상전시	氣象展示
meteorological district	기상관할구(역)	氣象管轄區(域)
meteorological dynamics	기상역학	氣象力學
meteorological education	기상교육	氣象敎育
meteorological element	기상요소	氣象要素
meteorological element series	기상요소계열	氣象要素系列

meteorological equator	기상적도	氣象赤道
meteorological equipment	기상장비	氣象裝備
meteorological factor	기상인자	氣象因子
meteorological forecast	기상예보	氣象豫報
meteorological imager	기상영상기	氣象映寫機
meteorological information	기상정보	氣象情報
meteorological information for aircraft in flight	볼멧기상정보	-氣象情報
meteorological institute	기상연구소	氣象研究所
meteorological instrument	기상장비, 기상기계	氣象裝備, 氣象機械
meteorological instruments and observations	기상측기 및 관측	氣象測器觀測
meteorological kinematics	기상운동학	氣象運動學
meteorological map	기상도	氣象圖
meteorological message	기상통보, 기상통보문	氣象通報, 氣象通報文
meteorological network	기상관측망	氣象觀測網
meteorological noise	기상잡음	氣象雜音
meteorological observation	기상관측	氣象觀測
meteorological observatory	기상관측소, 기상대	氣象觀測所, 氣象臺
meteorological observer	기상관측자	氣象觀測者
meteorological observing station	기상관측소	氣象觀測所
Meteorological Office	(영국)기상국	氣象局
Meteorological Operational Telecommunications Network Europe	유럽기상용통신망	-氣象用通信網
meteorological optical range	기상광학거리	氣象光學距離
meteorological optics	기상광학	氣象光學
meteorological organization	기상기구	氣象機構
meteorological phenomenon	기상현상	氣象現象
meteorological post	기상관측점	氣象觀測點
meteorological proverb	일기속담	日氣俗談
meteorological radar	기상레이더	氣象-
meteorological radar station	기상레이더관측소	氣象-觀測所
meteorological range	기상학적 등급	氣象學的等級
meteorological realm	기상영역	氣象領域
meteorological reconnaissance	기상정찰	氣象偵察
meteorological reconnaissance flight	기상정찰비행	氣象偵察飛行
meteorological region	기상구역	氣象區域
meteorological report	기상보고	氣象報告
meteorological resource	기상자원	氣象資源
meteorological rocket	기상로켓	氣象-
meteorological rocket network	기상로켓관측망	氣象-觀測網
meteorological satellite	기상위성	氣象衛星
meteorological season	기상계절	氣象季節
meteorological service	기상업무	氣象業務
meteorological service for electric power system	전력기상업무	電力氣象業務

M

Meteorological Service for International Air Navigation	국제항공항행을 위한 기상업무	國際航空航行-氣象業務
meteorological signal mark	기상신호표시	氣象信號表示
meteorological society	기상학회	氣象學會
meteorological special report	기상특보	氣象特報
meteorological statics	기상정역학	氣象靜力學
meteorological station	기상관측소	氣象觀測所
meteorological statistics	기상통계학	氣象統計學
meteorological survey	기상조사	氣象調査
meteorological symbols	기상기호	氣象記號
meteorological table	기상상용표	氣象常用表
meteorological target	기상목표물	氣象目標物
meteorological telecommunication	기상통신	氣象通信
meteorological telecommunication network	기상통신망	氣象通信網
meteorological telegraph	기상전보(유선)	氣象電報
meteorological teleprinter network	기상텔레타이프통신망	氣象-通信網
meteorological thermodynamics	기상열역학	氣象熱力學
meteorological tide	기상조석	氣象潮汐
meteorological transmission	기상정보송신	氣象情報送信
meteorological tropics	기상학적 열대	氣象學的熱帶
meteorological variable	기상변수	氣象變數
meteorological visibility	기상학적 시정	氣象學的視程
meteorological visibility at night	기상학적 야간시정	氣象學的夜間視程
meteorological warning message	기상경보	氣象警報
meteorological watch	기상감시	氣象監視
meteorological watch office	기상감시소	氣象監視所
meteorological yearbook	기상연감	氣象年鑑
meteorologist	기상학자	氣象學者
meteorology	기상학	氣象學
meteorology act	기상법	氣象法
meteorology for fisheries	어업기상	漁業氣象
meteorology of war	전쟁기상학	戰爭氣象學
meteoropathic reaction	기상생리반응	氣象生理反應
meteoropathology	기상병리학	氣象病理學
meteorotropic disease	기상병	氣象病
meteorotropism	기상병	氣象病
Meteosat	유럽 기상위성	-氣象衛星
meteor satellite	기상위성	氣象衛星
meteor shower	유성우	流星雨
Meteor sputnik	스푸트니크기상위성	-氣象衛星
meteor trail	유성흔적	流星痕迹
meterological acoustics	기상음향학	氣象音響學
meter-ton-second system of unit	mts단위계	-單位系
methane	메탄	

methane hydrate	메탄수화물, 메탄하이드레이트	-水化物
methanol	메탄올	
method for sampling stack gas	배기가스 시료채취 방법	排氣-試料採取方法
method of characteristics	특성법	特性法
method of continuity equation	연속방정식법	連續方程式法
method of equivalence	등치법	等値法
method of finite difference	차분법, 유한차분법	差分法, 有限差分法
method of least squares	최소제곱법, 최소자승법	最小自乘法
method of progressive average	점진평균방법	漸進平均方法
method of successive approximation	연차근사법	連次近似法
methylmercury	메틸수은	-水銀
MetOp	유럽 극궤도위성	-極軌道衛星
Metor satellites	기상위성	氣象衛星
metric acceleration	측도가속도	測度加速度
metric coefficient	도량계수	度量係數
metric system	미터법	-法
METROMEX	도시기상실험	都市氣象實驗
Michaelson actinograph	마이클슨자기일사계,	-自記日射計
	마이클슨일사기록계	-日射記錄計
microbar	마이크로바	
microbarogram	미기압기록(지)	微氣壓記錄(紙)
microbarograph	미기압기록계, 미압계	微氣壓記錄計, 微壓計
microbarometer	미기압계	微氣壓計
microbarovariograph	미기압기록계, 미압계	微氣壓記錄計, 微壓計
microburst	마이크로버스트(4km 이내	微細突風
	영역의 강한 하강류), 미세돌풍	
microburst alert	마이크로버스트경보,	微細突風警報
	미세돌풍경보	
microclimate	미기후	微氣候
microclimatology	미기후학	微氣候學
micro cloud modification model	미세구름조절모형	微細-調節模型
micro cloud physics	미세구름물리	微細-物理
microcyclone	미저기압	微低氣壓
microfilm	마이크로필름	
micrometeorograph	미기상기록계	微氣象記錄計
micrometeorology	미기상학	微氣象學
micrometer	마이크로미터(μm)	
micron	마이크론(길이의 단위, 10^{-6}m)	
microphotography	현미경사진술	顯微鏡寫眞術
microphysical model	미세물리모형	微細物理模型
microphysics	미세물리학	微細物理學
micropluviometer	미우량계	微雨量計
micro pulse lidar	마이크로 펄스라이다	
micro rain radar	마이크로 강수레이더	-降水-

M

microscale	미규모	微規模
microscale motion	미세규모운동	微細規模運動
microscopic	미시적	微視的
microseism	미진	微震
microseismograph	미진기록계	微震記錄計
microstructure	미세구조, 미구조	微細構造
microtherm	저온식생형	低溫植生型
microthermal climate	저온기후	低溫氣候
microthermal type	저온식생형	低溫植生型
microthermen	저온대	低溫帶
microturbulence	미세난류	微細亂流
microvariation of pressure	기압미세변화	氣壓微細變化
microwave	마이크로파	-波
microwave altimeter	마이크로파 고도계	-波高度計
microwave imager	마이크로파 영상계	-波映像計
microwave meteorology	마이크로파 기상학	-波氣象學
microwave probing	마이크로파 탐사	-波探査
microwave radiation	마이크로파 복사	-波輻射
microwave radiometer	마이크로파 라디오미터	-波-
microwave refractometer	마이크로파 굴절계	-波屈折計
microwave region	마이크로파 영역	-波領域
microwave remote sensing	마이크로파 원격탐사	-波遠隔探査
microwave satellite imagery	마이크로파 위성영상	-波衛星映像
microwave scattering	마이크로파 산란	-波散亂
microwave sounding unit	마이크로파 탐측계	-波探測計
microwave spectrum	마이크로파 스펙트럼	-波-
microwave weighting function	마이크로파 가중함수	-波加重函數
midday	정오, 한낮	正午
midday depression of photosynthesis	광합성의 일중저하현상	光合成日中低下現象
middle atmosphere	중층대기	中層大氣
middle atmosphere program	중층대기연구계획	中層大氣研究計畫
middle cloud	중층운	中層雲
middle latitude	중위도	中緯度
middle latitude air	중위도기단	中緯度氣團
middle latitude cyclone	중위도저기압	中緯度低氣壓
middle latitude westerlies	중위도편서풍	中緯度偏西風
middle latitude westerlies zone	중위도편서풍대	中緯度偏西風帶
middle-level cloud	중층운	中層雲
middle tropopause	중위도대류권계면	中緯度大流圈界面
midget	콩태풍	豆颱風
midget tropical cyclone	꼬마열대저기압	-熱帶低氣壓
midget tropical storm(= midget typhoon)	꼬마열대폭풍(지름 100km이내)	-熱帶颱風
mid-infrared region	중간 적외영역	中間赤外領域
midnight	자정, 한밤	子正

M

midnight sun	심야태양(극지방)	深夜太陽
mid-Pliocene warming period	중기선신세 온난기	中期鮮新世溫暖期
mid-range forecast	중기예보	中期豫報
mid-season month	계절중앙달(1, 4, 5, 11월)	季節中央-
mid-summer	한여름	
mid-term forecast	주간예보	週間豫報
mid-value	중앙값	中央-
Mie scattering	미산란	-散亂
Mie theory	미이론	-理論
migration	이동	移動
migratory	이동성	移動性
migratory anticyclone	이동성고기압	移動性高氣壓
migratory col	이동성안장부	移動性鞍裝部
migratory cyclone	이동성저기압	移動性低氣壓
migratory extratropical anticyclone	이동성온대저기압	移動性溫帶低氣壓
migratory high	이동성고기압	移動性高氣壓
MI Interface Unit	MI 접속장치	-接續裝置
mil	밀(1인치의 1/1000), 밀(원 둘레의 1/6400의 호에 대한 각)	
Milankovitch cycle	밀란코비치순환	-循環
Milankovitch oscillation	밀란코비치진동	-振動
Milankovitch Pleistocene climatic variation	밀란코비치홍적세 기후변동	-洪積世氣候變動
mild	온화한, 포근한	溫和
mile	마일	
military engineer operation weather	공병작전기상	工兵作戰氣象
military fire code	군사산불부호, 군사화재암호	軍事山-符號, 軍事火災暗號
military meteorology	군사기상학	軍事氣象學
military operation heat index	군사현업열지수	軍事現業熱指數
military operation icing condition standard	군사현업얼음기준	軍事現業-基準
military operation rainfall standard	군사현업강수량기준	軍事現業降水量基準
military operation snowfall standard	군사현업강설량기준	軍事現業降雪量基準
military operation weather	군사현업기상	軍事現業氣象
military operation wind standard	군사현업풍속기준	軍事現業風俗基準
military weather agency	군사기상기관	軍事氣象機關
military weather flash	군사기상특보	軍事氣象特報
military weather information network	군사기상정보망	軍事氣象情報網
military weather mobile maintenance	군사이동정비기상	軍事移動整備氣象
military weather mobile observation	군사기상이동관측	軍事氣象移動觀測
military weather modification	군사기상변조	軍事氣象變造
military weather observation environment	군사기상관측환경	軍事氣象觀測環境
milk ripe stage	유숙기	乳熟期
milky ice	우윳빛얼음	牛乳-
milky way	은하수	銀河水
milky weather	우윳빛하늘	牛乳-

M

millennial climate variability	천년단위 기후변동성	千年單位氣候變動性
milli-	밀리	
millibar	밀리바	
millibar-barometer	밀리바기압계	-氣壓計
millibar scale	밀리바눈금	
millimeter wave radar	밀리미터파 레이더	-波-
millipore filter	미세구멍거르개	微細-
min detectable signal	최소탐지신호	最小探知信號
mine atmosphere	광산대기	鑛山大氣
Minesweeping Operation weather	소해작전기상	掃海作戰氣象
mine watching weather	기뢰감시기상	機雷監視氣象
minimal media	최소배지	最少培地
minimum	최소, 최저, 극소	最小, 最低, 極小
minimum air temperature	최저온도	最低溫度
minimum altitude	최저고도	最低高度
minimum daily temperature	일최저기온	日最低氣溫
minimum detectable range	최소탐지거리	最小探知距離
minimum detectable signal	최소탐지신호	最小探知信號
minimum deviation	최소편향	最小偏向
minimum fetch	최소취송거리	最小吹送距離
minimum humidity	최저습도	最低濕度
minimum information method	최소정보추정법	最小情報推定法
minimum ionizing speed	최소이온화속도	最小-化速度
minimum safe altitude	최저안전고도	最低安全高度
minimum temperature	최저온도	最低溫度
minimum thermometer	최저온도계	最低溫度計
minimum time track	최단시간항로	最短時間航路
(= minimal flight path, least-time track)		
minimum value	최솟값, 극솟값	最小-, 極小-
minimum variance method	최소분산법	最小分散法
minimum visibility	최단시정	最短視程
minimum visible radiance	최소가시영역휘도	最小可視領域輝度
minor circulation	작은 순환, 부순환	副循環
minor cold	소한	小寒
minor heat	소서	小暑
minor planet	소행성	小行星
minor ridge	부(기압)마루	副(氣壓)-
minor snow	소설	小雪
minor trough	부(기압)골	副(氣壓)-
minor wave	부파	副波
mintra	민트라(특정기압에서 비행운이 생기는 최고기온)	
minuend	빼지는 수	-數
minute	분, 미소(의)	分, 微小
minute displacement	미소변위	微小變位

minute scale motion	미소규모운동	微小規模運動
mirage	신기루	蜃氣樓
mirror nephoscope	구름측정거울	-測定-
miscalibration	교정오류, 위 캘리브레이션	校定誤謬, 僞-
missile firing weather	미사일 발사기상	-發射氣象
mission to planet earth	행성지구연구	行星地球研究
mist	엷은 안개, 박무	薄霧
mistbow	흰 무지개	
mist droplet	박무우적, 박무방울	薄霧雨滴
mist interval	박무발생온도역	薄霧發生溫度域
mistral	미스트랄(프랑스 Rhone 계곡에서 Lions 만으로 부는 북풍)	
mixed cloud	혼합구름, 혼합무	混合-, 混合霧
mixed icing	혼합착빙	混合着氷
mixed layer	혼합층	混合層
mixed layer depth	혼합층깊이	混合層-
mixed layer height	혼합층높이	混合層-
mixed nucleus	혼합핵	混合核
mixed rain and snow	진눈깨비	
mixed Rossby-gravity wave	혼합로스비중력파	混合-重力波
mixed tide	혼합조	混合潮
mixed type cool summer damage	혼합형 여름냉해	混合型-障害
mixing	혼합	混合
mixing air mass	혼합기단	混合氣團
mixing condensation	혼합응결	混合凝結
mixing condensation level	혼합응결고도	混合凝結高度
mixing cooling	혼합냉각	混合冷却
mixing depth	혼합깊이	混合-
mixing fog	혼합안개	混合-
mixing height	혼합높이	混合-
mixing layer	혼합층	混合層
mixing length	혼합길이	混合-
mixing ratio	혼합비	混合比
mixing ratio line	등혼합비선	等混合比線
mixture	혼합물	混合物
Mizon method	미존법	-法
mizzle	는개	
M meter	M 미터	
Moazagotl	모아짜고틀(도이칠란드 남동쪽 수데텐 산맥과 그 풍하 측에 형성되는 구름 둑)	
mobile radar	이동형레이더	移動形-
mobile ship	이동선박	移動船舶
mobile ship station	이동선박관측소	移動船舶觀測所
mobile source	이동발생원	移動發生源

M

mobile weather observation system	이동식기상관측시스템	移動式氣象觀測-
mobile weather station	이동기상관측소	移動氣象觀測所
mobility	이동도, 이동성	移動度, 移動性
Moby Dick balloon	정고도기구	定高度氣球
mock fog	무리안개	
mock moon	무리달	
mock sun	무리해	
mock sun ring	무리해테	
mode	방식, 모드, 최빈값	方式, 最頻-
model	모형	模型
model atmosphere	모형대기	模型大氣
model expected tropical number	모형예상열대수	模型豫想熱帶數
model fluid	모형유체	模型流體
modeling criteria	모형기준	模型基準
model of numerical weather prediction	수치예보모형	數值豫報模型
model of photosynthesis of a community	군락광합성모형	群落光合成模型
model output statistics	모형출력통계	模型出力統計
model output statistics technique	모형산출통계기술	模型算出統計技術
modem	모뎀, 변복조장치	變復調裝置
mode radius	최빈값반지름, 최빈값반경	最頻-半徑
moderate breeze	건들바람(풍력계급 4)	
moderate fog	보통안개	普通-
moderate gale	센 바람(풍력계급 7)	
moderate rain	보통비	普通-
moderate sea	보통바다(파고 1~1.5m)	普通-
moderate visibility	보통시정	普通視程
modification	변조, 개조	變造, 改造
modified air-mass	변질기단	變質氣團
modified index of refraction	수정굴절률	修正屈折率
modified Korteweg-de Vries equation	변형 KdV 방정식	變形-方程式
modified refractive index	수정굴절률	修正屈折率
MODIS optical bands	모디스 광학파장역	-光學波長域
modular payload interface unit	위성탑재체 접속장치	衛星搭載體接續裝置
modulation	조절, 변조	調節, 變調
modulation transfer function	영상변환/변조 전달기능	映像變換/變造傳達機能
modulator	변조기	變調器
module	모듈	
modulus	율, 계수, 절댓값	率, 係數, 絶對-
Mögel-Dellinger effect	뫼겔-델린저 효과	-效果
moist adiabat	습윤단열선	濕潤斷熱線
moist adiabatic cooling	습윤단열냉각	濕潤斷熱冷却
moist adiabatic lapse rate	습윤단열감률	濕潤斷熱減率
moist adiabatic process	습윤단열과정	濕潤斷熱過程
moist air	습윤공기	濕潤空氣

moist climate	습윤기후	濕潤氣候
moist convection	습윤대류	濕潤對流
moist indifferent	습윤중립	濕潤中立
moist-labile energy	습윤불안정 에너지	濕潤不安定-
moist-lability	습윤불안정도	濕潤不安定度
moist snow-flake	습윤눈송이	濕潤-
moist static energy	습윤정적에너지	濕潤靜的-
moist subhumid climate	습윤아습기후	濕潤亞濕氣候
moist tongue	습윤혀	濕潤-
moisture	수분, 습기	水分, 濕氣
moisture adjustment	수분조절	水分調節
moisture boundary	수분경계값	水分境界-
moisture characteristic curve	수분특성곡선	水分特性曲線
moisture content	수분함량	水分含量
moisture continuity equation	수분연속방정식	水分連續方程式
moisture convergence	수분수렴	水分收斂
moisture equilibrium	수분평형	水分平衡
moisture equivalent	수분당량	水分當量
moisture factor	수분인자	水分因子
moisture index	수분지수	水分指數
moisture indicator	수분지시기	水分指示器
moisture inversion	수분역전	水分逆轉
moisture of rupture of capillary bond	모관단절수분점	毛管斷切水分點
moisture profile	수분연직분포	水分鉛直分布
moisture retrieval scheme	수분복원방안	水分復原方案
moisture-temperature index	수분-온도 지수	水分溫度指數
molar (specific) heat	몰비열	-比熱
molar heat of evaporation	몰증발열	-蒸發熱
molar heat of fusion	몰융해열	-融解熱
molar heat of fusion(melting)	몰녹음열	-熱
molar heat of liquefaction	몰액화열	-液化熱
molar heat of solidification	몰응고열	-凝固熱
molar heat of vaporization	몰기화열	-氣化熱
molarity	몰농도	-濃度
molar volume	몰부피	
mold rain	장맛비	
mole	몰	
molecular arrangement	분자배열	分子配列
molecular conduction	분자전도	分子傳導
molecular crystal	분자결정	分子結晶
molecular diffusion	분자확산	分子擴散
molecular diffusion coefficient	분자확산계수	分子擴散係數
molecular flow	분자류	分子流
molecular force	분자력	分子力

M

molecular formula	분자식	分子式
molecular mass	분자량	分子量
molecular model	분자모형	分子模型
molecular motion	분자운동	分子運動
molecular-scale temperature	분자규모온도	分子規模溫度
molecular scattering	분자산란	分子散亂
molecular structure	분자구조	分子構造
molecular theory	분자이론	分子理論
molecular viscosity	분자점성	分子粘性
molecular viscosity coefficient	분자점성계수	分子粘性係數
molecular weight	분자량	分子量
molecule	분자	分子
mole fraction	몰분율	-分率
mole fraction of saturation water vapour of moist air with respect to ice	습윤공기 빙면 포화수증기 몰비	濕潤空氣水面飽和水蒸氣 -比
mole fraction of saturation water vapour of moist air with respect to water	습윤공기 수면 포화수증기 몰비	濕潤空氣水面飽和水蒸氣 -比
mole fraction of water vapo(u)r	수증기몰비	水蒸氣-比
mole number	몰수	-數
mollisol	몰리솔(유기물이 풍부한 토양의 일종)	
Moll thermopile	몰열전퇴	-熱電堆
Molniya orbit	몰니냐궤도	-軌道
Molniya satellites	몰니냐위성	-衛星
Moltchanov board	몰차노프판	-板
moment	순간, 모멘트, 능률	瞬間, 能率
moment of inertia	관성능률, 관성모멘트	慣性能率
moment of momentum	운동량능률	運動量能率
moment of rotation	회전능률	回轉能率
momentum	운동량	運動量
momentum conservation law	운동량보존법칙	運動量保存法則
momentum equation	운동량방정식	運動量方程式
momentum flux	운동량플럭스	運動量-
momentum transfer	운동량전달	運動量傳達
Monge's phenomenon	몽게현상(북아프리가 사막에서 일어나는 광학현상)	-現象
Monin-Obukhov equation	모닌-오부코프 방정식	-方程式
Monin-Obukhov length	모닌-오부코프 길이	
Monin-Obukhov scaling length	모닌-오부코프 규모길이	-規模-
monitor	감시자, 모니터	監視者
monitoring	감시, 모니터링	監視
monitor of ultraviolet solar radiation	태양자외선복사감시기	太陽紫外線輻射監視器
monochromatic	단색(의)	單色
monochromatic light	단색광	單色光
monochromatic radiance	단색복사	單色輻射

M

monochromatic radiation	단색복사	單色輻射
monodispersed size distribution	균일크기분포, 단순크기분포	均一-分布, 單純-分布
monomolecular film	단분자두께의 필름	單分子-
monostatic radar	단안정 레이더	單安定-
monoterpene	모노터핀	
monsoon	계절풍, 몬순	季節風
monsoon air	계절풍기단, 계절풍공기	季節風氣團, 季節風空氣
monsoon air-mass	계절풍기단	季節風氣團
monsoon Asia	몬순아시아	
monsoon burst	계절풍버스트	季節風-
monsoon circulation	계절풍순환	季節風循環
monsoon climate	계절풍기후	季節風氣候
monsoon cluster	계절풍구름무리	季節風-
monsoon current	계절풍해류, 계절풍표류	季節風海流, 季節風漂流
monsoon depression	계절풍저기압	季節風低氣壓
monsoon drift	계절풍해류	季節風海流
monsoon fog	계절풍안개	季節風-
monsoon gyre	계절풍자이어, 몬순자이어	季節風-
monsoon index	계절풍지수	季節風指數
monsoon low	계절풍저기압	季節風低氣壓
monsoon rain	계절풍비	季節風-
monsoon rainfall	계절풍강우	季節風降雨
monsoon season	계절풍기	季節風期
monsoon surge	계절풍파도, 계절풍서지	季節風波濤
monsoon trough	계절풍기압골	季節風氣壓-
monsoon wind	계절풍	季節風
monster ice sheet	괴물빙상	怪物氷床
MONT code	MONT 전보식	-電報式
Monte Carlo method	몬테카를로방법	-方法
Montgomery stream function	몽고메리 유선함수	-流線函數
monthly bulletin	월보	月報
monthly forecast	월간예보	月間豫報
monthly maximum temperature	월최고기온	月最高氣溫
monthly mean	월평균	月平均
monthly minimum temperature	월최저기온	月最低氣溫
monthly precipitation	월강수량	月降水量
monthly record	월별기록	月別記錄
monthly total precipitation	월총강수량	月總降水量
monthly weather report	기상월보	氣象月報
Montreal Protocol	몬트리올의정서	-議定書
moonbow	달무지개	
moon dog	무리달	
moonlight	달빛	
moon pillar	달기둥	

M

moon's path	백도	白道
moor-gallop	황야폭풍	荒野暴風
moraine	빙퇴석	氷堆石
morbidity	이환율	罹患率
morning calm	아침고요	
morning glow	아침놀	
morning satellite	아침위성	-衛星
morning tide	아침조석	-潮汐
morphism	사상	寫像
Morse code	모스부호	-符號
mortality	사망률	死亡率
most frequent size	최빈크기	最頻-
most frequent wind direction	최다풍향	最多風向
mostly cloudy	구름많음	
mother-cloud	어미구름	
mother current	본류	本流
mother of pearl cloud	진주모운	眞珠母雲
motion	운동	運動
motion of falling body	낙체운동	落體運動
motion of free fall	자유낙하운동	自由落下運動
motion of projectile	포물선운동	抛物線運動
motor vehicle emission standard	자동차배출기준	自動車排出基準
mottling	모틀링(반점이 있는)	
mountain air	산악공기	山岳空氣
mountain and valley breeze	산곡풍	山谷風
mountain and valley wind	산골바람	山-
mountain area operation weather	산악지역작전기상	山岳地域作戰氣象
mountain barometer	산악기압계	山岳氣壓計
mountain barrier	산악장벽	山岳障壁
mountain breeze	산바람	山-
mountain climate	산악기후	山岳氣候
mountain fog	산안개	山-
mountain-gap wind	산사이바람	山-
mountain glacier	산악빙하, 고산빙하	山岳氷河, 高山氷河
mountain lee wave	산악풍하파	山岳風下波
mountain meteorology	산악기상학	山岳氣象學
mountain observation	산악관측	山岳觀測
mountainous area	산간, 산지	山間, 山地
mountainous terrain	산악지형	山岳地形
mountain sickness	고산병	高山病
mountain standard time	산악표준시(미국)	山岳標準時
mountain station	산악관측소	山岳觀測所
mountain torque	산악토크	山岳-
mountain tundra	고산툰드라	高山-

mountain turbulence	산악난류	山岳亂流
mountain uplift	산악융기	山岳隆起
mountain-valley wind	산곡풍	山谷風
mountain wave	산악파	山岳波
mountain wave cloud	산악파구름	山岳波-
mountain wave turbulence	산악파난류	山岳波亂流
mountain weather	산악기상	山岳氣象
mountain weather station	산악기상관측소	山岳氣象觀測所
mountain wind	산바람	山-
movable-scale barometer	이동자기압계	移動-氣壓計
moving average	이동평균	移動平均
moving fetch	이동취주거리	移動吹走距離
moving observation station	이동관측소	移動觀測所
moving target indicator	이동목표물지시기	移動目標物指示器
Mozambique current	모잠비크해류	-海流
Möller chart	묄러복사도	-輻射圖
M-region	M영역	-領域
mucin	점액	粘液
MU radar	MU레이더	
mud avalanche	진흙사태	-沙汰
mud flow	토사류	土砂流
mud rain	흙비	
muggy weather	무더운 날씨	
mulching	멀칭(표면 피복)	
multiangular observation	다각도관측	多角度觀測
multi-cell	다세포	多細胞
multi-cell type	다세포형	多細胞形
multi-cell convective storm	다세포대류폭풍	多細胞對流暴風
multi-cell storm	다세포폭풍, 다중세포스톰	多細胞暴風, 多重細胞-
multi-cell thunderstorm	다세포뇌우	多細胞雷雨
multi-cellular	다세포	多細胞
Multi-Channel Sea Surface Temperature	다중채널해수면온도	多重-海水面溫度
multi-color spin scan cloud camera	다색회전주사 구름카메라	多色回轉走査-
multi-dimensional warfare weather	입체전기상	立體戰氣象
multi-ensemble prediction model	다중앙상블예측모형	多重-豫測模型
multi-level model(= multilayer model)	다층모형	多層模型
multiparameter radar	다모수레이더	多母數-
multipath transmission	다중경로전송	多重經路傳送
multiple bond	다중결합	多重結合
multiple correlation	다중상관	多重相關
multiple correlation coefficient	다중상관계수	多重相關係數
multiple discharge	다중방전	多重放電
multiple discriminant analysis	다중판별분석	多重判別分析
multiple doppler analysis	다중도플러분석	多重-分析

M

multiple drift corrections	겹편류보정	-偏流補正
multiple equilibria	다중평형	多重平衡
multiple incursion theory	다중승강이론	多重昇降理論
multiple integral	중적분	重積分
multiple linear regression functions	다중선형회귀함수	多重線型回歸函數
multiple point	다중점	多重點
multiple-purpose reservoir	다목적저수지	多目的貯水池
multiple reflection	다중반사	多重反射
multiple register	다중기록계	多重記錄計
multiple regression	다중회귀	多重回歸
multiple regression analysis	다중회귀분석	多重回歸分析
multiple scattering	다중산란	多重散亂
multiple stroke	다중뇌격	多重雷擊
multiple trip return	다중왕복에코	多重往復-
multiple tropopause	겹대류권계면	-對流圈界面
multiple-window technique	다중창기법	多重窓技法
multiplexer	집중화장치	集中化裝置
multiport fuel injection	다중포트 연료분사	多重-燃料噴射
multiscale observation	다중규모관측	多重規模觀測
multiseasonal approach	다중계절접근	多重季節接近
multi-sensor	다중센서	多重-
multispectral analysis	다중분광분석	多重分光分析
multispectral camera	다중카메라	多重-
multispectral enhancement	다중분광강조	多重分光强調
multispectral imaging	다중분광영상	多重分光映像
multispectral scanner	다중스펙트럼 스캐너	多重-
multi-static radar	다중안테나 레이더	多重-
multitemporal analysis verification	다중시간 분석검증	多重時間分析檢證
multitemporal ratio	다중시간비율	多重時間比率
multitemporal regression analysis	다중시간 회귀분석	多重時間回歸分析
multivariate objective analysis	다변수객관분석	多變數客觀分析
M-unit	M-단위(수정굴절률측정단위)	-單位
mushroom cloud	버섯구름	
muslin	거즈, 모슬린	
mutagen	돌연변이(유발)원	突然變異(誘發)原
mutagenicity	돌연변이성	突然變異性
mutation theory	돌연변이설	突然變異說
mutatus	전화구름	轉化-
mutual coherence function	상호간섭함수	相互干涉函數
mutual inductance	상호인덕턴스	相互-
mutual induction	상호유도	相互誘導
muzzler	센 역풍	-逆風
mycotoxin	진균독	眞菌毒
Mylar balloon	마일라기구(폴리에스텔로 만든)	-氣球

N, n

영문	한글	한자
nabla operator(= del operator)	나블라연산자	-演算子
NACA standard atmosphere	NACA표준대기	-標準大氣
nacreous cloud	자개구름	
nadir	천저	天底
nadir angle	천저각	天底角
nadir sounding of temperature	천저온도탐측	天底溫度探測
nadir viewing	직하관측	直下觀測
name of TC	열대성저기압 이름	熱帶性低氣壓-
name of typhoon	태풍이름	颱風名
nano	나노(10^{-9})	
nanometer	나노미터	
nanotechnology	나노기술	-技術
Nansen bottle	난센병(해수채집용)	-瓶
nappe	냅(댐을 넘쳐흐르는 물)	
narboné(= narbonnasis)	나르보네(프랑스의 나르본으로부터 불어오는 바람)	
narrow angle camera	좁은 각 카메라	-角-
narrow-band radiation	좁은 띠 복사, 좁은 파장역 복사	-波長域輻射
narrow-band sensor	좁은 띠 센서, 좁은 파장역 센서	-波長域-
narrow beam radiogoniometer	좁은 빔무선방향 탐지기	-無線方向探知器
narrow frontal cloud band	좁은 전선구름대	-前線-帶
narrow-sector recorder	좁은 지역 기록계	-地域記錄計
NASA scatterometer	NASA 스캐터로미터	
nascent cyclone	초기저기압	初期低氣壓
nashi(= n'aschi)	나시(페르시아 만에서 겨울에 일어나는 북동풍의 아라비아말)	
National Acid Precipitation Assessment Program	(미국)국립산성비평가프로그램	國立酸性-評價-
National Adaptation Plans of Action	(미국)국가적응계획	國家適應計劃

National Advisory Committee for Aeronautics	미국항공자문위원회	美國航空諮問委員會
National Aeronautics and Space Administration	미국항공우주국	美國航空宇宙局
National Air Monitoring Stations	(미국)국립대기관측소	國立大氣觀測所
National Ambient Air Quality Standards	(미국)국립대기환경기준	國立大氣環境基準
National Atmospheric Emission Inventory	국가대기배출명세서	國家大氣排出明細書
National Center for Atmospheric Research	(미국)국립대기연구센터,	國立大氣研究,
	미국국립대기연구소	美國國立大氣研究所
National Center for Environmental Prediction	(미국)국립환경예측센터	國立環境豫測-
National Climate Center	국립기후센터	國立氣候-
National Climate Program Act	(미국)국가기후프로그램법	國家氣候-法
National Climatic Data Center	(미국)국립기후자료센터	國立氣候資料-
National Council on Radiation Protection	(미국)국립방사선방호측정	國立放射線防護測定
and Measurements	심의회	審議會
National Crop Loss Assessment Network	(미국)국가작물손실평가망	國家作物損失評價網
National Data Buoy Center	(미국)국립자료부이센터	國立資料浮漂-
National Digital Forecast Database	(미국)국립디지털예보	國立-豫報
	데이터베이스	
National Emissions Standards for	위해대기오염물질국가	危害大氣汚染物質國家
Hazardous Air Pollutants	배출기준	排出基準
National Environmental Satellite, Data,	(미국)국립환경위성-자료-	國立環境衛星資料情報-
and Information Service	정보서비스	
National Environmental Technology Centre	(영국)국립환경기술센터	國立環境技術-
National Fire Danger Rating System	(미국)국립산불위험률시스템	國立山-危險率-
National Flood Summary	(미국)국가홍수요약	國家洪水要約
National Food Insurance Program	(미국)국가식량보험프로그램	國家食糧保險-
National Hurricane Center	(미국)허리케인센터	
National Hurricane Operations Plan	(미국)국가허리케인운용계획	國家-運用計劃
National Ice Core Laboratory	(미국)국립아이스코어연구소	國立-研究所
National Materials Exposure Programme	(영국)국가물질노출프로그램	國家物質露出-
National Meteorological Centre	국립기상센터	國立氣象-
National Oceanic and Atmospheric Administration	(미국)국립해양대기청,	國立海洋大氣廳
	노아(NOAA)	
National Radiation Centre	국립방사선센터	國立放射線-
National Rainfall Index	(미국)국가강우지수	國家降雨指數
National Research Council	(미국)국립연구협의회	國立研究協議會
National Severe Storms Forecast Center	(미국)국립악뇌우예보센터	國立惡雷雨豫報-
National Snow and Ice Data Center	(미국)국립설빙자료센터	國立雪氷資料-
National Space Science Data Center	(미국)국립우주과학자료센터	國立宇宙科學資料-
National Standard Barometer	(미국)국가표준기압계	國家標準氣壓計
National Tidal Datum Epoch	(미국)국가조위기준년시대	國家潮位基準年時代
National Weather Association	(미국)국립기상협회	國立氣象協會
National Weather and Crop Summary	(미국)국가날씨와작물요약	國家氣象作物要約
National Weather Service	(미국)국립기상대	國立氣象臺
natural aerosol	자연에어로졸	自然-

natural cause	자연원인	自然原因
natural control	자연방제	自然防除
natural convection	자연대류	自然對流
natural coordinates	자연좌표	自然座標
natural coordinate system	자연좌표계	自然座標系
natural disaster	자연재해	自然災害
natural-draft cooling tower	자연통풍식 냉각탑	自然通風式冷却塔
natural flow	자연흐름, 자연유하량	自然流下量
natural frequency	고유진동수	固有振動數
natural gas	천연가스	天然-
natural hail embryo	자연우박눈	自然雨雹-
natural logarithm	자연대수	自然對數
natural mutation	자연돌연변이	自然突然變異
natural oscillation	고유진동, 자연진동	固有振動
natural period	고유주기, 자연주기	固有週期, 自然週期
natural phenomena	자연현상	自然現象
natural radioactivity	자연방사능, 천연방사능	自然放射能, 天然放射能
natural removal process	자연제거과정	自然除去過程
natural season	자연계절	自然季節
natural seasonal phenomenon	자연계절현상	自然季節現象
natural selection	자연도태	自然淘汰
natural snow enhancement	자연증설	自然增雪
natural source	자연오염원	自然汚染源
natural sulfur cycle	자연황순환	自然黃循環
natural synoptic period	자연종관기간	自然綜觀期間
natural synoptic region	자연종관영역	自然綜觀領域
natural synoptic season	자연종관계절	自然綜觀季節
natural variability of climate	기후자연변동성	氣候自然變動性
natural vegetation	자연식생	自然植生
nature	자연	自然
nautical dawn	항해여명	航海黎明
nautical dusk	항해황혼	航海黃昏
nautical mile	해리	海里
nautical system	항해시스템	航海-
nautical twilight	항해박명	航海薄明
naval meteorology	해양기상학	海洋氣象學
naval sea tactical information wing	해군해양전술정보단	海軍海洋戰術情報團
Naval Special Warfare weather	해군특수전기상	海軍特殊戰氣象
Navidad current	나비다드해류	-海流
Navier-Stokes equations	나비에-스토크스 방정식	-方程式
Navier-Stokes equations of motion	나비에-스토크스 운동방정식	-運動方程式
navigable semicircle	가항반원	可航半圓
navigation	항해, 항행	航海, 航行
navigation method	항법	航法

N

NAVTEX Forecast	NAVTEX 예보	-豫報
NBC Warfare weather	화생방전기상	化生放戰氣象
N-curve	N-곡선	-曲線
Neamtan solution	님탠해	-解
neap range	조금차	潮-差
neap tide	조금	潮-
near field	근거리장	近距離場
near gale	센 바람(풍력계급 7)	
near infrared	근적외선	近赤外線
Near Infrared Mapping Spectrometer	(갈릴레오 위성에 탑재된) 근적외선컴퓨터분광계	近赤外線-分光計
near-infrared radiation	근적외복사	近赤外輻射
near-infrared region	적외선영역부근	赤外線領域附近
near-infrared spectrum	적외선영역부근 스펙트럼	赤外線領域附近-
near-polar orbiting satellite	근극궤도위성	近極軌道衛星
near-real-time	근실시간	近實時間
nearshore forecast	연안예보	沿岸豫報
near-surface troposphere	지상부근대류권	地上附近對流圈
near-surface turbulence process	지상부근난류과정	地上附近亂流過程
neb	안개모양구름	-模樣-
nebewind(= fog wind)	안개바람(안데스산)	
Nebraskan glacial	네브래스카빙하기	-氷河期
NEBUL code	NEBUL 전보식	-電報式
nebula	성운	星雲
nebula hypothesis	성운가설	星雲假說
nebular theory	성운이론	星雲理論
nebule	네뷸(대기불투명도의 단위)	
nebulosus	안개모양(구름)	-模樣-
necessary ventilation amount	필요환기량	必要換氣量
necrosis	괴사	壞死
needle ice	서릿발	
Néel temperature(= Curie temperature)	닐온도	-溫度
negative area	음영역	陰領域
negative axis	음축	陰軸
negative buoyancy	음부력	陰浮力
negative charge	음전하	陰電荷
negative cloud-to-ground lightning (= negative ground flash)	역상구름-지면번개	逆狀-地面-
negative correlation	음상관관계	陰相關關係
negative electricity	음전기	陰電氣
negative electron	음전자	陰電子
negative feedback	음되먹임	陰-
negative feedback mechanism	음되먹임기작	陰-機作
negative feedback system	음되먹임시스템	陰-

negative ground flash	역상지면섬광	逆狀地面閃光
negative ion(= anion)	음이온	陰-
negative isothermal vorticity advection	음등온소용돌이도이류	陰等溫-度移流
negative pole	음극	陰極
negative rain	음대전비	陰帶電-
negative sign	음부호	陰符號
negative temperature	영하온도	零下溫度
negative-tilt trough	음경사골	陰傾斜-
negative viscosity	음점성	陰粘性
negative vorticity	음소용돌이도	陰-度
negative vorticity advection	음소용돌이도이류	陰-度移流
negatron	음전자	陰電子
nemere	네메레(헝가리의 차고 강한 치내리바람)	
Neogene climate	신생대기후	新生代氣候
Neoglacial	신빙하	新氷河
neoglaciation	신빙하작용	新氷河作用
neon	네온(물리량 측정비에 대한 지수단위)	
neper	네퍼	
nephanalysis	구름분석	-分析
neph chart	구름분석도	-分析圖
nephcurve	구름경계선	-境界線
nephelometer	네펠로미터, 탁도계, 비탁계	濁度計, 比濁計
nephelometry	구름측정법	-測定法
nepheloscope	제운기, 측운기	製雲器, 測雲器
nephology	구름학	-學
nephometer	운량계	雲量計
nephoscope	측운기, 구름측정기	測雲器, -測定器
nephsystem	구름시스템	
Neptune	해왕성	海王星
Nernst's theorem	네른스트정리	-定理
nested grids	둥지격자	-格子
net	순	純
net all-wave radiation	순전파동복사	純全波動輻射
net balance	순균형	純均衡
net condensation	순응결	純凝結
net energy storage	순에너지저장	純-貯藏
net evaporation	순증발	純蒸發
net force	순힘	純-
net outgoing IR	순방출적외선	純放出赤外線
net photosynthesis	순광합성	純光合成
net primary production	순일차생산	純一次生産
net pyranometer	순전천일사계	純全天日射計
net pyrgeometer	순전천장파복사계	純全天長波輻射計
net pyrradiometer	순전천복사계	純全天輻射計

N

net radiation	순복사	純輻射
net radiation flux	순복사속, 순복사플럭스	純輻射束
net radiative cooling	순복사냉각	純輻射冷却
net radiative flux density	순복사플럭스밀도, 순복사속밀도	純輻射束密度
net radiative forcing	순복사강제	純輻射强制
net radiometer	순복사계	純輻射計
net rainfall	순강우	純降雨
net solar radiation	순태양복사	純太陽輻射
net storm rain	순스톰비	純-
net terrestrial radiation	순지구복사	純地球輻射
net upward movement	순상향이동	純上向移動
network density	관측망밀도	觀測網密度
network of observation	관측망	觀測網
network of station	관측망	觀測網
Neuhoff diagram	노이호프선도	-線圖
Neumann problem	노이만문제	-問題
neutercane	중립열대저기압	中立熱帶低氣壓
neutral advection	중립이류	中立移流
neutral air	중립공기	中立空氣
neutral air mass	중립기단	中立氣團
neutral atmosphere	중립대기	中立大氣
neutral atmospheric boundary layer	중립대기경계층	中立大氣境界層
neutral boundary layer	중립경계층	中立境界層
neutral condition	중립조건	中立條件
neutral cyclone	중립저기압	中立低氣壓
neutral drag coefficient	중립항력계수	中立抗力係數
neutral equilibrium	중립평형	中立平衡
neutrality	중립, 중성	中立, 中性
neutralization	중화, 평형	中和, 平衡
neutral line	중립선	中立線
neutral mode	중립모드	中立-
neutral occlusion	중립폐색	中立閉塞
neutral oscillation	중립진동	中立振動
neutral particle	중성입자	中性粒子
neutral plane	중립면	中立面
neutral point	중립점	中立點
neutral stability	중립안정(도)	中立安定(度)
neutral stability wind	중립안정도바람	中立安定度-
neutral stratification	중립성층	中立成層
neutral surface	중립면	中立面
neutral temperature	중립온도	中立溫度
neutral temperature profile	중립온도연직분포	中立溫度鉛直分布
neutral wave	중립파	中立波

N

neutron	중성자	中性子
neutron moisture meter	중성자토양수분계	中性子土壤水分計
neutron probe	중성자탐측기	中性子探測器
neutron-scattering method	중성자산란법	中性子散亂法
neutropause	중성권계면	中性圈界面
neutrosphere	중성권	中性圈
nevada	네바다(산악의 설원 또는 빙원으로부터 불어내리는 찬바람의 스페인 말)	
nevados	네바도스(에콰도르 산맥과 높은 고원지대의 산맥을 따라서 부는 건조한 활강바람)	
névé(= firn)	만년설	萬年雪
névé line	영구설선, 만년설선	永久雪線, 萬年雪線
névé pénitent	만년적설	萬年積雪
new candle	신촉광(조도의 단위)	新燭光
New England seamounts	뉴잉글랜드해산	-海山
new equilibrium layer	신평형층	新平衡層
Newfoundland Ridge	뉴펀들랜드해령	-海嶺
New Guinea	뉴기니	
New Guinea Coastal Undercurrent	뉴기니연안잠류	-沿岸潛流
Newhall wind	(캐나다)뉴홀바람	
new moon	삭, 신월	朔, 新月
new snow	신적설	新積雪
newspaper forecast source	신문예보자료	新聞豫報資料
newton	뉴턴(힘의 SI 단위)	
Newtonian fluid	뉴턴유체	-流體
Newtonian friction law	뉴턴마찰법칙	-摩擦法則
Newtonian iteration	뉴턴반복	-反復
Newtonian mechanics	뉴턴역학	-力學
Newtonian speed of sound	뉴턴음속	-音速
Newtonian telescope	뉴턴식망원경	-式望遠鏡
Newtonian transformation	뉴턴변환	-變換
Newton-Raphson solution technique	뉴턴-래프슨 해기법	-解技法
Newton's attraction force	뉴턴인력	-引力
Newton's first law of motion	뉴턴운동제1법칙	-運動第一法則
Newton's hypothesis	뉴턴가설	-假說
Newton's law of attraction	뉴턴인력법칙	-引力法則
Newton's law of cooling	뉴턴냉각법칙	-冷却法則
Newton's law of friction	뉴턴마찰법칙	-摩擦法則
Newton's law of motion	뉴턴운동법칙	-運動法則
Newton's law of universal gravitation	뉴턴만유인력법칙	-萬有引力法則
Newton's law of viscosity	뉴턴점성법칙	-粘性法則
Newton's laws	뉴턴법칙	-法則
Newton's laws of motion	뉴턴운동법칙	-運動法則
Newton speed of sound	뉴턴음속	-音速

N

Newton's second law of motion	뉴턴운동제2법칙	-運動第二法則
New Zealand Meteorological Service	뉴질랜드기상청	-氣象廳
NEXRAD base data	NEXRAD 기초자료	-基礎資料
next generation doppler weather radar network	차세대도플러레이더망(미국)	次世代-網
next generation radar	차기레이더	-次期-
NHC method	엔에이치시방법	-方法
niche	생태적 지위	生態的地位
nieve penitente(= penitent ice)	니에베 페니텐테스(빙하지형을 가르키는 스페인어)	
Niger Delta	니제르삼각주	-三角洲
night	밤, 야간	夜間
night air-glow spectrum	야광스펙트럼	夜光-
night dew	밤이슬	
nightglow spectrum	야광, 야광스펙트럼	夜光
night jet	야간제트	夜間-
night-sky light	밤하늘빛	
night-sky luminescence	밤하늘발광	-發光
night-sky radiation	밤하늘복사	-輻射
night visibility	야간시정	夜間視程
night visual range	야간시계	夜間視界
night wind	밤바람	
Nile River	나일강(이집트)	
nimbostratus	난층운	亂層雲
nimbostratus pannus	조각난층운	-亂層雲
nimbostratus precipitation	강수난층운	降水亂層雲
nimbostratus virga	고리난층운	-亂層雲
nimbus	비구름, 적란운	積亂雲
nimbus cumuliformis	적운모양비구름	積雲模樣-
Nimbus N Microwave Spectrometer	Nimbus N 마이크로파분광계	-極超短波分光計
Nimbus satellite	님버스위성	-衛星
nine-light indicator	아홉등 풍향풍속지시기	-燈風向風速指示器
Ningaloo reef	닝갈루 사루(오스트레일리아)	
Niño region	니뇨지역	-地域
Nipher shield	니퍼막이	
nirta	니르타(북수마트라 토바호의 바람)	
nitrate	질산염	窒酸鹽
nitrate ion	질산염이온	窒酸鹽-
nitrate radical	질산염라디칼	窒酸鹽-
nitric acid	질산	窒酸
nitric acid trihydrate	질산삼수화물	窒酸三水化物
nitric oxide	일산화질소	一酸化窒素
nitrification	질화	窒化
nitrogen	질소	窒素
nitrogen assimilation	질소동화	窒素同化
nitrogen compound	질소화합물	窒素化合物

N

nitrogen cycle	질소순환	窒素循環
nitrogen deposition	질소침적	窒素沈積
nitrogen dioxide	이산화질소	二酸化窒素
nitrogen dioxide reduction	이산화질소 감소	二酸化窒素減少
nitrogen dioxide trend	이산화질소 경향	二酸化窒素傾向
nitrogen fixation	질소고정	窒素固定
nitrogen-fixing plant	질소고정식물	窒素固定植物
nitrogen isotope	질소동위원소	窒素同位元素
nitrogen molecule	질소분자	窒素分子
nitrogenous deposition	질소축적량	窒素蓄積量
nitrogen oxides	질소산화물	窒素酸化物
nitrogen oxides emission	질소산화물배출	窒素酸化物排出
nitrogen pentoxide	오산화질소	五酸化窒素
nitrogen saturation	질소포화	窒素飽和
nitrous acid	아질산	亞窒酸
nitrous oxide	아산화질소	亞酸化窒素
nival	니발(눈이 내리는 환경)	
nivation	눈의 침식, 설식	雪蝕
nivometer	설량계	雪量計
NOAA-class satellites	NOAA급 위성	-級衛星
NOAA satellite series	NOAA 위성시리즈	-衛星-
NOAA Weather Wire Service	NOAA날씨자료전달서비스	-資料傳達-
Noah's Ark	노아의 방주권운	-方舟卷雲
no-analog community	비유사집단	非類似集團
no-analog vegetation	비유사식생	非類似植生
Nobel Prize	노벨상	
noble gas(= inert gas)	불활성기체	不活性氣體
noctilucent cloud	야광운	夜光雲
nocturnal atmospheric boundary layer	야간대기경계층	夜間大氣境界層
nocturnal boundary layer	야간경계층	夜間境界層
nocturnal boundary layer depth	야간경계층깊이	夜間境界層-
nocturnal cooling	야간냉각	夜間冷却
nocturnal drainage flow	야간경사류	夜間傾斜流
nocturnal inversion	야간역전	夜間逆轉
nocturnal jet	야간제트	夜間-
nocturnal minimum temperature	야간최저온도	夜間最低溫度
nocturnal radiation	야간복사	夜間輻射
nocturnal ridge-top jet	야간능선꼭대기 제트	夜間-
nocturnal stable layer	야간안정층	夜間安定層
nocturnal thunderstorm	야간뇌우	夜間雷雨
nocturnal urban moisture level	야간도시습윤고도	夜間都市濕潤高度
nodal factor	마디점인자	-點因子
nodal increment	마디점증분	-點增分
nodal line	마디선	-線

N

nodal longitude increment	마디점경도증분	-點經度增分
nodal period	마디점주기	-點周期
node	마디, 교점	交點
nodule	단괴, 결절	團塊, 結節
no-emission equation	비방출방정식	非放出方程式
noise	잡음	雜音
noise bandwidth	잡음대역폭	雜音帶域幅
noise equivalent delta temperature	잡음상당온도변위(영상의)	雜音相當溫度變位
noise factor	잡음인자	雜音因子
noise figure	잡음지수	雜音指數
noise filtering	잡음여과	雜音濾過
noise guideline	잡음가이드라인	雜音-
noise level	잡음수준	雜音水準
noise level of atmospheric	공전잡음도	空電雜音度
noise power	잡음능, 잡음파워	雜音能
noise ratio	잡음비	雜音比
noise temperature	잡음온도	雜音溫度
noise threshold	잡음문턱값	雜音門-
nomenclature	명명법	命名法
nomogram	계산도표	計算圖表
nomograph	계산도표	計算圖表
nomographic chart(= nomogram)	계산도표	計算圖表
nomography	도표계산법	圖表計算法
non-adiabatic change	비단열변화	非斷熱變化
non-adiabatic cooling	비단열냉각	非斷熱冷却
non-adiabatic irreversible process	비단열비가역과정	非斷熱非可逆過程
non-adiabatic process	비단열과정	非斷熱過程
nonattainment	(환경기준)미달성	(環境基準)未達成
nonattainment area	비달성지역	非達成地域
nonattainment zone	(환경기준)미달성대	(環境基準)未達成帶
nonblackbody	비흑체	非黑體
nonbuoyant particle	비부력성입자	非浮力性粒子
nonclastic	비쇄설성	非碎屑性
noncoherent echo	비간섭에코	非干涉-
noncoherent integration	비정합적분	非整合積分
noncoherent radar	비간섭레이더	非干涉-
noncoherent target	비간섭표적	非干涉標的
non-compression	비압축성	非壓縮性
non-conductor	부도체	不導體
nonconformity	난정합	難整合
non-conservative property	비보존성	非保存性
non-convective flux	비대류성플럭스	非對流性-
non-degradation	불휘발성	不揮發性
non-deterministic	비결정론적	非決定論的

N

non-dimensional dissipation rate	무차원소산율	無次元消散率
non-dimensional equation	무차원방정식	無次元方程式
non-dimensional frequency	무차원진동수	無次元振動數
non-dimensional function	무차원함수	無次元函數
non-dimensional number	무차원수	無次元數
non-dimensional parameter	무차원매개변수	無次元媒介變數
non-dimensional quantity	무차원량	無次元量
non-dimensional specific humidity gradient	무차원비습경도	無次元比濕傾度
non-dimensional temperature gradient	무차원온도경도	無次元溫度傾度
non-dimensional temperature profile	무차원온도연직분포	無次元溫度鉛直分布
non-dimensional vertical flux	무차원연직플럭스	無次元鉛直-
non-dimensional wind shear	무차원윈드시어	無次元-
non-directivity	무지향성	無指向性
nondispersive infrared analyzer	비분산적외선 가스분석계	非分散赤外線-分析計
nondispersive infrared gas analyzer	비분산적외선 가스분석계	非分散赤外線-分析計
nondispersive infrared photometry	비분산적외선 광도측정법	非分散赤外線光度測定法
nondispersive infrared spectrometry	비분산적외선 분광법	非分散赤外線分光法
non-divergence	비발산	非發散
non-divergence layer	비발산층	非發散層
non-divergence level	무발산고도	無發散高度
non-divergent model	무발산모형	無發散模型
non-equilibrium statistical mechanics	비평형통계역학	非平衡統計力學
non-extensive entropy	비외연엔트로피	非外延-
non-frontal depression	비전선저기압	非前線低氣壓
non-frontal low	비전선성저기압	非前線性低氣壓
non-frontal squall line	비전선스콜선	非前線-線
non-frontal thunderstorm	비전선성뇌우	非前線性雷雨
non-frontal tidal analysis	비전선성조석분석	非前線性潮汐分析
non-Gaussian forcing	비가우스강제	非-强制
non-Gaussian turbulence	비가우스난류	非-亂流
non-geostationary orbit satellite	비정지궤도위성, 저궤도위성	非停止軌道衛星, 低軌道衛星
non-homogeneous boundary layer	비균질경계층	非均質境界層
non-hydrostatic	비정역학	非靜力學
non-hydrostatic model	비정역학모형	非靜力學模型
non-hydrostatic pressure	비정역학압력	非靜力學壓力
non-hygroscopic nucleus	비흡습성핵	非吸濕性核
noninductive charging mechanism	비유도충전메커니즘	非誘導充電-
non-isotropic turbulence	비등방성난류	非等方性亂流
nonlinear	비선형	非線形
nonlinear computational instability	비선형계산불안정	非線形計算不安定
nonlinear friction	비선형마찰	非線形摩擦
nonlinear instability	비선형불안정	非線形不安定
nonlinearity	비선형성	非線形性
nonlinear normal mode initialization	비선형 정상모드 초기화	非線形正常-初期化

N

nonlinear problem	비선형문제	非線形問題
nonlinear response	비선형반응	非線形反應
nonlinear stability	비선형안정성	非線形安定性
nonlinear turbulence	비선형난류	非線形亂流
nonlinear wave	비선형파	非線形波
non-local closure	비국지종결	非局地終結
non-local closure model	비국지종결모형	非局地終結模型
non-local closure scheme	비국지종결방식	非局地終結方式
non-local effect	비국지효과	非局地效果
non-local first order closure	비국지일차종결	非局地一次終結
non-local flux	비국지플럭스	非局地-
non-local mixing	비국지혼합	非局地混合
non-local static stability	비국지정적안정도	非局地靜的安定度
nonmeteorological scatter	비기상산란체	非氣象散亂體
nonmethane hydrocarbon	비메탄탄화수소	非-炭化水素
nonmethane organic gas	비메탄유기가스	非-有機-
non-Newtonian fluid	비뉴턴유체	非-流體
non-optimum method	비최적법	非最適法
nonorthogonal	비직교적	非直交的
nonpenetrative convection	비침투대류	非浸透對流
nonperiodic temperature change	비주기적 온도변화	非周期的溫度變化
nonradiative forcing	비복사강제	非輻射强制
non-real time	비실시간	非實時間
non-recording rain gauge	비기록우량계	非記錄雨量計
nonsaturated air	불포화공기	不飽和空氣
nonsaturated moist air	불포화습윤공기	不飽和濕潤空氣
nonscanner	비주사계	非走査計
non-scattering atmosphere	비산란대기	非散亂大氣
nonselective scattering	비선택산란	非選擇散亂
nonspherical particle	비구형입자	非球形粒子
non-static process	비정적과정	非靜的過程
non-stationary frontal surface	비정체전선면	非停滯前線面
nonsupercell tornado	비초대형세포 토네이도	非超大型細胞-
nonthreshold pollutant	비문턱오염물질,	非門-汚染物質,
	비임계치오염물질	非臨界值汚染物質
nonuniform beam filling	비균일빔채움	非均一-
nonuniform sky condition	비균일하늘상태	非均一-狀態
nonuniform visibility	비균일시정	非均一視程
nonviscous fluid	비점성유체	非粘性流體
nonvolatile	비휘발성	非揮發性
nonwetting liquid	확산하지 않는 액체,	非混合液體
	비혼합액체	
nonzero net force	비영순힘	非零-
noon	정오	正午

N

Nopsae	높새(푄의 한국명)	
Nopsae wind	높새바람	
Nordenskjöld line	노르덴스크욜드선	-線
nor'easter(= northeast storm)	노스이스터(북동쪽으로부터의 큰바람)	
Norfolk Island	노퍽섬(오스트레일리아)	
norm	평균, 정긋값	平均, 正規-
normal	정상, 정규, 평년값, 수직	正常, 正規, 平年-, 垂直
normal acceleration	법선가속도	法線加速度
normal aeration	정상환기	定常換氣
normal annual precipitation	정상연강수량	定常年降水量
normal atmosphere	정상대기	定常大氣
normal barometer	정상기압계	定常氣壓計
normal boiling point	기준끓는점	基準-點
normal chart(= normal map)	평년도	平年圖
normal circular distribution	정규원형분포	正規圓形分布
normal circulation	정상순환	定常循環
normal climate	평년기후	平年氣候
normal component	법선성분, 수직성분	法線成分, 垂直成分
normal crop	평년작	平年作
normal curvature	정상곡률	定常曲率
normal curve of error	오차정규곡선	誤差正規曲線
normal daily temperature	정상일평균기온	定常日平均氣溫
normal dispersion	정규분산	正規分散
normal distribution	정규분포	正規分布
normal distribution curve	정규분포곡선	正規分布曲線
normal distribution table	정규분포표	正規分布表
normal equation	정규방정식, 표준방정식	正規方程式, 標準方程式
normal fault	정단층	正斷層
normal force	법선력	法線力
normal function	정규함수	正規函數
normal gradient	정상경도, 법선경도	定常傾度, 法線傾度
normal gravity	정상중력	定常重力
normal incidence	수직입사, 연직입사	垂直入射, 鉛直入射
normalization	정규화	正規化
normalized difference vegetation index	정규식생지수	正規植生指數
normalized frequency	정규화진동수	正規化振動數
normalized radar cross section	정규화레이더반사면적	正規化-反射面積
normal lapse	정상감률	定常減率
normal lapse rate	정상기온감률	定常氣溫減率
normal law of error	정규오차법칙	定規誤差法則
normal map	정규도	正規圖
normal mode	정상모드, 정규모드	定常-, 正規-
normal mode initialization	정규모드초기화	正規-初期化
normal mode solution	정규모드해	正規-解

N

normal monthly mean weather chart	월별평년일기도	月別平年日氣圖
normal monthly temperature	정상월평균기온	定常月平均氣溫
normal operating condition	정상적 작동조건	定常的作動條件
normal-plate anemometer	직판풍속계	直板風速計
normal polarity	정상극성	定常極性
normal population	표준모집단	標準母集團
normal process	정규과정	正規過程
normal rainfall	평년강우	平年降雨
normal ratio method	정상년 강우량 비율법	定常年降雨量比率法
normal shear	정상시어	定常-
normal shock wave	수직충격파	垂直衝擊波
normal space mean chart	평년공간평균도	平年空間平均圖
normal state	표준상태, 정상상태	標準狀態, 定常狀態
normal stress	법선응력	法線應力
normal temperature	정상기온	定常氣溫
normal temperature and pressure	표준온도압력	標準溫度壓力
normal value	평년값, 정상값	平年-, 定常-
normal vector	법선벡터	法線-
normal water	표준해수	標準海水
normal year	예년, 평년	例年, 平年
Normand's theorem	노만드정리	-定理
nortada	노르타다(필리핀에서 부는 강한 북풍)	
North America	북아메리카	北-
North American anticyclone	북아메리카고기압	北-高氣壓
North American high	북아메리카고기압	北-高氣壓
North American ice sheet	북아메리카빙상	北-氷床
North American monsoon	북아메리카몬순	北-
North American steppe	북아메리카스텝	北-
North American summer monsoon	북아메리카여름몬순	北-
North Atlantic current	북대서양해류	北大西洋海流
North Atlantic deep water	북대서양심층수	北大西洋深層水
North Atlantic drift(= North Atlantic current)	북대서양편류	北大西洋偏流
North Atlantic gyre	북대서양환류	北大西洋還流
North Atlantic high	북대서양고기압	北大西洋高氣壓
North Atlantic mP air mass	북대서양해양한대기단	北大西洋海洋寒帶氣團
North Atlantic mT air mass	북대서양해양열대기단	北大西洋海洋熱帶氣團
North Atlantic Ocean	북대서양	北大西洋
North Atlantic oscillation	북대서양진동	北大西洋振動
North Atlantic oscillation record	북대서양진동기록	北大西洋振動記錄
northbound	북상	北上
North Brazil current	북브라질해류	北-海流
northeast	북동	北東
northeaster	북동강풍	北東強風
northeast storm	북동폭풍	北東暴風

northeast trades	북동무역풍	北東貿易風
northeast trade wind	북동무역풍	北東貿易風
north equatorial current	북적도해류	北赤道海流
norther	북풍	北風
northern annular mode	북반구환상모드	北半球環狀-
northern circuit	북쪽통과경로	北-通過經路
northern hemisphere	북반구	北半球
northern latitude	북위	北緯
northern lights	북극광	北極光
northern nanny	북극유모(잉글랜드의 우박을 동반하는 한랭한 북풍)	北極乳母
north föhn(foehn)	북푄	北-
North Frigid Zone(= artic zone)	북극한대	北極寒帶
North Island of New Zealand	뉴질랜드북섬	-北-
North Korea cold current	북한한류	北韓寒流
north magnetic pole	자북극	磁北極
north northeast	북북동	北北東
north northwest	북북서	北北西
North Pacific air mass	북태평양기단	北太平洋氣團
North Pacific current	북태평양해류	北太平洋海流
North Pacific high	북태평양고기압	北太平洋高氣壓
North Pacific intermediate water	북태평양중층수	北太平洋中層水
North Pacific mT air mass	북태평양해양성열대기단	北太平洋海洋性熱帶氣團
North Pacific Ocean	북태평양	北太平洋
North Pacific oscillation	북태평양진동	北太平洋振動
north point	북점	北點
north polar region	북극구역	北極區域
North Pole	북극	北極
North Sea	북해	北海
north star	북극성	北極星
north temperate zone	북온대	北溫帶
north tropic	북열대	北熱帶
north wall	북벽	北壁
northwest	북서	北西
northwest Australian cloud band	북서오스트레일리아구름대	北西-
northwester(= nor'wester)	노스웨스터(강한 북서풍)	
northwest monsoon	북서계절풍	北西季節風
northwest Queensland	북서퀸즐랜드(오스트레일리아)	北西-
Norway	노르웨이	
Norwegian current	노르웨이해류	-海流
Norwegian cyclone model	노르웨이저기압모형	-低氣壓模型
Norwegian School	노르웨이학교(학파)	-學校(學派)
Norwegian Sea	노르웨이해	-海
Norwegian Sea Deep Water	노르웨이해 심층수	-海深層水

N

nor'wester(= northwester)	노스웨스터(강한 북서풍)	
no-scattering equation	비산란방정식	非散亂方程式
notch width	노치폭	-幅
notice to airmen	항공고시보	航空告示報
Notos	노토스(그리스 남풍의 신)	
Nova Zemlya	노바젬랴(대기광학현상)	(大氣光學現象)
nowcast	실황예보(현재부터 1~2시간 앞까지)	實況豫報
nowcasting	실황예보, 현천예보	實況豫報, 現天豫報
NOx	질소산화물, 낙스	窒素酸化物
NOy(= odd nitrogen)	홀수질소산화물	-窒素酸化物
NRM wind scale	NRM 풍력등급	-風力等級
N-S coefficient	N-S 계수	-係數
nucleant(= nucleating agent)	핵생성제	核生成劑
nuclear autumn	핵가을	核-
nuclear binding energy	핵결합에너지	核結合-
nuclear energy	핵에너지	核-
nuclear exchange effect	핵교환효과	核交換效果
nuclear fireball	핵화염	核火焰
nuclear fission	핵분열	核分裂
nuclear force	핵력	核力
nuclear fusion	핵융합	核融合
nuclear isobars	핵동중체	核同重體
nuclear membrane	핵막	核膜
nuclear physics	핵물리학	核物理學
nuclear pore	핵공	核孔
nuclear power	원자력, 핵발전소	原子力, 核發電所
nuclear structure	핵구조	核構造
nuclear transmutation	핵변환	核變換
nuclear weapon	핵무기	核武器
nuclear winter	핵겨울	核-
nucleation	핵생성	核生成
nucleation threshold	핵생성문턱	核生成門-
nuclei counter	핵계수기	核計數器
nuclei mode	핵모드	核-
nuclepore filter	핵기공필터	核氣孔-
nucleus	핵	核
nucleus counter	핵계수기	核計數器
nudging	너징동화방법, 참여유도	-同化方法, 參與誘導
nudging coefficient	참여유도계수, 너징계수	參與誘導係數
null hypothesis	귀무가설	歸無假說
null layer(= zero layer)	영층	零層
number concentration	수농도	數濃度
number density	수밀도	數密度

number of consecutive wet days	강수지속일수, 강수계속일수	降水持續日數, 降水繼續日數
number of days with specified value	계급별일수	階級別日數
number of days with phenomena	현상일수	現象日數
number of days with precipitation	강수일수	降水日數
number of days with storm	폭풍일수	暴風日數
number of frost day	서리일수	-日數
number of sunless days	부조일수	不照日數
number of TC	열대저기압번호	熱帶低氣壓番號
number of vibration	진동수	振動數
numerator	분자	分子
numerical analysis	수치분석	數值分析
numerical dispersion	수치분산	數值分散
numerical dispersion model	수치분산모형	數值分散模型
numerical dynamical instability	수치역학불안정도	數值力學不安定度
numerical experiment	수치실험	數值實驗
numerical forecast	수치예보	數值豫報
numerical forecasting	수치예보	數值豫報
numerical forecasting model	수치예보모형	數值豫報模型
numerical instability	수치불안정	數值不安定
numerical integration	수치적분	數值積分
numerical model	수치모형	數值模型
numerical model analysis	수치모형분석	數值模型分析
numerical modeling	수치모델링	數值-
numerical notation	수치기호법	數值記號法
numerical simulation	수치모의, 수치시뮬레이션	數值模擬
numerical stability	수치안정	數值安定
numerical step by step method	수치순차법	數值順次法
numerical weather forecast	수치일기예보	數值日氣豫報
numerical weather prediction	수치일기예측	數值日氣豫測
numerical weather prediction model	수치일기예측모형	數值日氣豫測模型
numerical weather prediction output	수치일기예측산출물	數值日氣豫測産出物
N-unit	N-단위	-單位
Nusselt number	너셀트수	-數
nutation	장동	章動
N weather	N날씨	
NWS Cooperative Observer Network	NWS 협력관측자망	-協力觀測者網
n-year event(= T-year event)	n년사건	-年事件
Nyquist folding number	나이퀴스트수, 접힘횟수	-數, -回數
Nyquist frequency	나이퀴스트주파수	-周波數
Nyquist interval	나이퀴스트간격	-間隔
Nyquist velocity	나이퀴스트속도, 유효관측속도	有效觀測速度
Nyquist wavelength	나이퀴스트파장	-波長

N

O, o

영문	한글	한자
oak tree	오크나무	
oasis	오아시스	
oasis effect	오아시스효과	-效果
oasis situation	오아시스상황	-狀況
oberwind	오베르빈트(야간에 산맥 또는 고도가 높은 호수의 가장자리에서 부는 바람. 오스트리아 살츠캄메르구트에서 부는 바람)	
objective analysis	객관분석	客觀分析
objective forecast	객관예보	客觀豫報
objective method	객관방법	客觀方法
oblate spheroid	편평타원체	偏平楕圓體
oblates raindrop	편구형우적	偏球形雨滴
oblique Cartesian coordinate	사교카테시언좌표	斜交-座標
oblique circular cylinder	빗원기둥	-圓-
oblique (circular) cone	빗원뿔	-圓-
oblique coordinate	사교좌표	斜交座標
oblique incidence	빗입사	-入射
oblique shock wave	경사충격파	傾斜衝擊波
oblique visibility	빗시정	-視程
oblique visual range	빗가시거리	-可視距離
obliquity	경사	傾斜
obliquity of the ecliptic	황도경사	黃道傾斜
O'Brien cubic polynomial	오브리언3차다항식	-三次多項式
O'Brien polynomial	오브리언다항식	-多項式
obscuration	불명, 차폐	不明, 遮蔽
obscured sky cover	불명하늘가림	不明-
obscuring phenomenon	차폐현상	遮蔽現像
observation	관측	觀測
observational data	관측자료	觀測資料

observational day	관측일	觀測日
observational error	관측오차	觀測誤差
observational network	관측망	觀測網
observational validation	관측평가	觀測評價
observation angle	관측각	觀測角
observation balloon	관측기구	觀測氣球
observation equation	관측방정식	觀測方程式
observation error covariance	관측오차공분산	觀測誤差共分散
observation field	노장, 관측노장	露場, 觀測露場
observation frequency	관측주파수, 관측주기	觀測周波數, 觀測週期
observation increment	관측증분	觀測增分
observation instrument	관측기기	觀測器機
observation landplane	육상관측항공기	陸上觀測航空機
observation methods	관측방법	觀測方法
observation of visibility	시정관측	視程觀測
observation operator	관측연산자	觀測演算子
observation point	관측지점	觀測地點
observation post	관측소	觀測所
observation seaplane	해상관측항공기	海象觀測航空機
observation station	관측소	觀測所
observation time	관측시간	觀測時間
observation well	관측우물	觀測-
observatory	기상대, 관측소, 천문대	氣象臺, 觀測所, 天文臺
observatory-remote	원격관측소	遠隔觀測所
observed value	측정값	測定-
observer	관측자	觀測者
observing station	관측소	觀測所
observing system	관측시스템	觀測-
observing systems simulation experiment	관측시스템모의실험	觀測-模擬實驗
obstacle	장애물	障碍物
obstacle flow signature	장애흐름징후	障碍-徵候
obstruction to vision	시정장애	視程障碍
Obukhov length	오부코프길이	
occasional showers	한 번씩 소나기	
occluded cyclone	폐색저기압	閉塞低氣壓
occluded depression	폐색저기압	閉塞低氣壓
occluded front	폐색전선	閉塞前線
occluded frontal system	폐색전선계	閉塞前線系
occluded mesocyclone	폐색중규모저기압	閉塞中規模低氣壓
occlusion	폐색	閉塞
occlusion decay	폐색소멸	閉塞消滅
occlusion front	폐색전선	閉塞前線
occlusion process	폐색과정	閉塞過程
occultation	엄폐	掩蔽

occult deposition	눈에 보이지 않는 강하물, 엄폐침적	掩蔽沈積
occult precipitation	눈에 보이지 않는 강수	
occurrence	발생	發生
ocean	해양	海洋
ocean acidification	해양산성화	海洋酸性化
ocean air(mass)	해양기단, 해양대기	海洋氣團, 海洋大氣
ocean albedo	해양알베도	海洋-
ocean-atmosphere heat flux	해양-대기열플럭스	海洋大氣熱-
ocean-atmosphere system	해양-대기시스템	海洋大氣-
ocean basin	해양분지	海洋盆地
ocean carbon pump hypothesis	해양탄소 펌프가설	海洋炭素-假說
ocean chemistry	해양화학	海洋化學
ocean circulation	해양순환	海洋循環
ocean climate	해양성기후	海洋性氣候
ocean conveyor belt	해양컨베이어벨트	海洋-
ocean crust	해양지각	海洋地殼
ocean current	해류	海流
ocean deep	해연, 해구	海淵, 海溝
Ocean Drilling Program	해양시추프로그램, 해저지각시추프로그램	海洋試錐-, 海底地殼試錐-
Ocean Drilling Project	해양시추프로젝트	海洋試錐計劃
ocean floor	해저	海底
ocean floor spreading	해저확장	海底擴張
ocean fog	해양안개	海洋-
ocean gateway	해양통로	海洋通路
ocean general circulation	해양대순환	海洋大循環
ocean heat inertia	해양열관성	海洋熱慣性
ocean heat transport hypothesis	해양열수송가설	海洋熱輸送假說
ocean height variation	해양높이변동	海洋-變動
Oceania	오세아니아	
oceanic air mass	해양기단	海洋氣團
oceanic air traffic control	해양항공교통통제	海洋航空交通統制
oceanic anticyclone	해양고기압	海洋高氣壓
oceanic area	해양지역	海洋地域
oceanic circulation system	해양순환시스템	海洋循環-
oceanic climate	해양성기후	海洋性氣候
oceanic crust	해양지각	海洋地殼
oceanic front	해양전선	海洋前線
oceanic general circulation models	해양대순환모형	海洋大循環模型
oceanic heat transport	해양열수송	海洋熱輸送
oceanic hemisphere	해양반구	海洋半球
oceanic high(= subtropical high)	해양고기압(아열대고기압)	海洋高氣壓
oceanicity	해양도	海洋度

O

oceanic meteorological research	해양기상연구	海洋氣象研究
oceanic meteorology	해양기상학	海洋氣象學
oceanic mixed layer	해양혼합층	海洋混合層
oceanic moderate	해양성온난	海洋性溫暖
oceanic noise	해명, 해양잡음	海鳴, 海洋雜音
oceanic stratosphere	해양상공성층권	海洋上空成層圈
oceanic surface mixed layer	해양표층혼합층	海洋表層混合層
oceanic tide	해양조	海洋潮
oceanic weather station	해양기상관측소	海洋氣象觀測所
ocean influence	해양영향	海洋影響
ocean islands	해양섬	海洋島
oceanity(= oceanicity)	해양도	海洋度
ocean mixed layer model	해양혼합층모형	海洋混合層模型
ocean mixing	해양혼합	海洋混合
oceanographical observation	해양관측	海洋觀測
oceanography	해양학	海洋學
oceanophysics	해양물리학	海洋物理學
ocean paleocirculation	고해양순환	古海洋循環
ocean paleoproductivity	해양의 고생산력	海洋古生産力
ocean paleotemperature	고수온, 해양의 고온도	古水溫, 海洋古溫度
ocean planet	바다행성	-行星
ocean reservoir	해양저장소	海洋貯藏所
ocean ridges	해령	海嶺
ocean sediment	해양퇴적물	海洋堆積物
ocean station	해양관측소	海洋觀測所
ocean station vessel	해양관측선	海洋觀測船
ocean surface	해수면, 해양표층	海水面, 海洋表層
ocean surface wind	해양표면풍, 해상풍	海洋表面風, 海上風
ocean temperature	해수온도	海水溫度
Ocean Topography Experiment	해양지형실험	海洋地形實驗
ocean trench	해구	海溝
ocean type	해양형	海洋型
ocean wave	해파	海波
ocean weather ship	해양기상관측선	海洋氣象觀測船
ocean weather station	해양기상관측소	海洋氣象觀測所
ocean weather station vessel	해양기상관측선	海洋氣象觀測船
ocean wind	해양풍	海洋風
octa(= okta)	8분의 1의	
octane	옥탄	
octane number	옥탄가	-價
octane rating	옥탄가	-價
octant	8분위, 8방위	八分位, 八方位
octave band	옥타브밴드	
odd chlorine	홀수염소	-鹽素

Odderade interstadial	오더레이드간빙기	-間氷期
odd hydrogen	홀수수소	-水素
odd hydrogen species	기수수소입자	奇數水素粒子
odd nitrogen	홀수질소	-窒素
odd number	홀수	-數
odd oxygen	홀수산소	-酸素
odd oxygen system	홀수산소시스템	-酸素-
odd point	홀수점	-數點
Odintsovo interstadial	오딘초보 간빙기	-間氷期
odor	악취	惡臭
odor threshold	악취문턱값, 악취역치	惡臭閾値
oe	오(덴마크령 페로제도의 회오리바람)	
oecology(= ecology)	생태학	生態學
oersted	외르스테드(자기장강도의 단위)	
Oessn's equation	오센방정식	-方程式
off-airway	비정규항로	非定規航路
off-center PPI scope	편심PPI스코프	偏心-
Offensive Counter Air Operations weather	공세제공작전기상	攻勢制空作戰氣象
Offensive Mining weather	공격기뢰부설기상	攻擊機雷敷設氣象
off-ice wind	하빙풍	河氷風
official elevation of the aerodrome	비행장공식표고	飛行場公式標高
offing	앞바다	
offlevel	비규정면	非規定面
offline model	비연동모형	非聯動模型
offset policy	잔류편차정책	殘留偏差政策
offshore	연해	沿海
offshore and coastal dispersion model	이안류-연안류 분산모형	離岸流沿岸流分散模型
offshore bar	연해사주	沿海砂洲
offshore breeze	이안바람, 이안풍	離岸風
offshore current	이안류	離岸流
offshore patrol weather	외해초계기상	外海哨戒氣象
offshore route	먼항로	-航路
offshore trough	이안저기압골	離岸低氣壓-
offshore water	외해수	外海水
offshore wind	해안에서 바다로 부는 바람	
offtime	비정시	非定時
offtime report	비정시보고	非定時報告
Ogasawara high	오가사와라 고기압	-高氣壓
ogive	누적도수분포도	累積度數分布圖
ohm	옴(전기저항의 SI 단위)	
ohmic current	저항전류	抵抗電流
Ohm's law	옴법칙	-法則
oil	기름, 오일	
oil appropriate standard weather	오일상용기상, 오일표준기상	-常用氣象, -標準氣象

O

oil field	유전	油田
oil refinery	정유소	精油所
oil shale	유질혈암	油質頁岩
oil slick	유막(수면의)	油膜
oil spill	기름유출	-流出
oil substrate	오일기질	-基質
oil tanker	유조선	油槽船
Okhotsk high	오호츠크해 고기압	-海高氣壓
Okhotsk sea(= sea of Okhotsk)	오호츠크해	-海
Okhotsk sea air mass	오호츠크해 기단	-海氣團
okta	8분위	八分位
Older Dryas	올더 드리아스	
old snow	묵은 눈	
Old Wives' summer	노부인의 여름(유럽의 9월말	老夫人-
	부터 11월 초에 나타나는 따뜻하고 맑은 날씨)	
Olefins	올레핀류	-類
Oligocene	올리고세(신생대 제3기)	
Olland cycle	올랜드사이클	
Olland principle	올랜드원리	-原理
Oman	오만	
ombrology	측우학	測雨學
ombrometer	미우량계	微雨量計
ombroscope	강수지시기	降水指示器
omega equation	오메가방정식	-方程式
omega signal	오메가신호	-信號
omega sun	오메가모양태양	-模樣太陽
Omega-Wolff method	오메가울프법	-法
omega zonde	오메가존데	
omnidirectional	전방향성	全方向性
omnidirectional radiometer	전방향복사계	全方向輻射計
on and off instrument	온/오프 기기	-器機
onboard controls	기판제어	基板制御
one-and-a-half order closure	1.5차수종결	-次數終結
one-body reaction	동체반응	同體反應
one-cell Hadley circulation	해들리 단일대순환	-單一大循環
one-dimensional Bodin's model	1차원보딘모형	一次元-模型
one-dimensional cloud probe	1차원구름탐측기	一次元-探測機
one-dimensional humidity spectrum	1차원습도스펙트럼	一次元濕度-
one-dimensional model	1차원모형	一次元模型
one-dimensional spectral density tensor	1차원스펙트럼밀도텐서	一次元-密度-
one-dimensional temperature spectrum	1차원온도스펙트럼	一次元溫度-
one extinction thickness	한 소광두께	-消光-
O'Neill experiment	오닐실험(1953년)	-實驗
one-month forecast	월간예보	月間豫報

one-on-one climate consultation	일대일 기후상담	一對一氣候相談
one-on-one forecaster consultation	일대일 예보상담	一對一豫報相談
one-sided difference	한쪽차분	-差分
one-way entrainment process	한방향 유입과정	一方向流入課程
one-way interaction	한방향 상호작용	一方向相互作用
one-way mass transfer model	한방향 질량전달모형	一方向質量傳達模型
one-week forecast	주간예보	週間豫報
on instruments(= instrument flight)	계기비행	計器飛行
onion peeling	양파껍질벗기기	
online model	연동모형	聯動模型
Onsagar-Casimir reciprocity relation	온사거-카시미르 상반관계	-相反關係
onshore breeze	향안풍, 해풍	向岸風, 海風
onshore wind	바다바람	
on time	정시	定時
on top	구름위(비행)	(飛行)
oolite(= oolith)	어란석, 오얼라이트	魚卵石
Oort cloud	오르트구름	
opacity	불투명도	不透明度
opacus	불투명구름	不透明-
opalescence	젖빛광	-光
opalescent turbidity	젖빛혼탁도	-混濁度
opaline shell	단백석껍질	蛋白石殼
opal shell	오팔껍질	
opaque	불투명(의)	不透明
opaque ice	불투명얼음	不透明-
opaque layer	불투명층	不透明層
opaqueness	불투명도	-不透明度
opaque sky cover	하늘차폐	-遮蔽
open cell	열린세포	-細胞
open cell cumulus	열린세포적운	-細胞積雲
open channel flow	개수로흐름	開水路-
open circuit	열린회로	-回路
open circuit wind tunnel	개방형풍동, 개회로식 풍동	開放型風洞, 開回路式風洞
open curve	열린곡선	-曲線
open cylinder type	개원통형	開圓筒型
opening angle	열린각	-角
opening date of leaf	개엽일	開葉日
opening of leaf	개엽	開葉
open interval	열린구간	-區間
open scale barograph	열린눈금기압계	-氣壓計
open sea	부동해, 공해, 먼바다	不凍海, 公海
open set	열린집합	-集合
open system	열린계	-系
open-top chamber	상부개방형챔버	上部開放型-

O

open-top chamber experiment	상부개방형챔버실험	上部開放型-實驗
operating condition	작동상태	作動狀態
operation	연산	演算
operational calculus	연산자법	演算子法
operational cloud amount standard	운량조건, 작전운량조건	雲量條件, 作戰雲量條件
operational cloud layer visibility	작전운층시계	作戰雲層視界
operational linescan system	현업선주사시스템	現業線走査-
operational meteorological information	운영기상정보	運營氣象情報
operational numerical model	현업수치모형	現業數値模型
operational prediction	현업예보, 업무예보	現業豫報, 業務豫報
operational satellites	현업위성	現業衛星
operational support airlift weather	작전지원공수기상	作戰支援空輸氣象
operational use	현업용	現業用
operational weather limits	현업일기한계	現業日氣限界
operation center	관제센터	管制-
operation directive	업무지시	業務指示
operation schedule	운영일정	運營日程
operation weather	작전기상	作戰氣象
operative temperature	작용온도	作用溫度
operator	조작원, 연산자	操作員, 演算子
opposing wind	맞바람	
opposition	충, 반대	衝, 反對
oppressive	압제적인	壓制的
optical air mass	광학공기질량	光學空氣質量
optical air mass term	광학공기질량항	光學空氣質量項
optical amplification factor	광학증폭지수	光學增幅指數
optical axis	광축	光軸
optical condensation particle counter	광학응결입자계수기	光學凝結粒子計數器
optical counter	광학계수기	光學計數器
optical density	흡광도, 광밀도	吸光度, 光密度
optical density of a cloud	구름의 흡광도	-吸光度
optical depth	광학깊이	光學-
optical disdrometer	광학입경계	光學粒境界
optical-electronic sensor	광학적 전자센서	光學的電子-
optical haze	아지랑이	
optical hygrometer	광학습도계	光學濕度計
optical illusion	광학적 착각, 착시	光學的錯覺, 錯視
optical imaging probe	광학영상탐측기	光學映像探測機
optical length(= optical path)	광거리, 광길이	光距離
optical light-scattering instrument	광학광산란측기	光學光散亂測器
optically effective atmosphere	광학유효대기	光學有效大氣
optically homogeneous	광학적 균질	光學的均質
optically smooth	광학적 평활	光學的平滑
optical mass	광학질량	光學質量

optical meteor	대기광학현상	大氣光學現象
optical observation program	광학관측프로그램	光學觀測-
optical (aerosol) particle counter	광학입자계수기	光學粒子計數機
optical particle method	광학입자방법	光學粒子方法
optical particle probe	광학입자탐측기	光學粒子探測機
optical path(= optical pathlength)	광(행)로	光(行)路
optical path difference	광경로차	光經路差
optical phenomenon	광학현상	光學現像
optical precipitation particle counter (disdrometer)	광학강수입자계	光學降水粒子計
optical probing of the atmosphere	대기광학탐측	大氣光學探測
optical pyrometer	광학고온계	光學高溫計
optical rainfall particle counter	광학우적계	光學雨滴計
optical rain gauge	광학우량계	光學雨量計
optical refractive index	광굴절지수	光屈折指數
optical region	광학지역	光學地域
optical scattering probe	광학산란탐측기	光學散亂探測機
optical slant range	광학사선거리	光學斜線距離
optical snow cover meter	광학식 적설계	光學式積雪計
optical spectroscopy	광학분광술	光學分光術
optical thickness	광학두께	光學-
optical transfer function	광학전달함수,	光學傳達函數,
	광학적 전달함수	光學的傳達函數
optical transient detector	광과도현상탐지기	光過渡現象探知機
optics	광학	光學
optimal perturbation	최적섭동	最適攝動
optimal weight	최적가중치	最適加重値
optimal yield	최적양수량, 최적채수량	最適揚水量, 最適採水量
optimization	최적화	最適化
optimum climate	최적기후	最適氣候
optimum flight	최적비행	最適飛行
optimum humidity	최적습도	最適濕度
optimum interpolation	최적내삽	最適內揷
optimum interpolation sea surface temperature	최적내삽 해수면온도	最適內揷海水面溫度
optimum solution	최적해	最適解
optimum temperature	최적온도	最適溫度
ora(= aura)	오라(이탈리아 가르다호의 규칙적인 골바람)	
oraucan(→ hurricane)	열대성저기압	熱帶性低氣壓
orbit	궤도	軌道
orbital	궤도	軌道
orbital correction	궤도보정	軌道補正
orbital cycle	궤도주기	軌道週期
orbital eccentricity	궤도이심률	軌道離心率
orbital element	궤도요소	軌道要素
orbital forcing	궤도강제	軌道强制

O

orbital frequency	궤도진동수	軌道振動數
orbital monsoon hypothesis	궤도몬순가설	軌道-假說
orbital motion	궤도운동	軌道運動
orbital parameter	궤도인자	軌道因子
orbital-scale carbon transfer	궤도규모탄소전달	軌道規模炭素傳達
orbital-scale climate change	궤도규모기후변화	軌道規模氣候變化
orbital tuning	궤도동조	軌道同調
orbital variation	궤도변이, 궤도변동,	軌道變異, 軌道變動
orbital velocity	궤도속도	軌道速度
orbiting geophysical observatory	지구물리관측궤도위성	地球物理觀測軌道衛星
orbiting solar observatory	태양관측궤도위성	太陽觀測軌道衛星
orbit number	궤도번호	軌道番號
orbit period	궤도주기	軌道週期
orbit perturbation	궤도섭동	軌道攝動
orbit plane	궤도면	軌道面
orchard albedo	과수원알베도	果樹園反射率
orchard heater	과수서리방지가열기	果樹-防止加熱器
order of discontinuity	불연속차수	不連續次數
order of electrification	대전열	帶電列
order of magnitude	크기차수, 크기순서	-次數, -順序
ordinary agricultural meteorological station	일반농업기상관측소	一般農業氣象觀測所
ordinary agrometeorological station	일반농업기상관측소	一般農業氣象觀測所
ordinary cell	일반세포	一般-
ordinary climatological station	일반기후관측소	一般氣候觀測所
ordinary radiation station	일반복사선측정소	一般輻射線測定所
ordinary ray	정상광선	正常光線
ordinary wave	정상파	定常波
ordinary year	평년	平年
ordinate	세로좌표	-座標
Ordovician extinction	오르도비스기 멸종	-期滅種
Ordovician period	(고생대)오르도비스기	-期
Ordovician-Silurian event	오르도비스기-실루리아기 사건	-事件
organic acid	유기산	有機酸
organic carbon	유기탄소	有機炭素
organic carbon subcycle	유기탄소 부주기	有機炭素副週期
organic cloud seeding material	유기구름씨뿌리기물질	有機-物質
organic compound	유기화합물	有機化合物
organic geochemical proxy	유기지화학적 프록시	有機地化學的-
organic matter	유기물, 유기물질	有機物, 有機物質
organic nitrate	유기질산염	有機窒酸鹽
organic nitrate compound	유기질산염화합물	有機窒酸鹽化合物
organic peroxide	유기과산화물	-有機過酸化物
organization responsible for aviation forecast	항공예보관서	航空豫報官署

organizations to deliver massage	통보대상기관	通報對象機關
organized Cb cluster	적란운 무리	積亂雲-
organized large eddy	조직화 큰맴돌이	組織化-
organizes turbulent structure	난류구조구축	亂流構造構築
organochlorine compound	유기염소화합물	有機鹽素化合物
organochlorine	유기염소류	有機鹽素類
organophosphates	유기인제	有機燐劑
orientation	방향, 지향, 방위	方向, 指向, 方位
oriented overgrowth(= epitaxis)	방향성과성장	方向性過成長
orifice	물받이구멍, 수수구	受水口
origin	원점, 발원지, 기원	原點, 發源地, 起原
original record	원부	原簿
original vegetation	원식생	原植生
origin of coordinates	좌표원점	座標原點
Orion Nebula	오리온성운	-星雲
Ornstein-Uhlenbeck process	오른스타인-우렌벡 과정	-課程
orogeny	조산운동	造山運動
orographic	산악성, 지형성	山岳性, 地形性
orographic cloud	지형성구름	地形性-
orographic condition	지형조건	地形條件
orographic cyclogenesis	지형성저기압발생,	地形性低氣壓發生,
	지형저기압발생	地形低氣壓發生
orographic cyclone	산악저기압, 지형성저기압	山岳低氣壓, 地形性低氣壓
orographic depression	지형성저기압	地形性低氣壓
orographic effect	지형효과	地形效果
orographic flow	지형흐름	地形-
orographic high	지형성고기압	地形性高氣壓
orographic isobar	지형등압선	地形等壓線
orographic lifting	지형성상승	地形性上昇
orographic occlusion	지형성폐색	地形性閉塞
orographic precipitation	지형성강수, 산악형강수	地形性降水, 山岳形降水
orographic rain	지형성비	地形性-
orographic rainfall	지형강우(량)	地形降雨(量)
orographic rainfall enhancement	지형성강우증가	地形性降雨增加
orographic snow-line	지형설선	地形雪線
orographic storm	지형성스톰, 지형성폭풍	地形性暴風
orographic thunderstorm	산악뇌우, 지형뇌우	山岳雷雨, 地形雷雨
orographic uplift	지형성상승	地形性上昇
orographic vortex	지형성소용돌이	地形性-
orographic wave	지형성파	地形性波
orographic wind	지형성바람	地形性-
orographic wind flow	지형성바람흐름	地形性-
orography	산지지형학	山地地形學
orological observing station	기상관측소	氣象觀測所

O

Oroshi wind	(일본)오로치 바람(혼슈 중앙산맥으로부터 내려오는 건조한 북서풍계열의 보라)	
Orr-Sommerfeld equation	오르-좀메르펠트 방정식	-方程式
orsure	오르수르(프랑스 리옹만의 왕바람)	
orthogonal	직교(의)	直交
orthogonal antenna	직교안테나	直交-
orthogonal (Cartesian) coordinates (system)	직교좌표(계)	直角座標(系)
orthogonal curvilinear coordinates	직교곡선좌표	直交曲線座標
orthogonal expansion	직교확장	直交擴張
orthogonal function	직교함수	直交函數
orthogonality	직교성	直交性
orthogonal line	직교선	直交線
orthogonal polarization	직교편파	直交偏波
orthogonal polynomial	직교다항식	直交多項式
orthogonal projection	정사투영	正射投影
orthographic projection	정사투영	正射投影
orthomorphic map	정형도	正形圖
oscillating body	진동체	振動體
oscillating circuit	진동회로	振動回路
oscillating electric current	진동전류	振動電流
oscillation	진동	振動
oscillation series	진동급수	振動級數
oscillator	진동자	振動子
oscillator mechanism	진동자메커니즘	振動子-
oscillatory wave	진동파	振動波
oscillograph	오실로그래프	
oscilloscope	오실로스코프	
Oseen's approximation	오센근사	-近似
osmometer	삼투압계	滲透壓計
osmosis	삼투압	滲透壓
osmotic potential	삼투퍼텐셜	慘透-
osmotic pressure	삼투압(력)	滲透壓(力)
osmotic water	삼투수	滲透水
Osos wind	오소스바람(미국 캘리포니아주의 로스오소스골짜기로부터 샌루이스골짜기까지 부는 강한 북서풍)	
ostracodes	개형류	介形類
ostria(= auster)	오스트리아(불가리아 연해의 따뜻한 바람)	
ouari	오우아리(소말리아의 남풍)	
oued	갈수천	渴水川
outbreak	돌발	突發
outer atmosphere	바깥대기	-大氣
outer audible zone	외청역	外聽域
outer boundary layer	상부대기경계층	上部大氣境界層
outer core	외핵	外核

outer eyewall	바깥눈벽	-壁
outer halo	바깥무리	
outer ionosphere	외부전리권	外部電離圈
outer layer	바깥층	-層
outer layer scaling	바깥층 스케일링	-層-
outer product	외적	外積
outer zone of audibility	외청역	外聽域
outflow	유출(량)	流出(量)
outflow boundary	유출경계	流出境界
outflow channel	범람수로	氾濫水路
outflow from the cyclone	저기압유출	低氣壓流出
outflow hydrograph	유출수문곡선	流出水文曲線
outflow jet	유출제트	流出-
outflow layer	유출층	流出層
outflow spiral	유출나선	流出螺線
outflow wind	유출바람	流出-
outgassing	기체방출	氣體放出
outgoing longwave radiation(OLR)	방출장파복사	放出長波輻射
outgoing radiation	방출복사	放出輻射
outgoing radiation type	방출복사형, 방열형	放出輻射型, 放熱型
outlet	분수공, 취수구, 방류구	噴水孔, 取水口, 放流口
outlet glacier	분출빙하	噴出氷河
outlier	아웃라이어	
outline map	약도	略圖
outlook	전망	展望
outlook briefing	전망브리핑	展望-
outlook period	전망기	展望期
output	출력	出力
outside air temperature	외기온도	外氣溫度
outwash	융빙유수퇴적물	融氷流水堆積物
oven dried soil	건토	乾土
overbar	윗줄	
overcast	온흐림	
overcast circle	온흐림원	-圓
overcast day	온흐림날	
overcast sky	온흐림하늘	
overcompensate	과보상	過報償
overconstrained problem	과잉제약조건문제	過剩制約條件問題
overcooling	과냉각	過冷却
overexploitation	남획, 과잉개발	濫獲, 過剩開發
overflow	월류	越流
overhang	돌출, 오버행	突出
overhang echo	돌출에코, 오버행에코	突出-
overheat	과열	過熱

O

overheating layer	과열층	過熱層
overkill hypothesis	잔인가설	殘忍假設
overland flow	지면류, 지면흐름	地面流
overland runoff	지표면유출	地表面流出
overlap	중첩	重疊
overlapping average(mean)	중첩평균	重疊平均
overnormal wind	이상풍	異常風
overrelax	과완화	過緩和
over-relaxation	과완화	過緩和
overrunning	황폐화	荒廢化
overrunning cold front	추월한랭전선	追越寒冷前線
oversample	과표본, 과시료	過標本, 過試料
over sampling	과표본추출, 오버샘플링	過標本抽出
overseeding	과잉씨뿌리기	過剩-
overshoot	오버슛, 돌출	突出
overshooting cloud top	돌출구름꼭대기	突出-
overshooting thunderstorm	오버슈팅뇌우	-雷雨
overshooting thunderstorm top	돌출뇌우꼭대기	突出雷雨-
overshooting top	오버슛(돌출)꼭대기	-(突出)-
overspecified	과도명시된	過度明示-
overspeeding	배속	培速
overtone band	과색조 띠	
overtrades(= Krakatao wind)	고층무역풍(크라카타오 바람)	高層貿易風
overturn	뒤집힘	
overview	개요	槪要
overwater dispersion	유출수분산	流出水分散
Owens dust recorder	오웬 먼지기록계	-記錄計
oxbow lake	우각호	牛角湖
oxidant	산화제	酸化劑
oxidase	산화효소	酸化酵素
oxidation	산화	酸化
oxidation catalyst	산화촉매	酸化觸媒
oxidation reaction	산화반응	酸化反應
oxidation reductase	산화환원효소	酸化還元酵素
oxide	산화물	酸化物
oxides of nitrogen	질소산화물	窒素酸化物
oxidizing capacity of atmosphere	대기산화용량	大氣酸化容量
oxidizing flame	산화불꽃	酸化-
oxidizing power	산화력	酸化力
ox's eye	허리케인(기니에서 부르는 용어)	
oxygen	산소	酸素
oxygen absorption	산소흡수	酸素吸收
oxygenate	산소공급	酸素供給
oxygenated hydrocarbon	산화탄화수소	酸化炭化水素

oxygen band	산소띠	酸素-
oxygen free radical	활성산소	活性酸素
oxygen isotope	산소동위원소	酸素同位元素
oxygen isotope deep-sea record	산소동위원소 심해기록	酸素同位元素深海記錄
oxygen isotope method	산소동위원소방법	酸素同位元素方法
oxygen isotope ratio	산소동위원소비	酸素同位元素比
oxygen molecule	산소분자	酸素分子
oxygen-ozone system	산소-오존 시스템	酸素-
oxyhydrocarbon	산수소탄소	酸水素炭素
Oyashio	오야시오	
Oyashio current	오야시오해류	-海流
Ozmidov scale	오즈미도프 규모	-規模
ozone	오존	
ozone absorption	오존흡수	-吸收
ozone action day	오존활동일	-活動日
ozone advisory	오존주의보	-注意報
ozone alert	오존경보	-警報
ozone cloud	오존구름	
ozone concentration	오존농도	-濃度
ozone cutoff	오존분리	-分離
ozone-depleting potential	오존고갈퍼텐셜	-枯渴-
ozone-depleting substance	오존고갈물질	-枯渴物質
ozone depletion	오존감소, 오존고갈	-減少, -枯渴
ozone hole	오존구멍	
ozone isopleth plot	오존등치선도표	-等値線圖表
ozone layer	오존층	-層
ozone loss	오존손실	-損失
ozone molecule	오존분자	-分子
ozone monitoring	오존감시	-監視
ozone-oxygen cycle	오존-산소 주기	-酸素週期
ozone photodissociation rate	오존광해리율	-光解離率
ozone reaction	오존반응	-反應
ozone shape function	오존모양함수	-模樣函數
ozone shield	오존차폐	-遮蔽
ozone-sonde	오존존데	
ozone spectrophotometer	오존분광도계	-分光光度計
ozone transfer coefficient	오존전달계수	-傳達係數
ozone trend	오존경향	-傾向
ozonopause	오존권계면	-圈界面
ozonosphere	오존권	-圈

O

P, p

영문	한글	한자
Pacific air mass	태평양기단	太平洋氣團
Pacific and Indian-ocean common water	태평양 및 인도양 일반수	太平洋印度洋一般水
Pacific anticyclone(= Pacific high)	태평양고기압	太平洋高氣壓
Pacific decadal oscillation	태평양십년진동	太平洋十年振動
Pacific deep water	태평양심층수	太平洋深層水
Pacific high(= Pacific anticyclone)	태평양고기압	太平洋高氣壓
Pacific mP air masses	태평양해양한대기단	太平洋海洋寒帶氣團
Pacific-North American oscillation	태평양-북아메리카 진동	太平洋-北美振動
Pacific Ocean	태평양	太平洋
Pacific Plate	태평양판	太平洋板
Pacific ridge	태평양해령	太平洋海嶺
Pacific Standard Time	태평양표준시(미국)	太平洋標準時
Pacific subarctic current	태평양아북극해류	太平洋亞北極海流
pack(= pack ice)	부빙	浮氷
packet	속	束
pack ice(= ice pack)	부빙	浮氷
paddy field energy balance	벼논에너지평형(수지)	-平衡(收支)
Paekdusan high	백두산고기압	白頭山高氣壓
paesa	파에사(이탈리아 가르다 호에서 발생하는 격렬한 북북동풍)	
paesano	파에사노(이탈리아 가르다 호에서 밤에 북쪽 산에서부터 불어 내려오는 실바람)	
pagoscope	서리관측기	-觀測器
painter	페인터(페루 연해에 나타나는 안개)	
palaeoclimatic record	고기후기록	古氣候記錄
palaeodune system	고사구시스템	古砂丘-
palaeoecologists	고생태학자	古生態學者
palaeolatitude	고위도	古緯度
palaeosols(= paleosol)	고토양	古土壤

palaeotemperature record	고온도기록	古溫度記錄
Palaeozoic climate	고생대기후	古生代氣候
paleobotany	고식물학	古植物學
paleoceanography	고해양학	古海洋學
Paleocene	(신생대 제3기)팔레오세	
Paleocene-Eocene Thermal Maximum	피이티엠	-(-最高溫期)
	(팔레오세-에오세 최고온기)	
paleocirculation	고순환	古循環
paleoclimate(= geological climate)	고기후	古氣候
paleoclimate modeling	고기후모델링	古氣候-
paleoclimate proxy	고기후대리요소	古氣候代理要素
paleoclimatic change	고기후변화	古氣候變化
paleoclimatic sequence	고기후차례	古氣候次例
paleoclimatologist	고기후학자	古氣候學者
paleoclimatology	고기후학	古氣候學
paleocrystic ice	고결정얼음	古結晶-
paleoecology	고생태학	古生態學
Paleo-El Niño-Southern Oscillation record	고-엘니뇨-남방진동기록	古-南方振動記錄
paleoenvironmental proxy	고환경대리요소	古環境代理要素
paleogene	고제3기(의)	古第三紀
paleogene climate	고제3기기후	古第三紀氣候
paleoglaciology	고빙하학	古氷河學
paleohydrology	고수문학	古水文學
paleolimnology	고육수학	古陸水學
paleomagnetism	고지자기	古地磁氣
paleo-ocean modeling	고해양학모델링	古海洋學-
paleo-ocean pH	고해양학수소이온농도지수	古海洋學水素-濃度指數
paleo-precipitation indicator	고강수량지표	古降水量 指標
paleoproductivity	고생산력	古生産力
paleoscientists	고과학자	古科學者
paleosoil	고토양	古土壤
paleotemperature	고기온	古氣溫
paleotemperature data	고온도자료	古溫度資料
paleotempestology	고폭풍학(강한 허리케인의	古暴風學
	퇴적물 기록)	
Paleozoic Era	고생대	古生代
pallio-nimbus	층모양비구름	層模樣-
pallium	층모양비구름	層模樣-
Palmer Drought Severity Index(PDSI)	파머가뭄심도지수	-旱魃深度指數
Palmer Hydrological Drought Index(PHDI)	파머수문가뭄지수	-水文旱魃指數
palouser	팔로서(캐나다 래브라도 북서부의 먼지보라)	
Paluch diagram	팔루크다이어그램	
palynology	화분학	花粉學
pampas	팜파스(남아메리카 특히 아르헨티나의 스텝)	

pampero	팜페로(남아메리카 Andes 산맥에서 대서양으로 내리부는 찬바람)	
pan adjustment coefficient	증발계조정계수	蒸發計調整係數
Panama	파나마	
Panama Basin	파나마분지	-盆地
Panama Canal	파나마운하	-運河
panas oetara	파나스오에타라(인도네시아에서 2월에 부는 온난 건조한 폭풍)	
pancake ice	빈대떡얼음	
pan coefficient	증발계계수	蒸發計係數
pan evaporation	증발계증발	蒸發計蒸發
Pangaea	판게아(초대륙)	
Panhandle Hook	팬핸들 훅(폭풍)	
pan ice(= floe ice)	뜬 얼음	
pannus	토막구름, 파쇄구름	破碎-
panoramic distortion	파노라마비틀림	
papagayo	파파가요(중앙아메리카 니카라구아와 과테말라의 태평양 연안에서 부는 강한 북동풍)	
Papua New Guinea	파푸아뉴기니	
parabola	포물선	抛物線
parabolic antenna	포물면안테나, 접시안테나	抛物面-
parabolic orbit	포물선궤도	抛物線軌道
parabolic reflector	포물면반사기	抛物面反射器
paraboloid	포물면	抛物面
Paracas low-level Jet	파라카스 하층제트	-下層-
parachute radiosonde	낙하산라디오존데	落下傘-
paraffins	파라핀	
parafoveal vision	준중심시	準中心視
parallax	시차	時差
paralle lines	평행선	平行線
parallelism	평행, 대구법	平行, 對句法
parallelogram	평행사변형	平行四邊形
parameter	매개변수, 모수	媒介變數, 母數
parameter aggregation	매개변수집합	媒介變數集合
parameter estimation	매개변수추정	媒介變數推定
parameterization	매개변수화, 모수화	媒介變數化, 母數化
parameterization method	매개변수화방법, 모수화 방법	媒介變數化方法, 母數化方法
parameterization of turbulence statistics	난류통계학의 매개변수화	亂流統計學的媒介變數化
parameterization scheme	매개변수화방식	媒介變數化方式
parametric equation	매개변수방정식	媒介變數方程式
parametric function	매개변수함수	媒介變數函數
parametric model	매개변수모형, 모수모형	媒介變數模型, 母數模型
parametric resonance	매개변수공진	媒介變數共振

P

paranthelion	먼무리해	
Paranthropus	파란트로푸스(화석인류)	
parantiselena	먼무리달	
paraselene	무리달, 환월	幻月
paraselenic circle	무리달테	
parasite drag	유해항력	有害抗力
parcel	(공기)덩이	(空氣)-
parcel method	파슬법, 공기덩이법	空氣-法
parcel temperature	(공기)덩이온도	(空氣)-溫度
parent isotope	모원소	母元素
parhelia	환일	幻日
parhelia associated with the 22° halo	22도할로무리해	
parhelia associated with the 46° halo	46도할로무리해	
parhelic circle	무리해테	
parhelic ring	무리해고리	
parhelion	무리해	
Parry arcs	패리호	-弧
PAR scope	PAR 스코프	
Parseval's theorem	파시발정리	-定理
partial beam blocking	부분빔차폐	部分-遮蔽
partial charge	부분전하	部分電荷
partial correlation	부분상관, 편상관	部分相關, 偏相關
partial derivative	편도함수	偏導函數
partial differential coefficient	편미분계수	偏微分係數
partial differentiation	편미분(법)	偏微分(法)
partial drought	부분가뭄, 부분한발	部分旱魃
partial-duration series	부분-기간 계열	部分期間系列
partial eclipse	부분식	部分蝕
partial fraction	부분분수	部分分數
partially penetrating well	부분관통정	部分貫通井
partial melting	부분녹음	部分溶解
partial obscuration	부분가림	部分-
partial potential temperature	부분온위	部分溫位
partial pressure	분압	分壓
partial reflection	부분반사	部分反射
partial specific property	부분비성질	部分比性質
partial specific volume	부분비부피	部分比體積
partial specular reflection	부분거울면반사	部分-面反射
partial stability	부분안정도	部分安定度
partial tide	부분조석	部分潮汐
partial vacuum	부분진공	部分眞空
particle	입자, 질점	粒子, 質點
particle charge	입자전하	粒子電荷
particle collection system	입자채집시스템	粒子採集-

particle deposition velocity	입자침적속도	粒子沈積速度
particle derivative	입자도함수	粒子導函數
particle dispersion	입자분산	粒子分散
particle emission rate	입자방출률	粒子放出率
Particle Environment Monitor	입자환경모니터	粒子環境-
particle-height probability distribution	입자-높이확률분포	粒子-確率分布
particle matter	입자물질	粒子物質
particle model	입자모형	粒子模型
particle model of light	빛입자모형	-粒子模型
particle of cloud	구름입자	-粒子
particle-particle interaction	입자대입자 상호작용	粒子對粒子相互作用
particle size	입자크기, 입도	粒子-, 粒度
particle size distribution	입도분포	粒度分布
particle size parameter	입자크기매개변수(모수)	粒子-媒介變數(母數)
particle trajectory model	입자추적모형	粒子追跡模型
particle velocity	입자속도	粒子速度
particulate	부유미립자, 입자	浮遊微粒子, 粒子
particulate elemental carbon	원소탄소미세분진	元素炭素微細粉塵
particulate loading	입자하중, 미립자농도	粒子荷重, 微粒子濃度
particulate matter	부유물질, 입자상물질	浮遊物質, 粒子狀物質
particulate matter 10(= PM10)	10μ 미세입자	-微細粒子
particulate metal	금속성미세분진	金屬性微細粉塵
particulate organic matter	부유미립자유기물	浮遊微粒子有機物
partition of energy	에너지분배	-分配
partly cloudy	구름조금, 부분흐림	部分-
partly occluded	부분폐색(의)	部分閉塞
Pascal	파스칼(기압의 단위, N/m^2)	
Pascal's law	파스칼법칙	-法則
Pasquill category	파스퀼범주	-範疇
Pasquill-Gifford category	파스퀼-기퍼드 범주	-範疇
Pasquill-Gifford stability	파스퀼-기퍼드 안정도	-安定度
Pasquill-Gifford stability class	파스퀼-기퍼드 안정도등급	-安定度等級
Pasquill's beta	파스퀼베타	
Pasquill's stability	파스퀼 안정도	-安定度
Pasquill stability category	파스퀼 안정도범주	-安定度範疇
Pasquill stability class	파스퀼 안정도등급	-安定度等級
pass	태풍통과	颱風通過
passive anafrontal surface	수동상승활주면	受動上昇滑走面
passive cloud	비활동구름	非活動-
passive front	비활동전선	非活動前線
passive gas sampler	수동형가스채집기	手動型-採集器
passive instrument	수동형기기	受動型機器
passive katafrontal surface	수동하강활주면	受動下降滑走面
passive microwave sensing	수동마이크로파탐측	受動-波探測

P

passive microwave wind	수동마이크로파바람	受動-波-
passive permafrost(= fossil permafrost)	수동영구동토	受動永久凍土
passive radar	수동레이더	受動-
passive remote sensing	수동원격탐사	受動遠隔探査
passive sampling	수동시료채취	受動試料採取
passive scalar	수동스칼라	受動-
passive sensor	수동센서	受動-
passive solar collector	수동태양열집열기	受動太陽熱集熱器
passive system	수동시스템	受動-
pastagram	파스타그램	
past climate	과거기후	過去氣候
Past Global Changes(PGC)	과거지구변화	過去地球變化
past weather	과거일기	過去日氣
Patagonia	파타고니아(아르헨티나)	
patch	조각, 반점	斑點
patchy turbulence	부조화난류	不調和亂流
path	경로, 행로, 노선, 궤적	經路, 行路, 路線, 軌跡
path difference	경로차	經路差
path information entropy	경로정보엔트로피	經路情報-
path integrated attenuation	경로통합감쇠량	經路統合減衰量
path length	경로길이	路程-
path length of air	대기노정	大氣路程
path line	유적선, 경로선	流跡線, 經路線
path method	경로법	經路法
path of mixing	혼합경로	混合經路
pathway	경로	經路
patrol landplane	육상정찰기	陸上偵察機
patrol seaplane	해상정찰기	海上偵察機
pattern	모습, 유형, 패턴	類型
pattern correlation	패턴상관, 유형상관	類型相關
patterned ground	구조토	構造土
pattern of circulation	순환패턴, 순환유형	循環類型
pattern recognition	패턴인식, 유형인식	類型認識
pattern tropical number	PT(패턴열대)수	-(-熱帶)數
Patterson's rule	패터슨법칙	-法則
pause	권계면	圈界面
Pavlovsky's approximation	파블로프스키 근사	-近似
payload	유효하중	有效荷重
payload Interface Panel	탑재체접속판	搭載體接續板
P-D/N weather	공중요격기상	空中邀擊氣象
peak	꼭대기, 극대치	極大値
peak current	최대전류	最大電流
peak discharge	최대방전	最大放電
peak flow method	최대유량법	最大流量法

peak frequency	최대진동수, 최대주파수	最大振動數, 最大周波數
peak gust	절정돌풍, 최대돌풍	絶頂突風, 最大突風
peak instantaneous wind	최대순간바람	最大瞬間-
peak instantaneous wind speed	최대순간풍속	最大瞬間風速
peak power	절정공률, 절정전력,	絶頂工率, 絶頂電力,
	최대전력, 최대공률	最大電力, 最大工率
peak rate of runoff	첨두유출량	尖頭流出量
peak to average concentration	극대대평균 농도	極大對平均濃度
peak-to-mean ratio	극대대평균 비	極大對平均比
peak wind	최대풍, 절정바람	最大風 絶頂-
peak wind speed	최대풍속, 절정풍속	最大風速, 絶頂風速
pearl lightning(= beaded lightning)	구슬번개	
pearl-necklace lightning	진주목걸이번개	
Pearson distribution	피어슨분포	-分布
pea-soup fog	완두콩수프안개	
peat	토탄, 이탄	土炭, 泥炭
Péclet number	페클렛수	-數
pedestal	축받이대	
pedestal cloud(= wall cloud)	벽구름	壁-
pedestal rock	받침돌	
pedology	토양학	土壤學
peesash	피사시(인디아의 덥고 건조한 모래먼지 바람)	
peesweep storm	피스윕폭풍(잉글랜드와	-暴風
	스코틀랜드의 이른 봄 폭풍우)	
P-E index(precipitation-effectiveness index,	P-E지수(강수-유효지수,	-指數
precipitation-evaporation index)	강수-증발지수)	
pelagic organism	먼바다유기체	遠海有機體
Peléean eruption	폭발성분화	暴發性噴火
pellicular water	박막수, 흡착수	薄膜水, 吸着水
pendant cloud	깔때기구름	
pendant echo	깔때기에코	
pendulum	진자	振子
pendulum anemometer	흔들이풍속계	-風速計
pendulum day	진자일	振子日
peneplain	준평원	準平原
penetrating top	침투꼭대기	浸透頂上
penetration	침투	浸透
penetration depth	침투깊이	浸透深度
penetration frequency	돌발주파수	突發周波數
penetration range(→ night visual range)	침투거리	浸透距離
penetrative convection	침투대류, 관입대류	浸透對流, 貫入對流
penetrometer	투과도계	透過度計
peninsula	반도	半島
penitent ice (columnar shape of compacted ice)	압축얼음	壓縮-

P

penitent snow (columnar shape of compacted snow)	압축눈	壓縮-
penknife ice(= candle ice)	연필칼모양얼음	鉛筆-模樣-
Penman formula	펜만공식	-公式
Penman-Monteith equation	펜만-몬티스 방정식	-方程式
pennant	삼각깃(50노트의 풍속 표시)	三角-
Pennsylvanian Period	(고생대)펜실베이니아기	-期
penthemeron(= pentad)	반순	半旬
penumbra	반그림자	半-
P-E quotient(precipitation-evaporation quotient)	P-E몫(강수-증발몫)	
percentage error	백분오차	百分誤差
percentage of possible sunshine	가능일조율, 가조율	可能日照率, 可照率
percentile	백분위수	百分位數
percent of area of uncertainty	불확실영역퍼센트 (폭풍역에 들어갈 확률)	不確實領域-
percent reduction	백분율감소	百分率減少
perceptual climate	지각기후	知覺氣候
perched aquifer	주수대수층, 부유대수층	宙水帶水層, 浮遊帶水層
perched groundwater	주수지하수	宙水地下水
perched stream	주수하천	宙水河川
percolation	침루	浸漏
percolation path	침투경로	浸透經路
percolation zone	침투대	浸透帶
pereletok(= intergelisol)	상영구동토층	上永久凍土層
perennial stream	영구흐름	永久-
perennially frozen ground(= permafrost)	영구동토	永久凍土
perfect black body	완전흑체	完全黑體
perfect fluid	완전유체	完全流體
perfect forecast	완전예보	完全豫報
perfect gas(= ideal gas)	완전기체, 이상기체	完全氣體, 理想氣體
perfect gas law	완전기체법칙, 이상기체법칙	完全氣體法則, 理想氣體法則
perfectly diffuse radiator	완전확산복사체	完全擴散輻射體
perfectly diffuse reflector	완전확산반사체	完全擴散反射體
perfect mirror	완전반사거울	完全反射-
perfect mixing	완전혼합	完全混合
perfect prognostic	완전예단, 완전예상	完全豫斷, 完全豫想
perfect prognostic method	완전예단방법	完全豫斷方法
perfect radiation	완전복사	完全輻射
perfect radiator	완전복사체	完全輻射體
perfect storm	퍼펙트스톰	
perfluorocarbons	과플루오로탄소	過-炭素
performance standard	성과기준	成果基準
pergelation	영구동토화	永久凍土化
pergélisol(= permafrost)	영구동토〈프랑스어〉	永久凍土
pergélisol table(= permafrost table)	영구동결면	永久凍結面

perhumid climate	과습윤기후	過濕潤氣候
pericenter	가까운 중심점	近心點
perifocus	근점	近點
perigean range	근지점조차	近地點潮差
perigean tide	근지점조석	近地點潮汐
perigee	근지점	近地點
perigee passage	근지점통과	近地點通過
periglacial	빙하주변	氷河周邊
periglacial climate	빙하주변기후	氷河周邊氣候
periglacial geomorphology	주변빙하지형학	周邊氷河地形學
perihelion	근일점	近日點
period	기간, 주기	期間, 週期
period average (or mean)	기간평균(10년 이상)	期間平均
period forecasting	기간예보	期間豫報
periodical	주기운동을 하는, 주기적	週期的
periodic boundary condition	주기적 경계조건	週期的境界條件
periodic function	주기함수	週期函數
periodicity	주기성	週期性
periodic law	주기율	週期律
periodic motion	주기운동	週期運動
periodic oscillation	주기진동	週期振動
periodic property	주기적 성질	週期的性質
periodic variability	주기변동성	週期變動性
periodic variation	주기변동	週期變動
periodic wind	주기바람	週期-
period maxima	기간최대	期間最大
period of Changma	장마기간	-期間
period of drought	가뭄주기	-週期
period of growth	생장기간, 생장기	生長期間, 生長期
period of oscillation	진동주기	振動週期
period of record	기록주기	記錄週期
period of validity	유효기간	有效期間
periodogram	주기도	週期圖
periodogram analysis	주기도분석	週期圖分析
perlucidus	틈새(구름)	
permafrost(= pergéllisol)	영구동토층, 영구 언땅	永久凍土層
permafrost island	영구 언섬	永久-
permafrost melt	영구동토융해	永久凍土融解
permafrost table	영구 언면	永久-面
permanent anticyclone	영구고기압	永久高氣壓
permanent aurora	대기광	大氣光
permanent circulation	영구순환	永久循環
permanent community	영구군락	永久群落
permanent depression	영구저기압	永久低氣壓

P

permanent drought	영구가뭄	永久-
permanent echo	영구에코	永久-
permanent gas	영구기체	永久氣體
permanent high	영구고기압	永久高氣壓
permanent low	영구저기압	永久低氣壓
permanently frozen ground(= permafrost)	영구동토	永久凍土
permanent register	영구기록부	永久記錄簿
Permanent Service for Mean Sea Level	(영국)평균해수면상설서비스	平均海水面常設-
permanent threshold shift	영구임계값이동, 영구청력변위	永久臨界-移動, 永久聽力變位
permanent wave	영구파	永久波
permanent wilting	영구위조	永久萎凋
permanent wilting percentage	영구위조율	永久萎凋率
permanent wilting point	영구위조점	永久萎凋點
permanent wind	늘바람, 영구바람	永久-
permeability(= hydraulic conductivity)	투과성, 투과율, 투수율	透過性, 透過率, 透水率
permeability coefficient	투과성계수, 투수율계수	透過性係數, 透過率係數
permeable membrane	투과성막	透過性膜
permeameter	투과율계	透過率計
permeation tube	투과성 튜브	透過性-
Permian Period	(고생대)페름기	-期
Permian-Triassic extinction	페름기-트라이아스기멸종	-期-期滅種
per mille	퍼밀	
permittivity	유전율	誘電率
peroxides	과산화물	過酸化物
peroxyacetyl nitrate	질산과산화아세틸	窒酸過酸化-
peroxyalkyl radical	퍼옥시알킬라디칼	
peroxyl radical	퍼옥실라디칼	
peroxy radicals	퍼옥시라디칼, 과산화기라디칼	過酸化基-
perpendicular	수직(의)	垂直
perpendicular incidence	수직입사, 직사	垂直入射, 直射
perpetual frost climate	영구동결기후	永久凍結氣候
perry	페리(잉글랜드의 돌발성 강우)	
Pers sunshine recorder	퍼스일조계	-日照計
Persian Gulf	페르시아만	-灣
Persian Gulf Water	페르시아만수	-灣水
persistence	지속성	持續性
persistence forecast	지속성예보	持續性豫報
persistence method	지속법	連續法
persistence ratio	지속성비	持續性比
persistence tendency	지속경향	持續傾向
persistent oscillation	지속성진동	持續性振動
personal air pollution	개인대기오염	個人大氣汚染
personal air pollution exposure	개인대기오염노출	個人大氣汚染露出
personal emissions	개인배출	個人排出

personal error	개인오차	個人誤差
personal exposure	개인노출	個人露出
personal pollution exposure	개인오염피폭	個人汚染物質 被曝
pertinent area of special report	특보구역	特報區域
perturbation	요동, 섭동	搖動, 攝動
perturbation density	섭동밀도	攝動密度
perturbation equation	섭동방정식	攝動方程式
perturbation function	섭동함수	攝動函數
perturbation method	섭동법	攝動法
perturbation model	섭동모형	攝動模型
perturbation motion	섭동운동	攝動運動
perturbation pressure	섭동압력	攝動壓力
perturbation quantity	섭동량	攝動量
perturbation technique	섭동기법	攝動技法
perturbation temperature	섭동온도	攝動溫度
perturbation theory	섭동론	攝動論
Peru current	페루해류	-海流
Peruvian dew	페루비안이슬	
Peruvian paint	페루비안페인트	
pervious zone	투수대	透水帶
petit St. Bernard	페티 생 베르나르(프랑스 Haute Tarentaire에서 부는 산바람)	
Petterssen's development equation	페테르센발달방정식	-發達方程式
Petterssen's rule	페테르센규칙	-規則
Petterssen wave speed equation method	페터슨파속방정식법	-波速方程式法
Petterssen wave speed equation nomogram	페터슨파속방정식노모그램	-波速方程式-
Pfaffian differential form	파프미분형식	-微分型式
Pfaff's differential form	파프미분형식	-微分型式
pH	pH도	酸度
Phanerozoic Eon	현생이언, 현생누대	現生-, 現生累代
phantom echo	유령에코	幽靈-
phase	(위)상	(位)相
phase angle	위상각	位相角
phase array	위상배열	位相配列
phase array radar	위상배열레이더	位相配列-
phase averaging	위상평균	位相平均
phase calibration	위상보정	位相補正
phase change	상변화, 위상변화	相變化, 位相變化
phase constant	위상상수	位相常數
phased array	위상배열	位相配列
phased array antenna	위상배열안테나	位相配列-
phase delay	위상지연	位相遲延
phase detection	위상검파	位相檢波
phase detector	위상검출기, 위상감지기	位相檢出器, 位相感知器

P

phase diagram	위상도, 상(평형)선도	位相圖, 相(平衡)線圖
phase distortion	위상 이그러짐	位相-
phase diversity method	위상분산방식	位相分散方式
phase flow	위상흐름	位相-
phase fluid	위상유체	位相流體
phase front	위상전선(면)	位相前線(面)
phase function	위상함수	位相函數
phase instability	위상불안정	位相不安定
phase lag	위상지연	位相遲延
phase matrix	위상행렬	位相行列
phase modulation	위상변조	位相變調
phase noise	위상잡음	位相雜音
phase path	위상경로	位相經路
phase point	위상점	位相點
phase rule	상법칙	相法則
phase shift	위상변이	位相變移
phase space	위상공간	位相空間
phase spectrum	위상스펙트럼	位相-
phase speed(= phase velocity)	위상속력, 위상속도	位相速力, 位相速度
phase time modulation	위상시변조	位相時變調
phase transformation	상변환	相變換
phase velocity	위상속도	位相速度
phasor diagram	위상벡터선도	位相-線度
phenogram	생물계절도	生物季節圖
phenological observation	생물계절학적 관측	生物季節學的觀測
phenology	생물계절학, 생물기후학, 계절학	生物季節學, 生物氣候學, 季節學
phenomenological coefficient	현상계수	現象係數
phenomenological theory	현상학이론	現象學理論
phenomenon	현상	現象
Philip correction	필립보정	-補正
Philippine plate	필리핀판	-板
Philippines current	필리핀해류	-海流
Philippine sea	필리핀해	-海
Philips wind	필립스바람	
philosophy of pollution control	오염규제원리	汚染規制原理
phipigram	피피선도	-線圖
phosphorescence	인광	燐光
phosphorus cycle	인순환	燐循環
photochemical equilibrium	광화학평형	光化學平衡
Photochemical oxidants	광화학산화체	光化學酸化體
photochemical reaction	광화학반응	光化學反應
photochemical smog	광화학스모그, 광화학연무	光化學煙霧
photochemistry	광화학	光化學

photoconductive cell	광전도전지	光電導電池
photodiode	반도체광소자	半導體光素子
photodissociation	광분해, 광해리	光分解, 光解離
photoelectric cell	광전지	光電池
photoelectric effect	광전효과	光電效果
photoelectric hygrometer	광전습도계	光電濕度計
photoelectric photometer	광전광도계	光電光度計
photoelectric photometry	광전측광법	光電測光法
photoelectric transmittancemeter	광전투과계	光電透過計
photoelectron	광전자	光電子
photogrammeter	사진계측기	寫眞計測器
photographic barograph	사진기압계	寫眞氣壓計
photographic meteor	사진유성	寫眞流星
photographic projection	사진투영	寫眞投影
photographic sunshine recorder	사진일조계	寫眞日照計
photoionization	광전리	光電離
photology	촬영학, 광학	撮影學, 光學
photolysis	광분해	光分解
photometeor	대기빛현상	大氣光現像
photometer	광도계	光度計
photometry	측광(학), 측광법	測光(學), 測光法
photomicroscope	사진현미경	寫眞顯微鏡
photomultiplier tube	광전자증배관 (포토멀티플라이어관)	光電子增倍管
photon	광자	光子
photooxidation	광산화작용	光酸化作用
photoperiodism	일광효과, 광주기성	日光效果, 光週期性
photopolarimeter	광편광계	光偏光計
photorespiration	광호흡	光呼吸
photosensitivity	감광성	感光性
photosnow cover meter	광전식 적설계	光電式積雪計
photosphere	광구	光球
photosynthate	광합성산물	光合成産物
photosynthesis	광합성	光合成
Photosynthetically Active Radiation	광합성유효복사	光合成有效輻射
photo theodolite	사진경위(의)	寫眞經緯
photovoltaic effect	광기전력효과	光起電力效果
phreatic surface(= water table)	지하수면	地下水面
phreatic water(= ground water)	지하수	地下水
phreatic zone(= zone of saturation)	지하수층, 포화대	地下水層, 飽和帶
phreatophyte	깊은 뿌리 식물	-植物
pH scale	수소이온농도척도	水素-濃度尺度
pH sequence	수소이온농도서열	水素-濃度序列
physical change	물리변화	物理變化

P

physical chemistry	물리화학	物理化學
physical classification	물리분류	物理分類
physical climate	물리기후	物理氣候
physical climatic zone	물리기후대	物理氣候帶
physical climatology	물리기후학	物理氣候學
physical constant	물리상수	物理常數
physical enhancement	물리강화	物理强化
physical equation	물리방정식	物理方程式
physical forecasting	물리예보	物理豫報
physical hydrodynamics	물리유체역학	物理流體力學
physical meteorology	물리기상학	物理氣象學
physical mode	물리모드	物理-
physical model	물리모형	物理模型
physical oceanography	물리해양학	物理海洋學
physical optics	물리광학	物理光學
physical retrieval	물리추출	物理抽出
Physical Space Analysis System	물리적공간분석시스템	物理的空間分析-
physico-statistic prediction	물리통계예측	物理統計豫測
physics of front	전선물리	前線物理
physics of the atmosphere	대기물리	大氣物理
physiognomy	관상	觀相
physiography	지형학	地形學
physiological climatology	생리기후학	生理氣候學
physiological drought	생리적 가뭄	生理的-
physiological dryness	생리적 건조	生理的乾燥
physiological snow damage	생리적 설해	生理的雪害
phytoclimate	식물기후	植物氣候
phytoclimatology	식물기후학	植物氣候學
phytogeography	식물지리학	植物地理學
phytometer	증산계, 파이토미터	蒸散計
phytometric method of environmental measurement	식물이용환경측정법	植物利用環境測定法
phytotron	피토트론	
pibal	측풍기구, 파이발	測風氣球
Pich evaporimeter	피체증발계	-蒸發計
piedmont glacier	산기슭빙하	山-氷河
pieze	피즈(압력의 단위 104dyne/mm²)	
piezo coefficient	압성계수	壓性係數
piezometric equation	압성방정식	壓性方程式
piezometric surface	피압지하수면	被壓地下水面
piezotropy	압성	壓性
pigment layer	색소층	色素層
pile-up process	퇴적과정	堆積過程
pileus	두건구름	
pillow (cloud)	베개(구름)	

pilmer	필머(영국의 격렬한 소나기)	
pilot balloon	측풍기구	測風氣球
pilot-balloon ascent	측풍기구띄움	測風氣球-
pilot-balloon observation	측풍기구관측	測風氣球觀測
pilot-balloon plotting board	측풍기구기입판	測風氣球記入板
pilot-balloon self-recording theodolite	측풍기구자기경위(의)	測風氣球自記經緯
pilot-balloon slide rule	측풍기구계산자	測風氣球計算尺
pilot-balloon station	측풍기구관측소	測風氣球觀測所
pilot-balloon theodolite	측풍기구경위(의)	測風氣球經緯
pilot briefing	조종사브리핑	操縱士-
pilot chart	항로도	航路圖
pilot streamer	선두스트리머	先頭-
pilot weather report	조종사일기보고	操縱士日氣報告
pinched lightning	집게번개	
pingo	핑고(북극지방의 화산모양의 얼음언덕)	
pink snow	분홍눈	粉紅-
pin point weather	쪽집게기상	-氣象
pioneer fracture zone	파이어니어단열대	-斷裂帶
pioneer vegetation	선구식생	先驅植生
pip	핍, 돌아오는 에코	
piphigram	피피선도(압력-엔트로피도)	-線圖
pitch angle	피치각	-角
pitch axis	피치축	-軸
pi-theorem	파이정리	-定理
Pitot-static tube	피토정압관	-靜壓管
Pitot tube	피토관	-管
Pitot tube anemometer	피토관풍속계	-管風速計
Pivoting high	추축고기압	樞軸高氣壓
pixel	픽셀, 화소	畵素
pixel size	화소크기	畵素-
pixel value	화솟값	畵素-
plain	평원	平原
planar structure	평면구조	平面構造
Planckian radiator	플랑크의 복사체	-輻射體
Planck's constant	플랑크상수	-常數
Planck's function	플랑크함수	-函數
Planck's law	플랑크법칙	-法則
Planck's radiation law	플랑크복사법칙	-輻射法則
plane atmospheric wave	평면대기파	平面大氣波
plane curve	평면곡선	平面曲線
plane-dendritic crystal	평면나무가지모양결정	平面-模樣結晶
plane figure	평면도형	平面圖形
plane geometry	평면기하(학)	平面幾何(學)
plane of incidence	입사면	入射面

P

plane of the elliptic	황도면	黃道面
plane polarization	면편광	面偏光
plane polarization indicator	평면편광기	平面偏光器
plane-position indicator	평면위치지시기	平面位置指示器
plane shear indicator(PSI)	평면시어지시기	平面-指示器
plane source	면원	面源
planet	행성	行星
planetary albedo	행성반사율, 행성알베도	行星反射率
planetary atomic model	행성원자모형	行星原子模型
planetary atmosphere	행성대기	行星大氣
planetary boundary layer	행성경계층	行星境界層
planetary circulation	행성순환	行星循環
planetary motion	행성운동	行星運動
planetary scale	행성규모	行星規模
planetary scale system	행성규모시스템	行星規模-
planetary scale wind	행성규모바람	行星規模-
planetary system	행성계	行星系
planetary temperature	행성온도	行星溫度
planetary vorticity effect	행성소용돌이도효과	行星-度效果
planetary wave	행성파	行星波
planetary wave formula	행성파공식	行星波公式
planetary wind	행성바람	行星-
planetesimal hypothesis	미행성설	微行星說
planetoid	소행성	小行星
plane vector	평면벡터	平面-
plane wave	평면파	平面波
planimeter	면적계	面積計
plankton	플랑크톤, 부유생물	浮遊生物
planktonic foraminifera	부유성유공충	浮遊性有孔蟲
plant behavior	식물습성	植物習性
plant climate	식물기후	植物氣候
plant climatology	식물기후학	植物氣候學
plant community	식물군락	植物群落
plant damage	식물재해	植物災害
planting season	식수철	植樹-
plant injury	식물상해	植物傷害
plant phenology	식물계절학	植物季節學
plant temperature	식물온도	植物溫度
plant type	초형	草型
plasma	플라즈마	
plasmapause	플라즈마권계면	-圈界面
plasmasphere	플라즈마권	-圈
plasticity	가소성, 유연성	可塑性, 柔軟性
plastid	색소체	色素體

plate 363 pluviometry

plate	평판	平板
plate anemometer	평판풍속계	平板風速計
plateau	고원, 대지	高原, 大地
plateau glacier	고원빙하	高原氷河
plateau station	고원관측소	高原觀測所
plate boundaries	평판경계	平版境界
plate crystal	판모양결정	板模樣結晶
plate crystal frost	판모양결정서리	板模樣結晶-
platelet(= plate crystal)	판결정	板結晶
plate tectonics	판구조론	板構造論
plate tectonics theory	판구조이론	板構造理論
platform	관측대	觀測臺
platinum resistance thermometer	백금저항온도계	白金抵抗溫度計
platinum wire thermometer	백금선온도계	白金線溫度計
platykurtosis	저첨도	底尖度
playback	재현	再現
Pleistocene	빙하기(플라이스토세)	氷河期
Pleistocene climate	플라이스토세 기후	-氣候
Pleistocene epoch	플라이스토세 시대	-時代
Pleistocene glacial epoch	플라이스토신 빙하시대	-氷河時代
(= Quaternary Ice Age)	(제4빙하시대)	
Pleistocene glaciation	플라이스토신 빙하작용	-氷河作用
Pleistocene Ice Age	플라이스토신 빙하시대	-氷河時代
pleochroic halo	다색무리	多色-
plidar(= polychromatic lidar)	다색라이다	多色-
Pliocene climate	플라이오세기후, 선신세기후	-氣候, 鮮新世氣候
plotting	기입	記入
plotting board	기입판	記入板
plotting model	기입모형	記入模型
plotting symbol	기입부호	記入符號
ploughing season	갈이철(논, 밭)	
plow/plough wind	플라우바람(뇌우를 동반하고 직선적으로 이동하는 바람)	
plum rain	장마, 바이우	梅雨
plume	줄기흐름, 플룸	
plume rise	연기상승	煙氣上昇
plume rise height	연기상승높이	煙氣上昇-
plume rise model	연기상승모형	煙氣上昇模型
pluvial index	우량지수	雨量指數
pluvial period	다우기	多雨期
pluviograph	자기우량계	自記雨量計
pluviometer	우량계	雨量計
pluviometric coefficient	우량계수	雨量係數
pluviometric quotient	우량비	雨量比
pluviometry	우량측정법	雨量測定法

P

pluvioscope	우량계	雨量計
pluviothermic ratio	우온비	雨溫比
pocket anemometer	간이풍속계	簡易風速計
pocket aspiration psychrometer	간이통풍건습계	簡易通風乾濕計
pocket register	간이야장	簡易野帳
pocky cloud	유방구름	乳房-
poëchore	초지(기후)	草地
pogonip	얼음안개	
point	포인트(오스트레일리아의 우량 단위), 방위점	方位點
point charge	점전하	點電荷
point discharge	첨단방전	尖端放電
point discharge current	첨단방전류	尖端放電流
point frequency analysis	점빈도해석	點頻度解析
point of application	작용점	作用點
point of inflection	변곡점	變曲點
point of occlusion	폐색점	閉塞點
point of recurvature	전향점	轉向點
point of saturation	포화점	飽和點
point of symmetry	대칭점	對稱點
point rainfall	점우량(지점우량)	點雨量(地點雨量)
point source	점원	點源
point target	점표적	點標的
point-to-multipoint communication	점대다중점통신	點對多重點通信
point-to-point communication	점대점통신	點對點通信
point-to-point transmission	지점간전송	地點間傳送
poise	포이즈(점성계수의 단위)	
Poiseuille flow	푸아죄유흐름	
Poiseuille-Hagen law	푸아죄-하겐 법칙	-法則
Poisson constant	푸아송상수	-常數
Poisson distribution	푸아송분포	-分布
Poisson's equation	푸아송방정식	-方程式
polar air (mass)	한대기단	寒帶氣團
polar air haze	한대연무	寒帶煙霧
polar angle	극각	極角
polar anticyclone	한대고기압	寒帶高氣壓
polar aurora	극광	極光
polar axis	극축	極軸
polar band	한대, 오로라스펙트럼	寒帶
polar belt	한대	寒帶
polar blackout	극지암흑	極地暗黑
polar-cap	극관	極冠
polar-cap absorption	극관흡수	極冠吸收
polar-cap atmosphere	극관대기	極冠大氣

polar-cap ice	극관얼음	極冠-
polar cell	극세포	極細胞
polar circle	극권	極圈
polar circulation	극순환	極循環
polar climate	한대기후	寒帶氣候
polar continental air (mass)	한대대륙기단	寒帶大陸氣團
polar coordinates	극좌표	極座標
polar coordinates system	극좌표계	極座標系
polar current	한대기류	寒帶氣流
polar cyclone(= polar vortex)	극저기압, 극소용돌이	極低氣壓
polar distance	극거리	極距離
polar easterlies	극편동풍	極偏東風
polar easterlies index	극편동풍지수	極偏東風指數
polar equation	극방정식	極方程式
polar-front	한대전선	寒帶前線
polar-front jet	한대전선제트	寒帶前線-
polar-front jet stream	한대전선제트류	寒帶前線-流
polar-front model	한대전선모형	寒帶前線模型
polar-front theory	한대전선이론	寒帶前線理論
polar-front zone	한대전선대	寒帶前線帶
polar glacier	극빙하	極氷河
polar high	한대고기압, 극고기압	寒帶高氣壓, 極高氣壓
polar ice	극빙	極氷河
polar icecap	극빙모	極氷帽
polarization diversity	편파다양성(수신기)	偏波多樣性
polarimeter	편광계	偏光計
polar invasion	한기침입	寒氣侵入
polaris(= polar star)	북극성	北極星
polariscope	편광기	偏光器
polarity	극성	極性
polarizability	분극도	分極度
polarization	편광, 편극	偏光, 偏極
polarization agility	편파전환성(송신기)	偏波轉換性
polarization effect	편극효과	偏極效果
polarization isocline	편광등기울기선	偏光等傾斜線
polarization modulation factor	편광변조요소	偏光變造要素
polarization radar	편파레이더	偏波-
polarization signature	편파특징	偏波特徵
polarized radiation	편파복사	偏波輻射
polarizer	편광자	偏光子
polar jet	한대제트	寒帶-
polar jet stream	한대제트류	寒帶-流
polar low	극저기압	極低氣壓
polar magnetic storm	극지자기폭풍	極地磁氣暴風

P

polar marine air (mass)	한대해양기단	寒帶海洋氣團
polar maritime air mass	한대해양기단	寒帶海洋氣團
polar meteorology	한대기상학, 극지기상학	寒帶氣象學, 極地氣象學
polar night	극야	極夜
polar night jet	한대야간제트	寒帶夜間-
polar night jet stream	한대야간제트류, 극야제트류	寒帶夜間-流, 極夜-流
polar orbital meteorological satellite	극궤도기상위성	極軌道氣象衛星
polar orbiting geophysical observatory	극궤도지구물리관측위성	極軌道地球物理觀測衛星
polar orbiting satellite	극궤도위성	極軌道衛星
polar outbreak(= cold-air outbreak)	한기범람	寒氣氾濫
polar Pacific air mass	한대태평양기단	寒帶太平洋氣團
polar phenomenon	극현상	極現狀
polar projection	극투영	極投影
polar-radiometer	편광방사계	偏光放射計
polar region	극지	極地
polar star	북극성	北極星
polar stereographic projection	극평사도법	極平射圖法
polar stratospheric cloud	극성층운	極成層雲
polar tropopause	한대권계면	寒帶圈界面
polar trough	한대(기압)골	寒帶氣壓-
polar vortex	극소용돌이	極-
polar wave	극파	極波
polar westerlies	한대편서풍	寒帶偏西風
polar wind	극풍	極風
polar wind belt	극풍대	極風帶
polar wind divide	극풍분계	極風分界
polar wind zone	극풍대	極風帶
Polar Year	극관측년	極觀測年
polar zone	한대, 극권	寒帶, 極圈
pole	극	極
pole-star recorder	북극성기록계	北極星記錄計
pollen	꽃가루	
pollen analysis	꽃가루분석, 화분분석	花粉分析
pollen forecast	화분예보	花紛豫報
pollination	꽃가루받이	
pollutant	오염물질	汚染物質
polluted air	오염공기	汚染空氣
pollution	오염	汚染
pollution area	오염지역	汚染地域
polychlorinated biphenyls	폴리염화비페닐	
polychromatic lidar	다색광라이다	多色光-
polycyclic aromatic hydrocarbons	다환 방향족 탄화수소	多環芳香族炭化水素
polyethylene balloon	폴리에틸렌기구	-氣球
polygon	다각형	多角形

polymerization	중합	重合
polymetamorphism	복변성 작용	復變性作用
polymeter	다용도습도계	多用途濕度計
polymorphism	다형성, 동질이상	多形性, 同質異像
polynomial	다항식	多項式
polynomial expansions	다항식 확장	多項式-
polynomial function	다항함수	多項函數
polyn'ya	폴리니아(얼음틈, 얼음구멍)	
polytropic	다방상수	多方常數
polytropic atmosphere	다방대기	多方大氣
polytropic change	다방변화	多方變化
polytropic constant	다방상수	多方常數
polytropic Exner function	다방에크스너함수	多方-函數
polytropic potential temperature	다방온위	多方溫位
polytropic process	다방과정	多方過程
POMAR code	포마(POMAR) 전보식	-電報式
pondage	저수량	貯水量
ponding	담수	湛水
pool of cold air	한랭역	寒冷域
poorga(= purga)	퍼가	
poor spectral resolution	저스펙트럼분해능	低-分解能
poor visibility	나쁜 시정	惡視程
popcorn cumulonimbi	팝콘적란운군	-積亂雲群
population	모집단	母集團
population distribution	모집단분포	母集團分布
population growth	개체군성장	個體群成長
population mean	모평균	母平均
population standard deviation	모표준편차	母標準偏差
pop-up weather	저공기습기상	低空奇襲氣象
porosity	공극률	孔隙率
porous ice	다공성얼음	多孔性-
porphyroblast	반상변정	斑狀變晶
portable cup anemometer	휴대용컵풍속계	携帶式風向風速計
portable radiometers	휴대용복사계	携帶用輻射計
port meteorological liaison officer	항만기상연락관	港灣氣象連絡官
position	위치	位置
position indicator	위치지시기	位置指示器
positioning of satellite	위성위치잡기(배치)	衛星位置-(配置)
position report	위치통보	位置通報
position vector	위치벡터	位置-
positive	양(의), 정(의)	陽, 正
positive area	양영역, 정영역	陽領域, 正領域
positive axis	양축	陽軸
positive buoyancy	양부력	陽浮力

P

positive catalyst	정촉매	正觸媒
positive charge	양전하	陽電荷
positive cloud to ground lightning	양(의) 구름-지상 뇌전	陽-地上雷電
positive correlation	정상관	正相關
positive crankcase ventilation	자동차 통기장치	自動車 通氣裝置
positive direction	양방향	陽方向
positive feedback	양되먹임	陽-
positive feedback mechanism	양되먹임기작	陽-機作
positive feedback system	양되먹임시스템	陽-
positive integer	양정수	陽整數
positive ion	양이온	陽-
positive pole	양극	陽極
positive rain	양전기비	陽電氣-
positive rational number	양유리수	陽有理數
positive sign	양부호	陽符號
positive-tilt trough	양(의) 기울어짐기압골	陽-氣壓-
positive vorticity advection	양(의) 소용돌이도이류	陽-度移流
possible duration of sunshine	가조시간	可照時間
possible error	가능오차	可能誤差
post-analysis	사후분석	事後分析
post-analysis of special report	특보사후분석	特報事後分析
post-frontal fog	전선뒤안개	前線-
postglacial age	후빙기	後氷期
post-processing	후처리과정	後處理過程
potamology	하천학	河川學
potential	위치, 잠재, 가능, 전위	位置, 潛在, 可能, 電位
potential center of crystallization	잠재결정중심	潛在結晶中心
potential cyclogenesis	잠재저기압발생	潛在低氣壓發生
potential density	위치밀도	位置密度
potential dew-point temperature	이슬점위치온도, 이슬점온위	-點位置溫度, -點溫位
potential (voltage) drop	전압강하	電壓降下
potential energy	위치에너지, 퍼텐셜에너지	位置-
potential equivalent temperature	위치상당온도	位置相當溫度
potential evaporation	가능증발량	可能蒸發量
potential evapotranspiration	가능증발산(량)	可能蒸發散(量)
potential gradient	전위경도	電位傾度
potential index of refraction (= potential refractive index)	위치굴절률, 잠재굴절률	位置屈折率, 潛在屈折率
potential instability	잠재불안정도	潛在不安定度
potential productivity	잠재생산력	潛在生産力
potential pseudo-equivalent temperature	위치위상당온도	位置僞相當溫度
potential pseudo-wet-bulb temperature	위치위습구온도	位置僞濕球溫度
potential temperature	온위, 위치온도	溫位, 位置溫度
potential temperature coordinate system	온위좌표계	溫位座標系

potential transpiration	가능증산(량)	可能蒸散(量)
potential volume	위치체적	位置體積
potential vorticity	위치소용돌이도, 잠재소용돌이도	位置-度, 潛在-度
potential wet-bulb temperature	습구온위, 위치습구온도	濕球溫位, 位置濕球溫度
potentiation	상승작용	相乘作用
potentiometer	전위차계	電位差計
potentiometric surface	정수압면	靜水壓面
pothole	구혈	甌穴
potometer	흡수계, 증산계	吸收計, 蒸散計
poudrin	빙정	氷晶
pouring rain	호우, 폭우	豪雨, 暴雨
powdery snow(= powder snow)	가루눈	粉雪
power	공률, 율, 능률, 멱	工率, 率, 能率, 冪
power density	출력밀도, 속밀도	出力密度, 束密度
power density spectrum	출력밀도 스펙트럼	出力密度-
power dissipation	전력손실	電力損失
power divider	전력분리기	電力分離機
power law	멱법칙	冪法則
power law profile	멱법칙풍속분포	冪法則風速分布
power series	멱급수	冪級數
power source	전(력)원	電(力)源
power spectrum	멱스펙트럼	冪-
power splitter(= divider)	전력분배기	電力分配機
power supply	전원장치	電源裝置
praecipitatio	강수구름	降水-
PPI scope	PPI스코프	
practical unit	실용단위	實用單位
prairie	대초원(캐나다 남부와 미국 중서부의 대초원)	大草原
prairie climate	대초원기후	大草原氣候
Prandtl number	프란틀수	-數
preamplifier	전치증폭기	前置增幅器
preanalysis	사전분석	事前分析
prebaratic chart	지상예상일기도	地上豫想日氣圖
precast	예보	豫報
precession	세차운동	歲差運動
precipitable water	가강수량	可降水量
precipitable water depth	가강수수분	可降水水分
precipitable water diagram	가강수량선도	可降水量線圖
precipitable water vapour	가강수수증기	可降水水蒸氣
precipitation	강수	降水
precipitation aloft	상공강수	上空降水
precipitation area	강수역	降水域
precipitation attenuation	강수감쇠	降水減衰

P

precipitation causes	강수원인	降水原因
precipitation ceiling	강수실링	降水雲高
precipitation cell	강수세포	降水細胞
precipitation characteristics	강수특성	降水特性
precipitation current	강수전류	降水電流
precipitation day	강수일	降水日
precipitation density	강수밀도	降水密度
precipitation distribution	강수분포	降水分布
precipitation duration	강수시간	降水時間
precipitation echo	강수에코	降水-
precipitation effectiveness	강수과도, 강수유효도, 강수효과	降水果度, 降水有效度, 降水效果
precipitation effectiveness index	강수효과지수	降水效果指數
precipitation effectiveness ratio	강수효과비	降水效果比
precipitation efficiency	강수효율	降水效率
precipitation electricity	강수전기(학)	降水電氣(學)
precipitation-evaporation index	강수증발지수	降水蒸發指數
precipitation-evaporation quotient	강수증발몫	降水蒸發-
precipitation-evaporation ratio	강수증발비	降水蒸發比
precipitation fog	강수안개	降水-
precipitation gauge	강수량계	降水量計
precipitation-generating element	강수생성요소	降水生成要素
precipitation index	강수지수	降水指數
precipitation-induced downdraft	강수원인하강기류	降水原因下降氣流
precipitation intensity	강수강도	降水强度
precipitation inversion	강수역전	降水逆轉
precipitation loading	강수하중	降水荷重
precipitation mechanism	강수발달기작, 강수기구	降水發達機作, 降水器具
precipitation meter	강수량계	降水量計
precipitation phenomena	강수현상	降水現狀
precipitation physics	강수물리학	降水物理學
precipitation probability	강수확률	降水確率
precipitation radar	강수레이더	降水-
precipitation ratio	강수율	降水率
precipitation regime	강수영역	降水領域
precipitation rose	강수장미	降水薔薇
precipitation shadow	강수그늘	降水-
precipitation shield	강수차폐, 강수방패	降水遮蔽, 降水防牌
precipitation static	강수정전	降水靜電
precipitation station	강수관측소	降水觀測所
precipitation stimulation	인공강우	人工降雨
precipitation theory	강수이론	降水理論
precipitation trails	강수꼬리	降水-
precipitation trajectory	강수궤적	降水軌跡

precipitation type	강수형	降水形
precipitation water content	강수수량	降水水量
precipitation within sight	시계 내 강수	視界內降水
precipitus	강수구름	降水-
precision	정밀도	精密度
precision analysis	정밀분석	精密分析
precision aneroid barometer	정밀아네로이드기압계	精密-氣壓計
pre-cold-frontal squall line	한랭전선앞스콜선	寒冷前線-線
precursor	전조, 징조, 전조공전	前兆, 徵兆, 前兆空電
predictability	예측성, 예측가능성	豫測性, 豫測可能性
predictand	예측량	豫測量
prediction	예측	豫測
prediction diagram	예측도	豫測圖
prediction of the blooming day	개화예보	開花豫報
predictor	예측인자	豫測因子
preflight evaluation	비행 전 평가	飛行前評價
pre-frontal fog	전선앞안개	前線-
pre-frontal squall line	전선앞스콜선	前線-線
pre-frontal thunderstorm	전선앞뇌우	前線-雷雨
pre-hurricane squall line	허리케인앞스콜선	-線
Preliminary Air Operations weather	예비항공작전기상	豫備航空作戰氣象
preliminary phase	예비단계, 첫단계	豫備段階
preprocessing	전처리	前處理
preprocessor	전처리기, 전처리과정	前處理機, 前處理過程
Pre-Quaternary	제4기이전	第四紀以前
Pre-Quaternary Milankovitch cycle	제4기이전밀란코비치 주기	第四紀以前-週期
Pre-Quaternary monsoon	제4기이전몬순	第四紀以前-
present movement speed	현재이동속도	現在移動速度
present weather	현재날씨	現在天氣
present weather analysis	현재날씨분석	實況分析
pre-shipment review	선적적 검토	船積的檢討
pressure	압력	壓力
pressure altimeter	기압고도계, 압력고도계	氣壓高度計, 壓力高度計
pressure altitude	기압고도, 압력고도	氣壓高度, 壓力高度
pressure anemometer	풍압계	風壓計
pressure anomaly	기압편차, 압력편차	氣壓偏差, 壓力偏差
pressure broadening	기압넓어짐, 압력넓어짐	氣壓-, 壓力-
pressure capsule	기압계공합, 압력계공합	氣壓計空盒, 壓力計空盒
pressure center	기압중심, 압력중심	氣壓中心, 壓力中心
pressure change	기압변화, 압력변화	氣壓變化, 壓力變化
pressure change chart	기압변화도	氣壓變化圖
pressure chart	기압도	氣壓圖
pressure coefficient	압력계수	壓力係數
pressure coordinate system	기압좌표계	氣壓座標系

P

pressure correction	기압보정	氣壓補正
pressure depression	저기압, 저압부	低氣壓, 低壓部
pressure distribution	압력분포	壓力分布
pressure drag	기압항력	氣壓抗力
pressure equilibrium constant	기압평형상수	氣壓平衡常數
pressure-fall center	기압하강중심	氣壓下降中心
pressure field	기압장	氣壓場
pressure gauge	기압계, 압력계	氣壓計, 壓力計
pressure gradient	기압경도, 압력경도	氣壓傾度, 壓力傾度
pressure gradient cause	기압경도원인	氣壓傾度原因
pressure gradient force	기압경도력	氣壓傾度力
pressure height	기압고도, 압력고도	氣壓高度, 壓力高度
pressure ice	압력얼음	壓力-
pressure jump	기압비약	氣壓飛躍
pressure jump line	기압비약선	氣壓飛躍線
pressure level	기압면고도	氣壓面高度
pressure measurement	기압측정	氣壓測定
pressure melting	압력녹음	壓力-
pressure pattern	기압배치	氣壓配置
pressure pattern flight	기압배치비행	氣壓配置飛行
pressure plate anemometer	풍압판풍속계	風壓板風速計
pressure profile	기압연직분포	氣壓鉛直分布
pressure reduction	기압경정	氣壓更正
pressure ridge	기압마루	氣壓-
pressure rise center	기압상승중심	氣壓上昇中心
pressure sensor	기압센서, 기압감부	氣壓感部
pressure stress	기압응력	氣壓應力
pressure surge	기압급승	氣壓急昇
pressure surge line	기압급승선	氣壓急昇線
pressure system	기압계	氣壓系
pressure-temperature correlation	기압온도상관	氣壓溫度相關
pressure-temperature-humidity sounding	기압-온도-습도탐측	氣壓-溫度-濕度探測
pressure tendency	기압경향	氣壓傾向
pressure tendency chart	기압경향도	氣壓傾向圖
pressure tendency equation	기압경향방정식	氣壓傾向方程式
pressure topography	등압면높낮이	等壓面-
pressure trough	기압골	氣壓-
pressure tube anemograph	압력관풍속기록계	壓力管風速記錄計
pressure tube anemometer	압력관풍속계	壓力管風速計
pressure variation	기압변동	氣壓變動
pressure variograph	기압변화기록계	氣壓變化記錄計
pressure variometer	기압변화계	氣壓變化計
pressure wave	기압파	氣壓波
pressurization	여압	與壓

prestorm environment	폭풍전환경	暴風前環境
prevailing easterlies	탁월편동풍	卓越偏東風
prevailing visibility	우(세)시정	優(勢)視程
prevailing westerlies	탁월편서풍	卓越偏西風
prevailing wind	탁월풍	卓越風
prevailing wind direction	탁월풍향	卓越風向
prevailing wind system	탁월풍계	卓越風系
primary bow(= primary rainbow)	1차무지개	一次-
primary circulation	1차순환	一次循環
primary cycle	주순환	主循環
primary cyclone	주저기압	主低氣壓
primary depression	주저기압	主低氣壓
primary front	1차전선	一次前線
primary low	주저기압	主低氣壓
primary pollutant	1차오염물질	一次汚染物質
primary production	1차생산	一次生産
primary radar	주레이더	主-
primary rainbow	1차무지개	一次-
primary scattering	1차산란	一次散亂
primary standard	1차표준	一次標準
primary standard substance	1차표준물질	一次標準物質
primary succession	1차천이	一次遷移
primary trace atmospheric gases	1차대기미량가스	一次大氣微量-
primary tropopause	제1권계면	第一圈界面
Primary wave	P파	-波
primary wilting point	초기시듦점, 초기위조점	初期-點, 初期萎凋點
prime meridian	본초자오선	本初子午線
prime number	소수	素數
primitive air	원시대기	原始大氣
primitive equation	원시방정식	原始方程式
primitive equation model	원시방정식모형	原始方程式模型
primitive equation system	원시방정식계	原始方程式系
primitive man	원시인류	原始人類
primitive organism	원시생명체	原始生命體
primordial atmosphere	1차대기	一次大氣
principal agricultural meteorological station	주농업기상관측소	主農業氣象觀測所
principal climatological observation	영년기후관측	永年氣候觀測
principal climatological station	주기후관측소	主氣候觀測所
principal direction	주방향	主方向
principal front	주전선	主前線
principal land station	주육상관측소	主陸上觀測所
principal quantum number	주양자수	主量子數
principal station	기준관측소	基準觀測所
principal synoptic observation	주종관관측	主綜觀觀測

P

principle	원리, 원칙	原理, 原則
principle of Caratheodory	카라데오도리원리	-原理
principle of compensation	보상원리	補償原理
principle of conservation of energy	에너지보존원리	-保存原理
principle of increase of entropy	엔트로피증가원리	-增加原理
principle of superposition	중첩원리	重疊原理
principle of virtual displacement	가상변위원리	假想變位原理
principle of virtual work	가상일원리	假想日原理
prism	프리즘	
private weather agency	민간기상사업체	民間氣象事業體
private weather information agency	민간기상정보사업체	民間氣象情報事業體
private weather information service	민간기상정보서비스	民間氣象情報-
private weather organization	민간기상기구	民間氣象機構
private weather provider	민간기상제공자	民間氣象提供子
probability	확률	確率
probability density function	확률밀도함수	確率密度函數
probability distribution	확률분포	確率分布
probability distribution function	확률분포함수	確率分布函數
probability forecast	확률예보	確率豫報
probability function	확률함수	確率函數
probability integral	확률적분	確率積分
probability mass function	확률질량함수	確率質量函數
probability of occurrence	발생확률	發生確率
probability radius of typhoon location	태풍위치확률반경	颱風位置確率半徑
probability space	확률공간	確率空間
probability theory	확률론	確率論
probable error	확률오차	確率誤差
probable maximum precipitation	가능최대강수량	可能最大降水量
probable maximum storm	가능최대폭풍	可能最大暴風
probable rainfall amount	확률우량	確率雨量
probe	탐측장치	探測裝置
processed weather data	가공날씨자료	加工氣象資料
process lapse rate	과정감률	過程減率
processor	처리장치	處理裝置
productive structure	생산구조	生産構造
productivity	생산력	生産力
product-moment	승적률	乘積率
professional engineer weather forecaster	전문기상예보기술사	專門氣象豫報技術士
professional meteorological advice	전문기상학감정	專門氣象學鑑定
profile	단면, 프로파일	斷面
profile drag	프로파일항력	形狀抗力
profiler	연직단면관측기, 연직시어관측기	鉛直斷面觀測器, 鉛直-觀測器
prog	예상(prognostic의 약어)	豫想

prog chart	예상도	豫想圖
proglacial lacustrine sediment	빙하전면지형호수퇴적물	氷河前面地形湖水堆積物
prognosis	예상	豫想
prognostic analysis	예상분석	豫想分析
prognostic chart	예상도	豫想圖
prognostic contour chart	등압면예상도	等壓面豫想圖
prognostic equation	예상방정식	豫想方程式
prognostic upper level chart	예상상층도	豫想上層圖
prognostic weather chart	예상일기도	豫想日氣圖
prograde orbit	순행궤도	順行軌道
progression equation	진행방정식	進行方程式
progression of the monsoon	계절풍진행	季節風進行
progressive motion	전진운동, 진행운동	前進運動, 進行運動
progressive wave	전진파, 진행파	前進波, 進行波
projectile	포사체	抛射體
projection	투영	投影
prolate	편장(의)	扁長
prolonged frontal rain	지속성전선비	持續性前線-
prominence	홍염, 프로미넌스	紅焰
prontour chart	고층예상도	高層豫想圖
propagated wind	전파바람	傳播-
propagation	전파	傳播
propagation constant	전파상수	傳播常數
propagation forecasting	전파예보	傳播豫報
propagation velocity	전파속도	傳播速度
propeller anemometer	프로펠러풍속계	-風速計
propeller efficiency	프로펠러효율	-效率
proper vibration	고유진동	固有振動
proportion	비례	比例
proportional constant	비례상수	比例常數
proportional distribution	비례배분	比例配分
proportional expression	비례식	比例式
proportional part	비례부분	比例部分
protected reversing thermometer	방압전도온도계	防壓傳導溫度計
Proterozoic climate	원생대기후	原生代氣候
proton	양성자	陽性子
protonosphere	양성자권	陽性子圈
protothropic man	원시인류	原始人類
provisional observation	수시관측	隨時觀測
proxy	프록시	
pseudo-adiabat	위단열선	僞斷熱線
pseudo-adiabatic	위단열	僞斷熱
pseudo-adiabatic change	위단열변화	僞斷熱變化
pseudo-adiabatic chart	위단열선도	僞斷熱線圖

P

pseudo-adiabatic convection	위단열대류	僞斷熱對流
pseudo-adiabatic diagram	위단열선도	僞斷熱線圖
pseudo-adiabatic expansion	위단열팽창	僞斷熱膨脹
pseudo-adiabatic lapse rate	위단열감률	僞斷熱減率
pseudo-adiabatic process	위단열과정	僞斷熱過程
pseudo cold-front	위한랭전선	僞寒冷前線
pseudo-equivalent potential temperature	위상당온위	僞相當溫位
pseudo-equivalent temperature	위상당온도	僞相當溫度
pseudo front	위전선	僞前線
pseudo-geostrophic approximation	준지균근사	準地均近似
pseudo-instability	위불안정	僞不安定
pseudo-latent instability	위잠재불안정	僞潛在不安定
pseudo-occlusion	위폐색	僞閉塞
pseudo-potential temperature	위온위	僞溫位
pseudo wet-bulb potential temperature	위습구온위	僞濕球溫位
pseudo wet-bulb temperature	위습구온도	僞濕球溫度
psychrograph	자기건습계	自記乾濕計
psychrometer	건습구온도계, 건습계	乾濕球溫度計, 乾濕計
psychrometer equation	습도계방정식	濕度計方程式
psychrometer formula	건습계공식	乾濕計公式
psychrometric calculator	습도계산기	濕度計算器
psychrometric chart	습도계산도	濕度計算圖
psychrometric constant	건습계상수	乾濕計常數
psychrometric formula	건습계공식	乾濕計公式
psychrometric table	습도환산표	濕度換算表
psychrometry	습도측정술	濕度測定術
p system(= pressure system)	기압좌표계	氣壓座標系
public weather service	공공기상서비스	公共氣象-
puddle	얼음구멍	
puff	훅(부는)바람, 연기덩이	煙氣-
pulsation	맥동	脈動
pulse	펄스	
pulse actinometer	펄스일사계	-日射計
pulse-amplitude-code modulation	펄스진폭부호변조	-振幅符號變調
pulse amplitude modulation	펄스진폭변조	-振幅變調
pulse coded modulation	펄스부호변조	-符號變調
pulse compression	펄스압축	-壓縮
pulsed light	펄스빛	
pulse Doppler radar	펄스도플러레이더	
pulsed radar	펄스레이더	
pulse duration	펄스지속시간	-持續時間
pulse duration modulation	펄스지속변조	-持續變調
pulse frequency	펄스주파수	-周波數
pulse frequency modulation	펄스주파수변조	-周波數變調

pulse integrator	펄스집적기	-集積器
pulse integrator	펄스적분기	-積分器
pulse length	펄스길이	
pulse length modulation	펄스길이변조	-變調
pulse modulation	펄스변조	-變調
pulse modulator	펄스변조기	-變造器
pulse number modulation	펄스수변조	-數變調
pulse-pair processing	펄스쌍처리	-雙處理
pulse phase modulation	펄스위상변조	-位相變調
pulse position modulation	펄스위치변조	-位置變調
pulse recurrence frequency	펄스반복빈도	-反復頻度
pulse repetition frequency	펄스반복주파수	-反復周波數
pulse repetition rate	벌스반복률	-反復率
pulse repetition time	펄스반복시간	-反復時間
pulse resolution volume	펄스분해체적	-分解體積
pulse-time-modulated radiosonde	펄스시간변조라디오존데	-時間變調-
pulse time modulation	펄스시간변조	-時間變調
pulse-to-pulse coding	펄스간코딩	-間-
pulse volume	펄스부피	
pulse width	펄스너비	
pulse width modulation	펄스너비변조	-變調
pumping (of barometer)	진동(기압계수은면의)	振動(氣壓計水銀面)
punched card	펀치카드	
punching machine	펀치기	
pure air	순수공기	純粹空氣
pure brightness	청명	清明
pure imaginary number	순허수	純虛數
purga(= poorga)	퍼가(시베리아 툰드라지대의 겨울철 눈보라)	
purple glow	자광	紫光
purple light	보라놀, 보랏빛	
push-broom scanners	푸시브름 스캐너	
P-wave(= primary wave)	P파	-波
pyranogram	전천일사기록	全天日射記錄
pyranograph	자기전천일사계	自記全天日射計
pyranometer	전천일사계	全天日射計
pyrgeometer	지구복사계	地球輻射計
pyrheliogram	직달일사기록	直達日射記錄
pyrheliograph	자기직달일사계	自記直達日射計
pyrheliometer	직달일사계	直達日射計
pyrheliometric scale	직달일사스케일	直達日射-
pyrometer	고온계	高溫計
pyrometry	고온측정법	高溫測定法
pyrotechnic flare	구름씨 살포장치	-撒布裝置
pyrradiometer	전천복사계	全天輻射計

P

Q, q

영문	한글	한자
Q-band	Q띠	-帶
Q-code	큐코드, 큐약호	-略號
Q-factor	Q인자	-因子
quadrangle	사각형	四角形
quadrangular number	사각수	四角數
quadrangular prism	사각기둥	四角-
quadrangular pyramid	사각뿔	四角-
quadrant	사분(면), 사분(의)	四分(面), 四分
quadrant electrometer	사분전위계	四分電位計
quadrant visibility	사분시정	四分視程
quadratic curve	이차곡선	二次曲線
quadratic equation	이차방정식	二次方程式
quadratic expression	이차식	二次式
quadratic formula	근의 공식	根-公式
quadratic function	이차함수	二次函數
quadratic inequality	이차부등식	二次不等式
quadratic surface	이차곡면	二次曲面
quadratic surface approximation	이차곡면근사	二次曲面近似
quadratic term	이차항	二次項
quadrature component (Q-component)	90도위상차성분	-位相差成分
quadrature spectrum	사분스펙트럼	四分-
quadrilateral	사변형	四邊形
quadruple recorder	사중기록계	四重記錄計
qualitative observation	정성관측	定性觀測
quality	품질	品質
quality assurance	품질보증	品質保證
quality control	품질관리	品質管理
quality flags	품질플래그	品質-
quality indicator	품질계수	品質係數

quantile	분위수	分位數
quantitative analysis	정량분석	定量分析
quantitative forecast	양적예보, 정량예보	量的豫報, 定量豫報
quantitative precipitation estimation	정량적 강수추정	定量的降水推定
quantitative precipitation forecast	정량강수예보	定量降水豫報
quantity	양, 수량	量, 數量
quantity of direct solar radiation	직달일사량	直達日射量
quantity of electric charge	전하량	電荷量
quantity of illumination	광량	光量
quantity of radiant energy	복사에너지량	輻射-量
quantity of radiation	복사량	輻射量
quantity of solar radiation	일사량	日射量
quantum theory	양자론	量子論
quartic equation	사차방정식	四次方程式
quartile	사분위수	四分位數
quartile deviation	사분편차	四分偏差
quasi-	준, 유사	準, 類似
quasi-biennial oscillation	준2년주기진동	準二年週期振動
quasi-biennial periodicity	준2년주기성	準二年週期性
quasi-center of action	준작용중심	準作用中心
quasi-conservative	준보존	準保存
quasi-discontinuity	준불연속	準不連續
quasi-geostrophic adjustment	준지균조절	準地均調節
quasi-geostrophic approximation	준지균근사	準地均近似
quasi-geostrophic approximation model	준지균근사모형	準地均近似模型
quasi-geostrophic balance	준지균평형	準地均平衡
quasi-geostrophic equation	준지균방정식	準地均方程式
quasi-geostrophic equilibrium	준지균평형	準地均平衡
quasi-geostrophic wind system	준지균풍계	準地均風系
quasi-homogeneous	준균질	準均質
quasi-horizontal motion	준수평운동	準水平運動
quasi-hydrostatic approximation	준정역학근사	準靜力學近似
quasi-inverse method	준역산법	準逆算法
quasi-isobaric process	준등압과정	準等壓過程
quasi-Lagrangian coordinates	준라그랑지좌표	準-座標
quasi-laminar resistance	준층류저항	準層流抵抗
quasi-nondivergent	준비발산, 준무발산	準非發散, 準無發散
quasi-periodic	준주기(의)	準週期
quasi-permanent low	준영구저기압	準永久低氣壓
quasi-polar sun-synchronous orbit	태양동기준극궤도	太陽同期準極軌道
quasi-static process	준정적과정	準靜的過程
quasi-stationary	준정체(의)	準停滯
quasi-stationary front	준정체전선	準停滯前線
quasi-stationary perturbation	준정체섭동	準停滯攝動

quasi-stationary time series	준정체시계열	準停滯時系列
quasi-steady	준정상(의)	準定常
Quaternary	제4기	第四紀
Quaternary climate	제4기기후	第四紀氣候
Quaternary climate transition	제4기기후전이	第四紀氣候轉移
Quaternary Ice Age	제4빙하시대	第四氷河時代
Quaternary interglacial epoch	제4간빙기	第四間氷期
Quaternary monsoon	제4기몬순	第四紀-
Quaternary period	제4기	第四紀
Quaternary vegetation distribution	제4기식생분포	第四紀植生分布
Quenching	소광, 탈활성	消光, 脫活性
QuickBird satellite	퀵버드위성	-衛星
quicksilver	수은	水銀
Quiescent Prominence	휴면홍염	休眠紅焰
QuikSCAT satellite	QuiKSCAT위성	-衛星
Q-vector	큐벡터	
Q-vector forcing	큐벡터강제력	-强制力

Q

R, r

영문	한글	한자
rabal	레이벌(상층바람관측)	
Rackliff index	라클리프지수	-指數
radar	레이더	
radar altimeter	레이더고도계	-高度計
radar altitude	레이더고도	-高度
radar band	레이더파장대	-波長帶
radar beam	레이더빔	
radar calibration	레이더보정	-補正
radar climatology	레이더기후학	-氣候學
radar coded message	레이더코드신호	-信號
radar composite chart	레이더합성도	-合成圖
radar constant	레이더상수	-常數
radar cross-section	레이더단면	-斷面
radar data acquisition	레이더자료획득	-資料獲得
radar display	레이더표출	-表出
radar dome	레이더돔	
radar duct	레이더덕트	
radar echo	레이더에코	-反響
radar equation	레이더방정식	-方程式
radar frequency band	레이더주파수대	-周波數帶
radar ground echo	레이더지상에코	-地上-
radar horizon	레이더수평선	-水平線
radar indicator	레이더지시기	-指示器
radar meteorological observation	레이더기상관측	-氣象觀測
radar meteorology	레이더기상학, 레이더기상	-氣象學
radar microwave link	레이더마이크로파중계장치	-波中繼裝置
radar mile	레이더마일	
radar precipitation echo	레이더강수에코	-降水反響
radar product generator	레이더산출생성기	-産出物生成器

radar-rainfall relationship	레이더강수관계	-降水關係
radar-raingauge comparison	레이더우량계비교	-雨量計比較
radar range equation	레이더거리방정식	-距離方程式
radar reflectivity	레이더반사율	-反射率
radar reflectivity factor	레이더반사도인자	-反射圖因子
radar report	레이더보고	-報告
radar scatterometer	레이더산란체측정계	-散亂體測定計
radarscope	레이더스코프	
radar screen	레이더망	-網
radar shadow	레이더그림자	
radar signal control processor	레이더신호제어처리기	-信號制御處理機
radar signal spectrograph	레이더신호스펙트로그래프	-信號-圖表
radar signature	레이더신호	-信號
radarsonde	레이더존데	
radar sounding	레이더탐측	-探測
radar storm detection	레이더폭풍우탐지	-暴風雨探知
radar storm detection equation	레이더폭풍우탐지방정식	-暴風雨探知方程式
radar summary chart	레이더요약도	-要約圖
radar theodolite	레이더경위(의)	-經緯
radar triangulation	레이더삼각법	-三角法
radar upper band	레이더위파장대	-波長帶
radar volume	레이더부피	-體積
radar weather observation	레이더일기관측	-日氣觀測
radar wind sounding	레이더바람관측	-觀測
radar wind system	레이더바람시스템	
radial data	방사자료	放射資料
radial direction	방사방향	放射方向
radial drainage	방사유역	放射流域
radial inflow	방사유입	放射流入
radial symmetry	방사대칭	放射對稱
radial velocity	반지름방향속도, 시선속도	-方向速度, 視線速度
radial wind	방사풍	放射風
radian	라디안, 호도	弧度
radiance	복사휘도	輻射輝度
radiance background	복사휘도배경	輻射輝度背景
radiance ratio method	복사휘도비율방법	輻射輝度比率方法
radiance residual method	복사휘도잔차법	輻射輝度殘差法
radiance temperature	복사휘도온도	輻射輝度溫度
radiant density	복사밀도	輻射密度
radiant emittance	복사방출률	輻射放出率
radiant energy	복사에너지	輻射-
radiant energy density	복사에너지밀도	輻射-密度
radiant energy thermometer	복사에너지온도계	輻射-溫度計
radiant exitance	복사사출도	輻射射出度

R

radiant exposure	복사피폭	輻射被曝
radiant flux	복사플럭스	輻射-
radiant flux density	복사플럭스밀도	輻射-密度
radiant heat	복사열	輻射熱
radiant intensity	복사강도	輻射強度
radiant irradiance	복사조도	輻射照度
radiant point	복사점	輻射點
radiant power	복사능	輻射能
radiant ray	복사선	輻射線
radiation	복사	輻射
radiational cooling	복사냉각	輻射冷却
radiation balance	복사수지	輻射收支
radiation balance meter	복사수지계	輻射收支計
radiation belt	복사대	輻射帶
radiation budget	복사수지	輻射收支
radiation chart	복사도	輻射圖
radiation climate	복사기후	輻射氣候
radiation constant	복사상수	輻射常數
radiation cooling	복사냉각	輻射冷却
radiation damping	복사감쇠	輻射減衰
radiation equilibrium	복사평형	輻射平衡
radiation field	복사장	輻射場
radiation filter	복사필터	輻射-
radiation fog	복사안개	輻射-
radiation frost	복사서리	輻射-
radiation heating	복사가열	輻射加熱
radiation intensity	복사강도	輻射強度
radiation inversion	복사역전	輻射逆轉
radiation inversion fog	복사역전안개	輻射逆轉-
radiation law	복사법칙	輻射法則
radiation loss	복사손실	輻射損失
radiation model	복사모형	輻射模型
radiation night	복사밤	輻射-
radiation observation	복사관측	輻射觀測
radiation pattern	복사패턴	輻射-
radiation point	복사점	輻射點
radiation pressure	복사압력	輻射壓力
radiation process	복사과정	輻射過程
radiation recorder	복사기록계	輻射記錄計
radiation sensor	복사수감부	輻射受感部
radiation shield	복사선차폐, 복사가리개	輻射線遮蔽
radiation sonde	복사존데	輻射-
radiation station	복사측정소	輻射測定所
radiation summary chart	복사요약도	輻射要約圖

R

radiation thermometer	복사온도계	輻射溫度計
radiation transfer	복사전달	輻射傳達
radiative balance	복사평형	輻射平衡
radiative cooling	복사냉각	輻射冷却
radiative diffusivity	복사확산율	輻射擴散率
radiative equilibrium	복사평형	輻射平衡
radiative equilibrium temperature	복사평형온도	輻射平衡溫度
radiative forcing	복사강제력	輻射强制力
radiative heating	복사가열	輻射加熱
radiative heat transfer	복사열전달	輻射熱傳達
radiative heat transfer coefficient	복사열전달계수	輻射熱傳達係數
radiative interaction	복사상호작용	輻射相互作用
radiative transfer	복사전달	輻射傳達
radiative transfer equation	복사전달방정식	輻射傳達方程式
radiative transfer model	복사전달모형	輻射傳達模型
radiative tropopause	복사대류권계면	輻射對流圈界面
radiator	복사체	輻射體
radiatus	방사구름	放射-
radio	무선, 라디오	無線
radioactive collector	방사능집진기	放射能集塵機
radioactive emanation	방사능방출	放射能方出
radioactive fallout	방사성낙진	放射性落塵
radioactive fallout plot	방사능낙진기입	放射能落塵記入
radioactive gas	방사성기체	放射性氣體
radioactive precipitation	방사성강수	放射性降水
radioactive rain	방사성비	放射性-
radioactive ray	방사선	放射線
radioactive snow gauge	방사성설량계	放射性雪量計
radioactive tracer	방사능추적, 방사성추적자	放射能追跡, 放射能追跡者
radioactive washout	방사능세척	放射能洗滌
radioactivity	방사능	放射能
radioactivity weather	방사능기상	放射能氣象
radio acoustic sounding system	무선음파탐측계	無線音波探測系
radio altimeter	전파고도계	電波高度計
radioassay	방사능분석, 방사분석	放射能分析, 放射分析
radio atmometer	복사증발계	輻射蒸發計
radio aurora	전파극광	電波極光
radio autograph	무선자기기록계	無線自記記錄計
radio beacon	라디오비컨	
radio beam	전파빔	電波-
radio blackout	전파두절	電波杜絶
radiocarbon	방사성탄소	放射性炭素
radiocarbon dating	방사성탄소연대측정법	放射性炭素年代測定法
radio ceiling	전파실링, 전파실링계	電波-計

R

radio climatology	전파기후학	電波氣候學
radio direction finder	전파방향탐지기	電波方向探知器
radio direction finding	전파방향탐지	電波方向探知
radio distance finder	전파거리측정기	電波距離測定器
radio duct	전파관	電波管
radioelectric meteorology	전파기상학	電波氣象學
radio emission	전파방출	電波放出
radio energy	전파에너지	電波-
radio facsimile	전파팩시밀리	電波-
radio fadeout	전파소멸	電波消滅
radio fax	전파팩스	電波-
radio frequency	무선진동수(주파수)	無線振動數(周波數)
radio goniograph	전파방향기록계	電波方向記錄計
radio goniometer	전파방향탐지기	電波方向探知器
radio goniometry	전파방향탐지법	電波方向探知法
radio hole	전파구멍(난청)	電波-(難聽)
radio horizon	전파수평선	電波水平線
radio interference	전파간섭	電波干涉
radioisotope snow gauge	방사성동위원소적설계	放射性同位元素積雪計
radiolaria	방산충	放散蟲
radiolocator	전파위치측정기	電波位置測定器
radio maximograph	대기전기강도계	大氣電氣强度計
radio meteor	전파유성	電波流星
radio meteorograph	전파기상기록계	電波氣象記錄計
radio meteorology	전파기상학	電波氣象學
radiometer	복사계	輻射計
radiometer irradiance monitor	능동형공동복사계, 복사조도감시장치	能動型空洞輻射計, 輻射照度監視裝置
radiometric calibration	복사계보정	輻射計補正
radiometric contrast	복사계상수	輻射計常數
radiometric correction	복사계수정	輻射計修正
radiometric method	방사성연대측정법	放射性年代測定法
radiometric noise	복사계잡음	輻射計雜音
radiometric profile	복사계연직분포	輻射計鉛直分布
radiometric quantity	복사계량	輻射計量
radiometric resolution	복사계분해능	輻射計分解能
radiometry	방사분석, 복사측정	放射分析, 輻射測定
radio mirage	전파신기루	電波蜃氣樓
radio pilot balloon	전파측풍기구	電波測風氣球
radio refraction correction	전파굴절보정	電波屈折補正
radio refractive index	전파굴절률	電波屈折率
radio refractive index structure parameter	전파굴절률구조모수	電波屈折率構造母數
radio snow gauge	무선설량계	無線雪量計
radiosonde	라디오존데	

R

radiosonde balloon	라디오존데기구	-氣球
radiosonde commutator	라디오존데전환기	-轉換器
radiosonde modulator	라디오존데변조기	-變調器
radiosonde observation	라디오존데관측	-觀測
radiosonde-radiowind system	라디오존데-레윈관측시스템	-觀測-
radiosonde recorder	라디오존데기록계	-記錄計
radiosonde station	라디오존데관측소	-觀測所
radiosonde transmitter	라디오존데송신기	-送信機
radiosounding	전파관측	電波觀測
radio sounding equipment	무선탐측장비	無線探測裝備
radio telemeteograph	무선격측기상기록계	無線隔測氣象記錄計
radioteletype receiver	무선텔레타이프수신기	無線-受信機
radioteletype transmitter	무선텔레타이프송신기	無線-送信機
radioteletypewriter	무선전신타자기	無線電信打字機
radio theodolite	전파경위의	電波經緯儀
radio tracking	전파추적	電波追跡
radio wave	전파	電波
radio wave sounding	전파탐측	電波探測
radiowind	무선바람	無線-
radiowind observation	무선바람관측, 레윈관측	無線-觀測
radiowind station	무선바람관측소, 레윈관측소	無線-觀測所
radius	반경	半徑
radius of curvature	곡률반지름, 곡률반경	曲率半徑
radius of deformation	변형반지름, 변형반경	變形半徑
radius of maximum wind	최대풍속반지름, 최대풍속반경	最大風速半經
radius of protection	보호반지름(낙뢰), 보호반경	保護半徑
radius of strong wind area	강풍반지름	強風-
radius of strong wind area of typhoon	태풍강풍지역반경	颱風強風地域半徑
radius vector	반지름벡터, 반경벡터	半徑-
radix	기, 밑수	基, -數
radome	레이돔	
radon	라돈	
raffiche	라피체(지중해지방에서 산으로부터 불어오는 동풍)	
rafraichometer	생체온도계, 체온계	生體溫度計, 體溫計
rafted ice	뗏목얼음	
ragged ceiling	누더기실링	
ragged eye	태풍의 눈(직경이 40km 이내 크기의 눈)	颱風-
rain	비	
rain and snow mixed	진눈깨비	
rain area	강우역	降雨域
rain attenuation	강우감쇠	降雨減衰
rain band	강우대	降雨帶

rain belt	강우대	降雨帶
rainbow	무지개	虹
rain cloud	비구름	
rain crust	강우크러스트	降雨-
rain day	강우일	降雨日
rain detector	강우감지기	降雨感知器
raindrop	빗방울, 우적	雨滴
raindrop erosion	빗방울침식, 우적침식	雨滴浸蝕
rain droplet collector	작은빗방울채집기	-採集器
raindrop size distribution	빗방울크기분포	-分布
raindrop spectrograph	빗방울스펙트로그래프	
rain erosion	우식, 강우침식	雨蝕, 降雨浸蝕
rain factor	우량인자, 강우인자	雨量因子, 降雨因子
rainfall	강수량, 강우, 우량	降水量, 降雨, 雨量
rainfall amount	강우량	降雨量
rainfall amount of Changma	장마강우량	-降雨量
rainfall depth	강우량, 평균우량깊이	降雨量, 平均雨量-
rainfall depth-area relationship	평균우량깊이-유역면적관계	平均雨量-流域面積關係
rainfall duration	강우기간	降雨期間
rainfall effectiveness	강우효율	降雨效率
rainfall erosion	강우침식	降雨浸蝕
rainfall excess	초과강우량	超過降雨量
rainfall factor	강우인자	降雨因子
rainfall frequency	강우빈도	降雨頻度
rainfall hour	강우시간	降雨時間
rainfall intensity	강우강도	降雨強度
rainfall intensity-duration relationship	강우강도-지속기간관계	降雨強度持續時間關係
rainfall intensity-duration-frequency curve	강우강도-지속기간-빈도곡선	降雨強度持續期間頻度曲線
rainfall intensity meter	강우강도계	降雨強度計
rainfall inversion	강우역전	降雨逆轉
rainfall mass curve	강우누적곡선	降雨累積曲線
rainfall percentage	강우율	降雨率
rainfall probability	강우확률	降雨確率
rainfall rate	강우율	降雨率
rainfall recurrence interval	강우재현기간	降雨再現期間
rainfall regime	강우형	降雨型
rainfall station	우량관측소	雨量觀測所
rainfall totalizer	적산우량계	積算雨量計
rain field	우역	雨域
rainforest	우림	雨林
rainforest climate	우림기후	雨林氣候
rain frequency	강우빈도	降雨頻度
rain gauge	우량계	雨量計
rain-gauge float type	띄우개형우량계	-型雨量計

R

rain-gauge shield	우량계바람막이	雨量計-
rain-gauge tipping-bucket type	전도식우량계	轉倒式雨量計
rain-gauge weighing of balance type	칭량식우량계	秤量式雨量計
rain gush(= rain gust)	폭우	暴雨
rain hour	강우시간	降雨時間
raininess	강우특성	降雨特性
rain intensity	강우강도	降雨强度
rain intensity gauge	강우강도계	降雨强度計
rain making	인공강우	人工降雨
rain measuring glass	우량승, 우량되	雨量升
rain-out	성우제거	成雨除去
rain shadow	비그늘	
rainshaft	빗줄기	
rain shower	소나기	
rain spell	강우계속기간	降雨繼續期間
rain squall	강우스콜	降雨-
rain stage	비단계	-段階
rain storm	비보라, 폭풍우	暴風雨
rainstorm easing	폭풍우완화	暴風雨緩和
rain tree	비나무, 우림	雨林
rain washout	비세척제거	-洗滌除去
rain water	빗물, 우수	雨水
rain water content	빗물량, 우수량	雨水量
rainy climate	다우기후	多雨氣候
rainy day	강우일	降雨日
rainy period	우기	雨期
rainy season	우계, 우기, 강우철	雨季, 雨期, 降雨-
rainy spell	강수계속기간	降水繼續期間
rainy weather	장마날씨	
rainy wet season	장마철	
Raman effect	라만효과	-效果
Raman lidar	라만라이다	
ram penetrometer	램투과계, 빙설경도계	-透過計, 氷雪硬度計
ramsonde	램존데	
random	무작위	無作爲
random error	확률오차, 무작위오차	確率誤差, 無作爲誤差
random forecast	무작위예보	無作爲豫報
randomization	확률화, 무작위화	確率化, 無作爲化
randomized seeding trial	무작위씨뿌리기시험	無作爲-試驗
random medium	임의매체	任意媒體
random numbers	난수	亂數
random sample	무작위표본, 임의표본	無作爲標本, 任意標本
random sampling	무작위추출(법), 임의추출(법)	無作爲抽出(法), 任意抽出(法)
random variable	무작위변수	無作爲變數

R

random walk	확률보행	確率步行
range	교차, 거리, 범위, 주파수대	交差, 距離, 範圍, 周波數帶
range attenuation	거리감쇠	距離減衰
range bin	관측거리빈	觀測距離-
range delay	거리연장	距離延長
range distortion	거리뒤틀림	距離變形
range-elevation indicator	거리-고도지시기	距離高度指示器
range error	거리오차	距離誤差
range finder	거리계	距離計
range folding	관측거리접힘	觀測距離-
range gating	거리폐문	距離閉門
range-height indicator	거리-고도지시기	距離高度指示器
range marker	거리표지	距離標識
range of temperature	온도교차	溫度較差
range of tide	조차	潮差
range resolution	거리분해(능)	距離分解(能)
range strobe	거리지시표(레이더)	距離指示表
range tracking	거리추적	距離追跡
range unfolding	레인지 펼침	
range-velocity display	거리-속도표시기	距離速度表示器
range wind	사정풍	射程風
rank correlation	순위상관	順位相關
ranked reliability table	계급 신뢰도표	階級信賴圖表
Rankine	랭킨(온도의 단위)	
Rankine's combined vortex	랭킨복합소용돌이	-複合-
Rankine temperature scale	랭킨온도눈금	-溫度-
Rankine vortex	랭킨소용돌이	
Raoult's law	라울법칙	-法則
rapid deepening	급속하강	急速下降
rapid intensification	급속강화	急速强化
rapids	급경사수로, 급류	急傾斜水路, 急流
rapid scan	급속주사	急速走査
rapid update cycle	급속업데이트주기	急速-週期
rarefaction wave	희소파	稀疏波
rare gas	희유기체	稀有氣體
raster line	주사선	走査線
rasvodye	얼음면도랑	-面-
rate of change	변화율	變化率
rate-of-climb indicator	상승률지시계	上昇率指示計
rate of forecast error	오보율	誤報率
rate of increment	증분율	增分率
rate of rainfall	강우율	降雨率
rate of rainfall gauge	강우율계	降雨率計
rate of sunshine	일조율	日照率

R

rating curve	수위유량곡선	水位流量曲線
ratio	비, 비율	比, 比率
rational formula, rational runoff formula	합리식	合理式
rational horizon	지심수평, 합리수평	地心水平, 合理水平
ratio of clear line of sight	청천비율	晴天比率
ratio of expand	확대비	擴大比
ratio of the specific heat	비열비	比熱比
rattler	억수	
ravine wind	협곡바람	峽谷-
raw	냉습	冷濕
raw data	원자료	原資料
rawin (radiowind)	레윈	
rawin equipment	레윈장비	-裝備
rawin observation	레윈관측	-觀測
rawinsonde	레윈존데	
rawinsonde observation	레윈존데관측	-觀測
rawinsonde station	레윈존데관측소	-觀測所
rawin target	레윈표적	-標的
ray	선, 광선	線, 光線
Rayleigh	레일리	
Rayleigh absorption efficiency	레일리흡수효율	-吸收效率
Rayleigh atmosphere	레일리대기	-大氣
Rayleigh criterion	레일리기준점	-基準點
Rayleigh-Jeans approximation	레일리-진스근사	-近似
Rayleigh lidar	레일리라이다	
Rayleigh limit	레일리한계	-限界
Rayleigh number	레일리수	-數
Rayleigh reion	레이리영역	-領域
Rayleigh scattering	레일리산란	-散亂
Rayleigh's formular	레일리공식	-公式
Rayleigh's law	레일리법칙	-法則
Rayleigh's problem	레일리문제	-問題
Rayleigh wave	레일리파	-波
ray tracing	(광)선추적	(光)線追跡
reabsorption	재흡수	再吸收
reach	구간, 수로구간	區間, 水路區間
reaction	반응	反應
reaction mechanism	반응메커니즘	反應-
reaction model	반응모형	反應模型
reaction path	반응경로	反應經路
reaction rate	반응률	反應率
reactivation	재활성화	再活性化
reactive hydrocarbons	활성탄화수소류	活性炭化水素類
reactive oxygen	활성산소	活性酸素

R

reading	읽기(눈금)	
read out	읽어내기	
read-out station	기록수신소	記錄受信所
real air temperature	실제기온	實際氣溫
real horizon	참지평선	實地平線
realized energy	나타난 에너지	實-
real latent instability	실잠재불안정도, 진잠재불안정	實潛在不安定度, 眞潛在不安定
real of a depression	실제저기압	實際低氣壓
real time	실시간	實時間
real time liquid water observation equipment	실시간수액관측장비	實時間水液觀測裝備
real time monitoring	실시간감시	實時間監視
real time prediction	실시간예측	實時間豫測
real time transmission system	실시간송신시스템	實時間送信-
Réaumur thermometer	레오뮈르온도계, 열씨온도계	列氏溫度計
Réaumur thermometric scale	레오뮈르온도눈금,	列氏溫度-
	열씨온도눈금	
reboyo	레보요(브라질 연안의 폭풍)	
RECCO code	RECCO 전문형식,	-電文形式,
	기상정찰전문형식	氣象偵察電文形式
received power	수신공률	受信工率
receiver	수신기, 수집기	受信機, 收集機
receiver gain	수신기증폭	受信機增幅
receiving bucket	빗물받이통	-桶
receiving intensity	수신강도	受信强度
recent drizzle	관측시이슬비	觀測時-
recent freezing rain	관측시어는비	觀測時-
recent hail	관측시우박	觀測時雨雹
recent rain	관측시비	觀測時-
recent rain and snow	관측시진눈깨비	觀測時-
recent snow	관측시눈	觀測時-
recent snow shower	관측시소낙눈	觀測時-
recent thunderstorm	관측시뇌우	觀測時雷雨
receptor	수용체	受容體
recession curve	감쇠곡선	減衰曲線
recharge	함양	涵養
reciprocal equation	상반방정식	相反方程式
recombination	재결합	再結合
recombination coefficient	재결합계수	再結合係數
recombination energy	재결합에너지	再結合-
reconnaissance	정찰	偵察
reconstruction	복원	復元
recorder	기록계	記錄計
recording albedometer	자기알베도계	自記-計
recording anemometer	자기풍속계	自記風速計

R

recording arm	기록막대	記錄-
recording barometer	자기기압계	自記氣壓計
recording cylinder	기록원통	記錄圓筒
recording drum	기록원통	記錄圓筒
recording earth thermometer	자기지중온도계	自記地中溫度計
recording equipment	기록장비	記錄裝備
recording evaporation pan	자기증발계	自記蒸發計
recording evaporimeter	자기증발계	自記蒸發計
recording frigorimeter	자기냉각계	自記冷却計
recording instrument	기록장치	記錄裝置
recording lever	기록막대	記錄-
recording pen	기록펜	記錄-
recording pointer	기록펜촉	記錄-
recording potentiometer	자기전위계	自記電位計
recording rain gauge	자기우량계	自記雨量計
recording snow gauge	자기설량계	自記雪量計
recording theodolite	자기경위의	自記經緯儀
recording wind vane	자기풍향계	自記風向計
record observation	기록관측	記錄觀測
recovery factor	회수율	回收率
rectangle graph	사각형그래프	四角形-
rectangular coordinates	직교좌표	直交座標
rectangular coordinate system	직교좌표계	直交座標系
rectangular drainage pattern	장방형배수형	長方形排水形
rectangular waveguide	사각형도파관	四角形導波管
rectification	정류	整流
rectifier	정류기	整流器
rectifier circuit	정류회로	整流回路
rectifying action	정류작용	整流作用
rectifying tube	정류관	整流管
recurrence	재현, 반복	再現, 反復
recurrence formula	재현공식	再現公式
recurrence frequency	재현빈도	再現頻度
recurrence interval	재현기간	再現期間
recurrence of cold	꽃샘추위	
recurrent magnetic storm	재현자기폭풍	再現磁氣暴風
recurvature	전향	轉向
recurved occlusion	후굴폐색	後屈閉塞
red bed	적색층(붉은 퇴적암층)	赤色層
red coloring of leaves	홍엽	紅葉
red flag	적색깃발	赤色-
red flag warning	적색경보	赤色警報
red flash	붉은 섬광	-閃光
red noise	적색잡음	赤色雜音

redox reaction	산화환원반응	酸化還元反應
red rain	붉은 비	
red snow	붉은 눈	
red tide	적조	赤潮
reduced gravity	환산중력	換算重力
reduced pressure	환산기압	換算氣壓
reduced temperature	환산온도	換算溫度
reducing power	환원력	還元力
reduction	경정	更正
reduction catalysts	환원촉매	還元觸媒
reduction factor	경정계수, 환산계수	更正係數, 換算係數
reduction of pressure to a standard level	표준면기압경정	標準面氣壓更正
reduction of temperature to mean sea level	평균해면온도경정	平均海面溫度更正
reduction table	경정표	更正表
reduction to mean sea level	평균해면(기압)경정	平均海面(氣壓)更正
reduction to sea level	해면경정	海面更正
Reech number	리슈수	-數
re-entry	재진입	再進入
reference atmosphere	기준대기	基準大氣
reference climatological station	기준기후관측소	基準氣候觀測所
reference condition	기준상태	基準狀態
reference coordinate system	기준좌표계	基準座標系
reference data	참고자료	參考資料
reference document	참고문서	參考文書
reference frame	참고틀	參考-
reference level	기준고도	基準高度
reference radiosonde	기준라디오존데	基準-
reference sonde	기준존데	基準-
reference vacuum chamber	기준진공챔버	基準眞空-
reflectance	반사율	反射率
reflectance calculation	반사율산출	反射率算出
reflected global radiation	지구반사일사	地球反射日射
reflected light	반사광	反射光
reflected ray	반사광선	反射光線
reflected shortwave radiation	반사단파복사	反射短波輻射
reflected solar radiation	반사태양복사	反射太陽輻射
reflected terrestrial radiation	반사지구복사	反射地球輻射
reflected wave	반사파	反射波
reflecting nephoscope	반사측운기	反射測雲器
reflecting power	반사능	反射能
reflection	반사	反射
reflection coefficient	반사계수	反射係數
reflection factor	반사인자	反射因子
reflection of electromagnetic wave	전자기파반사	電磁氣波反射

R

reflection rainbow	반사무지개	反射-
reflective power	반사능	反射能
reflectivity	반사율	反射率
reflectivity factor	반사율인자	反射率因子
reflectometer	반사계	反射計
reflector	반사체, 반사기	反射體, 反射器
reflexion(= reflection)	반사	反射
reflexive law	반사법칙	反射法則
refracted global radiation	굴절전구복사	屈折全球輻射
refracted light	굴절광	屈折光
refracted radiation	굴절복사	屈折輻射
refracted ray	굴절광선	屈折光線
refraction	굴절	屈折
refraction coefficient	굴절계수	屈折係數
refraction diagram	굴절선도	屈折線圖
refraction factor	굴절인자	屈折因子
refraction index	굴절지수	屈折指數
refraction law	굴절법칙	屈折法則
refractive index	굴절지수	屈折指數
refractive index factor	굴절지수인자	屈折指數因子
refractive modulus	수정굴절률	修正屈折率
refractivity	굴절률, 굴절도	屈折率, 屈折度
refractometer	굴절계	屈折計
Refsdal chart	레프스달도	-圖
Refsdal diagram	레프스달선도	-線圖
regelation	되얼음, 복빙	復氷
regeneration	재생	再生
regeneration of a depression	저기압재생	低氣壓再生
regime	체계, 영역	體系, 領域
Regional Area Forecast Centre	지역예보센터	地域豫報-
regional basic synoptic network	지역기본종관네트워크	地域基本綜觀-
regional broadcast	지역방송	地域放送
regional code	지역코드, 지역기상전문형식	地域-, 地域氣象電文形式
regional flood frequency analysis	지역홍수빈도분석	地域洪水頻度分析
regional forecast	지역예보	地域豫報
regional forecast area	광역예보구역	廣域豫報區域
regional forecast over sea area	해상광역예보	海上廣域豫報
regional meteorological center	지역기상(통신)센터	地域氣象(通信)-
regional meteorological office	지방기상청	地方氣象廳
Regional Meteorological Telecommunication Network	지역기상통신네트워크	地域氣象通信-
regional model	지역모형	地域模型
Regional OPMET Bulletin Exchange	지역운영기상정보회보 교환	地域運營氣象情報會報交換
Regional Specialized Meteorological Centre	지역지정기상센터	地域指定氣象-

R

regional spectral model	지역스펙트럼모형	地域-模型
regional standard barometer	지역표준기압계	地域標準氣壓計
Regional Telecommunication Hub	지역통신허브	地域通信-
region for land forecast	육상기상경보구역	陸上氣象警報區域
region for land weather warning	육상특보구역	陸上特報區域
region of escape	이탈역	離脫域
region of frontolysis	전선소멸지역	前線消滅地域
region of thunderstorm activity	뇌우활동지역	雷雨活動地域
regions for marine forecast	해상예보구역	海上豫報區域
regions for marine special report	해상특보구역	海上特報區域
regions for sea area forecast	해상광역예보구역	海上廣域豫報區域
register	레지스터, 기록부	記錄簿
registrar	기록원	記錄員
regression	회귀	回歸
regression analysis	회귀분석	回歸分析
regression equation	회귀방정식, 회귀식	回歸方程式, 回歸式
regression estimation	회귀예측	回歸豫測
regression function	회귀함수	回歸函數
regressor	회귀변수	回歸變數
regular broadcast	정시방송	定時放送
regular dodecahedron	정십이면체	正十二面體
regular hexahedron	정육면체	正六面體
regular icosahedron	정이십면체	正二十面體
regular polygon	정다각형	正多角形
regular polyhedron	정다면체	正多面體
regular quadrilateral	정사각형	正四角形
regular reflection	정반사, 거울면반사	正反射, -面反射
regular reflector	정반사체	正反射體
regular tetrahedron	정사면체	正四面體
Regular World Days	정규세계일	正規世界日
regulating device	조정장치	調整裝置
regulator	조절기	調節器
regulatory factor	조절요인	調節要因
regulatory gene	조절유전자	調節遺傳子
Reid vapour pressure	레이드증기압	-蒸氣壓
reinforcement	강화, 증강	強化, 增强
rejection region	기각역	棄却域
relative	상대	相對
relative air density	상대공기밀도	相對空氣密度
relative angular momentum	상대각운동량	相對角運動量
relative charge	상대전하	相對電荷
relative concentration	상대농도	相對濃度
relative contour	상대등고선	相對等高線
relative cover	상대피도	相對皮度

R

relative coordinate system	상대좌표계	相對座標系
relative dating	상대연령측정	相對年齡測定
relative density	상대밀도	相對密度
relative dielectric constant	상대유전율	相對誘電率
relative divergence	상대발산	相對發散
relative earth rotation rate	상대지구자전율	相對地球自轉率
relative error	상대오차	相對誤差
relative evaporation	상대증발	相對蒸發
relative frequency	상대빈도	相對頻度
relative gradient current	상대경도류	相對傾度流
relative growth rate	상대성장률	相對成長率
relative height	상대고도	相對高度
relative helicity	상대헬리시티, 상대나선도	相對螺旋度
relative homogeneity	상대균질성	相對均質性
relative humidity	상대습도	相對濕度
relative humidity analysis	상대습도분석	相對濕度分析
relative humidity of moist air with respect to ice	빙면습윤공기상대습도	氷面濕潤空氣相對濕度
relative humidity of moist air with respect to water	수면습윤공기상대습도	水面濕潤空氣相對濕度
relative hypsography	상대측고(법)	相對測高(法)
relative ionospheric opacitymeter	상대이온층혼탁계	相對-層混濁計
relative isohypse	상대등고선	相對等高線
relative luminous efficiency	상대발광효율	相對發光效率
relative mass	상대질량	相對質量
relative moisture of the soil	토양상대수분	土壤相對水分
relative momentum	상대운동량	相對運動量
relative motion	상대운동	相對運動
relative pluviometric coefficient	상대우량계수	相對雨量係數
relative reduction	상대환산	相對換算
relative refractive index	상대굴절률	相對屈折率
relative scattering function	상대산란함수	相對散亂函數
relative scatter intensity	상대산란강도	相對散亂强度
relative stability	상대안정도	相對安定度
relative streamline	상대유선	相對流線
relative sunshine	상대일조	相對日照
relative sunspot number	상대태양흑점수	相對太陽黑點數
relative time	상대시간	相對時間
relative topography	상대지형학, 상대고도분포	相對地形學, 相對高度分布
relative velocity	상대속도	相對速度
relative visibility	상대시정	相對視程
relative vorticity	상대소용돌이도	相對-度
relative vorticity advection	상대소용돌이도이류	相對-度移流
relative wind	상대바람	相對-
relativistic mass	상대론적 질량	相對論的質量

relativity	상대성	相對性
relativity principle	상대성원리	相對性原理
relaxation method	완화법	緩和法
relaxation time	완화시간	緩和時間
releasable energy	방출가능에너지	放出可能-
release altitude weather	방출고도날씨	放出高度-
reliability	신뢰도	信賴度
relief	지형, 지세	地形, 地勢
remapping	리매핑, 복사상	複寫像
remnant low	잔존저기압	殘存低氣壓
remote control equipment	원격제어장치	遠隔制御裝置
remote mode	원격모드	遠隔-
remote observing system automation	자동원격측정장치	自動遠隔測定裝置
remote sensing	원격탐사	遠隔探査
remote sensing application	원격탐사응용	遠隔探査應用
remote sensing professional organization	원격탐사전문조직	遠隔探査專門組織
remote sensing satellite	원격탐사위성	遠隔探査衛星
remote sensing signal	원격탐사신호	遠隔探査信號
remote sensor	원격탐사센서	遠隔探査-
remote sounding	원격탐측	遠隔探測
removal correction	이동보정	移動補正
renewable	재생	再生
renewable energy	재생에너지	再生-
repeatability	반복성	反復性
repetition frequency	반복빈도	反復頻度
replenishment area weather	보급해역기상	補給海域氣象
reporting point	통보점	通報點
representative concentration pathway	대표농도경로	代表濃度經路
representative meteorological observation	대표기상관측점	代表氣象觀測點
representativeness	대표성	代表性
reproducibility	재생산성	再生産性
reproduction	재현, 재생	再現, 再生
reproductive growth	재생성장	再生成長
resampling	리샘플링	
réseau	관측망〈프랑스어〉	觀測網
réseau mondial	세계기상관측망〈프랑스어〉	世界氣象觀測網
resemblance	유사성	類似性
reservation	보류	保留
reservoir	저수지	貯水池
reset	되맞춤, 재설정	再設定
residence half-time	반체류시간	半滯留時間
residence time	체류시간	滯留時間
residual	나머지, 잔류, 잔차	殘留, 殘差
residual layer	나머지층, 잔류층	殘留層

R

residual mineral	잔류광물	殘留鑛物
residual moisture	나머지수분, 잔류수분	殘留水分
residual soil	잔적토	殘積土
residual variance	잔류분산	殘留分散
resilience	회복능, 탄력	回復能, 彈力
resinuous electricity	음전기	陰電氣
resistance	항력, 저항	抗力, 抵抗
resistance thermometer	저항온도계	抵抗溫度計
resistance to overhead flooding injury	내관수성	耐冠水性
resistivity	비저항	比抵抗
resolution	분해, 해상, 해, 해상도, 분해능	分解, 解像, 解, 解像度, 分解能
resolution cell	분해세포	分解細胞
resolution of force	힘의 분해	-分解
resolution of vector	벡터분해	-分解
resolution volume	분해부피	分解-
resonance	공명	共鳴
resonance box	공명상자	共鳴箱子
resonance frequency	공명진동수	共鳴振動數
resonance oscillation	공명진동	共鳴振動
resonance radiation	공명복사	共鳴輻射
resonance scattering	공명산란	共鳴散亂
resonance scattering lidar	공명산란라이다	共鳴散亂-
resonance structure	공명구조	共鳴構造
resonance theory	공명이론	共鳴理論
resonance trough	공명(기압)골	共鳴(氣壓)谷
resonant frequency	공명진동수	共鳴振動數
resonant lee wave	공명풍하파	共鳴風下波
resonator	공명장치, 공명기(체), 공진기	共鳴裝置, 共鳴氣(體), 共振器
respirable particle	호흡가능입자	呼吸可能粒子
respiration	호흡	呼吸
respiration quotient	호흡지수	呼吸指數
respiratory organ	호흡기관	呼吸器官
respiratory pigment	호흡색소	呼吸色素
respiratory substrate	호흡기질	呼吸基質
responder	응답기	應答器
response	반응	反應
response function	반응함수	反應函數
response time	반응시간	反應時間
rest energy	정지에너지	靜止-
restoring force	복원력	復元力
resultant	합, 합성, 합성량	合, 合成, 合成量
resultant force	합성력	合成力
resultant velocity	합성속도	合成速度
resultant wind	합성바람	合成-

resultant wind direction	합성풍향	合成風向
resultant wind velocity	합성풍속	合成風速
result validation	결과확인	結果確認
retardation	지연, 감속	遲延, 減速
retarding basin	유수지	留水池
retarding effect	지연효과	遲延效果
retention	체류	滯留
retreater	리트리터	
retrieval	복원, 회복, 환원, 산출	復原, 回復, 還元, 算出
retrieval method	환원방법	還元方法
retrieval module	환원모듈	還元-
retrieval problem	환원문제	還元問題
retrieved sounding	복원탐측	復原探測
retrograde depression	역행저기압	逆行低氣壓
retrograde orbit	역행궤도	逆行軌道
retrograde wave	역행파	逆行波
retrogression	역행	逆行
retrogressive of long wave	장파역행	長波逆行
retrogressive ridge	후진(기압)마루	後進(氣壓)稜
retrogressive trough	후진(기압)골	後進(氣壓)谷
retroreflection	역반사	逆反射
retroreflector	역반사체	逆反射體
return convection	되돌이대류	-對流
return(ed) flash	되돌이섬광	-閃光
return flow	되돌이흐름	
returning polar air	되돌이한대공기	-寒帶氣塊
return lightning	되돌이번개	
return period	재현기간	再現期間
return streamer	되돌이뇌격	-雷擊
return stroke	되돌이뇌격	-雷擊
reversal of the monsoon	계절풍반전	季節風反轉
reverse cell	역세포	逆細胞
reversed tide	역조	逆潮
reverse flow thermometer housing	역류온도계함	逆流溫度計函
reverse-oriented monsoon trough	역방향몬순기압골	逆方向-氣壓-
reversibility	가역성	可逆性
reversibility of light	빛역진성	-逆進性
reversible	가역적	可逆的
reversible change	가역변화	可逆變化
reversible cycle	가역순환	可逆循環
reversible engine	가역기관	可逆機關
reversible machine	가역기계	可逆機械
reversible phenomenon	가역현상	可逆現象
reversible process	가역과정	可逆過程

R

reversible reaction	가역반응	可逆反應
reversing current	역류	逆流
reversing layer	역변층	逆變層
reversing thermometer	전도온도계	轉倒溫度計
reversing tidal current	왕복조류	往復潮流
revival of Changma	장마부활	-復活
revolution	회전, 공전	回轉, 公轉
revolving fluid	회전유체	回轉流體
revolving storm	회전폭풍우	回轉暴風雨
Reynolds effect	레이놀즈효과	-效果
Reynolds number	레이놀즈수	-數
Reynolds stress	레이놀즈응력	-應力
RHI scope (range-height indicator scope)	RHI스코프	
rhumb line	등각선	等角線
ribbon ice(= ice ribbon)	리본얼음	
ribbon lightning	리본번개	
Richardson number	리처드슨수	-數
ridge	(기압)마루	(氣壓)-
ridge aloft	상층(기압)마루	上層(氣壓)-
ridge line	(기압)마루선	(氣壓)-線
ridge of high pressure	고기압마루	高氣壓-
riefne	리에프네(지중해에 있는 말타의 강한 폭풍)	
rift valley	열곡	裂谷
right angle	직각	直角
right ascension	적경	赤經
right circular polarization	우측원형편파	右側圓形偏波
right-handed rotation	오른손회전	-回轉
right-moving thunderstorm	우측(오른쪽)이동뇌우	右側移動雷雨
rigid system	강체계	剛體系
rime	상고대	
rime air hoar	나무서리	
rime fog	상고대안개	
rime ice	상고대얼음	
rime icing	상고대착빙	-着氷
riming	상고대화, 결착	-化, 結着
rind ice	껍질얼음	
ring current	고리흐름	
ring vortex	고리소용돌이	
Ringelmann chart	링겔만도	-圖
rip	거센물결	
rip current	역조, 이안류	逆潮, 離岸流
ripening	등숙	登熟
ripening period	등숙기간	登熟期間
ripple	(잔)물결에코, 잔물결, 급여울	急-

rip tide	역조	逆潮
rise time	상승시간	上昇時間
rising limb	상승부, 증수부	上昇部, 增水部
rising tide	밀물	
risk	위험도	危險度
risk assessment	위험도평가	危險度評價
river basin	(하천)유역	(河川)流域
river crossing assault weather	강습도하기상	强襲渡河氣象
rivercrossing operation weather	도하작전기상	渡河作戰氣象
river fog	하천안개, 강안개	河川-
river forecast	하천예보	河川豫報
river gauge	하천수위계	河川水位計
river stage	하천수위	河川水位
river system	수계	水系
river (surface) temperature	하천(면)온도	河川(面)溫度
R-meter	R-미터	
roaring forties	으르렁40도	-度
Robin Hood's wind	로빈후드바람(기온이 0℃에 가까운 습한 돌풍)	
Robinson's cup anemometer	로빈슨컵풍속계	-風速計
Robitzsch actinograph	로비치일사계	-日射計
robot station	무인관측소	無人觀測所
robot weather station	로봇기상관측소	-氣象觀測所
rocket balloon instrument	로켓기구관측장치	-氣球觀測裝置
rocket-grenade method	로켓유탄법(고층관측용)	-榴彈法
rocket-launching site	로켓발사지점	-發射地點
rocket lightning	로켓번개	
rocket meteorograph	로켓기상계기	-氣象計器
rocket projectile	로켓탄	-彈
rocketsonde	로켓존데	
rocket sounding	로켓탐측	-探測
rockfall	낙반	落盤
rock glacier	암석빙하	岩石氷河
rockoon sounding	로쿤탐측	-探測
roll axis	가로회전축	-回轉軸
roll cloud	두루마리구름	
roll cumulus	두루마리적운	-積雲
rondada	론다다(24시간의 짧은 시간 안에 풍향이 북서로부터 순차적으로 변하는 바람)〈스페인어〉	
room climate	실내기후	室內氣候
root mean square	2차제곱근	二次-根
root-mean-square error	2차제곱근오차	二次-根誤差
rope cloud	밧줄구름	
rope funnel	밧줄깔때기	
Rosemount temperature housing	로즈몬트온도계함	-溫度計函

R

Rossby deformation radius	로스비변형반경	-變形半徑
Rossby diagram	로스비선도	-線圖
Rossby factor	로스비인자	-因子
Rossby formula	로스비공식	-公式
Rossby-Haurwitz wave	로스비-호르비츠 파	-波
Rossby long wave equation	로스비장파방정식	-長波方程式
Rossby long wave formula	로스비장파공식	-長波公式
Rossby long wave nomogram	로스비장파노모그램	-長波-
Rossby number	로스비수	-數
Rossby parameter	로스비파라미터,	-媒介變數
	로스비매개변수	
Rossby radius of deformation	로스비변형반지름,	-變形半徑
	로스비변형반경	
Rossby regime	로스비영역	領域
Rossby term	로스비항	-項
Rossby wave	로스비파	-波
rotary current	회전류	回轉流
rotary tidal current	회전조류	回轉潮流
rotating annulus	회전수조	回轉水槽
rotating annulus experiment	회전수조실험	回轉水槽實驗
rotating band	회전대	回轉帶
rotating-beam ceilometer	회전빔운고계	回轉-雲高計
rotating ceilometer	회전운고계	回轉雲高計
rotating dishpan	회전수조	回轉水槽
rotating dishpan experiment	회전수조실험	回轉水槽實驗
rotating fluid	회전유체	回轉流體
rotating fluid dynamics	회전유체역학	回轉流體力學
rotating multicylinder	회전다중원통	回轉多重圓筒
rotating Reynolds number	회전레이놀즈수	回轉-數
rotating wind	회전풍	回轉風
rotation	회전, 자전	回轉, 自轉
rotational axis	회전축	回轉軸
rotational energy	회전에너지	回轉-
rotational field	회전장	回轉場
rotational instability	회전불안정(도)	回轉不安定(度)
rotational motion	회전운동	回轉運動
rotational transition	회전전이	回轉轉移
rotation anemometer	회전풍속계	回轉風速計
rotation Froude number	회전프루드수	回轉-數
rotation matrix	회전행렬	回轉行列
rotation moment	회전모멘트	回轉-
rotation rate of earth	지구자전율	地球自轉率
rotation-vibration spectrum	회전진동스펙트럼	回轉振動-
rotatory translation	회전이동	回轉移動

R

rotor cloud	두루마리구름, 풍동	風洞
rotor downburst	회전하강돌풍	回轉下降突風
rotoscope	로토스코프	
rotten ice	녹는 얼음	
rough air	악기류	惡氣流
roughness	거칠기, 조도	粗度
roughness coefficient	거칠기계수	-係數
roughness length	거칠기길이	
roughness parameter	거칠기파라미터	
rough sea	거친 바다	
rough surface	거친 면, 거친 지표	-面, -地表
rounding	어림(수의)	-(數-)
round number	어림수	-數
round-off error	반올림오차	-誤差
route	항공로, 경로	航空路, 經路
route component	항공로성분	航空路成分
route cross-section	항공로단면	航空路斷面
route forecast(= rofor)	항공로예보	航空路豫報
routine observation	정시관측	定時觀測
routine route forecast	정시항로예보	定時航路豫報
routine terminal aerodrome forecast	정시비행장예보	定時飛行場豫報
R scope	R스코프	
rubber balloon	고무기구	-氣球
rubber ice	고무얼음	
rudder	방향타	方向舵
rule	자, 규칙	規則
runaway greenhouse effect	이탈온실효과	離脫溫室效果
running mean	이동평균	移動平均
runoff	유출(량)	流出(量)
runoff cycle	유출순환	流出循環
runoff factor	유출인자	流出因子
runoff wind anemometer	풍정풍속계	風程風速計
run of wind	풍정	風程
runway	활주로	滑走路
runway elevation	활주로표고	滑走路標高
runway observation	활주로관측	滑走路觀測
runway temperature	활주로온도	滑走路溫度
runway visibility	활주로시정	滑走路視程
runway visual range	활주로가시거리	滑走路可視距離

R

S, s

영문	한글	한자
saddle	안장저압부	鞍裝低壓部
saddle-back	안장형 갠구역	鞍裝形區域
saddle point(= col)	안장점	鞍裝點
safety cultivation season	안전재배계절	安全栽培季節
safety period of cultivation	안전재배기간	安全栽培期間
Saffir-Simpson damage-potential scale	사피어-심프슨 위험잠재규모	-危險潛在規模
Saffir-Simpson hurricane intensity scale	사피어-심프슨 허리케인 강도규모	-强度規模
Saffir-Simpson scale	사피어-심프슨 규모	-規模
Saharan dust	사하라먼지	
Saint Elmo's fire	성엘모불	聖-
Sakagami's instability	사카가미불안정(도)	-不安定(度)
salinity	염분, 염도	鹽分, 鹽度
salinity bridge	염분측정기	鹽分測定器
salinization	염분축적	鹽分蓄積
salinometer	염분계	鹽分計
salt	소금, 염	鹽
saltation(= sandstorm)	모래폭풍	-暴風
salt breeze	소금바람	
salt concentration	염분농도	鹽分濃度
salt damage	소금피해, 염해	鹽害
salt extraction	염분추출	鹽分抽出
salt haze	소금연무	-煙霧
salt injury	소금피해, 염해	-被害, 鹽害
salt particle	소금입자, 염분입자	-粒子, 鹽分粒子
salt seeding	소금씨뿌리기	
salt tolerance	내염성	耐鹽性
salt water	소금물	
saltwater contamination	염수오염	鹽水汚染

salt-water invasion	염수침해	鹽水沈害
salty water damage	염수해	鹽水害
salty wind	소금바람	
salty wind damage	소금바람피해	-被害
salty wind tolerance	소금바람내성	-耐性
sample	표본	標本
sample mean	표본평균	標本平均
sample of precipitation	강수시료	降水試料
sampler	표본수집기, 표본채취기	標本收集機, 標本採取機
sample rate	표본비율	標本比率
sample space	표본공간	標本空間
sample standard deviation	표본표준편차	標本標準偏差
sample survey	표본조사	標本調査
sample time	표본시간	標本時間
sample variance	표본분산	標本分散
sample volume	표본부피	標本(體積)
sampling	표본채취	標本採取
sampling distribution	표본분포	標本分布
sampling error	표본오차	標本誤差
sampling frequency	표본수집주기, 표본채취주기	標本收集週期, 標本採取週期
sampling station	표본수집소, 표본채취소	標本收集所, 標本採取所
sand	모래	
sand auger	모래용오름	-龍-
sandbar	모래톱	
sand beach	모래해변	-海邊
sand devil(= dust devil)	모래회오리(먼지회오리)	
sand dune	모래언덕	
sand fall	모래바람	
sand haze	모래연무	-煙霧
sand mining	모래채굴	-採掘
sand mirage	모래신기루, 땅거울	-蜃氣樓
sand mist	모래실안개	
sand pillar	모래기둥	
sand sea	모래바다	
sand snow	모래눈	
sand soil	사토, 모래흙	沙土
sandstone	사암	砂岩
sandstorm	모래폭풍	-暴風
sand tornado	모래토네이도	
sand wall	모래벽	-壁
sand whirl	모래회오리	
sandy beach	모래해변	-海邊
sanitary clearance zone	위생제거지대	衛生除去地帶
Santa Ana	산타아나	

Santa Ana wind	산타아나바람	
Santa Rosa storm	산타로사폭풍	-暴風
sapropel(= organic-rich mud)	부니(유기물이 많은 진흙)	腐泥
saprophytism	사물기생	死物奇生
sastrugi(= zastrugi)	사스트루기(눈 위에 생긴 물결무늬)	
satan	먼지회오리	
satellite	위성	衛星
satellite appearance	위성출현	衛星出現
satellite band	위성대역, 위성밴드	衛星帶域
satellite bus	위성버스	衛星-
satellite cloud picture	위성구름사진	衛星-寫眞
satellite data distribution system	위성자료분배시스템	衛星資料分配-
satellite hydrology program	위성수문프로그램	衛星水文-
satellite image	위성영상, 위성이미지	衛星映像-
satellite image analysis system	위성영상분석시스템	衛星映像分析-
satellite imagery	위성영상	衛星映像
satellite infrared spectrometer	위성적외분광계	衛星赤外分光計
satellite mapping	위성도표화	衛星圖表化
satellite meteorological observation	위성기상관측	衛星氣象觀測
satellite meteorology	위성기상학	衛星氣象學
satellite observation	위성관측	衛星觀測
satellite orbit	위성궤도	衛星軌道
satellite picture	위성사진	衛星寫眞
satellite picture navigation	위성사진항법	衛星寫眞航法
satellite projection	위성투사	衛星投射
satellite relay of data	위성자료중계	衛星資料中繼
satellite sounding	위성탐측	衛星探測
satellite tracking antenna	위성추적안테나	衛星追跡-
satellite wind	위성바람	衛星-
satellite zenith angle	위성천정각	衛星天頂角
satin ice	비단얼음	緋緞-
saturated adiabat	포화단열선	飽和斷熱線
saturated-adiabatic lapse (rate)	포화단열감률	飽和斷熱減率
saturated air	포화공기	飽和空氣
saturated hydraulic conductivity	포화수리전도(도)	飽和水理傳導(度)
saturated soil	포화토양	飽和土壤
saturated solution	포화용액	飽和溶液
saturated vapo(u)r	포화수증기	飽和水蒸氣
saturated virtual temperature	포화가온도	飽和假溫度
saturated water capacity	포화용수량	飽和容水量
saturation	포화	飽和
saturation adiabat	포화단열선	飽和斷熱線
saturation adiabatic	포화단열	飽和斷熱
saturation adiabatic change	포화단열변화	飽和斷熱變化

S

saturation adiabatic cooling (rate)	포화단열냉각	飽和斷熱冷却
saturation adiabatic lapse (rate)	포화단열감률	飽和斷熱減率
saturation adiabatic process	포화단열과정	飽和斷熱過程
saturation adiabatics	포화단열선	飽和斷熱線
saturation curve	포화곡선	飽和曲線
saturation deficit (or deficiency)	포차	飽差
saturation mixing ratio	포화혼합비	飽和混合比
saturation mixing ratio line	포화혼합비선	飽和混合比線
saturation mixing ratio with respect to ice	얼음포화혼합비	-飽和混合比
saturation mixing ratio with respect to water	물포화혼합비	-飽和混合比
saturation point	포화점	飽和點
saturation signal	포화신호	飽和信號
saturation specific humidity	포화비습	飽和比濕
saturation vapo(u)r pressure	포화수증기압	飽和水蒸氣壓
saturation vapo(u)r pressure in the pure phase with respect to ice	얼음순수포화(수)증기압	-純粹飽和(水)蒸氣壓
saturation vapo(u)r pressure in the pure phase with respect to water	물순수포화(수)증기압	-純粹飽和(水)蒸氣壓
saturation vapo(u)r pressure of moist air with respect to ice	얼음습윤공기포화(수)증기압	-濕潤空氣飽和(水)蒸氣壓
saturation vapo(u)r pressure of moist air with respect to water	물습윤공기포화(수)증기압	-濕潤空氣飽和(水)蒸氣壓
saturation water vapo(u)r pressure	포화(수)증기압	飽和(水)蒸氣壓
savanna(h)	사바나	
savanna climate	사바나기후	-氣候
Savart polariscope	사바르편광기	-偏光器
S-band	S밴드	
S-band radar	S밴드레이더, S파레이더	-波-
scalar	스칼라	
scalar approximation	스칼라근사	-近似
scalar function	스칼라함수	-函數
scalar potential	스칼라퍼텐셜	
scalar product	스칼라곱	-積
scalar quantity	스칼라양	-量
scale	규모, 눈금, 축척, 척도	規模, 縮尺, 尺度
scale analysis	규모분석	規模分析
scale decision	규모결정	規模決定
scale factor	축척인자	縮尺因子
scale height	규모고도	規模高度
scale of atmospheric process	대기과정규모	大氣過程規模
scale of circulation	순환규모	循環規模
scale of motion	운동규모	運動規模
scale of turbulence	난류규모	亂流規模
scale of wind-force	풍력계급	風力階級

scale separation	규모분리	規模分離
scaling	규모화, 눈금매기기	規模化
scan	주사	走査
scan angle	주사각	走査角
scan circle	주사원	走査圓
Scandinavian ice sheet	스칸디나비아 빙상	-氷床
scan line	주사선	走査線
scanner	주사계	走査計
scanning	주사방식	走査方式
scanning electron microscope	주사전자현미경	走査電子顯微鏡
scanning microwave spectrometer	주사마이크로파분광계	走査-波分光計
scanning multifrequency microwave radiometer	주사다주파마이크로파복사계	走査多周波-波輻射計
scanning observation	주사관측	走査觀測
scanning radiometer	주사복사계	走査輻射計
scan pattern	주사유형	走査類型
scan spot	주사점	走査點
scarf cloud	스카프구름	
scatter	산란, 산포	散亂, 散布
scatter angle	산란각	散亂角
scatter communication	산란전달	散亂傳達
scatter diagram	산포도	散布圖
scattered	구름조금	
scattered light	산란광	散亂光
scattered power	산란공률	散亂工率
scattered radiation	산란복사	散亂輻射
scattered ray	산란광선	散亂光線
scattered shower	산발소나기	散發-
scatterer	산란체	散亂體
scattergram	산포도	散布圖
scatter graph	산포도	散布圖
scattering	산란	散亂
scattering amplitude	산란폭	散亂幅
scattering angle	산란각	散亂角
scattering area coefficient	산란면적계수	散亂面積係數
scattering area ratio	산란면적비	散亂面積比
scattering coefficient	산란계수	散亂係數
scattering cross-section	산란단면	散亂斷面
scattering efficiency	산란효율	散亂效率
scattering efficiency factor	산란효율인자	散亂效率因子
scattering function	산란함수	散亂函數
scattering index	산란지수	散亂指數
scattering matrix	산란행렬	散亂行列
scattering power	산란율	散亂率
scatterometer	산란계, 스캐트로미터	散亂計

scatter propagation	산란전파	散亂傳播
scatter-type visibility meter	산란시정계	散亂視程計
scavenging	정화	淨化
scavenging by precipitation	강수정화	降水淨化
scenario of future emission	미래배출시나리오	未來排出-
schistosity	편리	片離
schlieren method	굴절무늬법	屈折-法
Schumann-Runge band	슈만-룽에 띠	
Schumann-Runge continuum	슈만-룽에 연속체	-連續體
Schwarzchild's equation	슈와르츠차일드방정식	-方程式
scintillation	섬광, 반짝임	閃光
scintillometer	섬광계	閃光計
scirocco	시로코(북아프리카에서 남유럽으로 부는 열풍)	
scope	스코프	
scope indicator	스코프지시기	-指示器
scorer number	스코러수	-數
Scotch mist	스코틀랜드 박무	-薄霧
screen	백엽상, 영상막, 선발	百葉箱, 映像幕, 選拔
screened pan	그물증발계	-蒸發計
scrubber	세정기, 세척기	洗淨器, 洗滌器
scrubber dedustor	세정탈진장치	洗淨脫塵裝置
scud	조각구름, 토막구름	
S-curve method	S-곡선법	-曲線法
sea	바다, 바다상태	-狀態
sea air	바다공기	-空氣
sea-barometer	선박기압계	船舶氣壓計
sea area specially cared	특정관리해역	特定管理海域
sea breeze	해풍, 바다바람	海風
sea breeze circulation	해풍순환	海風循環
sea breeze convergence zone	해풍수렴대	海風收斂帶
sea breeze front	해풍전선	海風前線
sea chart	해도	海圖
sea clutter	해면에코, 파랑에코	海面-, 波浪-
sea echo	바다에코	
seafloor	해저	海底
seafloor sediment	해저퇴적	海底堆積
seafloor spreading	해저확장	海底擴張
sea fog	바다안개, 해무	海霧
sea ice	바다얼음, 해빙	海氷
sea ice thinning	해빙감소	海氷減少
sea level	해수면	海水面
sea level change	해수면변화	海水面變化
sea level chart	해수면도	海水面圖
sea level correction	해수면경정	海水面更正

sea level elevation	해발고도	海拔高度
sea level indicator	해수면고도계, 해수면지표	海水面高度計, 海水面指標
sea level pressure	해수면기압	海水面氣壓
sea level pressure chart	해수면기압도	海水面氣壓圖
sea level pressure pattern	해수면기압패턴	海水面氣壓-
sea level rise	해수면상승	海水面上昇
sealing	봉함	封緘
sea map	해도	海圖
sea mist	바다실안개	
sea of cloud	구름바다	
sea of fog	안개바다	
sea rainbow	바다무지개	
search and rescue	탐색구조	探索救助
search and rescue operation	탐색구조작전	探索救助作戰
search array	검색배열	檢索配列
search light	탐조등	探照燈
searchlighting	탐조, 탐색	探照, 探索
search radar	탐색레이더	探索-
sea return	바다반향, 바다에코	-反響
sea salt aerosol	해염에어로졸	海鹽-
sea salt nucleus	해염핵	海鹽核
sea salt particle	해염입자	海鹽粒子
Seasat	시새트(미국 해양탐사용 최초 지구궤도위성)	
sea smoke	바다증기안개	-蒸氣-
season	계절	季節
seasonal affective disorder	계절정서장애	季節情緒障碍
seasonal analysis method	계절분석법	季節分析法
seasonal change	계절변화	季節變化
seasonal change analysis	계절변화분석	季節變化分析
seasonal change observation	계절변화관측	季節變化觀測
seasonal character	계절특성	季節特性
seasonal disease	계절병	季節病
seasonal forecasting	계절예보	季節豫報
seasonal forecast model	계절예보모형	季節豫報模型
seasonal form	계절형	季節型
seasonal lag	계절지연	季節遲延
seasonal phenomenon	계절현상	季節現象
seasonal precipitation	계절강수	季節降水
seasonal rain front	계절강우전선	季節降雨前線
seasonal trend	계절추세	季節趨勢
seasonal variation	계절변동	季節變動
seasonal weather forecast	계절일기예보	季節日氣豫報
sea spray	바다물보라	
sea state	해면상태	海面狀態

S

sea station	해상관측소	海上觀測所
sea surface	해면	海面
sea surface temperature	해면온도	海面溫度
sea surface temperature anomaly chart	해면온도편차도	海面溫度偏差圖
sea surface temperature chart	해면온도분포도	海面溫度分布圖
sea swell	바다너울	
sea temperature	해수온도	海水溫度
seawall	방조제	防潮堤
sea water	바닷물, 해수	海水
sea water temperature	해수온도	海水溫度
sea water thermometer	해수온도계	海水溫度計
sea wave	해파	海波
sea wave model	해파모형	海波模型
Secchi disk(= Secchi disc)	투명도판	透明度板
seclusion	고립폐색	孤立閉塞
secondary aerosol	2차에어로졸	二次-
secondary air pollutant	2차대기오염물질	二次大氣汚染物質
secondary ambient air quality standard	2차표준대기질	二次標準大氣質
secondary atmosphere	2차대기	二次大氣
secondary circulation	2차순환	二次循環
secondary cold-front	2차한랭전선	二次寒冷前線
secondary cyclone	부저기압	副低氣壓
secondary depression	부저기압	副低氣壓
secondary diffraction	2차회절	二次回折
secondary emission	2차방출	二次放出
secondary flow	2차흐름	二次-
secondary front	2차전선	二次前線
secondary instrument	보조기기	補助機器
secondary low	부저기압	副低氣壓
secondary organic aerosol	2차유기에어로졸	二次有機-
secondary pollutant	2차오염물질	二次汚染物質
secondary production	2차생산	二次生産
secondary radar	2차레이더	二次-
secondary rainbow	2차무지개	二次-
secondary scattering	2차산란	二次散亂
secondary standard	2차표준기	二次標準器
secondary succession	2차천이	二次遷移
secondary trace atmospheric gases or particle	2차미량대기가스/입자	二次微量大氣-/粒子
secondary tropical tropopause	제2열대권계면	第二熱帶圈界面
secondary tropopause	제2권계면	第二圈界面
secondary warm front	2차온난전선	二次溫暖前線
secondary wave	2차에코	二次-
second law of thermodynamics	열역학제2법칙	熱力學第二法則
second moment	2차모멘트	二次-

S

second-order climatological station	2급기후관측소	二級氣候觀測所
second path retrieval	2차경로추출	二次經路抽出
second path sounding	2차경로탐측	二次經路探測
second priciple of thermodynamics	열역학제2법칙	熱力學第二法則
second purple light	제2보랏빛노을	第二-
second trip (radar) echo	2차(레이더)에코	二次-
second tropopause	제2대류권계면	第二對流圈界面
section	단면	斷面
sectoral breakdown	권역와해	圈域瓦解
sector boundary	부문경계	部門境界
sectorized hybrid scan	분할하이브리드주사	分割-走査
sectorized image	분할영상	分割映像
sector scan indicator	구획주사지시기	區劃走査指示器
sector scanning	구획주사	區劃走査
sector visibility	부분시정	部分視程
sector wind	구역바람	區域-
secular change	경년변화	經年變化
secular climatic change	경년기후변화	經年氣候變化
secular trend	경년경향	經年傾向
secular trend in climate	기후경년경향	氣候經年傾向
secular variation	경년변동	經年變動
sediment	침전물	沈澱物
sedimentary deposit	퇴적침전물	堆積沈澱物
sedimentary facies	퇴적상	堆積相
sedimentary indicator of climate change	기후변화퇴적물지표	氣候變化積物指標
sedimentary rock	퇴적암	堆積岩
sedimentation	퇴적작용, 침전작용	堆積作用, 沈澱作用
sediment storage capacity	퇴적물저장한계	堆積物貯藏限界
seedability	씨뿌림가능성	-可能性
seeder-feeder process	모이주기과정	-過程
seeding	씨뿌림	
seeding agent	씨뿌림사업자	-事業者
seeding method	씨뿌림방법, 파종법	播種法
seeding rate	씨뿌림율	-率
seeding subroutine	씨뿌림수순	-手順
seed of cloud	구름씨	
seed time	씨뿌림시기	-時期
seepage	누수	漏水
seiche	정진동, 세이시	靜振動
seism	지진	地震
seismic gap	지진공백	地震空白
seismic record	지진기록	地震記錄
seismic sea wave	지진해파	地震海波
seismic surveying	지진조사	地震調査

S

seismic tomography	지진단층촬영	地震斷層撮影
seismic wave	지진파	地震波
seismograph	지진계	地震計
seismological observation	지진관측	地震觀測
seismology	지진학	地震學
seismoscope	지진계	地震計
selected ship (station)	선발관측선지점	選拔觀測船地點
selected special observation	선특관측	選特觀測
selected special meteorological report	선특기상보고	選特氣象報告
selection rule	선택규칙	選擇規則
selective absorption	선택흡수	選擇吸收
selective chopper radiometer	선택절단복사계	選擇切斷輻射計
selective instability	선택불안정(도)	選擇不安定(度)
selective scattering	선택산란	選擇散亂
selective stability	선택안정(도)	選擇安定(度)
self-calibration	자체교정	自體矯正
self-consistency	자체일관성	自體一關性
self-finance forecasting service	자기금융예보서비스	自己金融豫報-
self-induction	자체유도	自體誘導
self recorder	자체기록계	自體記錄計
self-recording barometer	자기기압계	自記氣壓計
self-recording hygrometer	자기습도계	自記濕度計
self-recording rain gauge	자기우량계	自記雨量計
self-registering anemometer	자기풍력계	自記風力計
self-registering barometer	자기기압계	自記氣壓計
self-registering thermometer	자기온도계	自記溫度計
selsyn wind vane	셀신풍향계	-風向計
selvas	남미열대우림	南美熱帶雨林
semiarid	반건조	半乾燥
semiarid climate	반건조기후	半乾燥氣候
semiarid region	반건조구역	半乾燥區域
semiarid zone	반건조지대	半乾燥地帶
semiaverage method	반평균법	半平均法
semidesert climate	반사막기후	半沙漠氣候
semidiurnal period	반일주기	半日週期
semidiurnal pressure wave	반일기압파	半日氣壓波
semidiurnal tide	반일조석	半日潮汐
semidiurnal variation	반일변동	半日變動
semidiurnal wave	반일파	半日波
semi-implicit method	반함축법, 반암시법	半含蓄法, 半暗示法
semi-implicit scheme	반암시방안	半暗示方案
semi-interquartile range	반사분위범위	半四分位範圍
semi-Lagrangian method	반라그랑지법	半-法
semipermanent anticyclone	반영구고기압	半永久高氣壓

S

semipermanent depression	반영구저기압	半永久低氣壓
semipermanent high	반영구고기압	半永久高氣壓
semisphere	반구	半球
semitropical	아열대(의)	亞熱帶
semitropical cyclone	아열대저기압	亞熱帶低氣壓
sense of position	위치감각	位置感覺
sensible climate	체감기후	體感氣候
sensible heat	느낌열, 현열	顯熱
sensible heat flow	느낌열흐름	-熱-
sensible heating	느낌열가열	-熱加熱
sensible temperature	느낌온도, 체감온도	體感溫度
sensing element	수감부	受感部
sensing system	수감계	受感系
sensitive surveillance radar	고감도감시레이더	高感度監視-
sensitivity	민감도	敏感度
sensitivity time control	(민)감도시간제어기	(敏)感度時間制御器
sensitometer	감광도계	感光度計
sensitometry	감광도측정법	感光度測定法
sensor	센서, 수감부	受感部
sensor instrument	센서계기, 센서기기	-計器, -器機
sensor lag	센서지연	-遲延
sensor limitation	센서한계	-限界
sensor limitation error	센서한계오차	-限界誤差
sensor observation angle	센서관측각	-觀測角
sensor resolution	센서해상도	-解像度
sensor sensitivity	센서민감도	-敏感度
sensor viewing angle	센서시야각	-視野角
separation eddy	분리맴돌이	分離-
separation of charge	전하분리	電荷分離
separation of flow	분류	分流
separation of variable	변수분리	變數分離
sequence	차례, 순서, 수열	順序, 數列
sequence of weather	날씨순서	-順序
sequential analysis	순서분석	順序分析
sequential relaxation	순서완화	順序緩和
serein	하늘눈물(맑은 하늘에 내리는 이슬비)	
serial correlation	시계열상관	時系列相關
serial number	일련번호	一連番號
serial number of typhoon	태풍번호	颱風番號
service ceiling	실용비행고도	實用飛行高度
settling velocity	침강속도	沈降速度
severe cold	혹한	酷寒
severe cold damage	혹한해	酷寒害
severe cold wave	대한파	大寒波

S

severe convection storm	강대류폭풍우	强對流暴風雨
severe frost	된서리	
severe icing	악성착빙	惡性着氷
severe local storm	위험국지폭풍우	危險局地暴風雨
severe local storm unit	위험국지폭풍우단위	危險局地暴風雨單位
severe local storm watch	강국지폭풍우주의보	强局地暴風雨注意報
severe rain storm	위험폭풍우	危險暴風雨
severe right moving storm	강우편향폭풍우	强右偏向暴風雨
severe storm	강폭풍우	强暴風雨
severe storm precipitation	강폭풍우강수	强暴風雨降水
severe thunderstorm	위험뇌우	危險雷雨
severe thunderstorm outlook	위험뇌우전망	危險雷雨展望
severe thunderstorm warning	위험뇌우경보	危險雷雨警報
severe thunderstorm watch	위험뇌우감시보	危險雷雨監視報
severe tropical storm	위험열대폭풍우	危險熱帶暴風雨
severe warning	중대경보	重大警報
severe weather	위험기상, 악기상	危險氣象, 惡氣象
severe weather analysis	위험기상분석	危險氣象分析
severe weather element	위험기상요소	危險氣象要素
severe weather forecast	위험기상예보	危險氣象豫報
severe weather information	위험기상정보	危險氣象情報
severe weather potential statement	위험기상가능성보고서	危險氣象可能性報告書
severe weather probability	위험기상확률	危險氣象確率
severe weather signature	위험기상징후	危險氣象徵候
severe weather statement	위험기상보고서	危險氣象報告書
severe weather threat index	위험기상위협지수	危險氣象威脅指數
severe winter	엄동	嚴冬
SFAZI code	SFAZI 전문형식	-電文形式
SFAZU code	SFAZU 전문형식	-電文形式
sferic receiver	공전수신기	空電受信機
sferics	공전학, 공전	空電學, 空電
sferics fix	공전발원점	空電發源點
sferics network	공전관측망	空電觀測網
sferics observation	공전관측	空電觀測
sferics recorder	공전기록계	空電記錄計
shade ring	차광환	遮光環
shades of gray	회색도	灰色度
shade temperature	그늘온도	-溫度
shading disk	차광디스크	遮光-
shadow	그림자	
shadow band	차광띠	遮光-
shadow criteria	그늘범주	-範疇
shadowgraph	영회사진법	影繪寫眞法
shadow of the earth	지구그림자	地球-

shadow zone	그늘대	-帶
shaitan(= dust devil)	먼지회오리	
shallow assumption	얕은 근사	-近似
shallow fluid equation	천유방정식	淺流方程式
shallow fog	얕은 안개	
shallow low	얕은 저기압	-低氣壓
shallow moist zone	얕은 습윤대	-濕潤帶
shallow sea	천해	淺海
shallow water	천해	淺海
shallow water coral	천해산호	淺海珊瑚
shallow water effect	천수효과, 천해효과	淺水效果, 淺海效果
shallow water wave	천수파, 천해파	淺水波, 淺海波
shamal	샤말(페르시아 만과 티그리스 및 유프라테스 계곡의 열기와 먼지를 포함한 건조한 북서풍)	
shape factor	형상인자, 모양인자	形象因子, 模樣因子
shape of the sky	하늘모양	-模樣
sharp front	예리한 전선	-前線
sharp ridge	뾰족기압마루	-氣壓-
sharp trough	뾰족기압골	-氣壓-
shear	층밀림, 전단, 시어	剪斷
shear effect	전단효과, 시어효과	剪斷效果
shear-gravity wave	층밀림중력파, 시어중력파	-重力波
shearing deformation	층밀림변형, 시어변형	-變形
shearing gravity wave	시어중력파	-重力波
shearing instability	시어불안정도	-不安定度
shearing stress	층밀림응력, 시어응력	-應力
shearing wave	층밀림파, 시어파	-波
shear layer	층밀림층, 시어층	-層
shear line	층밀림선, 시어선	-線
shear pattern	시어패턴	
shear strain	층밀리기 변형	層-變形
shear strength	시어강도	-強度
shear stress	시어응력	-應力
shear vector	시어벡터	
shear vorticity	시어소용돌이도	-度
shear wave	시어파	-波
sheathed thermometer	이중관온도계	二重管溫度計
sheepgut hydrometer	양장습도계	羊腸濕度計
sheet cloud	판구름	板-
sheet erosion	면침식	面浸蝕
sheet flow	판흐름	板-
sheet frost	판서리	板-
sheet ice	판얼음	板-
sheet lightning	판번개	板-

S

shelf cloud	선반구름	
shelf ice	선반얼음	
shell ice	껍질얼음	
shelter	백엽상	百葉箱
shelterbelt	방풍대	防風帶
sheltering coefficient	차폐계수	遮蔽係數
shelter temperature	백엽상온도	百葉箱溫度
shelter thermometer	백엽상온도계	百葉箱溫度計
shield	바람막이, 차폐	遮蔽
shielding layer	가려막기층	-層
shield volcano	방패화산	防牌火山
shift	변위	變位
shifting level	변위고도	變位高度
shifting sand	흐르는 모래, 유사	流砂
shimmer	아지랑이	
ship-barometer	선박기압계	船舶氣壓計
ship icing	선박결빙	船舶結氷
ship observation	선박관측	船舶觀測
ship report	선박보고	船舶報告
ship synoptic code	선박종관전문형식	船舶綜觀電文形式
shock seeding	충격씨뿌리기	衝擊-
shock wave	충격파	衝擊波
shooting star	별똥별, 유성	流星
shore current	연안류	沿岸流
shore ice	연안얼음	沿岸-
shore wind	연안바람	沿岸-
short-crested wave	짧은 마루파	-波
short day plant	단일식물	短日植物
shortest visibility	최단시정	最短視程
short high	키작은 고기압	-高氣壓
short-interval image	단기영상	短期映像
short-period	단주기	短周期
short pulse	짧은 펄스	
short range ballistic missile operation weather	단거리미사일작전기상	短距離-作戰氣象
short range (weather) forecast	단기(일기)예보	短期(日氣)豫報
short term forecast	단기예보	短期豫報
short wave	단파	短波
short wave channel	단파채널	短波-
short wave fade	단파쇠퇴	短波衰退
short wave infrared spectrum	단파적외선스펙트럼	短波赤外線-
short wave irradiance	단파조도	短波照度
short wave radiation	단파복사	短波輻射
short wave trough	단파골	短波-
shot noise	산탄소음	散彈騷音

Showalter index	쇼월터지수	-指數
Showalter's stability index	쇼월터안정지수	-安定指數
shower	소나기	
shower cloud	소나기구름	
shower formula	소나기공식	-公式
shower of ash	재소나기	
shower of organic matter	유기물소나기	有機物-
showery precipitation	소낙성강수	-性降水
showery rain	소나기	
showery snow	소낙눈	
shred cloud	파편구름	破片-
shrieking sixties	울부짖는 60도(남위 60° 부근의 강한 서풍대)	
shrinking	움츠림, 수축	收縮
shroud	기구덮개	氣球-
shrub line	관목한계	灌木限界
shuidi	수적	水滴
S-hydrograph summation curve	S-수문곡선	-水文曲線
Siberian air mass	시베리아기단	-氣團
Siberian anticyclone	시베리아고기압	-高氣壓
Siberian continental air mass	시베리아대륙기단	-大陸氣團
Siberian high	시베리아고기압	-高氣壓
Siberian permafrost melt	시베리아동토융해	-凍土融解
side	연변	緣邊
side lobe	부돌출부	副突出部
sidelobe echo	부방사부에코	副放射部-
side-looking airbone radar	항공기탑재측방감시레이더	航空機搭載側方監視-
side-looking observation	측방관측	側方觀測
sidereal day	항성일	恒星日
sidereal hour	항성시	恒星時
sidereal year	항성년	恒星年
side wind	옆바람	
signal generator	신호생성기	信號生成機
sight observation	목측	目測
sigma coordinate	시그마좌표	-座標
sigma coordinate system	시그마좌표계	-座標系
sigma-t	시그마-티	
SIGMET information	위험기상정보	危險氣象情報
signal	신호	信號
signal conditioning	신호조절	信號調節
signal detectability	신호검출도	信號檢出度
signal generator	신호발생기	信號發生機
signal integration technique	신호적분기술	信號積分技術
signal interference	신호간섭	信號干涉
signal needle code	모스신호	-信號

S

signal processing	신호처리	信號處理
signal processor	신호처리기	信號處理機
signal quality index	신호품질지수	信號品質指數
signal strength	신호강도	信號强度
signal to noise ratio	신호잡음비	信號雜音比
signal velocity	신호속도	信號速度
significance	유의성, 유의도	有意性, 有意度
significance level	유의수준	有意水準
significance test	유의성검정	有意性檢定
significant cloud	특이구름	特異-
significant digit	유효숫자	有效數字
significant level	유의고도	有意高度
significant point	특이점	特異點
significant wave	유의파	有意波
significant wave height	유의파고	有意波高
significant weather	특이기상	特異氣象
significant weather chart	특이기상도	特異氣象圖
significant weather forecast	특이기상예보	特異氣象豫報
significant weather outlook	특이기상전망	特異氣象展望
significant weather prognostic chart	특이기상예상도	特異氣象豫想圖
silica sand	규사	硅沙
silicon detector	실리콘검출기	-檢出器
sill depth	문턱깊이	門-
silt-discharge rating	고운 모래방출률	-放出率
silver analysis	은량분석	銀量分析
silver-disk pyrheliometer	은반일사계	銀盤日射計
silver frost	은서리	銀-
silver iodide	요오드화은	-化銀
silver iodide seeding	요오드화은 씨뿌리기	-化銀-
silver storm	은폭풍	銀暴風
silver thaw	은상고대	銀-
similarity	상사(성), 유사(성)	相似(性), 類似(性)
similarity law	상사법칙	相似法則
similarity law of Monin-Obukhov	모닌-오부코프 상사법칙	-相似法則
similarity parameter	상사매개변수	相似媒介變數
similarity theory of turbulence	난류상사이론	亂流相似理論
similarity weather chart	유사일기도	類似日氣圖
similitude	유사성	類似性
Simon's earth thermometer	시몬 지중온도계	-地中溫度計
simoon	시문(아라비아사막의 모래폭풍)	
simple harmonic motion	단조화운동, 단진동운동	單調和運動, 單振動運動
simple harmonic wave	단진동파	單振動波
simple logistic curve	단순 로지스틱 곡선	單純-曲線
simple (harmonic) oscillation	단(조화)진동	單(調和)振動

simple pendulum	단진자	單振子
simplified fallout prediction weather	간이낙진예측기상	簡易落塵豫測氣象
Simpson's formula	심프슨공식	-公式
simulated rainfall	모의강우	模擬降雨
simulation	모의, 시뮬레이션	模擬
simultaneity	동시성	同時性
simultaneous equation	연립방정식	聯立方程式
simultaneous relaxation	동시완화	同時緩和
simultaneous transmission	동시송신	同時送信
sine curve	정현곡선, 사인곡선	正弦曲線
sine galvanometer	정현검류계, 사인검류계	正弦檢流計
sine wave	정현파, 사인파	正弦波
sine wave thermal pattern	사인파온도패턴	-波溫度-
Singapore index	싱가폴지수	-指數
single-cell circulation model	단세포대기순환모형	單細胞大氣循環模型
single-cell local thunderstorm	단세포국지뇌우	單細胞局地雷雨
single-cell thunderstorm	단세포뇌우	單細胞雷雨
single drift correction	단편류보정	單偏流補正
single observer forecast	단일관측소예보	單一觀測所豫報
single polarimetric radar	단편파레이더	單偏波-
single port fuel ignition	단일포트연료점화	單一-燃料點火
single-scatter albedo	단산란알베도	單散亂-
single-scattering approximation	단산란가정	單散亂假定
single-sideband	단측파대	單側波帶
single-station analysis	단일관측소분석	單一觀測所分析
single station forecast(ing)	단일관측소예보	單一觀測所豫報
singular advection	특이이류	特異移流
singular corresponding point	특이대응점	特異對應點
singularity	특이성	特異性
singular line	특이선	特異線
singular point	특이점	特異點
singular value decomposition	특이값분해	特異值分解
singular vector	특이벡터	特異-
sink	침강, 흡원, 싱크	沈降, 吸原
sinking	가라앉음, 침강	沈降
sinking stream	침강류	沈降流
sinuosity	굽이	
sinusoid	정현곡선	正弦曲線
siphon barograph	사이펀자기기압계	-自記氣壓計
siphon barometer	사이펀기압계	-氣壓計
siphon mercury barometer	사이펀수은기압계	-水銀氣壓計
siphon rainfall recorder	사이펀우량기록계	-雨量記錄計
siphon rain gauge	사이펀우량계	-雨量計
siphon recording rain gauge	사이펀자기우량계	-自記雨量計

S

siphon type rain gage	사이펀식우량계	-式雨量計
sirocco	시로코(사하라사막에서 지중해 연안으로 부는 열풍)	
SIRS-satellite infrared spectrometer	SIRS-위성적외분광계	-衛星赤外分光計
SI system	국제단위계	國際單位系
site of station	관측장소	觀測場所
site specific weather alert	장소특화날씨경보	場所特化氣象警報
siting	장소선정, 부지선정	場所選定, 敷地選定
SI units	SI-단위	-單位
Six's thermometer	식스온도계	-溫度計
	(최고, 최저 온도계)	
size distribution	크기분포	-分布
size parameter	크기매개변수	-媒介變數
skavler	스카블러(sastrugi의 노르웨이말)	
skew curve	기울어진 곡선	-曲線
skewness	왜도, 비대칭도	歪度, 非對稱度
skew T-log p diagram	스큐티로그피 선도	-線圖
skill score	기능평점, 숙련도	技能評點, 熟練度
skill score of forecast	예보숙련도	豫報熟練度
skin friction	표면마찰	表面摩擦
skin friction coefficient	표면마찰계수	表面摩擦係數
skin temperature	표면온도	表面溫度
skip bombing weather	저공비행폭격기상	低空飛行爆擊氣象
skip distance	도약거리	跳躍距離
skip effect	도약효과	跳躍效果
ski plane effect	썰매다리비행기효과	-飛行機效果
sky	하늘	
sky brightness temperature	하늘밝기온도	-溫度
sky clear	청천	晴天
sky condition	하늘상태	-狀態
sky cover	구름양	-量
skyhook balloon	고층등고도관측기구	高層等高度觀測氣球
skylight	하늘빛	
sky map	하늘지도	-地圖
sky radiance	하늘휘도	-輝度
sky radiation	하늘복사	-輻射
sky radiometer	하늘복사계	-輻射計
sky slightly clouded	조금 흐린하늘	
sky view factor	천공비	天空比
sky wave	공중파	空中波
slack water	정체수, 정체류	停滯水, 停滯流
slant convection	경사대류	傾斜對流
slant path	경사경로	傾斜經路
slant path molecular absorption	경사경로분자흡수	傾斜經路分子吸收
slant range	경사거리	傾斜距離

S

slant visibility	경사시정	傾斜視程
slant visual range	경사가시거리	傾斜可視距離
sleet	언비, 진눈깨비	
sleet warning	진눈깨비경보	-警報
slice method	슬라이스법	-法
sliding method	슬라이딩법	-法
sliding scale	슬라이딩척도	-尺度
slight air	약한 바람	
slight breeze	남실바람, 약한 바람	
slight haze	엷은 연무	-煙霧
slight rain	가랑비	
slight sea	잔물결바다	
slight thunderstorm	약뇌우	弱雷雨
sling psychrometer	휘돌이건습계	-乾濕計
sling thermometer	휘돌이온도계	-溫度計
slip stream	프로펠러후류	-後流
slob ice	덩이유빙	-流氷
slope creep	사면포복	斜面葡匐
slope current	경사류	傾斜流
slope of a front	전선기울기	前線-
slope of an isobaric surface	등압면기울기	等壓面-
slope of surface discontinuity	불연속면기울기	不連續面-
slope-Richardson number	경사-리처드수	傾斜-數
slope wind	활강풍	滑降風
sloping end walls	경사끝벽	傾斜-壁
slow ion	느린 이온	
sluff	눈흘러내림	
sluice(= flood gate)	수문	水門
slurry	슬러리, 현탁액	懸濁液
slush	질퍽눈	
small cloud	작은 구름	
small craft advisory	해안소형선박주의보	海岸小形船舶注意報
small evaporimeter	소형증발계	小型蒸發計
small hail	작은 우박	-雨雹
small ion	작은 이온	
small-scale lifting	소규모치올림	小規模-
small-scale structure	소규모구조	小規模構造
smaze(= smoke and haze)	스메이즈	
Smith and Woolf scheme	스미스-울프 방식	-方式
Smith's retrieval scheme	스미스복귀방식	-復歸方式
smog(= smoke and fog)	스모그	
smog horizon	연개지평, 스모그선	煙-地平, -線
smoke	연기	煙氣
smoke cloud	연기구름	煙氣-

S

smoke damage	연기피해	煙氣被害
smoke dispersal	연기분산	煙氣分散
smoke fog(= smog)	연무, 스모그	煙霧
smoke horizon	연기지평	煙氣地平
smoke management	연기관리	煙氣管理
smoke pall	연기보	煙氣褓
smokes	스모크스(아프리카의 기니아 해안에서 건조한 계절에 발생하는 짙은 흰 안개)	
smoke screen	연막	煙幕
smoke screen operation weather	연막작전기상	煙幕作戰氣象
smokestack	연돌	煙突
smoke wire method	연기선법	煙氣線法
smoothed sunspot number	평활태양흑점수	平滑太陽黑點數
smoothing	고르기, 평활	平滑
smoothing process	평활과정	平滑過程
smooth sea	남실바다	
smooth sea water region	평수구역	平水區域
smooth surface	매끈한 면	-面
smudging	연기뿜기(서리방지용)	煙氣-
snap	갑작스러운 추위	
Snell's law	스넬법칙	-法則
snotel(= snow telemeter)	자동적설측정망	自動積雪測定網
snow	눈	
snow accretion	착설	着雪
snow accumulation	적설(량)	積雪(量)
snow accumulation and ablation model	적설-삭설 모형	積雪-朔雪模型
snow advisory	눈주의보	-注意報
snow and ice verification service	눈얼음검증서비스	-檢證-
snowball	눈덩이	
snowball earth	눈덩이지구	-地球
snowball earth hypothesis	눈덩이지구 가설	-地球假說
snowbank	눈더미	
snow banner	눈깃발	
snow bed	눈발	
snowbelt	강설대	降雪帶
snow bin	눈상자	-箱子
snow blindness	설맹(증)	雪盲(症)
snow blink	눈반짝임	
snow board	적설판	積雪板
snowbreak	눈막이	
snow bridging	눈다리	
snowburn	눈그을음	
snow camera	스노카메라	
snow cap	관설	冠雪

snow-capped crater	눈덮힌 분화구	-噴火口
snow cap profile	적설윤곽	積雪輪廓
snow climate	눈기후	-氣候
snow cloud	눈구름	
snow concrete	눈콘크리트	
snow core	설심, 눈핵	雪心, -核
snow cornice	눈처마	
snow course	눈관측길	-觀測-
snow cover	적설, 눈덮임	積雪
snowcover chart	적설도	積雪圖
snowcover line	적설선	積雪線
snow cover melting	적설융해	積雪融解
snow cover meter	적설계	積雪計
snow creep	적설기어내림	積雪-
snow crust	눈크러스트	
snow crystal	눈결정	-結晶
snow cutter	눈절단기	-切斷機
snow damage	눈피해, 설해	-被害, 雪害
snow day	강설일, 눈 온 날	降雪日
snow density	눈밀도	-密度
snow density meter	눈밀도계	-密度計
snow depth	적설, 눈깊이	積雪
snow devil	눈회오리	
snowdrift	날려쌓인 눈	
snowdrift glacier	날려쌓인 눈빙하, 설계	-氷河, 雪溪
snowdrift ice	날려쌓인 눈얼음	
snow eater	눈녹이(바람, 안개)	
snow endurance	내설성	耐雪性
snowfall	신적설, 강설(량)	新積雪, 降雪(量)
snowfall meter	강설계	降雪計
snowfall totalizer	적산설량계	積算雪量計
snow fence	눈울타리	
snowfield	눈벌판	
snow flake	설편, 눈송이	雪片
snow flurries	가벼운 눈발	(積雪量記錄)
snow flurry	눈소나기	
snow forest climate	설림기후	雪林氣候
snow garland	눈고리밧줄, 눈화환	-花環
snow gauge	설량계	雪量計
snow geyser	눈치솟음	
snow grain	쌀알눈	
snow ice	눈얼음	
snow level	눈고도	-高度
snow line	설선	雪線

S

snow lying	눈깊이	
snow mat	눈방석	-方席
snow measuring plate	적설판	積雪板
snow melt	눈녹음	
snowmelt flooding	융설홍수	融雪洪水
snow meter	적설계	積雪計
snow mist	눈실안개	
snow pack	눈쌓임	
snow particle	눈입자	-粒子
snow patch	눈밭	
snow pellet	눈싸라기, 싸락눈	
snow pillow	눈베개	
snow plume(= snow banner)	눈플룸	
snow precipitation line	강설선	降雪線
snowquake(= snow tremor)	설진	雪震
snow recorder	적설기록계	積雪記錄計
snow regime	눈영역	-領域
snow roller	두루마리눈	
snow sampler	채설기	採雪器
snow server	스노우서버	
snow settling force	적설침강력	積雪沈降力
snow shed	눈사태막이	-沙汰-
snow-shelter forest	방설림	防雪林
snow shield	눈막이	
snow shower	소낙눈	
snow sky(= snow blink)	눈하늘	
snowslide	눈사태	-沙汰
snow slush	질펀눈	
snow smoke	눈연기	-煙氣
snow squall	눈스콜	
snow stage	눈단계	-段階
snow stick	설고막대	雪高-
snow storm	눈보라	
snow storm warning	눈보라경보	-警報
snow survey	눈측정	-測定
snow thunderstorm	뇌설	雷雪
snow tremor	눈요동	-搖動
snow tube	눈채집관	-採集管
snow virga	눈꼬리구름	
snow water	눈녹은 물	
snow water equivalent	수상당량비, 설상당우량	水相當量比, 雪相當雨量
soft air(= moist air)	습윤공기	濕潤空氣
soft hail	연우박	軟雨雹
soft radiation	연방사선	軟放射線

S

soft rime	연상고대	軟-
soil	토양	土壤
soil aeration	토양통기	土壤通氣
soil air	토양공기	土壤空氣
soil air temperature	토양공기온도	土壤空氣溫度
soil atmosphere	토양온도	土壤溫度
soil borne diseases	토양기원질병	土壤起源疾病
soil climate	토양기후	土壤氣候
soil climatology	토양기후학	土壤氣候學
soil conductivity	토양전도도	土壤傳導度
soil creep	토양기어내림	土壤-
soil drought	토양가뭄	土壤-
soil erosion	토양침식	土壤浸蝕
soil evaporation	토양증발	土樓蒸發
soil evaporimeter	토양증발계	土壤蒸發計
soil flow	토양흐름	土壤-
soil gas	토양기체	土壤氣體
soil hardness	토양경도	土壤硬度
soil horizon	토양수평선	土壤水平線
soil moisture	토양수분	土壤水分
soil moisture content	토양수분함유량	土壤水分含有量
soil moisture deficit	토양수분부족(량)	土壤水分不足(量)
soil moisture meter	토양수분계	土壤水分計
soil moisture tensiometer	토양수분응력계	土壤水分應力計
soil moisture tension	토양수분응력	土壤水分應力
soil pore	토양공극	土壤孔隙
soil property	토양특성	土壤特性
soil reflectance properties	토양반사특성	土壤反射特性
soil respiration	토양호흡	土壤呼吸
soil salt injury	토양염류장해	土壤鹽類障害
soil structure	토양구조	土壤構造
soil temperature	토양온도, 지온	土壤溫度, 地溫
soil texture	토양질감	土壤質感
soil thermograph	자기지온계,	自記地溫計,
	자기토양온도계	自記土壤溫度計
soil thermometer	지온계, 토양온도계	地溫計, 土壤溫度計
soil water	토양수분	土壤水分
soil water balance	토양물수지	土壤-收支
soil water potential	토양수분퍼텐셜	土樓-水分-
solar activity	태양활동	太陽活動
solar air mass	태양공기질량	太陽空氣質量
solar altitude	태양고도	太陽高度
solar atmosphere	태양대기	太陽大氣
solar atmospheric tide	태양대기조석	太陽大氣潮汐

S

solar barometric variation	태양기압변화	太陽氣壓變化
solar battery	태양전지	太陽電池
solar cell	태양전지	太陽電池
solar climate	태양기후	太陽氣候
solar constant	태양상수	太陽常數
solar coordinates	태양좌표계	太陽座標系
solar corona	해코로나	
solar corpuscle	태양미립자	太陽微粒子
solar corpuscular theory	태양미립자설	太陽微粒子說
solar cycle	태양주기	太陽週期
solar day	태양일	太陽日
solar declination	태양경사	太陽傾斜
solar distance	태양거리	太陽距離
solar eclipse	일식	日蝕
solar elevation angle	태양고도각	太陽高度角
solar energy	태양에너지	太陽-
solar flare	태양플레어	太陽-
solar flare disturbance	태양플레어요란	太陽-擾亂
solar halo	햇무리	
solarigram	자기전천일사곡선	自記全天日射曲線
solarigraph	자기일사계	自記日射計
solarimeter	일사계	日射計
solar maximum	최고태양활동	最高太陽活動
solar minimum	최소태양활동	最小太陽活動
solar noon	정오	正午
solar occultation	태양엄폐	太陽掩蔽
solar paddle	태양전지판	太陽電池板
solar particle environment	태양입자환경	太陽粒子環境
solar power	태양동력	太陽動力
solar prominence	태양홍염	太陽紅焰
solar proton event	태양양자소동	太陽陽子騷動
solar proton monitor	태양양자감시기	太陽陽子監視器
solar radiant exitance	태양복사발산도	太陽輻射發散度
solar radiation	태양복사, 일사	太陽輻射, 日射
solar radiation environment	태양복사환경	太陽輻射環境
solar radiation observation	태양복사관측	太陽輻射觀測
solar radiation shield	태양복사차폐	太陽輻射遮蔽
solar radiation sonde	일사존데	日射-
solar radiation spectrum	태양복사스펙트럼	太陽輻射-
solar radiation thermometer	태양복사온도계	太陽輻射溫度計
solar radio emission	대기복사방출	大氣輻射放出
solar spectral irradiance	태양스펙트럼조도	太陽-照度
solar spectrum	태양분광, 태양광스펙트럼	太陽分光, 太陽光-
solar system	태양계	太陽界

solar temperature	태양온도	太陽溫度
solar term	절기	節氣
solar terrestrial physics	태양지구물리학	太陽地球物理學
solar terrestrial relationship	태양지구관계	太陽地球關係
solar thermal energy	태양열에너지	太陽熱-
solar tide	태양조석	太陽潮汐
solar time	태양시	太陽時
solar ultraviolet radiation	태양자외복사	太陽紫外輻射
solar variability	태양변동성	太陽變動性
solar variation	태양변동	太陽變動
solar wind	태양풍	太陽風
solar year	태양년	太陽年
solar zenith angle	태양천정각	太陽天頂角
sole mark	저흔, 발바닥흔적	底痕, -痕迹
solenoid	솔레노이드	
solenoidal field	솔레노이드장	-場
solenoidal index	솔레노이드지수	-指數
solid	입체	立體
solid angle	입체각	立體角
solid boundary	고체경계	固體境界
solidification	응고	凝固
solidifying point	응고점	凝固點
solid of revolution	회전체	回轉體
solid phase	고체상	固體相
solid precipitation	고체강수, 고형강수	固體降水, 固形降水
solid rotation	고체회전	固體回轉
solid state	고체상태	固體狀態
solid state transmitter	고체송신기	固體送信機
solitary wave	고립파	孤立波
solstice	지점(하지, 동지)	至點
solstitial colure	지점시간권	至點時間圈
solstitial tide	지점조석	至點潮汐
solubility	용해도	溶解度
solubility product	용해도곱	溶解度-
solute	용질	溶質
solution	용액	溶液
solvent	용매, 용제	溶媒, 溶劑
Somali current	소말리아해류	-海流
Somali jet	소말리아제트	
sonar detection weather	소나탐지기상	-探知氣象
sonar range weather	소나탐지기상	-探知氣象
sonde	존데	
sonic anemometer	음파풍속계	音波風速計
sonic anemometer thermometer	음파풍속온도계	音波風速溫度計

S

sonic energy	소리에너지	
sonic speed	음속	音速
sonic thermometer	음파온도계	音波溫度計
sonic wave	음파	音波
sonora	소노라(캘리포니아 및 멕시코 북서부 산지와 사막의 여름철 뇌우)	
sonora weather	소노라날씨	
soot	검댕이, 그을음	
soot carbon aerosol	검댕탄소에어로졸	-炭素-
sorbents	흡수제, 흡착제	吸收劑, 吸着劑
soroche	고산병〈스페인어〉	高山病
sorption	흡착	吸着
sound	소리, 음	音
sound absorption	소리흡수	-吸收
sound detection and ranging(SODAR)	소다	
sound frequency	소리주파수	-周波數
sounding	탐측	探測
sounding balloon	탐측기구	探測氣球
sounding by satellite	위성탐측	衛星探測
sounding chart	탐측도	探測圖
sounding lapse rate	환경감률	環境減率
sounding meteorograph	탐측기상기록계	探測氣象記錄計
sounding rocket	탐측로켓	探測-
sound intensity	소리강도	-强度
sound power	음향출력	音響出力
sound pressure level	음압수준	音壓水準
sound pressure level meter	음압수준계	音壓水準計
sound propagation	소리전파	-傳播
sound ray	소리선	-線
sound shadow	소리그늘	
sound speed	음속	音速
sound velocity	음속	音速
sound virtual temperature	음파가온도	音波假溫度
sound wave	음파	音波
source	발원, 샘, 원천, 소스	發源, 源泉
source category	발원범주	發源範疇
source function	원천함수	源泉函數
source of an atmospherics	공전원	空電源
source of thunderstorm activity	뇌우활동원	雷雨活動源
source property	발원지특성	發源地特性
source region	발원지	發源地
South Atlantic current	남대서양해류	南大西洋海流
South Atlantic high	남대서양고기압	南大西洋高氣壓
South China Sea	남중국해	南中國海

S

southeaster	남동편풍	南東偏風
southeast trades	남동무역풍	南東貿易風
southeastern monsoon	남동계절풍	南東季節風
south equatorial current	남적도해류	南赤道海流
southern annular mode	남방연간모드	南方年間-
southern hemisphere	남반구	南半球
southern oscillation	남방진동	南方振動
southern oscillation index	남방진동지수	南方振動指數
South Pacific convergence zone	남태평양수렴대	南太平洋收斂帶
South Pacific current	남태평양해류	南太平洋海流
South Pole	남극	南極
south tropical disturbance	남열대요란	南熱帶擾亂
southwester(= sou'wester)	남서편풍	南西偏風
southwest monsoon	남서계절풍	南西季節風
Soviet F-2 rocket	소련F-2로켓	蘇聯-
space	공간, 우주(공간)	空間, 宇宙(空間)
space-based subsystem	우주근거하위계	宇宙根據下位系
space change	공간변화	空間變化
space charge	공간전하	空間電荷
space control operations weather	우주통제작전기상	宇宙統制作戰氣象
spacecraft computer unit	위성메인컴퓨터	衛星-
space environmental monitor	우주환경감시기	宇宙環境監視器
space environment center	우주환경센터	宇宙環境-
spaceflight	우주비행체	宇宙飛行體
space lidar	우주라이다	宇宙-
spacelink	우주도킹	宇宙-
space mean	공간평균	空間平均
space mean anomaly chart	공간평균평년편차도	空間平均平年偏差圖
space mean vorticity advection method	공간평균소용돌이도이류법	空間平均-度移流法
space medicine	우주의학	宇宙醫學
space meteorology	우주기상학	宇宙氣象學
space power	우주전력	宇宙電力
space probe	우주탐측기	宇宙探測器
space research	우주연구	宇宙研究
space shuttle	우주왕복선	宇宙往復船
space support	우주지원	宇宙支援
space support operations weather	우주지원작전기상	宇宙支援作戰氣象
space-time sampling	시공표본채취	時空標本採取
space tracking	우주추적	宇宙追跡
space vector	공간벡터	空間-
space warfare	우주전쟁	宇宙戰爭
space weather	우주기상	宇宙氣象
spark	불꽃, 스파크	
spark discharge	불꽃방전	-放電

S

spark ignition engine	불꽃점화기관	-點火機關
spark retardation	불꽃지연	-遲延
sparse data area	희소자료지역	稀少資料地域
spatial coherence	공간정합	空間整合
spatial dendrite	나뭇가지모양결정	-模樣結晶
spatial resolution	공간해상도	空間解像度
spatial sampling	공간샘플링, 공간표본수집	空間標本收集
spatial scale	공간규모	空間規模
spearhead echo	창끝에코	槍-
special avalanche warning	특별눈사태경보	特別-沙汰警報
special climate change fund	특별기후변화기금	特別氣候變化基金
special fire weather	특별화재기상	特別火災氣象
special marine warning	특별해상경보	特別海上警報
special meteorological report	특별기상보고	特別氣象報告
special observation	특별관측	特別觀測
special report for aviation weather	항공기상특보	航空氣象特報
special report for tsunami	지진해일특보, 쓰나미특보	地震海溢特報
special report regions for aviation weather	항공기상특보구역	航空氣象特報區域
special station	특수관측소	特殊觀測所
special tropical disturbance statement	특별열대요란발표문	特別熱帶擾亂發表文
special weather report	특별일기통보, 특별기상통보	特別日氣通報, 特別氣象通報
special world interval	특별세계간격	特別世界間隔
species of clouds	구름종	-種
specific	비(의), 특수한	比, 特殊-
specific conductivity	비전도율	比傳導率
specific differential phase	비차등위상	比差等位相
specific energy	비에너지	比-
specific forecast	특수예보	特殊豫報
specific gas constant	비기체상수	比氣體常數
specific gravity	비중	比重
specific heat	비열	比熱
specific heat at constant pressure	정압비열	定壓比熱
specific heat at constant volume	정적비열	定積比熱
specific heat capacity	비열용량	比熱容量
specific humidity	비습	比濕
specific humidity line	비습선	比濕線
specific leaf area	비엽면적	比葉面積
specific radiation intensity	비복사강도	比輻射强度
specific tension	비장	比張
specific volume	비체적	比體積
specific volume anomaly	비체적이상	比體積異常
specific yield	비산출량, 비추수량	比産出量, 比秋收量
speckle	(레이더영상의)반점	班點
spectra	스펙트럼	

S

spectral	스펙트럼(의), 분광(의), 단색광(의)	分光, 單色光
spectral channel	스펙트럼채널	
spectral concentration	스펙트럼농도	-濃度
spectral density	스펙트럼밀도	-密度
spectral detectivity	스펙트럼검출률	-檢出率
spectral element method	분광요소법	分光要素法
spectral function	스펙트럼함수	-函數
spectral hygrometer	분광습도계	分光濕度計
spectral line	스펙트럼선	-線
spectral method	스펙트럼법, 분광법	分光法
spectral model	스펙트럼모형	-模型
spectral numerical analysis	스펙트럼수치분석	-數值分析
spectral numerical prediction	스펙트럼수치예측	-數值豫測
spectral photoconductivity	스펙트럼광전도율	-光傳導率
spectral pyranometer	분광전천일사계	分光全天日射計
spectral pyrheliometer	분광직달일사계	分光直達日射計
spectral radiance	분광휘도	分光輝度
spectral range	분광영역	分光領域
spectral reflectance	분광반사율	分光反射率
spectral resolution	스펙트럼분해능	-分解能
spectral solar radiation	태양분광복사	太陽分光輻射
spectral width	스펙트럼폭	-幅
spectrobologram	분광복사기록	分光輻射記錄
spectrograph	분광사진기	分光寫眞機
spectroheliograph	분광일광계	分光日光計
spectrometer	분광계	分光計
spectrometry	분광분석(법)	分光分析(法)
spectrophotometer	분광광도계	分光光度計
spectropyrheliometer	분광일사계	分光日射計
spectroscope	분광기	分光器
spectroscopy	분광법	分光法
spectroscopic hygrometer	분광습도계	分光濕度計
spectrum	스펙트럼	
spectrum analysis	스펙트럼분석	-分析
spectrum function	스펙트럼함수	-函數
spectrum intensity	스펙트럼강도	-强度
spectrum of global radiation	지구복사스펙트럼	地球輻射-
spectrum series	스펙트럼계열	-系列
spectrum width	스펙트럼폭	-幅
specular reflection	정반사, 거울면반사	正反射, -面反射
specular reflector	정반사체, 거울반사체	正反射體, -反射體
speed	속력	速力
speed of light	광속	光速

S

speed of sound	음속	音速
speedometer	속도계	速度計
speleo-meteorology	동굴기상학	洞窟氣象學
speleothem	스펠레오뎀	
sphere	구, 권	球, 圈
spheric aerostat	구형기구	球形氣球
spherical albedo	구면반사도, 구면알베도	球面反射度
spherical coordinates	구면좌표	球面座標
spherical coordinate system	구면좌표계	球面座標系
spherical curvature	구면곡률	球面曲率
spherical function	구면함수	球面函數
spherical harmonic analysis	구면조화분석	球面調和分析
spherical harmonics	구면조화함수	球面調和函數
spherical projection	구면투영	球面投影
spherical pyranometer	구형전천일사계	球型全天日射計
spherical pyrgeometer	구형지구복사계	球型地球輻射計
spherical pyrradiometer	구형전천일사계	球型全天日射計
spherical shell apparatus	구면효과측정기	球面效果測定器
spherical symmetry	구면대칭	球面對稱
spherical wave	구면파	球面波
spherics	공중전기	空中電氣
spheroid	회전타원체	回轉楕圓體
spicule	얼음침, 빙침	氷針
spillover	날림비, 날림눈	
spin	선회, 회전	旋回, 回轉
spindle nephoscope	선회측운기	旋回測雲器
spin down	선회감소	旋回減少
spin down time	선회감소시간	旋回減少時間
spin scan cloud camera	자전주사구름카메라	自轉走査-
spin stabilization	선회안정	旋回安定
spin up	선회증가	旋回增加
spiral band	나선띠	螺旋-
spiral band cloud	나선띠구름	螺旋-
spiral cirrus	나선권운	螺旋卷雲
spiral cloud band	나선구름띠	螺旋-
spiral layer	나선층	螺旋層
spiral pattern	나선패턴	螺旋-
spiral scanning	나선주사	螺旋走査
spirit thermometer	알코올온도계	-溫度計
spissatus	농밀구름	濃密-
splashdown	수면착수(우주선 등의)	水面着水
splintering	파열	破裂
split front	파열전선	破裂前線
split method	분리법	分離法

S

splitting method	분리법	分離法
splitting of supercells	초대형세포분리	超大型細胞分離
split window method	창분리법	窓分離法
spongy hail	스펀지우박	-雨雹
spongy ice	스펀지얼음	
spongy zone	스펀지구역	-區域
spontaneous condensation	자발응결	自發凝結
spontaneous convection	자발대류	自發對流
spontaneous freezing	자발결빙	自發結氷
spontaneous nucleation	자발핵화	自發核化
spontaneous sublimation	자발승화	自發昇華
sporadic E-layer	돌발E층	突發E層
sporadic E-region(= sporadic E-layer)	돌발E영역	突發E地域
sporadic meteor	산발유성	散發流星
spotting(= plotting)	기입	記入
spot wind	정점바람	定點-
spout	용오름	龍-
spray	물보라	
spray electrification	물보라대전	-帶電
spray region	분무역	噴霧域
spray tower	분무탑	噴霧塔
spread F	퍼진 F층	-層
spring	봄	
spring equinox(= vernal equinox)	춘분(점)	春分(點)
springkler irrigation	살수관개	撒水灌漑
springness	봄기운	-氣運
spring snow	봄눈	
spring tide	사리	
sprinkle	후두두비	
Sprung barograph	슈프룽자기기압계	-自記氣壓計
Sputnik	스푸트니크(인공위성)	
spy satellite	첩보위성	諜報衛星
squall	스콜	
squall cloud	스콜구름	
squall line	스콜선	-線
squall line thunderstorm	스콜선뇌우	-線雷雨
squall surface	스콜면	-面
squall wave	스콜파	-波
square	제곱	
square number	제곱수	-數
square plane	사각평면	四角平面
square pyramid	사각뿔	四角-
square root	제곱근	-根
staff gauge	수위측정기	水位測定器

S

stability	안정도	安定度
stability chart	안정도도	安定度圖
stability class	안정등급	安定等級
stability condition	안정도조건	安定度條件
stability diagram	안정도도표	安定度圖表
stability index	안정도지수	安定度指數
stability of flow	흐름안정도	-安定度
stability parameter	안정도매개변수	安定度媒介變數
stability ratio	안정비	安定比
stabilization	안정화	安定化
stabilization system	안정화시스템	安定化-
stabilizing factor	안정인자	安定因子
stabilizing force	안정력	安定力
stable air	안정공기	安定空氣
stable air layer	안정기층	安定氣層
stable air mass	안정기단	安定氣團
stable equilibrium	안정평형	安定平衡
stable flow	안정흐름	安定-
stable isotope analysis	안정동위원소분석	安定同位元素分析
stable layer	안정층	安定層
stable local oscillator	안정국부발진기	安定局部發振器
stable motion	안정운동	安定運動
stable oscillation	안정진동	安定振動
stable state	안정상태	安定狀態
stable stratification	안정성층	安定成層
stable type	안정형	安定型
stable wave	안정파	安定波
stack effluent	굴뚝배출물	-排出物
stack sampling	굴뚝배기표본채취	-排氣標本採取
staff gauge	수위계	水位計
stage	단계, 수위	段階, 水位
stage-discharge relation	수위유량관계	水位流量關係
stage relation	수위상관	水位相關
stage wind	무대바람	舞臺-
stagewise regression procedure	축차회귀법	逐次回歸法
staggered grid	엇갈림격자	-格子
staggering of horizontal grid	수평격자엇갈림	水平格子-
stagger scheme	엇갈림방식	-方式
stagnant air	정체공기	停滯空氣
stagnant glacier	정체빙하	停滯氷河
stagnant ground water	정체지하수	停滯地下水
stagnation area	정체지역	停滯地域
stagnation pressure	정체압	停滯壓
stalling mach number	실속마하수	失速-數

stalling velocity	실속속도	失速速度
stamp map	우표분포도	郵票分布圖
	(앙상블 예측결과 표현 방법의 일종)	
standard artillery atmosphere	표준탄도대기	標準彈道大氣
standard artillery zone	표준탄도구역	標準彈道區域
standard atmosphere	표준대기	標準大氣
standard atmosphere altitude	표준대기고도	標準大氣高度
standard atmosphere lapse rate	표준대기기온감률	標準大氣氣溫減率
standard atmosphere pressure	표준대기압	標準大氣壓
standard atmospheric lapse rate	표준대기기온감률	標準大氣氣溫減率
standard barometer	표준기압계	標準氣壓計
standard briefing	표준브리핑	標準-
standard density	표준밀도	標準密度
standard density altitude	표준밀도고도	標準密度高度
standard depth	표준깊이	標準-
standard deviation	표준편차	標準偏差
standard error	표준오차	標準誤差
standard error of estimate	추정표준오차	推定標準誤差
standard(normal) form	표준형	標準形
standard glass beads	표준유리구슬	標準琉璃-
standard gravity	표준중력	標準重力
standard instrument	표준계기	標準計器
standard isobaric surface	표준등압면	標準等壓面
standardization	표준화	標準化
standard meridian	표준자오선	標準子午線
standard method	표준방법	標準方法
standard normal distribution	표준정규분포	標準正規分布
standard operative temperature	표준작업온도	標準作業溫度
standard point	기준점	基準點
standard pressure	표준기압	標準氣壓
standard pressure altitude	표준기압고도	標準氣壓高度
standard pressure level	표준기압면	標準氣壓面
standard project flood	표준계획홍수량	標準計劃洪水量
standard propagation	표준전파	標準傳播
standard raingauge	표준우량계	標準雨量計
standard refraction	표준굴절	標準屈折
standard sea level temperature	표준해면온도	標準海面溫度
standard sea water	표준해수	標準海水
standard state	표준상태	標準狀態
standard target	표준표적	標準標的
standard temperature	표준온도	標準溫度
standard thermometer	표준온도계	標準溫度計
standard time	표준시	標準時
standard time of observation	표준관측시각	標準觀測時刻

S

standard unit	표준단위	標準單位
standard visibility	표준시정	標準視程
standard visual range	표준가시거리	標準可視距離
standard weather observatory	표준기상관측소	標準氣象觀測所
standing cloud	정체구름	停滯-
standing crop	현존곡물량	現存穀物量
standing eddy	정체에디	停滯-
standing wave	정립파	定立波
Stanton number	스탠턴수	-數
star	별, 항성	恒星
Stark effect	슈타르크효과	-效果
starshine recorder	별빛기록계	-記錄計
starting wind velocity	기동풍속	起動風速
start of autumn	입추	立秋
start of spring	입춘	立春
start of summer	입하	立夏
start of winter	입동	立冬
startscope	스타트스코프	
starute(= star and parachute)	스타류트	
state	상태, 국가	狀態, 國家
state and local air monitoring stations	국가지역대기질관측소	國家地域大氣質觀測所
state curve	상태곡선	狀態曲線
state function	상태함수	狀態函數
state implementation plan	국가시행계획	國家施行計劃
state of equilibrium	평형상태	平衡狀態
state of ground	지면상태	地面狀態
state of sea scale	파고자	波高-
state of sky	하늘상태	-狀態
state of the sea	바다상태	-狀態
state parameter	상태매개변수	狀態媒介變數
state variable	상태변수	狀態變數
static	공전, 정적	空電, 靜的
statical meteorology	정기상학	靜氣象學
static ceiling	기구평형고도	氣球平衡高度
static climatology	정기후학	靜氣候學
static cloud seeding	정적구름씨뿌리기	靜的-
static current	정지전류	靜止電流
static electricity	정전기	靜電氣
static equilibrium	정적평형	靜的平衡
static friction	정지마찰	靜止摩擦
static instability	정적불안정(도)	靜的不安定(度)
static level	정적수위	靜的水位
static pressure	정압	定壓
static process	정적과정	靜的過程

statics	정역학	靜力學
static stability	정적안정(도)	靜的安定(度)
station	관측소, 기상대	觀測所, 氣象臺
stationary	정체, 정지	停滯, 靜止
stationary atmosphere	정지대기	靜止大氣
stationary cyclone	정체저기압	停滯低氣壓
stationary electric current	정상전류	定常電流
stationary emission source	고정배출원	固定排出源
stationary flow	정상류	定常流
stationary front	정체전선	停滯前線
stationary Gaussian process	정체가우스과정	停滯-過程
stationary Gaussian time series	정체가우스시계열	停滯-時系列
stationary motion	정체운동	停滯運動
stationary orbit	정지궤도	靜止軌道
stationary phase	정체기	停滯期
stationary Rossby wave	정체로스비파	停滯-波
stationary satellite	정지위성	靜止衛星
stationary source	고정(점)오염원	固定(點)汚染源
stationary state	정체상태	定滯狀態
stationary time series	정체시계열	停滯時系列
stationary wave	정체파	停滯波
stationary wave length	정체파길이	停滯波-
station atmospheric pressure	관측소기압	觀測所氣壓
station circle	지점원	地點圓
station continuity chart	관측소연속도	觀測所連續圖
station designator	관측지점번호	觀測地點番號
station elevation	관측소고도	觀測所高度
station index	지점색인	地點索引
station location	관측소위치	觀測所位置
station mercury barometer	관측소수은기압계	觀測所水銀氣壓計
station network	관측소망	觀測所網
station (index) number	지점(색인)번호	地點(索引)番號
station pressure	관측소기압	觀測所氣壓
statistic	통곗값	統計-
statistical assimilation method	통계동화법	通計同化法
statistical climatology	통계기후학	統計氣候學
statistical dependence	통계의존성	統計依存性
statistical dynamic prediction	통계동역학예측	統計動力學豫測
statistical enhancement	통계강조	統計強調
statistical forecast	통계예보	統計豫報
statistical independence	통계독립성	統計獨立性
statistical method	통계법	統計法
statistical methods of forecasting	통계예보법	統計豫報法
statistical parameter	통계파라미터, 통계매개변수	統計媒介變數

S

statistical probability	통계확률	統計確率
statistical significance test	통계유의성검정	統計有意性檢定
statistical table	통계표	統計表
statistical thermodynamics	통계열역학	統計熱力學
statistical weight	통계가중	統計加重
statistics	통계, 통계학	統計, 統計學
statistics of successive years	누년통계	累年統計
statoscope	자기미압계	自記微壓計
staubosphere	먼지권, 먼지층	-圈, -層
steadiness	정상도	定常度
steady field	정상장	定常場
steady flow	정상류	定常流
steady motion	정상운동	定常運動
steady state	정상상태	定常狀態
steady velocity	정상속도	定常速度
steady wave	정상파	定常波
steady wave regime	정상파영역	定常波領域
steam	김, 증기	蒸氣
steam engine	증기기관	蒸氣機關
steam fog	김안개	
steam mist	김실안개	
steam turbine	증기터빈	蒸氣-
steepness	가파르기	
steering	지향	指向
steering current	지향류	指向流
steering flow	지향흐름	指向-
steering law	지향법칙	指向法則
steering level	지향고도	指向高度
steering line	지향선	指向線
steering method	지향법	指向法
steering surface	지향면	指向面
Stefan-Boltzmann constant	스테판-볼츠만 상수	-常數
Stefan-Boltzmann law	스테판-볼츠만 법칙	-法則
Stefan's law	스테판법칙	-法則
stellar crystal	별모양결정	-模樣結晶
stellar lightning	별모양번개	-模樣-
stellar scintillation	별반짝임	
St. Elmo's fire	성엘모불	聖-
stem-and-leaf diagram	줄기잎선도	-線圖
stem thermometer	막대온도계	-溫度計
step-by-step (or successive) approximation	차례어림, 축차어림	逐次-
step(ped) function	계단함수	階段函數
steppe	스텝, 초원	草原
steppe climate	스텝기후, 초원기후	草原氣候

stepped leader	계단선도	階段先導
stepwise regression analysis	단계회귀분석	段階回歸分析
stepwise sequential test	단계순차검정	段階順次檢定
steradian	스테라디안(입체각의 단위)	
stereo	스테레오, 입체	立體
stereogram	입체도	立體圖
stereographic projection	입체투영	立體投影
stereophotogrammetry	입체사진측량법	立體寫眞測量法
steric anomaly	비체적이상	比體積異常
Stevenson screen	스티븐슨백엽상	-百葉箱
Stevenson's shelter	스티븐슨백엽상	-百葉箱
stilb	스틸브(밝기의 단위)	
still-water level	정수위	靜水位
still well	정수통	靜水桶
stimulated emission	유도방출	誘導放出
stirring	휘젓기	
St. Luke's summer	성루크여름(10월 18일 전후의 맑고 따뜻한 기간)	聖-
St. Martin's summer	성마틴여름(11월 11일 전후의 맑고 따뜻한 기간)	聖-
stochastic	확률(적)	確率(的)
stochastic dynamic prediction	확률역학예보	確率力學豫報
stochastic process	확률과정	確率過程
stoichiometric coefficient	화학량계수	化學量係數
stoichiometric number	화학량수	化學量數
stoichiometric ratio	화학량비	化學量比
stoke	스토크(점성단위)	
Stokesian wave	스토크스파	-波
Stokes vector	스토크스벡터	
Stokes's approximation	스토크스근사	-近似
Stokes's law	스토크스법칙	-法則
Stokes's law of resistance	스토크스저항법칙	-抵抗法則
Stokes's stream function	스토크스유선함수	-流線函數
Stokes's theorem	스토크스정리	-定理
stomata	기공	氣孔
stomatal aperture	기공저항	氣孔抵抗
stomatal conductance	기공전도도	氣孔傳導度
stomatal resistance	기공저항	氣孔低抗
stone ice(= ground ice)	지면얼음	地面-
stooping	위축	萎縮
storage	저장, 저류	貯藏, 貯留
storage equation	저류방정식	貯留方程式
storage raingauge(→ streamflow routing)	저장우량계	貯藏雨量計
storage routing	저류경로	貯留經路

S

storm	폭풍, 폭풍우, 노대바람(풍력계급 10)	暴風, 暴風雨
storm alert service	폭풍경보서비스	暴風警報-
storm area	폭풍(우)역	暴風(雨)域
storm cell	폭풍(우)세포	暴風(雨)細胞
storm cellar	폭풍(우)대피소	暴風(雨)待避所
storm cone	폭풍(우)표지	暴風(雨)標識
storm feeder band	폭풍(우)발생대	暴風(雨)發生帶
storm model	폭풍(우)모형	暴風(雨)模型
storm path	폭풍(우)경로	暴風(雨)經路
storm period	폭풍(우)기간	暴風(雨)期間
storm sudden commencement	폭풍(우)돌연시작	暴風(雨)突然始作
storm surge	폭풍(우)해일	暴風(雨)海溢
storm surge advisory	폭풍해일주의보	暴風海溢注意報
storm surge warning	폭풍해일경보	暴風海溢警報
storm tide	폭풍조석	暴風潮汐
storm tide warning	폭풍조석경보	暴風潮汐警報
storm track	폭풍(우)경로	暴風(雨)經路
storm transposition	폭풍(우)이동	暴風(雨)移動
storm warning	폭풍(우)경보	暴風(雨)警報
storm wave	폭풍(우)파	暴風(雨)波
stormy westerly wind	격렬서풍	激烈西風
straight line	직선	直線
straight line motion	직선운동	直線運動
strain	(응력)변형	(應力)變形
strain tensor	응력변형텐서	應力變形-
stratification	성층, 성층화	成層, 成層化
stratification curve	성층곡선	成層曲線
stratified fluid	성층유체	成層流體
stratiform cloud	층모양구름	層模樣-
stratiform echo	층모양에코	層模樣-
stratiformis	층모양구름	層-
stratiform precipitation	층모양강수	層模樣-降水
stratigraphy	층서학	層序學
stratocumulus	층적운	層積雲
stratocumulus castellanus	탑층적운	塔層積雲
stratocumulus duplicatus	겹층적운	-層積雲
stratocumulus floccus	송이층적운	-層積雲
stratocumulus lacunosus	벌집층적운	-層積雲
stratocumulus lenticularis	렌즈층적운	-層積雲
stratocumulus mammatus	유방층적운	乳房層積雲
stratocumulus opacus	불투명층적운	不透明層積雲
stratocumulus perlucidus	틈새층적운	-層積雲
stratocumulus radiatus	방사층적운	放射層積雲

stratocumulus stratiformis	층모양층적운	層模樣層積雲
stratocumulus translucidus	반투명층적운	半透明層積雲
stratocumulus undulatus	파도층적운	波濤層積雲
stratocumulus vesperalis	석양층적운	夕陽層積雲
stratocumulus virga	꼬리층적운	-層積雲
stratopause	성층권계면	成層圈界面
stratosphere	성층권	成層圈
stratosphere radiation	성층권복사	成層圈輻射
stratosphere-troposphere exchange	성층권-대류권 교환	成層圈-對流圈交換
stratospheric circulation index	성층권순환지수	成層圈循環指數
stratospheric compensation	성층권보상	成層圈補償
stratospheric coupling	성층권접합, 성층권결합	成層圈接合, 成層圈結合
stratospheric fallout	성층권낙진	成層圈落塵
stratospheric jet stream	성층권제트류	成層圈-流
stratospheric monsoon	성층권계절풍	成層圈季節風
stratospheric ozone	성층권오존	成層圈-
stratospheric polar vortex	성층권극소용돌이	成層圈極-
stratospheric steering	성층권지향	成層圈指向
stratospheric sudden warming	성층권돌연승온	成層圈突然昇溫
stratospheric turbidity	성층권혼탁도	成層圈混濁度
stratospheric warming	성층권승온	成層圈昇溫
stratostat	성층권기구	成層圈氣球
stratovolcano	성층화산	成層火山
stratus	층운	層雲
stratus communis(= common stratus)	보통층운	普通層雲
stratus fractus	조각층운	-層雲
stratus lenticularis	렌즈층운	-層雲
stratus maculosus	얼룩층운	-層雲
stratus nebulosus	베일층운	-層雲
stratus opacus	불투명층운	不透明層雲
stratus translucidus	반투명층운	半透明層雲
stratus undulatus	파도층운	波濤層雲
stray(= atmospherics)	공전	空電
streak	줄무늬, 스트리크	
streak lightning	줄무늬번개	
streak line	줄무늬선	-線
stream devil	흐름회오리	
streamer	스트리머, 코로나광휘	-光輝
streamflow	하천유수	河川流水
streamflow routing	하천유수추적	河川流水追跡
stream function	유선함수	流線函數
stream gauge	유량계	流量計
streamline	유선	流線
streamline amplitude	유선진폭	流線振幅

S

streamline analysis	유선분석	流線分析
streamline chart	유선도	流線圖
streamline curvature	유선곡률	流線曲率
streamlined shape	유선형	流線形
streamline field	유선장	流線場
streamline flow	유선류	流線流
streamline pattern	유선형, 유선패턴	流線型
streamlining	쏠림	
streamlining effect	쏠림효과	-效果
stream order	하천차수	河川次數
stream piracy	지하수이동	地下水移動
stream tube	유관	流管
strength of magnetic pole	자기극세기	磁氣極-
stress	응력	應力
stress tensor	응력텐서	應力-
stretching	신장, 늘림	伸長
stretching deformation	늘림변형	-變形
striation	줄무늬	
stroke	전격, 벼락, 낙뢰	電擊, 落雷
stroke density	벼락밀도	-密度
stroke of lightning	벼락, 낙뢰	落雷
strong breeze	된바람(풍력계급 6)	
strong gale	큰센바람(풍력계급 9)	
strong wind	강풍	強風
strong wind advisory	강풍주의보	強風注意報
strong wind warning	강풍경보	強風警報
strontium isotope	스트론튬동위원소	-同位元素
Strouhal number	스트로할수	-數
structural classification	구조분류	構造分類
structural icing	구조착빙	構造着氷
structure function	구조함수	構造函數
strut thermometer	버팀목온도계	-木溫度計
Student's 't' test	't' 검증	't' 檢證
sturmpause(= foehn pause)	푄멈춤	
Stüve diagram	스튜브선도	-線圖
subarcric current	아(북)극해류	亞(北)極海流
subarctic climate	아(북)극기후	亞(北)極氣候
subarctic zone	아(북)극대	亞(北)極帶
subarid	아건조	亞乾燥
subcloud	구름아래, 구름아래층	-層
subcooling(= supercooling)	과냉각	過冷却
subcritical instability	아임계불안정	亞臨界不安定
subequatorial belt	아적도대	亞赤道帶
subfield	부분장	部分場

subfrigid zone	아한대	亞寒帶
subfrontal cloud	아전선구름	亞前線-
subgeostrophic wind	아지균풍	亞地均風
subgradient wind	아경도풍	亞傾度風
subgrid-scale process	아격자규모과정	亞格子規模過程
subhumid	아습윤(의)	亞濕潤
subhumid climate	아습윤기후	亞濕潤氣候
subinversion layer	역전아래층	逆轉-層
subjective forecast	주관예보	主觀豫報
sublimation	승화	昇華
sublimation adiabat	승화단열선	昇華斷熱線
sublimation curve	승화곡선	昇華曲線
sublimation nucleus	승화핵	昇華核
submergence	관수, 침수	冠水, 沈水
subnormal refraction	반정상굴절	半正常屈折
subpolar	아한대(의)	亞寒帶
subpolar air mass	아한대기단	亞寒帶氣團
subpolar anticyclone	아한대고기압	亞寒帶高氣壓
subpolar belt	아한대	亞寒帶
subpolar cap	아한대공기덩이	亞寒帶空氣-
subpolar climate	아한대기후	亞寒帶氣候
subpolar glacier	아한대빙하	亞寒帶氷河
subpolar high	아한대고기압	亞寒帶高氣壓
subpolar low	아한대저기압	亞寒帶低氣壓
subpolar low pressure belt	아한대저압대	亞寒帶低壓帶
subpolar westerlies	아한대편서풍(대)	亞寒帶偏西風(帶)
subpolar zone	아한대	亞寒帶
subrefraction	아굴절	亞屈折
subregional broadcast	준광역방송	準廣域放送
subsatellite point	위성직하점	衛星直下點
subsatellite point track	위성직하점궤도	衛星直下點軌道
subsequence	부분수열	部分數列
subset	부분집합	部分集合
subsidence	침강	沈降
subsidence (temperature) inversion	침강(온도)역전	沈降(溫度)逆轉
subsoil ice	땅속얼음	
subsonic	아음속(의)	亞音速
substance	물질	物質
substandard propagation	아표준전파	亞標準傳播
substandard refraction	아표준굴절	亞標準屈折
substantial derivative	실질도함수	實質導函數
substantial sheet	물질박층	物質薄層
substantial surface	물질면	物質面
substitutional vegetation	대용식생	代用植生

S

substratosphere	아성층권	亞成層圈
subsurface	아지표	亞地表
subsurface drainage	지하배수	地下排水
subsurface flow	지하류, 복류	地下流, 伏流
subsurface runoff	지하유출	地下流出
subsurface water	복류수	伏流水
subsynoptic	아종관	亞綜觀
subsynoptic scale	아종관규모	亞綜觀規模
subtarget area	부목표역	副目標域
subterranean ice	땅속얼음	
subterranean water	지하수	地下水
subtraction	뺄셈	
subtropical	아열대(의)	亞熱帶
subtropical air mass	아열대기단	亞熱帶氣團
subtropical anticyclone	아열대고기압	亞熱帶高氣壓
subtropical belt	아열대	亞熱帶
subtropical calm	아열대무풍대	亞熱帶無風帶
subtropical cell	아열대세포	亞熱帶細胞
subtropical climate	아열대기후	亞熱帶氣候
subtropical cyclone	아열대저기압	亞熱帶低氣壓
subtropical depression	아열대저압부	亞熱帶低壓部
subtropical easterlies	아열대편동풍(대)	亞熱帶偏東風(帶)
subtropical easterlies index	아열대편동풍지수	亞熱帶偏東風指數
subtropical front	아열대전선	亞熱帶前線
subtropical high	아열대고기압	亞熱帶高氣壓
subtropical high-pressure belt	아열대고압대	亞熱帶高壓帶
subtropical high-pressure zone	아열대고압대	亞熱帶高壓帶
subtropical jet stream	아열대제트류	亞熱帶-流
subtropical ridge	아열대기압마루	亞熱帶氣壓-
subtropical storm	아열대폭풍우, 아열대스톰	亞熱帶暴風雨
subtropical trade-wind inversion	아열대무역풍역전	亞熱帶貿易風逆轉
subtropical westerlies	아열대편서풍(대)	亞熱帶偏西風(帶)
subtropical zone	아열대지역	亞熱帶地域
subtropics	아열대	亞熱帶
subwarning	주의보, 준경보	注意報, 準警報
succession	천이	遷移
successive correction method	천이수정법	遷移修正法
suction anemometer	흡입풍속계	吸入風速計
suction pressure	흡입압	吸入壓
suction tube	흡입관	吸入管
suction vortices	흡입소용돌이	吸入-
Sudan type	수단형(Hann의 기후 분휴형의 하나)	-型
sudden change report	돌연변화보고	突然變化報告

sudden enhancement of atmospherics	돌연공전증강	突然空電增强
sudden ionospheric disturbance	돌연전리층요란	突然電離層擾亂
sudden warming	돌연승온	突然昇溫
Suess effect	서스효과	-效果
Suez Canal	수에즈운하	-運河
sufficient condition	충분조건	充分條件
sugar berg	곰보빙산	-氷山
sugar snow	곰보눈	
sukhovei	수크호베이	
sulfate	황산염	黃酸鹽
sulfur dioxide	아황산가스	亞黃酸-
sulphuric acid	황산	黃酸
sulphuric mist	황산미스트, 황산실안개	黃酸-
sulfur isotope	황동위원소	黃同位元素
sulfur oxide	산화황	酸化黃
sulfur oxide densitometer	산화황농도계	酸化黃濃度計
sulfuric acid	황산	黃酸
sulfuric mist	황산미스트, 황산실안개	黃酸-
sulphur dioxide	아황산가스	亞黃酸-
sulphur rain	황산비	黃酸-
sultriness	무더위	
sumatra	수마트라(돌풍의 일종)	
summation principle	합산원칙	合算原則
summer	여름	
summer crop	여름작물	-作物
summer day	여름날	
summer depression	더위 시듦	
summer half year	여름반년	-半年
summer hemisphere	여름반구	-半球
summer monsoon	여름계절풍	-季節風
summer solstice	하지	夏至
sun	태양	太陽
sunbeam	일광, 태양광	日光, 太陽光
sunburn	볕탐, 볕타기	
sun-climate connection	태양-기후연관성	太陽氣候聯關性
sun cross(= cross)	십자햇무리	十字-
sundial	해시계	-時計
sun dog	해, 환일	幻日
sun drawing water	구름틈새햇살	
sun-earth geometry	태양-지구기하학	太陽地球幾何學
sun glint	태양반짝임 (위성구름사진 위의)	太陽-
sun leaf	태양잎새	太陽-
sunlight	일광	日光

S

sunlight zone	일광지대	日光地帶
sunlit aurora	일조극광	日照極光
sunny	양지바른	陽地-
sunny interval	햇빛기간	-其間
sun-photometer	썬포토미터, 태양광도계	太陽光度計
sun pillar	해기둥	
sun plant	양지식물	陽地植物
sun pointing	썬포인팅	
sun protection factor	자외선차단지수	紫外線遮斷指數
sun proton monitor	태양양자감시기	太陽陽子監視器
sunrise	일출	日出
sun's altitude	태양고도	太陽高度
sunscald	볕뎀	
sun sensor	태양수감부	太陽受感部
sunset	일몰	日沒
sunshine	일조	日照
sunshine duration	일조시간	日照時間
sunshine hour	일조시간	日照時間
sunshine integrator	일조적분계	日照積分計
sunshine record	일조기록(지)	日照記錄(紙)
sunshine recorder	일조계	日照計
sunshine time	일조시간	日照時間
sunspot	태양흑점	太陽黑點
sunspot cycle	태양흑점주기	太陽黑點週期
sunspot number	태양흑점수	太陽黑點數
sunspot periodicity	태양흑점주기	太陽黑點週期
sunspot relative number	태양흑점상대수	太陽黑點相對數
sun strobe echo	태양섬광에코	太陽閃光-
sunstroke	일사병	日射病
sun synchronous correlation	태양동기상관	太陽同期相關
sun synchronous orbit	태양동기궤도	太陽同期軌道
sun synchronous satellite	태양동기위성	太陽同期衛星
Supan's climatic zone	수판기후대	-氣候帶
superadiabat	초단열	超斷熱
superadiabatic lapse rate	초단열감률	超斷熱減率
superadiabatic state	초단열상태	超斷熱狀態
supercooled cloud	과냉각구름	過冷却-
supercell	거대세포	巨大細胞
supercell thunderstorm	거대세포뇌우	巨大細胞雷雨
supercell type	거대세포형	巨大細胞型
supercooled cloud	과냉각구름	過冷却-
supercooled drop	과냉각수적	過冷却水滴
supercooled fog	과냉각안개	過冷却-
supercooled rain	과냉각비	過冷却-

S

supercooled water	과냉각수	過冷却水
supercooled water droplet	과냉각물방울	過冷却-
supercooling	과냉각	過冷却
supercooling cloud partical	과냉각운립	過冷却雲粒
superensemble	슈퍼앙상블	
superfusion	과융해	過融解
supergeostrophic wind	초지균풍	超地均風
supergradient	초경도	超傾度
supergradient wind	초경도풍	超傾度風
superheated vapour	과열(수)증기	過熱(水)蒸氣
superheating	과열	過熱
superheterodyne	초헤테로다인	超-
superior air	상층공기	上層空氣
superior air mass	상층기단	上層氣團
superior mirage	상층신기루	上層蜃氣樓
superior planet	외행성	外行星
supernatant liquid	상청액	上淸液
supernormal refraction	초정상굴절	超定常屈折
supernumerary bow	과잉무지개	過剩-
supernumerary rainbow	과잉무지개	過剩-
superobservation	슈퍼관측	-觀測
superposition	중첩	重疊
superpressure balloon	초압기구	超壓氣球
superrefraction	초굴절	超屈折
supersaturated air	과포화공기	過飽和空氣
supersaturation	과포화	過飽和
supersaturation cloud	과포화구름	過飽和-
supersaturation with respect to ice	얼음과포화	-過飽和
supersaturation with respect to water	물과포화	-過飽和
supersonic	초음속(의)	超音速
supersonic anemometer	초음파풍속계	超音波風速計
superstandard propagation	초표준전파	超標準傳播
superstandard refraction	초표준굴절	超標準屈折
superstratosphere	초성층권	超成層圈
superturbulent flow	초난류	超亂流
supplementary feature	부속특징	附屬特徵
supplementary land station	보조지상관측소	補助地上觀測所
supplementary meteorological office	보조기상대	補助氣象臺
supplementary observation	보조관측	補助觀測
supplementary ship	보조관측선박	補助觀測船舶
supplementary ship station	보조선박관측소	補助船舶觀測所
supplementary station	보조관측소	補助觀測所
supply current	보급류	補給流
surf	쇄파	碎波

S

surface	면, 표면, 지면, 지상	面, 表面, 地面, 地上
surface air observation	지상관측	地上觀測
surface air temperature	지상기온	地上氣溫
surface albedo	표면반사율	表面反射率
surface analysis chart	지상분석도	地上分析圖
surface area	표면적	表面積
surface avalanche	표면설붕	表面雪崩
surface backscatter variable	표면후방산란변수	表面後方散亂變數
surface-based convection	지면기원대류	地面起源對流
surface boundary layer	표면(경계)층, 접지(경계)층, 지표(경계)층	表面(境界)層, 接地(境界)層, 地表(境界)層
surface charge	표면전하	表面電荷
surface chart	지상(일기)도	地上(日氣)圖
surface composition mapping radiometer	표면조성화상복사계	表面造成畵像輻射計
surface condensation	표면응결	表面凝結
surface density	지상밀도	地上密度
surface detention	지면체류수	地面滯留水
surface disturbance	지상요란	地上擾亂
surface energy	지상에너지	地上-
surface flow	지상흐름	地上-
surface forecast chart	지상예보도	地上豫報圖
surface free energy	지상자유에너지	地上自由-
surface freezing index	지상동결지수	地上凍結指數
surface friction	지상마찰	地上摩擦
surface front	지상전선	地上前線
surface hoar	지상서리	地上-
surface humidity	지상습도	地上濕度
surface integral	면적분	面積分
surface inversion	접지역전	接地逆轉
surface layer	지표층	地表層
surface map	지상일기도	地上日氣圖
surface observation	지상관측	地上觀測
surface of discontinuity	지상불연속면	地上不連續面
surface of separation	분계면	分界面
surface of subsidence	침강면	沈降面
surface ozone	지상오존	地上-
surface pressure	지상기압	地上氣壓
surface prognostic chart	지상예상도	地上豫想圖
surface radiation budget	지면복사수지	地面輻射收支
surface reflection	지표반사	地表反射
surface resistance	지표저항	地表抵抗
surface retention	지면정체수	地面停滯水
surface Rossby number	지표로스비수	地表-數
surface roughness	지면거칠기	地面-

surface runoff	표면유출	表面流出
surface spectral reflectance	표면분광반사율	表面分光反射率
surface storage	지면저류(수)	地面貯留(水)
surface streamflow	지표수흐름	地表水-
surface synoptic station	지상종관관측소	地上綜觀觀測所
surface temperature	지상기온, 지면온도	地上氣溫, 地面溫度
surface temperature calculation	지상온도계산	地上溫度計算
surface temperature distribution	지상기온분포	地上氣溫分布
surface temperature of water	수면온도	水面溫度
surface temperature test	표면온도검정	表面溫度檢定
surface tension	표면장력	表面張力
surface thawing index	지면융해지수	地面融解指數
surface thermal resistance	표면열저항	表面熱抵抗
surface thermometer	표면온도계, 지면온도계	表面溫度計, 地面溫度計
surface topography	지표지형	地表地形
surface trough	지상기압골	地上氣壓-
surface turbulence	지상난류	地上亂流
surface visibility	지상시정	地上視程
surface water	지표수	地表水
surface wave	표면파, 지상파	表面波, 地上波
surface weather chart	지상일기도	地上日氣圖
surface weather map	지상일기도	地上日氣圖
surface weather observation	지상기상관측	地上氣象觀測
surface wind	지상바람, 지표바람	地上-, 地表-
surface wind speed	지상풍속	地上風速
surface wind stress	지표바람응력	地表-應力
surf beat	쇄파맥놀이	碎波-
surge	기압파, 서지	氣壓波
surge current	서지흐름	
surge lightning	서지번개	
surge line	서지선	-線
surplus area(= accumulation area)	(빙하의)누적지역	累積地域
survey	실측	實測
suspended particulate matter	부유입자물질	浮遊粒子物質
suspended phase	현탁상, 현탁단계	懸濁相, 懸濁段階
suspension	현탁	懸濁
suspensoid	현탁질	懸濁質
sustainability science	지속가능과학	持續可能科學
sustained speed	지속속도	持續速度
sustained wind	지속풍	持續風
sustained wind speed	지속풍속	持續風速
S-value	S값	
swallow storm(= peesweep storm)	제비폭풍	-暴風
swamp	늪, 습지	濕地

S

swash	처오름파	-波
swath width	관측폭, 주사폭	觀測幅, 走査幅
S wave(= secondary wave)	S파	-波
sway	스웨이	
sweat	응결수	凝結水
SWEAT index	스웨트지수	-指數
sweep	쓸기	
sweep length	쓸기길이	
swell	너울	
swelling	증대	增大
swinging plate anemometer	진동판풍속계	振動板風速計
symbol	부호, 기호, 전조	符號, 記號, 前兆
symbol for weather analysis	일기분석기호	日氣分析記號
symbol of weather	일기기호	日氣記號
symmetrical center	대칭중심	對稱中心
symmetric component	대칭성분	對稱成分
symmetric instability	대칭불안정(도)	對稱不安定(度)
symmetric regime	대칭영역	對稱領域
symmetric tensor	대칭텐서	對稱-
symmetric transformation	대칭변환	對稱變換
symmetry point	대칭점	對稱點
sympathetic development	감응발달	感應發達
sympathetic discharge	감응방전	感應放電
sympiesometer	심피조미터	
synapsis stage	접합기	接合期
synchronization	동조화, 동시화	同調化, 同時化
synchronous change	동시변화	同時變化
synchronous meteorological satellite	정지기상위성	靜止氣象衛星
synergism	상승작용	相乘作用
synodic month	삭망월	朔望月
synodic period	회합주기	會合週期
SYNOP code	SYNOP 코드, 지상종관기상실황전문형식	地上綜觀氣象實況電文型式
SYNOP message	지상종관기상실황전문	地上綜觀氣象實況電文
synoptic	종관(의)	綜觀
synoptic aerology	종관고층기상학	綜觀高層氣象學
synoptic analysis	종관분석	綜觀分析
synoptic chart	종관(일기)도	綜觀(日氣)圖
synoptic climatology	종관기후학	綜觀氣候學
synoptic code	종관전문형식	綜觀電文型式
synoptic forecast	종관예보	綜觀豫報
synoptic forecasting	종관예보	綜觀豫報
synoptic hour	종관시간	綜觀時間
synoptic map	종관(일기)도	綜觀(日氣)圖

synoptic meteorology	종관기상학	綜觀氣象學
synoptic model	종관모형	綜觀模型
synoptic observation	종관관측	綜觀觀測
synoptic report	종관(기상)보고	綜觀(氣象)報告
synoptic scale	종관규모	綜觀規模
synoptic-scale circulation	종관규모순환	綜觀規模循環
synoptic-scale forcing	종관규모강제	綜觀規模强制
synoptic situation	종관개황	綜觀槪況
synoptic station	종관관측소	綜觀觀測所
synoptic type	종관형태	綜觀形態
synoptic wave chart	종관파랑도	綜觀波浪圖
synoptic weather chart	종관일기도	綜觀日氣圖
synoptic weather code	종관일기부호	綜觀日氣符號
synoptic weather observation system	종관기상관측시스템	綜觀氣象觀測-
synoptic weather situation	종관일기개황	綜觀日氣槪況
synthetic aperture radar	합성개구레이더	合成開口-
synthetic hydrograph	합성수문도	合成水文圖
synthetic polarimetric algorithm	합성편파알고리즘	合成偏波-
synthetic unit hydrograph	합성단위수문도	合成單位水文圖
syphon	빨대	
system	시스템, 계	系
systematic approach	체계적 접근	體系的接近
systematic error	계통오차	系統誤差
systematics	분류학	分類學
Système International d'Unités	국제단위계〈프랑스어〉	國際單位系
systemic circulation	온몸순환	-循環
system of winds	풍계	風系
syzygy	삭망	朔望

T, t

영문	한글	한자
tabetisol	부동지〈러시아어〉	不凍地
tablecloth	책상보구름	册床褓雲
table iceberg	책상빙산	册床氷山
table of square root	제곱근표	二乘根表
table of standard atmosphere	표준대기표	標準大氣表
tabular crystal	평판결정	平板結晶
tabular difference	표차	表差
tabular iceberg	평판빙산	平板氷山
tabulating machine	도표작성기기	圖表作成器機
tabulation	도표작성	圖表作成
tachometer	회전속도계	回轉速度計
t'aifung(= typhoon)	태풍	颱風
taiga climate	타이가기후	北方針葉林氣候
tailored products weather services	맞춤예보업무	-豫報業務
tail wind	뒷바람	
taino	타이노(대〈大〉안틸레스제도의 태풍)	
take-off and landing weather	이착륙기상	離着陸氣象
take-off distance	이륙활주거리	離陸滑走距離
take-off forecast	이륙예보	離陸豫報
take-off weather	이륙기상	離陸氣象
talik	융해땅	融解-
tall high	키큰 고기압	参天高氣壓
tampering	부당변경	不當變更
tangent arcs	접호	接弧
tangent arcs 22° halo	접호 22도 무리	接弧-度-
tangent arcs 46° halo	접호 46도 무리	接弧-度-
tangential acceleration	접선가속도	接線加速度
tangential component	접선성분	接線成分
tangential force	접선력	接線力

tangential stress	접선응력	接線應力
tangential velocity	접선속도	接線速度
tangential wind	접선바람	接線-
tangent line	접선	接線
tangent linear model	접선선형모형	接線線形模型
tangent (tangential) plane	접평면	接平面
tapioca snow(= snow pellet)	눈싸라기, 싸락눈	
target	목표	目標
target area	목표지역	目標地域
targeting observation	목표관측	目標觀測
target position indicator	목표위치지시기	目標位置指示器
target signal	표적신호	標的信號
target volume	표적부피	標的體積
taryn	장기육지동결〈러시아어〉	長期陸地凍結
tau-value	타우값	
Taylor column	테일러기둥	
Taylor diagram	테일러선도	-線圖
Taylor effect	테일러효과	-效果
Taylor number	테일러수	-數
Taylor-Proudman theorem	테일러-프라우드맨 정리	-定理
Taylor series	테일러급수	-級數
Taylor's theorem	테일러정리	-定理
Taylor's vortex	테일러소용돌이	
T-E index(Temperature-Efficiency Index)	T-E지수	-指數
T-E ratio	T-E비	-比
teardrop balloon	눈물방울기구	-氣球
technoclimatology	기술기후학	技術氣候學
technology-based standard	기술기반표준	技術基盤標準
telecommand	원격명령	遠隔命令
telecommunication	장거리통신, 원격통신	長距離通信, 遠隔通信
telecommunication hub	통신중추	通信中樞
telecommunication operations weather	원거리통신작전기상	遠距離通信作戰氣象
teleconnection	원격상관	遠隔相關
teleconnection index	원격상관지수	遠隔相關指數
telecontrol	원격제어	遠隔制御
telegraph exchange	전보교신	電報交信
telegraphic equation	전신방정식	電信方程式
telemeteorograph	격측자기기상계	隔測自記氣象計
telemeteorography	격측기상측기학	隔測氣象測器學
telemeteorometry	격측기상관측법	隔測氣象觀測法
telemeter	격측기	隔測器
telemetering	격측	隔測
telemetry	격측법, 격측자료	隔測法, 隔測資料
telemetry command and ranging	원격명령거리측정	遠隔命令距離測定

telephotometer	격측광도계	隔測光度計
telephotometry	원격광도측정법	遠隔光度測定法
teleprinter	전신인쇄기, 텔레프린터	電信印刷機
telepsychrometer	격측건습계	隔測乾濕計
telerecording	격측기록	隔測記錄
telescope	망원경	望遠鏡
telescoped ice(= rafted ice)	뗏목얼음	浮木氷
telescope pointing	망원조준	望遠照準
telethermometer	격측온도계	隔測溫度計
telethermoscope	격측온도측정기	隔測溫度測定器
teletype(= teleprinter)	전신타자기, 텔레타이프	電信打字機
television and infrared observing satellite (TIROS)	텔레비전적외관측위성 (타이로스)	-赤外觀測衛星
television weather broadcasting	텔레비전기상방송	-氣象放送
telex(= telegraph exchange)	텔렉스	
telluric line	대기흡수선	大氣吸收線
TEMPAL code	TEMPAL전문형식	-電文形式
temperate belt	온대	溫帶
temperate climate	온대기후	溫帶氣候
temperate climate with summer rain	온대하우기후	溫帶夏雨氣候
temperate climate with winter rain	온대동우기후	溫帶冬雨氣候
temperate deciduous forest	온대낙엽수림	溫帶落葉樹林
temperate glacier	온대빙하	溫帶氷河
temperate humid climate	온대습윤기후	溫帶濕潤氣候
temperate monsoon climate	온대계절풍기후	溫帶季節風氣候
temperate rainforest	온대우림	溫帶雨林
temperate rainy climate	온대다우기후	溫帶多雨氣候
temperate westerlies	온대편서풍	溫帶偏西風
temperate westerlies index	온대편서풍지수	溫帶偏西風指數
temperate zone	온대	溫帶
temperature	온도	溫度
temperature analysis	온도분석	溫度分析
temperature anomaly	온도이상	溫度異常
temperature average	평균온도	平均溫度
temperature belt	온도대	溫度帶
temperature coefficient	온도계수	溫度係數
temperature condition	온도조건	溫度條件
temperature conductivity	온도전도도	溫度傳導度
temperature control	온도조절, 온도제어	溫度調節, 溫度制御
temperature correction	온도보정	溫度補正
temperature departure	온도편차	溫度偏差
temperature dewpoint spread	기온이슬점차, 기온노점차	氣溫露點差
temperature difference	온도차	溫度差
temperature distribution	온도분포	溫度分布

temperature effect	온도효과	溫度效果
temperature efficiency	온도효율	溫度效率
temperature efficiency index	온도효율지수	溫度效率指數
temperature efficiency ratio	온도효율비	溫度效率比
temperature extreme	온도극값	溫度極-
temperature field	온도장	溫度場
temperature gradient	온도경도	溫度傾度
temperature humidity index	온습지수	溫濕指數
temperature humidity infrared radiometer	온습도적외복사계	溫濕度赤外輻射計
temperature inversion	기온역전	氣溫逆轉
temperature lapse rate	기온감률	氣溫減率
temperature layer	온도층, 기온층	溫度層, 氣溫層
temperature measurement	온도측정	溫度測定
temperature moisture diagram	온습선도	溫濕線圖
temperature moisture index	온습지수	溫濕指數
temperature of the soil surface	토양표면온도	土壤表面溫度
temperature pressure curve	온도기압곡선	溫度氣壓曲線
temperature province	온도영역	溫度領域
temperature range	온도교차, 기온교차	溫度較差, 氣溫較差
temperature reduction	온도보정	溫度補正
temperature rising	승온	昇溫
temperature-salinity curve	온도-염도곡선	溫度-鹽度曲線
temperature-salinity diagram	온도-염도선도	溫度-鹽度線圖
temperature scale	온도눈금	溫度-
temperature sensing element	온도수감부	溫度受感部
temperature variation	온도변동	溫度變動
temperature wave	온도파	溫度波
temperature zone	온도대	溫度帶
template	형판	型板
temporal change	시간변화, 일시변화	時間變化, 一時變化
temporales	템포랄레스	
temporal resolution	시간해상도, 일시해상도	時間解像度, 一時解像度
temporary blindness	일시시력상실	一時視力喪失
temporary threshold shift	일시역치변위, 일시문턱변위	一時閾値變位, 一時門-變位
TEMPSHIP message	해상고층기상실황전문	海上高層氣象實況電文
tendency	경향	傾向
tendency chart	경향선도	傾向線圖
tendency equation	경향방정식	傾向方程式
tendency interval	경향간격	傾向間隔
tendency method	경향법	傾向法
tendency profile	경향단면	傾向斷面
tensiometer	텐시오미터	
tension	장력	張力
tensor	텐서	張量

tensor quadric	텐서 2차곡면	張量二次曲面
tephigram	테피그램	
tephrochronology	화산재연대학, 테프라연대학	火山災年代學, -年代學
tercile	백분위점	百分位點
terminal aerodrome forecast(TAF)	비행장예보	飛行場豫報
terminal infiltration rate	말기침투율	末期浸透率
terminal point	종점	終點
terminal (fall) velocity	종단(낙하)속도	終端(落下)速度
terminator	명암경계선	明暗境界線
terrain analysis	지형분석	地形分析
terrain clearance weather	가시침투기상	可視浸透氣象
terrestrial broadcast	영역방송	領域放送
terrestrial magnetic field	지구자기장	地球磁氣場
terrestrial magnetism	지자기	地磁氣
terrestrial (surface) radiation	지구복사, 지면복사	地球輻射, 地面輻射
terrestrial radiation balance	지구복사평형	地球輻射平衡
terrestrial radiation thermometer	지구복사온도계	地球輻射溫度計
terrestrial refraction	지구굴절, 대기굴절	地球屈折, 大氣屈折
terrestrial scintillation	아지랑이	
test flight weather	시험비행기상	試驗飛行氣象
test of significance(= significance test)	유의성검정	有意性檢定
tethered balloon	계류기구	繫留氣球
tetrahedron	사면체	四面體
tetroon	사면체기구	四面體氣球
thalweg	수로중앙선	水路中央線
thaw	해빙(기)	解氷(期)
thawing index	해빙지수	解氷指數
thawing season	해빙계절	解氷季節
thaw water	해빙수	海氷水
themopile	열전퇴	熱電堆
theodolite	경위의	經緯儀
theoretical meteorology	이론기상학	理論氣象學
theory of cyclone	저기압이론	低氣壓理論
theory of relativity	상대론	相對論
theory of similarity	상사이론	相似理論
theory of the general circulation of the atmosphere	대기대순환론	大氣大循環論
thermal	열기포, 열(의)	熱氣泡, 熱
thermal advection	열이류	熱移流
thermal belt	온난대	溫暖帶
thermal capacity	열용량	熱容量
thermal circulation	열순환	熱循環
thermal climate	열기후	熱氣候
thermal conductivity	열전도성, 열전도도	熱傳導性, 熱傳導度

T

thermal constant	열상수	熱常數
thermal contrast	열대비	熱對比
thermal convection	열대류	熱對流
thermal current	열기류	熱氣流
thermal depression	열저압부	熱低壓部
thermal diffusivity	열확산율	熱擴散率
thermal distortion	열변형	熱變形
thermal efficiency	열효율	熱效率
thermal efficiency index	열효율지수	熱效率指數
thermal efficiency ratio	열효율비	熱效率比
thermal emission spectrum	열방출스펙트럼	熱放出-
thermal energy	열에너지	熱能量
thermal equator	열적도	熱赤道
thermal equilibrium	열평형	熱平衡
thermal equivalent	열당량	熱當量
thermal expansion	열팽창	熱膨脹
thermal gradient	열경도	熱傾度
thermal high	열고기압	熱高氣壓
thermal inertia	열관성	熱慣性
thermal infrared domain	열적외선 영역	熱赤外線領域
thermal instability	열불안정	熱不安定
thermal island(= heat island)	열섬	熱島
thermal jet	열제트	熱-
thermal low	열저기압	熱低氣壓
thermally driven local wind	열국지풍	熱局地風
thermal motion(= heat motion)	열운동	熱運動
thermal mountain effect	열산악효과	熱山岳效果
thermal neutrality	열중립	熱中立
thermal oxidation	열산화	熱酸化
thermal pollution	열오염	熱汚染
thermal potential	열퍼텐셜	熱-
thermal power generation	화력발전	火力發電
thermal precipitator	열집진기	熱集塵器
thermal radiation	열복사	熱輻射
thermal regime	열영역	熱領域
thermal ridge	온도마루	溫度-
thermal Rossby number	열로스비수	熱-數
thermal roughness	열거칠기	熱-
thermal slope	열경사	熱傾斜
thermal steering	열지향	熱指向
thermal stratification	열성층	熱成層
thermal tide	열조석	熱潮汐
thermal trough	온도골	溫度-
thermal turbulence	열난류	熱亂流

thermal unit	열단위	熱單位
thermal vacuum	열진공	熱眞空
thermal vorticity	열소용돌이도	熱-度
thermal vorticity advection	열소용돌이도이류	熱-度移流
thermal wind	온도풍	溫度風
thermal wind equation	온도풍방정식	溫度風方程式
thermal zone	온난지대	溫暖地帶
thermic cumulus	열적운	熱積雲
thermionic tube	열이온관	熱-管
thermistor	서미스터	
thermistor anemometer	서미스터 풍속계	-風速計
thermistor bolometer	서미스터 열복사계	-熱輻射計
thermistor thermometer	서미스터 온도계	-溫度計
thermocline	수온약층	水溫躍層
thermocouple	열전(기)쌍	熱電(氣)雙
thermocyclogenesis	열저기압발생	熱低氣壓發生
thermodynamical function	열역학함수	熱力學函數
thermodynamical potential	열역학퍼텐셜	熱力學-
thermodynamic chart	열역학선도	熱力學線圖
thermodynamic dew-point temperature	열역학이슬점온도	熱力學-點溫度
thermodynamic diagram	열역학선도	熱力學線圖
thermodynamic efficiency	열역학효율	熱力學效率
thermodynamic energy equation	열역학에너지방정식	熱力學-方程式
thermodynamic frost-point temperature	열역학서리점온도	熱力學-點溫度
thermodynamic function of state	열역학상태함수	熱力學狀態函數
thermodynamic ice-bulb temperature	열역학빙구온도	熱力學氷球溫度
thermodynamic potential	열역학퍼텐셜	熱力學-
thermodynamic process	열역학과정	熱力學過程
thermodynamic profiler	열역학탐측기, 열역학프로파일러	熱力學探測器
thermodynamics	열역학	熱力學
thermodynamic solenoid	열역학솔레노이드	熱力學-
thermodynamic stratification	열역학성층	熱力學成層
thermodynamic temperature	열역학온도(절대온도)	熱力學溫度(絶對溫度)
thermodynamic temperature scale	열역학온도눈금	熱力學溫度-
thermodynamic variable	열역학변수	熱力學變數
thermodynamic wet-bulb temperature	열역학습구온도	熱力學濕球溫度
thermoelectric actinograph	열전자기일사계	熱電自記日射計
thermoelectric actinometer	열전일사계	熱電日射計
thermoelectric pyrheliometer	열전직달일사계	熱電直達日射計
thermoelectric thermometer	열전온도계	熱電溫度計
thermo-element	열전소자	熱電素子
thermogram	자기온도기록	自記溫度記錄
thermograph	자기온도계	自記溫度計
thermograph correction card	자기온도계보정표	自記溫度計補正表

T

thermohaline circulation	열염분순환	熱鹽分循環
thermohygrogram	자기온습도기록	自記溫濕度記錄
thermohygrograph	자기온습계	自記溫濕計
thermohygrometer	온습계	溫濕計
thermo-integrator	열적산계	熱積算計
thermoisopleth	등온선	等溫線
thermo-junction	열전대	熱電對
thermo-junction thermometer	열전대온도계	熱電對溫度計
thermolysis	열분해	熱分解
thermometer	온도계	溫度計
thermometer bulb	온도계구부	溫度計球部
thermometer screen	백엽상	百葉箱
thermometer shelter	백엽상	百葉箱
thermometric conductivity	온도전도율	溫度傳導率
thermometric constant	온도상수	溫度常數
thermometric scale	온도눈금	溫度-
thermometry	온도측정법	溫度測定法
thermoneutrality	열중립	熱中立
thermopause	열권계면	熱圈界面
thermopile	열전기더미	熱電氣-
thermoscope	온도측정기	溫度測定器
thermoscreen	백엽상	百葉箱
thermosensitivity	감온성	感溫性
thermosonde	온도존데	溫度-
thermosphere	열권	熱圈
thermostat	온도조절기	溫度調節器
thermosteric anomaly	비부피편차	比-偏差
thermotropic model	향열모형	向熱模型
thetagram	세타그램	
theta (coordinate) system	세타(θ)좌표계	-(座標)系
thickness	층후	層厚
thickness advection	층후류	層厚流
thickness anomaly	층후편차	層厚偏差
thickness change	층후변화	層厚變化
thickness change advection	층후변화이류	層厚變化移流
thickness chart	층후선도	層厚線圖
thickness line	층후선	層厚線
thickness relative vorticity	층후상대소용돌이도	層厚相對-度
thickness scale	층후척도	層厚尺度
Thiessen (polygon) method	티센(다각형)법	-(多角形)法
third law of thermodynamics	열역학제3법칙	熱力學第三法則
Thornthwaite heat index	손스웨이트 열지수	-熱指數
Thornthwaite moisture index	손스웨이트 수분지수	-水分指數
Thornthwaite's classification of climate	손스웨이트 기후구분	-氣候區分

thoron	토론(radon의 방사성 동위원소)	
threat score	위협점수	威脅點數
three-cup anemometer	3배풍속계	三盃風速計
threshold illuminance	문턱조도	門-照度
threshold of audibility	문턱가청값	門-可聽-
threshold temperature	문턱온도	門-溫度
thrust-anemometer	추력풍속계	推力風速計
thrust horse power	추력마력	推力馬力
thunder	천둥	
thunder and lightning	천둥번개, 뇌전	雷電
thunderbolt	벼락, 낙뢰	落雷
thunderbolt observation system	낙뢰관측시스템	落雷觀測-
thundercloud	뇌운	雷雲
thunderhead	모루구름	
thunder shower	천둥소나기, 뇌우	雷雨
thundersquall	뇌우스콜	雷雨-
thunderstorm (rain)	뇌우	雷雨
thunderstorm cell	뇌우세포	雷雨細胞
thunderstorm charge separation	뇌우전하분리	雷雨電荷分離
thunderstorm cirrus	뇌우권운	雷雨卷雲
thunderstorm cloud	뇌우구름	雷雨-
thunderstorm cluster	뇌우클러스터, 뇌우무리	雷雨-
thunderstorm complex	뇌우복합체	雷雨複合體
thunderstorm cyclone	뇌우저기압	雷雨低氣壓
thunderstorm day	뇌우일	雷雨日
thunderstorm dynamics	뇌우역학	雷雨力學
thunderstorm electricity	뇌우전기(학)	雷雨電氣(學)
thunderstorm electrification	뇌우대전	雷雨帶電
thunderstorm environment	뇌우환경	雷雨環境
thunderstorm high	뇌우고기압	雷雨高氣壓
thunderstorm line	뇌우선	雷雨線
thunderstorm observing station	뇌우관측소	雷雨觀測所
thunderstorm outflow	뇌우유출류	雷雨流出流
thunderstorm recorder	뇌우기록계	雷雨記錄計
thunderstorm turbulence	뇌우난류	雷雨亂流
thunderstorm type	뇌우형태	雷雨形態
thunderstorm wind	뇌우바람	雷雨-
thunderstroke	뇌격	雷擊
thundery cloud system	뇌운시스템	雷雲-
thundery precipitation	뇌우강수	雷雨降水
Tibetan high (pressure)	티벳고기압	-高氣壓
tidal bore	조석해일, 고조	潮汐海溢, 高潮
tidal breeze	조석바람	潮汐-
tidal current	조석류	潮汐流

T

tidal day	조석일	潮汐日
tidal excursion	조석노정	潮汐路程
tidal power	조류발전	潮流發電
tidal prism	조석프리즘	潮汐-
tidal range	조차, 조석차	潮差, 潮汐差
tidal rip	조석격류	潮汐激流
tidal stream	조석류	潮汐流
tidal wave	조석파	潮汐波
tidal wind	조석바람	潮汐-
tide	조석	潮汐
tide amplitude	조석진폭	潮汐振幅
tide crack	조석균열	潮汐龜裂
tide gauge	검조기	檢潮器
tide-generating force	기조력	起潮力
tide level	조위	潮位
tide-predicting machine	조석예상장치	潮汐豫想裝置
tide prediction	조석예측	潮汐豫測
tide-producing potential	기조퍼텐셜	起潮-
tide table	조석표	潮汐表
tide range	조차, 조석차	潮差, 潮汐差
tilted trough	기운 (기압)골	-(氣壓)-
tilting bucket rain gauge	전복형우량계	顚覆形雨量計
tilting effect	기울기효과	-效果
tilting term	기울기항	-項
timber line	수목한계선	樹木限界線
time-area-depth curve	시간면적깊이곡선, 시면심곡선, TAD곡선	時間面積-曲線, 時面深曲線
time cross-section	시간단면도	時間斷面圖
time-dependent flow	시간종속류	時間從屬流
time development	시간발전	時間發展
time-interval radiosonde	시간간격라디오존데	時間間隔-
time lag	시간지연	時間遲延
time lapse	시간경과	時間經過
time-mean chart	기간평균도	期間平均圖
time of concentration	도달시간	到達時間
time of removal	해제시각	解除時刻
time scale	시간규모, 시간척도	時間規模, 時間尺度
time section	시간단면, 시간단면도	時間斷面圖
time series	시계열	時系列
time series analysis	시계열분석	時系列分析
tipping bucket	전도되	轉倒-
tipping bucket rain gauge	전도형우량계	轉倒形雨量計
tipping term	기울기항	-項
TIROS operational vertical sounder	타이로스연직탐측계	-鉛直探測計

tissue	조직	組織
titration	적정	滴定
titration curve	적정곡선	滴定曲線
titration error	적정오차	滴定誤差
T-number(= Tropical number)	T-수, 열대수	熱帶數
tolerance	허용오차	許容誤差
TOPEX Poseidon satellite	해양지형학실험위성	海洋地形學實驗衛星
topoclimatology	지형기후학	地形氣候學
top of cloud	구름꼭대기, 운정	雲頂
topographic frontogenesis	지형전선발생	地形前線發生
topography	지형(학), 높낮이	地形(學)
topside sounder	이온탐측기	-探測器
tornadic thunderstorm	토네이도뇌우	-雷雨
tornado	토네이도, 용오름	龍-
tornado belt	토네이도대, 용오름대	龍-帶
tornado cave	토네이도대피소	-待避所
tornado cycone	토네이도저기압	-低氣壓
tornado echo	토네이도에코	
tornado intensity category	토네이도강도분류	-強度分類
tornado outbreak	토네이도돌발	-突發
tornado vortex signature	토네이도 소용돌이 신호	-信號
toroid	토로이드, 원추곡선회전면	圓錐曲線回轉面
torque	토크	
torr	토르(수은주 1mm의 압력)	
Torricelli	토리첼리	
Torricellian vacuum	토리첼리진공	-眞空
Torricelli's tube	토리첼리관	-管
torrid zone	열대	熱帶
torsion	비틀림	
torsion hygrometer	비틀림습도계	-濕度計
total cloud cover	전운량	全雲量
total derivative	전도함수	全導函數
total differential	전미분	全微分
total eclipse	개기식	皆旣蝕
total index	전체파수, 전체지수(구면조화함수)	全體波數, 全體指數
totalizer(= totalizator)	적산계	積算計
totalizer raingauge	적산우량계	積算雨量計
totalizer snow gauge	적산설량계	積算雪量計
total liquid water path	총수액량	總水液量
total ozone	전오존량	全-量
total sky cover	전운량	全雲量
total totals index	총대류지수	總對流指數
total variance	총분산	總分散

Toussaint's formula	토우사공식	-公式
towering cumulus	탑적운	塔積雲
tower visibility	(관제)탑시정	(管制)塔視程
town fog	도시안개	都市-
Townsend support	타운센드받침대	-臺
toxic air pollutants	독성공기오염물	毒性空氣汚染物
toxicant	독성물	毒性物
toxicity	독성	毒性
trace	흔적, 기록곡선, 미량	痕迹, 記錄曲線, 微量
trace element	미량요소	微量要素
trace gas	미량기체	微量氣體
trace of precipitation	강수흔적	降水痕跡
trace of rainfall	강우흔적	降雨痕跡
tracer	추적자	追跡子
trace recorder	(강수)흔적기록계	(降水)痕跡記錄計
tracer method	추적자법	追跡子法
track	경로, 진로, 궤적	經路, 進路, 軌跡
tracking	추적	追跡
tracking system	추적장치	追跡裝置
track wind	항적바람	航跡-
traction	견인력	牽引力
trade air	무역풍공기	貿易風空氣
trade cumulus	무역풍적운	貿易風積雲
trade inversion	무역풍역전	貿易風逆轉
trade (wind)	무역풍	貿易風
trade-wind belt	무역풍대	貿易風帶
trade-wind cumulus	무역풍적운	貿易風積雲
trade-wind desert	무역풍사막	貿易風砂漠
trade-wind equatorial trough	무역풍적도기압골	貿易風赤道氣壓-
trade-wind front	무역풍전선	貿易風前線
trade-wind inversion	무역풍역전	貿易風逆轉
trade-wind zone	무역풍대	貿易風帶
trail	항적, 궤적, 미류	航跡, 軌跡, 尾流
trailing front	꼬리전선	-前線
trails of precipitation	미류구름, 강수줄기	尾流-, 降水-
trails of rain	미류구름, 강수줄기	尾流-, 降水-
training data set	연습자료집	演習資料集
trainment ratio	유인비	誘引比
trajectory	궤적, 유적	軌跡, 流跡
trajectory analysis	궤적분석, 유적분석	軌跡分析, 流跡分析
trajectory curvature	궤적곡률, 유적곡률	軌跡曲率, 流跡曲率
transcribed weather broadcast	녹음일기방송	錄音日氣放送
transducer	변환기	變換器
transfer function	전이함수	轉移函數

T

transfer of atmospheric radiation	대기방사전달	大氣輻射傳達
transfer orbit	전이궤도	轉移軌道
transformation of an air mass	기단변질	氣團變質
transform boundary	변환경계	變換境界
transformed air-mass	변질기단	變質氣團
transform fault	변환단층	變換斷層
transient climate response	일시기후반응	一時氣候反應
transient eddy	일시에디	一時-
transient variation	일시변동, 잠시변동	一時變動, 暫時變動
transient wave	일시파	一時波
transitional air-mass	전이기단	轉移氣團
transitional climate	전이기후	轉移氣候
transitional flow	전이흐름	轉移-
transitional surface	전이표면	轉移表面
transition altitude	전환고도	轉換高度
transition element	전이원소	轉移元素
transition state	전이상태	轉移狀態
transition temperature	전이온도	轉移溫度
translation	병진, (평행)이동	立進, (平行)移動
translational motion	병진운동	立進運動
translatory field	병진장	立進場
translatory velocity	병진속도	立進速度
translatory wave	병진파	立進波
translucency(= translucence)	반투명도	半透明度
translucent	반투명(의)	半透明
translucidus	반투명구름	半透明-
transmissibility	투과성	透過性
transmissiometry	투과측정술	透過測定術
transmission	전송, 투과	傳送, 透過
transmission coefficient	투과계수	透過係數
transmission curve	투과곡선	透過曲線
transmission function	투과함수	透過函數
transmission loss	투과손실	透過損失
transmission range	투과거리	透過距離
transmissivity	투과율	透過率
transmissometer	투과율계	透過率計
transmittance	투과도	透過度
transmittance meter	투과율계	透過率計
transmittancy	투과율	透過率
transmitted power	송신출력	送信出力
transmitter	송신기	送信器
transonic	전이음속(의)	轉移音速
transosonde	송신존데	送信-
transparency	투명도	透明度

T

transparent body	투명체	透明體
transpiration	증산작용	蒸散作用
transpiration coefficient	증산계수	蒸散係數
transpiration ratio	증산비	蒸散比
transplanting	이식, 이앙(벼의 경우)	移植, 移秧
transplanting period	이식기, 이앙기	移植期, 移秧期
transponder(= transmitter-responder)	송수신기	送受信器
transponder ranging	송수신폭	送受信幅
transported soil	객토, 운적토	客土, 運積土
transport wind	수송풍	輸送風
transversal wave	가로파, 횡파	橫波
transversal vibration	가로진동	-振動
transverse bands	가로밴드	
transverse cloud line	가로구름선	-線
transverse divergence	가로발산, 횡발산	橫發散
transverse dune	가로사구	-沙丘
transverse force	가로압력, 횡압력	橫壓力
transverse rolls	가로파	-波
trapezoid	사다리꼴	梯形
trapped lee wave	갇힌 산악파	-山岳波
travelling anticyclone	이동성고기압	移動性高氣壓
travelling ionospheric disturbance	이온층진행파	-層進行波
travelling wave	진행파	進行波
travelling wave tube	진행파통로	進行波通路
tree climate	수목기후	樹木氣候
tree deformation	수목변형	樹木變形
tree line(= timber line)	수목한계선	樹木限界線
tree-ring climatology	나이테기후학	-氣候學
trellis drainage	창살하계, 격자상하계	窓-河系, 格子狀河系
triangular truncation	삼각절단	三角切斷
triangulation	삼각측량	三角測量
Triassic	트라이아스기(삼첩기)(의)	-期(三疊期)
Triassic extinction	트라이아스기소산	-期消散
Triassic period	트라이아스기	-期
triboelectrification	마찰대전	摩擦帶電
trigger action	방아쇠작용, 격발작용	激痛作用
trigger effect	방아쇠효과, 격발효과	激痛效果
triggering mechanism	유발기구	誘發機構
trigonometric function	삼각함수	三角函數
trigonometric ratio	삼각비	三角比
triple Doppler	3중도플러	三重-
triple Doppler radar	3중도플러 레이더	三重-
triple point	삼중점	三重點
triple point temperature	삼중점온도	三重點溫度

T

triple register	삼중기록계	三重記錄計
triple root	삼중근	三重根
triple state	삼중상태	三重狀態
trivial solution	평범해	平凡解
tropical	열대	熱帶
tropical advisory	열대주의보	熱帶注意報
tropical air (mass)	열대기단	熱帶氣團
tropical air current	열대(기)류	熱帶(氣)流
tropical air fog	열대기단안개	熱帶氣團-
tropical anticyclone	열대고기압	熱帶高氣壓
tropical arid climate	열대건조기후	熱帶乾燥氣候
tropical belt	열대	熱帶
tropical calm zone	열대무풍대	熱帶無風帶
tropical climate	열대기후	熱帶氣候
tropical climatology	열대기후학	熱帶氣候學
tropical continental air mass	열대대륙기단	熱帶大陸氣團
tropical cyclone	열대저기압, 사이클론(인도양태풍)	熱帶低氣壓, -(印度洋颱風)
tropical cyclone position estimate	열대저기압위치 추정	熱帶低氣壓位置推定
tropical day	열대일	熱帶日
tropical depression	열대저압부	熱帶低壓部
tropical desert climate	열대사막기후	熱帶沙漠氣候
tropical discontinuity line	열대불연속선	熱帶不連續線
tropical disturbance	열대요란	熱帶擾亂
tropical dry climate	열대건조기후	熱帶乾燥氣候
tropical easterlies	열대편동풍	熱帶偏東風
tropical easterly jet	열대편동풍제트	熱帶偏東風-
tropical easterly zone	열대편동풍대	熱帶偏東風帶
tropical front	열대전선	熱帶前線
tropical maritime air mass	열대해양기단	熱帶海洋氣團
tropical meteorology	열대기상학	熱帶氣象學
tropical monsoon climate	열대계절풍기후, 열대몬순기후	熱帶季節風氣候, 熱帶-氣候
tropical night	열대야	熱帶夜
tropical number	열대수	熱帶數
tropical plume	열대플룸	熱帶-
tropical rain climate	열대다우기후	熱帶多雨氣候
tropical rainfall measuring mission	열대강우측정임무, 열대강우측정위성	熱帶降雨測定任務, 熱帶降雨測定衛星
tropical rainforest	열대우림	熱帶雨林
tropical rainforest climate	열대우림기후	熱帶雨林氣候
tropical rainy climate	열대다우기후	熱帶多雨氣候
tropical savanna climate	열대사바나기후	熱帶-氣候
tropical storm	열대폭풍(우)	熱帶暴風(雨)
tropical storm warning	열대폭풍(우)경보	熱帶暴風(雨)警報

T

tropical storm watch	열대폭풍(우)감시보	熱帶暴風(雨)監視報
tropical tropopause	열대(대류)권계면	熱帶(對流)圈界面
tropical upper-tropospheric trough(TUTT)	열대상부대류권(기압)골	熱帶上部對流圈(氣壓)-
tropical wave	열대파	熱帶波
tropical weather	열대일기, 열대기상	熱帶日氣, 熱帶氣象
tropical weather outlook	열대일기전망, 열대기상전망	熱帶日氣展望, 熱帶氣象展望
tropical wet climate	열대습윤기후	熱帶濕潤氣候
tropical zone	열대	熱帶
Tropic of Cancer	북회귀선	北回歸線
Tropic of Capricorn	남회귀선	南回歸線
tropics	열대, 회귀선	熱帶, 回歸線
tropogram	트로포그램	
tropopause	대류권계면, 권계면	對流圈界面, 圈界面
tropopause break	권계면단절	圈界面斷絶
tropopause break-line	권계면단절선	圈界面斷絶線
tropopause chart	권계면도	圈界面圖
tropopause discontinuity	권계면불연속	圈界面不連續
tropopause dome	권계면돔	圈界面-
tropopause folding	권계면접힘	圈界面-
tropopause funnel	권계면깔대기	圈界面-
tropopause high	권계면고기압	圈界面高氣壓
tropopause inversion	권계면역전	圈界面逆轉
tropopause jet (stream)	권계면제트(류)	圈界面-(流)
tropopause layer	권계면층	圈界面層
tropopause low	권계면저기압	圈界面低氣壓
tropopause wave	권계면파	圈界面波
troposcatter	대류권산란	對流圈散亂
troposphere	대류권	對流圈
tropospheric aerosol	대류권에어로졸	對流圈-
tropospheric fallout	대류권낙진	對流圈落塵
tropospheric ozone	대류권오존	對流圈-
trough	기압골, 골	氣壓-
trough aloft	상층기압골	上層氣壓-
trough line	기압골선	氣壓-線
trowal(= trowell)	상층온난혀	上層溫暖-
true azimuth	진방위	眞方位
true solar day	진태양일	眞太陽日
truncated spur	절단산각	切斷山脚
truncation error	절단오차	切斷誤差
Tse method	체방법	-方法
Tse nomogram	체노모그램	
T-sonde	T-존데	
tsunami	지진해일, 쓰나미	地震海溢
tsunami warning	지진해일경보, 쓰나미경보	地震海溢警報

T

tsunami watch	쓰나미주의보	地震海溢注意報
Tsushima current	쓰시마해류	-海流
Tsuyu	매우, 장마, 바이우	梅雨
t-test	t-검정	-檢定
tuba	깔때기구름	
tube anemometer	관풍속계	管風速計
tuff	응회석	凝灰石
tuft method	타래법	-法
tundra	툰드라, 동토지대	凍土地帶
tundra climate	툰드라기후	凍土氣候
tundra desert	툰드라사막	凍土沙漠
turbidimeter	혼탁도계	混濁度計
turbidity	혼탁도	混濁度
turbidity coefficient	혼탁계수	混濁係數
turbidity factor	혼탁인자	混濁因子
turbid medium	혼탁매질	混濁媒質
turbopause	난류권계면	亂流圈界面
turbosphere	난류권	亂流圈
turbulence	난류	亂流
turbulence body	난류체	亂流體
turbulence cloud	난류구름	亂流-
turbulence coefficient	난류계수	亂流係數
turbulence component	난류성분	亂流成分
turbulence condensation level	난류응결고도	亂流凝結高度
turbulence energy	난류에너지	亂流-
turbulence intensity	난류강도	亂流强度
turbulence inversion	난류역전	亂流逆轉
turbulence level	난류도, 난류수준	亂流度, 亂流水準
turbulence measure	난류측정	亂流測定
turbulence penetration speed	난류침투속도	亂流浸透速度
turbulence shear stress	난류층밀림응력	亂流層-應力
turbulence spectrum	난류스펙트럼	亂流-
turbulent boundary layer	난류경계층	亂流境界層
turbulent diffusion	난류확산	亂流擴散
turbulent diffusion coefficient	난류확산계수	亂流擴散係數
turbulent dissipation	난류소산	亂流消散
turbulent eddy	난류맴돌이	亂流-
turbulent energy	난류에너지	亂流-
turbulent exchange	난류교환	亂流交換
turbulent flow	난류	亂流
turbulent flux	난류속, 난류플럭스	亂流束
turbulent gust	난류돌풍	亂流突風
turbulent inversion	난류역전	亂流逆轉
turbulent medium	난류매질	亂流媒質

T

turbulent mixing	난류혼합	亂流混合
turbulent motion	난류운동	亂流運動
turbulent shear stress	난류시어응력	亂流-應力
turbulent transfer	난류전달	亂流傳達
turbulent wake	난류후류	亂流後流
turbulent zone	난류역	亂流域
turning point	전향점	轉向點
turnover frequency	반전주파수	反轉周波數
turreted cloud	탑구름	塔-
TVOC theory	총휘발성유기화합물 계산이론	總揮發性有機化合物計算理論
twilight	박명	薄明
twilight airglow	박명대기광	薄明大氣光
twilight arch	박명호	薄明弧
twilight color	박명색	薄明色
twilight correction	박명보정	薄明補正
twilight flash	박명광	薄明光
twilight phenomenon	박명현상	薄明現象
twilight spectrum	박명스펙트럼	薄明-
twilight zone	박명지역	薄明地域
twister	토네이도, 용오름, 회오리	龍-
twisting term	뒤틀림항	-項
two-dimensional stereo imager	2차원 입자영상기	二次元粒子映像機
two-dimensional turbulence	2차원 난류	二次元亂流
two-way attenuation	양방향감쇠	兩方向減衰
two-way interaction	양방향상호작용	兩方向相互作用
Tyndallometer	틴달로미터	
typhoon	태풍	颱風
typhoon advisory	태풍주의보	颱風主意報
typhoon bar	태풍구름벽	颱風-壁
typhoon bogussing	태풍보거싱	颱風-
typhoon committee	태풍위원회	颱風委員會
typhoon eye	태풍눈	颱風-
typhoon forecast	태풍예보	颱風豫報
typhoon forecast system	태풍예보시스템	颱風豫報-
typhoon information	태풍정보	颱風情報
typhoon intensity	태풍강도	颱風强度
typhoon intensity forecast	태풍강도예보	颱風强度豫報
typhoon intensity number	태풍강도지수	颱風强度指數
typhoon model	태풍모형	颱風模型
typhoon season	태풍계절	颱風季節
typhoon squall	태풍스콜	颱風-
typhoon track	태풍경로	颱風經路
typhoon track forecast	태풍경로예보	颱風經路豫報
typhoon warning	태풍경보	颱風警報

T

U, u

영문	한글	한자
U burst	U파열, 돌발, 위상신호	-破裂, 突發, 位相信號
UDT operation weather	수중파괴작전기상	水中破壞作戰氣象
U figure	U형태, U지수	-形態, -指數
U index	U지수	-指數
Ulloa's circle(→ Bouguer's halo)	울로아고리	
Ulloa's ring(→ Bouguer's halo)	울로아고리	
ultrafine particle	초미립자, 초미세먼지	超微粒子, 超微細-
ultra high frequency	초고주파수	超高周波數
ultra long wave	초장파	超長波
ultrasonic	초음파(의)	超音波
ultrasonic anemometer	초음파풍속계	超音波風速計
ultrasonic snow cover meter	초음파적설계	超音波積雪計
ultra-upper atmosphere	초고층대기	超高層大氣
ultraviolet	자외(선)	紫外(線)
ultraviolet index	자외(선)지수	紫外(線)指數
ultraviolet light	자외광	紫外光
ultraviolet pyrheliometer	자외일사계	紫外日射計
ultraviolet radiation	자외복사	紫外輻射
ultraviolet radiation index	자외복사지수	紫外輻射指數
ultraviolet ray	자외선	紫外線
ultraviolet ray hygrograph	자외선습도계	紫外線濕度計
ultraviolet region	자외역	紫外域
ultraviolet spectrum	자외스펙트럼	紫外-
ultraviolet wavelength	자외파장	紫外波長
Umkehr effect	움케르효과	-效果
uncinus	갈퀴구름	
undercooling(= supercooling)	과냉각	過冷却
undercurrent	하층류	下層流
underground ice	지하얼음, 지하빙	地下氷

underground runoff	지하유출	地下流出
underground water	지하수	地下水
undermelting	밑녹음	
underrun	밑흐름, 지하류	地下流
undershoot	고압편향	高壓偏向
undersun	아래무리해	
undertows	해양저류	海洋底流
undulation	파동	波動
undulatus	파도구름	波濤-
unfiltered model	비여과모형	非濾過模型
unfree water	갇힌 물	
unfreezing	녹음	
unifilar electrometer	단선전위계	單線電位計
uniform acceleration	등가속도	等加速度
uniform circular motion	등속원운동	等速圓運動
uniform flow	균질류, 균일흐름	均質流, 均一-
uniformitarianism	동일과정설	同一過程說
uniformly accelerated motion	등가속운동	等加速運動
unimodal distribution	단봉분포	單峰分布
unipolar electrical charge	단극전하	單極電荷
unique solution	단일해	單一解
unit cell	단위세포	單位細胞
unit control position	대수제어지점,	帶水制御地點,
	단일제어지점	單一制御地點
United Nations Conference on Environment and Development	국제연합환경개발회의	國際聯合環境開發會議
United Nations Framework Convention on Climate Change	국제연합기후변화기획회의	國際聯合氣候變化企劃會議
unit hydrograph	단위유량도	單位流量圖
unit lattice	단위격자	單位格子
unit normal distribution	단위정규분포	單位正規分布
unit volume	단위부피	單位-
universal day	세계일(0도 자오선의 날)	世界日
universal decimal classification	국제십진분류법	國際十進分類法
universal equilibrium hypothesis	보편평형가설	普遍平衡假說
universal gas constant	보편기체상수	普遍氣體常數
universal geographic code	범용지리코드	汎用地理-
universal gravitation	만유인력	萬有引力
universal gravitational constant	만유인력상수	萬有引力常數
universal raingauge	보편우량계	普遍雨量計
universal time	세계시 (그리니치 자오선의 시각)	世界時
Universal Time Coordinated(UTC)	세계협정시	世界協定時
universal transmission function	보편투과함수	普遍透過函數

U

University Corporation for Atmospheric Research (UCAR)	대기연구 대학연맹	大氣研究大學聯盟
unpredictable	비예측	非豫測
unrestricted visibility	무한시정	無限視程
unsaturated bond	불포화결합	不飽和結合
unsaturated condition	불포화상태	不飽和狀態
unsaturated hydraulic conductivity	불포화수리전도도	不飽和水理傳導度
unsaturated percolation	불포화투수	不飽和透水
unsaturated solution	불포화용액	不飽和溶液
unsaturation	불포화	不飽和
unstable	불안정	不安定
unstable air	불안정공기, 불안정기단	不安定空氣, 不安定氣團
unstable air layer	불안정공기층	不安定空氣層
unstable air mass	불안정기단	不安定氣團
unstable atmosphere	불안정대기	不安定大氣
unstable equilibrium	불안정평형	不安定平衡
unstable flow	불안정흐름	不安定-
unstable oscillation	불안정진동	不安定振動
unstable wave	불안정파	不安定波
updraft(= updraught)	상승(기)류	上昇(氣)流
updraft base	상승(기)류바탕	上昇(氣)流-
updraft strength	상승류강도	上昇流强度
updraft velocity	상승(기)류속도	上昇(氣)流速度
upglide cloud	활승운	滑昇雲
upgradient flux	역경도속, 역경도플럭스	逆傾度束
upper air	상층대기	上層大氣
upper air analysis	상층(대기)분석	上層(大氣)分析
upper air anticyclone	상층고기압	上層高氣壓
upper air chart	상층일기도	上層日氣圖
upper air circulation	상층대기순환	上層大氣循環
upper air climatology	상층(대기)기후학	上層(大氣)氣候學
upper air current	상층(기)류	上層(氣)流
upper air cyclone	상층저기압	上層低氣壓
upper air disturbance	상층요란	上層擾亂
upper air front	상층전선	上層前線
upper air observation	고층관측	高層觀測
upper air ridge	상층기압마루	上層氣壓-
upper air sounding	고층대기탐측	高層大氣探測
upper air (synoptic) station	고층대기(종관)관측소	高層大氣(綜觀)觀測所
upper air temperature	상층기온	上層氣溫
upper air trough	상층기압골	上層氣壓-
upper air weather chart	상층일기도	上層日氣圖
upper air wind	상층풍	上層風
upper atmosphere	고층대기	高層大氣

U

upper atmosphere chart	고층대기선도	高層大氣線圖
upper atmospheric boundary layer	상부대기경계층	上部大氣境界層
upper circumazenithal arc	상단천정호	上端天頂弧
upper cloud	상층운, 상층구름	上層雲
upper cold front	상층한랭전선	上層寒冷前線
upper cyclone	상층저기압	上層低氣壓
upper flight information region	상층비행정보구역	上層飛行情報區域
upper frictional region	상부마찰구역	上部摩擦區域
upper front	상층전선	上層前線
upper frontal surface	상층전선면	上層前線面
upper high	상층고기압	上層高氣壓
upper layer inversion	상층역전	上層逆轉
upper level	상층	上層
upper-level anticyclone	상층고기압	上層高氣壓
upper-level chart	상층일기도	上層日氣圖
upper-level cyclone	상층저기압	上層低氣壓
upper-level disturbance	상층요란	上層擾亂
upper-level flow	상층(기)류	上層(氣)流
upper-level front	상층전선	上層前線
upper-level high	상층고기압	上層高氣壓
upper-level low	상층저기압	上層低氣壓
upper-level map	상층일기도	上層日氣圖
upper-level ridge	상층(기압)마루	上層(氣壓)陵
upper-level system	상층계, 상층시스템	上層系
upper-level trough	상층(기압)골	上層(氣壓)谷
upper-level weather chart	상층일기도	上層日氣圖
upper-level wind	상층바람	上層-
upper limit of atmosphere	대기상한	大氣上限
upper low	상층저기압	上層低氣壓
upper mirage	상층신기루	上層蜃氣樓
upper mixing layer	상부혼합층	上部混合層
upper precipitation part	상층강수역	上層降水域
upper ridge	상층마루	上層-
upper ridge(= upper-level ridge)	상층(기압)마루	上層(氣壓)-
upper trough(= upper-level trough)	상층(기압)골	上層(氣壓)-
upper stratosphere	상부성층권	上部成層圈
upper troposphere	상부대류권	上部對流圈
upper trough(= upper-level trough)	상층골	上層-
upper warm front	상층온난전선	上層溫暖前線
upper wind	상층바람	上層-
upper wind observation	상층풍관측	上層風觀測
uprush	급상승류	急上昇流
upshear	역시어	逆-
upslide surface	활승면	滑昇面

upslope effect	활승효과	滑昇效果
upslope flow	활승류	滑昇流
upslope fog	활승안개	滑昇霧
upslope wind	활승바람	滑昇風
upstream	풍상쪽, 흘러오는쪽	風上-
upstream slope	흘러오는쪽 경사면, 풍상경사면	風上傾斜面
up-valley wind	계곡역풍	溪谷逆風
upward atmospheric radiation	상향대기복사	上向大氣輻射
upward motion	상향운동	上向運動
upward radiation	상향복사	上向輻射
upward terrestrial radiation	상향지구복사	上向地球輻射
upward total radiation	상향총복사	上向總輻射
upward vertical velocity	상승연직속도	上昇鉛直速度
upwelling	용승(류)	湧昇(流)
upwind	풍상	風上
upwind effect	풍상효과	風上效果
uranium-series dating	우라늄연대측정법	-年代測定法
urban and small stream flood advisory	도시소하천홍수주의보	都市小河川洪水注意報
urban and small stream flooding	도시소하천홍수	都市小河川洪水
urban climate	도시기후	都市氣候
urban climatology	도시기후학	都市氣候學
urban flash flood guidance	도시돌발홍수정보	都市突發洪水情報
urban flooding	도시홍수	都市洪水
urban heat island	도시열섬	都市熱島
urbanisation	도시화	都市化
urban plume	도시플룸	都市-
USDA soil classification system	미국농무부 토양분류체계	美國農務部土壤分類體系
U.S. Standard Atmosphere	미국표준대기	美國標準大氣
UV-A	자외선-A	紫外線-
UV-B	자외선-B	紫外線-
UV biometer	자외선측정계	紫外線測定計
UV-C	자외선-C	紫外線-
UV dosimeter	자외선측정계	紫外線測定計
UV photometry	자외선광도측정법	紫外線光度測定法

U

V, v

영문	한글	한자
vacuum	진공	眞空
vacuum correction	진공보정	眞空補正
vacuum discharge	진공방전	眞空放電
vacuum ga(u)ge	진공계	眞空計
vacuum-tube electrometer	진공관전위계	眞空管電位計
vacuum ultraviolet ray	진공자외선	眞空紫外線
vadose zone	통기대	通氣帶
VAD wind profile	연직바람단면, VAD바람단면, 연직바람프로파일	鉛直-斷面
valid time	유효시, 유효시간	有效時, 有效時間
valley breeze	골바람, 계곡풍	溪谷風
valley breeze circulation	골바람순환, 계곡풍순환	溪谷風循環
valley fog	골안개, 계곡안개	溪谷-
valley glacier	골빙하, 계곡빙하	溪谷氷河
valley wind	골바람, 계곡풍	溪谷風
Vallot heliothermometer	발롯일사온도계	-日射溫度計
Van Allen (radiation) belt	반알렌(복사)대	-(輻射)帶
Van der Waals' equation	반데르발스방정식	-方程式
vane	풍향계	風向計
vane anemometer	풍향풍속계	風向風速計
van't Hoff's factor	반트호프인자	-因子
van't Hoff's law	반트호프법칙	-法則
vapo(u)r	수증기, 증기	水蒸氣, 蒸氣
vapo(u)r concentration	(수)증기농도	(水)蒸氣濃度
vapo(u)r density	(수)증기밀도	(水)蒸氣密度
vaporization	기화	氣化
vaporizing ice	기화얼음	氣化-
vapo(u)r line	(수)증기선	(水)蒸氣線
vapo(u)r pressure	(수)증기압	(水)蒸氣壓

vapo(u)r pressure curve	(수)증기압곡선	(水)蒸氣壓曲線
vapo(u)r pressure deficit	(수)증기압편차, 포차	(水)蒸氣壓偏差, 飽差
vapo(u)r pressure lowering	(수)증기압 강하	(水)蒸氣壓降下
vapo(u)r tension	(수)증기장력	(水)蒸氣張力
variable of state	상태변수	狀態變數
variance	분산	分散
variance analysis	분산분석	分散分析
variance ratio	분산비	分散比
variational assimilation	변분동화	變分同化
variational objective analysis	변분객관분석	變分客觀分析
variational quality control	변분품질관리	變分品質管理
variation curve	변동곡선	變動曲線
variograph	자기미변계	自記微變計
variometer	미변계	微變計
varve	계절적 퇴적	季節的堆積
varved sediment	빙호퇴적물	氷縞堆積物
V-band	V-띠	
vectopluviometer(= vector gauge)	방향별우량계	方向別雨量計
vector amount	벡터량	-量
vector component	벡터성분	-成分
vector gauge	벡터우량계	-雨量計
vector mean chart	벡터평균도	-平均圖
vector potential	벡터퍼텐셜	
vector rain gauge	벡터우량계	-雨量計
vector space	벡터공간	-空間
veering	순전(풍향)	順轉(風向)
veering wind	순전바람	順轉-
vegetation	식생	植生
vegetation biomass application	식생 바이오매스 응용	植生-應用
vegetation canopy	식생수관	植生樹冠
vegetation monitoring	식생모니터링, 식생감시	植生監視
vegetation period	식생기간	植生期間
vegetation reflectance	식생반사율	植生反射率
vegetative growth	식생생장, 영양생장	植生生長, 營養生長
vegetative period	생육기	生育期
vegetative stage	영양생장기	營養生長期
velium	면사포구름	面紗布-
velocity aliasing	속도접힘	速度-
velocity azimuth display	속도방위표시, 시선속도-방위각 표출	速度方位表示, 視線速度方位角表出
velocity cross section	유속단면	流速斷面
velocity defect law	속도손실법칙	速度損失法則
velocity distribution	속도분포	速度分布
velocity divergence	속도발산	速度發散

velocity field	속도장	速度場
velocity folding	속도접힘	速度-
velocity governor	속도조절기	速度調節器
velocity indication coherent integrator	속도지시간섭적산계	速度指示干涉積算計
velocity of light	광속(도)	光速(度)
velocity potential	속도퍼텐셜	速度-
velocity pressure	속도압, 풍압	速度壓, 風壓
velocity profile	속도연직분포	速度鉛直分布
velocity-radius vortex	속도반지름소용돌이	速度-
velocity selector	속도선택기	速度選擇器
velocity unfolding	속도 펼침	速度-
velopause	무속도면	無速度面
velum	면사포구름	面紗布-
ventifact	풍식돌	風蝕-
ventilated psychrometer	통풍건습계	通風乾濕計
ventilated thermometer	통풍온도계	通風溫度計
ventilation	환기, 통풍	換氣, 通風
ventilation index	환기지수, 통풍지수	換氣指數, 通風指數
venturi effect	벤투리효과	-效果
venturi scrubber	벤투리스크러버	
venturi tube	벤투리관	-管
verdant zone(= frostless zone)	무서리지대	-地帶
Verderman method	버더만법	-法
verglas	베르글라(바위 표면을 얇게 덮은 얼음)	
verification	검증	檢證
verification sample	검증표본	檢證標本
vernal equinox	춘분	春分
vernier	부척	副尺
vernier scale	부척눈금	副尺-
vertebratus	늑골구름	肋骨-
vertical advection	연직이류	鉛直移流
vertical airflow	연직류	鉛直流
vertical air motion	연직공기운동	鉛直空氣運動
vertical anemometer	연직풍속계	鉛直風速計
vertical anemoscope	연직풍향계	鉛直風向計
vertical angle	앙각	仰角
vertical beam radar	연직빔레이더, 연직빛살레이더	鉛直-
vertical circle	연직원	鉛直圓
vertical circulation	연직순환	鉛直循環
vertical component	연직성분	鉛直成分
vertical convection	연직대류	鉛直對流
vertical coordinate	연직좌표	鉛直座標
vertical cross section	연직단면(도)	鉛直斷面(圖)
vertical current recorder	연직류기록계	鉛直流記錄計

V

vertical differential chart	연직차이도	鉛直差異圖
vertical distribution	연직분포	鉛直分布
vertical flow	연직류	鉛直流
vertical gustiness	연직돌풍(성)	鉛直突風(性)
vertical instability	연직불안정(도)	鉛直不安定(度)
vertical integrated liquid water content	연직적분액체수량	鉛直積分液體水量
vertical jet	연직제트	鉛直-
vertical layer	연직층	鉛直層
vertical leaved canopy	수직엽형군락, 연직엽형수관	垂直葉形群落, 鉛直葉形樹冠
vertically integrated liquid	연직적분액체수량, 연직적산수액	鉛直積分液體水量, 鉛直積算水液
vertical mixing	연직혼합	鉛直混合
vertical motion	연직운동	鉛直運動
vertical motion equation	연직운동방정식	鉛直運動方程式
vertical pointing radar	연직지향레이더	鉛直指向-
vertical polarization	연직편광	鉛直偏光
vertical pressure gradient	연직기압경도	鉛直氣壓傾度
vertical profile	연직프로파일, 연직단면	鉛直斷面
vertical p-velocity	연직기압속도	鉛直氣壓速度
vertical rain radar	연직강우레이더	鉛直降雨-
vertical ray	연직광선, 연직복사선	鉛直光線, 鉛直輻射線
vertical resolution	연직해상도	鉛直解像度
vertical scale	연직규모	鉛直規模
vertical section	연직단면(도)	鉛直斷面(圖)
vertical shear	연직시어	鉛直-
vertical sounding	연직측심, 연직탐측	鉛直測深, 鉛直探測
vertical speed indicator	연직속력지시기	鉛直速力指示器
vertical stability	연직안정(도)	鉛直安定(度)
vertical stretching	연직늘림	鉛直-
vertical temperature gradient	연직온도경도	鉛直溫度傾度
vertical temperature profile	연직온도분포, 연직온도단면	鉛直溫度分布, 鉛直溫度斷面
vertical temperature profile radiometer	연직온도분포복사계	鉛直溫度分布輻射計
vertical velocity	연직속도	鉛直速度
vertical velocity chart	연직속도(분포)도	鉛直速度(分布)圖
vertical velocity ditribution	연직속도분포	鉛直速度分布
vertical velocity prognostic chart	연직속도예상도	鉛直速度豫想圖
vertical visibility	연직시정	鉛直視程
vertical vorticity	연직소용돌이도	鉛直-度
vertical wind profile	연직풍속분포, 연직바람단면, 연직바람프로파일	鉛直風速分布, 鉛直-斷面
vertical wind shear	연직바람시어	鉛直-
vertical wind velocity	연직바람속도	鉛直-速度
very high frequency	초고주파	超高周波
very high resolution radiometer	초고분해복사계	超高分解輻射計

very low frequency	초저주파	超低周波
very-short-range forecast	초단기예보	超短期豫報
VETH chart	VETH 선도	-線圖
VFR flight weather information system	시계비행기상정보시스템	視界飛行氣象情報-
VFR terminal minimum	시계비행공항최저값	視界飛行空港最低-
VFR weather	시계비행기상	視界飛行氣象
vibrational motion	진동운동	振動運動
vibrational transition	진동전이	振動轉移
video frequency(= vision frequency)	비디오주파수	-周波數
video gain	비디오증폭, 영상증폭	映像增幅
video integrator and processor	비디오통합처리기	-統合處理器
video integrator processor level	비디오통합처리수준	-統合處理水準
video mapping	영상묘화	映像描畵
video signal	영상신호	映像信號
vidicon	비디콘	
vidicon camera system	비디콘카메라시스템	
viento roterio	비엔토로테리오	
viewing angle	시야각	視野角
viewing angle error	시야각오류	視野角誤謬
village forecast	동네예보	鄕村豫報
Vince's phenomenon	빈스현상(북아프리카 사막의 빛현상의 일종)	-現象
VIP level	VIP수준, 비디오통합처리수준	-水準, 統合處理水準
virga	꼬리구름	
Virginia high	버지니아 고기압	-高氣壓
virtual displacement	가상변위	假想變位
virtual gravity	가중력, 가상중력, 겉보기중력	假重力, 假想重力
virtual height	가고도	假高度
virtual impactor	가상충격기, 가상임팩터	假想衝擊器
virtual potential temperature	가온위	假溫位
virtual pressure	가압력	假壓力
virtual stress	가응력	假應力
virtual temperature	가온도	假溫度
virtual work	겉보기일	
viscosity	점성, 점도	粘性, 粘度
viscosity coefficient	점성계수	粘性係數
viscosity resistance	점성저항	粘性抵抗
viscosity term	점성항	粘性項
viscous body	점성체	粘性體
viscous dissipation	점성소산	粘性消散
viscous drag	점성항력	粘性抗力
viscous flow	점성류	粘性流
viscous fluid	점성유체	粘性流體
viscous force	점성력	粘性力

V

viscous stress	점성응력	粘性應力
visibility	시정	視程
visibility index	시정지수	視程指數
visibility marker	시정목표물	視程目標物
visibility meter	시정계	視程計
visibility object	시정목표물	視程目標物
visibility ratio(= luminosity)	가시율, 가시도	可視率, 可視度
visibility sensor	시정감지기	視程感知器
visible	가시(의)	可視
visible and infrared spin scan radiometer	가시적외회전주사복사계	可視赤外回轉走査輻射計
visible band	가시대역	可視帶域
visible flight weather condition	시계비행기상조건	視界飛行氣象條件
visible horizon	가시지평(선)	可視地平(線)
visible image	가시영상, 가시이미지	可視映像
visible imagery	가시영상	可視映像
visible infrared spin scan radiometer atmospheric sounder	가시적외회전주사사대기 탐측기	可視赤外回轉走査輻射大氣探測器
visible light	가시광(선)	可視光線
visible (light) radiation	가시(광)복사	可視(光)輻射
visible ray	가시광(선)	可視光線
visible region	가시역	可視域
visible satellite imagery	가시역 위성영상	可視域衛星映像
visible spectrum	시정분광, 가시역 분광	視程分光, 可視域分光
visible wavelength	가시파장	可視波長
visual analysis element	시각분석요소	視角分析要素
visual and instrument weather condition	시계계기기상조건	視界計器氣象條件
visual angle	시각	視角
visual/aural range	가시가청거리	可視可聽距離
visual contact height	시계접촉높이	視界接觸-
visual field	가시구역, 보임구역	可視區域
visual flight rule	시계비행규칙	視界飛行規則
visual meteorological condition	시계(비행)기상조건	視界(飛行)氣象條件
visual observation	목측	目測
visual photometer	맨눈광도계	-光度計
visual photometry	맨눈측광법	-測光法
visual range	가시거리	可視距離
visual range formula	가시거리공식	可視距離公式
visual spectrum	가시스펙트럼	可視-
visual storm signal	폭풍표시신호	暴風表示信號
visual storm warning	폭풍경보표지	暴風警報標識
vital temperature	생존온도	生存溫度
VLF emission(= very low frequency emission)	초저주파방출	超低周波放出
voice frequency	음성주파수	音聲周波數
volatile organic compound	휘발성 유기화합물	揮發性有機化合物

V

volatility	휘발성	揮發性
volatilization	휘발	揮發
volcanic activity	화산활동	火山活動
volcanic aerosol	화산에어로졸	火山-
volcanic ash	화산재	火山灰
volcanic ash cloud	화산재구름	火山灰-
volcanic bomb	화산탄	火山彈
volcanic cinder	화산분석	火山噴石
volcanic dust	화산먼지	火山-
volcanic eruption cloud	화산폭발구름, 화산분출구름, 분화운	火山爆發-, 火山噴出-, 噴火雲
volcanic eruption	화산폭발, 화산분출	火山爆發, 火山噴出
volcanic gas	화산가스	火山-
volcanic island chain	화산섬고리	火山-
volcanic lightning	화산번개	火山-
volcanic sand	화산모래	火山-
volcanic storm	화산폭풍	火山暴風
volcanic thunder	화산천둥	火山天動
volcanic thunderstorm	화산뇌우	火山雷雨
volcanic wind	화산바람	火山-
volcanism	화산작용	火山作用
volcano	화산	火山
volcanogenic massive sulfide deposit	화산대량 황화합물침적	火山大量黃化合物沈積
VOLMET broadcast	비행 중 항공기 기상정보 방송, VOLMET방송	飛行中航空機氣象情報放送
voltage	전압	電壓
voltage regulator	전압조정기	電壓調整器
voltmeter	전압계	電壓計
volume absorption coefficient	부피흡수계수	-吸收係數
volume expansion	부피팽창, 체적팽창	體積膨脹
volume median diameter	부피중앙지름	-中央-
volume scan	볼륨스캔, 부피주사	-走査
volume scattering function	부피산란함수	-散亂函數
volumetric density	체적밀도	體積密度
volumetric rain rate	용적강우율	容積降雨率
volumetric water content	용적수분함량	容積水分含量
volume velocity processing	부피속도처리, 용적속도처리	-速度處理, 容積速度處理
voluntary observing ship	관측지원선박	觀測支援船舶
von Kármán's constant	폰카르만상수	-常數
von Kármán's law	폰카르만법칙	-法則
V point(= radiant point)	V점, 방사점	放射點
vortex	소용돌이, 와	
vortex cloud street	소용돌이구름길	
vortex filament	소용돌이필라멘트	

V

vortex line	소용돌이선	-線
vortex motion	소용돌이운동	-運動
vortex rain	소용돌이비	
vortex ring	소용돌이고리	
vortex sheet	소용돌이판(줄)	-板
vortex stretching	소용돌이늘임	
vortex theorem	소용돌이정리, 와정리	渦定理
vortex thermometer	소용돌이온도계	-溫度計
vortex trail(= vortex street)	소용돌이꼬리	
vortex train	소용돌이줄	
vortex tube	소용돌이관	-管
vortex wake turbulence	소용돌이후류난류	-後流亂流
vorticity	소용돌이도, 와도	渦度
vorticity advection	소용돌이도이류, 와도이류	渦度移流
vorticity advection method	소용돌이도이류법, 와도이류법	渦度移流法
vorticity effect	소용돌이도효과, 와도효과	渦度效果
vorticity equation	소용돌이도방정식, 와도방정식	渦度方程式
vorticity equation of quasi-geostrophic wind	준지균풍와도방정식, 준지균풍소용돌이도방정식	準地均風渦度方程式
vorticity flux	소용돌이도플럭스, 소용돌이도속, 와도플럭스, 와도속	渦度束
vorticity maximum	소용돌이도최대, 와도최대	渦度最大
vorticity transfer	소용돌이도전이, 와도전이	渦度轉移
vorticity transport hypothesis	소용돌이도수송가설, 와도수송가설	渦度輸送假說
vorticity transport theory	소용돌이도수송이론, 와도수송이론	渦度輸送理論
Voss polariscope	보스편광기	-偏光器
V-R vortex	속도반지름소용돌이	速度-
V-shaped depression	V모양저압부	-模樣低壓部
V-shaped isobar	V모양등압선	-模樣等壓線

V

W, w

영문	한글	한자
wake	후류, 꼬리흐름(정지유체 속을 운동하는 물체를 뒤따르는 흐름)	後流
wake capture	후류포착	後流捕捉
wake depression	후류저기압	後流低氣壓
wake turbulence	후류난류, 항적난류	後流亂流, 航跡亂流
Walker circulation	워커순환	-循環
Wallace-Kousky wave	월리스-코스키파	-波
wall cloud	벽구름, 벽운	壁雲
warm advection	온난이류	溫暖移流
warm air advection	난기이류	暖氣移流
warm air drop(→ warm pool)	난기웅덩이	暖氣-
warm air mass	온난기단	溫暖氣團
warm anticyclone	온난고기압	溫暖高氣壓
warm conveyor belt	온난컨베이어벨트	溫暖-
warm-core anticyclone	온난핵고기압	溫暖核高氣壓
warm-core cyclone	온난핵저기압	溫暖核低氣壓
warm-core high	온난핵고기압	溫暖核高氣壓
warm-core low	온난핵저기압	溫暖核低氣壓
warm current	온난류, 난류	溫暖流, 暖流
warm cyclone	온난저기압	溫暖低氣壓
warm downslope wind	온난하강풍	溫暖下降風
warm fog	온난안개	溫暖-
warm front	온난전선	溫暖前線
warm frontal surface	온난전(선)면	溫暖前(線)面
warm front occlusion	온난전선폐색	溫暖前線閉塞
warm front precipitation	온난전선강수	溫暖前線降水
warm front rain	온난전선비	溫暖前線雨
warm front surface	온난전선면	溫暖前線面
warm front thunderstorm	온난전선뇌우	溫暖前線雷雨

warm-front-type occlusion	온난전선형폐색	溫暖前線型閉塞
warm-front wave	온난전선파(동)	溫暖前線波(動)
warm high(= warm anticyclone)	온난고기압	溫暖高氣壓
warming	온난화	溫暖化
warm low(= warm cyclone)	온난저기압	溫暖低氣壓
warm occluded front	온난폐색전선	溫暖閉塞前線
warm occlusion	온난폐색	溫暖閉塞
warm pool	온난웅덩이	溫暖-
warm rain	온난비	溫暖-
warm sector	온난역	溫暖域
warm tongue	온난혀	溫暖-
warm trough	온난(기압)골	溫暖(氣壓)-
warm-type occluded front	온난형 폐색전선	溫暖型閉塞前線
warm-type occlusion	온난형 폐색	溫暖型閉塞
warm wave	온난파(동)	溫暖波(動)
warning	경보	警報
warning area	(기상)경보구역	(氣象)警報區域
warning stage	경보단계	警報段階
warning for ground-loosening	지면약화경보, 지면연화경보	地面弱化警報, 地面軟化警報
warping	뒤틀림, 개흙(거르기)	歪曲
washout	세척제거	洗滌除去
watch	감시, 주의보	監視, 注意報
watch area	감시구역, 주의보구역	監視區域, 注意報區域
water balance	수평	水平
water-borne disease	수인성전염병	水因性傳染病
water budget	물수지	-收支
water circulation coefficient	물순환계수	-循環係數
water cloud	물구름	
water constant	물상수, 수분상수	水分常數
water consumption	물소비량	-消費量
water content	함수량	含水量
water content of cloud	구름수분량	-水分量
water cycle	물순환	-循環
water divide(= water parting, watershed)	분수계	分水界
water drop	물방울	
water droplet	물방울	
water dropper	물방울 채집기	-採集器
water dropping collector	물방울 채집기	-採集器
water equivalent	물당량	-當量
water equivalent of snow	눈 물당량	-當量
water equivalent of snow cover	적설물당량, 적설상당수량	積雪-當量, 積雪相當水量
water erosion	물침식, 수식	-浸蝕, 水蝕
waterfall	폭포	瀑布
waterfall effect	폭포효과	瀑布效果

W

water-flow pyrheliometer	물일사계	-日射計
water fog	물안개	
water gap	물틈, 수극	水隙
water gas	수성기체	水性氣體
water hemisphere	수반구, 해양반구	水半球, 海洋半球
water level	수위	水位
water loss	물손실	-損失
water mass	수단	水團
water molecule	물분자	-分子
water optical properties	물광학특성	-光學特性
water phase	물상, 수상	-相
water pollution	물오염, 수질오염	水質汚染
water potential of soil	토양수분퍼텐셜	土壤水分-
water power	수력	水力
water recycling	물재활용	-再活用
water requirement	물소요량	-所要量
water resource	수자원	水資源
water retentivity	보수성	保水性
water saturation	수면포화	水面飽和
water-saving technology	절수기술	節水技術
water scrubber	워터 스크러버	
watershed	분수령, 집수역	分水嶺, 集水域
water sky	물하늘	
water smoke(= steam fog)	물연기, 김안개	-煙氣
water snow	습설	濕雪
water-solubility	수용성	水溶性
water spectral reflection	물분광반사	-分光反射
waterspout	용오름	龍-
water spray seeding	물(살포)씨뿌리기	-(撒布)-
water-stage recorder	수위기록계	水位記錄計
water supply outlook	물공급전망	-供給展望
water table	지하수위	地下水位
water temperature	수온	水溫
water thermometer	수온계	水溫計
watertight stratum	불투수층	不透水層
water turbidity	물혼탁도	-混濁度
water type	물유형	-類型
water use efficiency	물사용효율	-使用效率
water-use ratio(= transpiration ratio)	증산율	蒸散率
water vapo(u)r	수증기, 증기	水蒸氣, 蒸氣
water vapo(u)r absorption	수증기흡수	水蒸氣吸收
water vapo(u)r band	수증기띠	水蒸氣-
water vapo(u)r content	수증기함량	水蒸氣含量
water vapo(u)r density	수증기밀도	水蒸氣密度

W

water vapo(u)r image	수증기영상	水蒸氣映像
water vapo(u)r pressure	수증기압	水蒸氣壓
water vapo(u)r radiation	수증기복사	水蒸氣輻射
water vapo(u)r spectroscope	수증기분광광도계	水蒸氣分光光度計
water warming channel	온수로	溫水路
water warming pond	온수못, 온수지	溫水池
water wave	물결	
wave	파, 파동	波, 波動
wave age	파동나이	波動年齡
wave amplitude	파진폭	波振幅
wave base	파바닥	波-
wave breaking	쇄파	碎波
wave celerity	파민첩성, 파속도	波敏捷性, 波速度
wave cloud	파도구름	波濤-
wave crest	파봉, 파꼭대기	波峰
wave cycle	물결주기, 파주기	波週期
wave cyclone	파동저기압	波動低氣壓
wave depression	파동저기압	波動低氣壓
wave direction	물결방향, 파향	-方向, 波向
wave disturbance	파동요란	波動搖亂
wave dynamometer	파력계	波力計
wave equation	파동방정식	波動方程式
wave forecasting	파랑예보	波浪豫報
waveform	파형	波形
wave frequency	파동주파수	波動周波數
wave front	파동전선	波動前線
wave front method	파동전선법	波動前線法
wave gauge	파고계	波高計
waveguide	도파관	導波管
wave height	파고	波高
wave height meter	파고계	波高計
wavelength	파장	波長
wavelet	파엽, 파, 작은 파, 잔물결	波葉, 波
wavelet analysis	파엽분석	小波分析
wave function	파동함수	波動函數
wave mechanics	파동역학	波動力學
wave model	파동모형	波動模型
wave motion	파동	波動
wave number	파수	波數
wave of condensation and rarefaction	소밀파	疏密波
wave of oscillation	진동파	振動波
wave of translation	병진파	竝進波
wave optics	파동광학	波動光學
wave packet	파속, 파다발	波束

wave period	파주기	波周期
wave pole	파고측정기둥	波高測定棒
wave power	파력	波力
wave propagation	파동전파	波動傳播
wave recorder	파기록계	波記錄計
wave regime	파동양식	波動樣式
wave solution	파동해	波動解
wave source	파원	波源
wave spectrum	파스펙트럼	波分光
wave speed	파속력	波速力
wave staff(= wave pole)	파고측정기둥	波高測定棒
wave steepness	파동가파르기, 파동경사	波動傾斜
wave system	파동체계	波動体系
wave theory	파동이론	波動理論
wave theory of cyclogenesis	저기압발생파동설	低氣壓發生波動說
wave train	파동열	波動列
wave trough	파동골	波谷
wave velocity	파속도	波速度
weak echo region	약에코역, 약반향역	弱反響域
weak interaction	약한 상호작용	弱-相互作用
weak-line approximation	약한 선근사	弱-線近似
weapon to ground weather	공대지(사격)기상	空對地(射擊)氣象
weather ability	내후성	耐候性
weather advisory	기상주의보	氣象注意報
weather analysis	일기분석, 날씨분석	日氣分析
weather analysis data	일기분석자료	日氣分析資料
weather and climate impact appraisal	기상·기후 영향 평가	氣象氣候影響評價
weather anomaly	일기이상, 날씨이상	日氣異常
weather appraisal	기상감정	氣象鑑定
weather appraisal business	기상감정업	氣象鑑定業
weather appraiser	기상감정사	氣象鑑定士
weather briefing	기상브리핑	氣象報告
weather broadcast	기상방송	氣象放送
weather bureau	기상국	氣象局
weather business agent	기상사업자	氣象事業者
weather certification	기상증명	氣象證明
weather chart	일기도, 기상도	日氣圖, 氣象圖
weather chart analysis	일기도분석	日氣圖分析
weather cock	바람개비	
weather code	기상전문형식	氣象電文形式
weather communication	기상통신	氣象通信
weather condition	기상(조건)	氣象(條件)
weather consultant	기상상담사	氣象相談士
weather consulting business	기상상담업	氣象相談業

W

weather contingency insurance	날씨사고보험, 기상사고보험	氣象事故保險
weather control	일기조절, 날씨조절	日氣調節
weather cycle	날씨순환, 일기순환	日氣循環
weather data processing	기상자료처리	氣象資料處理
weather decision-analytic model	날씨결정분석모형	-決定分析模型
weather depiction chart	일기표시도	日氣標示圖
weather derivative	날씨파생상품	-派生商品
weather diagnosis	일기진단	日氣診斷
weather divide	일기경계	日氣境界
weather effect	날씨효과, 일기효과	日氣效果
weather element	일기요소, 기상요소	日氣要素, 氣象要素
weather equipment business	기상장비업	氣象裝備業
weather evaluation process	일기평가과정	日氣評價過程
weather fax	일기모사전송, 일기팩스	日氣模寫電送
weather finance and insurance	기상금융보험	氣象金融保險
weather flight	기상관측비행	氣象觀測飛行
weather forecast	일기예보	日氣豫報
weather forecaster	일기예보자, 일기예보관	日氣豫報者, 日氣豫報官
weather forecasting	일기예보	日氣豫報
weather gauge(= barometer)	청우계, 기압계	晴雨計, 氣壓計
weather glass	청우계	晴雨計
weather goods	날씨상품	-商品
weather group	기상전대	氣象戰隊
weather hazard	위험일기, 위험기상, 악기상	危險日氣, 危險氣象, 惡氣象
weather impact variables	날씨영향변수, 기상영향변수	氣象影響變數
weather index	기상지수	氣象指數
weather industry	기상산업	氣象産業
weather information	기상정보	氣象情報
weather-information-sensitive	기상정보민감도	氣象情報敏感度
weathering	풍화	風化
weather instrument	기상측기	氣象測器
weather insurance solution	기상보험솔루션	氣象保險-
weather intelligence	기상정찰	氣象偵察
weather litigation	날씨소송, 기상소송	氣象訴訟
weather lore	일기속담	日氣俗談
weather maneuver space	기상기동공간	氣象機動空間
weather map	일기도	日氣圖
weather-map scale	일기도규모, 종관규모, 일기도축척	日氣圖規模, 綜觀規模, 日氣圖縮尺
weather map type	일기도형	日氣圖型
weather market	기상시장	氣象市場
weather maxim	일기속담	日氣俗談
weather message	기상전문	氣象電文
weather minimum	최저기상조건	最低氣象條件

W

weather modification	기상개조, 기상조절	氣象改造, 氣象調節
weather modification law	기상조절법	氣象調節法
weather observation	기상관측	氣象觀測
weather observation plane	기상관측비행기	氣象觀測飛行機
weather observer	기상관측자	氣象觀測者
weather outlook	기상전망	氣象展望
weather patrol ship	기상정찰선	氣象偵察船
weather phenomena	기상현상	氣象現像
weather platoon	기상소대	氣象小隊
weather prediction	일기예측, 기상예측, 날씨예측	日氣豫測, 氣象豫測
weather prediction business	기상예보업	氣象豫報業
weather predictor	기상예보사	氣象豫報士
weather prognosis	날씨예상	-豫想
weather prospect	기상전망	氣象展望
weather proverb	날씨속담	-俗談
weather radar	기상레이더	氣象-
weather radar data processor and analyzer	기상레이더자료처리분석기	氣象-資料處理分析器
weather recall	기상소환	氣象召還
weather reconnaissance (flight)	기상정찰(비행)	氣象偵察(飛行)
weather recurrence	날씨반복	-反復
weather report	기상통보	氣象通報
weather routing	항로기상정보이용	航路氣象情報利用
weather satellite	기상위성	氣象衛星
weather-sensitive business	날씨민감사업	-敏感事業
weather-sensitive decision	날씨민감도결정	-敏感度決定
weather-sensitive user	날씨민감소비자, 날씨민감사용자	-敏感消費者, -敏感使用者
weather service	기상업무	氣象業務
weather service agent	기상서비스대리인	氣象業務代理人
weather ship	기상관측선	氣象觀測船
weather sign	일기기호	日氣記號
weather situation	날씨개황	-槪況
weather symbol	일기기호	日氣記號
weather squad	기상분대	氣象分隊
weather squadron	기상대대	氣象大隊
weather staff officer	기상참모	氣象參謀
weather station	기상관측소, 기상대	氣象觀測所, 氣象臺
weather surveillance radar	기상감시레이더	氣象監視-
weather system	일기계, 일기시스템, 날씨시스템	日氣系
weather technology	기상기술	氣象技術
weather tide	기상조석	氣象潮汐
weather type	날씨형	-型
weather value of information	기상정보가치	氣象情報價値
weather vane	풍향계	風向計
weather warfare	날씨전쟁	-戰爭

W

weather warning	기상경보	氣象警報
weather warning bulletin	기상경보발표문	氣象警報發表文
weather watch	기상감시, 기상주의보	氣象監視, 氣象注意報
Weber-Fechner law	웨버-페히너 법칙	-法則
Weber number	웨버수	-數
Weber's law	웨버법칙	-法則
wedge	쐐기	
wedge high	쐐기고기압, 설상고기압	舌狀高氣壓
wedge isobar	쐐기등압선	-等壓線
wedge line	쐐기선	-線
wedge tornado	쐐기토네이도	
Wegener-Bergeron process	베게너-베르게론 과정	-過程
Wegener-Bergeron-Findeisen theory	베게너-베르게론-핀다이젠 이론	-理論
Weger aspirator	베거통풍기, 베거통풍통	-通風器, -通風桶
weighing gauge	저울계수기, 저울강수계	-計數器, -降水計
weighing rain-gauge	저울우량계	-雨量計
weighing snow-gauge	무게식적설계, 저울설량계	-積雪計, -雪量計
weighing-type precipitation gauge	저울형 강수계	-型降水計
weight barograph	중력자기기압계	重力自記氣壓計
weight barometer	중력기압계	重力氣壓計
weight chart	(기주)중량도	(氣柱)重量圖
weighted average	가중평균	加重平均
weighted mean	가중평균	加重平均
weighted mean velocity	하중평균속도, 가중평균속도	荷重平均速度, 加重平均速度
weighting factor	가중인자	加重因子
weighting function	가중함수	加重函數
weight measuring gauge	무게식우량계	-雨量計
weight measuring snow-recorder	무게식자기설량계	-自記雪量計
weir	보, 둑	洑
Werenskiold diagram	웨렌스키올드선도	-線圖
West African disturbance line	서아프리카 요란대	西-擾亂帶
west coast climate	서안기후	西岸氣候
west coast subtropical	서안아열대	西岸亞熱帶
west coast subtropical desert	서안아열대사막	西岸亞熱帶沙漠
west-east transport	동서수송	東西輸送
westerlies	편서풍	偏西風
westerly belt	편서풍대	偏西風帶
westerly rain belt	편서풍다우대	偏西風多雨帶
westerly trough	편서풍(기압)골	偏西風(氣壓)-
westerly wave	편서풍파	偏西風波
westerly wind burst	(편)서풍돌진	(偏)西風突進
westerly zone	편서풍대	偏西風帶
Western North Pacific monsoon	서북태평양계절풍, 서북태평양몬순	西北太平洋季節風

west wind drift	서풍해류	西風海流
wet adiabat	습윤단열선	濕潤斷熱線
wet adiabatic	습윤단열(의)	濕潤斷熱
wet-adiabatic change	습윤단열변화	濕潤斷熱變化
wet-adiabatic cooling	습윤단열냉각	濕潤斷熱冷却
wet-adiabatic lapse rate	습윤단열감률	濕潤斷熱減率
wet-adiabatic process	습윤단열과정	濕潤斷熱過程
wet-adiabatic rate	습윤단열률	濕潤斷熱率
wet-adiabatic temperature difference	습윤단열온도차	濕潤斷熱溫度差
wet air	습윤공기	濕潤空氣
wet-and-dry-bulb hygrometer	건습구습도계, 건습계	乾濕球濕度計, 乾濕計
wet bulb	습구	濕球
wet-bulb depression	건습구온도차	乾濕球溫度差
wet-bulb effect	습구효과	濕球效果
wet-bulb globe thermometer	습구온도계	濕球溫度計
wet-bulb potential temperature	습구온위	濕球溫位
wet-bulb pseudo-potential temperature	위습구온위	僞濕球溫位
wet-bulb temperature	습구온도	濕球溫度
wet-bulb thermometer	습구온도계	濕球溫度計
wet-bulb zero	습구영점, 습구빙점	濕球零點, 濕球氷點
wet-bulb zero height	습구영점고도, 습구빙점고도	濕球零點高度, 濕球氷點高度
wet climate	습윤기후	濕潤氣候
wet day	습윤일	濕潤日
wet deposition	습성침착	濕性沈着
wet downburst	습윤하강돌풍	濕潤下降突風
wet electrostatic dust precipitator	습식전기집진장치	濕式電氣集塵裝置
wet endurance	내습성	耐濕性
wet flood proofing	습윤홍수방지	濕潤洪水防止
wet fog	습윤안개	濕潤-
wet growth	습윤생장	濕潤生長
wet injury	습해	濕害
wet instability	습윤불안정(도)	濕潤不安定(度)
wet katathermometer	습윤카타온도계	濕潤-溫度計
wetland	습지	濕地
wet number	습윤수	濕潤數
wet period	습윤기간, 우기	濕潤期間, 雨期
wet scavenging(= wet deposition)	습식 포집	濕式捕集
wet scrubber	습식 스크러버	濕式-
wet season	습윤계절	濕潤季節
wet snow	습한 눈	濕-
wet spell	습윤지속기간	濕潤持續期間
Wetterlage	기상상태, 기상개황〈독일어〉	氣象狀態, 氣象槪況
wetting front	침윤전선, 습윤전선	浸潤前線, 濕潤前線
wet tongue	습설	濕舌

W

wet tropics	습윤열대	濕潤熱帶
wet unstable	습윤불안정	濕潤不安定
whaleback cloud	고래등구름	
whale-tail-type pressure pattern	고래꼬리형 기압분포	-型氣壓分布
wheel barometer	바퀴기압계	-氣壓計
whip-poor-will storm	개구리폭풍	-暴風
whirling psychrometer	휘돌이건습계	-乾濕計
whirling thermometer	휘돌이온도계	-溫度計
whirl wind(= whirls)	선풍, 회오리바람	旋風
whistling meteor	휘파람유성	-流星
whistling wind	휘파람바람	
white band	흰 얼음층	-層
white body	백체	白體
whitecap	백파	白波
white dew	언 이슬	
white frost	흰 서리	
white head	백수	白穗
white horizontal circle	무리해테	
white ice	흰 얼음	
white light(WL)	백색광	白色光
white light flare	백색광 불꽃	白色光-
white noise	백색잡음	白色雜音
whiteout	화이트아웃	
white rainbow	흰 무지개, 흰 파도	
white squall	흰 스콜	
white wave	백파	白波
whiting stability index	화이팅안정(도)지수	-安定(度)指數
whole gale	노대바람(풍력계급 10)	
whole gale warning	노대바람경보	-警報
Wien' displacement law	비인변위법칙	-變位法則
Wien' distribution law	비인분포법칙	-分布法則
Wien' law	비인법칙	-法則
Wigand visibility meter	위갠드시정계	-視程計
wild fence	와일드바람막이(우량계의)	
wildfire	산불	山-
Wild's evaporimeter	와일드증발계	-蒸發計
wild snow	거친 눈	
willy-willy	윌리-윌리('오스트레일리아 부근의 열대저기압(태풍)' 으로 잘못 알려진 용어)	
Wilson cloud-chamber	윌슨구름상자	-箱子
Wilson cycle	윌슨주기	-周期
wilting	시듦	
wilting coefficient	시듦계수	-係數
wilting point	시듦점	-點

W

wind	바람	
wind action	풍작용, 바람작용	風作用
wind advisory	바람주의보	-注意報
wind aloft	상층바람	上層-
wind analysis	바람분석	-分析
wind and drift chart	바람편류도	-偏流圖
wind and flood damage	풍수해	風水害
wind and temperature chart	바람기온(분포)도	-氣溫(分布)圖
wind angle	풍향각	風向角
wind arrow	바람화살표	-表
wind avalanche	바람눈사태	-沙汰
wind axis	바람축	-軸
wind backing	풍향반전	風向反轉
wind barb	바람깃	
wind belt	바람띠	
windbreak	바람막이	防風
windbreak forest	방풍림	防風林
windburn	바람그을림	
wind chart	바람(분포)도	-(分布)圖
wind-chill	바람냉각	-冷却
wind-chill advisory	바람냉각주의보	-冷却注意報
wind-chill factor	바람냉각인자	-冷却因子
wind-chill index	바람냉각지수	-冷却指數
wind-chill warning	바람냉각경보	-冷却警報
wind cloud	바람구름, 파도구름	波濤-
wind column	바람기둥	
wind component	바람성분	-成分
wind cone	바람자루	
wind corrosion	바람침식	-浸蝕
wind countess	백작부인바람	伯爵夫人-
wind couplet	바람연결	-連結
wind crust	바람크러스트	
wind current	기류	氣流
wind damage	바람피해, 풍해	-被害, 風害
wind deflection	바람쏠림	
wind direction	풍향	風向
wind-direction shaft	풍향살(기호)	風向-(記號)
wind direction indicator	풍향지시기	風向指示器
wind direction recorder	풍향기록계	風向記錄計
wind distribution	바람분포	-分布
wind divide	풍계	風界
wind drift	취송류, 바람표류	吹送流, -漂流
wind drift distance	취송거리, 바람표류거리	吹送距離, -漂流距離
wind-driven circulation	풍성순환, 바람몰이순환	風成循環

W

wind-driven current	취송류, 바람몰이(해)류	吹送流, -(海)流
wind-driven oceanic circulation	취송해양순환, 바람몰이해양순환	吹送海洋循環
wind duration	취송기간, 바람기간	吹送期間
wind eddy	바람소용돌이, 바람맴돌이	
wind energy	바람에너지	
wind erosion	바람침식	-浸蝕
W-index	바람지수	-指數
wind factor	바람인자	-因子
wind field	바람장	-場
wind flow	기류, 바람흐름	氣流
wind flurry	일진광풍	一陣狂風
wind force	풍력	風力
wind force scale	풍력계급	風力階級
wind gap	바람틈새	
wind generator	풍력발전기	風力發電機
wind gradient	바람경도	-傾度
wind gust	돌풍	突風
wind layer	바람층	-層
wind load	풍하중	風荷重
wind lull	바람휴식	-休息
windmill anemometer	풍차풍속계	風車風速計
wind mixing	바람혼합	-混合
window channel test	대기창역시험	大氣窓域試驗
window frost	성에, 창서리	窓-
window function	창함수	窓函數
window hoar	성에	窓白霜
window ice	창얼음	窓-
windowing	성에, 창얼음	窓-
window region	창(영)역	窓(領)域
window thermometer	창온도계	窓溫度計
wind passage	바람길	
wind power	풍력	風力
wind power station	풍력발전소	風力發電所
wind pressure	풍압	風壓
wind profile	바람프로파일, 바람연직분포	-鉛直分布
wind profiler	연직측풍장비	鉛直測風裝備
wind protection	방풍, 바람막이	防風
wind radius	바람반경	-半徑
wind recorder	바람기록계	-記錄計
wind resistance of soil	토양내풍식성	土壤耐風蝕性
wind resource map	풍력자원(지)도	風力資源(地)圖
wind reversal	바람역전	-逆轉
wind ridge	바람마루, 바람이랑	

wind ripple	바람무늬	
wind rose	바람장미, 풍배도	-薔薇, 風配圖
wind run	풍정	風程
winds-aloft observation	상층바람관측	上層-觀測
winds-aloft plotting board	상층바람기록판	上層-記錄板
winds-aloft report	상층바람보고	上層-報告
winds- and temperatures- aloft forecast	상층바람기온예보	上層-氣溫豫報
wind scale	풍력계급	風力階級
wind scoop	바람주걱	
wind sea	풍랑	風浪
wind shadow	바람그늘	
wind shaft	바람깃대	
wind shear	윈드시어	
wind shear alert	윈드시어경계, 윈드시어경보	-警戒, -警報
wind shear profile	윈드시어연직분포	-鉛直分布
wind shear warning	윈드시어경보	-警報
wind shield	바람막이	
wind shift	바람급변, 풍향급변	風向急變
wind shift line	바람급변선, 풍향급변선	風向急變線
wind slab	바람슬랩	
wind sleeve	바람자루	
wind sock	바람자루	
wind speed	풍속	風速
wind speed counter	풍속계수기	風速計數器
wind-spun vortex	바람회전소용돌이	-回轉-
wind squall	바람스콜	
wind storm	폭풍, 바람폭풍(우)	暴風, -暴風(雨)
wind stratum	바람층	-層
wind stress	바람응력	-應力
wind stress curl	바람응력컬	-應力-
wind structure	바람구조	-構造
wind swath	바람자췻길	
wind system	풍계	風系
wind tee	T꼴바람개비	
wind tunnel	풍동	風洞
wind tunnel velocity	풍동풍속	風洞風速
wind turbine	풍차	風車
wind vane	풍향계	風向計
wind vane and anemometer	풍향풍속계	風向風速計
wind vector	바람벡터	
wind veering	바람순전, 풍향순전	風向順轉
wind velocity	풍속	風速
wind velocity counter	풍속계수기	風速計數器
wind velocity equation	풍속방정식	風速方程式

W

wind velocity indicator	풍속지시기	風速指示器
wind velocity profile	풍속연직분포	風速鉛直分布
windward	풍상(의)	風上
wind wave	풍랑, 풍파	風浪, 風波
wind wave advisory	풍랑주의보	風浪注意報
windy	바람 부는	
windy cirrus(= cirrus ventosus)	바람권운	-卷雲
wind zone	바람구역, 풍역	風域
wing-tip contrail	날개끝비행운	-飛行雲
winter	겨울	冬
winter crop	겨울작물	-作物
winter hemisphere	겨울반구	-半球
winter ice	동빙	冬氷
winterization	겨울준비	-準備
winter monsoon	겨울계절풍, 겨울몬순	-季節風
winter season	동계	冬季
winter solstice	동지	冬至
winter storm	겨울폭풍(우), 겨울스톰	-暴風(雨)
winter storm warning	겨울폭풍(우)경보	-暴風(雨)警報
winter storm watch	겨울폭풍(우)감시	-暴風(雨)監視
winter weather advisory	겨울날씨주의보	-注意報
wintry blast	겨울폭풍	-暴風
wiresonde	계류존데	繫留-
wire-weight gauge	유선저울우량계	有線-雨量計
Wisconsinian (Weichselian, Wurm) glaciation	위스콘신빙하작용	-氷河作用
Wolf number	볼프수(상대태양흑점수)	-數
Wolf-Wolfer number	볼프-볼퍼수	-數
woolpack cloud	뭉게구름	
work function	일함수	-函數
Workman-Reynolds effect	워크먼-레이놀즈 효과	-效果
World Area Forecast Centre	세계지역예보센터	世界地域豫報-
World Area Forecast System	세계공역예보시스템	世界空域豫報-
world climate	세계기후	世界氣候
world climate programme	세계기후프로그램	世界氣候-
World Health Organization	세계보건기구	世界保健機構
World Meteorological Congress(WMC)	세계기상대회	世界氣象大會
world meteorological day	세계기상의 날	世界氣象-
World Meteorological Organization(WMO)	세계기상기구	世界氣象機構
World Radiation Center(WRC)	세계복사센터	世界輻射-
World Radiometric Reference(WRR)	세계복사분석기준	世界輻射分析基準
World Weather Watch(WWW)	세계기상감시	世界氣象監視
World Weather Watch Program	세계기상감시프로그램, 세계기상감시계획	世界氣象監視計劃

X, x

영문	한글	한자
X-band	X띠	
xenon	크세논	
xerochore	사막지역	砂漠地域
xerophyte	건생식물	乾生植物
xerothermal index	건열지수	乾熱指數
X-Interdiction weather	공중비상대기항공차단기상	空中非常待機航空遮斷氣象
x-ray	X선	-線
x-ray burst	엑스선분리, 엑스선폭발	-線分離, -線爆發
x-ray diffraction	엑스선회절	-廻折

Y, y

영문	한글	한자
Yagi antenna	야기안테나	
Yanai-Maruyama wave	야나이-마루야마 파	-波
Yangtze-River air mass	양쯔강기단	-江氣團
Yangtze-River high	양쯔강고기압	-江高氣壓
yellow coloring of leaves	황엽	黃葉
yellow ripe stage	황숙기	黃熟期
yellow sand	황사	黃砂
yellow sand advisory	황사주의보	黃砂注意報
yellow sand warning	황사경보	黃砂警報
yellow snow	노랑눈	
yellow wind	황사바람	黃砂-
yield component	수확량구성요소, 수확성분	收穫量構成要素, 收穫成分
yield forecast test	작황예보시험	作況豫報試驗
yield point	산출점, 항복점	産出點, 降伏點
yield prediction	수확량예측	收穫量豫測
young ice	새얼음	
younger Dryas	신드리아스기	新-期

Z, z

영문	한글	한자
zastrugi(= sastrugi)	사스트루기	
zenith	천정	天頂
zenith angle	천정각	天頂角
zenith distance	천정거리	天頂距離
zephyr	산들바람	微風
Zephyros	제피로스(서풍의 그리스 옛말)	
zero absolute vorticity	영절대소용돌이도, 영절대와도	零絶對渦度
zero ceiling	영실링	零-
zero curtain	영도층(토양)	零度層(土壤)
zero datum	영점기준	零點基準
zero gravity	무중력	無重力
zero-isodop	영시선속도선	零視線速度線
zero isotherm	영(도)등온선	零(度)等溫線
zero layer(= null layer)	영층	零層
zero-order discontinuity	영차불연속	零次不連續
zero-plane displacement	영면변위	零面變位
zeroth law of thermodynamics	열역학제0법칙	熱力學第零法則
zero visibility	영시정	零視程
zeta potential	제타퍼텐셜	
zodiac	황도	黃道
zodiacal band	황도대	黃道帶
zodiacal cone	황도원뿔	黃道圓-
zodiacal counterglow	황도대일조	黃道對日照
zodiacal light	황도광	黃道光
zodiacal pyramid	황도피라미드	黃道-
zonal airflow pattern	동서기류형태, 동서기류패턴	東西氣流形態
zonal available potential energy	동서유효위치에너지	東西有效位置-
zonal circulation	동서순환, 띠순환	東西循環
zonal flow	동서(기)류, 띠흐름	東西(氣)流

zonal flow pattern	동서(기)류패턴, 띠기류패턴,	東西(氣)流-, -氣流形態
	띠기류형태	
zonal frontal surface	띠전선면	-前線面
zonal high	띠고기압	-高氣壓
zonal index	동서지수, 띠지수	東西指數
zonality	띠특성, 띠분포	-特性, -分布
zonal kinetic energy	동서운동에너지	東西運動-
zonal motion	동서운동	東西運動
zonal precipitation pattern	동서강수패턴, 띠강수패턴	東西降水-
zonal wave number	동서파수	東西波數
zonal westerlies	띠편서풍	-偏西風
zonal wind	동서풍, 띠바람	東西風
zonal wind profile	동서풍분포	東西風分布
zonal wind-speed profile	동서풍속단면분포,	東西風速斷面分布
	동서풍속프로파일	
zone of ablation	융삭영역, 융삭대	融削領域, 融削帶
zone of abnormal audibility	이상청역	異常聽域
zone of accumulation	집적구역	集積區域
zone of aeration	통기구역	通氣區域
zone of audibility	가청구역	可聽區域
zone of constant temperature	등온구역	等溫區域
zone of discontinuity	불연속구역	不連續區域
zone of evaporative regulation	증발조정역	蒸發調整域
zone of fracture	파쇄구역	破碎區域
zone of leaching	침출구역	浸出區域
zone of maximum precipitation	최다강수역	最多降水域
zone of saturation	포화구역	飽和區域
zone of silence	무성역	無聲域
zone of trade wind	무역풍대	貿易風帶
zone of transition	전이역	轉移域
zone of variable wind	가변바람구역, 바람변동구역	可變-區域, -變動區域
zoophenology	동물계절학	動物季節學
Z-R relationship(radar reflectivity(Z)-rain	Z-R관계(레이더반사도인자(Z)와 -關係	
rates(R) relationship)	강수강도(R)의 관계)	
Z-system	Z시스템	
Zulu(Z) Time	Zulu타임(0도 자오선시간)	-(零度子午線時間)

Z

한영편

ㄱ

한글	영문	한자
가간섭성목표물	coherent target	可干涉性目標物
가강수량	precipitable water	可降水量
가강수량선도	precipitable water diagram	可降水量線圖
가강수수분	precipitable water depth	可降水水分
가강수수증기	precipitable water vapour	可降水水蒸氣
가고도	virtual height	假高度
가공날씨자료	processed weather data	加工-資料
가까운 중심점	pericenter	近心點
가끔(때때로)소나기	isolated shower	
가능	potential	可能
가능오차	possible error	可能誤差
가능일조율	percentage of possible sunshine	可能日照率
가능증발량	potential evaporation	可能蒸發量
가능증발산(량)	potential evapotranspiration	可能蒸發散(量)
가능증산(량)	potential transpiration	可能蒸散(量)
가능최대강수량	probable maximum precipitation	可能最大降水量
가능최대폭풍	probable maximum storm	可能最大暴風
가라앉음	sinking	
가랑비	slight rain	
가랑이번개	forked lightning	
가래톳 흑사병	bubonic plague	-黑死病
가렛-뭉크 모형	Garrett-Munk model	-模型
가려막기층	shielding layer	-層
가로구름선	transverse cloud line	-線
가로발산	transverse divergence	-發散
가로밴드	transverse bands	
가로사구	transverse dune	-沙丘
가로압력	transverse force	-壓力
가로좌표	abscissa	-座標

가로진동	transversal vibration	-振動
가로파	transverse rolls, transversal wave	-波
가로회전축	roll axis	-回轉軸
가루눈	powdery snow, powder snow	粉雪
가루아	garua, camanchaca	
가르니에기후분류	Garnier's climate classification	-氣候分類
가뭄	drought	
가뭄감시	drought monitoring	-監視
가뭄빈도	drought frequency	-頻度
가뭄스트레스	drought stress	
가뭄저항성	drought resistance	-抵抗性
가뭄주기	period of drought	-週期
가뭄지속기간	drought duration	-持續期間
가뭄지수	drought index	-指數
가뭄피해	drought damage	-被害
가벼운 눈발	snow flurries	
가벼운 이온	light ion	
가변격자	adaptive grid	可變格子
가변바람구역	zone of variable wind	可變-區域
가변수	dummy variable	假變數
가상변위	virtual displacement	假想變位
가상변위원리	principle of virtual displacement	假想變位原理
가상일원리	principle of virtual work	假想日原理
가상임팩터	virtual impactor	假想-
가상전구기후	hypothetical global climate	假想全球氣候
가상중력	virtual gravity	假想重力
가상충격기	virtual impactor	假想衝擊器
가색법	additive color	加色法
가설	hypothesis	假說
가소성	plasticity	可塑性
가속기	accelerator	加速器
가속기류	accelerated airflow	加速氣流
가속도	acceleration	加速度
가속도계	accelerometer	加速度計
가속도 길이규모	acceleration length scale	加速度-規模
가속도법칙	law of acceleration	加速度法則
가속운동	accelerated motion	加速運動
가속침식	accelerated erosion	加速浸蝕
가수분해	hydrolysis	加水分解
가스크로마토그래프	gas chromatograph	
가스하이드레이트	clathrate hydrate	
가습기발열	humidifier fever	加濕器發熱
가시(의)	visible	可視
가시가청거리	visual/aural range	可視可聽距離

가시거리	visual range	可視距離
가시거리공식	visual range formula	可視距離公式
가시광(선)	visible light, visible ray	可視光(線)
가시구역	visual field	可視區域
가시대역	visible band	可視帶域
가시도	visibility ratio, luminosity	可視度
가시(광)복사	visible (light) radiation	可視(光)輻射
가시스펙트럼	visual spectrum	可視-
가시역	visible region	可視域
가시역 분광	visible spectrum	可視域分光
가시역 위성영상	visible satellite imagery	可視域衛星映像
가시영상	visible image, visible imagery	可視映像
가시영상의 강조	highlight on visible imagery	可視影像强調
가시율	visibility ratio, luminosity	可視率
가시이미지	visible image	可視-
가시적외회전주사복사계	visible and infrared spin scan radiometer	可視赤外回轉走査輻射計
가시적외회전주사대기 탐측기	visible infrared spin scan radiometer atmospheric sounder	可視赤外回轉走査輻射大氣 探測器
가시지평(선)	visible horizon	可視地平(線)
가시침투기상	terrain clearance weather	可視浸透氣象
가시파장	visible wavelength	可視波長
가압력	virtual pressure	假壓力
가압층	confining bed	加壓層
가역과정	reversible process	可逆過程
가역기계	reversible machine	可逆機械
가역기관	reversible engine	可逆機關
가역반응	reversible reaction	可逆反應
가역변화	reversible change	可逆變化
가역성	reversibility	可逆性
가역순환	reversible cycle	可逆循環
가역적	reversible	可逆的
가역현상	reversible phenomenon	可逆現象
가연원소	combustible element	可燃元素
가온도	virtual temperature	假溫度
가온위	virtual potential temperature	假溫位
가용수두	available head	可用水頭
가용수분	available water	可用水分
가용에너지	available energy	可用-
가용위치에너지	available potential energy	可用位置-
가용저장용량	available storage capacity	可用貯藏容量
가용태양복사	available solar radiation	可用太陽輻射
가용토양수분	available moisture of the soil, available soil moisture	可用土壤水分
가우스	gauss	

가우스격자	Gaussian grid	-格子
가우스곡선	Gauss curve, Gaussian curve	-曲線
가우스과정	Gaussian process	-過程
가우스난류	Gaussian turbulence	-亂流
가우스모형	Gaussian model	-模型
가우스발산정리	Gauss's divergence theorem	-發散定理
가우스방정식	Gaussian equation	-方程式
가우스분산모형	Gaussian dispersion model	-分散模型
가우스분포	Gaussian distribution	-分布
가우스위도	Gaussian latitude	-緯度
가우스잡음	Gaussian noise	-雜音
가우스정리	Gauss's theorem	-定理
가우스퍼프모형	Gaussian puff model	-模型
가우스퍼프확산	Gaussian puff diffusion	-擴散
가우스플룸방정식	Gaussian plume equation	-方程式
가우스플룸분산	Gaussian plume dispersion	-分散
가우스해	Gaussian year	
가우스형태	Gaussian shape	-形態
가우스확률함수	Gaussian probability function	-確率函數
가우스확산모형	Gaussian diffusion model	-擴散模型
가을	fall	
가을더위	after summer	
가을얼음	autumn ice	
가을장마	autumn changma	
가응력	virtual stress	假應力
가이거계수기	Geiger counter	-計數器
가이아가설	Gaia hypothesis	-假說
가장자리얼음	border ice	
가조시간	possible duration of sunshine	可照時間
가조율	percentage of possible sunshine	可照率
가중력	virtual gravity	假重力
가중인자	weighting factor	加重因子
가중평균	weighted average, weighted mean	加重平均
가중평균속도	weighted mean velocity	加重平均速度
가중함수	weighting function	加重函數
가짜전선	fictitious front	-前線
가짜태양	fictitious sun	-太陽
가청구역	zone of audibility	可聽區域
가청도	audibility	可聽度
가청역	audibility zone	可聽域
가청주파수변조라디오존데	audio-modulated radiosonde	可聽周波數變調-
가청진동수역	audio frequency band	可聽振動數域
가청한계	limit of audibility	可聽限界
가파르기	steepness	

가항반원	navigable semicircle	可航半圓
각가속도	angular acceleration	角加速度
각도너비	angular width	角度-
각면적	angular area	角面積
각변위	angular displacement	角變位
각분해능	angular resolution	角分解能
각속도	angular velocity	角速度
각운동량	angular momentum	角運動量
각운동량균형	angular momentum balance	角運動量均衡
각운동량보존	conservation of angular momentum	角運動量保存
각운동량보존식	conservation equation of angular momentum	角運動量保存式
각운동량불변성	constancy of angular momentum	角運動量不變性
각진동수	angular frequency	角振動數
각초	arcsecond	角秒
각퍼짐	angular spreading	角-
각퍼짐인자	angular spreading factor	角-因子
각폭	angular width	角暴
각표류	angular drift	角漂流
간격	interval	間隔
간디스형	Ganges type	-型
간빙기	interglacial period, interglacial phase	間氷期
간빙기기후	interglacial climate	間氷期氣候
간섭	interference	干涉
간섭계	interference instrument, interferometer	干涉計
간섭광거리탐지기	coherent light detection and ranging	干涉光距離探知器
간섭광학레이더	coherent optical radar	干涉光學-
간섭기억필터	coherent memory filter	干涉記憶-
간섭무늬	interference fringe	干涉-
간섭발진기	coherent oscillator	干涉發振器
간섭복사	coherent radiation	干涉輻射
간섭색	interference colour	干涉色
간섭성	coherence	干涉性
간섭성레이더	coherent radar	干涉性-
간섭성산란	coherent scattering	干涉性散亂
간섭성오차 적분회로	coherent error integrator	干涉性誤差積分回路
간섭송신기	coherent transmitter	干涉送信機
간섭수신기	coherent receiver	干涉受信機
간섭신호	coherent signal	干涉信號
간섭역	interference region	干涉域
간섭잎	interference lobe	干涉-
간섭적분	coherent integration	干涉積分
간섭후방산란 레이더	coherent backscatter radar	干涉後方散亂-
간수	bittern	-水
간우기(의)	interpluvial	間雨期

간이낙진예측기상	simplified fallout prediction weather	簡易落塵豫測氣象
간이습도계	hygroscope	簡易濕度計
간이야장	pocket register	簡易野帳
간이통풍건습계	pocket aspiration psychrometer	簡易通風乾濕計
간이풍속계	pocket anemometer	簡易風速計
간접고층기상학	indirect aerology	間接高層氣象學
간접고층분석	indirect aerological analysis	間接高層分析
간접세포	indirect cell	間接細胞
간접순환	indirect circulation	間接循環
간접영향	indirect hit (or effect)	間接影響
간접오염원	indirect source	間接汚染源
간조	low tide	干潮
간헐적 시료채취	intermittent sampling	間歇的試料採取
간헐적인	periodical	間歇的
간헐천	intermittent stream	間歇川
갇힌 물	unfree water	
갇힌 산악파	trapped lee wave	-山岳波
갈고리계기	hook gauge	-計器
갈고리구름유형	hook cloud pattern	-類型
갈고리에코	hook echo	
갈고리에코 레이더 신호	hook echo radar signature	-信號
갈라진 틈새	crack	
갈변	browning	褐變
갈비권운	cirrus vertebratus	-卷雲
갈색구름	brown cloud	褐色-
갈색눈	brown snow	褐色-
갈색스모그	brown smog	褐色-
갈색연무	brown cloud	褐色煙霧
갈색탄소에어로졸	brown carbon aerosol	褐色炭素-
갈수기	droughty season	渴水期
갈수천	oued	渴水川
갈이철	ploughing season	
갈퀴구름	uncinus	
갈퀴권운	cirrus uncinus, cirrus uncinus cloud	-卷雲
감광도계	sensitometer	感光度計
감광도측정법	sensitometry	感光度測定法
감광성	photosensitivity	感光性
감극	depolarization	減極
감률	lapse rate	減率
감률선	lapse line	減率線
감마복사	gamma radiation	-輻射
감마선	gamma ray	-線
감마선 적설계	gamma ray snow gauge	-線積雪計
감상관시간	decorrelation time	減相關時間

감소	decrease	減少
감소 가용성 물균형 모형	decreasing availability water balance model	減少可用性-均衡模型
감속	decelerate, retardation	減速
감쇠	attenuation, damping, decay	減衰
감쇠거리	decay distance	減衰距離
감쇠계수	attenuation coefficient	減衰係數
감쇠곡선	recession curve	減衰曲線
감쇠기	attenuator	減衰器
감쇠길이	attenuation length	減衰-
감쇠깊이	damping depth	減衰-
감쇠모드	damped mode, decaying mode	減衰-
감쇠보정	attenuation correction	減衰補正
감쇠비	damping ratio	減衰比
감쇠상수	attenuation constant	減衰常數
감쇠오류	attenuation error	減衰誤謬
감쇠인자	attenuation factor, damping factor	減衰因子
감쇠지역	decay area	減衰地域
감쇠진동	damped oscillation	減衰振動
감쇠파	damped wave, evanescent wave	減衰波
감수곡선	depletion curve	減水曲線
감습저항	humistor	感濕抵抗
감습콘덴서	humicap	感濕-
감시	monitoring, watch	監視
감시구역	watch area	監視區域
감시자	monitor	監視者
감온성	thermosensitivity	感溫性
감응발달	sympathetic development	感應發達
감응방전	sympathetic discharge	感應放電
갑작스러운 추위	snap	
갑판	deck, lid	甲板
강결합수	firmly bound water	强結合水
강관수은온도계	mercury-in-steel thermometer	鋼管水銀溫度計
강교점	descending node	降交點
강국지폭풍우주의보	severe local storm watch	强局地暴風雨注意報
강대류폭풍우	severe convection storm	强對流暴風雨
강도	intensity	强度
강도변조	intensity modulation	强度變調
강도-변조 지시기	intensity-modulated indicator	强度變調指示器
강도-시간 공식	intensity-duration formula	强度時間公式
강설(량)	snowfall	降雪(量)
강설계	snowfall meter	降雪計
강설깊이	depth of snow fall	降雪-
강설대	snowbelt	降雪帶
강설선	snow precipitation line	降雪線

강설일	snow day	降雪日
강설입자영상기	ice particle imager	降雪粒子映像機
강수	precipitation	降水
강수감쇠	precipitation attenuation	降水減衰
강수강도	precipitation intensity	降水强度
강수계속기간	rainy spell	降水繼續期間
강수계속일수	number of consecutive wet days	降水繼續日數
강수고층운	altostratus precipitation, altostratus precipitus	降水高層雲
강수과도	precipitation effectiveness	降水果度
강수관측소	precipitation station	降水觀測所
강수구름	praecipitatio, precipitus	降水-
강수궤적	precipitation trajectory	降水軌跡
강수궤적에코	mares' tails	江水軌跡-
강수그늘	precipitation shadow	降水-
강수기구	precipitation mechanism	降水器具
강수꼬리	precipitation trails	降水-
강수난층운	nimbostratus precipitation	降水亂層雲
강수대륙도	hygrometric continentality	降水大陸度
강수도표	hyetograph	降水圖表
강수띠	band of precipitation	降水-
강수량	amount of precipitation, catch	降水量
강수량계	precipitation gauge, precipitation meter	降水量計
강수레이더	precipitation radar	降水-
강수물리학	precipitation physics	降水物理學
강수밀도	precipitation density	降水密度
강수발달기작	precipitation mechanism	降水發達機作
강수방패	precipitation shield	降水防牌
강수백분율	isomeric value	降水百分率
강수분포	precipitation distribution	降水分布
강수분포곡선	hyetographic curve	降水分布曲線
강수생성요소	precipitation-generating element	降水生成要素
강수세포	precipitation cell	降水細胞
강수수량	precipitation water content	降水水量
강수시간	precipitation duration	降水時間
강수시료	sample of precipitation	降水試料
강수실링	precipitation ceiling	降水雲高
강수안개	precipitation fog	降水-
강수에코	precipitation echo	降水-
강수역	precipitation area	降水域
강수역전	inversion of precipitation, precipitation inversion	降水逆轉
강수영역	precipitation regime	降水領域
강수원인	precipitation causes	降水原因
강수원인하강기류	precipitation-induced downdraft	降水原因下降氣流
강수유효도	precipitation effectiveness	降水有效度

강수-유효지수	P-E index(precipitation-effectiveness index, precipitation-evaporation index)	降水有效指數
강수율	precipitation ratio	降水率
강수이론	precipitation theory	降水理論
강수일	day of precipitation, precipitation day	降水日
강수일수	number of days with precipitation	降水日數
강수장미	precipitation rose	降水薔薇
강수적란운	cumulonimbus praecipitatio	降水積亂雲
강수적운	cumulus praecipitatio	降水積雲
강수전기	electricity of precipitation	降水電氣
강수전기(학)	precipitation electricity	降水電氣(學)
강수전류	precipitation current	降水電流
강수정전	precipitation static	降水靜電
강수정화	scavenging by precipitation	降水淨化
강수줄기	trails of precipitation, trails of rain	降水-
강수증가	enhancement of precipitation	降水增加
강수증가촉매제	agent of precipitation enhancement	降水增加觸媒劑
강수증발몫	precipitation-evaporation quotient, P-E quotient	降水蒸發-
강수증발비	precipitation-evaporation ratio	降水蒸發比
강수증발지수	precipitation-evaporation index, P-E index	降水蒸發指數
강수지속일수	number of consecutive wet days	降水持續日數
강수지수	precipitation index	降水指數
강수지시기	ombroscope	降水指示器
강수차단	interception of precipitation	降水遮斷
강수차폐	precipitation shield	降水遮蔽
강수특성	precipitation characteristics	降水特性
강수하중	precipitation loading	降水荷重
강수학	hyetology	降水學
강수(분포)학	hyetography	降水(分布)學
강수현상	precipitation phenomena	降水現狀
강수형	precipitation type	降水形
강수화학조성	chemical composition of precipitation	降水化學組成
강수확률	precipitation probability	降水確率
강수효과	precipitation effectiveness	降水效果
강수효과비	precipitation effectiveness ratio	降水效果比
강수효과지수	precipitation effectiveness index	降水效果指數
강수효율	effectiveness of precipitation, precipitation efficiency	降水效率
강수흔적	trace of precipitation	降水痕跡
강습도하기상	river crossing assault weather	強襲渡河氣象
강안개	river fog	
강우	rainfall	降雨
강우(의)	hyetal	降雨
강우감쇠	rain attenuation	降雨減衰

강우감지기	rain detector	降雨感知器
강우강도	rain intensity, rainfall intensity	降雨强度
강우강도계	rain intensity gauge, rainfall intensity meter	降雨强度計
강우강도-지속기간관계	rainfall intensity-duration relationship	降雨强度持續時間關係
강우강도-지속기간-빈도곡선	rainfall intensity-duration-frequency curve	降雨强度持續期間頻度曲線
강우강조	enhancement of rainfall	降雨强調
강우계속기간	rain spell	降雨繼續期間
강우기간	rainfall duration	降雨期間
강우누적곡선	rainfall mass curve	降雨累積曲線
강우대	rain band, rain belt	降雨帶
강우량	rainfall, rainfall amount, rainfall depth	降雨量
강우량극값	extreme rainfall	降雨量極-
강우량측정	measurement of rainfall amount	降雨量測定
강우빈도	rain frequency, rainfall frequency	降雨頻度
강우스콜	rain squall	降雨-
강우시간	rain hour, rainfall hour	降雨時間
강우역	rain area	降雨域
강우역전	inversion of rainfall, rainfall inversion	降雨逆轉
강우율	rainfall percentage, rainfall rate, rate of rainfall	降雨率
강우율계	rate of rainfall gauge	降雨率計
강우인자	rainfall factor, rain factor	降雨因子
강우일	day of rain, rain day, rainy day	降雨日
강우재현기간	rainfall recurrence interval	降雨再現期間
강우지대(의)	hyetal	降雨地帶
강우차단	interception of rainfall	降雨遮斷
강우철	rainy season	降雨-
강우침식	rainfall erosion, rain erosion	降雨浸蝕
강우크러스트	rain crust	降雨-
강우특성	raininess	降雨特性
강우편향폭풍우	severe right moving storm	强右偏向暴風雨
강우형	rainfall regime	降雨型
강우확률	rainfall probability	降雨確率
강우효율	rainfall effectiveness	降雨效率
강우흔적	trace of rainfall	降雨痕跡
강제(력)	forcing	强制(力)
강제강하	forced dropping	强制降下
강제기류	forced airflow	强制氣流
강제대류	forced convection	强制對流
강제력이류	forcing advection	强制力移流
강제력인자	forcing factor	强制力因子
강제력함수	forcing function	强制力函數
강제력항	forcing term	强制力項
강제응결과정	forced condensation process	强制凝結過程

강제직교화	forced orthogonalization	强制直交化
강제진동	forced oscillation, forced vibration	强制振動
강제파	forced wave	强制波
강제확산	forced diffusion	强制擴散
강제환기	forced ventilation	强制換氣
강조곡선	enhancement curves	强調曲線
강조사항	enhanced wording	强調事項
강조이미지	enhanced image	强調-
강조테이블	enhancement tables	强調-
강체계	rigid system	剛體系
강추위	bitter cold	强-
강추위바람	biting wind	强-
강폭풍우	severe storm	强暴風雨
강폭풍우강수	severe storm precipitation	强暴風雨降水
강풍	high wind, strong wind	强風
강풍감시	high wind watch	强風監視
강풍경보	extreme wind warning, high wind warning, strong wind warning	强風警報
강풍반지름	radius of strong wind area	强風-
강풍주의보	high wind advisory, strong wind advisory	强風注意報
강한 대류현상	deep convection	-對流現象
강한 연직운	cloud with great vertical development	-鉛直雲
강화	intensification, reinforcement	强化
개구리도약법	leapfrog method	-跳躍法
개구리뜀방법	leapfrog method	-方法
개구리폭풍	frog storm, whip-poor-will storm	-暴風
개기식	total eclipse	皆旣蝕
개념도구	conceptual tool	槪念道具
개념모형	conceptual model	槪念模型
개량TIROS운영위성	improved TIROS operational satellite	改良-運營衛星
개방형풍동	open circuit wind tunnel	開放型風洞
개별도함수	individual derivative	個別導函數
개별방울	individual droplet	個別-
개별수적	individual droplet	個別水滴
개별핵	individual nucleus	個別核
개수로흐름	open channel flow	開水路-
개엽	opening of leaf	開葉
개엽일	opening date of leaf	開葉日
개요	overview	槪要
개원통형	open cylinder type	開圓筒型
개인노출	personal exposure	個人露出
개인대기오염	personal air pollution	個人大氣汚染
개인대기오염노출	personal air pollution exposure	個人大氣汚染露出
개인배출	personal emissions	個人排出

개인오염피폭	personal pollution exposure	個人汚染物質 被曝
개인오차	personal error	個人誤差
개조	modification	改造
개체군성장	population growth	個體群成長
개형류	ostracodes	介形類
개화	flowering	開花
개화기	flowering time	開花期
개화시기	beginning of flowering	開花始期
개화예보	prediction of the blooming day	開花豫報
개화일	flowering date	開花日
개화전선	isochronous line of the blooming day	開花前線
개회로식 풍동	open circuit wind tunnel	開回路式風洞
개흙(거르기)	warping	
객관방법	objective method	客觀方法
객관분석	objective analysis	客觀分析
객관예보	objective forecast	客觀豫報
객토	transported soil	客土
갠	fair	
갠날(씨)	fair-weather	
갠날 권운	fair-weather cirrus	-卷雲
갠날씨 계	fair-weather system	-系
갠날 적운	fair-weather cumulus	-積雲
갠날 전기장	fair-weather electric field	-電氣場
갠날 전류	fair-weather current	-電流
갤러킨방법	Galerkin method	-方法
갬	fairness	
거대뇌우발달지수	energy helicity index	巨大雷雨發達指數
거대세포	supercell	巨大細胞
거대세포뇌우	supercell thunderstorm	巨大細胞雷雨
거대세포형	supercell type	巨大細胞型
거대입자	coarse particle	巨大粒子
거대입자분율	coarse particle fraction	巨大粒子分率
거대핵	giant nucleus	巨大核
거르개	filter	
거리	range	距離
거리감쇠	range attenuation	距離減衰
거리계	range finder	距離計
거리-고도단면도	distance-altitude cross section	距離高度斷面圖
거리-고도지시기	range-elevation indicator, range-height indicator	距離高度指示器
거리뒤틀림	range distortion	距離變形
거리분해(능)	range resolution	距離分解(能)
거리-속도표시기	range-velocity display	距離速度表示器
거리연장	range delay	距離延長
거리오차	range error	距離誤差

거리의 역자승법칙	inverse square law of distance	距離逆自乘法則
거리-이웃그래프	distance-neighbor graph	距離-
거리지시표(레이더)	range strobe	距離指示表
거리추적	range tracking	距離追跡
거리폐문	range gating	距離閉門
거리표지	range marker	距離標識
거센물결	rip	
거스트너파	Gerstner wave	-波
거시물리학	macro-physics	巨視物理學
거시적	macro-scopic	巨視的
거시점성	macro-viscosity	巨視粘性
거울면반사	regular reflection, specular reflection	-面反射
거울반사체	specular reflector	-反射體
거울얼음	glare ice	
거점지역기상	Lodgement Area weather	據點地域氣象
거즈	muslin	
거짓권운	false cirrus	-卷雲
거짓난역	false warm sector	-暖域
거짓반사	false reflection	-反射
거짓색깔	false color	-色-
거짓색깔 영상	false color image	-色-映像
거짓전선	false front	-前線
거짓흰무지개	false white rainbow	
거친 눈	wild snow	
거친 면	rough surface	-面
거친 바다	high sea, rough sea	
거친 알갱이눈	coarse-grained snow	
거친 지표	rough surface	-地表
거친 파도	heavy surf	-波濤
거친 파도주의보	heavy surf advisory	-波濤注意報
거칠기	roughness	
거칠기계수	roughness coefficient	-係數
거칠기길이	roughness length	
거칠기파라미터	roughness parameter	
거품	bubble	
거품고기압	bubble high	-高氣壓
거품대류	bubble convection	-對流
거품수위계	bubble gauge	-水位計
거품얼음	bubble ice, bubbly ice	
거품정책	bubble policy	-政策
거품핵	bubble nucleus	-核
건구	dry bulb	乾球
건구온도	dry bulb temperature	乾球溫度
건구온도계	dry bulb thermometer	乾球溫度計

건기	dry spell, dry period	乾期
건들바람	moderate breeze	
건물량	dry matter	乾物量
건물생산	dry matter production	乾物生産
건생식물	xerophyte	乾生植物
건성침착	dry deposition	乾性沈着
건습계	psychrometer, wet-and-dry-bulb hygrometer	乾濕計
건습계공식	psychrometer formula, psychrometric formula	乾濕計公式
건습계상수	psychrometric constant	乾濕計常數
건습구습도계	dry-and-wet-bulb hygrometer, hygrodeik, wet-and-dry-bulb hygrometer	乾濕球濕度計
건습구온도계	dry-and-wet-bulb thermometer, psychrometer	乾濕球溫度計
건습구온도차	depression of the wet bulb, wet-bulb depression	乾濕球溫度差
건식꽃받침	dry sepal	
건식배연탈황법	dry method of exhaust gas desulfurization	乾式排煙脫黃法
건식세정기	dry scrubbers	乾式洗淨器
건열지수	xerothermal index	乾熱指數
건조	desiccation	乾燥
건조경보	dryness warning	乾燥警報
건조계수	aridity coefficient, coefficient of aridity	乾燥係數
건조계절	dry season	乾燥季節
건조공기	dry air	乾燥空氣
건조공기밀도	density of dry air	乾燥空氣密度
건조공기피해	injury due to dry air	乾燥空氣披害
건조관입	dry intrusion	乾燥貫入
건조구역	arid region	乾燥區域
건조기	altithermal period	乾燥期
건조기간	dry period	乾燥期間
건조기류	dry airstream	乾燥氣流
건조기후	arid climate, dry climate	乾燥氣候
건조능	drying power	乾燥能
건조단계	dry stage	乾燥段階
건조단열	dry adiabat	乾燥斷熱
건조단열(의)	dry adiabatic	乾燥斷熱
건조단열감률	dry adiabatic lapse rate	乾燥斷熱減率
건조단열곡선	dry adiabatic curve	乾燥斷熱曲線
건조단열과정	dry adiabatic process	乾燥斷熱過程
건조단열냉각	dry adiabatic cooling	乾燥斷熱冷却
건조단열대기	dry adiabatic atmosphere	乾燥斷熱大氣
건조단열률	dry adiabatic rate	乾燥斷熱率
건조단열변화	dry adiabatic change	乾燥斷熱變化
건조대류	dry convection	乾燥對流
건조대류조정	dry convective adjustment	乾燥對流調定
건조대류층	dry convection layer	乾燥對流層

건조대 수문학	arid zone hydrology	乾燥帶水文學
건조도	aridity, dryness	乾燥度
건조도지수	dryness index	乾燥度指數
건조마이크로버스트	dry microburst	乾燥-
건조메커니즘	drying mechanism	乾燥-
건조바람	dry wind	乾燥-
건조분자입사	dry impingement	乾燥分子入射
건조불안정	dry instability	乾燥不安定
건조선	dry line	乾燥線
건조선전선	dry line front	乾燥線前線
건조선폭풍	dry line storm	乾燥線暴風
건조성장	dry growth	乾燥成長
건조순간돌풍	dry microburst	乾燥瞬間突風
건조슬롯	dry slot	乾燥-
건조아습윤기후	dry subhumid climate	乾燥亞濕潤氣候
건조여름아열대기후	dry summer subtropical climate	乾燥-亞熱帶氣候
건조역전	dry inversion	乾燥逆轉
건조염호	dry salt lake	乾燥鹽湖
건조영구 언땅	dry permafrost	乾燥永久-
건조인자	aridity factor	乾燥因子
건조주기	arid cycle	乾燥週期
건조주의보	dryness advisory	乾燥注意報
건조지대	arid zone, drylands	乾燥地帶
건조지대붕괴	dryland degradation	乾燥地帶崩壞
건조지속기	dry spell	乾燥持續期
건조지수	aridity index, index of aridity	乾燥指數
건조지역	dry region	乾燥地域
건조하강돌연풍	dry downburst	乾燥下降突然風
건조한	arid	乾燥-
건조한계	arid boundary	乾燥限界
건조해	dry damage, injury due to dryness	乾燥害
건조혀	dry tongue	乾燥-
건축	architecture	建築
건축기상학	architectural meteorology	建築氣象學
건축기하학	building geometry	建築幾何學
건축기후학	architectural climatology, building climatology	建築氣候學
건축설계	building design	建築設計
건축자재효과	building material effect	建築資材效果
건토	oven dried soil	乾土
걸린 빙하	hanging glacier	-氷河
검댕이	soot	
검댕탄소에어로졸	soot carbon aerosol	-炭素-
검력계주기	dynamometer cycles	檢力計週期
검류계	galvanometer	檢流計

검사	inspection	檢查
검사분석	inspectional analysis	檢查分析
검색배열	search array	檢索配列
검색부호	interrogation sign	檢索符號
검습기	hygroscope	檢濕器
검은 먼지보라	black blizzard	
검은 번개	black lightning, dark lightning	
검은 북동풍	black northeaster	
검은 비	black rain	
검은 서리	black frost	
검은 스콜	black squall	
검은 안개	black fog	
검은 얼음	black ice	
검은 연기	black smoke	-煙氣
검은 층운	black stratus	黑層雲
검은 폭풍	black storm	-暴風
검전기	electroscope	檢電器
검정	inspection	檢定
검정탄소에어로졸	black carbon aerosol	-炭素-
검조계	marigraph	檢潮計
검조기	tide gauge	檢潮器
검증	verification	檢證
검증표본	verification sample	檢證標本
검출	detection	檢出
검출한계	limit of detection	檢出限界
검파	demodulation, detection	檢波
겉보기바람	apparent wind	
겉보기밝기	apparent brightness	
겉보기비중	apparent specific gravity	-比重
겉보기속도	apparent velocity	-速度
겉보기수평선	apparent horizon	-水平線
겉보기어는점	apparent freezing point	-點
겉보기온도	apparent temperature	-溫度
겉보기응력	apparent stress	-應力
겉보기일	virtual work	
겉보기정오	apparent noon	-正午
겉보기중력	apparent gravity, virtual gravity	-重力
겉보기지름	apparent diameter	
겉보기지하수속도	apparent groundwater velocity	-地下水速度
겉보기태양시	apparent solar time	-太陽時
겉보기태양일	apparent solar day	-太陽日
겉보기팽창	apparent expansion	-膨脹
겉보기휘도	apparent luminance	-輝度
겉보기힘	apparent force	

계각도	crab angle	-角度
게그	geg	
게르디엔통풍기	Gerdien aspriator	-通風器
게이뤼삭법칙	Gay-Lussac's law	-法則
게이뤼삭실험	Gay-Lussac experiment	-實驗
게인	gain	
겨울	winter	
겨울(의)	hibernal	
겨울계절풍	winter monsoon	-季節風
겨울날씨주의보	winter weather advisory	-注意報
겨울몬순	winter monsoon	
겨울반구	winter hemisphere	-半球
겨울스톰	winter storm	
겨울작물	winter crop	-作物
겨울잠	hibernation	
겨울준비	winterization	-準備
겨울폭풍	wintry blast	-暴風
겨울폭풍(우)	winter storm	-暴風(雨)
겨울폭풍(우)감시	winter storm watch	-暴風(雨)監視
겨울폭풍(우)경보	winter storm warning	-暴風(雨)警報
격렬서풍	stormy westerly wind	激烈西風
격리	isolation	隔離
격발작용	trigger action	激痛作用
격발효과	trigger effect	激痛效果
격변설	catastrophism	激變說
격자	grating, grid	格子
격자간격	mesh scale	格子間隔
격자계	grid system	格子系
격자모양측운기	grid nephoscope	格子模樣測雲器
격자방향	grid heading	格子方向
(회절)격자분광기	grating spectrometer	(回折)格子分光器
격자법	grid method	格子法
격자북	grid north	格子北
격자상수	grid constant	格子常數
격자상하계	trellis drainage	格子狀河系
격자압축	grid telescoping	格子壓縮
격자자오선	grid meridian	格子子午線
격자점	grid-point, lattice point	格子點
격자점값	grid-point value	格子點-
격자점표현	grid-point representation	格子點表現
격자항법	grid navigation	格子航法
격자화	gridding	格子化
격측	telemetering	隔測
격측건습계	telepsychrometer	隔測乾濕計

격측광도계	telephotometer	隔測光度計
격측기	telemeter	隔測器
격측기록	telerecording	隔測記錄
격측기상관측법	telemeteorometry	隔測氣象觀測法
격측기상측기학	telemeteorography	隔測氣象測器學
격측법	telemetry	隔測法
격측온도계	telethermometer	隔測溫度計
격측온도측정기	telethermoscope	隔測溫度測定器
격측자기기상계	telemeteorograph	隔測自記氣象計
격측자료	telemetry	隔測資料
견인기구	balloon drag	牽引氣球
견인력	traction	牽引力
견지성	consistency	堅持性
결과확인	result validation	結果確認
결로	dew condensation, dewfall, dewing	結露
결빙	freezing	結氷
결빙경보	freeze warning	結氷警報
결빙고도	freezing level	結氷高度
결빙고도도	freezing level chart	結氷高度圖
결빙도일	freezing degree-day	結氷度日
결빙물보라	freezing spray	結氷-
결빙물보라 주의보	freezing spray advisory	結氷-注意報
결빙열	heat of freezing	結氷熱
결빙일	freeze-up date	結氷日
결빙자료	freeze data	結氷資料
결빙점	freezing point	結氷點
결빙점선	freezing point line	結氷點線
결빙지수	freezing index	結氷指數
결빙지시기	ice accretion indicator	結氷指示器
결빙핵	freezing nucleus	結氷核
결빙핵 스펙트럼	freezing nucleus spectrum	結氷核-
결우성	anhyetism	缺雨性
결절	nodule	結節
결정	crystal	結晶
결정격자	crystal lattice	結晶格子
결정고도	decision height	決定高度
결정구조	crystal structure	結晶構造
결정나무	decision tree	決定-
결정론적 카오스	deterministic chaos	決定論的-
결정론적 항	deterministic term	決定論的項
결정서리	crystalline frost	結晶-
결정성(도)	crystallinity	結晶性(度)
결정습성	crystal habit	結晶習性
결정열	heat of crystallization	結晶熱

결정조건	determination condition	決定條件
결정트리	decision tree	決定-
결정프레임워크	decision framework	決定-
결정학적	crystallographic	結晶學的
결정핵	crystallization nucleus	結晶核
결정형	crystal form	結晶形
결정화	crystallization	結晶化
결집	agglomeration, coagulation	結集
결착	accretion, riming	結着
결착률	accretion efficiency	結着率
결착방정식	accretion equation	結着方程式
결핍증	deficiency	缺乏症
결합	bond, coupling	結合
결합계수	combination coefficient	結合係數
결합극성	bond polarity	結合極性
결합기후계	coupled climate system	結合氣候系
결합기후시스템	coupled climate system	結合氣候-
결합길이	bond length	結合-
결합모형	combination model, coupled model, coupling model	結合模型
결합모형계	coupled model system	結合模型系
결합모형시스템	coupled model system	結合模型-
결합방정식	combination equation, coupled equation	結合方程式
결합밴드	combination band	結合-
결합법칙	associative law	結合法則
결합수	bound water	結合水
결합에너지	binding energy, bond energy	結合-
결합용도	conjunctive use	結合用度
결합원리	combination principle	結合原理
결합주기	bond cycle	結合週期
겹고적운	altocumulus duplicatus	-高積雲
겹고층운	altostratus duplicatus	-高層雲
겹구름	duplicatus	
(두)겹권운	cirrus duplicatus	-卷雲
(두)겹권층운	cirrostratus duplicatus	-卷層雲
겹대류권계면	multiple tropopause	-對流圈界面
겹층적운	stratocumulus duplicatus	-層積雲
겹편류보정	multiple drift corrections	-偏流補正
경각	dip, inclination	傾角
경계	boundary	境界
경계구분	demarcation	境界區分
경계농도	alert concentration	警戒濃度
경계류	boundary current	境界流
경계면	boundary surface, interface	境界面

경계부-약에코역	bounded weak echo region	境界部弱-域
경계샘	border spring	境界-
경계역	border zone	境界域
경계조건	boundary condition	境界條件
경계층	boundary layer	境界層
경계층가설	boundary layer hypothesis	境界層假設
경계층기후학	boundary layer climatology	境界層氣候學
경계층기후학자	boundary layer climatologist	境界層氣候學者
경계층높이	boundary layer height	境界層-
경계층레이더	boundary layer radar	境界層-
경계층롤	boundary layer roll	境界層-
경계층매개변수화	boundary layer parameterization	境界層媒介變數化
경계층방정식	boundary layer equation	境界層方程式
경계층분리	boundary layer separation	境界層分離
경계층수렴선	boundary layer convergence line	境界層收斂線
경계층스케일링	boundary layer scaling	境界層-
경계층이론	boundary layer theory	境界層理論
경계층펌핑	boundary layer pumping	境界層-
경계층효과	boundary layer effect	境界層效果
경계파	boundary wave	境界波
경계표지	landmark	境界標識
경계혼합	boundary mixing	境界混合
경곗값문제	boundary value problem	境界-問題
경관기후학	landscape climatology	景觀氣候學
경관복잡성 오차	landscape complexity errors	景觀複雜性誤差
경관생태계	landscape ecology	景觀生態系
경년경향	secular trend	經年傾向
경년기후변화	secular climatic change	經年氣候變化
경년변동	interannual oscillation, interannual variation, secular variation	經年變動
경년변동성	interannual variability	經年變動性
경년변화	secular change	經年變化
경도	longitude	經度
경도	gradient	傾度
경도류	gradient current	傾度流
경도리처드슨수	gradient Richardson number	傾度-數
경도방정식	gradient equation	傾度方程式
경도방향바람	longitudinal wind	經度方向-
경도수송이론	gradient transport theory	傾度輸送理論
경도풍	gradient wind	傾度風
경도풍고도	gradient level, gradient wind level	傾度風高度
경도풍방정식	gradient wind equation	傾度風方程式
경도풍속	gradient wind speed	傾度風速
경도풍속도	gradient velocity	傾度風速度

경도확산근사	gradient diffusion approximation	傾度擴散近似
경도효과	longitude effect	經度效果
경도흐름	gradient flow	傾度-
경로	channel, path, pathway, route, track	經路
경로길이	path length	路程-
경로법	path method	經路法
경로선	path line	經路線
경로정보엔트로피	path information entropy	經路情報-
경로차	path difference	經路差
경로통합감쇠량	path integrated attenuation	經路統合減衰量
경보	warning	警報
(기상)경보구역	warning area	(氣象)警報區域
경보단계	warning stage	警報段階
경보수위	alarm level	警報水位
경사	inclination, obliquity	傾斜
경사가시거리	slant visual range	傾斜可視距離
경사각	angle of inclination, canting angle, inclination angle	傾斜角
경사거리	slant range	傾斜距離
경사경로	slant path	傾斜經路
경사경로분자흡수	slant path molecular absorption	傾斜經路分子吸收
경사끝벽	sloping end walls	傾斜-壁
경사대류	slant convection	傾斜對流
경사류	slope current	傾斜流
경사-리처드슨수	slope-Richardson number	傾斜-數
경사면을 따라서 부는 바람시스템	along-slope wind system	傾斜面-
경사방위	aspect	傾斜方位
경사시정	slant visibility	傾斜視程
경사충격파	oblique shock wave	傾斜衝擊波
경선	meridian	經線
경시변동성	interhourly variability	經時變動性
경압	barocline	傾壓
경압(의)	baroclinic	傾壓
경압골구름패턴	baroclinic trough cloud pattern	傾壓-
경압구역	baroclinic zone	傾壓區域
경압대	baroclinic zone	傾壓帶
경압대기	baroclinic atmosphere	傾壓大氣
경압매개변수	baroclinity parameter	傾壓媒介變數
경압매질	baroclinic medium	傾壓媒質
경압모형	baroclinic model	傾壓模型
경압모형대기	baroclinic model atmosphere	傾壓模型大氣
경압발달모형	baroclinic development model	傾壓發達模型
경압벡터	baroclinity vector	傾壓-

경압불안정	baroclinic instability	傾壓不安定
경압상태	barocline state	傾壓狀態
경압성	baroclinicity, baroclinity, barocliny	傾壓性
경압성구름	baroclinic leaf cloud	傾壓性-
경압성대류 되먹임고리	baroclinic convection feedback loop	傾壓性對流-
경압성조건	baroclinicity condition	傾壓性條件
경압예보	baroclinic forecast	傾壓豫報
경압요란	baroclinic disturbance	傾壓搖亂
경압운동	baroclinic motion	傾壓運動
경압유체	baroclinic fluid	傾壓流體
경압잎새구름	baroclinic leaf cloud	傾壓-
경압장	baroclinic field	傾壓場
경압조건	baroclinic condition	傾壓條件
경압토크벡터	baroclinic torque vector	傾壓-
경압파	baroclinic wave	傾壓波
경압항	baroclinic term	傾壓項
경압행성경계층	baroclinic planetary boundary layer	傾壓行星境界層
경압효과	baroclinic effect	傾壓效果
경압흐름	baroclinic flow	傾壓-
경월변동	inter-monthly variation	經月變動
경월변동성	inter-monthly variability	經月變動性
경월압력변동	inter-monthly pressure variation	經月壓力變動
경월온도변동	inter-monthly temperature variation	經月溫度變動
경위의	theodolite	經緯儀
경일변동성	interdiurnal variability	經日變動性
경일압력변동	interdiurnal pressure variation	經日壓力變動
경일온도변동	interdiurnal temperature variation	經日溫度變動
경작	cultivation	耕作
경정	reduction	更正
경정계수	reduction factor	更正係數
경정표	reduction table	更正表
경제적 정미방사계	economical net radiometer	經濟的正味放射計
경질방사선	hard radiation	硬質放射線
경칩	awakening of insects	驚蟄
경풍계	anemoclinometer, anemogram	傾風計
경향	tendency	傾向
경향간격	tendency interval	傾向間隔
경향단면	tendency profile	傾向斷面
경향방정식	tendency equation	傾向方程式
경향법	tendency method	傾向法
경향선도	tendency chart	傾向線圖
경험기후분류계	empirical climatic classification system	經驗氣候分類系
경험기후분류시스템	empirical climatic classification system	經驗氣候分類-
경험식	empirical formula	經驗式

경험적 접근	empirical approach	經驗的接近
경험적 확률	empirical probability	經驗的確率
경험직교함수	empirical orthogonal function	經驗直交函數
경화	hardening	硬化
경화계수	coefficient of consolidation	硬化係數
계	system	系
계곡빙하	valley glacier	溪谷氷河
계곡안개	valley fog	溪谷-
계곡역풍	up-valley wind	溪谷逆風
계곡을 따라서 부는 바람	along-valley wind	溪谷-
계곡을 따라서 부는 바람시스템	along-valley wind system	溪谷-
계곡풍	valley breeze, valley wind	溪谷風
계곡풍순환	valley breeze circulation	溪谷風循環
계곡횡단바람	cross-valley wind	溪谷橫斷-
계곡횡단순환	cross-valley circulation	溪谷橫斷循環
계급	class	階級
계급값	class mark	階級-
계급구간	class interval	階級區間
계급별일수	number of days with specified value	階級別日數
계급 신뢰도표	ranked reliability table	階級信賴圖表
계기	apparatus, gauge	計器
계기날씨	instrument weather	計器-
계기노출	instrument exposure	計器露出
계기보정	instrument correction	計器補定
계기비행	instrument flight, on instruments	計器飛行
계기비행규칙	instrument flight rule	計器飛行規則
계기비행 기상	IFR weather	計器飛行氣象
계기비행 기상조건	instrument meteorological condition	計器飛行氣象條件
계기압력	gauge pressure	計器壓力
계기오차(기차)	instrumental error, instrument error	計器誤差(器差)
계기오차표	instrument error table	計器誤差表
계기잡음	instrument noise	計器雜音
계기착륙장치	instrument landing system	計器着陸裝置
계기착빙	instrument icing	計器着氷
계단	cascade	階段
계단선도	stepped leader	階段先導
계단적 변화	jump	階段的變化
계단함수	step(ped) function	階段函數
계류기구	captive balloon, kytoon, kite balloon, tethered balloon	繫留氣球
계류기구탐측	captive balloon sounding	繫留氣球探測
계류존데	wiresonde	繫留-
계면장력	interfacial tension	界面張力

계산기	counter	計算器
계산도표	alignment chart, nomogram, nomograph, nomographic chart	計算圖表
계산모드	computational mode	計算-
계산방식	computational mode	計算方式
계산분산	computational dispersion	計算分散
계산불안정(성)	computational instability	計算不安定(性)
계산적 분산	computational dispersion	計算的分散
계산정확도	computational accuracy	計算正確度
계산좌표	enumeration coordinate	計算座標
계산확산	computational diffusion	計算擴散
계수	coefficient, modulus	係數
계수기	counter	計數器
계수컵풍속계	cup counter anemometer	計數-風速計
계수풍속계	counting anemometer	計數風速計
계절	season	季節
계절강수	seasonal precipitation	季節降水
계절강우전선	seasonal rain front	季節降雨前線
계절변동	seasonal variation	季節變動
계절변화	seasonal change	季節變化
계절변화관측	seasonal change observation	季節變化觀測
계절변화분석	seasonal change analysis	季節變化分析
계절병	seasonal disease	季節病
계절분석법	seasonal analysis method	季節分析法
계절사이변동성	interseasonal variability	季節-變動性
계절안변동	intraseasonal variation	季節內變動
계절안변동성	intraseasonal variability	季節內變動性
계절안진동	intraseasonal oscillation	季節內振動
계절예보	seasonal forecasting	季節豫報
계절예보모형	seasonal forecast model	季節豫報模型
계절일기예보	seasonal weather forecast	季節日氣豫報
계절적 퇴적	varve	季節的堆積
계절정서장애	seasonal affective disorder	季節情緒障碍
계절중앙달	mid-season month	季節中央-
계절지연	seasonal lag	季節遲延
계절추세	seasonal trend	季節趨勢
계절특성	seasonal character	季節特性
계절풍	monsoon, monsoon wind	季節風
계절풍강우	monsoon rainfall	季節風降雨
계절풍공기	monsoon air	季節風空氣
계절풍구름무리	monsoon cluster	季節風-
계절풍기단	monsoon air, monsoon air-mass	季節風氣團
계절풍기압골	monsoon trough	季節風氣壓-
계절풍기후	monsoon climate	季節風氣候

계절풍반전	reversal of the monsoon	季節風反轉
계절풍버스트	monsoon burst	季節風-
계절풍비	monsoon rain	季節風-
계절풍서지	monsoon surge	季節風-
계절풍순환	monsoon circulation	季節風循環
계절풍안개	monsoon fog	季節風-
계절풍자이어	monsoon gyre	季節風-
계절풍저기압	monsoon depression, monsoon low	季節風低氣壓
계절풍지수	monsoon index	季節風指數
계절풍진행	progression of the monsoon	季節風進行
계절풍기	monsoon season	季節風期
계절풍파도	monsoon surge	季節風波濤
계절풍표류	monsoon current	季節風漂流
계절풍해류	monsoon current, monsoon drift	季節風海流
계절학	phenology	季節學
계절현상	seasonal phenomenon	季節現象
계절형	seasonal form	季節型
계차	difference	階差
계통오차	systematic error	系統誤差
고감도감시레이더	sensitive surveillance radar	高感度監視-
고강수량지표	paleo-precipitation indicator	古降水量 指標
고강수초대형세포	high precipitation supercell	高降水超大型細胞
(북아메리카의)고건다 빙하작용	Gowganda glaciation	-氷河作用
고결정얼음	paleocrystic ice	古結晶-
고고학자료	archeological data	考古學資料
고공	high altitude	高空
고공강하날씨	high altitude low opening weather	高空降下-
고공병	aeroembolism	高空病
고공연무	haze aloft	高空煙霧
고공전위기록계	alti-electrograph	高空電位記錄計
고공폭격날씨	high altitude bombing weather	高空爆擊-
고공폭발기상	high air burst weather	高空爆發氣象
고공 푄	free air foehn	高空-
고과학자	paleoscientists	古科學者
고기압	anticyclone, barometric high, high, high atmospheric pressure, high pressure	高氣壓
고기압계	high pressure system	高氣壓系
고기압능	high pressure ridge	高氣壓-
고기압마루	ridge of high pressure	高氣壓-
고기압발생	anticyclogenesis	高氣壓發生
고기압블로킹	anticyclonic blocking	高氣壓-
고기압성	anticyclonic, anticyclonicity, cum sole	高氣壓性
고기압성곡률	anticyclonic curvature	高氣壓性曲率

고기압성기류	anticyclonic flow	高氣壓性氣流
고기압성날씨	anticyclonic weather	高氣壓性日氣
고기압성날씨형	anticyclonic weather type	高氣壓性日氣型
고기압성맴돌이	anticyclonic eddy	高氣壓性-
고기압성발산	anticyclonic divergence	高氣壓性發散
고기압성보라	anticyclonic bora	高氣壓性-
고기압성소용돌이	anticyclonic vortex	高氣壓性-
고기압성소용돌이도	anticyclonic vorticity	高氣壓性-度
고기압성순환	anticyclonic circulation	高氣壓性循環
고기압성시어	anricycloniic shear	高氣壓性-
고기압성위상	anticyclonic phase	高氣壓性位相
고기압성침강	anticyclone subsidence	高氣壓性沈降
고기압성회전	anticyclonic rotation	高氣壓性回轉
고기압성흐름	anticyclonic flow	高氣壓性-
고기압성흐림	anticyclonic gloom	高氣壓性-
고기압소멸	anticyclolysis	高氣壓消滅
고기압역	area of high pressure	高氣壓域
고기압의 축	axis of anticyclone	高氣壓軸
고기압이동	anticyclone movement	高氣壓移動
고기압저지	anticyclonic blocking	高氣壓沮止
고기압중심	high pressure center	高氣壓中心
고기압지역	anticyclonic region, high pressure area	高氣壓地域
고기압축경사	inclination of the axis of an anticyclone	高氣壓軸傾斜
고기온	paleotemperature	古氣溫
고기준신호	high reference signal	高基準信號
고기후	paleoclimate, geological climate	古氣候
고기후기록	palaeoclimatic record	古氣候記錄
고기후대리요소	paleoclimate proxy	古氣候代理要素
고기후모델링	paleoclimate modeling	古氣候-
고기후변화	paleoclimatic change	古氣候變化
고기후차례	paleoclimatic sequence	古氣候次例
고기후학	paleoclimatology	古氣候學
고기후학의 역사	history of paleoclimatology	古氣候學歷史
고기후학자	paleoclimatologist	古氣候學者
고더드우주과학연구소(미국)	Goddard Institute of Space Sciences	-宇宙科學硏究所(美國)
(해발)고도	height, altitude	(海拔)高度
고도각	angel of elevation	高度角
고도각주사	elevation scan	高度角走査
고도경도	height gradient	高度傾度
고도계	altimeter	高度計
고도계방정식	altimeter equation	高度計方程式
고도계보정	altimeter corrections	高度計補正
고도계산	altitude calculation, calculation of altitude, height calculation	高度計算

고도계산척	height nomogram	高度計算尺
고도권	almucantar	高度圈
고도등변화선	contour change line	高度等變化線
고도면	level surface, geopotential surface	高度面
고도-방위-거리 위치지시기	height-azimuth-range position indicator	高度方位距離位置指示器
고도변화	changes with altitude	高度變化
고도변화도	height-change chart	高度變化圖
고도변화도법	allohypsography	高度變化圖法
고도변화선	height-change line	高度變化線
고도보정	altitude correction	高度補正
고도분석법	height analysis method	高度分析法
고도설정(값)	altimeter setting	高度設定
고도설정지시기	altimeter setting indicator	高度設定指示器
고도오차	elevation error	高度誤差
고도정보	altitude information	高度情報
고도좌표	height coordinate	高度座標
고도차측정기	cathetometer	高度差測定器
고도탐지레이더	height-finding radar	高度探知-
고도패턴	height pattern	高度-
고드름	icicle	
고등어바람	mackerel breeze	
고래꼬리형기압분포	whale-tail-type pressure pattern	-型氣壓分布
고래등구름	whaleback cloud	
고랭지	highland	高冷地
고르기	smoothing	
고리난층운	nimbostratus virga	-亂層雲
고리모양	looping	-模樣
고리모양 플룸	looping plume	-模樣-
고리소용돌이	ring vortex	
고리흐름	ring current	
고립	isolation	孤立
고립계	isolated system	孤立系
고립소나기	isolated showers	孤立-
고립에코	isolated echo	孤立-
고립파	solitary wave	孤立波
고립폐색	seclusion	孤立閉塞
고무기구	rubber balloon	-氣球
고무얼음	rubber ice	
고문서	archives	古文書
고반사면	high reflectance surface	高反射面
고분해엑스선회절기	high resolution X-ray diffractometer	高分解-線回折器
고분해적외복사계	high resolution infrared radiometer	高分解赤外輻射計
고분해적외탐측계	high resolution infrared sounder	高分解赤外探測計
고비사막	Gobi desert	-砂漠

고빙하학	paleoglaciology	古氷河學
고사구시스템	palaeodune system	古砂丘-
고사리서리	fern frost	
고산관측소	high-altitude observatory	高山觀測所
고산광	alpenglow, Alpenglühen, alpine glow	高山光
고산기후	alpine climate	高山氣候
고산병	mountain sickness, soroche	高山病
고산빙하	alpine glacier, mountain glacier	高山氷河
고산식생	alpine vegetation	高山植生
고산실험	alpine experiment	高山實驗
고산툰드라	alpine tundra, mountain tundra	高山-
고생대	Paleozoic Era	古生代
고생대기후	Palaeozoic climate	古生代氣候
고생대말 기후	Late Paleozoic climate	古生代末氣候
고생대전기기후 (캄브리아기-데본기)	early Paleozoic climates (Cambrian-Devonian)	古生代前期氣候
고생산력	paleoproductivity	古生産力
고생태학	paleoecology	古生態學
고생태학자	palaeoecologists	古生態學者
고속류	high speed stream	高速流
고속반응센서	fast-response sensor	高速反應-
고속얼음	fast ice	高速-
고속이온	fast ion	高速-
고속정보전송	High Rate Information Transmission	高速情報電送
고속투하기상	high velocity drop weather	高速投下氣象
고속팩시밀리	high speed facsimile	高速-
고속푸리에변환	fast Fourier transform	高速-變換
고쇄파	high surf	高碎波
고쇄파경보	high surf warning	高碎波警報
고쇄파주의보	high surf advisory	高碎波注意報
고수문학	paleohydrology	古水文學
고수온	ocean paleotemperature	古水溫
고수위	high water level	高水位
고수위방수	high water discharge	高水位放水
고수위표지	high water mark	高水位標識
고순환	paleocirculation	古循環
고식물학	paleobotany	古植物學
고(기)압	barometric maximum	高(氣)壓
고압	high pressure	高壓
고압대	high belt	高壓帶
고압편향	undershoot	高壓偏向
고-엘니뇨-남방진동기록	Paleo-El Niño-Southern Oscillation record	古-南方振動記錄
고영향날씨	high-impact weather	高影響-
고온계	pyrometer	高溫計

고온기간	megathermal period, Climatic Optimum	高溫期間
고온기후	hot climate, megathermal climate	高溫氣候
고온다습	high temperature and humidity	高溫多濕
고온대	hot belt, megathermen	高溫帶
고온도기록	palaeotemperature record	古溫度記錄
고온도자료	paleotemperature data	古溫度資料
고온식생	megatherm	高溫植生
고온식생형	megathermal type	高溫植生形
고온장해	high temperature injury	高溫障害
고온측정법	pyrometry	高溫測定法
고요	calm	
고용량에어샘플러	high volume sampler	高容量-
고운 모래방출률	silt-discharge rating	-放出率
고원	plateau	高原
고원관측소	plateau station	高原觀測所
고원빙하	plateau glacier	高原氷河
고원안개	high land fog	高原-
고위도	high latitudes	高緯度
고위도	palaeolatitude	古緯度
고유곡선	eigencurve	固有曲線
고유방향	eigendirection	固有方向
고유벡터	eigenvector, charcteristic vector, latent vector	固有-
고유벡터분석	eigenvector analysis	固有-分析
고유분석	eigenanalysis	固有分析
고유종-면적 관계	endemics-area relationship	固有種-面積關係
고유주기	eigenperiod, natural period	固有週期
고유진동	natural oscillation, proper vibration	固有振動
고유진동수	eigenfrequency, natural frequency	固有振動數
고유함수	eigenfunction, characteristic function	固有函數
고육수학	paleolimnology	古陸水學
고윳값	eigenvalue, charcteristic value, latent value	固有値
고윳값문제	eigenvalue problem	固有値問題
고윳값방정식	eigenvalue equation	固有値方程式
고의적 기상조절	intentional weather modification	故意的氣象調節
고이더(강우)경계선	Goyder Line	-(降雨)境界線
고적운	altocumulus, altocumulus cloud	高積雲
고전유체역학	classical hydrodynamics	古典流體力學
고전응결이론	classical condensation theory	古典凝結理論
고전적 초대형세포	classic supercell	古典的超大型細胞
고정기구	fixed balloon	固定氣球
고정바다플랫폼	fixed sea platform	固定-
고정배출원	stationary emission source	固定排出源
고정빔운고계	fixed-beam ceilometer	固定-雲高計
고정(점)오염원	stationary source	固定(點)汚染源

고정점	fixed point	固定點
고정파	fixed wave	固定波
고제3기(의)	paleogene	古第三紀
고제3기기후	paleogene climate	古第三紀氣候
고조	high tide, tidal bore	高潮
고조간격	high water interval	高潮間隔
고조경보	high tide warning	高潮警報
고조위	high water level	高潮位
고조위표지	high water mark	高潮位標識
고조주의보	high tide advisory	高潮注意報
고주파	high frequency	高周波
고주파발진기	high frequency oscillator	高周波發振器
고주파증폭기	high frequency amplifier	高周波增幅器
고지기후	highland climate	高地氣候
고지빙하	highland glacier	高地氷河
고지얼음	highland ice	高地-
고지자기	paleomagnetism	古地磁氣
고진공	high vacuum	高眞空
고체강수	solid precipitation	固體降水
고체경계	solid boundary	固體境界
고체상태	solid state	固體狀態
고체상	solid phase	固體相
고체송신기	solid state transmitter	固體送信機
고체회전	solid rotation	固體回轉
고초열	hay fever	枯草熱
고층고기압	high-level anticyclone	高層高氣壓
고층(기압)골	high-level trough	高層(氣壓)-
고층관측	upper air observation	高層觀測
고층관측의 날	aerological days	高層觀測-
고층구름	high-level cloud	高層-
고층권계면	aeropause	高層圈界面
고층기상관측	aerological observation	高層氣象觀測
고층기상관측기	aerological instrument	高層氣象觀測器
고층기상관측소	aerological station	高層氣象觀測所
고층기상관측시스템	aerological observation system	高層氣象觀測-
고층기상기록계	aerometeorograph	高層氣象記錄計
고층기상도	aerographical chart	高層氣象圖
고층기상선도	aerological diagram	高層氣象線圖
고층기상탐측	aerological sounding	高層氣象探測
고층기상표	aerological table	高層氣象表
고층기상학	aerology	高層氣象學
고층기후학	aeroclimatology	高層氣候學
고층뇌우	high-level thunderstorm	高層雷雨
고층대기	upper atmosphere	高層大氣

고층대기(종관)관측소	upper air (synoptic) station	高層大氣(綜觀)觀測所
고층대기물리학	aeronomy	高層大氣物理學
고층대기선도	upper atmosphere chart	高層大氣線圖
고층대기탐측	upper air sounding	高層大氣探測
고층등고도관측기구	skyhook balloon	高層等高度觀測氣球
고층(기압)마루	high-level ridge	高層(氣壓)-
고층무역풍(크라카타오 바람)	overtrades, Krakatao wind	高層貿易風
고층분석	aerological analysis, high-level analysis	高層分析
고층예보	high-level forecast(HLFOR)	高層豫報
고층예상도	prontour chart	高層豫想圖
고층운	altostratus, altostratus cloud	高層雲
고층저기압	high-level cyclone	高層低氣壓
고층적운	altostratocumulus	高層積雲
고층전류	electrojet	高層電流
고층중요기상예상도	high-level significant weather prognostic chart	高層重要氣象豫想圖
고층탐측	aerological ascent	高層探測
고토양	palaeosols, paleosoil	古土壤
고폭풍학	paleotempestology	古暴風學
고해양순환	ocean paleocirculation	古海洋循環
고해양학	paleoceanography	古海洋學
고해양학모델링	paleo-ocean modeling	古海洋學-
고해양학수소이온농도지수	paleo-ocean pH	古海洋學水素-濃度指數
고환경대리요소	paleoenvironmental proxy	古環境代理要素
고형강수	solid precipitation	固形降水
곡관온도계	bent stem thermometer	曲管溫度計
곡관지중온도계	bent stem earth thermometer	曲管地中溫度計
곡률반경	radius of curvature	曲率半徑
곡률반지름	radius of curvature	曲率-
곡률방향	direction of curvature	曲率方向
곡률소용돌이도	curvature vorticity	曲率-
곡률중심	center of curvature	曲率中心
곡률텐서	curvature tensor	曲律-
곡률효과	curvature effect	曲率效果
곡선맞춤	curve fitting	曲線-
곡선운동	curved motion	曲線運動
곡선좌표	curvilinear coordinates	曲線座標
곡우	grain rain	穀雨
곤드와나빙하시대	Gondwana Ice Age	-氷河時代
곤충에코	insect echo	昆蟲-
골	trough	
골드계산척	Gold slide	-計算尺
골바람	valley breeze, valley wind	
골바람순환	valley breeze circulation	-循環
골빙하	valley glacier	-氷河

골안개	valley fog	
곰보눈	sugar snow	
곰보빙산	sugar berg	-氷山
공간	space	空間
공간고분해능	high spatial resolution	空間高分解能
공간고분해능감부	high spatial resolution sensor	空間高分解能感部
공간규모	spatial scale	空間規模
공간등온선	choroisotherm	空間等溫線
공간벡터	space vector	空間-
공간변화	space change	空間變化
공간샘플링	spatial sampling	空間-
공간전하	space charge	空間電荷
공간정합	spatial coherence	空間整合
공간평균	space mean	空間平均
공간평균소용돌이도이류법	space mean vorticity advection method	空間平均-度移流法
공간평균평년편차도	space mean anomaly chart	空間平均平年偏差圖
공간표본수집	spatial sampling	空間標本收集
공간해상도	spatial resolution	空間解像度
공격개시선기상	line of departure weather	攻擊開始線氣象
공격기뢰부설기상	Offensive Mining weather	攻擊機雷敷設氣象
공공기상서비스	public weather service	公共氣象-
공군기상단	Air Force Weather Wing	空軍氣象團
공극률	porosity	孔隙率
공급자구름	feeder cloud	供給者-
공기	air	空氣
공기(의)	airborne	空氣
공기격류	air torrent	空氣激流
공기광	airlight	空氣光
공기광공식	airlight formula	空氣光公式
공기교환	massflow of air	空氣交換
공기궤적	air trajectory	空氣軌跡
공기기둥	air column	空氣-
공기기둥의 평균온도	mean temperature of air column	空氣-平均溫度
공기덩이	air parcel	空氣-
공기덩이가속도	acceleration of air parcel	空氣-加速度
공기덩이궤적	airparcel trajectory	空氣-軌跡
공기덩이법	parcel method	空氣-法
공기덩이역학방법	airparcel dynamic method	空氣塊力學方法
공기독소	air toxin	空氣毒素
공기먼지분석	air dust analysis, airborne dust analysis	空氣-分析
공기밀도	air density	空氣密度
공기범프	air bumps	空氣-
공기분자	air molecule	空氣分子
공기비율	airborne fraction	空氣比率

공기산도	air acidity	空氣酸度
공기상승	ascent of air	空氣上昇
공기샘	air fountain	空氣-
(외연)공기샤워	air shower	(外延)空氣-
공기소나기	air shower	空氣-
공기수송	air transport	空氣輸送
공기순환	air circulation, circulation of air	空氣循環
공기압축성	air compressibility	空氣壓縮性
공기여과기	air filter	空氣濾過機
공기역학	aerodynamics	空氣力學
공기역학(의)	aerodynamic	空氣力學
공기역학감쇠	aerodynamic damping	空氣力學減衰
공기역학강제	aerodynamic forcing	空氣力學强制
공기역학 거친 흐름	aerodynamic rough flow	空氣力學-
공기역학거칠기	aerodynamic roughness	空氣力學-
공기역학거칠기길이	aerodynamic roughness length	空氣力學-
공기역학규모	aerodynamic scale	空氣力學規模
공기역학균형	aerodynamic balance	空氣力學均衡
공기역학 매끄러움	aerodynamic smoothness	空氣力學-
공기역학 매끈한 흐름	aerodynamic smooth flow	空氣力學-
공기역학방법	aerodynamic method	空氣力學方法
공기역학불안정	aerodynamic instability	空氣力學不安定
공기역학비행운	aerodynamic contrail, aerodynamic trail	空氣力學飛行雲
공기역학상당직경	aerodynamic equivalent diameter	空氣力學相當直徑
공기역학성질	aerodynamic property	空氣力學性質
공기역학저항	aerodynamic resistance	空氣力學抵抗
공기역학적 거친 면	aerodynamically rough surface	空氣力學的-面
공기역학적 매끄러운 면	aerodynamically smooth surface	空氣力學的-面
공기역학접근법	aerodynamic approach	空氣力學接近法
공기역학항력	aerodynamic drag	空氣力學抗力
공기역학힘	aerodynamic force	空氣力學-
공기연령	age-of-air	空氣年齡
공기연료비	air-fuel ratio	空氣燃料比
공기오염물질	air pollutant	空氣汚染物質
공기온도경	air thermoscope	空氣溫度鏡
공기용량	air capacity	空氣容量
공기운동	air motion	空氣運動
공기이슈	air issue	空氣-
공기입자	air particle	空氣粒子
공기입자지수	air particle index	空氣粒子指數
공기입자크기	airborne particle size	空氣粒子-
공기저울	aerostatic balance, air poise	空氣-
공기저항	air resistance	空氣抵抗
공기정역학	aerostatics	空氣靜力學

공기조절	air conditioning	空氣調節
공기청정	air cleaning	空氣淸淨
공기측온계	air thermoscope	空氣測溫計
공기침입	invasion of air	空氣侵入
공기트랩	air trap	空氣-
공기파	air wave	空氣波
공기파열	air burst	空氣破裂
공기포집기	air sampler	空氣捕集器
공기폭포	air cascade	空氣瀑布
공기하강	descent of air	空氣下降
공기흐름	air flow	空氣-
공기흡원	air sink	空氣吸源
공누적스펙트럼	co-cumulative spectrum	共累積-
공대지(사격)기상	weapon to ground weather	空對地(射擊)氣象
공동	cavity	空洞
공동복사계	cavity radiometer	空洞輻射計
공동직달일사계	cavity pyrheliometer	空洞直達日射計
공동체	collective	共同體
공동크리깅	cokriging	共同-
공동화작용	cavitation	空洞化作用
공률	power	工率
공명	resonance	共鳴
공명(기압)골	resonance trough	共鳴(氣壓)谷
공명구조	resonance structure	共鳴構造
공명기(체)	resonator	共鳴氣(體)
공명복사	resonance radiation	共鳴輻射
공명산란	resonance scattering	共鳴散亂
공명산란라이다	resonance scattering lidar	共鳴散亂-
공명상자	resonance box	共鳴箱子
공명이론	resonance theory	共鳴理論
공명장치	resonator	共鳴裝置
공명진동	resonance oscillation	共鳴振動
공명진동수	resonance frequency, resonant frequency	共鳴振動數
공명풍하파	resonant lee wave	共鳴風下波
공변수	covariable	共變數
공병작전기상	military engineer operation weather	工兵作戰氣象
공분산	covariance	共分散
공분산곡률텐서	covariant curvature tensor	共分散曲率-
공분산시간연산자	covariant time operator	共分散時間演算子
공분산텐서	covariance tensor, covariant tensor	共分散-
공분산행렬	covariance matrix	共分散行列
공세제공작전기상	Offensive Counter Air Operations weather	攻勢制空作戰氣象
공스펙트럼	co-spectrum	共-
공스펙트럼유사성	cospectral similarity	共-類似性

공스펙트럼 피크	co-spectrum peak	共-
공식	formula	公式
공액좌표	conjugate coordinate	共軛座標
공역	airspace, flight area	空域
공역	codomain	共域
공유결정	covalent crystal	共有結晶
공유결합	covalent bond	共有結合
공유결합길이	covalent bond length	共有結合-
공전	atmospherics, sferics, static, stray	空電
공전	revolution	公轉
공전관측	sferics observation	空電觀測
공전관측망	sferics network	空電觀測網
공전기록계	sferics recorder	空電記錄計
공전발원점	sferics fix	空電發源點
공전빈도기록계	atmoradiograph	空電頻度記錄計
공전수신기	sferic receiver	空電受信機
공전원	source of an atmospherics	空電源
공전잡음도	noise level of atmospheric	空電雜音度
공전파형분류	classification of wave forms of atmospherics	空電波形分類
공전학	atmospherics, sferics	空電學
공중	aerial	空中
공중방전	air discharge	空中放電
공중부유(의)	airborne	空中浮遊
공중부유입자	airborne particles	空中浮遊粒子
공중비상대기항공차단기상	X-Interdiction weather	空中非常待機航空遮斷氣象
공중생물학	aerobiology	空中生物學
공중요격기상	P-D/N weather	空中邀擊氣象
공중위치지시기	air-position indicator	空中位置指示器
공중전기	spherics	空中電氣
공중파	sky wave	空中波
공중플랑크톤	aerial plankton, air plankton	空中-
공지전도전류	air-earth conduction current	空地傳導電流
공지전류	air-earth current	空地電流
공진기	resonator	共振器
공차	common difference	公差
공축상관법	coaxial correlation method	共軸相關法
공통요인분석	common factor analysis	共通要因分析
공학자기상학	engineer meteorology	工學子氣象學
공합	bellow	空盒
공합기압계	holosteric barometer	空盒氣壓計
공항	airport	空港
공항고도	airport height	空港高度
공항모형	airport model	空港模型
공항예보	airport forecast	空港豫報

공항표고	airport elevation	空港標高
공해	open sea	公海
공해배출부과금	emission charges, effluent charges	公害排出賦課金
과거기후	past climate	過去氣候
과거일기	past weather	過去日氣
과거지구변화	Past Global Changes(PGC)	過去地球變化
과냉각	overcooling, subcooling, supercooling undercooling	過冷却
과냉각구름	supercooled cloud	過冷却-
과냉각물방울	supercooled water droplet	過冷却-
과냉각비	supercooled rain	過冷却-
과냉각수	supercooled water	過冷却水
과냉각수적	supercooled drop	過冷却水滴
과냉각안개	supercooled fog	過冷却-
과냉각운립	supercooling cloud partical	過冷却雲粒
과다구름씨뿌리기	excessive cloud seeding	過多-
과도명시된	overspecified	過度明示-
과민성폐렴	hypersensitivity pneumonitis	過敏性肺炎
과보상	overcompensate	過報償
과산화기라디칼	peroxy radicals	過酸化基-
과산화물	peroxides	過酸化物
과색조 띠	overtone band	
과수서리방지가열기	orchard heater	果樹-防止加熱器
과수원알베도	orchard albedo	果樹園反射率
과습윤기후	perhumid climate	過濕潤氣候
과시료	oversample	過試料
과열	overheat, superheating	過熱
과열도	degree of superheat	過熱度
과열(수)증기	superheated vapour	過熱(水)蒸氣
과열층	overheating layer	過熱層
과완화	over-relaxation, overrelax	過緩和
과융해	superfusion	過融解
과잉	excess	過剩
과잉개발	overexploitation	過剩開發
과잉계수	coefficient of excess	過剩係數
과잉공기율	excess air ratio	過剩空氣率
과잉무지개	supernumerary bow, supernumerary rainbow	過剩-
과잉씨뿌리기	overseeding	過剩-
과잉제약조건문제	overconstrained problem	過剩制約條件問題
과정감률	process lapse rate	過程減率
과포화	supersaturation	過飽和
과포화공기	supersaturated air	過飽和空氣
과포화구름	supersaturation cloud	過飽和-
과표본	oversample	過標本

과표본추출	over sampling	過標本抽出
과플루오로탄소	perfluorocarbons	過-炭素
관개시설	irrigation facilities	灌漑施設
관계식	expression of relation	關係式
관목한계	shrub line	灌木限界
관상	physiognomy	觀相
관상방전	crown flash	冠狀放電
관설	snow cap	冠雪
관성	inertia	慣性
관성계	inertial system	慣性(座標)系
관성능률	moment of inertia	慣性能率
관성력	inertial force	慣性力
관성모멘트	moment of inertia	慣性-
관성반경	inertial radius	慣性半徑
관성반지름	inertial radius	慣性-
관성법칙	law of inertia	慣性法則
관성불안정(도)	inertial instability	慣性不安定(度)
관성아구간	inertial subrange	慣性亞區間
관성아층	inertial sublayer	慣性亞層
관성안정(도)	inertial stability	慣性安定(度)
관성영역	inertial range	慣性領域
관성예보	inertial forecast	慣性豫報
관성원	circle of inertia, inertia circle	慣性圓
관성저항	inertial resistance	慣性抵抗
관성좌표계	inertial coordinate system	慣性座標系
관성주기	inertial period	慣性週期
관성중력파	inertial gravity wave, inertio-gravitational wave	慣性重力波
관성지연	inertial lag	慣性遲延
관성진동	inertial oscillation	慣性振動
관성진동수	inertial frequency	慣性振動數
관성질량	inertial mass	慣性質量
관성파	inertia wave, inertial wave	慣性波
관성풍속	inertia velocity	慣性風速
관성항	inertia term	慣性項
관성흐름	inertial flow	慣性-
관수	submergence	冠水
관입	intrusion	貫入
관입대류	penetrative convection	貫入對流
관입법칙	cross cutting relationships	貫入法則
관제공역	controlled airspace	管制空域
관제센터	operation center	管制-
관제탑시정	control tower visibility	管制塔視程
관측	observation	觀測
관측각	observation angle	觀測角

관측거리빈	range bin	觀測距離-
관측거리접힘	range folding	觀測距離-
관측기구	observation balloon	觀測氣球
관측기기	observation instrument	觀測器機
관측노장	observation field	觀測露場
관측대	platform	觀測臺
관측망	network of observation, network of station, observational network, réseau	觀測網
관측망밀도	network density	觀測網密度
관측방법	observation methods	觀測方法
관측방정식	observation equation	觀測方程式
관측선	instrumentation ship	觀測船
관측소	observation post, observation station, observatory, observing station, station	觀測所
관측소고도	station elevation	觀測所高度
관측소기압	station atmospheric pressure, station pressure	觀測所氣壓
관측소망	station network	觀測所網
관측소수은기압계	station mercury barometer	觀測所水銀氣壓計
관측소연속도	station continuity chart	觀測所連續圖
관측소위치	station location	觀測所位置
관측시간	observation time	觀測時間
관측시뇌우	recent thunderstorm	觀測時雷雨
관측시눈	recent snow	觀測時-
관측시비	recent rain	觀測時-
관측시소낙눈	recent snow shower	觀測時-
관측시스템	observing system	觀測-
관측시스템모의실험	observing systems simulation experiment	觀測-模擬實驗
관측시어는비	recent freezing rain	觀測時-
관측시우박	recent hail	觀測時雨雹
관측시이슬비	recent drizzle	觀測時-
관측시진눈깨비	recent rain and snow	觀測時-
관측야장	field book	觀測野帳
관측연산자	observation operator	觀測演算子
관측오차	observational error	觀測誤差
관측오차공분산	observation error covariance	觀測誤差共分散
관측우물	observation well	觀測-
관측역	area of converage	觀測域
관측일	observational day	觀測日
관측자	observer	觀測者
관측자료	observational data	觀測資料
관측장소	site of station	觀測場所
관측주기	observation frequency	觀測週期
관측주파수	observation frequency	觀測周波數
관측증분	observation increment	觀測增分

관측지원선박	voluntary observing ship	觀測支援船舶
관측지점	observation point	觀測地點
관측지점번호	station designator	觀測地點番號
관측평가	observational validation	觀測評價
관측폭	swath width	觀測幅
관풍속계	tube anemometer	管風速計
관할공역기상	Jurisdiction Airspace weather	管轄空域氣象
관할구역	coverage	管轄區域
관할수역기상	Jurisdiction Sea Zone weather	管轄水域氣象
광거리	optical length, optical path	光距離
광경로차	optical path difference	光經路差
광과도현상탐지기	optical transient detector	光過渡現象探知機
광구	photosphere	光球
광굴절지수	optical refractive index	光屈折指數
광기전력효과	photovoltaic effect	光起電力效果
광기후	light climate, illumination climate	光氣候
광길이	optical length, optical path	光-
광년	light year	光年
광대역	broadband	廣帶域
광대역방출	broadband emissivity	廣帶域放出
광대역복사	broadband radiation	廣帶域輻射
광대역수감	broadband sensor	廣帶域隨感
광대한 범위(의)	macroscopic	廣大-範圍
광도	luminosity, candle power	光度
광도계	photometer	光度計
광도계급	luminosity class	光度階級
광량	quantity of illumination	光量
광(행)로	optical path, optical pathlength	光(行)路
광밀도	luminous density, optical density	光密度
광보상점	light compensation point	光補償點
광복사고온측정계	ardometer	光輻射高溫測定計
광분산	light dispersion	光分散
광분해	photodissociation, photolysis	光分解
광사용효율	light use efficiency	光使用效率
광산대기	mine atmosphere	鑛山大氣
광산화작용	photooxidation	光酸化作用
광선	light rays, ray	光線
(광)선추적	ray tracing	(光)線追跡
광속	luminance flux, speed of light	光束
광속(도)	velocity of light	光速(度)
광압	light pressure	光壓
광역예보구역	regional forecast area	廣域豫報區域
광자	photon	光子
광전광도계	photoelectric photometer	光電光度計

광전도전지	photoconductive cell	光電導電池
광전리	photoionization	光電離
광전습도계	photoelectric hygrometer	光電濕度計
광전식 적설계	photosnow cover meter	光電式積雪計
광전자	photoelectron	光電子
광전자증배관	photomultiplier tube	光電子增倍管
광전지	photoelectric cell	光電池
광전측광법	photoelectric photometry	光電測光法
광전투과계	photoelectric transmittancemeter	光電透過計
광전효과	photoelectric effect	光電效果
광정웨어	broad-crested weir	廣頂-
광주	luminous pillar	光柱
광주기성	photoperiodism	光週期性
광축	optical axis	光軸
광파	light wave	光波
광편광계	photopolarimeter	光偏光計
광포화	light saturation	光飽和
광포화점	light saturation point	光飽和點
광플럭스	luminous flux	光-
광학	optics, photology	光學
광학강수입자계	optical precipitation particle counter (disdrometer)	光學降水粒子計
광학계수기	optical counter	光學計數器
광학고온계	optical pyrometer	光學高溫計
광학공기질량	optical air mass	光學空氣質量
광학공기질량항	optical air mass term	光學空氣質量項
광학관측프로그램	optical observation program	光學觀測-
광학광산란측기	optical light-scattering instrument	光學光散亂測器
광학깊이	optical depth	光學-
광학두께	optical thickness	光學-
광학분광술	optical spectroscopy	光學分光術
광학사선거리	optical slant range	光學斜線距離
광학산란탐측기	optical scattering probe	光學散亂探測機
광학센서의 시선경로	line of sight	光學-視線經路
광학습도계	optical hygrometer	光學濕度計
광학식 적설계	optical snow cover meter	光學式積雪計
광학영상탐측기	optical imaging probe	光學映像探測機
광학우량계	optical rain gauge	光學雨量計
광학우적계	optical rainfall particle counter	光學雨滴計
광학유효대기	optically effective atmosphere	光學有效大氣
광학응결입자계수기	optical condensation particle counter	光學凝結粒子計數器
광학입경계	optical disdrometer	光學粒境界
광학입자계수기	optical (aerosol) particle counter	光學粒子計數機
광학입자방법	optical particle method	光學粒子方法

광학입자탐측기	optical particle probe	光學粒子探測機
광학적 균질	optically homogeneous	光學的均質
광학적 전달함수	optical transfer function	光學的傳達函數
광학적 전자센서	optical-electronic sensor	光學的電子-
광학적 착각	optical illusion	光學的錯覺
광학적 착시	optical illusion	光學的錯視
광학적 평활	optically smooth	光學的平滑
광학전달함수	optical transfer function	光學傳達函數
광학증폭지수	optical amplification factor	光學增幅指數
광학지역	optical region	光學地域
광학질량	optical mass	光學質量
광학현상	optical phenomenon	光學現像
광합성	photosynthesis	光合成
광합성산물	photosynthate	光合成産物
광합성유효복사	Photosynthetically Active Radiation	光合成有效輻射
광합성의 일중저하현상	midday depression of photosynthesis	光合成日中低下現象
광해리	photodissociation	光解離
광행차	aberration	光行差
광호흡	photorespiration	光呼吸
광화학	photochemistry	光化學
광화학반응	photochemical reaction	光化學反應
광화학반응(의)	actinic	光化學反應
광화학복사	actinic radiation	光化學輻射
광화학산화체	Photochemical oxidants	光化學酸化體
광화학선	actinic rays	光化學線
광화학스모그	photochemical smog	光化學-
광화학연무	photochemical smog	光化學煙霧
광화학저울	actinic balance	光化學-
광화학평형	photochemical equilibrium	光化學平衡
광화학플럭스	actinic flux	光化學-
광화학흡수	actinic absorption	光化學吸收
광환	corona	光環
광흡수 에어로졸	light absorbing aerosol	光吸收-
괴물빙상	monster ice sheet	怪物氷床
괴사	necrosis	壞死
괴틀러소용돌이	Görtler vortex	
괴틀러수	Görtler number	-數
교대단위텐서	alternating unit tensor	交代單位-
교대텐서	alternating tensor	交代-
교란토양시료	disturbed soil sample	攪亂土壤試料
교류	alternating current	交流
교점	node	交點
교정	calibration	較正
교정방정식	calibration equation	較正方程式

교정오류	miscalibration	校定誤謬
교정오차	calibration error	較正誤差
교정주기	calibration interval	較正週期
교집합	intersection	交集合
교차	intersect, range	交差
교차공분산	cross covariance	交叉共分散
교차미분	cross differentiation	交叉微分
교차상관	cross correlation	交叉相關
교차상관계수	cross correlation coefficient	交差相觀係數
교차상관방법	cross correlation method	交叉相關方法
교차소용돌이	crossed vortex	交叉-
교차송신	alternate transmission	交叉送信
교차스펙트럼	cross spectrum	交叉-
교차실험	cross-over experiment	交叉實驗
교차총량지수	cross totals index	交叉總量指數
교차편광	cross polarization	交叉偏光
교차편파성분	cross polar component	交叉偏波成分
교차편파송수신상관계수	cross polar correlation coefficient	交叉偏波送受信相關係數
교차편파신호	cross polarization signal	交叉偏波信號
교차현상	cross-over phenomenon	交叉現象
교착	agglutination	膠着
교환	exchange, Austausch	交換
교환계수	Austausch coefficient, coefficient of exchange, exchange coefficient, interchange coefficient	交換係數
교환계수가설	exchange coefficient hypothesis	交換係數假設
교환법칙	commutative law	交換法則
교환적분	exchange integral	交換積分
교환층	exchange layer	交換層
구	sphere	球
구간	interval, reach	區間
구간규모영상	interval scale image	區間規模映像
구간추정	interval estimation	區間推定
구경	aperture, caliber	口徑
구경각	angle of aperture	口徑角
구름	cloud	
구름가족	family of clouds	-家族
구름감쇠	cloud attenuation	-減衰
구름거울	cloud mirror	
구름경계	cloud boundary	-境界
구름경계선	nephcurve	-境界線
구름경로바람	cloud-track winds	-經路-
구름계	cloud system	-系
구름고도	cloud level	-高度
구름고도역	cloud étage	-高度域

구름과 땅 사이의 방전	cloud-to-ground discharge	-放電
구름과 수증기 추적	cloud and vapo(u)r tracking	-水蒸氣追跡
구름관측	cloud observation	-觀測
구름광학깊이	cloud optical depth	-光學-
구름광학두께	cloud optical depth	-光學-
구름구역	cloud sector	-區域
구름구조	cloud structure	-構造
구름그림자	cloud shadow	
구름(속)기류	cloud current	-氣流
구름기호	cloud symbol	-記號
구름기후	cloud climatology	-氣候
구름기후학	cloud climatology	-氣候學
구름깃발	cloud banner	
구름깊이	cloud depth	
구름꼬리표	cloud tags	
구름꼭대기	cloud top, top of cloud	
구름낀 경계층	cloud-topped boundary layer	-境界層
구름나선	cloud spiral	-螺線
구름난류	cloud turbulence	-亂流
구름높이	cloud height, height of cloud	
구름높이지시기	cloud height indicator	-指示器
구름단계	cloud stage	-段階
구름대	cloudy zone	-帶
구름대류	cloudy convection	-對流
구름대전	cloud electrification	-帶電
구름덩이	cloud mass	
구름도	cloud chart	-圖
구름도감	cloud atlas	-圖鑑
구름동영상	cloud image animation	-動映像
구름되먹임	cloud feedback	
구름두께	cloud depth, cloud thickness	
구름둑	cloud bank	
구름띠	band of cloud, cloud band, cloud streaks	
구름레이더	cloud radar	
구름막	cloud veil	-幕
구름많음	mostly cloudy	
구름(망원사진)편광계	cloud photopolarimeter	-(望遠寫眞)偏光計
구름매개변수화	cloud parameterization	-媒介變數化
구름머리	cloud head	
구름모양	cloud form	-模樣
구름모자	cloud cap, cap cloud	-帽子
구름모형	cloud model	-模型
구름모형기법	cloud model technique	-模型技法
구름무리	cloud cluster	

구름무지개	cloudbow	
구름물리관측장비	cloud physics observation instrument	-物理觀測裝備
구름물리학	cloud physics	-物理學
구름물리학적함축	cloud physical implication	-物理學的含蓄
구름미세구조	cloud microstructure	-微細構造
구름미세물리	cloud microphysics	-微細物理學
구름미세물리학모형	cloud microphysical model	-微細物理學模型
구름미풍	cloud-breeze	-微風
구름밑면	cloud base	-面
구름밑면고도	height of cloud base	-面高度
구름밑면기록계	cloud base recorder	-面記錄計
구름밑면온도	cloud base temperature	-面溫度
구름바다	sea of cloud, cloud sea	
구름바람	cloud winds	
구름반사거울	cloud reflector	-反射-
구름반사도	cloud reflectance	-反射度
구름반사율	cloud albedo	-反射率
구름발달	cloud development	-發達
구름방울	cloud drop	
구름방울병합설	drop coalescence theory	-倂合說
구름방울 크기	cloud droplet size	
(작은)구름방울성장	cloud droplet growth	-成長
구름방울수집기	drop collector	-收集器
구름방울스펙트럼	drop spectrum	
구름방울채집기	cloud drop sampler	-採集器
구름방울크기분광계	drop-size spectrometer	-分光計
구름방울크기분포	drop-size distribution	-分布
구름방울크기측정계	drop-size meter	-測定計
구름방울포착기	cloud droplet collector	-捕捉器
구름방전	cloud discharge	-放電
구름방출도	cloud emittance	-放出度
구름방패	cloud shield	-防牌
구름방향	cloud direction	-方向
구름배기	cloud detrainment	-排氣
구름변조	cloud modification	-變造
구름변종	cloud variety	-變種
구름복사강제력	cloud radiative forcing	-輻射强制力
구름분류	classification of clouds, cloud classification	-分類
구름분류방법	cloud classification scheme	-分類方法
구름분석	nephanalysis	-分析
구름분석도	neph chart	-分析圖
구름분해능모형	cloud resolving model	-分解能模型
구름불형성층	forbidden zone of cloud formation	-不形成層
구름빈도	cloud frequency	-頻度

구름빙정핵	cloud icing nucleus	-氷晶核
구름사이방전	cloud-to-cloud discharge, intercloud discharge	-放電
구름사진	cloud picture	-寫眞
구름사진측량기법	cloud photogrammetry	-寫眞測量技法
구름상도	cloud phase chart	-相圖
구름상부	cloud head	-上部
구름상부면	cloud deck	-上部面
구름상자	cloud chamber	-箱子
구름생성	cloud formation	-生成
구름선	cloud line, line of cloud	-線
구름성장	cloud growth	-成長
구름세포	cloud cells	-細胞
구름소산	cloud dispersal, cloud dissipation	-消散
구름소산기	cloudburster	-消散器
구름 속	inside of cloud	
구름속도	cloud speed, cloud velocity	-速度
구름 속 방전	cloud flash, cloud discharge, intracloud discharge	-放電
구름 속 번개	in-cloud lightning	
구름 속 비행	cloud flying	-飛行
구름수분량	cloud liquid water, water content of cloud	-水分量
구름수분함량	cloud water content, cloud liquid water content	-水分含量
구름숲	cloud forest	
구름시뮬레이터	cloud simulator	
구름시스템	nephsystem	
구름시트	cloud sheet	
구름식별	cloud identification	-識別
구름씨	seed of cloud	
구름씨뿌리기	cloud seeding	
구름씨뿌리기물질	cloud seeding agent	-物質
구름씨 살포장치	pyrotechnic flare	-撒布裝置
구름씨 투하기구	dropping flare	-投下器具
구름아래층	subcloud	-層
구름알갱이	cloud particle	
구름알고리즘	cloud algorithm	
구름알베도	cloud albedo	
구름양	amount of clouds, cloudage, cloud amount, cloudiness, cover of cloud, sky cover	-量
구름양 복사계	cloud cover radiometer	-量輻射計
구름 없는 시선	cloud-free line of sight	-視線
구름에코	cloud echo	
구름역학	cloud dynamics	-力學
구름요소	cloud element	-要素
구름용카메라	cloud camera	-用-

구름위(비행)	on top	(飛行)
구름위상	cloud phase	-位相
구름유출	cloud detrainment	-流出
구름응결고도	Cloud Condensation Level(CCL)	-凝結高度
구름응결핵	cloud condensation nucleus	-凝結核
구름응결핵계수기	cloud condensation nuclei counter	-凝結核計數機
구름의 종류	genera of cloud	-種類
구름의 종	cloud species	-種
구름의 흡광도	optical density of a cloud	-吸光度
구름이동	cloud movement	-移動
구름이동벡터	cloud motion vector	-移動-
구름일지	cloud diary	-日誌
구름입자	cloud particle, particle of cloud	-粒子
구름입자영상기	cloud particle imager	-粒子映像機
구름입자카메라	cloud particle camera	-粒子-
구름조각	cloud leaf, cloudlet	
구름조금	scattered, partly cloudy	
구름종	species of clouds	-種
구름종류	cloud genera, cloud genus	-種類
구름줄	cloud bar	
구름-지면 사이 번개	cloud-to-ground lightning	-地面-
구름-지면 사이 섬광	cloud-to-ground flash	-地面-閃光
구름지수화	cloud indexing	-指數化
구름총	cloud gun	-銃
구름촬영술	cloud photography	-撮影術
구름추정	cloud estimation	-推定
구름측정거울	mirror nephoscope	-測定-
구름측정기	nephoscope	-測定器
구름측정법	nephelometry	-測定法
구름층	cloud layer, cloud level	-層
구름캐노피	cloud canopy	
구름탐색기	cloud tracker	-探索機
구름탐조등	cloud searchlight	-探照燈
구름탐지레이더	cloud-detection radar	-探知-
구름투과도	cloud transmissivity	-透過度
구름틈새햇살	sun drawing water	
구름패턴	cloud pattern	
구름표류바람	cloud drift winds	-漂流-
구름필터링	cloud filtering	
구름학	nephology	-學
구름함양수	cloud water content	-涵養水
구름핵	cloud nucleus	-核
구름형	cloud type	-型
구름형성	cloud formation	-形成

구름-환경계	cloud-environment system	-環境系
구름휘도	cloud luminance	-輝度
구름흡수	cloud absorption	-吸收
구름흡입	cloud entrainment	-吸入
구리하라 격자	Kurihara grid	-格子
구멍	cavity	
구면곡률	spherical curvature	球面曲率
구면대칭	spherical symmetry	球面對稱
구면반사도	spherical albedo	球面反射度
구면알베도	spherical albedo	球面-
구면조화분석	spherical harmonic analysis	球面調和分析
구면조화함수	spherical harmonics	球面調和函數
구면좌표	spherical coordinates	球面座標
구면좌표계	spherical coordinate system	球面座標系
구면투영	spherical projection	球面投影
구면파	spherical wave	球面波
구면함수	spherical function	球面函數
구면효과측정기	spherical shell apparatus	球面效果測定器
구부	bulb	球部
구상번개	ball lightning, globe lightning	球狀-
구상투영	globular projection	球狀投影
구서훈련기상	Kuso weather	驅鼠訓練氣象
구성호흡	constructive respiration	構成呼吸
구속전하	bound charge	拘束電荷
구슬번개	beaded lightning, pearl lightning	
구심	centripetal	求心
구심가속도	centripetal acceleration	求心加速度
구심력	centripetal force	求心力
구역바람	sector wind	區域-
구역번호	block number	區域番號
구역예보	district forecast	區域豫報
구역최고제한고도	highest minimum sector altitude	區域最高制限高度
구역코드	block code	區域-
구역평균	block averaging	區域平均
구조분류	structural classification	構造分類
구조착빙	structural icing	構造着氷
구조토	patterned ground	構造土
구조함수	structure function	構造函數
구트라	guttra	
구티날씨	Guti weather	
구혈	kettle, pothole	甌穴
구형기구	spheric aerostat	球形氣球
구형일사계	ball pyranometer	球形日射計
구형전천일사계	spherical pyranometer, spherical pyrradiometer	球型全天日射計

구형지구복사계	spherical pyrgeometer	球型地球輻射計
구형화	balling	球形化
구획주사	sector scanning	區劃走査
구획주사지시기	sector scan indicator	區劃走査指示器
궤도변이	orbital variation	軌道變異
국가	state	國家
(미국)국가강우지수	National Rainfall Index	國家降雨指數
(미국)국가기후프로그램법	National Climate Program Act	國家氣候-法
(미국)국가날씨와작물요약	National Weather and Crop Summary	國家氣象作物要約
국가대기배출명세서	National Atmospheric Emission Inventory	國家大氣排出明細書
(영국)국가물질노출프로그램	National Materials Exposure Programme	國家物質露出-
국가시행계획	state implementation plan	國家施行計劃
(미국)국가식량보험프로그램	National Food Insurance Program	國家食糧保險-
(미국)국가작물손실평가망	National Crop Loss Assessment Network	國家作物損失評價網
(미국)국가적응계획	National Adaptation Plans of Action	國家適應計劃
(미국)국가조위기준년시대	National Tidal Datum Epoch	國家潮位基準年時代
국가지역대기질관측소	state and local air monitoring stations	國家地域大氣質觀測所
(미국)국가표준기압계	National Standard Barometer	國家標準氣壓計
(미국)국가허리케인운용계획	National Hurricane Operations Plan	國家-運用計劃
(미국)국가홍수요약	National Flood Summary	國家洪水要約
국내비행장예보	domestic aerodrome forecast	國內飛行場豫報
(미국)국립기상대	National Weather Service	國立氣象臺
국립기상센터	National Meteorological Centre	國立氣象-
(미국)국립기상협회	National Weather Association	國立氣象協會
국립기후센터	National Climate Center	國立氣候-
(미국)국립기후자료센터	National Climatic Data Center	國立氣候資料-
(미국)국립대기관측소	National Air Monitoring Stations	國立大氣觀測所
(미국)국립대기연구센터	National Center for Atmospheric Research	國立大氣研究
(미국)국립대기환경기준	National Ambient Air Quality Standards	國立大氣環境基準
(미국)국립디지털예보 데이터베이스	National Digital Forecast Database	國立-豫報
(미국)국립방사선방호측정 심의회	National Council on Radiation Protection and Measurements	國立放射線防護測定審議會
국립방사선센터	National Radiation Centre	國立放射線-
(미국)국립산불위험률시스템	National Fire Danger Rating System	國立山-危險率-
(미국)국립산성비평가프로그램	National Acid Precipitation Assessment Program	國立酸性-評價-
(미국)국립설빙자료센터	National Snow and Ice Data Center	國立雪氷資料-
(미국)국립아이스코어연구소	National Ice Core Laboratory	國立-研究所
(미국)국립악뇌우예보센터	National Severe Storms Forecast Center	國立惡雷雨豫報-
(미국)국립연구협의회	National Research Council	國立研究協議會
(미국)국립우주과학자료센터	National Space Science Data Center	國立宇宙科學資料-
(미국)국립자료부이센터	National Data Buoy Center	國立資料浮漂-
(미국)국립해양대기청	National Oceanic and Atmospheric Administration	國立海洋大氣廳
(영국)국립환경기술센터	National Environmental Technology Centre	國立環境技術-

(미국)국립환경예측센터	National Center for Environmental Prediction	國立環境豫測-
(미국)국립환경위성-자료-정보서비스	National Environmental Satellite, Data, and Information Service	國立環境衛星資料情報-
국부진동자	local oscillator	局部振動子
국소등방성	local isotropy	局所等方性
국제관측기간	international period	國際觀測期間
국제구름관측년	international cloud year	國際-觀測年
국제구름도감	international cloud atlas	國際-圖鑑
국제극관측년	international polar year	國際極觀測年
국제기상기구	International Meteorological Organization(IMO)	國際氣象機構
국제기상전문부호	international weather code	國際氣象電文符號
국제기상전문형식	international code, international weather code	國際氣象電文型式
국제기상코드	international code	國際氣象-
국제기상텔레프린터망	International Meteorological Teleprinter Network	國際氣象-網
국제기상통신망	International Meteorological Telecommunication Network(IMTN)	國際氣象通信網
국제기준대기	international reference atmosphere	國際基準大氣
국제기호	international symbol	國際記號
국제날짜선	International Date Line	國際-線
국제단위	International Units	國際單位
국제단위계	SI system, Système International d'Unités	國際單位系
국제민간항공기구	International Civil Aviation Organization	國際民間航空機構
국제민간항공조약	International Civil Aviation Agreement	國際民間航空條約
국제민간항공조약부속서	International Civil Aviation Agreement Annex	國際民間航空條約附屬書
국제부호	international symbol, international signs	國際符號
국제분석전문부호	international analysis code	國際分析電文符號
국제비행장예보	international aerodrome forecast	國際飛行場豫報
국제시간국	International Time Bureau	國際時間局
국제십진분류법	universal decimal classification	國際十進分類法
국제연합기후변화기획회의	United Nations Framework Convention on Climate Change	國際聯合氣候變化企劃會議
국제연합환경개발회의	United Nations Conference on Environment and Development	國際聯合環境開發會議
국제온도척도	international temperature scale	國際溫度尺度
국제운형약어	International Cloud Abbreviations	國際雲形略語
국제위성통신체제	International Satellite Broadcast	國際衛星通信體制
국제정온태양관측년	International Quiet Sun Year, International Years of the Quiet Sun	國際靜穩太陽觀測年
국제종관기상전문형식	international synoptic code	國際綜觀氣像電文型式
국제종관전문부호	international synoptic code	國際綜觀電文符號
국제종관지상관측전문부호	international synoptic surface observation code	國際綜觀地上觀測電文符號
국제지구물리관측년	International Geophysical Year	國際地球物理觀測年
국제지구물리협력	International Geophysical Cooperation	國際地球物理協力
국제직달일사척도	international pyrheliometric scale	國際直達日射尺度

국제촉광	international candle	國際燭光
국제측지학지구물리학연합	International Union of Geodesy and Geophysics	國際測地學地球物理學聯合
국제테이블칼로리	international table calorie	國際-
국제폭풍경보신호	international visual storm warning signal	國際暴風警報信號
국제표준대기	international standard atmosphere	國際標準大氣
국제항공로화산감시	International Airways Volcano Watch	國際航空路火山監視
국제항공항행을 위한 기상업무	Meteorological Service for International Air Navigation	國際航空航行-氣象業務
국제해사기구	International Maritime Organization	國際海事機構
국지(의)	local	局地
국지가속도	local acceleration	局地加速度
국지가열	local heating	局地加熱
국지각운동량	local angular momentum	局地角運動量
국지고기압	local anticyclone	局地高氣壓
국지기상	local weather	局地氣象
국지기후	local climate	局地氣候
국지기후학	local climatology	局地氣候學
국지낙진	local fallout	局地落塵
국지도함수	local derivative	局地導函數
국지마흐수	local Mach number	局地-數
국지면적	local area	局地面積
국지바람	local wind	局地-
국지번개계수기	local lightning counter	局地-計數器
국지번개섬광계수기	local lightning-flash counter	局地-閃光計數器
국지변동	local variation	局地變動
국지변화	local change, local variation	局地變化
국지비행공역기상	local flight area weather	局地飛行空域氣象
국지성강수	local precipitation	局地性降水
국지성뇌우	local thunderstorm	局地性雷雨
국지성스톰	local storm	局地性-
국지소용돌이도	local vorticity	局地-度
국지수송기상	local haul weather	局地輸送氣象
국지수평선	local horizon	局地水平線
국지순환	local circulation	局地循環
국지열역학적 평형	local thermodynamic equilibrium	局地熱力學的平衡
국지예보	local forecast	局地豫報
국지유입	local inflow	局地流入
국지유사가설	local similarity hypothesis	局地類似假說
국지임시관측	local extra observation	局地臨時觀測
국지작용	local action	局地作用
국지저기압	local cyclone	局地低氣壓
국지적 불연속선	local discontinuity line	局地的不連續線
국지축	local axis	局地軸
국지폭풍(우)	local storm	局地暴風(雨)

국지풍	local wind	局地風
국지효과	local effect	局地效果
군락	canopy	群落
군락광합성	canopy photosynthesis	群落光合成
군락광합성모형	canopy photosynthesis model, model of photosynthesis of a community	群落光合成模型
군락구조	canopy structure	群落構造
군사기상관측환경	military weather observation environment	軍事氣象觀測環境
군사기상기관	military weather agency	軍事氣象機關
군사기상변조	military weather modification	軍事氣象變造
군사기상이동관측	military weather mobile observation	軍事氣象移動觀測
군사기상정보망	military weather information network	軍事氣象情報網
군사기상특보	military weather flash	軍事氣象特報
군사기상학	military meteorology	軍事氣象學
군사산불부호	military fire code	軍事山-符號
군사이동정비기상	military weather mobile maintenance	軍事移動整備氣象
군사현업강설량기준	military operation snowfall standard	軍事現業降雪量基準
군사현업강수량기준	military operation rainfall standard	軍事現業降水量基準
군사현업기상	military operation weather	軍事現業氣象
군사현업얼음기준	military operation icing condition standard	軍事現業-基準
군사현업열지수	military operation heat index	軍事現業熱指數
군사현업풍속기준	military operation wind standard	軍事現業風俗基準
군사화재암호	military fire code	軍事火災暗號
군속도	group velocity	群速度
군집	community	群集
군집기후모형	community climate model	群集氣候模型
군집잡음표준	community noise standard	群集雜音標準
굳은 결빙	hard freeze	-結氷
굳은 상고대	hard rime, kernel ice	-霜高臺
굴곡기류	curved airflow	屈曲氣流
굴곡흐름	curved flow	屈曲-
굴드버그-몬 바람	Guldberg-Mohn wind	
굴뚝구름	chimney cloud	
굴뚝기류	chimney current	-氣流
굴뚝높이	chimney height	
굴뚝배기표본채취	stack sampling	-排氣標本採取
굴뚝배출물	stack effluent	-排出物
굴절	refraction	屈折
굴절각	angle of refraction	屈折角
굴절계	refractometer	屈折計
굴절계수	refraction coefficient	屈折係數
굴절광	refracted light	屈折光
굴절광선	refracted ray	屈折光線
굴절도	refractivity	屈折度

굴절률	index of refraction, refractivity	屈折率
굴절무늬법	schlieren method	屈折-法
굴절법칙	refraction law	屈折法則
굴절복사	refracted radiation	屈折輻射
굴절선도	refraction diagram	屈折線圖
굴절인자	refraction factor	屈折因子
굴절전구복사	refracted global radiation	屈折全球輻射
굴절지수	refraction index, refractive index	屈折指數
굴절지수인자	refractive index factor	屈折指數因子
굽이	sinuosity	
굽힘방식	bending mode	-方式
권	sphere	圈
권계면	pause, tropopause	圈界面
권계면고기압	tropopause high	圈界面高氣壓
권계면깔대기	tropopause funnel	圈界面-
권계면단절	tropopause break	圈界面斷絶
권계면단절선	tropopause break-line	圈界面斷絶線
권계면도	tropopause chart	圈界面圖
권계면돔	tropopause dome	圈界面-
권계면불연속	tropopause discontinuity	圈界面不連續
권계면역전	tropopause inversion	圈界面逆轉
권계면저기압	tropopause low	圈界面低氣壓
권계면접힘	tropopause folding	圈界面-
권계면제트(류)	tropopause jet (stream)	圈界面-(流)
권계면층	tropopause layer	圈界面層
권계면파	tropopause wave	圈界面波
권곡	cirques	圈谷
권곡빙하	cirque glacier	圈谷氷河
권역와해	sectoral breakdown	圈域瓦解
권운	cirrus, cirrus cloud	卷雲
권운덮개	cirrus canopy	卷雲-
권운사이구역	intercirrus region	卷雲-區域
권운시트	cirrus sheet	卷雲-
권운형	cirriform	卷雲形
권운형구름	cirriform cloud, cirrus-type cloud	卷雲形-
권적운	cirrocumulus, cirrocumulus cloud	卷積雲
권층운	cirrostratus, cirrostratus cloud	卷層雲
궤도	orbit, orbital	軌道
궤도강제	orbital forcing	軌道强制
궤도규모기후변화	orbital-scale climate change	軌道規模氣候變化
궤도규모탄소전달	orbital-scale carbon transfer	軌道規模炭素傳達
궤도동조	orbital tuning	軌道同調
궤도를 따라서 주사하는 복사계	along track scanning radiometer	軌道走査輻射計

궤도를 따른 주사	along track scanners	軌道走査
궤도면	orbit plane	軌道面
궤도몬순가설	orbital monsoon hypothesis	軌道季節風假說
궤도방향	along track direction	軌道方向
궤도방향해상도	along track resolution	軌道方向解像度
궤도번호	orbit number	軌道番號
궤도변동	orbital variation	軌道變動
궤도변이	orbital variation	軌道變異
궤도보정	orbital correction	軌道補正
궤도섭동	orbit perturbation	軌道攝動
궤도속도	orbital velocity	軌道速度
궤도요소	orbital element	軌道要素
궤도운동	orbital motion	軌道運動
궤도이심률	orbital eccentricity	軌道離心率
궤도인자	orbital parameter	軌道因子
궤도주기	orbital cycle, orbit period	軌道週期
궤도진동수	orbital frequency	軌道振動數
궤적	locus, path, track, trail, trajectory	軌跡
궤적곡률	trajectory curvature	軌跡曲率
궤적분석	trajectory analysis	軌跡分析
궤적제어	locus control	軌跡制御
귀납법	induction	歸納法
귀무가설	null hypothesis	歸無假說
규모	scale	規模
규모결정	scale decision	規模決定
규모고도	scale height	規模高度
규모분리	scale separation	規模分離
규모분석	scale analysis	規模分析
규모축소	downscaling	規模縮小
규모축소자료	downscaled data	規模縮小資料
규모화	scaling	規模化
규사	silica sand	硅沙
규조류	Aulacoseira granulata, diatoms	硅藻類
규칙	rule	規則
균시차	equation of time	均時差
균열생성	fracturing	龜裂生成
균일크기분포	monodispersed size distribution	均一-分布
균일흐름	uniform flow	均一-
균질결빙	homogeneous freezing	均質結氷
균질계	homogeneous system	均質系
균질계열	homogeneous series	均質系列
균질권	homosphere	均質圈
균질권계면	homopause	均質圈界面
균질난류	homogeneous turbulence, homologous turbulence	均質亂流

균질대기	homogeneous atmosphere	均質大氣
균질류	uniform flow	均質流
균질매체	homogeneous medium	均質媒體
균질성	homogeneity	均質性
균질승화	homogeneous sublimation	均質昇華
균질역	homogeneous area	均質域
균질유체	homogeneous fluid	均質流體
균질응결	homogeneous condensation	均質凝結
균질핵화	homogeneous nucleation	均質核化
균질핵화물질	homogeneous nucleating material	均質核化物質
균질화	homogenization	均質化
균형	balance	均衡
(기구의)균형고도	balance height(of aerostat)	均衡高度
균형계	balance system	均衡系
균형기	balancer	均衡器
균형년	balance year	均衡年
균형력	balanced force	均衡力
균형모형	balanced model	均衡模型
균형바람	balanced wind	均衡-
균형방정식	balance equation	均衡方程式
균형힘	balanced force	均衡-
그늘대	shadow zone	-帶
그늘범주	shadow criteria	-範疇
그늘온도	shade temperature	-溫度
그라쇼프 수	Grashof number	-數
그라스만규칙	Grassmann rule	-規則
그래프분석	graphical analysis	-分析
그래프판	graphing board	-板
그래프표시	graphic representation	-表示
그램당량	gram equivalent	-當量
그램분자량	gram molecular weight	-分子量
그램원자량	gram atomic weight	-原子量
그램원자부피	gram atomic volume	-原子-
그레이코드	Gray code	
그레이트 베이슨 고기압	Great Basin high	-高氣壓
그레이트해양 컨베이어벨트	great Ocean Conveyor Beat	-海洋-
그룹화	grouping	-化
그리니치상용시	Greenwich civil time	-常用時
그리니치시(태양시)	Greenwich apparent time, Greenwich time	-時
그리니치자오선	Greenwich meridian	-子午線
그리니치평균태양시	Greenwich mean time	-平均太陽時
그리니치항성시	Greenwich sidereal time	-恒星時
그린랜드고기압	Greenland anticyclone, Greenland high	-高氣壓
그린정리	Green's theorem	-定理

그린함수	Green's function	-函數
그림	figure	
그림자	shadow	
그림자광륜	glory	-光輪
그물전류	mesh current	-電流
그물증발계	screened pan	-蒸發計
그을리기	fumigation	
그을음	soot	
극	pole	極
극각	polar angle	極角
극값	extremes, extreme value	極-
극거리	polar distance	極距離
극고기압	polar high	極高氣壓
극고주파	extremely high frequency	極高周波
극관	polar-cap	極冠
극관대기	polar-cap atmosphere	極冠大氣
극관얼음	polar-cap ice	極冠-
극관측년	Polar Year	極觀測年
극관흡수	polar-cap absorption	極冠吸收
극광	aurora, polar aurora	極光
극광대	aurora bands, auroral bands, auroral zone	極光帶
극광등빈도선	isaurore	極光等頻度線
극광선	aurora ray, auroral rays	極光線
극광스톰	auroral storm	極光-
극광스트리머	aurora streamer	極光-
극광스펙트럼	aurora spectrum	極光-
극광음	auroral hiss	極光音
극광장막(커튼)	aurora draperies, auroral draperies, auroral curtains	極光帳幕
극광전자류	auroral electrojet	極光電子流
극광지대	aurora zone	極光地帶
극광초록선	auroral green line	極光草錄線
극광코로나	aurora corona, auroral corona	極光-
극광타원	auroral oval	極光楕圓
극광호	aurora arc	極光弧
극권	polar circle, polar zone	極圈
극궤도기상위성	polar orbital meteorological satellite	極軌道氣象衛星
극궤도위성	polar orbiting satellite	極軌道衛星
극궤도지구물리관측위성	polar orbiting geophysical observatory	極軌道地球物理觀測衛星
극단기후	extreme climate	極端氣候
극단날씨	extreme weather	極端-
극단날씨사건	extreme weather event	極限-事件
극대	maximum	極大
극대대평균 농도	peak to average concentration	極大對平均濃度

극대대평균 비	peak-to-mean ratio	極大對平均比
극대치	peak	極大値
극댓값	maximum value	極大-
극둘레기압골	circumpolar trough	極-氣壓-
극둘레소용돌이	circumpolar vortex	極-
극둘레저기압	circumpolar cyclone	極-低氣壓
극둘레지도	circumpolar map	極-地圖
극둘레편서풍	circumpolar westerlies	極-偏西風
극둘레회오리	circumpolar whirl	極-
극둘레흐름	circumpolar flow	極-
극방정식	polar equation	極方程式
극빙	polar ice	極氷河
극빙모	polar icecap	極氷帽
극빙하	polar glacier	極氷河
극상	climax	極相
극상식생	climax vegetation	極相植生
극성	polarity	極性
극성층운	polar stratospheric cloud	極成層雲
극세포	polar cell	極細胞
극소	minimum	極小
극소용돌이	polar vortex	極-
극솟값	minimum value	極小-
극순환	polar circulation	極循環
극심가뭄	devastatiing drought	極甚旱魃
극야	polar night	極夜
극야제트류	polar night jet stream	極夜-流
극자외복사	extreme ultraviolet radiation	極紫外輻射
극자외선	extreme ultraviolet	極紫外線
극저기압	polar cyclone, polar low, polar vortex	極低氣壓
극저주파	extremely low frequency	極低周波
극좌표	polar coordinates	極座標
극좌표계	polar coordinates system	極座標系
극지	polar region	極地
극지기상학	polar meteorology	極地氣象學
극지암흑	polar blackout	極地暗黑
극지자기폭풍	polar magnetic storm	極地磁氣暴風
극지화	arcticization	極地化
극초음속	hypersonic	極超音速
극축	polar axis	極軸
극투영	polar projection	極投影
극파	polar wave	極波
극편동풍	polar easterlies	極偏東風
극편동풍지수	polar easterlies index	極偏東風指數
극평사도법	polar stereographic projection	極平射圖法

극풍	polar wind	極風
극풍대	polar wind belt, polar wind zone	極風帶
극풍분계	polar wind divide	極風分界
극한	limit	極限
극한대	cold cap	極寒帶
극한점	limiting point	極限點
극한지작전	cold weather operation	極寒地作戰
극현상	polar phenomenon	極現狀
근거리장	near field	近距離場
근극궤도위성	near-polar orbiting satellite	近極軌道衛星
근사	approximation	近似
근사식	approximate expression	近似式
근사온도풍방정식	approximate thermal wind equation	近似溫度風方程式
근사절대온도눈금	approximate absolute temperature scale	近似絶對溫度-
근사좌표계	approximate coordinate system	近似座標系
근사확대	approximation spread	近似擴大
근삿값	approximate value	近似-
근설	continuous snow cover	根雪
근실시간	near-real-time	近實時間
근의 공식	quadratic formula	根-公式
근일점	perihelion	近日點
근적외복사	near-infrared radiation	近赤外輻射
근적외선	near infrared	近赤外線
근적외선컴퓨터분광계	Near Infrared Mapping Spectrometer	近赤外線-分光計
근점	perifocus	近點
근접공중지원작전기상	close air support operation weather	近接支援作戰氣象
근접낙진	close-in fallout	近接落塵
근지점	perigee	近地點
근지점의 편각	argument of perigee	近地點偏角
근지점조석	perigean tide	近地點潮汐
근지점조차	perigean range	近地點潮差
근지점통과	perigee passage	近地點通過
글라이스베르크 주기	Gleissberg cycle	-週期
글로방전	glow discharge	-放電
글로브온도계	glove thermometer	-溫度計
금박막습도계	goldbeater's skin hygrometer	金薄膜濕度計
금속관액체 자기온도계	liquid(-in-metal) thermograph	金屬管液體自記溫度計
금속기압계	metal barometer, metallic barometer	金屬氣壓計
금속박편	foil	金屬薄片
금속부식	metal corrosion	金屬腐蝕
금속성미세분진	particulate metal	金屬性微細粉塵
금속온도계	metallic thermometer	金屬溫度計
금속조각	chaff	金屬-
금속조각살포	chaff seeding	金屬-撒布

금환식	annual eclipse	金環蝕
급격기후변화	abrupt climate change	急激氣候變化
급경사수로	rapids	急傾斜水路
급류	rapids	急流
급상승류	uprush	急上昇流
급속강화	rapid intensification	急速強化
급속업데이트주기	rapid update cycle	急速-週期
급속주사	rapid scan	急速走査
급속하강	rapid deepening	急速下降
급여울	ripple	急-
급첨(의)	leptokurtic	急尖
급첨도	leptokurtosis	急尖度
기	era	紀
기	radix	基
기각역	rejection region	棄却域
기간	period	期間
기간예보	period forecasting	期間豫報
기간최대	period maxima	期間最大
기간통계학	duration statistics	期間統計學
기간평균	period average (or mean)	期間平均
기간평균도	time-mean chart	期間平均圖
기계난류	mechanical ventilation	機械亂流
기계적 내부경계층	mechanical internal boundary layer	機械的內部境界層
기계적 설해	mechanical snow damage	機械的雪害
기계적 환기	mechanical ventilation	機械的換氣
기공	stomata	氣孔
기공저항	stomatal aperture, stomatal resistance	氣孔抵抗
기공전도도	stomatal conductance	氣孔傳導度
기관지염	bronchitis	氣管支炎
기구	aerostat, balloon	氣球
기구관측	balloon observation	氣球觀測
기구껍질	envelope	氣球-
기구덮개	balloon shroud, shroud	氣球-
기구바구니	balloon basket	氣球-
기구상승한계	ballonet ceiling	氣球上昇限界
기구실링	balloon ceiling	氣球-
기구용기상기록계	aerostat meteorograph	氣球用氣象記錄計
기구추적	balloon trajectory	氣球追跡
기구측정	balloon measurement	氣球測定
기구커버	balloon cover	氣球-
기구탐측	balloon sounding	氣球探測
기구평형고도	static ceiling	氣球平衡高度
기권	air-sphere	氣圈
기(체)권	gas-sphere, air-sphere	氣(體)圈

기기보정	calibration of an instrument	器機補正
기기오차보정	correction for instrument error	器機誤差補正
기능검사실험장치	functional validation bench	機能檢查實驗裝置
기능평점	skill score	技能評點
기니아해류	Guinea current	-海流
기단	air-mass	氣團
기단강수	air-mass precipitation	氣團降水
기단기상학	air-mass meteorology	氣團氣象學
기단기후학	air-mass climatology	氣團氣候學
기단날씨	air-mass weather	氣團日記
기단뇌우	air-mass thunderstorm	氣團雷雨
기단도	air-mass chart	氣團圖
기단력	air-mass calendar	氣團歷
기단발원지	air-mass source	氣團發源地
기단발원지역	air-mass source region	氣團發源地域
기단변질	air-mass modification, air-mass transformation, transformation of an air mass	氣團變質
기단변질실험	air-mass transformation experiment	氣團變質實驗
기단분류	air-mass classificationm, classification of air mass	氣團分類
기단분류시스템	air-mass classification system	氣團分類-
기단분석	air-mass analysis	氣團分析
기단불연속	air-mass discontinuity	氣團不連續
기단빈도	air-mass frequency	氣團頻度
기단선도	air-mass diagram	氣團線圖
기단성질	air-mass property	氣團性質
기단소나기	air-mass shower	氣團-
기단식별	air-mass identification	氣團識別
기단안개	air-mass fog	氣團-
기단온도	air-mass temperature	氣團溫度
기단윈드시어	air-mass wind shear	氣團-
기단유형도표	air-mass-type diagram	氣團類型圖表
기단의 특성	air-mass characteristic	氣團特性
기단이동	air-mass transport	氣團移動
기단정체모형	air stagnation model	氣團停滯模型
기댓값	expectation, expected value	期待-
기동조종	maneuvering	機動操縱
기동풍속	starting wind velocity	起動風速
기둥(빙정의)	column	
기둥눈결정	columnar snow crystals	-結晶
기둥모양결정	columnar crystal	-模樣結晶
기둥모양소용돌이	columnar vortex	-模樣-
기둥모형	column model	-模型
기둥재결합	columnar recombination	-再結合
기둥저항	columnar resistance	-抵抗

기둥함량비	column abundance	-含量比
기록계	recorder	記錄計
기록곡선	trace	記錄曲線
기록관측	record observation	記錄觀測
기록막대	recording arm, recording lever	記錄-
기록보관소	archives	記錄保管所
기록보관소자료	archiving data	記錄保管所資料
기록부	register	記錄簿
기록수신소	read-out station	記錄受信所
기록원	registrar	記錄員
기록원통	recording cylinder, recording drum	記錄圓筒
기록장비	recording equipment	記錄裝備
기록장치	recording instrument	記錄裝置
기록적 홍수	flood of record	記錄的洪水
기록주기	period of record	記錄週期
기록펜	recording pen	記錄-
기록펜촉	recording pointer	記錄-
기뢰감시기상	mine watching weather	機雷監視氣象
기류	air current, air flow, air stream, wind current, wind flow	氣流
(소규모의)기류	draft, draught	氣流
기류개념	air stream concept	氣流槪念
기류퇴적(현상)	banking of the current	氣流堆積
기름	oil	
기름유출	oil spill	-流出
기본관측시각	main standard time	基本觀測時刻
기본단위	fundamental unit	基本單位
기본방정식	basic equation	基本方程式
기본벡터	basis vector	基本-
기본상태	basic state	基本狀態
기본상호작용	fundamental interaction	基本相互作用
기본소용돌이	elemental vortex	基本-
기본수준면	datum level	基本水準面
기본시스템	basic system	基本-
기본유체역학방정식	fundamental equation of hydrodynamics	基本流體力學方程式
기본입자	elementary particle	基本粒子
기본전하량	elementary electric charge	基本電荷量
기본조직계	fundamental tissue system	基本組織系
기본지표면복사네트워크	basic surface radiation network	基本地表面輻射觀測網
기본진동	fundamental vibration	基本振動
기본침투율	basic intake rate	基本浸透率
기본컴퓨터모형	basic computer model	基本-模型
기본텐서	fundamental tensor	基本-
기본함수	basis function	基本函數

기본활주로기상	basic runway weather	簡易滑走路氣象
기본흐름	basic flow	基本-
기본흡입률	basic intake rate	基本吸入率
기브스기본방정식	Gibbs' fundamental equation	-基本方程式
기브스-뒤엥 관계	Gibbs-Duhem relation	-關係
기브스-뒤엥 방정식	Gibbs-Duhem equation	-方程式
기브스분포	Gibbs distribution	-分布
기브스상규칙	Gibbs' phase rule	-相規則
기브스에너지	Gibbs energy	
기브스자유에너지	Gibbs free energy	-自由-
기브스함수	Gibbs function	-函數
기브스-헬름홀츠 방정식	Gibbs-Helmholtz equation	-方程式
기브스현상	Gibbs phenomenon	-現象
기상(조건)	weather condition	氣象(條件)
기상감시	meteorological watch, weather watch	氣象監視
기상감시레이더	weather surveillance radar	氣象監視-
기상감시소	meteorological watch office	氣象監視所
기상감정	weather appraisal	氣象鑑定
기상감정사	weather appraiser	氣象鑑定士
기상감정업	weather appraisal business	氣象鑑定業
기상개조	weather modification	氣象改造
기상개황	Wetterlage	氣象槪況
기상결과보고	meteorological debriefing	氣象結果報告
기상경보	meteorological warning message, weather warning	氣象警報
기상경보발표문	weather warning bulletin	氣象警報發表文
기상계절	meteorological season	氣象季節
기상관측	meteorological observation, weather observation	氣象觀測
기상관측기구	meteorological balloon	氣象觀測氣球
기상관측망	meteorological network	氣象觀測網
기상관측부이	meteorological buoy	氣象觀測-
기상관측비행	weather flight	氣象觀測飛行
기상관측비행기	weather observation plane	氣象觀測飛行機
기상관측선	weather ship	氣象觀測船
기상관측소	meteorological observing station,	氣象觀測所
	meteorological station,	
	orological observing station,	
	meteorological observatory, weather station	
기상관측자	meteorological observer, weather observer	氣象觀測者
기상관측점	meteorological post	氣象觀測點
기상관측환경	environment of meteorological observation	氣象觀測環境
기상관할구(역)	meteorological district	氣象管轄區(域)
기상광학	meteorological optics	氣象光學
기상광학거리	meteorological optical range	氣象光學距離
기상교육	meteorological education	氣象敎育

기상구역	meteorological region	氣象區域
기상국	weather bureau	氣象局
(영국)기상국	Meteorological Office	氣象局
(이집트)기상국	Meteorological Authority	氣象局
(기상)국제지점번호	international index numbers	(氣象)國際地點番號
기상금융보험	weather finance and insurances	氣象金融保險
기상기계	meteorological instrument	氣象機械
기상기구	meteorological organization	氣象機構
기상기기 검증	calibration of meteorological instruments	氣象器機檢證
기상기동공간	weather maneuver space	氣象機動空間
기상기록(지)	meteorogram	氣象記錄(紙)
기상기록계	meteorograph	氣象記錄計
기상기술	weather technology	氣象技術
기상기호	meteorological symbols	氣象記號
기상 · 기후 영향 평가	weather and climate impact appraisal	氣象氣候影響評價
기상대	meteorological observatory, observatory, station, weather station	氣象臺
기상대대	weather squadron	氣象大隊
기상도	meteorological chart, meteorological map, weather chart	氣象圖
기상레이더	meteorological radar, weather radar	氣象-
기상레이더관측소	meteorological radar station	氣象-觀測所
기상레이더자료처리분석기	weather radar data processor and analyzer	氣象-資料處理分析器
기상로켓	meteorological rocket	氣象-
기상로켓관측망	meteorological rocket network	氣象-觀測網
기상목표물	meteorological target	氣象目標物
기상 및 기후 요소	elements of weather and climate	氣象氣候要素
기상방송	meteorological broadcast, weather broadcast	氣象放送
기상법	meteorology act	氣象法
기상변수	meteorological variable	氣象變數
기상병	meteorological disease, meteorotropic disease, meteorotropism	氣象病
기상병리학	meteoropathology	氣象病理學
기상보고	meteorological report	氣象報告
기상보험솔루션	weather insurance solution	氣象保險-
기상분대	weather squad	氣象分隊
기상분석	meteorological analysis	氣象分析
기상브리핑	weather briefing	氣象報告
기상사고보험	weather contingency insurance	氣象事故保險
기상사업자	weather business agent	氣象事業者
기상산업	weather industry	氣象産業
기상상담사	weather consultant	氣象相談士
기상상담업	weather consulting business	氣象相談業
기상상용표	meteorological table	氣象常用表

기상상태	Wetterlage	氣象狀態
기상생리반응	meteoropathic reaction	氣象生理反應
기상서비스대리인	weather service agent	氣象業務代理人
기상소대	weather platoon	氣象小隊
기상소송	weather litigation	氣象訴訟
기상소환	weather recall	氣象召還
기상시장	weather market	氣象市場
기상신호표시	meteorological signal mark	氣象信號表示
기상업무	meteorological service, weather service	氣象業務
기상역학	dynamic meteorology, meteorological dynamics	氣象力學
기상연감	meteorological yearbook	氣象年鑑
기상연구소	meteorological institute	氣象研究所
기상연보	annual climatological report	氣象年報
기상열역학	meteorological thermodynamics	氣象熱力學
기상영상기	meteorological imager	氣象映寫機
기상영역	meteorological realm	氣象領域
기상영향변수	weather impact variables	氣象影響變數
기상예보	meteorological forecast	氣象豫報
기상예보사	weather predictor	氣象豫報士
기상예보업	weather prediction business	氣象豫報業
기상예측	weather prediction	氣象豫測
기상요소	meteorological element, weather element	氣象要素
기상요소계열	meteorological element series	氣象要素系列
기상요소등치선	isometeoric line	氣象要素等值線
기상운동학	meteorological kinematics	氣象運動學
기상월보	monthly weather report	氣象月報
기상위성	meteor satellite, meteorogical satellite, meteorological satellite, Metor satellites, weather satellite	氣象衛星
기상음향학	meteorological acoustics, meterological acoustics	氣象音響學
기상인자	meteorological factor	氣象因子
기상자료	meteorological data	氣象資料
기상자료센터	meteorological data center	氣象資料-
기상자료처리	weather data processing	氣象資料處理
기상자원	meteorological resource	氣象資源
기상잡음	meteorological noise	氣象雜音
기상장비	meteorological equipment, meteorological instrument	氣象裝備
기상장비업	weather equipment business	氣象裝備業
기상재해	meteorological disaster	氣象災害
기상적도	meteorological equator	氣象赤道
기상전기학	atmospheric electricity	氣象電氣學
기상전대	weather group	氣象戰隊
기상전망	weather outlook, weather prospect	氣象展望

기상전문	weather message	氣象電文
기상전문부호	meteorological code	氣象電文符號
기상전문부호형식	meteorological code form	氣象電文符號形式
기상전문형식	weather code	氣象電文形式
기상전보(유선)	meteorological telegraph	氣象電報
기상전보식	meteorological code form	氣象電報式
기상전보식시스템	FM system	氣象電報式-
기상전시	meteorological display	氣象展示
기상정보	meteorological information, weather information	氣象情報
기상정보가치	weather value of information	氣象情報價値
기상정보민감도	weather-information-sensitive	氣象情報敏感度
기상정보송신	meteorological transmission	氣象情報送信
기상정역학	meteorological statics	氣象靜力學
기상정찰	meteorological reconnaissance, weather intelligence	氣象偵察
기상정찰(비행)	weather reconnaissance (flight)	氣象偵察(飛行)
기상정찰비행	meteorological reconnaissance flight	氣象偵察飛行
기상정찰전문형식	RECCO code	氣象偵察電文形式
기상정찰선	weather patrol ship	氣象偵察船
기상조사	meteorological survey	氣象調査
기상조석	meteorological tide, weather tide	氣象潮汐
기상조절	meteorological control, weather modification	氣象調節
기상조절법	weather modification law	氣象調節法
기상조절연소탄	flare for weather modification	氣象調節燃燒炭
기상주의보	weather advisory, weather watch	氣象注意報
기상증명	weather certification	氣象證明
기상지수	weather index	氣象指數
기상참모	weather staff officer	氣象參謀
(한국)기상청	Korea Meteorological Administration	氣象廳
(일본)기상청	Meteorological Agency	氣象廳
(오스트레일리아)기상청	Bureau of Meteorology	氣象廳
기상측기	weather instrument	氣象測器
기상측기 및 관측	meteorological instruments and observations	氣象測器觀測
기상텔레타이프통신망	meteorological teleprinter network	氣象-通信網
기상통계학	meteorological statistics	氣象統計學
기상통보	meteorological message, weather report	氣象通報
기상통보문	meteorological message	氣象通報文
기상통신	meteorological communications, meteorological telecommunication, weather communication	氣象通信
기상통신망	meteorological telecommunication network	氣象通信網
기상특보	meteorological special report	氣象特報
기상학	meteorology	氣象學
기상학자	meteorologist	氣象學者

기상학적 등급	meteorological range	氣象學的等級
기상학적 시정	meteorological visibility	氣象學的視程
기상학적 야간시정	meteorological visibility at night	氣象學的夜間視程
기상학적 열대	meteorological tropics	氣象學的熱帶
기상학회	meteorological society	氣象學會
기상항공보고서	METAR	氣象航空報告書
기상해설	meteorological briefing	氣象解說
기상현상	meteorological phenomenon, weather phenomena	氣象現象
기상회보	meteorological bulletin	氣象會報
기수	brackish water, esturine water	汽水
기수-수소입자	odd hydrogen species	奇數水素粒子
기수역	esturine zone	汽水域
기술기반표준	technology-based standard	技術基盤標準
기술기상학	aerography	記述氣象學
기술기후학	technoclimatology	技術氣候學
기압	air pressure, barometric pressure	氣壓
기압(의)	barometric	氣壓
기압감부	pressure sensor	氣壓感部
기압감소	katabaric	氣壓減少
기압검정상자	altichamber	氣壓檢定箱子
기압경도	barometric gradient, pressure gradient	氣壓傾度
기압경도력	pressure gradient force	氣壓傾度力
기압경도원인	pressure gradient cause	氣壓傾度原因
기압경정표	barometric reduction table	氣壓更正表
기압경정	barometer reduction, pressure reduction	氣壓更正
기압경향	air pressure tendency, baric tendency, barometric tendency, pressure tendency	氣壓傾向
기압경향도	pressure tendency chart	氣壓傾向圖
기압경향방정식	pressure tendency equation	氣壓傾向方程式
기압계	air gauge, atmospheric barometer, barometer, pressure gauge, weather gauge	氣壓計
기압계	pressure system	氣壓系
기압계공합	pressure capsule	氣壓計空盒
기압계기차보정표	barometric correction table	氣壓計器差補正表
기압계보정	barometer corrections	氣壓計補正
기압계상자	barometer box, barometer case	氣壓計箱子
기압계수은조	barometer cistern	氣壓計水銀槽
기압계수은주	barometer column	氣壓計水銀柱
기압계영점고도	elevation of zero point of a barometer	氣壓計零點高度
기압계읽기	barometer reading	氣壓計-
기압계중심	center of pressure system	氣壓系中心
기압계표고	barometer elevation	氣壓計標高
기압고도	barometric height, pressure altitude, pressure height	氣壓高度

기압고도계	barometric altimeter, pressure altimeter	氣壓高度計
기압고도공식	barometric height formula	氣壓高度公式
기압골	pressure trough, trough	氣壓-
기압골선	trough line	氣壓-線
기압골축	axis of trough	氣壓-軸
기압구역	baric area	氣壓區域
기압급승	pressure surge	氣壓急昇
기압급승선	pressure surge line	氣壓急昇線
기압기둥	barometric column	氣壓柱
기압기록	barogram	氣壓記錄
기압기록계	barograph	氣壓記錄計
기압넓어짐	pressure broadening	氣壓-
기압높낮이	baric topography	氣壓-
기압도	pressure chart	氣壓圖
기압마루	pressure ridge	氣壓-
기압마루축	axis of ridge	氣壓-軸
기압면고도	pressure level	氣壓面高度
기압미세변화	microvariation of pressure	氣壓微細變化
기압미파	barometric ripple	氣壓微波
기압바람법칙	baric wind law	氣壓-法則
기압방정식	barometric equation	氣壓方程式
기압배치	pressure pattern	氣壓配置
기압배치비행	pressure pattern flight	氣壓配置飛行
기압배치형	isobar type	氣壓配置型
기압법칙	barometric law	氣壓法則
기압변동	barometric fluctuation, pressure variation	氣壓變動
기압변화	pressure change	氣壓變化
기압변화계	pressure variometer	氣壓變化計
기압변화기록계	pressure variograph	氣壓變化記錄計
기압변화도	pressure change chart	氣壓變化圖
기압변화율	barometric rate	氣壓變化率
기압보정	barometric correction, pressure correction	氣壓補正
기압분석	baric analysis	氣壓分析
기압분포	distribution of air pressure	氣壓分布
기압비약	pressure jump	氣壓飛躍
기압비약선	pressure jump line	氣壓飛躍線
기압상수	barometric constant	氣壓常數
기압상승(의)	anallobaric	氣壓上昇
기압상승중심	anallobaric center, isallobaric high, pressure rise center	氣壓上昇中心
기압센서	pressure sensor	氣壓-
기압스위치	barometric switch, baroswitch	氣壓-
기압실험실	altichamber	氣壓實驗室
기압연직분포	pressure profile	氣壓鉛直分布

기압오차	barometric error	氣壓誤差
기압온도상관	pressure-temperature correlation	氣壓溫度相關
기압-온도-습도탐측	pressure-temperature-humidity sounding	氣壓-溫度-濕度探測
기압온습계	barothermohygrometer	氣壓溫濕計
기압응력	pressure stress	氣壓應力
기압장	field of pressure, pressure field	氣壓場
기압좌표계	p system, pressure system, pressure coordinate system	氣壓座標系
기압중심	pressure center	氣壓中心
기압증가(의)	anabaric, ananllobaric	氣壓增加
기압최곳값	barometer maximum, barometric maximum	氣壓最高-
기압최저값	barometric minimum	氣壓最低-
기압측고법	barometric altimetry, barometric hypsometry	氣壓測高法
기압측정	pressure measurement	氣壓測定
기압측정법	barometry	氣壓測定法
기압특성	barometric characteristics	氣壓特性
기압파	barometric wave, pressure wave, surge	氣壓波
기압편차	pressure anomaly	氣壓偏差
기압평형상수	pressure equilibrium constant	氣壓平衡常數
기압하강선	katallobar	氣壓下降線
기압하강중심	isallobaric low, katallobaric center, pressure-fall center	氣壓下降中心
기압항력	pressure drag	氣壓抗力
기압효과	barometric effect	氣壓效果
기억	memory	記憶
기억장치	memory	記憶裝置
기여지역	contributing region	寄與地域
기여함수	contribution function	寄與函數
기온	air temperature, atmospheric temperature	氣溫
기온감률	temperature lapse rate	氣溫減率
기온감부	air temperature sensor	氣溫感部
기온교차	temperature range	氣溫較差
기온극값	extreme temperature	氣溫極-
기온노점차	depression of the dew point, dew point deficit, dew point spread, temperature dewpoint spread	氣溫露點差
기온역전	air temperature inversion, temperature inversion	氣溫逆轉
기온연직분포	air temperature profile, atmospheric temperature profile	氣溫鉛直分布
기온이슬점차	depression of the dew point, dew point deficit, dew point spread, temperature dewpoint spread	氣溫-點差
기온장	air temperature field	氣溫場
기온층	temperature layers	氣溫層
기운 (기압)골	tilted trough	-(氣壓)-
기울기	gradient	

기울기벡터	lamellar vector	
기울기항	tilting term, tipping term	-項
기울기효과	tilting effect	-效果
기울어진 곡선	skew curve	-曲線
기원	genesis, origin	起原
기입	plotting, spotting	記入
기입모형	plotting model	記入模型
기입부호	plotting symbol	記入符號
기입자	chart plotter	記入者
기입판	plotting board	記入板
기작	mechanism	機作
기저시간	base length, base period	基底時間
기저유출	base flow, base runoff	基底流出
기저유출저장	baseflow storage	基底流出貯藏
기저유출후퇴	baseflow recession	基底流出後退
기전력	electromotive force	起電力
기조력	tide-generating force	起潮力
기조퍼텐셜	tide-producing potential	起潮-
기주	air column	氣柱
(기주)중량도	weight chart	(氣柱)重量圖
기준고도	reference level	基準高度
기준관측소	base station, principal station	基準觀測所
기준기후관측소	reference climatological station	基準氣候觀測所
기준끓는점	normal boiling point	基準-點
기준대기	reference atmosphere	基準大氣
기준등압면	mandatory level	基準等壓面
기준라디오존데	reference radiosonde	基準-
기준량	basic(base) quantity	基準量
기준면	datum level	基準面
기준상태	reference condition	基準狀態
기준선	baseline	基準線
기준선감시	baseline monitoring	基準線監視
기준선점검	baseline check	基準線點檢
기준오염물질	criteria pollutant	基準汚染物質
기준온도	fiducial temperature	基準溫度
기준점	fiducial point, standard point	基準點
기준존데	reference sonde	基準-
기준좌표계	reference coordinate system	基準座標系
기준지침서	criteria document	基準指針書
기준진공챔버	reference vacuum chamber	基準眞空-
기준표지	fiducial mark	基準標識
기차보정	correction for instrument error	器差補正
기체	gas	氣體
기체계량기	gasometer	氣體計量器

기체고정화	fixing of atmospheric gas	氣體固定化
기체교환	gas exchange	氣體交換
기체농도측정장치	gas chromatograph	
기체론	gas theory	氣體論
기체반응법칙	law of gaseous reaction	氣體反應法則
기체방전	gas electric discharge, gaseous electric discharge	氣體放電
기체방정식	gas equation	氣體方程式
기체방출	outgassing	氣體放出
기체배출물질	gaseous effluent material	氣體排出物質
기체법칙	gas law	氣體法則
기체보정분광법	gas correction spectroscopy	氣體補正分光法
기체분자운동론	kinetic theory of gas	氣體分子運動論
기체상	gas phase	氣體相
기체상관분광법	gas correlation spectroscopy	氣體相關分光法
기체상반응	gas phase reaction	氣體相反應
기체상수	gas constant	氣體常數
기체상태	gaseous state	氣體狀態
기체성분	gaseous constituent	氣體成分
기체온도계	gas thermometer	氣體溫度計
기체유출	gaseous discharge	氣體流出
기체의 용해도	dissolutions of gas	氣體-溶解度
기체화	gasification	氣體化
기체확산	gaseous diffusion	氣體擴散
기체흡입	gas uptake	氣體吸入
기초대사	basal metabolism	基礎代謝
기초도	base map	基礎圖
기초방정식계	basic equational system	基礎方程式系
기초자료	base data	基礎資料
기판제어	onboard controls	基板制御
기포수위계	bubble gauge	氣泡水位計
기포터짐	bubble bursting	氣泡-
기포핵	bubble nucleus	氣泡核
기폭	air cataract	氣瀑
기하광선추적	geometrical ray tracing	幾何光線追跡
기하광학	geometric optics, geometrical optics	幾何光學
기하규모	geometric scale	幾何規模
기하미분	geometric differential	幾何微分
기하분포	geometric distribution	幾何分布
기하상사	geometric similarity	幾何相似
기하수평선	geometrical horizon	幾何水平線
기하(학)적 확률	geometric(geometrical) probability	幾何(學)的確率
기하평균	geometric mean, geometrical mean	幾何平均
기하학	geometry	幾何學
기호	symbol	記號

기화	gasification, vaporization	氣化
기화기착빙	carburetor icing	氣化器着氷
기화얼음	vaporizing ice	氣化-
기화열	heat of vaporization	氣化熱
기화잠열	latent heat of vaporization	氣化潛熱
기후	climate	氣候
기후각본	climate scenario	氣候脚本
기후감시	climate monitoring, climate watch	氣候監視
기후감시진단연구소	climate monitoring and diagnostics laboratory	氣候監視診斷研究所
기후강제	climatic forcing	氣候强制
기후강제력	climate forcing	氣候强制力
기후개량	climatic amelioration	氣候改良
기후개황	climatological summary	氣候概況
기후경계	climatic boundary	氣候境界
기후경관	climatic landscape	氣候景觀
기후경년경향	secular trend in climate	氣候經年傾向
기후경향	climatic trend	氣候傾向
기후계	climate system	氣候系
기후계절	climatic season	氣候季節
기후계절관측	climatic seasonal observation	氣候季節觀測
기후공식	climatic formula	氣候公式
기후과학	climate science	氣候科學
기후관측	climatological observation	氣候觀測
기후관측망	climatological network	氣候觀測網
기후관측분소	climatological substation	氣候觀測分所
기후관측소	climatological station	氣候觀測所
기후관측소기압	climatological station pressure	氣候觀測所氣壓
기후관측소망	climatological station network	氣候觀測所網
기후관측소표고	climatological station elevation	氣候觀測所標高
기후구	climate region, climatic province, climatic region	氣候區
기후구분	classification of climate, climate division, climatic division, climate classification, climatic classification	氣候區分
기후극값	climatic extremes	氣候極-
기후기록	climatic record	氣候記錄
기후기호	climatic symbol	氣候記號
기후난민	climate refugee	氣候亂民
기후내성	climatic tolerances	氣候耐性
기후년	climatic year, climatological year	氣候年
기후노이즈	climatic noise	氣候-
기후단위	climatic unit	氣候單位
기후대	climate zone, climatic belt, climatic zone	氣候帶
기후대리요소	climate proxy	氣候代理要素
기후대조	climatic contrast	氣候對照

기후데이터베이스	climate database	氣候-
기후도	climate atlas, climate diagram, climatic chart, climatic map, climatogram, climatological chart	氣候圖
기후도감	climatological atlas	氣候圖鑑
기후도표	climagram, climagraph, climatic diagram, climatograph	氣候圖表
기후되먹임	climate feedback	氣候-
기후되먹임기구	climate feedback mechanism	氣候-機構
기후-레저상호작용	climate-leisure interactions	氣候-相互作用
기후리듬	climatic rhythm	氣候-
기후모델링	climate modelling	氣候-
기후모의	climate simulation	氣候模擬
기후모형	climate model, climatic model	氣候模型
기후민감도	climate sensitivity, climatic sensitivity	氣候敏感度
기후변동	climate variability, climatic fluctuation, climatic variation	氣候變動
기후변동성	climatic variability	氣候變動性
기후변형	climatic alteration	氣候變形
기후변화	climate change, climatic change	氣候變化
기후변화과학프로그램	climate change science program	氣候變化科學-
기후변화되먹임	climate change feedback	氣候變化-
기후변화시나리오	climate change scenario	氣候變化-
기후변화에 관한 정부 간 협의체	Intergovernmental Panel on Climate Change	氣候變化-政府-協議體
기후변화영향분석그룹	climate change impacts review group	氣候變化影響分析-
기후변화 커뮤니케이션	communicating climate change	氣候變化-
기후변화탐지	climate change detection	氣候變化探知
기후변화퇴적물지표	sedimentary indicator of climate change	氣候變化積物指標
기후변화투영	climate change projection	氣候變化投影
기후변화협약	climate change convention, climatic change convention, Framework Convention on Climate Change	氣候變化協約
기후병리학	climatic pathology	氣候病理學
기후분류(법)	climate classification, climatic classification	氣候分類(法)
기후분산모형	climatological dispersion model	氣候分散模型
기후분할	climate divide, climatic divide	氣候分割
기후불변성	constancy of climate	氣候不變性
기후불안정(성)	climatic instability	氣候不安定(性)
기후불연속	climatic discontinuity	氣候不連續
기후비약	climatic jump	氣候飛躍
기후사고	climate accident	氣候事故
기후-사회 연구	climate-society study	氣候社會研究
기후생리학	climatic physiology	氣候生理學
기후생산력지수	climatic productivity index	氣候生産力指數

기후생성	climatogenesis	氣候生成
기후서비스	climate services	氣候-
기후선도	climatological diagram	氣候線圖
기후설선	climate snow line, climatic snowline	氣候雪線
기후순환	climatic cycle	氣候循環
기후스트레스	climatic stress	氣候-
기후시나리오	climate scenario	氣候-
기후시뮬레이션	climate simulation	氣候-
기후시스템	climatic system	氣候-
기후시험	climatic test	氣候試驗
기후신호	climate signal, climatic signal	氣候信號
기후실험실	climatizer	氣候實驗室
기후심리학	climatic psychology	氣候心理學
기후역학	climate dynamics	氣候力學
기후연구	climate research	氣候研究
기후연구소	climate research unit	氣候研究所
기후연표	climate clonological table	氣候年表
기후-엽다변량분석계획	climate-leaf analysis multivariate program	氣候葉多變量分析計劃
기후영향모형	climate impact models	氣候影響模型
기후영향연구	climate impact study	氣候影響研究
기후영향평가	climate impact assessment	氣候影響評價
기후예보	climate forecast, climate forecasting, climatic forecast	氣候豫報
기후예측	climate prediction, climatic prediction	氣候豫測
기후예측센터	climate prediction center	氣候豫測-
기후요동	climate fluctuation, climatic vacillation	氣候搖動
기후요법	climatic therapy, climatotherapy	氣候療法
기후요소	climate element, climatic element	氣候要素
기후위기	climatic risk	氣候危機
기후유사	climate analogs, climatic analogue	氣候類似
기후응용 및 자료에 관한 자문위원회	advisory committee on climate applications and data	氣候應用資料諮問委員會
기후이상	climate anomaly, climatic anomaly	氣候異常
기후-인간 되먹임	climate-human feedback	氣候-人間-
기후인자	climatic factor	氣候因子
기후자료	climate data, climatic data, climatological data	氣候資料
기후자료연보	climatological data annual report	氣候資料年報
기후자료출력	climate data output	氣候資料出力
기후자연변동성	natural variability of climate	氣候自然變動性
기후자원	climatic resources	氣候資源
기후잡음	climate noise, climatic noise	氣候雜音
기후장기변동	climatic revolution	氣候長期變動
기후장애	climatic barrier	氣候障碍
기후쟁점	climate issue	氣候爭點

기후저장소	climatological archives	氣候貯藏所
기후적응	acclimatization, climatization	氣候適應
기후전도도	climatological conductance	氣候傳導度
기후전망	climate outlook, climatic projection	氣候展望
기후전선	climatological front	氣候前線
기후점	climate point	氣候點
기후점프	climatic jump	氣候-
기후정보예측서비스	climate information and prediction service	氣候情報豫測-
기후조절	climate modification	氣候調節
기후조절(인자)	climatic control	氣候調節(因子)
기후주기	climate cycle	氣候週期
기후주기성	climate periodicity, climatic periodicity	氣候週期性
기후지	climatography	氣候誌
기후지도작성 및 예측계획	climatic mapping and prediction project	氣候地圖作成豫測計劃
기후지도작성예측계획	climatic mapping and prediction project	氣候地圖作成豫測計劃
기후지수	climate index, climatic index	氣候指數
기후지연	climatic retardation	氣候遲延
기후진동	climatic oscillation	氣候振動
기후체제	climatic regime	氣候體制
기후최적기	climatic optimum	氣候最適期
기후층서학	climatostratigraphy	氣候層序學
기후쾌적	climate comfort	氣候快適
기후통계(학)	climatological statistics	氣候統計(學)
기후통계학	climate statistics	氣候統計學
기후투영	climate projection	氣候投影
기후편이	climatic shift	氣候偏異
기후편차	climate anomaly, climatic anomaly	氣候偏差
기후평균값	climatic normal	氣候平均-
기후표류	climate drift	氣候漂流
기후표준평년값	climatological standard normals	氣候標準平年-
기후프록시자료	climate proxy data	氣候-資料
기후학	climamatology, climatology	氣候學
기후학자	climatologist	氣候學者
기후학적	climatological	氣候學的
기후학적 계절간진동	climatological intra-seasonal oscillations	氣候學的季節間振動
기후학적 규모	climatological scale	氣候學的規模
기후학적 방법	climatological method	氣候學的方法
기후학적 분할	climatological division	氣候學的分割
기후학적 예보	climatological forecast	氣候學的豫報
기후학적 저항	climatological resistance	氣候學的抵抗
기후학적 전망	climatological outlook	氣候學的展望
기후학적 접근	climatological approach	氣候學的接近
기후학적 평균연직분포	climatological mean profile	氣候學的平均鉛直分布
기후한계	climatic limit	氣候限界

기후형	climate type, climatic type	氣候型
기후형성분류(법)	genetic classification of climate	氣候形成分流(法)
기후환경	climatic environment	氣候環境
기후-환경반응	climate-environment response	氣候-環境反應
긴급구조대	emergency services	緊急救助隊
긴급항공지원기상	Immediate Air Support weather	緊急航空支援氣象
긴 방전	long discharge	-放電
긴 불꽃	long spark	
긴츠부르그-란다우 이론	Ginzburg-Landau theory	-理論
긴 펄스	long pulse	
긴 행렬공중전기	long-train atmospherics	-行列空中電氣
길베르규칙	Guilbert's rule	-法則
김	steam	
김실안개	steam mist	
김안개	steam fog, water smoke	
(바람기호의)깃	feather	
깃가지(풍속기호)	barb	
깃구름	collar cloud	
깃발구름	banner cloud	
깃털결정	feathery crystal	-結晶
깃털권운	cirrus plume	-卷雲
깊은 뿌리 식물	phreatophyte	-植物
깊은 편동풍	deep easterlies	-偏東風
깊이표시기	depth marker	-表示器
깔때기구름	funnel, funnel cloud, funnel column, pendant cloud, tuba	
깔때기대	funnel stand	-臺
깔때기에코	pendant echo	
깔때기적란운	cumulonimbus tuba	-積亂雲
깔때기적운	cumulus tuba	-積雲
깔때기효과	funnel effect, funneling effect, funneling	-效果
껍질딱정벌레	barker beetle	
껍질얼음	rind ice, shell ice	
꼬리고적운	altocumulus virga	-高積雲
꼬리고층운	altostratus virga	-高層雲
꼬리구름	fall streaks, Fallstreifen, virga	
꼬리권운	cirrus caudatus	-卷雲
꼬리권적운	cirrocumulus virga	-卷積雲
꼬리적란운	cumulonimbus virga	-積亂雲
꼬리적운	cumulus virga	-積雲
꼬리전선	trailing front	-前線
꼬리층적운	stratocumulus virga	-層積雲
꼬리흐름	wake	
꼬마열대저기압	midget tropical cyclone	-熱帶低氣壓

꼬마열대폭풍	midget tropical storm, midget typhoon	-熱帶颱風
꼬마저기압	cyclonette	-低氣壓
꼭대기	peak	
꽃가루	pollen	
꽃가루받이	pollination	
꽃가루병	hay fever	-病
꽃가루분석	pollen analysis	-分析
꽃샘추위	cold weather in blooming season, recurrence of cold	
꽃양배추구름	cauliflower cloud	-洋-
꽃잎(벚꽃)소나기	blossom shower	
꽃줄구름	festoon cloud	
끌개	attractor	
끓는점	boiling point	-點
끓는점 오름	boiling point elevation	-點-
끓는점 온도계	boiling point thermometer	-點溫度計
끓음	boil, boiling	
끝서리	latest frost	
끝속도	final velocity	-速度

한글	영문	한자
나날기압변동	interdiurnal pressure variation	-壓力變動
나날변동성	interdiurnal variability	-變動性
나날온도변동	interdiurnal temperature variation	-溫度變動
나노	nano	
나노기술	nanotechnology	-技術
나노미터	nanometer	
나르보네	narboné, narbonnasis	
나머지	residual	
나머지수분	residual moisture	-水分
나머지층	residual layer	-層
나무서리	air hoar, rime air hoar	
나뭇가지모양	dendrite	-模樣
나뭇가지모양결정	dendritic crystal, spatial dendrite	-模樣結晶
나뭇가지모양눈결정	dendritic snow crystal	-模樣-結晶
나블라연산자	nabla operator, del operator	-演算子
나비다드해류	Navidad current	-海流
나비에-스토크스 방정식	Navier-Stokes equations	-方程式
나비에-스토크스 운동방정식	Navier-Stokes equations of motion	-運動方程式
나비효과	butterfly effect	-效果
나쁜 시정	poor visibility	惡視程
나사선주사	helical scanning	螺絲線走査
나선구름띠	spiral cloud band	螺旋-
나선권운	spiral cirrus	螺旋卷雲
나선도	helicity	螺線度
나선띠	spiral band	螺旋-
나선띠구름	spiral band cloud	螺旋-
나선안테나	helical antenna	螺旋-
나선주사	spiral scanning	螺旋走査
나선층	spiral layer	螺旋層

(태풍의)나선팔	feeder band	螺線-
나선패턴	spiral pattern	螺旋-
나선풍속계	helicoidal anemometer	螺旋風速計
나선형 구름	cloud of spiral pattern	螺旋形-
나시	nashi, n'aschi	
나이퀴스트간격	Nyquist interval	-間隔
나이퀴스트속도	Nyquist velocity	-速度
나이퀴스트수	Nyquist folding number	-數
나이퀴스트주파수	Nyquist frequency	-周波數
나이퀴스트파장	Nyquist wavelength	-波長
나이테기후학	tree-ring climatology	-氣候學
나일강	Nile River	
나지	bare soil	裸地
나침반	compass	羅針盤
나타난 에너지	realized energy	實-
낙구법	falling sphere method	落球法
낙뢰	lightning strike, stroke, stroke of lightning, thunderbolt	落雷
낙뢰관측시스템	thunderbolt observation system	落雷觀測-
낙반	rockfall	落盤
낙스	NOx	
낙엽기	defoliation time	落葉期
낙엽층	duff	落葉層
낙진	dust fall, fallout	落塵
낙진도	fallout plot	落塵圖
낙진바람	fallout wind	落塵-
낙진바람벡터표시	fallout wind vector plot	落塵-標示
낙진영역기상	fallout area weather	落塵領域氣象
낙진유형	fallout pattern	落塵類型
낙진전선	fallout front	落塵前線
낙진채집조	dust fall jar	落塵採集槽
낙체운동	motion of falling body	落體運動
낙하	fall	落下
낙하구	falling sphere	落下球
낙하바람존데	dropwindsonde	落下-
낙하산라디오존데	parachute radiosonde	落下傘-
낙하속도	fall velocity	落下速度
낙하속력	falling speed	落下速力
낙하입자	falling particle	落下粒子
낙하존데	dropsonde	落下-
낙하존데관측	dropsonde observation	落下-觀測
낙하존데자료	dropsonde data	落下-資料
낙하존데투하장치	dropsonde dispenser	落下-投下裝置
낙하탐측	drop sounding	落下探測

낙회	ash fall	落灰
난기웅덩이	warm air drop, warm pool	暖氣-
난기이류	warm air advection	暖氣移流
난동해	injury by warm temperature in winter	暖冬害
난류	turbulence, turbulent flow	亂流
난류	warm current	暖流
난류강도	intensity of turbulence, turbulence intensity	亂流强度
난류경계층	turbulent boundary layer	亂流境界層
난류계수	turbulence coefficient	亂流係數
난류교환	turbulent exchange	亂流交換
난류구름	turbulence cloud	亂流-
난류구조구축	organizes turbulent structure	亂流構造構築
난류권	turbosphere	亂流圈
난류권계면	turbopause	亂流圈界面
난류규모	scale of turbulence	亂流規模
난류도	turbulence level	亂流度
난류돌풍	turbulent gust	亂流突風
난류매질	turbulent medium	亂流媒質
난류맴돌이	turbulent eddy	亂流-
난류상사이론	similarity theory of turbulence	亂流相似理論
난류성공기	bumpy air	亂流性空氣
난류성분	turbulence component	亂流成分
난류소산	turbulent dissipation	亂流消散
난류속	turbulent flux	亂流束
난류수준	turbulence level	亂流水準
난류스펙트럼	turbulence spectrum	亂流-
난류시어응력	turbulent shear stress	亂流-應力
난류에너지	turbulence energy, turbulent energy	亂流-
난류역	turbulent zone	亂流域
난류역전	turbulence inversion, turbulent inversion	亂流逆轉
난류영역	irregular regime	亂流領域
난류운동	turbulent motion	亂流運動
난류응결고도	turbulence condensation level	亂流凝結高度
난류전달	turbulent transfer	亂流傳達
난류체	turbulence body	亂流體
난류측정	turbulence measure	亂流測定
난류층밀림응력	turbulence shear stress	亂流層-應力
난류침투속도	turbulence penetration speed	亂流浸透速度
난류통계학의 매개변수화	parameterization of turbulence statistics	亂流統計學的媒介變數化
난류플럭스	turbulent flux	亂流-
난류혼합	turbulent mixing	亂流混合
난류확산	turbulent diffusion	亂流擴散
난류확산계수	turbulent diffusion coefficient	亂流擴散係數
난류후류	turbulent wake	亂流後流

난반사	diffuse reflection, irregular reflection	亂反射
난반사체	diffuse reflector	亂反射體
난방기후	heating climate	暖房氣候
난방도일	heating degree-day	暖房度日
난방지수	heating index	暖房指數
난산란	diffuse scattering	亂散亂
난센병	Nansen bottle	-甁
난수	random numbers	亂數
난역저기압	low with warm sector	暖域低氣壓
난정합	nonconformity	難整合
난층운	nimbostratus	亂層雲
난투수대	confining zone	亂透水帶
난투수층	aquiclude, confining unit	難透水層
날개끝비행운	wing-tip contrail	-飛行雲
날개단면	airfoil section	-斷面
날려쌓인 눈	driven snow, snowdrift	
날려쌓인 눈빙하	snowdrift glacier	-氷河
날려쌓인 눈얼음	snowdrift ice	
(바람에)날린	blowby	
날린 물보라	blowing spray	
날림눈	spillover	
날림먼지	drifting-blowing dust	
날림비	spillover	
날씨개황	weather situation	-槪況
날씨결정분석모형	weather decision-analytic model	-決定分析模型
날씨민감도결정	weather-sensitive decision	-敏感度決定
날씨민감사업	weather-sensitive business	-敏感事業
날씨민감사용자	weather-sensitive user	-敏感使用者
날씨민감소비자	weather-sensitive user	-敏感消費者
날씨반복	weather recurrence	-反復
날씨분석	weather analysis	-分析
날씨사고보험	weather contingency insurance	-事故保險
날씨상품	weather goods	天氣商品
날씨소송	weather litigation	-訴訟
날씨속담	weather proverb	-俗談
날씨순서	sequence of weather	-順序
날씨순환	weather cycle	-循環
날씨시스템	weather system	
날씨영향변수	weather impact variables	-影響變數
날씨예상	weather prognosis	-豫想
날씨예측	weather prediction	-豫測
날씨오보	false weather forecast	-誤報
날씨이상	weather anomaly	-異常
날씨전쟁	weather warfare	-戰爭

날씨조절	weather control	-調節
날씨파생상품	weather derivative	-派生商品
날씨형	weather type	-型
날씨효과	weather effect	-效果
날짜선	date line	-線
남극	Antarctic Pole, South Pole	南極
남극(의)	Antarctic	南極
남극계	Antarctic system	南極系
남극고기압	Antarctic anticyclone, Antarctic high	南極高氣壓
남극광	aurora australis	南極光
남극권	Antarctic Circle	南極圈
남극기단	Antarctic air (mass)	南極氣團
남극기후	Antarctic climate	南極氣候
남극대	Antarctic zone	南極帶
남극대륙	Antarctica	南極大陸
남극바다김안개	Antarctic sea smoke	南極-
남극바다얼음	Antarctic sea ice	南極-
남극발산	Antarctic divergence	南極發散
남극빙상	Antarctic ice sheets	南極氷床
남극빙하역사	Antarctic glaciation history	南極氷河歷史
남극성층권소용돌이	Antarctic stratospheric vortex	南極成層圈-
남극수렴	Antarctic convergence	南極收斂
남극순환수	Antarctic circumpolar water	南極循環水
남극순환파동	Antarctic circumpolar wave	南極循環波動
남극순환해류	Antarctic circumpolar current	南極循環海流
남극시스템	Antarctic system	南極-
남극심층수	Antarctic deep water	南極深層水
남극오존구멍	Antarctic ozone hole	南極-
남극오존층	Antarctic ozone layer	南極-層
남극저층수	Antarctic bottom water	南極底層水
남극전선	Antarctic front	南極前線
남극중층수	Antarctic intermediate water	南極中層水
남극진동	Antarctic oscillation	南極振動
남극추위반전	Antarctic cold reversal	南極-反轉
남극표층수	Antarctic surface water	南極表層水
남극한대소용돌이	Antarctic polar vortex	南極寒帶-
남극한대해류	Antarctic polar current	南極寒帶海流
남극한랭반전	Antarctic cold reversal	南極寒冷反轉
남극한랭역전	Antarctic cold reversal	南極寒冷逆轉
남극해빙	Antarctic sea ice	南極海氷
남극해빙역사	Antarctic sea ice history	南極海氷歷史
남대서양고기압	South Atlantic high	南大西洋高氣壓
남대서양해류	South Atlantic current	南大西洋海流
남동계절풍	southeastern monsoon	南東季節風

남동무역풍	southeast trades	南東貿易風
남동편풍	southeaster	南東偏風
남미열대우림	selvas	南美熱帶雨林
남반구	southern hemisphere	南半球
남반구아열대무풍대	calms of Capricorn	南半球亞熱帶無風帶
남방연간모드	southern annular mode	南方年間-
남방진동	southern oscillation	南方振動
남방진동지수	southern oscillation index	南方振動指數
남북교환	meridional exchange	南北交換
남북기압골형 저기압발생	meridional trough cyclogensis	南北氣壓谷型低氣壓發生
남북류	meridional flow	南北流
남북변동	latitudinal fluctuation	南北變動
남북순환	meridional circulation	南北循環
남북지수	meridional index	南北指數
남빙해	Antarctic Ocean	南氷海
남서계절풍	southwest monsoon	南西季節風
남서편풍	southwester, sou'wester	南西偏風
남실바다	smooth sea	
남실바람	light breeze, slight breeze	
남열대요란	south tropical disturbance	南熱帶擾亂
남적도해류	south equatorial current	南赤道海流
남중국해	South China Sea	南中國海
남중시	meridian transit time	南中時
남태평양수렴대	South Pacific convergence zone	南太平洋收斂帶
남태평양해류	South Pacific current	南太平洋海流
남풍	maparam	南風
남회귀선	Tropic of Capricorn	南回歸線
남획	overexploitation	濫獲
납	lead	鉛
납작지붕권운	cirrus deck	-卷雲
낭떠러지맴돌이	cliff eddy	
낮	daytime	
낮대기광	dayglow	-大氣光
낮미풍	day breeze	-微風
낮시정	daytime visual range	-視程
낮은 땅눈보라	drifting snow	
낮은 지수	low index	-指數
낮은 풍진	drifting dust	-風塵
낮활승류	daytime upslope	-滑昇流
내관수성	resistance to overhead flooding injury	耐冠水性
내냉수성	cold water resistance	耐冷水性
내도복성	lodging resistance	耐倒伏性
내동성	freezing hardiness	耐凍性
내륙	inland	內陸

내륙기후	inland climate	內陸氣候
내륙담수습지	inland freshwater wetlands	內陸淡水濕地
내륙도	inlandity	內陸度
내륙빙하	inland ice	內陸氷河
내륙사막	inland desert	內陸沙漠
내륙수	inland water	內陸水
내륙안개	inland fog	內陸-
내륙얼음	inland ice	內陸-
내리바람	fall wind	
내리바람기압골	lee trough	-氣壓-
내리바람쪽	lee side, leeward	
내리바람파	lee wave	-波
내림각	depression angle	-角
내병성	disease tolerance	耐病性
내부경계면	internal boundary	內部境界面
내부경계층	internal boundary layer	內部境界層
내부마찰	inner friction, internal friction	內部摩擦
내부물순환	internal water circulation	內部-循環
내부보정목표	internal calibration target	內部補正目標
내부 숨은 열	internal latent heat	內部-熱
내부에너지	internal energy	內部-
내부중력파	internal atmospheric gravity wave, internal gravity wave	內部重力波
내부채널 회귀검사	interchannel regression test	內部-回歸檢查
내부파	internal wave	內部波
내부혼합	internal mixture	內部混合
내삽(법)	interpolation	內揷(法)
내삽공식	interpolation formula	內揷公式
내삽방안	interpolation scheme	內揷方案
내상성	frost hardiness	耐霜性
내서성	heat tolerance	耐暑性
내설성	snow endurance	耐雪性
내수면변화	lake level variation	內水面變化
내습성	wet endurance	耐濕性
내연량	intensive quantity	內延量
내열성	heat tolerance	耐熱性
내염성	salt tolerance	耐鹽性
내인성과정	endogenetic processes	內因性過程
내적	inner product, dot product, scalar product	內積
내청역	inner audible zone	內聽域
내충성	insect resistance	耐蟲性
내한성	cold hardiness, cold resistance	耐寒性
내행성	inner planet	內行星
내후성	weather ability	耐候性

냅	nappe	
냉각	chilling, cooling	冷却
냉각거울습도계	chilled mirror hygrometer	冷却-濕度計
냉각거울이슬점온도계	chilled mirror dew-point thermometer	冷却-點濕度計
냉각계	coolometer	冷却計
냉각곡선	cooling curve	冷却曲線
냉각과정	cooling process	冷却過程
냉각능풍속계	cooling power anemometer	冷却能風速計
냉각단위	chill unit	冷却單位
냉각력기록계	frigorigraph	冷却力記錄計
냉각력측정계	frigorimeter	冷却力測定計
냉각률	cooling power, cooling rate	冷却率
냉각바람인자	chill wind factor	冷却-因子
냉각상자	cold box	冷却箱子
냉각시간	chill hour	冷却時間
냉각온도	cooling temperature	冷却溫度
냉각타워플룸	cooling tower plume	冷却-
냉기내성	cold weather resistance	冷氣耐性
냉기댐	cold air dam	冷氣-
냉기류	cold air drainage, cold air flow	冷氣流
냉기배기	cold air drainage	冷氣排氣
냉기하강류(의)	katabatic	冷氣下降流
냉기호	cold air lake	冷氣湖
냉방도일	cooling degree day	冷房度日
냉습	raw	冷濕
냉파	cool wave	冷波
냉하	cool summer	冷夏
냉하해	cold summer damage	冷夏害
냉해	cold summer damage, chilling injury, cold damage, cool summer damage	冷害
너셀트수	Nusselt number	-數
너울	swell	
너징계수	nudging coefficient	-係數
너징동화방법	nudging	-同化方法
(바람으로)넘어진	blowdown	
넘친 물 빙결	flood icing	-氷結
넘친 물 얼음	flooding ice	
넙적적운	cumulus humilis	-積雲
네른스트정리	Nernst's theorem	-定理
네메레	nemere	
네바다	nevada	
네바도스	nevados	
네뷸	nebule	
네브래스카빙하기	Nebraskan glacial	-氷河期

네온	neon	
네퍼	neper	
네펠로미터	nephelometer	
노대바람	storm, whole gale	
노대바람경보	whole gale warning	-警報
노랑눈	yellow snow	
노력	endeavour	努力
노르덴스크욜드선	Nordenskjöld line	-線
노르웨이	Norway	
노르웨이저기압모형	Norwegian cyclone model	-低氣壓模型
노르웨이학교(학파)	Norwegian School	-學校(學派)
노르웨이해	Norwegian Sea	-海
노르웨이해류	Norwegian current	-海流
노르웨이해 심층수	Norwegian Sea Deep Water	-海深層水
노르타다	nortada	
노만드정리	Normand's theorem	-定理
노바젬랴	Nova Zemlya	
노벨상	Nobel Prize	
노부인의 여름	Old Wives' summer	老夫人-
노선	path	路線
노스웨스터	northwester, nor'wester	
노스이스터	nor'easter, northeast storm	
노아	National Oceanic and Atmospheric Administration(NOAA)	
노아의 방주권운	Noah's Ark	-方舟卷雲
노을	glow	
노을구름	fiery cloud	
노이만문제	Neumann problem	-問題
노이호프선도	Neuhoff diagram	-線圖
노장	observation field	露場
노점	dew point	露點
노점감률	dew point lapse rate	露點減率
노점계	dew point instrument	露點計
노점공식	dew point formula	露點公式
노점기록계	dew point recorder	露點記錄計
노점불연속선	dew point line	露點不連續線
노점습도계	dew point hygrometer	露點濕度計
노점온도	dew point temperature	露點溫度
노점온도감률	dew point temperature lapse rate	露點溫度減率
노점전선	dew point front	露點前線
노점차	dew point depression	露點差
노점측정계	dew point apparatus	露點測定計
노출	exposure	露出
노출광	exposure	露出光
노출도	exposure	露出度

노치폭	notch width	-幅
노트	knot	
노토스	Notos	
노퍽섬	Norfolk Island	
녹는 고도	melting level	-高度
녹는 띠	melting band	
녹는 얼음	rotten ice	
녹는 온도	melting temperature	-溫度
녹는점	melting point	-點
녹는점내림	melting point depression	-點-
녹색선	green ray	綠色線
녹색섬광	green flash	綠色閃光
녹색성장	green growth	綠色成長
녹음	fusion, melting, unfreezing	
녹음곡선	melting curve	-曲線
녹음일기방송	transcribed weather broadcast	錄音日氣放送
녹지대	green belt, frostless zone	綠地帶
놀치는 바다	choppy sea	
농도	concentration, loading	濃度
농도경도	concentration gradient	濃度傾度
농도구덩이	concentration basin	濃度-
농도대류	concentration convection	濃度對流
농도분산	concentration variance	濃度分散
농도분포	concentration distribution	濃度分布
농무기	dense fog seasons	濃霧期
농밀고층운	altostratus densus	濃密高層雲
농밀구름	spissatus	濃密-
농사계절	agricultural season	農事季節
농업	agriculture	農業
농업가뭄	agricultural drought	農業-
농업기상관측소	agricultural meteorological station, agrometeorological station	農業氣象觀測所
농업기상예보	agrometeorological forecast	農業氣象豫報
농업기상학	agricultural meteorology, agrometeorology	農業氣象學
농업기후	agricultural climate	農業氣候
농업기후구	agroclimatic region	農業氣候區
농업기후구분	agroclimatic classification	農業氣候區分
농업기후지수	agroclimatic index	農業氣候指數
농업기후학	agricultural climatology, agroclimatology	農業氣候學
농업오염원	agricultural source	農業汚染源
농업일기예보	agricultural weather forecast	農業日氣豫報
농업지형기후학	agrotopoclimatology	農業地形氣候學
농업혁명	agricultural revolution	農業革命
농작물산출량	crop yield	農作物産出量

농작물산출량모형	crop yield model	農作物産出量模型
농작물습윤지수	crop moisture index	農作物濕潤指數
농작물열단위	crop heat unit	農作物熱單位
높날림눈	blowing snow	
높날림눈주의보	blowing snow advisory	-注意報
높날림먼지	blowing dust	
높날림모래	blowing sand	
높낮이	topography	
높새	Nopsae	
높새바람	Nopsae wind	
높은	high	
높은 권운	cirrus excelsus	-卷雲
높은 안개	high fog	
높은 에코	elevated echo	
높은 역전안개	high-inversion fog	-逆轉-
높은 지수	high index	-指數
높은 혼합층	elevated mixed layer	-混合層
높은 흐림	high overcast	
높이	altitude, height, level	
높이경도	height gradient	-傾度
높이이득	height gain	高度利得
뇌격	thunderstroke	雷擊
뇌격전류계	fulchronograph	雷擊電流計
뇌설	snow thunderstorm	雷雪
뇌우	electric storm, thunderstorm (rain), thunder shower	雷雨
뇌우강수	thundery precipitation	雷雨降水
뇌우계	brontometer	雷雨計
뇌우고기압	thunderstorm high	雷雨高氣壓
뇌우관측소	thunderstorm observing station	雷雨觀測所
뇌우구름	thunderstorm cloud	雷雨-
뇌우권운	thunderstorm cirrus	雷雨卷雲
뇌우기록계	brontograph, thunderstorm recorder	雷雨記錄計
뇌우난류	thunderstorm turbulence	雷雨亂流
뇌우대전	thunderstorm electrification	雷雨帶電
뇌우무리	thunderstorm cluster	雷雨-
뇌우바람	thunderstorm wind	雷雨-
뇌우복합	cluster of thunderstorms	雷雨複合
뇌우복합체	thunderstorm complex	雷雨複合體
뇌우선	line of thunderstorm, thunderstorm line	雷雨線
뇌우세포	thunderstorm cell	雷雨細胞
뇌우스콜	thundersquall	雷雨-
뇌우역학	thunderstorm dynamics	雷雨力學
뇌우유출류	thunderstorm outflow	雷雨流出流

뇌우일	thunderstorm day	雷雨日
뇌우저기압	thunderstorm cyclone	雷雨低氣壓
뇌우전기(학)	thunderstorm electricity	雷雨電氣(學)
뇌우전하분리	thunderstorm charge separation	雷雨電荷分離
뇌우코	crochet d'orage, crochet degrain	
뇌우클러스터	cluster of thunderstorms, thunderstorm cluster	雷雨-
뇌우형태	thunderstorm type	雷雨形態
뇌우환경	thunderstorm environment	雷雨環境
뇌우활동원	source of thunderstorm activity	雷雨活動源
뇌우활동지역	region of thunderstorm activity	雷雨活動地域
뇌운	thundercloud	雷雲
뇌운시스템	thundery cloud system	雷雲-
뇌전	thunder and lightning	雷電
뇌전기록계	keraunograph, ceraunograph	雷電記錄計
뇌 크기	brain size	
누가곡선	cumulative curve, mass curve	累加曲線
누년통계	statistics of successive years	累年統計
누더기실링	ragged ceiling	
누수	seepage	漏水
누적	accumulation, cumulation	累積
누적가능물손실	accumulated potential water loss	累積可能-損失
누적강수	accumulated precipitation	累積降水
누적강수량	cumulative precipitation	累積降水量
누적곡선	mass curve	累積曲線
누적냉각	accumulated cooling	累積冷却
누적도수	cumulative frequency	累積度數
누적도수그래프	cumulative frequency graph	累積度數-
누적도수분포도	ogive	累積度數分布圖
누적률	cumulant	累積率
누적모드	accumulation mode	累積-
누적분포함수	cumulative distribution function	累積分布函數
누적빈도분포	cumulative frequency distribution	累積頻度分布
누적상대도수	cumulative relative frequency	累積相對度數
누적오차	cumulative error	累積誤差
누적온도	cumulative temperature	累積溫度
(빙하의)누적지역	surplus area, accumulation area	累積地域
누출곡선	breakthrough curve	漏出曲線
눈	snow	
눈결정	snow crystal	-結晶
눈고도	snow level	-高度
눈고리밧줄	snow garland	
눈관측길	snow course	-觀測-
눈구름	snow cloud	
눈그을음	snowburn	

눈금	scale	
눈금매기기	scaling	
눈금매김	graduation	
눈금실린더	graduated glass cylinder	
눈기후	snow climate	-氣候
눈깃발	snow banner	
눈깊이	snow depth, snow lying	
눈꼬리구름	snow virga	
눈녹은 물	snow water	
눈녹은 홍수	freshet	-洪水
눈녹음	snow melt	
눈녹이	snow eater	
눈다리	snow bridging	
눈단계	snow stage	-段階
눈더미	snowbank	
눈덩이	snowball	
눈덩이지구	snowball earth	-地球
눈덩이지구 가설	snowball earth hypothesis	-地球假說
눈덮임	snow cover	
눈덮힌 분화구	snow-capped crater	-噴火口
눈막이	snowbreak, snow shield	
눈 물당량	water equivalent of snow	-當量
눈물방울기구	teardrop balloon	-氣球
눈밀도	density of snow, snow density	-密度
눈밀도계	density of snow gauge, snow density meter	-密度計
눈반짝임	snow blink	
눈발	snow bed	
눈방석	snow mat	-方席
눈밭	snow patch	
눈벌판	snowfield	
눈베개	snow pillow	
눈벽	eye wall	
눈벽굴뚝	eye-wall chimney	
눈보라	blizzard, snow storm	
눈보라경보	blizzard warning, snow storm warning	-警報
눈사태	avalanche, snowslide	-沙汰
눈사태막이	snow shed	-沙汰-
눈사태바람	flurry	-沙汰-
눈상자	snow bin	-箱子
눈소나기	snow flurry	
눈송이	flakes, snow flake	
눈스콜	snow squall	
눈실안개	snow mist	
눈싸라기	snow pellet, tapioca snow	

눈쌓임	snow pack	
눈얼음	snow ice	
눈얼음검증서비스	snow and ice verification service	-檢證-
눈에 보이지 않는 강수	occult precipitation	-降水
눈에 보이지 않는 강하물	occult deposition	-降下物
눈연기	snow smoke	-煙氣
눈영역	snow regime	-領域
눈 온 날	snow day	
눈요동	snow tremor	-搖動
눈울타리	snow fence	
눈을 좋아하는	chinophile	
눈의 변형	metamorphosis of snow	-變形
눈의 침식	nivation	
눈입자	snow particle	-粒子
눈절단기	snow cutter	-切斷機
눈주의보	snow advisory	-注意報
눈채집관	snow tube	-採集管
눈처마	cornice, snow cornice	
눈측정	snow survey	-測定
눈치솟음	snow geyser	
눈콘크리트	snow concrete	
눈크러스트	snow crust	
눈패턴	eye pattern	
눈플룸	snow plume, snow banner	
눈피해	snow damage	-被害
눈하늘	snow sky, snow blink	
눈핵	snow core	-核
눈화환	snow garland	-花環
눈회오리	snow devil	
눈흘러내림	sluff	
뉴기니	New Guinea	
뉴기니연안잠류	New Guinea Coastal Undercurrent	-沿岸潛流
뉴잉글랜드해산	New England seamounts	-海山
뉴질랜드기상청	New Zealand Meteorological Service	-氣象廳
뉴질랜드북섬	North Island of New Zealand	-北-
뉴턴	newton	
뉴턴가설	Newton's hypothesis	-假說
뉴턴냉각법칙	Newton's law of cooling	-冷却法則
뉴턴-래프슨 해기법	Newton-Raphson solution technique	-解技法
뉴턴마찰법칙	Newtonian friction law, Newton's law of friction	-摩擦法則
뉴턴만유인력법칙	Newton's law of universal gravitation	-萬有引力法則
뉴턴반복	Newtonian iteration	-反復
뉴턴법칙	Newton's laws	-法則
뉴턴변환	Newtonian transformation	-變換

뉴턴식망원경	Newtonian telescope	-式望遠鏡
뉴턴역학	Newtonian mechanics	-力學
뉴턴운동법칙	Newton's law of motion,	-運動法則
	Newton's laws of motion	
뉴턴운동제1법칙	Newton's first law of motion	-運動第一法則
뉴턴운동제2법칙	Newton's second law of motion	-運動第二法則
뉴턴유체	Newtonian fluid	-流體
뉴턴음속	Newton speed of sound,	-音速
	Newtonian speed of sound	
뉴턴인력	Newton's attraction force	-引力
뉴턴인력법칙	Newton's law of attraction	-引力法則
뉴턴점성법칙	Newton's law of viscosity	-粘性法則
뉴펀들랜드해령	Newfoundland Ridge	-海嶺
(캐나다)뉴홀바람	Newhall wind	
느낌열	sensible heat	-熱
느낌열가열	sensible heating	-熱加熱
느낌열흐름	sensible heat flow	-熱-
느낌온도	sensible temperature	-溫度
느린 이온	slow ion	
늑골구름	vertebratus	肋骨-
는개	mizzle	
늘림	stretching	
늘림변형	stretching deformation	-變形
늘바람	permanent wind	
능동계	active system	能動系
능동감지	active sensor	能動感知
능동면	active surface	能動面
능동시스템	active system	能動-
능동원격감지부	active remote sensor	能動遠隔感知部
능동적 상승활주면	active anafrontal surface	能動的上昇滑走面
능동측정기	active instrument	能動測定器
능동태양열집열기	active solar collector	能動太陽熱集熱器
능동형계기	active instrument	能動型計器
능동형공동복사계	active cavity radiometer irradiance monitor	能動型空洞輻射計
능동형극초단파장치	active microwave instrument	能動型極超短波裝置
능동형레이더	active radar	能動形-
능동형원격탐사	active remote sensing	能動形遠隔探査
능률	moment, power	能率
늦더위	afterheat	
늦서리	last frost, late hoarfrost	
늪	marsh, swamp	
늪가스	marsh gas	
니뇨지역	Niño region	-地域
니르타	nirta	

L

니발	nival	
니에베 페니텐테스	nieve penitente, penitent ice	
니제르삼각주	Niger Delta	-三角洲
니퍼막이	Nipher shield	
닐온도	Néel temperature, Curie temperature	-溫度
님버스위성	Nimbus satellite	-衛星
님탠해	Neamtan solution	-解
닝갈루 사루	Ningaloo reef	

ㄷ

한글	영문	한자
다각도관측	multiangular observation	多角度觀測
다각형	polygon	多角形
다공성얼음	porous ice	多孔性-
다단과정	cascade process	多段過程
다단샤워	cascade shower	多段-
다단충돌채집기	cascade impactor	多段衝突採集器
다단폭포	cascade	多段瀑布
다모수레이더	multiparameter radar	多母數-
다목적저수지	multiple-purpose reservoir	多目的貯水池
다방계수	coefficient of polytropy	多方係數
다방과정	polytropic process	多方過程
다방대기	polytropic atmosphere	多方大氣
다방변화	polytropic change	多方變化
다방상수	polytropic, polytropic constant	多方常數
다방에크스너함수	polytropic Exner function	多方-函數
다방온위	polytropic potential temperature	多方溫位
다변수객관분석	multivariate objective analysis	多變數客觀分析
다색광라이다	polychromatic lidar	多色光-
다색라이다	plidar	多色-
다색무리	pleochroic halo	多色-
다색회전주사 구름카메라	multi-color spin scan cloud camera	多色回轉走査-
다세포	multi-cell, multi-cellular	多細胞
다세포뇌우	multi-cell thunderstorm	多細胞雷雨
다세포대류폭풍	multi-cell convective storm	多細胞對流暴風
다세포폭풍	multi-cell storm	多細胞暴風
다세포형	multi-cell type	多細胞形
다시법칙	Darcy's law	-法則
다시속도	Darcian velocity	-速度
다양성	diversification	多樣性

다용도습도계	polymeter	多用途濕度計
다우기	pluvial period	多雨期
다우기후	rainy climate	多雨氣候
다운버스트	downburst	
다운스케일링	downscaling	
다운컨버터	downconverter	
다이나모이론	dynamo theory	-理論
다이나믹미터	dynamic meter	
다이나믹킬로미터	dynamic kilometer	
다이메틸설파이드	dimethyl sulfide	
다이메틸아세트아마이드	dimethyl acetamide	
다이아몬드먼지	diamond dust	
다이애드곱	dyadic product	
다이오드	diode	
다이옥산	dioxane	
다이옥신	dioxins	
다인	dyne	
다인스보상	Dines compensation	-補償
다인스보상원리	Dines compensation theorem	-補償原理
다인스복사계	Dines radiometer	-輻射計
다인스풍속계	Dines anemometer	-風速計
다인스풍압풍속기록계	Dines pressure anemograph	-風壓風速記錄計
다중결합	multiple bond	多重結合
다중경로전송	multipath transmission	多重經路傳送
다중계절접근	multiseasonal approach	多重季節接近
다중규모관측	multiscale observation	多重規模觀測
다중기록계	multiple register	多重記錄計
다중뇌격	multiple stroke	多重雷擊
다중도플러분석	multiple doppler analysis	多重-分析
다중반사	multiple reflection	多重反射
다중방전	multiple discharge	多重放電
다중분광강조	multispectral enhancement	多重分光強調
다중분광분석	multispectral analysis	多重分光分析
다중분광영상	multispectral imaging	多重分光映像
다중산란	multiple scattering	多重散亂
다중상관	multiple correlation	多重相關
다중상관계수	coefficient of multiple correlation, multiple correlation coefficient	多重相關係數
다중선형회귀함수	multiple linear regression functions	多重線型回歸函數
다중세포스톰	multi-cell storm	多重細胞-
다중센서	multi-sensor	多重-
다중스펙트럼 스캐너	multispectral scanner	多重-
다중승강이론	multiple incursion theory	多重昇降理論
다중시간 분석검증	multitemporal analysis verification	多重時間分析檢證

다중시간비율	multitemporal ratio	多重時間比率
다중시간 회귀분석	multitemporal regression analysis	多重時間回歸分析
다중안테나 레이더	multi-static radar	多重-
다중앙상블예측모형	multi-ensemble prediction model	多重-豫測模型
다중왕복에코	multiple trip return	多重往復-
다중점	multiple point	多重點
다중창기법	multiple-window technique	多重窓技法
다중채널해수면온도	Multi-Channel Sea Surface Temperature	多重-海水面溫度
다중카메라	multispectral camera	多重-
다중판별분석	multiple discriminant analysis	多重判別分析
다중평형	multiple equilibria	多重平衡
다중포트 연료분사	multiport fuel injection	多重-燃料噴射
다중회귀	multiple regression	多重回歸
다중회귀분석	multiple regression analysis	多重回歸分析
다층모형	multi-level model, multilayer model	多層模型
다항식	polynomial	多項式
다항식 확장	polynomial expansions	多項式-
다항함수	polynomial function	多項函數
다형성	polymorphism	多形性
다환 방향족 탄화수소	polycyclic aromatic hydrocarbons	多環芳香族炭化水素
단거리미사일작전기상	short range ballistic missile operation weather	短距離-作戰氣象
단계	stage	段階
단계순차검정	stepwise sequential test	段階順次檢定
단계회귀분석	stepwise regression analysis	段階回歸分析
단괴	nodule	團塊
단극전하	unipolar electrical charge	單極電荷
단기영상	short-interval image	短期映像
단기예보	short term forecast	短期豫報
단기(일기)예보	short range (weather) forecast	短期(日氣)豫報
단면	cross-section, section, profile	斷面
단면검사	cross-section test	斷面檢查
단면도	cross-section diagram	斷面圖
단면분석	cross-sectional analysis	斷面分析
단백석껍질	opaline shell	蛋白石殼
단봉분포	unimodal distribution	單峰分布
단분자두께의 필름	monomolecular film	單分子-
단산란가정	single-scattering approximation	單散亂假定
단산란알베도	single-scatter albedo	單散亂-
단색(의)	monochromatic	單色
단색광	monochromatic light	單色光
단색광(의)	spectral	單色光
단색복사	monochromatic radiance, monochromatic radiation	單色輻射
단선전위계	unifilar electrometer	單線電位計
단세포국지뇌우	single-cell local thunderstorm	單細胞局地雷雨

단세포뇌우	single-cell thunderstorm	單細胞雷雨
단세포대기순환모형	single-cell circulation model	單細胞大氣循環模型
단속성비	intermittent rain	斷續性-
단순 로지스틱 곡선	simple logistic curve	單純-曲線
단순크기분포	monodispersed size distribution	單純-分布
단스고-외슈거 사건	Dansgaard-Oeschger event	-事件
단스고-외슈거 주기	Dansgaard-Oeschger(D-O) cycle	-週期
단안정 레이더	monostatic radar	單安定-
단열(의)	adiabatic	斷熱
단열가열	adiabatic heating	斷熱加熱
단열감률	adiabatic lapse rate	斷熱減率
단열고체수 함량	adiabatic solid water content	斷熱固體水含量
단열곡선	adiabatic curve	斷熱曲線
단열과정	adiabatic process	斷熱過程
단열구역	adiabatic region	斷熱區域
단열근사	adiabatic approximation	斷熱近似
단열기준상태	adiabatic reference state	斷熱基準狀態
단열냉각	adiabatic cooling	斷熱冷却
단열닫힌계	adiabatically enclosed system	斷熱-系
단열대기	adiabatic atmosphere	斷熱大氣
단열도	adiabatic chart	斷熱圖
단열미분방정식	adiabatic differential equation	斷熱微分方程式
단열방정식	adiabatic equation	斷熱方程式
단열법	adiabatic method	斷熱法
단열벽	insulation wall	斷熱壁
단열변화	adiabatic change	斷熱變化
단열불변량	adiabatic invariant	斷熱不變量
단열상당온도	adiabatic equivalent temperature	斷熱相當溫度
단열선	adiabat, adiabatics	斷熱線
단열선도	adabatic chart, adiabatic diagram	斷熱線圖
단열습구온도	adiabatic wet-bulb temperature	斷熱濕球溫度
단열승온	adiabatic warming	斷熱昇溫
단열시행	adiabatic trial	斷熱施行
단열압축	adiabatic compression	斷熱壓縮
단열액체수	adiabatic liquid water content	斷熱液體水
단열영역	adiabatic region	斷熱領域
단열온도경도	adiabatic temperature gradient	斷熱溫度傾度
단열온도변화	adiabatic temperature change	斷熱溫度變化
단열운동	adiabatic motion	斷熱運動
단열응결	adiabatic condensation	斷熱凝結
단열응결압력	adiabatic condensation pressure	斷熱凝結壓力
단열응결온도	adiabatic condensation temperature	斷熱凝結溫度
단열응결점	adiabatic condensation point	斷熱凝結點
단열조건	adiabatic condition	斷熱條件

단열층	adiabatic layer	斷熱層
단열팽창	adiabatic expansion	斷熱膨脹
단열평형	adiabatic equilibrium	斷熱平衡
단열폐쇄계	adiabatically enclosed system	斷熱閉鎖系
단열포화압력	adiabatic saturation pressure	斷熱飽和壓力
단열포화온도	adiabatic saturation temperature	斷熱飽和溫度
단열포화점	adiabatic saturation point	斷熱飽和點
단열효과	adiabatic effect	斷熱效果
단열흔적	adiabatic trail	斷熱痕跡
단위	cell	單位
단위격자	unit lattice	單位格子
단위부피	unit volume	單位-
단위세포	unit cell	單位細胞
단위유량도	unit hydrograph	單位流量圖
단위정규분포	unit normal distribution	單位正規分布
단일관측소분석	single-station analysis	單一觀測所分析
단일관측소예보	single observer forecast,	單一觀測所豫報
	single station forecast(ing)	
단일식물	short day plant	短日植物
단일제어지점	unit control position	單一制御地點
단일포트연료점화	single port fuel ignition	單一-燃料點火
단일해	unique solution	單一解
단절	break	中斷
단절기	chopper	斷截機
단절선	break line	斷絶線
단조화운동	simple harmonic motion	單調和運動
단주기	short-period	短周期
단(조화)진동	simple (harmonic) oscillation	單(調和)振動
단진동운동	simple harmonic motion	單振動運動
단진동파	simple harmonic wave	單振動波
단진자	simple pendulum	單振子
단측파대	single-sideband	單側波帶
단파	short wave	短波
단파골	short wave trough	短波-
단파복사	short wave radiation	短波輻射
단파쇠퇴	short wave fade	短波衰退
단파적외선스펙트럼	short wave infrared spectrum	短波赤外線-
단파조도	short wave irradiance	短波照度
단파채널	short wave channel	短波-
단편구름양	fractional cloud cover	斷片-量
단편류보정	single drift correction	單偏流補正
단편캐노피덮개	fractional canopy cover	斷片-
단편파레이더	single polarimetric radar	單偏波-
단풍전선	maple front	丹楓前線

닫힌 계	closed system	-系
닫힌 고기압	closed high	-高氣壓
닫힌 곡선	closed curve	-曲線
닫힌 구간	closed interval	-區間
닫힌 날씨	closed weather	-氣象
닫힌 등압선	closed isobar	-等壓線
닫힌 만	closed basin	
닫힌 (구름)세포	closed cells	-細胞
닫힌 세포층적운	closed-cell stratocumulus	閉細胞層積雲
닫힌 순환	closed circulation	-循環
닫힌 유역	closed drainage	-流域
닫힌 저기압	closed low	-低氣壓
닫힌 호수	closed lake	-湖水
달거리	lunar distance	-距離
달광환	lunar corona	-光環
달기둥	moon pillar	
달대기조석	lunar atmospheric tide	-大氣潮汐
달랑베르역설	d'Alembert's paradox	-逆說
달랑베르해	d'Alembert's solution	-解
달링소나기	darling shower	
달무리	lunar halo	
달무지개	lunar rainbow, moonbow	
달빛	moonlight	
달조석	lunar tide	-潮汐
달중력조	lunar gravitational tide	-重力潮
달탐측기	lunar probe	-探測器
닭 말라리아	avian malaria	
담배흡연	cigarette smoking	-吸煙
담수	ponding	湛水
담수계	fresh-water system	淡水系
담수렌즈	fresh-water lens	淡水-
담쾰러수	Damköhler number	-數
당연한 결과	consequent	當然結果
당화혈색소	carboxyhemoglobin	糖化血色素
대	era	代
대구법	parallelism	對句法
대규모	large scale, macroscale	大規模
대규모난류	macroturbulence	大規模亂流
대규모대기과정	large scale atmospheric processes	大規模大氣過程
대규모무리	macroscale cluster	大規模-
대규모상승	large scale lifting	大規模上昇
대규모소용돌이교환	gross-austausch	大規模-交換
대규모요란	large scale disturbance	大規模搖亂
대규모운동	large scale motion	大規模運動

대규모일기상황	large scale weather situation	大規模日氣狀況
대규모일기패턴	macroscale weather pattern	大規模日氣-
대규모클러스터	macroscale cluster	大規模-
대규모혼합	large scale mixing	大規模混合
대규모흐름	large scale flow	大規模-
대기	air, atmosphere	大氣
대기(의)	atmospheric	大氣
대기각	atmospheric shell	大氣殼
대기간섭	atmospheric interference	大氣干涉
대기감률	atmospheric lapse rate	大氣減率
대기감쇠	atmosphere attenuation, atmospheric attenuation	大氣減衰
대기감쇄	atmospheric extinction	大氣減殺
대기건조	atmospheric drought	大氣乾燥
대기경계층	atmospheric boundary layer	大氣境界層
대기경로길이	atmospheric path length	大氣經路-
대기공전강도	activity of a foyer of atmospherics	大氣空電强度
대기공전원	foyer of atmospherics	大氣空電源
대기과정규모	scale of atmospheric process	大氣過程規模
대기과제	air issue	大氣課題
대기과학	atmospheric sciences	大氣科學
대기과학(기상학)온도	atmospheric scientific(meteorological) temperature	大氣科學(氣象學)溫度
대기관측	aeroscopy	大氣觀測
대기광	airglow, permanent aurora	大氣光
대기광학	atmospheric optics	大氣光學
대기광학탐측	optical probing of the atmosphere	大氣光學探測
대기광학현상	low parry arc, optical meteor	大氣光學現象
대기구름	atmospheric cloud	大氣-
대기구조	atmospheric structure	大氣構造
대기굴절	atmospheric refraction, terrestrial refraction	大氣屈折
대기권	aerosphere, atmosphere	大氣圈
대기규모	atmospheric scales	大氣規模
대기난류	atmospheric turbulence	大氣亂流
대기난류풍동	atmospheric-turbulence wind tunnel	大氣亂流風洞
대기노정	path length of air	大氣路程
대기대순환	general circulation of atmosphere	大氣大循環
대기대순환론	theory of the general circulation of the atmosphere	大氣大循環論
대기대순환모형	atmospheric general circulation model	大氣大循環模型
대기도시열섬	atmospheric urban heat island	大氣都市熱島
대기독성	air toxic	大氣毒性
대기로켓관측	atmospherics rocket observation	大氣-觀測
대기맴돌이	atmospheric eddy	大氣-
대기먼지	atmospheric dust	大氣-
대기먼지현상	lithometeor	大氣-現象

대기모형	atmospheric model	大氣模型
대기모형상호비교프로젝트	atmospheric model intercomparison project	大氣模型相互比較-
대기물결	atmospheric billow	大氣-
대기물리	physics of the atmosphere	大氣物理
대기물리학	atmospheric physics	大氣物理學
대기물수지	atmospheric water budget	大氣-收支
대기물현상	hydrometeor	大氣-現象
대기미량원소	atmospheric tracer	大氣微量元素
대기밀도	atmospheric density	大氣密度
대기방사	atmospheric emission	大氣放射
대기방사전달	transfer of atmospheric radiation	大氣輻射傳達
대기방출률	atmospheric emissivity, emittance of the atmosphere	大氣放出率
대기변수	atmospheric variables	大氣變數
대기변질	atmospheric metamorphism	大氣變質
대기보정	atmospheric correction	大氣補正
대기복사	atmospheric radiation	大氣輻射
대기복사모듈	atmospheric radiance module	大氣輻射-
대기복사방출	solar radio emission	大氣輻射放出
대기복사수지	atmospheric radiation budget	大氣輻射收支
대기복원문제	atmospheric retrieval problem	大氣復元問題
대기복잡성	atmospheric complexity	大氣複雜性
대기분극	atmospheric polarization	大氣分極
대기분산	atmospheric dispersion	大氣分散
대기분수계	air shed	大氣分水系
대기불순물	atmospheric impurities	大氣不純物
대기불안정(도)	atmospheric instability	大氣不安定(度)
대기불투명도	atmospheric opacity	大氣不透明度
대기블로킹	atmospheric blocking	大氣-
대기빛현상	photometeor	大氣光現像
대기산란	atmospheric scattering	大氣散亂
대기산란체	atmospheric scattering body	大氣散亂體
대기산화	atmospheric oxidation	大氣酸化
대기산화용량	oxidizing capacity of atmosphere	大氣酸化容量
대기상학	macrometeorology	大氣象學
대기상한	upper limit of atmosphere	大氣上限
대기성질	atmospheric properties	大氣性質
대기성층	atmospheric stratification	大氣成層
대기세정효과	atmospheric scavening effect	大氣洗淨效果
대기세포	atmospheric cell	大氣細胞
대기소산	atmospheric extinction	大氣消散
대기소용돌이	atmospheric vortex	大氣-
대기속력	airspeed	大氣速力
대기수	meteoric water	大氣水

대기수명	atmospheric lifetime	大氣壽命
대기수분	atmospheric moisture	大氣水分
대기수상 비디오존데	hydrometeor videosonde	大氣水象-
대기수송	atmospheric transport	大氣輸送
대기수요량	atmospheric demand	大氣需要量
대기순환	atmospheric circulation, circulation of atmosphere	大氣循環
대기순환모형	atmospheric circulation model	大氣循環模型
대기시스템	atmospheric system	大氣-
대기안정도	atmospheric stability	大氣安定度
대기압	atmosphere, atmospheric pressure	大氣壓
대기야광	light-of-the-night-sky, airglow	大氣夜光
대기에너지균형	atmospheric energy balance	大氣-均衡
대기에너지론	atmospheric energetics	大氣-論
대기에어로졸	atmospheric aerosol	大氣-
대기역학	atmospheric dynamics, dynamics of the atmosphere	大氣力學
대기연구 대학연맹	University Corporation for Atmospheric Research (UCAR)	大氣硏究大學聯盟
대기연구 및 환경프로그램	Atmospheric Research and Environment Program	大氣硏究環境-
대기연직구조	atmospheric vertical structure	大氣鉛直構造
대기연직구조적외탐측기	atmospheric infrared sounder	大氣鉛直構造赤外探測機
대기열기관	atmospheric engine	大氣熱機關
대기열수지	atmospheric heat budget	大氣熱收支
대기열역학	atmospheric thermodynamics	大氣熱力學
대기영향	atmospheric influences	大氣影響
대기오염	atmospheric pollution, air pollution	大氣汚染
대기오염경보	air pollution alert	大氣汚染警報
대기오염관측소	air pollution observation station	大氣汚染觀測所
대기오염기상학	air pollution meteorology	大氣汚染氣象學
대기오염기준	air quality criteria	大氣汚染基準
대기오염물질	air pollutant, atmospheric pollutants	大氣汚染物質
대기오염방지	air pollution prevention	大氣汚染防止
대기오염방지법	air pollution control act	大氣汚染防止法
대기오염사건	air pollution episode	大氣汚染事件
대기오염통제	air pollution control	大氣汚染統制
대기오염퍼텐셜	air pollution potential	大氣汚染-
대기오존	atmospheric ozone	大氣-
대기온도	atmospheric temperature	大氣溫度
대기요란	atmospheric disturbance	大氣擾亂
대기운동	atmospheric motion	大氣運動
대기원격상관	atmospheric teleconnection	大氣遠隔相關
대기위험	atmospheric hazard	大氣危險
대기음향학	atmospheric acoustics	大氣音響學
대기의 화학적 조성	chemical composition of air	大氣化學的組成

대기이온	atmospheric ion	大氣-
대기이온화	atmospheric ionization	大氣-化
대기입자	atmospheric particle	大氣粒子
대기자료	atmospheric data	大氣資料
대기자원관리	atmospheric resource management	大氣資源管理
대기자원지역 오염평가모형	Air Resources Regional Pollution Assessment (ARRPA) Model	大氣資源地域汚染評價模型
대기잡음	atmospheric noise	大氣雜音
대기적도	atmospheric equator	大氣赤道
대기전기	atmospheric electricity	大氣電氣
대기전기강도계	radio maximograph	大氣電氣強度計
대기전기장	atmospheric electric field	大氣電氣場
대기전기현상	electrometeor	大氣電氣現象
대기전도도	atmospheric conductivity, air conductivity	大氣傳導度
대기전류	electric currents in the atmosphere	大氣電流
대기정역학	atmospheric statics	大氣靜力學
대기조석	atmospheric tide	大氣潮汐
대기조성	atmospheric composition, composition of atmosphere	大氣組成
대기종의 수지	budgets of atmospheric species	大氣種收支
대기중규모 우기 캠페인	atmospheric mesoscale wet season campaign	大氣中規模雨期-
대기중력파	atmospheric gravity wave	大氣重力波
대기중력풍	atmospheric gravity wind	大氣重力風
대기지	aerography	大氣誌
대기지역	atmospheric region	大氣地域
대기지질학	atmospheric geology	大氣地質學
대기지표층	atmospheric surface layer	大氣地表層
대기진동	atmospheric oscillation	大氣振動
대기진화	atmospheric evolution	大氣進化
대기질	air quality	大氣質
대기질관련조례	air quality act	大氣質關聯條例
대기질관리	air quality management	大氣質管理
대기질디스플레이모형	air quality display model	大氣質-模型
대기질량	atmospheric mass, mass of atmosphere	大氣質量
대기질모형화	air quality modeling	大氣質模型化
대기질범주	air quality criteria	大氣質範疇
대기질법률제정	air quality legislation	大氣質法律制定
대기질병	atmospheric disaster	大氣疾病
대기질순환	air quality cycle	大氣質循環
대기질제어지역	air quality control region	大氣質制御地域
대기질지수	air quality index	大氣質指數
대기질표준	air quality standard	大氣質標準
대기창역시험	window channel test	大氣窓域試驗
대기창	atmospheric window	大氣窓

대기천	atmospheric river	大氣川
대기청정법	clean air act	大氣淸淨法
대기추적분자형광분광술	atmospheric trace molecular spectroscopy	大氣追跡分子螢光分光術
대기층	atmospheric layer, atmospheric shell	大氣層
대기침적	atmospheric deposition	大氣沈積
대기침적망	atmospheric deposition networks	大氣沈積網
대기탐측로켓	atmospheric sounding projectile	大氣探測-
대기투과	atmospheric transmission	大氣透過
대기파	atmospheric wave	大氣波
대기파동	atmospheric wave motion	大氣波動
대기편광	atmospheric polarization	大氣偏光
대기-해양 대순환결합모형	coupled atmospheric and oceanic general circulation model	大氣海洋大循環結合模型
대기-해양 대순환모형	atmosphere-ocean general circulation model	大氣海洋大循環模型
대기-해양 상호작용	air-sea interaction, atmosphere-ocean interaction	大氣海洋相互作用
대기현상	atmospheric phenomenon	大氣現象
대기혼탁도	atmospheric turbidity	大氣混濁度
대기혼합	atmospheric mixing	大氣混合
대기-혼합층 해양모형	atmosphere-mixed layer ocean model	大氣混合層海洋模型
대기화학	atmospheric chemistry	大氣化學
대기확산	atmospheric diffusion	大氣擴散
대기확산방정식	atmospheric diffusion equation	大氣擴散方程式
대기효과	atmospheric effect	大氣效果
대기후	macroclimate	大氣候
대기후학	macroclimatology	大氣候學
대기흐름	atmospheric flow	大氣-
대기흡수	atmosphere absorption, atmospheric absorption	大氣吸收
대기흡수선	telluric line	大氣吸收線
대돌연풍	macroburst	大突然風
대량멸종	mass extinction	大量滅種
대류	convection	對流
대류가속	convective acceleration	對流加速
대류가열	convective heating	對流加熱
대류가용잠재에너지	convective available potential energy	對流可用潛在-
대류간격	convective interval	對流間隔
대류거품	convective bubble	對流-
대류경계층	convective boundary layer	對流境界層
대류경계층두께	convective boundary layer depth	對流境界層-
대류계	convective system	對流系
대류고도	ceiling of convection	對流高度
대류구름	convection cloud, convective cloud	對流-
대류구름높이선도	convective cloud height diagram	對流-線圖
대류구름무리	convective complex	對流-
대류구역	convective region	對流區域

대류권	troposphere	對流圈
대류권계면	lapse limit, tropopause	對流圈界面
대류권계면접힘현상	folding tropopause	對流圈界面-現象
대류권낙진	tropospheric fallout	對流圈落塵
대류권산란	troposcatter	對流圈散亂
대류권에어로졸	tropospheric aerosol	對流圈-
대류권오존	tropospheric ozone	對流圈-
대류규모	convective scale	對流規模
대류규모상호작용	convective scale interaction	對流規模相互作用
대류기류	convective current	對流氣流
대류난류	convective turbulence	對流亂流
대류뇌우	convective thunderstorm	對流雷雨
대류리처드슨수	convective Richardson number	對流-數
대류무리	convective cluster	對流-
대류발생점	burst point of convection	對流發生點
대류방정식	convection equation	對流方程式
대류불안정	convectively unstable	對流不安定
대류불안정(도)	convective instability	對流不安定(度)
대류불안정선	convectively unstable line	對流不安定線
대류성강수	convectional precipitation, convective precipitation	對流性降水
대류성기류	convection current	對流性氣流
대류성대기	convective atmosphere	對流性大氣
대류성두루마리구름	convective roller	對流性-
대류성분	convective component	對流成分
대류성비	convectional rain, convective rain	對流性-
대류성소나기	convective showers	對流性-
대류성토네이도	convective tornado	對流性-
대류성폭풍기폭메커니즘	convective storm initiation mechanism	對流性暴風起爆-
대류세포	cell of convection, convection cell, convective cell	對流細胞
대류속도	convection velocity	對流速度
대류수송	convective transport	對流輸送
대류수송이론	convective transport theory	對流輸送理論
대류순환	convectional circulation, convective circulation	對流循環
대류스케일링	convective scaling	對流-
대류스톰	convective storms	對流-
대류시간규모	convective time scale	對流時間規模
대류시그멧	convective SIGMET	對流-
대류안정	convectively stable	對流安定
대류안정(도)	convective stability	對流安定(度)
대류억제	convective inhibition	對流抑制
대류에너지	energy of convection	對流-
대류에코	convective echo	對流-
대류역전	convective inversion	對流逆轉

대류온도	convective temperature	對流溫度
대류요소	convective element	對流要素
대류응결고도	convection condensation level, convective condensation level	對流凝結高度
대류이론	convectional theory	對流理論
대류임펄스	convective impulse	對流-
대류적합층	convective matching layer	對流適合層
대류전류	convection current	對流轉流
대류전복	convective overturn	對流顚覆
대류조절	convective adjustment	對流調節
대류조절방안	convective adjustment scheme	對流調節方案
대류줄	convection street	對流-
대류지수	convective index	對流指數
대류질량플럭스	convective mass flux	對流質量-
대류층	convective layer	對流層
대류층운형기법	convective stratiform technique	對流層雲形技法
대류치올림	convective lifting	對流-
대류탱크	convection tank	對流-
대류파열점	burst point of convection	對流破裂點
대류평형	convective equilibrium	對流平衡
대류폭풍	convective storms	對流暴風
대류플럭스	convective flux	對流-
대류플룸	convective plume	對流-
대류항	convective term	對流項
대류현상	convective phenomenon	對流現象
대류혼합층	convective mixed layer	對流混合層
대류활동	convective activity	對流活動
대륙고기압	continental anticyclone, continental high	大陸高氣壓
대륙구름	continental cloud	大陸-
대륙권	continental sphere	大陸圈
대륙기단	continental air (mass)	大陸氣團
대륙(성)기후	continental climate	大陸(性)氣候
대륙날씨계	continental weather system	大陸日氣系
대륙내륙체제	continental interior regime	大陸內陸體制
대륙대지	continental platform	大陸臺地
대륙도	continentality	大陸度
대륙도계수	coefficient of continentality	大陸度係數
대륙도인자	continentality factor	大陸度因子
대륙도지수	continentality index, index of continentality	大陸度指數
대륙도효과	continentality effect	大陸度效果
대륙(성)바람	continental wind	大陸(性)-
대륙반구	continental hemisphere	大陸半球
대륙방송	continental broadcast	大陸放送
대륙붕	continental shelf	大陸棚

대륙붕파	continental shelf wave	大陸棚波
대륙빙상	continental ice sheet	大陸氷床
대륙빙하	continental glacier	大陸氷河
대륙사면	continental slope	大陸斜面
대륙성열대기단	continental tropical air (mass)	大陸性熱帶氣團
대륙성한대기단	continental polar air (mass)	大陸性寒帶氣團
대륙아북극기후	continental subarctic climate	大陸亞北極氣候
대륙얼음	continental ice	大陸-
대륙에어로졸	continental aerosol	大陸-
대륙연변땅	continental borderland	大陸緣邊-
대륙열저기압	continental thermal low	大陸熱低氣壓
대륙이동	continental drift	大陸移動
대륙(성)저기압	continental cyclone	大陸(性)低氣壓
대륙적운	continental cumulus	大陸積雲
대륙지각	continental crust	大陸地殼
대륙충돌	continental collision	大陸衝突
대머리구름	calvus	
대머리적란운	cumulonimbus calvus	-積亂雲
대머리적운	cumulus calvus	-積雲
대방사능대	great radiation belt	大放射能帶
대비	contrast	對備
대비단계	conditions of readiness	對備段階
대서	major heat	大暑
대서양고기압	Atlantic high	大西洋高氣壓
대서양분지	Atlantic basin	大西洋盆地
대서양블로킹에피소드	Atlantic blocking episode	大西洋-
대서양시간	Atlantic time	大西洋時間
대서양십년단위진동	Atlantic multidecadal oscillation	大西洋十年單位振動
대서양저기압	Atlantic depression	大西洋低氣壓
대서양컨베이어	Atlantic conveyor	大西洋-
대서양허리케인	Atlantic hurricane	大西洋-
대설	heavy snow, major snow	大雪
대설경보	heavy snow warning	大雪警報
대설주의보	heavy snow advisory	大雪注意報
대수성시험	aquifer test	帶水性試驗
대수정규분포	log-normal distribution	對數正規分布
대수제어지점	unit control position	帶水制御地點
대수층시스템	aquifer system	帶水層-
대수풍속분포	logarithmic velocity profile	對數風速分布
(대기)대순환	general circulation	(大氣)大循環
(대기)대순환모형	general circulation model	(大氣)大循環模型
대역분석	band analysis	帶域分析
대역 통과 여과기	band pass filter	帶域通過濾過器
대역 통과 필터	band pass filter	帶域通過-

대용식생	substitutional vegetation	代用植生
대원	great circle	大圓
대원경로	great-circle course	大圓經路
대원운항	great-circle navigation	大圓運航
대원항해	great-circle navigation	大圓航海
대응	correspondence	對應
대응점	corresponding point	對應點
대일점	anti-solar point	對日點
대일조	counterglow, gegenschein	對日照
대전	charge, electrification	帶電
대전구름	charged cloud	帶電-
대전빙정핵	electrification ice nucleus	帶電氷晶核
대전열	order of electrification	帶電列
대전판	charged plate	帶電板
대조	contrast	對照
대조구름	control cloud	對照-
대조구역	control area, contrast area	對照區域
대조문턱값	contrast threshold	對照門-
대조우물	control well	對照-
대조일	control day	對照日
대중매체예보	media forecast	大衆媒體豫報
대중매체예보자료	media forecast source	大衆媒體豫報資料
대지	plateau	大地
대지방전	flash-to-ground	對地放電
대지속도	ground speed	對地速度
대진폭	large amplitude	大振幅
대체공항	alternate airport	代替空港
대체에너지	alternative energy	代替-
대초원	prairie	大草原
대초원기후	prairie climate	大草原氣候
대칭변환	symmetric transformation	對稱變換
대칭불안정(도)	symmetric instability	對稱不安定(度)
대칭성분	symmetric component	對稱成分
대칭영역	symmetric regime	對稱領域
대칭점	point of symmetry, symmetry point	對稱點
대칭중심	center of symmetry, symmetrical center	對稱中心
대칭텐서	symmetric tensor	對稱-
대평원실험	Great Plain Experiment(GPE)	大平原實驗
대표기상관측점	representative meteorological observation	代表氣象觀測點
대표농도경로	representative concentration pathway	代表濃度經路
대표성	representativeness	代表性
대표성오차	error of representativeness	代表性誤差
대한	major cold	大寒
대한파	severe cold wave	大寒波

대한해류	Daehan current	大韓海流
대형일후형세	Grosswetterlage	大型日候形勢
대형증발계	large evaporimeter	大型蒸發計
대홍수	deluge	大洪水
더스트돔	dust dome	
더스트보울	dust bowl	
더위 시듦	summer depression	
더위탈진	heat exhaustion	-脫盡
덕트	duct	
덩어리(구름)	glomeratus	
덩어리 얼음	cake ice	
(공기)덩이	parcel	(空氣)-
(공기)덩이온도	parcel temperature	(空氣)-溫度
덩이유빙	slob ice	-流氷
데네캄프간빙기	Denekamp interstrade	-間氷期
데본기말멸종	end-Devonian extinction	-期末滅種
데시벨	decibel(db)	
데이비드슨-브라이언트 방정식	Davidson-Bryant equation	-方程式
데이비드슨해류(해)	Davidson current	-海流
데이비스수	Davies number	-數
데이터세트	dataset	
데이터세트명부	dataset directory	-名簿
데이터세트목록	dataset catalog	-目錄
데이터시스템시험	data system test	-試驗
데이터통신시스템	data communication system	-通信-
데카르트광선	Descartes ray	-光線
데판트모형	Defant's model	-模型
데판트법	Defant method	-法
데피그램	depegram	
덴디방법	Dendy method	-方法
델린저효과	Delinger effect	-效果
델연산자	del-operator	-演算子
도	degree	度
도	chart	圖
도나우-귄츠 간빙기	Donau-Günz interglacial	間氷期
도나우빙기	Donau Glacial	-氷期
도노라사건	Donora episode	-事件
도달각	angle of arrival	到達角
도달시간	time of concentration	到達時間
도량계수	metric coefficient	度量係數
도마뱀기구	lizard balloon	-氣球
도법	map projection	圖法
도배법칙	Dove's law	-法則

도복	lodging	倒伏
도브슨기기	Dobson instrument	-器機
도브슨단위	Dobson unit	-單位
도브슨분광광도계	Dobson spectrophotometer	-分光光度計
도브슨오존분광광도계	Dobson ozone spectrophotometer	-分光光度計
도수분포곡선	frequency distribution curve	度數分布曲線
도수분포다각형	frequency distribution polygon	度數分布多角形
도수분포표	frequency distribution table	度數分布表
도시	degree hour	度時
도시기상실험	METROMEX	都市氣象實驗
도시기후	city climate, urban climate	都市氣候
도시기후학	urban climatology	都市氣候學
도시돌발홍수정보	urban flash flood guidance	都市突發洪水情報
도시미터	dosimeter	
도시소하천홍수	urban and small stream flooding	都市小河川洪水
도시소하천홍수주의보	urban and small stream flood advisory	都市小河川洪水注意報
도시안개	city fog, town fog	都市-
도시열섬	urban heat island	都市熱島
도시플룸	urban plume	都市-
도시홍수	urban flooding	都市洪水
도시화	urbanisation	都市化
도식가법	graphical addition	圖式加法
도식감법	graphical subtraction	圖式減法
도식계산	graphical calculation	圖式計算
도식분석	graphical analysis	圖式分析
도약	jump	跳躍
도약거리	skip distance	跳躍距離
도약효과	skip effect	跳躍效果
도열병	blast disease	稻熱病
도일	degree day	度日
도체	conductor	導體
도파관	waveguide	導波管
도표	chart	圖表
도표계산법	nomography	圖表計算法
도표작성	tabulation	圖表作成
도표작성기기	tabulating machine	圖表作成器機
도플러	Doppler	
도플러넓어짐	Doppler broadening	
도플러딜레마	Doppler dilemma	
도플러라이다	Doppler lidar	
도플러레이더	Doppler radar	
도플러레이저 레이더	Doppler laser radar	
도플러바람관측	Doppler wind measurements	-觀測
도플러방정식	Doppler equation	-方程式

도플러변이	Doppler shift	-變移
도플러소다	Doppler sodar	
도플러속도	Doppler velocity	-速度
도플러스펙트럼	Doppler spectrum	
도플러오차	Doppler error	-誤差
도플러원리	Doppler principle	-原理
도플러진동수 변이	Doppler frequency shift	-振動數變異
도플러측정	Doppler measurement	-測定
도플러형상	Doppler shape	-形象
도플러효과	Doppler effect	-效果
도하작전기상	rivercrossing operation weather	渡河作戰氣象
도함수	derivative	導函數
도해	graphical solution	圖解
도해기후학	cartographical climatology	圖解氣候學
도해법	graphic method, graphical method	圖解法
도해역학	graphical statics	圖解力學
독립변수	independent variable	獨立變數
독립사건	independent event	獨立事件
독립표본	independent sample	獨立標本
독성	toxicity	毒性
독성공기오염물	toxic air pollutants	毒性空氣汚染物
독성물	toxicant	毒性物
돌발	burst, outbreak, U burst	突發
돌발서리	flash frost	突發-
돌발주파수	penetration frequency	突發周波數
돌발홍수	flash flood	突發洪水
돌발홍수경보	flash flood warning	突發洪水警報
돌발홍수주의보	flash flood watch	突發洪水注意報
돌발E영역	sporadic E-region	突發E地域
돌발E층	sporadic E-layer	突發E層
돌아오는 에코	pip	
돌연공전증강	sudden enhancement of atmospherics	突然空電增强
돌연변이설	mutation theory	突然變異說
돌연변이성	mutagenicity	突然變異性
돌연변이(유발)원	mutagen	突然變異(誘發)原
돌연변화	jump	突然變化
돌연변화보고	sudden change report	突然變化報告
돌연승온	sudden warming	突然昇溫
돌연전리층요란	sudden ionospheric disturbance	突然電離層擾亂
돌출	overhang, overshoot	突出
돌출구름꼭대기	overshooting cloud top	突出-
돌출뇌우꼭대기	overshooting thunderstorm top	突出雷雨-
돌출에코	overhang echo	突出-
돌턴법칙	Dalton's law	-法則

돌턴수	Dalton number	-數
돌턴접근법	Dalton approach	-接近法
돌풍	blast, crochet d'orage, crochet degrain, flurry, gust, wind gust	突風
돌풍감쇠시간	gust decay time	突風減衰時間
돌풍경도거리	gust-gradient distance	突風傾度距離
돌풍관측존데	gustsonde	突風觀測-
돌풍기간	gust duration	突風期間
돌풍도(성)	gustiness	突風度(性)
돌풍률	gustiness factor	突風率
돌풍벡터	gust vector	突風-
돌풍부하	gust load	突風負荷
돌풍빈도	gust frequency	突風頻度
돌풍빈도간격	gust frequency interval	突風頻度間隔
돌풍성분	gustiness component	突風成分
돌풍전선	gust front	突風前線
돌풍진폭	gust amplitude	突風振幅
돌풍최대풍속	gust peak speed	突風最大風速
돌풍토네이도	gustnado	突風-
돌풍형성시간	gust formation time	突風形成時間
돌효과	Dole effect	-效果
동결건조	freeze drying	凍結乾燥
동결깊이선	frost line	凍結-線
동결난류	frozen turbulence	凍結亂流
동결난류가설	frozen turbulence hypothesis	凍結亂流假說
동결띠	frost belt, frost dam	凍結-
동결쐐기	frost wedging	凍結-
동결열	heat of congelation	凍結熱
동결자기장	frozen magnetic field	凍結磁氣場
동결작용	frost action	凍結作用
동결측정계	cryopedometer	凍結測定計
동결파동가설	frozen wave hypothesis	凍結波動假說
동결파동근사	frozen wave approximation	凍結波動近似
동결풍화작용	congelifraction	凍結風化作用
동결학	cryopedology	凍結學
동결-해빙 패턴	freeze-thaw pattern	凍結-解氷-
동계	winter season	冬季
동굴기상학	speleo-meteorology	洞窟氣象學
동기검파	coherent detection	同期檢波
동기후학	dynamic climatology	動氣候學
동기후학기후대	climatic zone by dynamic climatology	動氣候學氣候帶
동네예보	village forecast, Dong-Nae Forecast	(鄕村)豫報
동대서양형	Eastern Atlantic pattern	東大西洋型
동마찰	kinetic friction	動摩擦

동면	hibernation	冬眠
동면기	hibernating period	冬眠期
동물계절학	zoophenology	動物季節學
동물군	fauna	動物群
동물기후	animal climate	動物氣候
동물생산	animal production	動物生産
동물안개	animal fog	動物-
동빙	winter ice	冬氷
동상	acrocyanosis, chilblains, cold stroke, frostbite, freezer burn, frostbite, frostburn	凍傷
동상	frost heaving, heaving	凍上
동상해	frost injury	凍霜害
동상해	frost heaving damage	凍上害
동서강수패턴	zonal precipitation pattern	東西降水-
동서기류패턴	zonal airflow pattern	東西氣流-
동서기류형태	zonal airflow pattern	東西氣流形態
동서(기)류	zonal flow	東西(氣)流
동서(기)류패턴	zonal flow pattern	東西(氣)流-
동서수송	west-east transport	東西輸送
동서순환	zonal circulation	東西循環
동서운동	zonal motion	東西運動
동서운동에너지	zonal kinetic energy	東西運動-
동서유효위치에너지	zonal available potential energy	東西有效位置-
동서파수	zonal wave number	東西波數
동서지수	zonal index	東西指數
동서풍	zonal wind	東西風
동서풍분포	zonal wind profile	東西風分布
동서풍속단면분포	zonal wind-speed profile	東西風速斷面分布
동서풍속프로파일	zonal wind-speed profile	東西風速-
동시변화	synchronous change	同時變化
동시상관	coincidental correlation	同時相關
동시생성수	connate water	同時生成水
동시성	simultaneity	同時性
동시성오류	coincidence error	同時性誤謬
동시송신	simultaneous transmission	同時送信
동시완화	simultaneous relaxation	同時緩和
동시출현선	isopipteses	同時出現線
동시화	synchronization	同時化
동심눈벽	concentric eyewall	同心眼壁
동심눈벽순환	concentric eyewall cycle	同心眼壁循環
동심원군	consentric circle group	同心圓群
동아시아여름몬순	East Asian summer monsoon	東-
동아프리카제트	East African jet	東-
동아프리카하층제트	East African low-level jet	東-下層-

동안경계류	east coast boundary current	東岸境界流
동안기후	eastern coastal climate	東岸氣候
동압	dynamic pressure	動壓
동역학	kinematics, kinetics	動力學
동역학적 유사법칙	law of dynamical similarity	動力學的類似法則
동역학파라우팅모형	dynamic wave routing model	動力學波-
동역학파전달모형	dynamic wave routing model	動力學波傳達模型
동영상	animation of imagery	動映像
동오스트레일리아해류	East Australia current	東-海流
동위원소	isotope	同位元素
동위원소분별	isotopic fractionation	同位元素分別
동위원소분석	isotopic analysis	同位元素分析
동위원소원자	isotopic atom	同位元素原子
동위원소추적자	isotopic tracer	同位元素追跡子
동일과정설	uniformitarianism	同一過程說
동일기후	homoclime	同一氣候
동일면기법	coplane technique	同一面技法
동일면주사	coplane scanning	同一面走査
동일편광신호	copolarized signal	同一偏光信號
동일편파성분	copolar component	同一偏波成分
동일편파송수신상관계수	copolar correlation coefficeint	同一偏波送受信相關係數
동적구름씨뿌리기	dynamic cloud seeding	動的-
동적범위	dynamic range	動的範圍
동적비상	dynamic soaring	動的飛上
동적선형모형	dynamic linear model	動的線型模型
동적성능	dynamic performance	動的性能
동점성	kinematic viscosity	動粘性
동조	coherence	同調
동조구조	coherent structure	同調構造
동조시간	coherence time	同調時間
동조에코	coherent echo	同調-
동조요소	coherence element	同調要素
동조적분	coherent integration	同調積分
동조화	synchronization	同調化
동족원소	homologous element	同族元素
동중국해	East China Sea	東中國海
동중국해 저기압	depression in East China Sea	東中國海低氣壓
동중체	isobars	同重體
동지	winter solstice	冬至
동질성	ergodicity	同質性
동질이상	polymorphism	同質異像
동체반응	one-body reaction	同體反應
동축상관법	coaxial correlation method	同軸相關法
동축쌍극	coaxial collinear dipole	同軸雙極

동축케이블	coaxial cable	同軸-
동케이프해류	East Cape current	東-海流
동토지대	tundra	凍土地帶
동파지수	freezing burst index	凍破指數
동풍	Apheliotes	東風
동해	East Sea	東海
동해	freezing damage	凍害
동해안기후	east coast climate	東海岸氣候
동해안저기압	east coast cyclone	東海岸低氣壓
동해안폐색	east-coast occlusion	東海岸閉塞
동해저기압	East Sea cyclone	東海低氣壓
동화작용	anabolism, assimilation	同化作用
돛대	mast	
되돌이뇌격	return streamer, return stroke	-雷擊
되돌이대류	return convection	-對流
되돌이번개	return lightning	
되돌이섬광	return(ed) flash	-閃光
되돌이한대공기	returning polar air	-寒帶空氣
되돌이흐름	return flow	
되맞춤	reset	
되먹임	feedback	
되먹임과정	feedback process	-過程
되먹임기구	feedback mechanism	-機構
되먹임루프	feedback loop	
되먹임인자	feedback factor	-因子
되얼음	regelation	
되추적	backtracking	-追跡
된바람	strong breeze	
된서리	hard frost, killing frost, severe frost	
된추위	killing freeze	
두갈래치기지점	bifurcation point	-地點
두개골 크기	braincase size	
두건구름	pileus	
두건적란운	cumulonimbus pileus	頭巾積亂雲
두건적운	cumulus pileus	頭巾積雲
두꺼운 극빙하	high polar glacier	-極氷河
두꺼운 무역풍	deep trades	-貿易風
두꺼운 얼음	heavy ice	
두루마리구름	roll cloud, rotor cloud	
두루마리눈	snow roller	
두루마리적운	roll cumulus	-積雲
두방향풍향계	bivane	-方向風向計
두보이스지역	DuBois area	-地域
두줄전위계	bifilar electrometer	-電位計

둑	bank, embankment, weir	
둔체	bluff body	鈍體
둥근 얼음	ball ice	
둥지격자	nested grids	-格子
뒤집힘	overturn	
뒤틀림	warping	歪曲
뒤틀림항	twisting term	-項
뒷문한랭전선	backdoor cold front	後門寒冷前線
뒷바람	tail wind, following wind	
듀티주기	duty cycle	
드라이스크러버	dry scrubbers	
드라이아이스	dry ice	
드라이아이스씨뿌리기	dry ice seeding	
드레이쇼	drecho	
드보락분석법	Dvorak analysis	-分析法
드셍베낭방정식	de Saint-Venant equation	-方程式
드와이	dwigh, dwey, dwoy	
들뜸	excitation	
(오스트레일리아)들불	bushfire	
등가속도	uniform acceleration	等加速度
등가속운동	uniformly accelerated motion	等加速運動
등가	equivalent	等價
등가정리	equivalence theorem	等價定理
등각나선	equiangular spiral	等角螺線
등각도법	conformal projection	等角圖法
등각류	isopods	等脚類
등각선	isogonal line, rhumb line	等角線
등각원뿔투영(법)	conformal conic projection	等角圓-投影(法)
등각지도	conformal map, isogonal map	等角地圖
등각투영	conformal projection, equal-angle projection	等角投影
등강수량선	equipluve, isohyet, isohyets	等降水量線
등강수량선(의)	isohyetal	等降水量線
등강압중심	isokatabaric center	等降壓中心
등개화선	isanthesic line	等開花線
등결빙일선	isopectrics	等結氷日線
등경도선	isogradient	等傾度線
등계수선	iso-coefficient	等係數線
등고도면도	fixed-level chart	等高度面圖
등고도변화선	isallohypse	等高度變化線
등고면	isohypsic surface	等高面
등고선	contour line, isoheight, isohypse, contour	等高線
등고선간격	contour interval	等高線間隔
등고선경도	contour gradient	等高線傾度
등고선고도	contour height	等高線高度

등고선곡률	contour curvature	等高線曲率
등고선도	contour chart, contour map, isohypsic chart	登高線圖
등고선분석	contour analysis	等高線分析
등고선오차	contour height error	等高線誤差
등고선재배	contour tillage	等高線栽培
등고선코드	contour code	等高線-
등고선횡단흐름	cross contour flow	等高線橫斷-
등광도선	isophote	等光度線
등급	magnitude	等級
등기상도선	isometeorograde	等氣象度線
등기압교차선	isanakatabar, isokatanabar	等氣壓較差線
등기압교차선도	isanakatabaric chart	等氣壓較差線圖
등기압변화(의)	isallobaric	等氣壓變化
등기압변화경도	isallobaric gradient	等氣壓變化傾度
등기압변화도	isallobaric chart	等氣壓變化圖
등기압변화분석	isallobaric analysis	等氣壓變化分析
등기압변화선	isallobar	等氣壓變化線
등기압변화최대	isallobaric maximum	等氣壓變化最大
등기압변화최소	isallobaric minimum	等氣壓變化最小
등기압상승선	anallobar	等氣壓上昇線
등기압상승중심	isoanabaric center	等氣壓上昇中心
등기압평형선	isothene	等氣壓平衡線
등기압하강선	kata-isallobar	等氣壓下降線
등기울기선	isocline, isoclinal line, isoclinic line	等-線
등기준선	isobase	等基準線
등기후선	isoclimatic line	等氣候線
등넘기차분	leapfrog differencing	-差分
등노점선	isodrosotherm	等露點線
등노점차선	isoline of dewpoint deficit	等露點差線
등뇌우빈도선	isoceraunic line, isokeraunic line	等雷雨頻度線
등뇌우선	isobront	等雷雨線
등뇌우시선	homobront	等雷雨時線
등뇌전빈도	isokeraunic	等雷電頻度
등뇌전빈도선	isokeraunic line	等雷電頻度線
등대선	lightship	燈臺船
등대선관측소	lightship station	燈臺船觀測所
등대선전보식	lightship code	燈臺船電報式
등더위선	isothere, isotheral	等-線
등도플러속도선	isodop	等-速度線
등동결기선	isopag	等凍結期線
등동력면	isodynamical surface	等動力面
등두께선	isothickness line	等-線
등두께이류선	isothickness advection line	等-移流線
등D값선	iso-D	等-線

등력선	isodynamic	等力線
등면적도	equal-area chart	等面積圖
등면적변환	equal-area transformation	等面積變換
등면적지도	equal-area map	等面積地圖
등면적투영	equal-surface projection	等面積投影
등물질면	equisubstantial surface	等物質面
등밀도(선)	isopycnic	等密度(線)
등밀도면	isopycnic surface	等密度面
등밀도선	isodense, isopycnic line	等密度線
등밀도혼합	cabelling	等密度混合
등발현선	isophane	等發現線
등방성	isotropy	等方性
등방성(의)	isotropic	等方性
등방성난류	isotropic turbulence	等方性亂流
등방성반사	isotropic reflection	等方性反射
등방성복사	isotropic radiation	等方性輻射
등방성복사체	isotropic radiator	等方性輻射體
등방성산란	isotropic scattering	等方性散亂
등방성안테나	isotropic antenna	等方性-
등방성전파파발사기	isotropic radiator	等方性電波波發射機
등변화	isallo-	等變化
등복각선	isoclinal line, isoclinic line	
등분배	equipartition	等分配
등비부피	isosteric	等比-
등비부피면	isosteric surface	等比-面
등비부피선	isostere	等比-線
등비수열	geometric sequence (progression)	等比數列
등비율선	isomer	等比率線
등비적면	isostere surface	等比積面
등사습곡	isoclinal fold	等斜褶曲
등상관선	isocorrelation	等相關線
등설선	isochion	等雪線
등설선고도선	isohion	等積線高度線
등설일수선	isohion	等積日數線
등속원운동	uniform circular motion	等速圓運動
등속표본채취	isokinetic sampling	等速標本採取
등수심선	isochore line	等水深線
등수온선	isothermobath	等水溫線
등숙	ripening	登熟
등숙기간	ripening period	登熟期間
등습도선	isohume, isohygrometric line	等濕度線
등습수선	iso-dewpoint deficit line	等濕數線
등시선	isochrone	等時線
등시성	isochronism	等時性

등시어선	isoshear	等-線
등심선	fathom curve, isobath	等深線
등심온도선	isobathytherm	等深溫度線
등압(의)	isobaric	等壓
등압고도도	isobaric contour chart	等壓高度圖
등압과정	isobaric process	等壓過程
등압관	isobaric tube	等壓管
등압냉각	isobaric cooling	等壓冷却
등압높낮이	isobaric topography	等壓-
등압도	isobaric chart	等壓圖
등압두께선도	isobaric thickness chart	等壓-線圖
등압등적관	isobaric isosteric tube	等壓等積管
등압등적 솔레노이드	isobaric isosteric solenoid	等壓等積-
등압면	isobaric surface	等壓面
등압면기울기	slope of an isobaric surface	等壓面-
등압면높낮이	pressure topography	等壓面-
등압면분석	isobaric analysis	等壓面分析
등압면예상도	prognostic contour chart	等壓面豫想圖
등압발산	isobaric divergence	等壓發散
등압상당온도	isobaric equivalent temperature	等壓相當溫度
등압상당온위	isobaric equivalent potential temperature	等壓相當溫位
등압선	isobar, isobaric line, isobarometric line	等壓線
등압선교각	cross isobar angle	等壓線交角
등압선교차흐름각	cross isobar flow angle	等壓線交叉-角
등압선묘화	isobar drawing	等壓線描畵
등압소용돌이도	isobaric vorticity	等壓-度
등압습구온도	isobaric wet-bulb temperature	等壓濕球溫度
등압습구온위	isobaric wet-bulb potential temperature	等壓濕球溫位
등압온도변화	isobaric temperature change	等壓溫度變化
등압운동	isobaric motion	等壓運動
등압조건	isobaric condition	等壓條件
등압통로	isobaric channel	等壓通路
등에코선	iso echo, isoecho line	等-線
등엔트로피과정	isentropic process	等-過程
등연온도교차선	iseoric line, isoparallage, isotalant	等年溫度較差線
등연직속도선	isanabat, isanabation	等鉛直速度線
등염분선	isohaline	等鹽分線
등오로라빈도선	isochasm	等-頻度線
등오로라선	isoaurora	等極光線
등온	isothermal	等溫
등온과정	isothermal process	等溫過程
등온구역	zone of constant temperature	等溫區域
등온대기	isothermal atmosphere	等溫大氣
등온도변화선	isallotherm	等溫度變化線

등온면	isothermal surface	等溫面
등온변화	isothermal change	等溫變化
등온선	isotherm, isothermohypse, thermoisopleth	等溫線
등온선밀집대	isotherm ribbon	等溫線密集帶
등온압축	isothermal compression	等溫壓縮
등온위	equipotential temperature	等溫位
등온위과정	isentropic process	等溫位過程
등온위뇌우	isentropic thunderstorm	等溫位雷雨
등온위대기	isentropic atmosphere	等溫位大氣
등온위덮개	isentropic cap	等溫位-
등온위도	isentropic chart	等溫位圖
등온위두께선도	isentropic thickness chart	等溫位-線圖
등온위면분석	isentropic analysis, isentropic surface	等溫位面分析
등온위무게선도	isentropic weight chart	等溫位-線圖
등온위발산	isentropic divergence	等溫位發散
등온위변화	isentropic change	等溫位變化
등온위선	isentrope	等溫位線
등온위선도	isentropic thickness chart	等溫位線圖
등온위수렴	isentropic convergence	等溫位收斂
등온위운동	isentropic motion	等溫位運動
등온위응결고도	isentropic condensation level	等溫位凝結高度
등온위판	isentropic sheet	等溫位板
등온위혼합	isentropic mixing	等溫位混合
등온층	isothermal layer	等溫層
등온팽창	isothermal expansion	等溫膨脹
등온평형	isothermal equilibrium	等溫平衡
등와도선	isovorticity line	等渦度線
등우량선	isopluvial	等雨量線
등우선법	isohyetal method	等雨線法
등운량선	isoneph	等雲量線
등월변화선	isodiaphore	等月變化線
등이상선	isabnormal, isoabnormal	等異常線
등이슬점선	isodrosotherm	等-點線
등이슬점차선	isoline of dewpoint deficit	等-點差線
등일사강도선	isactine	等日射強度線
등일사선	isalea	等日射線
등일조선	isohel, isohelic	等日照線
등장력선	isotonicity	等張力線
등장액	isotonic solution	等張液
등재해선	isophtor	等災害線
등적과정	isochoric process	等積過程
등적면	isasteric surface	等積面
등적변형	equiareal transform	等積變形
등적변화	isochoric change	等積變化

등적설깊이선	isohion	等積雪-線
등전자이온	isoelectronic ion	等電子-
등조선	cotidal line	等潮線
등조시	cotidal hour	等潮時
등조차선	corange line	等潮差線
등증발량선	isothyme	等蒸發量線
등증발선	isoatmic, isoombre	等蒸發線
등지오퍼텐셜면	equigepotential surface	等-面
등지온선	geoisotherm, isogeotherm	等地溫線
등진도선	isoseismal line	等震度線
등진폭선	iso-amplitude	等振幅線
등첨도곡선	isokurtic curve	等尖度曲線
등추선	isometropal	等秋線
등춘선	isoeral	等春線
등치면	equiscalar surface	等値面
등치법	method of equivalence	等値法
등치선	contour, equiscalar line, isarithm, isogram, isoline, isopleth	等値線
등치선꺾임	kinks on isopleths	等値線-
등퍼텐셜면	equipotential surface	等-面
등퍼텐셜선	equipotential line, isopotential	等-線
등편차	equideparture	等偏差
등편차(의)	isanomalous	等偏差
등편차선	isabnormal line, isametric, isanomal, isanomalous line, isanomaly	等偏差線
등풍속선	isanemone, isokinetic, isotach, isovel	等風速線
등풍속선도	isotach chart	等風速線圖
등풍속선분석	isotach analysis	等風速線分析
등풍향선	isogon	等風向線
등한선	isocheim	等寒線
등할	equal cleavage	等割
등해동선	isotac	等解凍線
등향선	isogonic line	等向線
등호	equal(equality) sign	等號
등혼합비선	mixing ratio line	等混合比線
등화기선	isoanth	等花期線
디랙델타	Dirac's delta	
디랙델타함수	Dirac delta function	-函數
디리클레안정도정리	Dirichlet's stability theorem	-安定度定理
디리클레조건	Dirichlet condition	-條件
디리클레프로그램	Dirichlet program	
디벤시아빙기	Devensian glacial	-氷期
디어도프속도	Deardorff velocity	-速度
디에일리어싱	dealiasing	

디옵터	diopter	
디지털비디오통합처리기	digital video integrator processor	-統合處理器
디지털샘플링	digital sampling	
디지털수	digital count	-數
디지털-아날로그 변환	digital-to-analog conversion	-變換
디지털-아날로그 변환기	digital-to-analog converter	-變換器
디지털영상처리	digital image processing	-映像處理
디지털예보	digital forecast	-豫報
디지털이미지	digital image	
디지털자료전송	digital data transmission	-資料電送
디지털자료처리기	digital data processor	-資料處理器
디지털컴퓨터	digital computer	
디지털프리필터링	digital prefiltering	
디지털필터	digital filter	
디카르복실산	dicarboxylic acids	-酸
디콘바람프로파일인자	Deacon wind profile parameter	-因子
디콘볼루션	deconvolution	
디트로이트-윈저 지역	Detroit-Windsor area	-地域
딕시길목	Dixie Alley	
딸동위원소	daughter isotope	-同位元素
딸림방정식	adjoint equation	-方程式
땅거미	crepuscule, dusk, twilight	
땅거울	sand mirage	
땅-구름방전	ground-to-cloud discharge	-放電
땅날림눈	drifting snow	
땅날림먼지	drifting-blowing dust, drifting dust	
땅날림모래	drifting sand	
땅너울	ground swell	
땅방전	ground discharge	-放電
땅속얼음	subsoil ice, subterranean ice	
땅스트리머	ground streamer	
땅안개	ground fog	
땅얼음	ground ice	
땅얼음언덕	ground ice mound	
땅울림	brontides	
땅전류	earth current	-電流
땅전류폭풍	earth current storm	-電流暴風
땅하늘	land sky	
떠보임	looming	
뗏목얼음	rafted ice, telescoped ice	
뚜렷한 CDO	distinct CDO	
뚝구름	barrage cloud	
뚝모양안개	bank fog	
뜬 얼음	pan ice, floe ice	

띄우개형우량계	rain-gauge float type	-型雨量計
띠	band	
띠간격	band gap	-間隔
띠강수	banded precipitation	-降水
띠강수패턴	zonal precipitation pattern	-降水-
띠고기압	zonal high	-高氣壓
띠구조	banded structure	-構造
띠권운	band cirrus	-卷雲
띠권층운	cirrostratus vittatus	-卷層雲
띠그래프	band graph	
띠기류패턴	zonal flow pattern	-氣流-
띠기류형태	zonal flow pattern	-氣流形態
띠너비	band width	
띠모양	band pattern	-模樣
띠바람	zonal wind	
띠번개	band lightning, fillet lightning	
띠분포	zonality	-分布
띠순환	zonal circulation	-循環
띠스펙트럼	band spectrum	
띠전선면	zonal frontal surface	-前線面
띠지수	zonal index	-指數
띠특성	zonality	-特性
띠편서풍	zonal westerlies	-偏西風
띠폭	band width	帶幅
띠효과	banding effect	-效果
띠흐름	zonal flow	
띠흡수	band absorption	-吸收

ㄹ

한글	영문	한자
라그랑지미분	Lagrangian differentiation	-微分
라그랑지방법	Lagrangian method	-法
라그랑지방정식	Lagrangian equation	-方程式
라그랑지상관	Lagrangian correlation	-相關
라그랑지유사	Lagrangian similarity	-類似
라그랑지좌표	Lagrangian coordinates	-座標
라그랑지파	Lagrangian wave	-波
라노	llanos, savanna	
라니냐	La Nina	
라돈	radon	
라돈측정기	emanometer	-測定器
라디안	radian	
라디오	radio	
라디오비컨	radio beacon	
라디오존데	radiosonde	
라디오존데관측	radiosonde observation	-觀測
라디오존데관측소	radiosonde station	-觀測所
라디오존데기구	radiosonde balloon	-氣球
라디오존데기록계	radiosonde recorder	-記錄計
라디오존데-레윈관측시스템	radiosonde-radiowind system	-觀測-
라디오존데변조기	radiosonde modulator	-變調器
라디오존데송신기	radiosonde transmitter	-送信機
라디오존데전환기	radiosonde commutator	-轉換器
라르센 빙붕	Larsen ice shelf	-氷棚
라마마 현상	La Mama	-現像
라만라이다	Raman lidar	
라만효과	Raman effect	-效果
라브라도해류	Labrador current	-海流

라설페	la serpe	
라울법칙	Raoult's law	-法則
라이다	lidar	
라이다-광 탐지분류	lidar-light detection and ranging	-光 探知分類
라이다기상학	lidar meteorology	-氣象學
라이먼알파방출선	Lyman alpha emission line	-放出線
라클리프지수	Rackliff index	-指數
라테라이트	laterite	
라퐁드표	Lafond's table	-表
라플라스방정식	Laplace equation	-方程式
라플라스변환	Laplace transform	-變換
라플라스연산자	Laplace operator, Laplacian operator	-演算子
라플라스음속	Laplacian speed of sound	-音速
라피체	raffiche	
람베르트	Lambert	
람베르트공식	Lambert's formula	-公式
람베르트등각	Lambert conformal	-等角
람베르트등각도법	Lambert conformal projection	-等角圖法
람베르트등각원뿔 투영법	Lambert conformal conic projection	-等角圓-投影法
람베르트반사경	Lambertian reflector	-反射鏡
람베르트반사체	Lambertian reflector	-反射體
람베르트(코사인)법칙	Lambert's (cosine) law	-法則
람베르트흡수법칙	Lambert's law of absorption	-吸收法則
람브레흐트 다용도습도계	Lambrecht's polymeter	-多用途濕度計
랑주뱅이온	Langevin ion	
래리	larry	
랙스동등이론	Lax's equivalence theorem	-同等理論
랙스-리히르마이어 정리	Lax-Richtmyer theorem	-定理
랙스방안	Lax's scheme	-方案
랙스-웬돌프 차분방안	Lax-Wendorff differencing scheme	-差分方案
랜드래시	land-lash	
랜드새트	LANDSAT	
랜드새트프로그램	Landsat program	
램존데	ramsonde	
램투과계	ram penetrometer	-透過計
랭글리	Langley	
랭뮤어연쇄반응	Langmuir chain reaction	-連鎖反應
랭뮤어탐지기	Langmuir probe	-探知機
랭킨	Rankine	
랭킨복합소용돌이	Rankine's combined vortex	-複合-
랭킨소용돌이	Rankine vortex	
랭킨온도눈금	Rankine temperature scale	-溫度-
러브파	love wave	-波
런던스모그	London smog	

런던안개	London fog	
레나르트효과	Lenard effect	-效果
레보요	reboyo	
레오뮈르온도계	Réaumur thermometer	-溫度計
레오뮈르온도눈금	Réaumur thermometric scale	-溫度-
레윈	rawin (radiowind)	
레윈관측	radiowind observation, rawin observation	-觀測
레윈관측소	radiowind station	-觀測所
레윈장비	rawin equipment	-裝備
레윈존데	rawinsonde	
레윈존데관측	rawinsonde observation	-觀測
레윈존데관측소	rawinsonde station	-觀測所
레윈표적	rawin target	-標的
레이놀즈수	Reynolds number	-數
레이놀즈응력	Reynolds stress	-應力
레이놀즈효과	Reynolds effect	-效果
레이더	radar	
레이더강수관계	radar-rainfall relationship	-降水關係
레이더강수에코	radar precipitation echo	-降水反響
레이더거리방정식	radar range equation	-距離方程式
레이더경위(의)	radar theodolite	-經緯
레이더고도	radar altitude	-高度
레이더고도계	radar altimeter	-高度計
레이더그림자	radar shadow	
레이더기상	radar meteorology	-氣象
레이더기상관측	radar meteorological observation	-氣象觀測
레이더기상학	radar meteorology	-氣象學
레이더기후학	radar climatology	-氣候學
레이더단면	radar cross-section	-斷面
레이더덕트	radar duct	
레이더돔	radar dome	
레이더마이크로파중계장치	radar microwave link	-波中繼裝置
레이더마일	radar mile	
레이더망	radar screen	-網
레이더바람관측	radar wind sounding	-觀測
레이더바람시스템	radar wind system	
레이더반사도인자	radar reflectivity factor	-反射圖因子
레이더반사도인자(Z)와 강수강도(R)의 관계	radar reflectivity(Z)-rain rates(R) relationship	
레이더반사율	radar reflectivity	-反射率
레이더방정식	radar equation	-方程式
레이더보고	radar report	-報告
레이더보정	radar calibration	-補正
레이더부피	radar volume	-體積

레이더빔	radar beam	
레이더 빔 아래	below radar beam	
레이더산란체측정계	radar scatterometer	-散亂體測定計
레이더산출생성기	radar product generator	-産出物生成器
레이더삼각법	radar triangulation	-三角法
레이더상수	radar constant	-常數
레이더수평선	radar horizon	-水平線
레이더스코프	radarscope	
레이더신호	radar signature	-信號
레이더신호스펙트로그래프	radar signal spectrograph	-信號-圖表
레이더신호제어처리기	radar signal control processor	-信號制御處理機
레이더에코	radar echo	-反響
레이더요약도	radar summary chart	-要約圖
레이더우량계비교	radar-raingauge comparison	-雨量計比較
레이더위파장대	radar upper band	-波長帶
레이더일기관측	radar weather observation	-日氣觀測
레이더자료획득	radar data acquisition	-資料獲得
레이더존데	radarsonde	
레이더주파수대	radar frequency band	-周波數帶
레이더지상에코	radar ground echo	-地上-
레이더지시기	radar indicator	-指示器
레이더코드신호	radar coded message	-信號
레이더탐측	radar sounding	-探測
레이더파장대	radar band	-波長帶
레이더폭풍우탐지	radar storm detection	-暴風雨探知
레이더폭풍우탐지방정식	radar storm detection equation	-暴風雨探知方程式
레이더 표본체적	bin	-標本體積
레이더표출	radar display	-表出
레이더합성도	radar composite chart	-合成圖
레이돔	radome	
레이드증기압	Reid vapour pressure	-蒸氣壓
레이리영역	Rayleigh reion	-領域
레이벌	rabal	
레이저	laser	
레이저레이더	laser radar	
레이저운고계	laser ceilometer, laser pulse ceilometer	-雲高計
레이저측정거리	Laser Detection and Ranging	-測定距離
레인지 펼침	range unfolding	
레일리	Rayleigh	
레일리공식	Rayleigh's formular	-公式
레일리기준점	Rayleigh criterion	-基準點
레일리대기	Rayleigh atmosphere	-大氣
레일리라이다	Rayleigh lidar	
레일리문제	Rayleigh's problem	-問題

레일리법칙	Rayleigh's law	-法則
레일리산란	Rayleigh scattering	-散亂
레일리수	Rayleigh number	-數
레일리-진스근사	Rayleigh-Jeans approximation	-近似
레일리파	Rayleigh wave	-波
레일리한계	Rayleigh limit	-限界
레일리흡수효율	Rayleigh absorption efficiency	-吸收效率
레지스터	register	
레프스달도	Refsdal chart	-圖
레프스달선도	Refsdal diagram	-線圖
렌즈고적운	altocumulus lenticularis,	-高積雲
	lenticular altocumulus cloud	
렌즈고층운	altostratus lenticularis	-高層雲
렌즈구름	lenticular cloud, lenticularis	
렌즈구름띠	lenticular cloud band	
렌즈권적운	cirrocumulus lenticularis	-卷積雲
렌즈적운	cumulus lenticularis	-積雲
렌즈층운	stratus lenticularis	-層雲
렌즈층적운	stratocumulus lenticularis	-層積雲
로그수신기	logarithmic receiver	-受信機
로렌츠변환	Lorentz transformation	-變換
로렌츠선모양	Lorentz lineshape	-線模樣
로렌츠인자	Lorentz factor	-因子
로렌타이드빙상	Laurentide ice sheet	-氷床
로봇기상관측소	robot weather station	-氣象觀測所
로비츠호	arcs of Lowitz, Lowitz arcs	-弧
로비치일사계	Robitzsch actinograph	-日射計
로빈슨컵풍속계	Robinson's cup anemometer	-風速計
로빈후드바람	Robin Hood's wind	
로슈미트수	Loschmidt's number	-數
로스비공식	Rossby formula	-公式
로스비매개변수	Rossby parameter	-媒介變數
로스비변형반경	Rossby deformation radius,	-變形半徑
	Rossby radius of deformation	
로스비변형반지름	Rossby radius of deformation	-變形半-
로스비선도	Rossby diagram	-線圖
로스비수	Rossby number	-數
로스비영역	Rossby regime	領域
로스비인자	Rossby factor	-因子
로스비장파공식	Rossby long wave formula	-長波公式
로스비장파노모그램	Rossby long wave nomogram	-長波-
로스비장파방정식	Rossby long wave equation	-長波方程式
로스비파	Rossby wave	-波
로스비파라미터	Rossby parameter	

로스비항	Rossby term	-項
로스비-호르비츠 파	Rossby-Haurwitz wave	-波
로즈몬트온도계함	Rosemount temperature housing	-溫度計函
로지스틱곡선	logistic curve	-曲線
로켓기구관측장치	rocket balloon instrument	-氣球觀測裝置
로켓기상계기	rocket meteorograph	-氣象計器
로켓발사지점	rocket-launching site	-發射地點
로켓번개	rocket lightning	
로켓유탄법	rocket-grenade method	-榴彈法
로켓존데	rocketsonde	
로켓탄	rocket projectile	-彈
로켓탐측	rocket sounding	-探測
로쿤탐측	rockoon sounding	-探測
로토스코프	rotoscope	
론다다	rondada	
루멘	lumen	
루밍	looming	
루버백엽상	louvered screen	-百葉箱
루파	Loofah	
루프안테나	loop antenna	
룩스	lux	
르장드르함수	Legendre function	-函數
리만해류	Liman Current	-海流
리매핑	remapping	
리밴터	levanter	
리본번개	ribbon lightning	
리본얼음	ribbon ice, ice ribbon	
리빙스턴구	Livingstone sphere	-球
리샘플링	resampling	
리슈수	Reech number	-數
리아푸노프벡터	Lyapunov vector	
리아푸노프지수	Lyapunov exponent	-指數
리에프네	riefne	
리처드슨수	Richardson number	-數
리트리터	retreater	
리틀브라더	little brother	
리틀시스터	little sister	
링겔만도	Ringelmann chart	-圖
링케척도	Linke-scale, blue sky scale	-尺度
링케혼탁인자	Linke turbidity factor	-混濁因子

ㅁ

한글	영문	한자
마개역전	capping inversion	-逆轉
마그네트론	magnetron	
마그마물	magmatic water, juvenile water	
마디	node	
마디선	nodal line	-線
마디점경도증분	nodal longitude increment	-點經度增分
마디점인자	nodal factor	-點因子
마디점주기	nodal period	-點周期
마디점증분	nodal increment	-點增分
(기압)마루	ridge	(氣壓)-
마루단계	crest stage	-段階
(기압)마루선	ridge line	(氣壓)-線
마루종단면	crest profile	-縱斷面
마루표고	crest elevation	-標高
마르굴레스공식	Margules formula	-公式
마르굴레스방정식	Margules's equation	-方程式
마르굴레스전선기울기	Margules frontal slope	-前線-
마르스덴일기도	Marsden chart	-日氣圖
마르코프사슬	Markov chain	
마른 결빙	dry freeze	-結氷
마른 뇌우	dry thunderstorm	-雷雨
마른 눈	dry snow	
마른 눈사태	dust avalanche, loose avalanche	-沙汰
마른 안개	dry fog	
마른 연무	dry haze	-煙霧
마른 장마	dry Baiu, dry Changma	
(식물의)마름병	blight	-病
마리오테법칙	Mariotte's law, Boyle-Mariotte law	-法則
마린형 수은기압계	marine mercury barometer	-水銀氣壓計

마빈일조계	Marvin sunshine recorder	-日照計
마샬-팔머 레이더-강우 함수	Marshall-Palmer radar-rainfall function	-降雨函數
마샬-팔머 분포	Marshall-Palmer distribution	-分布
마우루	maulu	
마운더 극소기	Maunder minimum	-極小期
마이크로 강수레이더	micro rain radar	-降水-
마이크로미터(μm)	micrometer	
마이크로바	microbar	
마이크로버스트	microburst	
마이크로버스트경보	microburst alert	-警報
마이크로파	microwave	-波
마이크로파 가중함수	microwave weighting function	-波加重函數
마이크로파 고도계	microwave altimeter	-波高度計
마이크로파 굴절계	microwave refractometer	-波屈折計
마이크로파 기상학	microwave meteorology	-波氣象學
마이크로파 라디오미터	microwave radiometer	-波-
마이크로파 복사	microwave radiation	-波輻射
마이크로파 산란	microwave scattering	-波散亂
마이크로파 스펙트럼	microwave spectrum	-波-
마이크로파 영상계	microwave imager	-波映像計
마이크로파 영역	microwave region	-波領域
마이크로파 원격탐사	microwave remote sensing	-波遠隔探査
마이크로파 위성영상	microwave satellite imagery	-波衛星映像
마이크로파 탐사	microwave probing	-波探査
마이크로파 탐측계	microwave sounding unit	-波探測計
마이크로 펄스라이다	micro pulse lidar	
마이크로필름	microfilm	
마이크론	micron	
마이클슨일사기록계	Michaelson actinograph	-日射記錄計
마이클슨자기일사계	Michaelson actinograph	-自記日射計
마일	mile	
마일라기구	Mylar balloon	-氣球
마찰	friction	摩擦
마찰각	angle of friction	摩擦角
마찰강제력	frictional forcing	摩擦强制力
마찰계수	coefficient of friction, frictional coefficient	摩擦係數
마찰고도	frictional height	摩擦高度
마찰대전	triboelectrification	摩擦帶電
마찰력	frictional force	摩擦力
마찰발산	frictional divergence	摩擦發散
마찰불연속	frictional discontinuity	摩擦不連續
마찰소멸	frictional dissipation	摩擦消滅
마찰속도	friction velocity, frictional velocity	摩擦速度
마찰수렴	frictional convergence	摩擦收斂

마찰온도	friction temperature	摩擦溫度
마찰응력	frictional stress	摩擦應力
마찰2차흐름	frictional secondary flow	摩擦二次-
마찰층	friction layer, frictional layer	摩擦層
마찰토크	frictional torque	摩擦-
마찰풍	antitriptic wind	摩擦風
마찰항력	frictional drag	摩擦抗力
마찰효과	frictional effect	摩擦效果
마토그로소	matto grosso	
마파람	maparam	
마하수	Mach number	-數
막내리바람	downrush	
막대온도계	stem thermometer	-溫度計
막힌 배수	blind drainage	-排水
만곡도복	bending type lodging	彎曲倒伏
만년설	firn, névé	萬年雪
만년설선	firn line, névé line	萬年雪線
만년설층	firn layer	萬年雪層
만년설화	firnification	萬年雪化
만년적설	névé pénitent	萬年積雪
만상	late frost	晚霜
만상해	late frost damage	晚霜害
만성기관지염	chronic bronchitis	慢性氣管支炎
만성노출	chronic exposure	慢性露出
만성산성화	chronic acidification	慢性酸性化
만성알레르기비염	chronic allergic rhinitis	慢性-鼻炎
만성절여름	All-hallown summer, All Saints' summer	萬聖節-
만성질환	chronic disease	慢性疾患
만성폐쇄성폐질환	chronic obstructive lung disease	慢性閉鎖性肺疾患
만성피폭	chronic exposure	慢性被爆
만유인력	gravitation, gravitational force, universal gravitation, force of gravitation	萬有引力
만유인력법칙	law of universal gravitation	萬有引力法則
만유인력상수	universal gravitational constant	萬有引力常數
만조	flood, flood tide	滿潮
만조수위	bankful stage	滿槽水位
말기침투율	terminal infiltration rate	末期浸透率
말꼬리구름	mares' tails	
말위도	horse latitude	-緯度
맑은	clear	
맑은 공기	clean air	-空氣
맑은 구역	cloud-free area	-區域
맑은 구역복사휘도	clear column radiance	-區域輻射輝度
맑은 날	clear day	

맑은 밤	clear night	
맑은 얼음	clear ice	
맑은 착빙	clear icing	-着氷
맑음	clear, clear sky, cloudless, fine	
망고소나기	mango shower	
망원경	telescope	望遠鏡
망원조준	telescope pointing	望遠照準
망종	grain in ear	芒種
맞무리달	antiselena	
맞무리해	antihelion, anthelion, counter sun	
맞무리호	anthelic arcs	-弧
맞바람	head wind, opposing wind	
맞추기	fitting	
맞추기모드	fitting mode	
맞춤관측	adaptive observation	-觀測
맞춤예보업무	tailored products weather services	-豫報業務
맞춤형 예보	customized forecast	-形豫報
매(황사)	bai	霾
매개변수	parameter	媒介變數
매개변수공진	parametric resonance	媒介變數共振
매개변수모형	parametric model	媒介變數模型
매개변수방정식	parametric equation	媒介變數方程式
매개변수집합	parameter aggregation	媒介變數集合
매개변수추정	parameter estimation	媒介變數推定
매개변수함수	parametric function	媒介變數函數
매개변수화	parameterization	媒介變數化
매개변수화방법	parameterization method	媒介變數化方法
매개변수화방식	parameterization scheme	媒介變數化方式
매그너스효과	Magnus effect	-效果
매끈한 면	smooth surface	-面
매든-줄리안 진동	Madden-Julian oscillation	-振動
매몰	filling up	埋沒
매몰저기압	filling cyclone	埋沒低氣壓
매몰플럭스	burial flux	埋沒-
매시관측	hourly observation	每時觀測
매연농도	density of flue gas pollutants	煤煙濃度
매우	maeu, Tsuyu, Baiu	梅雨
매질	medium	媒質
매체	medium	媒體
매키효과	Macky effect	-效果
매트릭 퍼텐셜	matric potential	
맥간괴사	interveinal necrosis	脈間壞死
맥놀이주파수발진기	beat frequency oscillator	-周波數發振器
맥놀이횟수	beat frequency	-回數

맥동	pulsation	脈動
맥스웰공식	Maxwell's equation	-公式
맥스웰방정식	Maxwell's equation	-方程式
맥스웰법칙	Maxwell's law	-法則
맨눈광도계	visual photometer	-光度計
맨눈측광법	visual photometry	-測光法
맨땅	bare soil	
맨얼음	bare ice	
맴돌이	eddy	
맴돌이경사	eddy slope	-傾斜
맴돌이계수	eddy coefficients	-係數
맴돌이공분산	eddy covariance	-共分散
맴돌이공분산법	eddy covariance method	-共分散法
맴돌이공분산소프트웨어	eddy covariance software	-共分散-
맴돌이난류	eddy turbulence	-亂流
맴돌이마찰	eddy friction	-摩擦
맴돌이반전시간규모	eddy turnover time scale	-反轉時間規模
맴돌이분해모형	eddy resolving model	-分解模型
맴돌이상관법	eddy correlation method	-相關法
맴돌이섭동법	eddy fluctuation method	-攝動法
맴돌이속도	eddy velocity	-速度
맴돌이수송	eddy transport	-輸送
맴돌이스펙트럼	eddy spectrum	
맴돌이시어응력	eddy shearing stress	-應力
맴돌이에너지	eddy energy	
맴돌이열속	eddy heat flux	-熱束
맴돌이열전도	eddy heat conduction	-熱傳導
맴돌이운동	eddy motion	-運動
맴돌이운동량	eddy momentum	-運動量
맴돌이운동에너지	eddy kinetic energy	-運動-
맴돌이유효퍼텐셜에너지	eddy available potential energy	-有效-
맴돌이응력	eddy stress	-應力
맴돌이응력텐서	eddy stress tensor	-應力-
맴돌이이류	eddy advection	-移流
맴돌이저항	eddy resistance	-抵抗
맴돌이전도	eddy conduction	-傳導
맴돌이전도계수	coefficient of eddy conduction	-傳導係數
맴돌이전도도	eddy conductivity	-傳導度
맴돌이점성	eddy viscosity	-粘性
맴돌이점성계수	coefficient of eddy viscosity, eddy viscosity coefficient	-粘性係數
맴돌이지름	eddy diameter	-直徑
맴돌이축적모형	eddy accumulation method	-蓄積模型
맴돌이파	eddy wave	-波

맴돌이플럭스	eddy flux	
맴돌이확산	eddy diffusion	-擴散
맴돌이확산계수	coefficient of eddy diffusion	-擴散係數
맴돌이확산도	eddy diffusivity	-擴散度
맴돌이흐름	eddy current	
먼거리방전	distant flash	-距離放電
먼무리달	parantiselena	
먼무리해	paranthelion	
먼바다	open sea	
먼바다유기체	pelagic organism	-有機體
먼지	dust	
먼지경보	dust warnig	-警報
먼지계	coniscope, koniscope	-計
먼지계수기	dust counter	-計數器
먼지구름	dust cloud	
먼지권	konisphere, staubosphere	-圈
먼지나르기	dust burden	
먼지농도	dust loading	-濃度
먼지덤불	bush	
먼지덩이	dust bowl	
먼지돔	dust dome	
먼지베일	dust veil	
먼지베일지수	dust veil index	-指數
먼지벽	dust wall	-壁
먼지보라	duster, dust storm	
먼지비	dust rain	
먼지소나기	dust shower	
먼지수송	dust transport	-輸送
먼지안개	dust fog	
먼지알갱이	dust particle	
먼지 에어로졸	dust aerosol	
먼지연무	dust haze	-烟霧
먼지지시기	koniscope, coniscope	-指示機
먼지진드기	dust mite	
먼지짐	dust burden	
먼지채집기	conimeter, dust collector, konimeter	-採集器
먼지측정계	conimeter	-測定計
먼지층	staubosphere	-層
먼지층선	dust horizon	-層線
먼지투성이	dustiness	
먼지폭풍	dust storm	-暴風
먼지플룸	dust plume	
먼지학	coniology, koniology	-學
먼지함량	dust content	-含量

먼지혼탁도	dust turbidity	-混濁度
먼지회오리	dancing devil, dancing dervish, desert devil, devil, dust devil, dust whirl, satan, shaitan	
먼항로	offshore route	-航路
멀칭	mulching	
메가	mega-	
메가바리에	megabarye	
메가사이클	megacycle	
메니스커스	meniscus	
메르카토르투영(도법)	Mercator projection	-投影(圖法)
메시니아 염분위기	Messinian salinity crisis	-鹽分危機
메시지	message	
메시지형식시스템	FM system	-形式-
메이유	mei-yu	
메이유전선	mei-yu front	-前線
메탄	methane	
메탄가스	marsh gas	
메탄수화물	methane hydrate	-水化物
메탄올	methanol	
메탄하이드레이트	methane hydrate	
메틸수은	methylmercury	-水銀
멕시코만류	Gulf Stream	-灣流
멕시코만류계	Gulf Stream System	-灣流系
멱	power	冪
멱급수	power series	冪級數
멱법칙	power law	冪法則
멱법칙풍속분포	power law profile	冪法則風速分布
멱스펙트럼	power spectrum	冪-
면	surface	面
면사포고적운	altocumulus nebulosus	面紗布高積雲
면사포구름	velium, velum	面紗布-
면사포권운	cirro-nebula, cirro-velum, cirrus nebula	面紗布卷雲
면사포권적운	cirrocumulus nebulosus	面紗布卷積雲
면사포권층운	cirrostratus nebulosus	面紗布卷層雲
면사포적란운	cumulonimbus velum	面紗布積亂雲
면사포적운	cumulus velum	面紗布積雲
면오염원	area source	面汚染源
면오염원해법	area source solution	面汚染源解法
면원	plane source	面源
면적감소인자	areal reduction factor	面積減少因子
면적강수	areal precipitation	面積降水
면적계	planimeter	面積計
면적-고도 곡선	area-elevation curve	面積高度曲線
면적변환인자	area conversion factor	面積變換因子

면적분	surface integral	面積分
면적속도	areal velocity	面積速度
면적(강)우량	areal rainfall	面積(降)雨量
면적평균	area average, area averaging	面積平均
면적평균문제	area averaging problem	面積平均問題
면침식	sheet erosion	面浸蝕
면편광	plane polarization	面偏光
멸종	extinction of species	滅種
멸종사건	extinction event	滅種事件
멸종에피소드	extinction episode	滅種-
멸종위기	extinction risk	滅種危機
멸종위기(의)	endangered	滅種危機
멸종위기종	endangered species	滅種危機種
(미국)멸종위기종보호법	Endangered Species Act	滅種危機種保護法
명료성	clarity(of gas)	明瞭性
명명법	nomenclature	命名法
명시방안	explicit scheme	明示方案
명시평활	explicit smoothing	明示平滑
명암경계선	terminator	明暗境界線
명암대비스트레칭	contrast stretching	明暗對比-
명주실구름	fibratus, filosus	明紬-
모관단절수분점	moisture of rupture of capillary bond	毛管斷切水分點
모관수두	capillary head	毛管水頭
모노터핀	monoterpene	
모니터	monitor	
모니터링	monitoring	
모닌-오부코프 규모길이	Monin-Obukhov scaling length	-規模-
모닌-오부코프 길이	Monin-Obukhov length	
모닌-오부코프 방정식	Monin-Obukhov equation	-方程式
모닌-오부코프 상사법칙	similarity law of Monin-Obukhov	-相似法則
모뎀	modem	
모듈	module	
모드	mode	
모디스 광학파장역	MODIS optical bands	-光學波長域
모래	sand	
모래기둥	sand pillar	
모래눈	sand snow	
모래바다	sand sea	
모래바람	sand fall	
모래벽	sand wall	-壁
모래신기루	sand mirage	-蜃氣樓
모래실안개	sand mist	
모래언덕	dunes, sand dune	
모래연무	sand haze	-煙霧

모래용오름	sand auger	-龍-
모래채굴	sand mining	-採掘
모래토네이도	sand tornado	
모래톱	sandbar	
모래폭풍	saltation, sandstorm	-暴風
모래해변	sand beach, sandy beach	-海邊
모래회오리	sand devil, sand whirl, dust devil	
모래흙	sand soil	
모루	anvil	
모루구름	anvil cloud, incus, thunderhead	
모루돔	anvil dome	
모루적란운	cumulonimbus incus	-積亂雲
모멘트	moment	
모발습도계	hair hygrometer	毛髮濕度計
모사도	facsimile chart	模寫圖
모사전송	facsimile	模寫電送
모서리반사경	corner reflector	-反射鏡
모서리효과	corner effect	-效果
모서리흐름	corner stream	
모세관강하	capillary depression	毛細管降下
모세관공극	capillary interstice	毛細管空隙
모세관물띠	capillary fringe	毛細管-
모세관보정	capillarity correction	毛細管補正
모세관상승	capillary rise	毛細管上昇
모세관수	capillary water	毛細管水
모세관압력	capillary pressure	毛細管壓力
모세관용수량	capillary moisture capacity	毛細管用水量
모세관이력현상	capillary hysteresis	毛細管履歷現象
모세관작용	capillary action	毛細管作用
모세관전도도	capillary conductivity	毛細管傳導度
모세관전위계	capillary electrometer	毛細管電位計
모세관채수기	capillary collector	毛細管採水器
모세관층	capillary layer	毛細管層
모세관파	capillary ripple, capillary wave	毛細管波
모세관퍼텐셜	capillary potential	毛細管-
모세관현상	capillarity, capillary phenomenon	毛細管現象
모세관확산	capillary diffusion	毛細管擴散
모세관흡입	capillary suction	毛細管吸入
모세관힘	capillary force	毛細管-
모수	parameter	母數
모수모형	parametric model	母數模型
모수화	parameterization	母數化
모수화 방법	parameterization method	母數化方法
모순	conflict	矛盾

모스부호	Morse code	-符號
모스신호	signal needle code	-信號
모슬린	muslin	
모습	pattern	
모아짜고틀	Moazagotl	
모양인자	shape factor	模樣因子
모원소	parent isotope	母元素
모의	simulation	模擬
모의강우	simulated rainfall	模擬降雨
모의복사방출에 의한 레이저광 증폭	laser-light amplification by stimulated emission of radiation	模擬輻射放出-光增幅
모이주기과정	seeder-feeder process	-過程
모자구름	cap cloud	帽子-
모자 쓴 기둥	capped column	帽子-
모자안정층	capping stable layer	帽子安定層
모잠비크해류	Mozambique current	-海流
모조관측	bogus observation	模造觀測
모조소용돌이	bogus vortex	模造-
모조자료	bogus data	模造資料
모조태풍	bogus typhoon	模造颱風
모집단	population	母集團
모집단분포	population distribution	母集團分布
모틀링(반점이 있는)	mottling	
모평균	population mean	母平均
모표준편차	population standard deviation	母標準偏差
모형	model	模型
모형기준	modeling criteria	模型基準
모형대기	model atmosphere	模型大氣
모형산출통계기술	model output statistics technique	模型算出統計技術
모형예상열대수	model expected tropical number	模型豫想熱帶數
모형유체	model fluid	模型流體
모형자료원본출력	direct model output	模型資料原本出力
모형출력통계	model output statistics	模型出力統計
모호성함수	ambiguity function	模糊性函數
목성	Jupiter	木星
목성형행성	Jovian planet	木星型行星
목측	eye observation, sight observation, visual observation	目測
목측(어림)실링	estimated ceiling	目測-
목측(어림)연직시정	estimated vertical visibility	目測鉛直視程
목표	target	目標
목표관측	targeting observation	目標觀測
목표위치지시기	target position indicator	目標位置指示器
목표지역	target area	目標地域

목화띠기후	cotton belt climate	木花-氣候
목화지역백엽상	cotton region shelter	木花地域百葉箱
몬순	monsoon	
몬순돌발	burst of the monsoon	-突發
몬순아시아	monsoon Asia	
몬순자이어	monsoon gyre	
몬순휴식	break of the monsoon	-休息
몬테카를로방법	Monte Carlo method	-方法
몬트리올의정서	Montreal Protocol	-議定書
몰	mole	
몰기화열	molar heat of vaporization	-氣化熱
몰녹음열	molar heat of fusion(melting)	-熱
몰농도	molarity	-濃度
몰니냐궤도	Molniya orbit	-軌道
몰니냐위성	Molniya satellites	-衛星
몰리솔	mollisol	
몰부피	molar volume	
몰분율	mole fraction	-分率
몰비열	molar (specific) heat	-比熱
몰수	mole number	-數
몰액화열	molar heat of liquefaction	-液化熱
몰열전퇴	Moll thermopile	-熱電堆
몰융해열	molar heat of fusion	-融解熱
몰응고열	molar heat of solidification	-凝固熱
몰증발열	molar heat of evaporation	-蒸發熱
몰차노프판	Moltchanov board	-板
몸화학	body chemistry	-化學
몽게현상	Monge's phenomenon	-現象
몽고메리 유선함수	Montgomery stream function	-流線函數
뫼겔-델린저 효과	Mögel-Dellinger effect	-效果
묄러복사도	Möller chart	-輻射圖
무강수계속일수	continuous dry spell days	無降水繼續日數
무게식우량계	weight measuring gauge	-雨量計
무게식자기설량계	weight measuring snow-recorder	-自記雪量計
무게식적설계	weight snow-gauge	-積雪計
무결빙기간	freeze-free period	無結氷期間
무늬고층운	altostratus maculosus	-高層雲
무늬권운	cirro-macula	-卷雲
무대바람	stage wind	舞臺-
무더운 날씨	muggy weather	
무더위	sultriness	
무리	cluster, halo	
무리달	mock moon, moon dog, paraselene	
무리달테	paraselenic circle	

무리수	irrational number	無理數
무리식	irrational expression	無理式
무리안개	mock fog	
무리짓기	clustering	
무리함수	irrational function	無理函數
무리해	false sun, mock sun, parhelion	
무리해고리	parhelic ring	
무리해테	mock sun ring, parhelic circle, white horizontal circle	
무발산고도	level of non-divergence, non-divergence level	無發散高度
무발산모형	non-divergent model	無發散模型
무산소물	anoxic water	無酸素-
무상기간	duration of frost-free period	無霜期間
무서리	light frost	無-
무서리지대	verdant zone, frostless zone	-地帶
무선	radio	無線
무선격측기상기록계	radio telemeteograph	無線隔測氣象記錄計
무선바람	radiowind	無線-
무선바람관측	radiowind observation	無線-觀測
무선바람관측소	radiowind station	無線-觀測所
무선설량계	radio snow gauge	無線雪量計
무선음파탐측계	radio acoustic sounding system	無線音波探測系
무선자기기록계	radio autograph	無線自記記錄計
무선전신타자기	radioteletypewriter	無線電信打字機
무선진동수	radio frequency	無線振動數
무선주파수	radio frequency	無線周波數
무선탐측장비	radio sounding equipment	無線探測裝備
무선텔레타이프송신기	radioteletype transmitter	無線-送信機
무선텔레타이프수신기	radioteletype receiver	無線-受信機
무성역	zone of silence	無聲域
무속도면	velopause	無速度面
무에코원형덮개	echo free vault	無-圓形-
무역풍	trade (wind)	貿易風
무역풍공기	trade air	貿易風空氣
무역풍대	trade-wind belt, trade-wind zone, zone of trade wind	貿易風帶
무역풍사막	trade-wind desert	貿易風砂漠
무역풍역전	trade inversion, trade-wind inversion	貿易風逆轉
무역풍적도기압골	trade-wind equatorial trough	貿易風赤道氣壓-
무역풍적운	trade cumulus, trade-wind cumulus	貿易風積雲
무역풍전선	trade-wind front	貿易風前線
무인관측소	robot station	無人觀測所
무인비행기	drone	無人飛行機
무작위	random	無作爲

무작위변수	random variable	無作爲變數
무작위씨뿌리기시험	randomized seeding trial	無作爲-試驗
무작위오차	random error	無作爲誤差
무작위예보	random forecast	無作爲豫報
무작위추출(법)	random sampling	無作爲抽出(法)
무작위표본	random sample	無作爲標本
무작위화	randomization	無作爲化
무정위(의)	astatic	無定位
무정형	amorphous	無定形
무정형고적운	altocumulus informis	無定形高積雲
무정형구름	amorphous cloud	無定形-
무정형구름무리	amorphous cloud cluster	無定形-
무정형구름하늘	amorphous sky	無定形-
무정형눈	amorphous snow	無定形-
무정형서리	amorphous frost	無定形-
무조역	amphidromic region	無潮域
무조점	amphidrome, amphidromic point	無潮點
무중력	gravity-free, zero gravity	無重力
무지개	bow, rainbow	
무지갯빛	iridescence	
무지갯빛 고적운	iridescent altocumulus	-高積雲
무지갯빛 구름	iridescent cloud	
무지갯빛 현상	irisation	-現象
무지향성	non-directivity	無指向性
무차원높이	dimensionless height	無次元高度
무차원량	dimensionless quantity, non-dimensional quantity	無次元量
무차원매개변수	dimensionless parameter, non-dimensional parameter	無次元媒介變數
무차원바람시어	dimensionless wind shear, non-dimensional wind shear	無次元-
무차원방정식	non-dimensional equation	無次元方程式
무차원비습경도	non-dimensional specific humidity gradient	無次元比濕傾度
무차원소산율	non-dimensional dissipation rate	無次元消散率
무차원수	dimensionless number, non-dimensional number	無次元數
무차원수문곡선	dimensionless hydrograph	無次元水文曲線
무차원연직플럭스	non-dimensional vertical flux	無次元鉛直-
무차원온도경도	dimensionless temperature gradient, non-dimensional temperature gradient	無次元溫度傾度
무차원온도연직분포	non-dimensional temperature profile	無次元溫度鉛直分布
무차원주파수	dimensionless frequency	無次元周波數
무차원진동수	non-dimensional frequency	無次元振動數
무차원함수	non-dimensional function	無次元函數
무편각선	agonic line	無偏角線
무풍	calm	無風

무풍대	calm zone	無風帶
무풍띠	calm belt	無風-
무풍중심눈	calm central eye	無風中心-
무풍층	calm layer	無風層
무하지역	arheic region	無河地域
무한	infinite, infinity	無限
무한급수	infinite series	無限級數
무한등비급수	infinite geometric(geometrical) series	無限等比級數
무한등비수열	infinite geometric(geometrical) sequence	無限等比數列
무한소수	infinite(nonterminating) decimal	無限小數
무한시정	unrestricted visibility	無限視程
묵은 눈	old snow	
문자숫자데이터	alphanumeric data	文字數字-
문턱가청값	threshold of audibility	門-可聽-
문턱깊이	sill depth	門-
문턱대조	liminal contrast	門-對照
문턱온도	threshold temperature	門-溫度
문턱조도	threshold illuminance	門-照度
묻힌 순환	embedded circulation	-循環
물결	water wave	
물결구름	billow cloud	
물결구름파동	billow wave	-波動
물결방향	wave direction	-方向
(잔)물결에코	ripple	
물결주기	wave cycle	-供給週期
물공급전망	water supply outlook	-供給展望
물과포화	supersaturation with respect to water	-過飽和
물광학특성	water optical property	-光學特性
물구름	water cloud	
물길조절	channel control	-調節
물당량	water equivalent	-當量
물대기	irrigation	
물뜀	hydraulic jump	
물리강화	physical enhancement	物理强化
물리광학	physical optics	物理光學
물리기상학	physical meteorology	物理氣象學
물리기후	physical climate	物理氣候
물리기후대	physical climatic zone	物理氣候帶
물리기후학	physical climatology	物理氣候學
물리모드	physical mode	物理-
물리모형	physical model	物理模型
물리방정식	physical equation	物理方程式
물리변화	physical change	物理變化
물리분류	physical classification	物理分類

물리상수	physical constant	物理常數
물리예보	physical forecasting	物理豫報
물리유체역학	physical hydrodynamics	物理流體力學
물리적공간분석시스템	Physical Space Analysis System	物理的空間分析-
물리추출	physical retrieval	物理抽出
물리통계예측	physico-statistic prediction	物理統計豫測
물리해양학	physical oceanography	物理海洋學
물리화학	physical chemistry	物理化學
물매	gradient	
물받이구멍	orifice	
물방울	water drop, water droplet	
물방울 관개	drip irrigation	-灌漑
물방울 분열이론	breaking-drop theory	-分裂理論
물방울 채집기	water dropper, water dropping collector	-採集器
물보라	spray	
물보라대전	spray electrification	-帶電
물분광반사	water spectral reflection	-分光反射
물분자	water molecule	-分子
물사용효율	water use efficiency	-使用效率
물상	water phase	-相
물상수	water constant	-常數
물소비량	water consumption	-消費量
물소요량	water requirement	-需要量
물손실	water loss	-損失
물수지	water budget	-收支
물순수포화(수)증기압	saturation vapo(u)r pressure in the pure phase with respect to water	-純粹飽和(水)蒸氣壓
물순환	hydrologic cycle, water cycle	-循環
물순환계수	water circulation coefficient	-循環係數
물습윤공기포화(수)증기압	saturation vapo(u)r pressure of moist air with respect to water	-濕潤空氣飽和(水)蒸氣壓
물(살포)씨뿌리기	water spray seeding	-(撒布)-
물안개	water fog	
물연기	water smoke, steam fog	-煙氣
물오염	water pollution	-汚染
물유형	water type	-類型
물의 비열	kilogram calorie	-比熱
물의 산도	acidity of water	-酸度
물일사계	water-flow pyrheliometer	-日射計
물재활용	water recycling	-再活用
물질	material, substance	物質
물질도함수	material derivative	物質導函數
물질면	material surface, substantial surface	物質面
물질박층	substantial sheet	物質薄層

물질부피	material volume	物質-
물질요동현상	congeliturbation	物質搖動現象
물질좌표	material coordinates	物質座標
물침식	water erosion	-浸蝕
물통기압계	cistern barometer	-桶氣壓計
물통법	bucket method	-桶法
물통온도	bucket temperature	-桶溫度
물통온도계	bucket thermometer	-桶溫度計
물틈	water gap	
물평형	hydrologic balance (budget)	-平衡
물포화혼합비	saturation mixing ratio with respect to water	-飽和混合比
물하늘	water sky	
물혼탁도	water turbidity	-混濁度
뭉게구름	woolpack cloud	
미구조	microstructure	微構造
미국국립대기연구소	National Center for Atmospheric Research	美國國立大氣研究所
미국기상학회	American Meteorological Society(AMS)	美國氣象學會
미국농무부 토양분류체계	USDA soil classification system	美國農務部土壤分類體系
미국지구물리연합	American Geophysical Union	美國地球物理聯合
미국표준대기	U.S. Standard Atmosphere	美國標準大氣
미국항공우주국	National Aeronautics and Space Administration	美國航空宇宙局
미국항공자문위원회	National Advisory Committee for Aeronautics	美國航空諮問委員會
미규모	microscale	微規模
미기상기록계	micrometeorograph	微氣象記錄計
미기상학	micrometeorology	微氣象學
미기압계	microbarometer	微氣壓計
미기압기록(지)	microbarogram	微氣壓記錄(紙)
미기압기록계	microbarograph, microbarovariograph	微氣壓記錄計
미기후	microclimate	微氣候
미기후학	microclimatology	微氣候學
(환경기준)미달성	nonattainment	(環境基準)未達成
(환경기준)미달성대	nonattainment zone	(環境基準)未達成帶
미디어예보	media forecast	-豫報
미래배출시나리오	scenario of future emission	未來排出-
미량	trace	微量
미량기체	trace gas	微量氣體
미량요소	trace element	微量要素
미류	trail	尾流
미류구름	trails of precipitation, trails of rain	尾流-
미립자농도	particulate loading	微粒子濃度
미립자복사	corpuscular radiation	微粒子輻射
미립자선	corpuscular ray	微粒子線
미립자설	corpuscular theory	微粒子說
미변계	variometer	微變計

미분가능(한)	differentiable	微分可能
미분계수	differential coefficient	微分係數
미분기하(학)	differential geometry	微分幾何(學)
미분방정식	differential equation	微分方程式
미분분석기	differential analyser	微分分析機
미분연산자	differential operator	微分演算子
미분운동학	differential kinematics	微分運動學
미사일 발사기상	missile firing weather	-發射氣象
미산란	Mie scattering	-散亂
미세	fine	微細
미세구름물리	micro cloud physics	微細-物理
미세구름조절모형	micro cloud modification model	微細-調節模型
미세구멍거르개	millipore filter	微細-
미세구조	fine structure, microstructure	微細構造
미세규모운동	microscale motion	微細規模運動
미세난류	microturbulence	微細亂流
미세돌풍	microburst	微細突風
미세돌풍경보	microburst alert	微細突風警報
미세망	fine-mesh	微細網
미세망격자	fine-mesh grid	微細網格子
미세망모형	fine-mesh model	微細網模型
미세물리모형	microphysical model	微細物理模型
미세물리학	microphysics	微細物理學
미세입자	aerosol, fine particle	微細粒子
미소(의)	minute	微小
미소규모운동	minute scale motion	微小規模運動
미소변위	minute displacement	微小變位
미스트랄	mistral	
미시적	microscopic	微視的
미압계	microbarograph, microbarovariograph	微壓計
미우량계	micropluviometer, ombrometer	微雨量計
미이론	Mie theory	-理論
미저기압	microcyclone	微低氣壓
미존법	Mizon method	-法
미주진로	erratic track	迷走進路
미지화합물	compound-X	未知化合物
미진	microseism	微震
미진기록계	microseismograph	微震記錄計
미터법	metric system	-法
미풍	breeze	微風
미풍계	airmeter	微風計
미풍전선	breeze front	微風前線
미행성설	planetesimal hypothesis	微行星說
민간기상기구	private weather organization	民間氣象機構

민간기상사업체	private weather agency	民間氣象事業體
민간기상정보사업체	private weather information agency	民間氣象情報事業體
민간기상정보서비스	private weather information service	民間氣象情報-
민간기상제공자	private weather provider	民間氣象提供子
민간항공관리국	Civil Aeronautical Administration	民間航空管理局
민감도	sensitivity	敏感度
(민)감도시간제어기	sensitivity time control	(敏)感度時間制御器
민물	fresh water	
민첩속도	celerity	敏捷速度
민트라	mintra	
밀	mil	
밀도	density	密度
밀도고도	density altitude	密度高度
밀도류	density current	密度流
밀도보정	density correction	密度補正
밀도불변고도	density constant height	密度不變高度
밀도비	density ratio	密度比
밀도성층	density stratification	密度成層
밀도이류	density advection	密度移流
밀도(두께)이류	density (thickness) advection	密度移流
밀도채널	density channel	密度-
밀도함수	density function	密度函數
밀도행렬	density matrix	密度行列
밀란코비치순환	Milankovitch cycle	-循環
밀란코비치 홍적세 기후변동	Milankovitch Pleistocene climatic variation	-洪積世氣候變動
밀란코비치진동	Milankovitch oscillation	-振動
밀리	milli-	
밀리미터파 레이더	millimeter wave radar	-波-
밀리바	millibar	
밀리바기압계	millibar-barometer	-氣壓計
밀리바눈금	millibar scale	
밀물	flood current, flood tide, rising tide	
밀유빙	close pack ice	密遊氷
밀집고적운	altocumulus glomeratus	密集高積雲
밑녹음	undermelting	
밑면가열	basal heating	-面加熱
밑면미끄럼	basal slip	-面-
밑면사태	basal sliding	-面沙汰
밑수	radix	-數
밑흐름	underrun	

ㅂ

한글	영문	한자
바	bar	
바기오	baguio, bagio, bagyo, vaguio, vario	
바깥눈벽	outer eyewall	-壁
바깥대기	outer atmosphere	-大氣
바깥무리	outer halo	
바깥층	outer layer	-層
바깥층 스케일링	outer layer scaling	-層-
바늘얼음	acicular ice, frazil ice	
바다	sea	
바다공기	sea air	-空氣
바다나리류	crinoids	-流
바다너울	sea swell	
바다무지개	marine rainbow, sea rainbow	
바다물보라	sea spray	
바다바람	sea breeze	
바다반향	sea return	-反響
바다상태	sea, state of the sea	-狀態
바다실안개	sea mist	
바다안개	sea fog	
바다얼음	basal ice, sea ice	
바다바람	onshore wind	
바다에코	sea echo, sea return	
바다증기안개	sea smoke	-蒸氣-
바다행성	ocean planet	-行星
바닥경계조건	bottom boundary condition	-境界條件
바닥높낮이	bottom topography	
바닥마찰	bottom friction	-摩擦
바닥얼음	anchor ice, bottom ice	
바닥온도	bottom temperature	-溫度

바닥층	bottom layer	-層
바닷가얼음	beach ice	
바닷물	sea water	
바라이	barye	
바람	wind	
바람가락	aeolian tones	
바람가속도	acceleration of wind	-加速度
바람감속	deceleration of wind	-減速
바람개비	weather cock	
바람경각	inclination of the wind	-傾角
바람경도	wind gradient	-傾度
바람관측탑	anemometer mast (or tower)	-觀測塔
바람구름	wind cloud	
바람구역	wind zone	-區域
바람구조	wind structure	-構造
바람권운	cirrus ventosus, windy cirrus	-卷雲
바람그늘	wind shadow	
바람그을림	windburn	
바람급변	wind shift	-急變
바람급변선	wind shift line	-急變線
바람기간	wind duration	-期間
바람기둥	wind column	
바람기록계	wind recorder	-記錄計
바람기온(분포)도	wind and temperature chart	-氣溫(分布)圖
바람길	wind passage	
바람깃	wind barb	
바람깃대	wind shaft	
바람냉각	wind-chill	-冷却
바람냉각경보	wind-chill warning	-冷却警報
바람냉각인자	wind-chill factor	-冷却因子
바람냉각주의보	wind-chill advisory	-冷却注意報
바람냉각지수	wind-chill index	-冷却指數
바람눈사태	wind avalanche	-沙汰
바람(분포)도	wind chart	-(分布)圖
바람띠	wind belt	
바람마루	wind ridge	
바람막이	shield, windbreak, wind protection, wind shield	
바람맞이의	aweather	
바람맴돌이	wind eddy	
바람몰이(해)류	wind-driven current	-(海)流
바람몰이순환	wind-driven circulation	-循環
바람몰이해양순환	wind-driven oceanic circulation	-海洋循環
바람무늬	wind ripple	
바람반경	wind radius	-半徑

바람벡터	wind vector	
바람변동구역	zone of variable wind	-變動區域
바람 부는	windy	
바람분석	wind analysis	-分析
바람분포	wind distribution	-分布
바람성분	wind component	-成分
바람소리	aeolian sound	
바람소리풍속계	aeolian anemometer	-風速計
바람소용돌이	wind eddy	
바람순전	wind veering	-順轉
바람스콜	wind squall	
바람슬랩	wind slab	
바람쏠림	wind deflection	
바람에너지	wind energy	
바람재앙	catastrophic windthrow	-災殃
바람역전	wind reversal	-逆轉
바람연결	wind couplet	-連結
바람연직분포	wind profile	-鉛直分布
바람응력	wind stress	-應力
바람응력컬	wind stress curl	-應力-
바람의 고형물수송력	competence of the wind	-固形物輸送力
바람의 숨	gustiness	
바람의 일정성	constancy of wind	-一定性
바람이랑	wind ridge	
바람인자	wind factor	-因子
바람자루	wind cone, wind sleeve, wind sock	
바람자췻길	wind swath	
바람작용	wind action	-作用
바람장	wind field	-場
바람장미	wind rose	-薔薇
바람주걱	wind scoop	
바람주의보	wind advisory	-注意報
바람지수	W-index	-指數
바람축	wind axis	-軸
바람층	wind layer, wind stratum	-層
바람침식	corrasion, eolation, wind corrosion, wind erosion	-浸蝕
바람크러스트	wind crust	
바람틈새	wind gap	
바람편류도	wind and drift chart	-偏流圖
바람폭풍(우)	wind storm	-暴風(雨)
바람표류	wind drift	-漂流
바람표류거리	wind drift distance	-漂流距離
바람표시화살	barbed arrow	-表示-
바람프로파일	wind profile	

바람피해	wind damage	-被害
바람혼합	wind mixing	-混合
바람화살표	wind arrow	-表
바람회전소용돌이	wind-spun vortex	-回轉-
바람휴식	wind lull	-休息
바람흐름	wind flow	
바렌츠빙상	Barents ice sheet	-氷床
바로밀	baromil	
바르칸	barkhan, barchane, barchan	
바르한	barchan	
바르한사구	barchan dune	-砂丘
바리	barih	
바베이도스해양기상실험	Barbados oceanographic and meteorological experiment	-海洋氣象實驗
바비네원리	Babinet's principle	-原理
바비네중립점	Babinet's point	-中立點
바스카라 위성	Bhaskara satellite	-衛星
바이램풍속계	Byram anemometer	-風速計
바이메탈온도계	bimetal thermometer	-溫度計
바이메탈일사계	bimetallic actinometer	-日射計
바이메탈일조계	bimetal sunshine sensor	-日照計
바이메탈자기온도계	bimetallic thermograph	-自記溫度計
바이메탈자기일사계	bimetallic actinograph	-自記日射計
바이메탈판	bimetal strip	-板
바이스발롯 규칙	Buys-Ballot rule	-規則
바이스발롯 법칙	Buys-Ballot's law	-法則
바이스태틱	bistatic	
바이오가스	biogas	
바이오매스	biomass	
바이오매스 연소	biomass burning	-燃燒
바이오매스 조리기구	biomass cookers	-調理器具
바이오매스 향상비	biomass enhancement ratio	-向上比
바이오연료	biofuel	-燃料
바이오온도	biotemperature	-溫度
바이우	Baiu, plum rain, Tsuyu	
바이우전선	Baiu front	-前線
바퀴기압계	wheel barometer	-氣壓計
바퀴오스	Baquios	
바탕복사	background radiation	-輻射
바탕수준	background level	-水準
바탕오염	background pollution	-汚染
박막수	pellicular water	薄膜水
박명	twilight	薄明
박명광	twilight flash	薄明光

박명대기광	twilight airglow	薄明大氣光
박명보정	twilight correction	薄明補正
박명색	twilight color	薄明色
박명스펙트럼	twilight spectrum	薄明-
박명지역	twilight zone	薄明地域
박명현상	twilight phenomenon	薄明現象
박명호	arch twilight, crepuscular arch, twilight arch	薄明弧
박무	mist	薄霧
박무방울	mist droplet	薄霧-
박무발생온도역	mist interval	薄霧發生溫度域
박무우적	mist droplet	薄霧雨滴
박스-젠킨스 방법	Box-Jenkins method	-方法
박스카함수	boxcar function	-函數
반감기	half-life, half-value period	半減期
반건조	semiarid	半乾燥
반건조구역	semiarid region	半乾燥區域
반건조기후	semiarid climate	半乾燥氣候
반건조지대	semiarid zone	半乾燥地帶
반걸	ban-gull	
반경	radius	半徑
반경도전달	countergradient transfer	反傾度傳達
반경도풍	countergradient wind	反傾度風
반경도플럭스	countergradient flux	反傾度-
반경도확산	countergradient diffusion	反傾度擴散
반경도흐름	countergradient flow	反傾度-
반경벡터	radius vector	半徑-
반구	hemisphere, semisphere	半球
반구교환센터	hemisphere exchange center	半球交換-
반구모형	hemispheric model	半球模型
반구방송	hemisphere broadcast	半球放送
반구제트	hemispheric jet	半球-
반구파수	hemispheric wave number	半球波數
반그림자	penumbra	半-
반다이나모	antidynamo	反-
반다해수	Banda Sea Water	-海水
반달	half moon	半-
반대	counter, opposition	反對
반대계곡풍	antivalley wind	反對溪谷風
반대계절풍	antimonsoon	反對季節風
반대광환	anticorona, Brocken bow	反對光環
반대무역풍	antitrade wind, antitrades, countertrades	反對貿易風
반대박명	anti-twilight, countertwilight	反對薄明
반대박명광	anticrepuscular ray	反對薄明光
반대박명호	anticrepuscular arch, anti-twilight arch	反對薄明弧

반대산풍	anti-mountain wind	反對山風
반대해풍	antisea breeze	反對海風
반데르발스방정식	Van der Waals' equation	-方程式
반도	peninsula	半島
반도체광소자	photodiode	半導體光素子
반라그랑지법	semi-Lagrangian method	半-法
반류	counter flow, countercurrent	反流
반류열교환	counter current heat exchange	反流熱交換
반무리해	counter parhelion	反-
반변미분	contravariant differentiation	反變微分
반변벡터	contravariant vector	反變-
반복	iteration, recurrence	反復
반복법	iteration	反復法
반복빈도	repetition frequency	反復頻度
반복성	repeatability	反復性
반사	reflection, reflexion	反射
반사각	angle of reflection	反射角
반사계	reflectometer	反射計
반사계수	reflection coefficient	反射係數
반사광	reflected light	反射光
반사광선	reflected ray	反射光線
반사기	reflector	反射器
반사능	reflecting power, reflective power	反射能
반사단파복사	reflected shortwave radiation	反射短波輻射
반사막기후	semidesert climate	半沙漠氣候
반사무지개	reflection rainbow	反射-
반사법칙	reflexive law	反射法則
반사분위범위	semi-interquartile range	半四分位範圍
반사율	albedo, reflectance, reflectivity	反射率
반사율계	albedometer	反射率計
반사율기록계	albedograph	反射率記錄計
반사율산출	reflectance calculation	反射率算出
반사율인자	reflectivity factor	反射率因子
반사인자	reflection factor	反射因子
반사지구복사	reflected terrestrial radiation	反射地球輻射
반사체	reflector	反射體
반사측운기	reflecting nephoscope	反射測雲器
반사태양복사	reflected solar radiation	反射太陽輻射
반사파	reflected wave	反射波
반상변정	porphyroblast	斑狀變晶
반송주파수	carrier frequency	搬送周波數
반송파	carrier wave	搬送波
반순	penthemeron, pentad	半旬
반순환	countercirculation	反循環

반시계(방향)	anticlockwise	反時計(方向)
반시계방향	counterclockwise	反時計方向
반시계방향바람	counterclockwise wind	反時計方向-
반시계방향풍향전환	counterclockwise wind shift	反時計方向風向轉換
반알렌(복사)대	Van Allen (radiation) belt	-(輻射)帶
반암시방안	semi-implicit scheme	半暗示方案
반암시법	semi-implicit method	半暗示法
반압류	antibaric flow	反壓流
반양성자	antiproton	反陽性子
반염수	brackish water, esturine water	半鹽水
반영구고기압	semipermanent anticyclone, semipermanent high	半永久高氣壓
반영구저기압	semipermanent depression	半永久低氣壓
반올림오차	round-off error	-誤差
반응	reaction, response	反應
반응경로	reaction path	反應經路
반응률	reaction rate	反應率
반응메커니즘	reaction mechanism	反應-
반응모형	reaction model	反應模型
반응시간	response time	反應時間
반응체적차원	dimension of reaction volume	反應體積次元
반응함수	response function	反應函數
반일기압파	semidiurnal pressure wave	半日氣壓波
반일변동	semidiurnal variation	半日變動
반일조석	semidiurnal tide	半日潮汐
반일주기	semidiurnal period	半日週期
반일파	semidiurnal wave	半日波
반전	backing	反轉
반전력점	half-power point	半電力點
반전바람	backing wind	反轉-
반전주파수	turnover frequency	反轉周波數
반점	patch	斑點
(레이더영상의)반점	speckle	班點
반정상굴절	subnormal refraction	半正常屈折
반주기역	half-period zone	半週期域
반지름방향속도	radial velocity	-方向速度
반지름벡터	radius vector	
반지속성	antipersistence	反持續性
반짝빛	glare	
반짝임	blink, scintillation	
반쪽반응	half reaction	-反應
반체류시간	residence half-time	半滯留時間
반투명(의)	translucent	半透明
반투명고적운	altocumulus translucidus	半透明高積雲
반투명고층운	altostratus translucidus	半透明高層雲

반투명구름	translucidus	半透明-
반투명도	translucency, translucence	半透明度
반투명얼음	glime	半透明-
반투명층운	stratus translucidus	半透明層雲
반투명층적운	stratocumulus translucidus	半透明層積雲
반트호프법칙	van't Hoff's law	-法則
반트호프인자	van't Hoff's factor	-因子
반편차중심	antipleion	反偏差中心
반평균법	semiaverage method	半平均法
반폭	half-width	半幅
반호각	half-arc angle	半弧角
반ENSO사건	anti-ENSO event	反-事件
받침돌	pedestal rock	
발광	luminescence	發光
발광강도	luminous intensity	發光强度
발광노출	luminous exposure	發光露出
발광능	luminous power	發光能
발광도	luminous emittance	發光度
발광률	luminous exitance	發光率
발광비행운	luminous vapour-trail	發光飛行雲
발광속밀도	luminous flux density	發光束密度
발광에너지	luminous energy	發光-
발광운	luminous cloud	發光雲
발광체	luminous body	發光體
발광효율	luminous efficiency	發光效率
발달단계	developing stage	發達段階
발달성전선	frontogenetical front	發達性前線
발달지수	development index	發達指數
발달파	development wave	發達波
발롯일사온도계	Vallot heliothermometer	-日射溫度計
발류트	ballute(= ball + parachute)	
발리	bali	
발리바람	Bali wind	
발바닥흔적	sole mark	-痕迹
발사	launching	發射
발산	divergence	發散
발산구역	divergence region	發散區域
발산방정식	divergence equation	發散方程式
발산선	divergence line	發散線
발산순압대기	divergent barotropic atmosphere	發散順壓大氣
발산장	divergence field, field of divergence	發散場
발산점	divergent point	發散點
발산점근선	asymptote of divergence	發散漸近線
발산정리	divergence theorem	發散定理

발산주변부	divergent margin	發散周邊部
발산징후	divergence signature	發散徵候
발산효과	divergence effect	發散效果
발생	genesis, occurrence	發生
발생기	incipient stage	發生期
발생기저기압	incipient low	發生期低氣壓
발생단계	incipient stage	發生段階
발생빈도	frequency of occurrence	發生頻度
발생확률	probability of occurrence	發生確率
발아	bud burst	發芽
발아구	colpus	發芽口
발아일	germination date	發芽日
발암물질	carcinogen	發癌物質
발암성	carcinogenicity	發癌性
발암성물질	carcinogenic compound	發癌性物質
발열량	heating value	發熱量
발열반응	exothermic reaction	發熱反應
발원	source	發源
발원범주	source category	發源範疇
발원지	source region, origin	發源地
발원지특성	source property	發源地特性
발자국	footprint	
발자국모형	footprint model	-模型
발트지역	Baltic region	-地域
발트해	Baltic Sea	-海
발포탐사	gunprobe	發砲探査
밝기	brightness	
밝기값	brightness value	
밝기기준	brightness criteria	-基準
밝기등치선	isopleth of brightness	-等值線
밝기수준	brightness level	-水準
밝기온도	brightness temperature	-溫度
밝은 띠	bright band	
밝은 띠 에코	bright band echo	
(채층의) 밝은 망상조직	bright network	-網狀組織
밝은 부분	bright segment	-部分
밤	night	
밤바람	night wind	
밤이슬	night dew	
밤하늘발광	night-sky luminescence	-發光
밤하늘복사	night-sky radiation	-輻射
밤하늘빛	night-sky light	
밧줄구름	rope cloud	
밧줄깔때기	rope funnel	

방류구	outlet	放流口
방류하천	effluent stream	放流河川
방사강도	intensity of radiation	放射强度
방사고적운	altocumulus radiatus	放射高積雲
방사고층운	altostratus radiatus	放射高層雲
방사구름	radiatus	放射-
방사권운	cirrus radiatus	放射卷雲
방사능	radioactivity	放射能
방사능기상	radioactivity weather	放射能氣象
방사능낙진기입	radioactive fallout plot	放射能落塵記入
방사능방출	radioactive emanation	放射能放出
방사능분석	radioassay	放射能分析
방사능세척	radioactive washout	放射能洗滌
방사능집진기	radioactive collector	放射能集塵機
방사능추적	radioactive tracer	放射能追跡
방사대칭	radial symmetry	放射對稱
방사방향	radial direction	放射方向
방사분석	radioassay, radiometry	放射分析
방사선	radioactive ray	放射線
방사성강수	radioactive precipitation	放射性降水
방사성기체	radioactive gas	放射性氣體
방사성낙진	radioactive fallout	放射性落塵
방사성동위원소적설계	radioisotope snow gauge	放射性同位元素積雪計
방사성비	radioactive rain	放射性-
방사성설량계	radioactive snow gauge	放射性雪量計
방사성연대측정법	radiometric method	放射性年代測定法
방사성추적자	radioactive tracer	放射性追跡者
방사성탄소	radiocarbon	放射性炭素
방사성탄소연대측정법	radiocarbon dating	放射性炭素年代測定法
방사유역	radial drainage	放射流域
방사유입	radial inflow	放射流入
방사자료	radial data	放射資料
방사적운	cumulus radiatus	放射積雲
방사점	V point, radiant point	放射點
방사층적운	stratocumulus radiatus	放射層積雲
방사풍	radial wind	放射風
방사형(의)	actiniform	放射形
방산충	radiolaria	放散蟲
방설림	snow-shelter forest	防雪林
방송	broadcast	放送
방송기상학	broadcast meteorology	放送氣象學
방수	discharge	防水
방식	mode	方式
방아쇠작용	trigger action	-作用

방아쇠효과	trigger effect	-效果
방압전도온도계	protected reversing thermometer	防壓傳導溫度計
방열형	outgoing radiation type	放熱型
방울	drop	
방위	bearing, cardinal direction, orientation	方位
방위각	azimuth, azimuth angle, horizontal angle	方位角
방위각파수	azimuth wavenumber	方位角波數
방위거리	azran	方位距離
방위고도각	azimuth elevation	方位高度角
방위-고도지시기	azel-scope	方位高度指示器
방위기상위성계획	defense meteorological satellite program	防衛氣象衛星計劃
방위눈금	azimuth scale	方位-
방위분해능	azimuthal resolution, azimuth resolution	方位分解能
방위섭동	azimuthal perturbation	方位攝動
방위오차	azimuth error	方位誤差
방위왜곡	azimuth distortion	方位歪曲
방위점	point	方位點
방위표지	azimuth marker	方位標識
방위해상도	azimuth resolution	方位解像度
방재기간	disaster prevention period	防災期間
방전	discharge, electric discharge	放電
방전관	discharge tube	放電管
방전식풍속계	condenser discharge anemometer	放電式風速計
방조제	seawall	防潮堤
방출	emission, exhalation	放出
방출가능에너지	releasable energy	放出可能-
방출강도	emission intensity	放出强度
방출고도날씨	release altitude weather	放出高度-
방출능	emissive power	放出能
방출률	emission rate, emissivity, emittance	放出率
방출률수정	emissivity correction	放出率修整
방출복사	emission radiation, outgoing radiation	放出輻射
방출복사형	outgoing radiation type	放出輻射型
방출선	emission line	放出線
방출스펙트럼	emission spectrum	放出-
방출인자	emission factor	放出因子
방출장파복사	outgoing longwave radiation(OLR)	放出長波輻射
방출체	emitter	放出體
방출층	emission layer	放出層
방파제	breakwater	防波堤
방패권운	cirrus shield	防牌卷雲
방패화산	shield volcano	防牌火山
방풍	wind protection	防風
방풍대	shelterbelt	防風帶

방풍림	windbreak forest	防風林
방한	cold protection	防寒
방한복착용기상	heavy winter clothing wear weather	防寒服着用氣象
(항행 또는 비행)방향	heading	方向
방향	attitude, orientation	方向
방향각	direction angle	方向角
방향반구반사율	directional hemispherical reflectance	方向半球反射率
방향반사율	directional reflectance	方向反射率
방향발산	directional divergence	方向發散
방향변화발산	divergence due to directional change	方向變化發散
방향별우량계	vectopluviometer, vector gauge	方向別雨量計
방향복사휘도	directional radiance	方向輻射輝度
방향분포	directional distribution	方向分布
방향성결합기	directional coupler	方向性結合器
방향성과성장	oriented overgrowth, epitaxis	方向性過成長
방향수	direction number	方向數
방향수리전도도	directional hydraulic conductivity	方向水理傳導度
방향시어	directional shear	方向-
방향안테나	directional antenna	方向-
방향족 탄화수소	aromatic hydrocarbon	芳香族炭化水素
방향족 화합물	aromatic compound, aromatics	芳香族化合物
방향타	rudder	方向舵
방향탐지	direction finding	方向探知
방향탐지기	direction finder	方向探知器
배경대기오염관측망	background air pollution monitoring network	背景大氣汚染觀測網
배경대비	background contrast	背景對比
배경복사	background radiation	背景輻射
배경수준	background level	背景水準
배경신호	background signal	背景信號
배경오염	background pollution	背景汚染
배경오차공분산	background error covariance	背景誤差共分散
배경장	background field	背景場
배경휘도	background luminance	背景輝度
배기	air drainage	排氣
배기가스	exhaust gas	排氣-
배기가스 시료채취 방법	method for sampling stack gas	排氣-試料採取方法
배기가스 재순환	exhaust gas recirculation	排氣-再循環
배기구름(비행운)	engine-exhaust trail, exhaust trail	排氣-(飛行雲)
배기비행운	exhaust contrail	排氣飛行雲
배기흔적	exhaust trail	排氣痕跡
배분	allocation	配分
배속	overspeeding	培速
배수	drainage	排水
배수량계	drainage gauge	排水量計

배수력	drainage force	排水力
배수밀도	drainage density	排水密度
배심문제	jury problem	陪審問題
배연탈질	flue gas denitrogenization	排煙脫窒
배연탈황	flue gas desulfurization	排煙脫黃
배증법	doubling method	倍增法
배출	emission	排出
배출감축	emission reduction	排出減縮
배출검사주기	emission test cycle	排出檢查週期
배출권거래제	emission trading	排出權去來制
배출권거래제도	emission trading system, marketable permit	排出權去來制度
배출규제	emission control	排出規制
배출기준	emission standard	排出基準
배출량거래체계	emission trading scheme	排出量去來體系
배출량평가	emissions assessment	排出評價
배출류	drainage flow	排出流
배출물	effluent	排出物
배출물부력	effluent buoyancy	排出物浮力
배출물속도	effluent velocity	排出物速度
배출물운동량	effluent momentum	排出物運動量
배출반영	emission reflection	排出反影
배출부과금	emission tax	排出賦課金
배출시나리오	emission scenario	排出-
배출역	drainage area	排出域
배출이론	emission theory	放出理論
배출지역	discharge area	排出地域
배출풍	drainage wind	排出風
배치	allocation	配置
배타적 홍수조절저류량	exclusive flood control storage capacity	排他的洪水調節貯流量
배핀아일랜드	Baffin Island	
배핀해류	Baffin current	-海流
백금선온도계	platinum wire thermometer	白金線溫度計
백금저항온도계	platinum resistance thermometer	白金抵抗溫度計
백도	moon's path	白道
백두산고기압	Paekdusan high	白頭山高氣壓
백반	faculae	白斑
백분도	centigrade	百分度
백분도눈금	centigrade scale	百分度-
백분도온도계	centigrade thermometer	百分度溫度計
백분도온도눈금	centigrade temperature scale	百分度溫度-
백분오차	percentage error	百分誤差
백분위수	percentile	百分位數
백분위점	tercile	百分位點
백분율감소	percent reduction	百分率減少

백색광	white light (WL)	白色光
백색광 불꽃	white light flare	白色光-
백색잡음	white noise	白色雜音
백수	white head	白穗
(중생대)백악기	Cretaceous period	白堊期
백악기기후	Cretaceous climate	白堊期氣候
백악기온난기후	Cretaceous warm climate	白堊期溫暖氣候
백악기-제3기 경계층	Cretaceous-Tertiary(K-T) boundary	白堊期第三期境界層
백악기-제3기 멸종	Cretaceous-Tertiary(K-T) extinction	白堊期第三期滅種
백엽상	instrument shelter, shelter, thermometer screen, thermometer shelter, thermoscreen, screen	百葉箱
백엽상온도	shelter temperature	百葉箱溫度
백엽상온도계	shelter thermometer	百葉箱溫度計
백작부인바람	wind countess	伯爵夫人-
백중사리	flood tide in Baekjung	百中-
백지도	base map	白地圖
백체	white body	白體
백파	white wave, whitecap	白波
백필터	bag filter	
백하우스	baghouse	
백화현상(식물의)	bleaching, chlorosis	白化現象
밴드모형	band model	-模型
밴드방법	band method	-方法
밴드지정	band designation	-指定
밴드패턴	band pattern	
버거수	Burger number	-數
버거스물질	Burgers material	-物質
버거스방정식	Burgers equation	-方程式
버거스벡터	Burgers vector	
버거의 소용돌이	Buger's vortex	
버널-파울러 규칙	Bernal-Fowler rule	-規則
버더만법	Verderman method	-法
버드버스트	bird burst	
버드스트라이크	bird strike	
버뮤다고기압	Bermuda high	-高氣壓
버뮤다-아조레스 고기압	Bermuda-Azores high	-高氣壓
버블터짐	bubble bursting	
버섯구름	mushroom cloud	
버스트	burst	
버지니아 고기압	Virginia high	-高氣壓
버캔냉난기	Buchan spells	-冷暖期
버크랜드해류	Birkeland current	-海流
버킷법	bucket method	-法
버킹엄 파이 정리	Buckingham pi theorem	-定理

버팀목온도계	strut thermometer	-木溫度計
버퍼링마하수	buffering Mach number	-數
번개	bolt, lightning	
번개계수기	lightning counter	-計數器
번개기록계	lightning recorder	-記錄計
번개방전	lightning discharge	-放電
번개보라	lightning storm	
번개섬광	lightning flash	-閃光
번개스톰	lightning storm	
번개스펙트럼	lightning spectrum	
번개억제	lightning suppression	-抑制
번개에코	lightning echo	
번개영상센서	lightning imaging sensor	-映像-
번개전류	lightning current	-電流
번개지도작성자	lightning mapper	-地圖作成者
번개지도화시스템	lightning mapping system	-地圖化-
번개탐지	lightning detection	-探知
번개피해	lightning damage	-被害
번개확산	lightning diffusion	-擴散
번갯길	lightning channel	
벌스반복률	pulse repetition rate	-反復率
벌집(구름)	lacunaris, lacunosus	
벌집고적운	altocumulus lacunosus	-高積雲
벌집권적운	cirrocumulus lacunosus	-卷積雲
벌집층적운	stratocumulus lacunosus	-層積雲
벌크법	bulk method	-法
벌틴크-말레 안정도	Bultink-Malet stability	-安定度
범람수로	outflow channel	氾濫水路
범람평원	flood plain	氾濫平原
범용지리코드	universal geographic code	汎用地理-
범위	coverage, range	範圍
범위도	coverage diagram	範圍圖
범죄기상학	forensic meteorology	犯罪氣象學
범주예보	categorical forecast, category forecast	範疇豫報
법선가속도	normal acceleration	法線加速度
법선경도	normal gradient	法線傾度
법선력	normal force	法線力
법선방정식	equation of normal	法線方程式
법선벡터	normal vector	法線-
법선성분	normal component	法線成分
법선응력	normal stress	法線應力
벚꽃전선	cherry-blossom front	-前線
베개(구름)	pillow (cloud)	
베거통풍기	Weger aspirator	-通風器

베거통풍통	Weger aspirator	-通風桶
베게너-베르게론 과정	Wegener-Bergeron process	-過程
베게너-베르게론-핀다이젠이론	Wegener-Bergeron-Findeisen theory	-理論
베나르-레일리 대류	Bénard-Rayliegh convection	-對流
베나르문제	Bénard problem	-問題
베나르세포	Bénard cell	-細胞
베르게론	Bergeron	
베르게론과정	Bergeron process	-過程
베르게론기후구분시스템	Bergeron Climatic Classification System	-氣候區分-
베르게론-핀다이젠 과정	Bergeron-Findeisen process	-過程
베르게론-핀다이젠 메커니즘	Bergeron-Findeisen mechanism	
베르게론-핀다이젠 이론	Bergeron-Findeisen theory	-理論
베르게론합류	Bergeron confluence	-合流
베르게론효과	Bergeron effect	-效果
베르겐지구물리연구소	Bergen Geophysical Institute	-地球物理研究所
베르겐학교(학파)	Bergen School	-學校(學派)
베르글라	verglas	
베르누이메커니즘	Bernoulli mechanism	
베르누이방정식	Bernoulli's equation	-方程式
베르누이분포	Bernoulli distribution	-分布
베르누이원리	Bernoulli principle	-原理
베르누이정리	Bernoulli's theorem	-定理
베르누이효과	Bernoulli effect	-效果
베르데곶형	cape Verde type	-型
베르법칙	Beer's law	-法則
베르-부게 법칙	Beer-Bouguer law	-法則
베르손 서풍대	Berson winds (or westerlies)	-西風帶
베르그바람	berg wind	
베릴륨	beryllium(Be)	
베릴륨구리	beryllium copper	
베릴륨중독증	berylliosis	-中毒症
베링경사류	Bering slope current	-傾斜流
베링해	Bering Sea	-海
베링해류	Bering current	-海流
베링해협	Bering Strait	-海峽
베셀방정식	Bessel equation	-方程式
베셀함수	Bessel function	-函數
베송 빗모양측운기	Besson comb nephoscope	-模樣測雲器
베이지안 정리	Baysian theorem	-定理
베일층운	stratus nebulosus	-層雲
베크렐	becquerel(Bq)	
베크만온도	Beckmann thermometer	-溫度
베타감쇠	beta decay	-減衰
베타나선	beta spiral	-螺線

베타면근사	beta-plane approximation	-面近似
베타붕괴	beta decay	-崩壞
베타선	beta ray(β-ray)	-線
베타유효	beta effectiveness	-有效
베타입자	beta particle	-粒子
베타자이어	beta gyre	
베타평면	beta plane	-平面
베타평면근사	beta plane approximation	-平面近似
베타표류	beta drift	-漂流
베타환류	beta gyre	-還流
베타효과	beta effect	-效果
베타효과성	beta effectiveness	-效果性
벡터공간	vector space	-空間
벡터량	vector amount	-量
벡터분해	resolution of vector	-分解
벡터성분	vector component	-成分
벡터우량계	vector gauge, vector rain gauge	-雨量計
벡터퍼텐셜	vector potential	
벡터평균도	vector mean chart	-平均圖
벡터합성	composition of vector	-合成
벤젠	benzene	
벤조피렌	benzopyrene	
벤투리관	venturi tube	-管
벤투리스크러버	venturi scrubber	
벤투리효과	venturi effect	-效果
벤틀리 해구	Bentley Subglacial Trench	-海溝
벨(음향강도측정단위, 10데시벨)	bel	
벨라미 방법	Bellamy method	-方法
벨탄-리	beltane-ree	
벨 테이퍼	bell taper	
벨트라미방정식	Beltrami equation	-方程式
벨트라미흐름	Beltrami flow	
벰포라드공식	Bemporad's formula	-公式
벵골만 해수	Bay of Bengal Water	-灣海水
벵구엘라해류	Benguela current	-海流
벼논에너지평형(수지)	paddy field energy balance	-平衡(收支)
벼락	cloud-to-ground discharge, flash-to-ground, ground discharge, lightning strike, stroke, stroke of lightning, thunderbolt	
벼락동반소나기	light shower	-同件-
벼락밀도	stroke density	-密度
벽구름	pedestal cloud, wall cloud	壁-
벽력	lightning strike	霹靂

벽운	wall cloud	壁雲
변고풍	allohypsic wind, isallohypsic wind	變高風
변곡점	inflexion point, point of inflection	變曲點
변덕권운	inconstant cirrus	變德卷雲
변덕날씨	blirty	變德-
변동곡선	variation curve	變動曲線
변동-소산정리	fluctuation-dissipation theorem	變動-消散定理
변동중심	center of variation	變動中心
변동플룸모형	fluctuating plume model	變動-模型
변복조장치	modem	變復調裝置
변분객관분석	variational objective analysis	變分客觀分析
변분동화	variational assimilation	變分同化
변분품질관리	variational quality control	變分品質管理
변수분리	separation of variable	變數分離
변압(의)	allobaric	變壓
변압구역	allobar	變壓區域
변압풍	allobaric wind, isallobaric wind	變壓風
변압풍수렴	isallobaric wind convergence	變壓風收斂
변위	displacement, shift	變位
변위거리	displacement distance	變位距離
변위고도	displacement height, shifting level	變位高度
변위두께	displacement thickness	變位-
변위벡터	displacement vector	變位-
변위전류	displacement current	變位電流
변조	modification	變造
변조	modulation	變調
변조기	modulator	變調器
변질기단	modified air-mass, transformed air-mass	變質氣團
변형	deformation, diversification	變形
변형구역	deformation zone	變形區域
변형기압계	elastic barometer	變形氣壓計
변형된 빗방울	distorted raindrop	變形-
변형면	deformation plane	變形面
변형반경	deformation radius, radius of deformation	變形半徑
변형반지름	deformation radius, radius of deformation	變形-
변형벡터	deformation vector	變形-
변형온도계	deformation thermometer	變形溫度計
변형온도기록계	deformation thermograph	變形溫度記錄計
변형장	deformation field, field of deformation	變形場
변형축	deformation axis	變形軸
변형항	deformation term	變形項
변형흐름	deformational flow	變形-
변형 KdV 방정식	modified Korteweg-de Vries equation	變形-方程式
변화	change	變化

변화도	change chart	變化圖
변화문턱	change threshold	變化門-
변화벡터분석	change vector analysis	變化-分析
변화율	rate of change	變化率
변환경계	transform boundary	變換境界
변환기	converter, transducer	變換器
변환단층	transform fault	變換斷層
별	star	
별똥별	shooting star	
별모양결정	stellar crystal	-模樣結晶
별모양번개	stellar lightning	-模樣-
별반짝임	stellar scintillation	
별빛기록계	starshine recorder	-記錄計
병목	bottleneck	瓶-
병온도계	bottle thermometer	瓶溫度計
병진	translation	竝進
병진속도	translatory velocity	竝進速度
병진운동	translational motion	竝進運動
병진장	translatory field	竝進場
병진파	translatory wave, wave of translation	竝進波
병참작전기상	logistics operation weather	兵站作戰氣象
병합	coalescence	倂合
병합계수	coalescence coefficient	倂合係數
병합과정	coalescence process	倂合過程
병합설	coalescence theory	倂合說
병합효과	coalescence effect	倂合效果
병합효율	coalescence efficiency	倂合效率
병해저항성	disease resistance	病害抵抗性
볕뎀	sunscald	
볕타기	sunburn	
볕탐	sunburn	
보	weir	洑
보가스해법	Borgas solution	-解法
보강재배	cultivation with supplementary light	補强栽培
보그	bog	
보급류	supply current	補給流
보급해역기상	replenishment area weather	補給海域氣象
보네빌 소금 평원	Bonneville Salt Flats	-平原
보드	baud	
보드율	baud rate	-率
보라	bora	
보라놀	purple light	
보라바람	bora wind	
보라스	borras	

보라스코	borasco, borasca, bourrasque	
보라안개	bora fog	
보랏빛	purple light	
보레아스	Boreas, borras	
보류	reservation	保留
보름달	full moon	
보먼트기간	Beaumont period	-期間
보상	compensation	補償
보상기류	compensation current	補償氣流
보상깊이	depth of compensation	補償-
보상원리	principle of compensation	補償原理
보상점	compensation point	補償點
보상층	layer of compensation	補償層
보상해류	compensation current	補償海流
보색	complementary color	補色
보색광	complementary light	補色光
보수성	water retentivity	保水性
보스편광기	Voss polariscope	-偏光器
보스포루스해협	Bosporus Strait	-海峽
보어	bore, eagre	
보어가설	Bohr's hypothesis	-假說
보어모형	Bohr's model	-模型
보엔-루들럼 과정	Bowen-Ludlam process	-過程
보엔비	Bowen ratio	-比
보엔비상사	Bowen ratio similarity	-比相似
보이든지수	Boyden index	-指數
보이스카메라	Boys camera	
보일-마리오트 법칙	Boyle-Mariotte law	-法則
보일법칙	Boyle's law	-法則
보일-샤를 법칙	Boyle-Charles' law	-法則
보임구역	visual field	-區域
보전협동작전기상	Infantry Tank Combined Arms Operation weather	步戰協同作戰氣象
보정고도	corrected altitude	補正高度
보정공기속도	corrected airspeed	補正空氣速度
보정눈금기압계	compensated scale barometer	補正-氣壓計
보정문제	calibration problem	補正問題
보정반사도	corrected reflectivity	補正反射度
보정수조	calibration tank	補正水曹
보정온도계	compensated air thermometer	補正溫度計
보정직달일사계	compensated pyrheliometer, compensating pyrheliometer	補正直達日射計
보조관측	supplementary observation	補助觀測
보조관측선박	supplementary ship	補助觀測船舶
보조관측소	auxiliary station, supplementary station	補助觀測所

보조기기	secondary instrument	補助機器
보조기낭	ballonet	補助氣囊
보조기상대	dependent meteorological office,	補助氣象臺
	supplementary meteorological office	
보조농업기상관측소	auxiliary agricultural meteorological station	補助農業氣象觀測所
보조선	auxiliary(adjoint) line	補助線
보조선박	auxiliary ship	補助船舶
보조선박관측소	auxiliary ship station, supplementary ship station	補助船舶觀測所
보조온도계	auxiliary thermometer	補助溫度計
보조이주	assisted migration	補助移住
보조일기도	auxiliary chart	補助日氣圖
보조자료	auxiliary data	補助資料
보조지상관측소	supplementary land station	補助地上觀測所
보존량	conservative quantity, conserved quantity	保存量
보존력	conservative force	保存力
보존매개변수선도	conserved parameter diagram	保存媒介變數線圖
보존멱지수	conservation exponent	保存冪指數
보존방식	conservative scheme	保存方式
보존법칙	conservation law, law of conservation	保存法則
보존변수	conservation variable, conserved variable	保存變數
보존변수선도	conserved variable diagram	保存變數線圖
보존성	conservatism, conservative	保存性
보존성질	conservative property	保存性質
보존오염물질	conservative pollutant	保存汚染物質
보존요소	conservative element	保存要素
보존장	conservative field	保存場
보통권층운	cirrostratus communis	普通卷層雲
보통바다	moderate sea	普通-
보통변수	crisp variable	普通變數
보통비	moderate rain	普通-
보통시정	moderate visibility	普通視程
보통안개	moderate fog	普通-
보통층운	stratus communis, common stratus	普通層雲
보틀링거고리	Bottlinger's ring	
보퍼트(풍력)계급	Beaufort (wind) scale	-(風力)階級
보퍼트기상표기	Beaufort weather notation	-氣象表記
보퍼트수	Beaufort number	-數
보퍼트일기기호	Beaufort notation	-日氣記號
보퍼트풍력	Beaufort force	-風力
보편기체상수	universal gas constant	普遍氣體常數
보편우량계	universal raingauge	普遍雨量計
보편투과함수	universal transmission function	普遍透過函數
보편평형가설	universal equilibrium hypothesis	普遍平衡假說
보호반경	radius of protection	保護半徑

보호반지름	radius of protection	保護-
복각	dip, inclination, magnetic dip, magnetic inclination	伏角
복각적도	dip equator	伏角赤道
복굴절	birefringence	復屈折
복류	subsurface flow	伏流
복류수	subsurface water	伏流水
복변성 작용	polymetamorphism	復變性作用
복비례	compound proportion	複比例
복빙	regelation	復氷
복사	radiation	輻射
복사가리개	radiation shield	輻射-
복사가열	radiation heating, radiative heating	輻射加熱
복사감쇠	radiation damping	輻射減衰
복사강도	radiant intensity, radiation intensity	輻射強度
복사강제력	radiative forcing	輻射強制力
복사계	radiometer	輻射計
복사계량	radiometric quantity	輻射計量
복사계보정	radiometric calibration	輻射計補正
복사계분해능	radiometric resolution	輻射計分解能
복사계상수	radiometric contrast	輻射計常數
복사계수정	radiometric correction	輻射計修正
복사계연직분포	radiometric profile	輻射計鉛直分布
복사계잡음	radiometric noise	輻射計雜音
(복사)공동	hohlraum	(輻射)空洞
복사과정	radiation process	輻射過程
복사관측	radiation observation	輻射觀測
복사기록	bologram	輻射記錄
복사기록계	bolograph, radiation recorder	輻射記錄計
복사기후	radiation climate	輻射氣候
복사냉각	radiation cooling, radiational cooling, radiative cooling	輻射冷却
복사능	radiant power	輻射能
복사능지수	actinothermic index	輻射能指數
복사대	radiation belt	輻射帶
복사대류권계면	radiative tropopause	輻射對流圈界面
복사도	radiation chart	輻射圖
복사량	quantity of radiation	輻射量
복사모형	radiation model	輻射模型
복사밀도	radiant density	輻射密度
복사밤	radiation night	輻射-
복사방지	antiradiance	輻射防止
복사방출	emission of radiation	輻射放出
복사방출률	radiant emittance	輻射放出率

복사법칙	radiation law	輻射法則
복사사출도	radiant exitance	輻射射出度
복사상	remapping	複寫像
복사상수	radiation constant	輻射常數
복사상호작용	radiative interaction	輻射相互作用
복사서리	radiation frost	輻射-
복사선	radiant ray	輻射線
복사선차폐	radiation shield	輻射線遮蔽
복사손실	radiation loss	輻射損失
복사수감부	radiation sensor	輻射受感部
복사수지	economy of radiation, radiation balance, radiation budget	輻射收支
복사수지계	balance meter, radiation balance meter	輻射收支計
복사안개	radiation fog	輻射-
복사압력	radiation pressure	輻射壓力
복사에너지	radiant energy	輻射-
복사에너지량	quantity of radiant energy	輻射-量
복사에너지밀도	radiant energy density	輻射-密度
복사에너지온도계	radiant energy thermometer	輻射-溫度計
복사역전	radiation inversion	輻射逆轉
복사역전안개	radiation inversion fog	輻射逆轉-
복사열	radiant heat	輻射熱
복사열전달	radiative heat transfer	輻射熱傳達
복사열전달계수	radiative heat transfer coefficient	輻射熱傳達係數
복사온도계	radiation thermometer	輻射溫度計
복사요약도	radiation summary chart	輻射要約圖
복사장	radiation field	輻射場
복사전달	radiation transfer, radiative transfer	輻射傳達
복사전달모형	radiative transfer model	輻射傳達模型
복사전달방정식	equation of radiative transfer(ERT), radiative transfer equation	輻射傳達方程式
복사점	radiant point, radiation point	輻射點
복사조도	irradiance, radiant irradiance	輻射照度
복사조도감시장치	radiometer irradiance monitor	輻射照度監視裝置
복사조사	irradiation	輻射照射
복사존데	radiation sonde	輻射-
복사주파수	frequency of radiation	輻射周波數
복사증발계	radio atmometer	輻射蒸發計
복사체	radiator	輻射體
복사측정	radiometry	輻射測定
복사측정소	radiation station	輻射測定所
복사패턴	radiation pattern	輻射-
복사평형	radiation equilibrium, radiative balance, radiative equilibrium	輻射平衡

복사평형온도	radiative equilibrium temperature	輻射平衡溫度
복사플럭스	flux of radiation, radiant flux	輻射-
복사플럭스밀도	radiant flux density	輻射-密度
복사피폭	radiant exposure	輻射被曝
복사필터	radiation filter	輻射-
복사확산율	radiative diffusivity	輻射擴散率
복사휘도	radiance	輻射輝度
복사휘도배경	radiance background	輻射輝度背景
복사휘도비율방법	radiance ratio method	輻射輝度比率方法
복사휘도온도	radiance temperature	輻射輝度溫度
복사휘도잔차법	radiance residual method	輻射輝度殘差法
복사흡수	absorption of radiation	輻射吸收
복소수	complex number	複素數
복소수변수	complex variable	複素數變數
복소수승법	complex multiplication	複素數乘法
복소수장	complex number field	複素數場
복소평면	complex plane	複素平面
복슬적란운	cumulonimbus capillatus	-積亂雲
복슬적운	cumulus capillatus	-積雲
복원	reconstruction	復元
복원	retrieval	復原
복원력	restoring force	復元力
복원탐측	retrieved sounding	復原探測
복잡교차스펙트럼	complex cross spectrum	複雜交叉-
복잡성	complexity	複雜性
복잡수문곡선	complex hydrograph	複雜水文曲線
복잡지형	complex terrain	複雜地形
복잡지형대기과학프로그램	atmospheric sciences in complex terrain program	複雜地形大氣科學-
복잡지형대기연구	atmospheric studies in complex terrain	複雜地形大氣研究
복잡품질관리	complex quality control	複雜品質管理
복조	demodulation	復調
복중	dog day	伏中
복합계	complex system	複合系
복합굴절률	complex index of refraction, complex refractive index	複合屈折率
복합굴절률지수	complex refractive index	複合屈折率指數
복합기후학	complex climatology	複合氣候學
복합방전	composite flash	複合放電
복합수문곡선	compound hydrograph	複合水文曲線
복합습윤안정도	composite moisture stability	複合濕潤安定度
복합시스템	complex system	複合-
복합신호	complex signal	複合信號
복합원심력	compound centrifugal force	複合遠心力

복합저기압	complex low	複合低氣壓
복합특이성	complex singularity	複合特異性
복합확률	compound probability	複合確率
본류	mother current	本流
본초자오선	prime meridian	本初子午線
볼기아노스케일링	Bolgiano scaling	
볼륨스캔	volume scan	
볼멧기상정보	meteorological information for aircraft in flight	-氣象情報
볼스터 맴돌이	bolster eddy	
볼츠만방정식	Boltzmann equation	-方程式
볼츠만분포	Boltzmann distribution	-分布
볼츠만상수	Boltzmann constant, Boltzmann's constant	-常數
볼츠만인자	Boltzmann factor	-因子
볼프-볼퍼수	Wolf-Wolfer number	-數
볼프수(상대태양흑점수)	Wolf number	-數
봄	spring	
봄기운	springness	-氣運
봄눈	spring snow, corn snow	
봄바람(중국)	chun fung	
봄비(중국)	chunyuh	
봉쇄기상	blockade weather	封鎖氣象
봉함	sealing	封緘
뵐링-알레뢰드 아간빙기	Bølling-Allerød interstadial	-亞間氷期
뵐링-알레뢰드 온난기	Bølling-Allerød interstadial	-溫暖期
부게법칙	Bouguer's law	-法則
부게의 무리	Bouguer's halo	
부(기압)골	minor trough	副(氣壓)-
부니	sapropel, organic-rich mud	腐泥
부당변경	tampering	不當變更
부도체	non-conductor	不導體
부돌출부	side lobe	副突出部
부동점	floating point	浮動點
부동지	tabetisol	不凍地
부동해	open sea	不凍海
부등식	inequality	不等式
부디코기후분류시스템	Budyko Climatic Classification System	-氣候分類-
부디코분류	Budyko classification	-分類
부디코비	Budyko ratio	-比
부디코수	Budyko number	-數
부라스크	bourrasque	
부력	buoyancy, buoyancy force, buoyant force	浮力
부력가속도유사성	buoyancy acceleration analogy	浮力加速度類似性
부력곡률유사성	buoyancy curvature analogy	浮力曲率類似性
부력규모	buoyancy scale	浮力規模

부력길이규모	buoyancy length scale	浮力-規模
부력대류	buoyant convection	浮力對流
부력매개변수	buoyancy parameter	浮力媒介變數
부력불안정도	buoyant instability	浮力不安定度
부력상승	buoyancy lift	浮力上昇
부력생산	buoyant production	浮力生産
부력생산율	buoyant production rate	浮力生産率
부력속도	buoyancy velocity	浮力速度
부력아영역	buoyancy subrange	浮力亞領域
부력에너지	buoyancy energy, buoyant energy	浮力-
부력인자	buoyancy factor	浮力因子
부력중심	center of buoyancy	浮力中心
부력증발계	floating pan	浮力蒸發計
부력진동	buoyancy oscillation	浮力振動
부력진동수	buoyancy frequency	浮力振動數
부력파	buoyancy wave	浮力波
부력파괴율	buoyancy destruction rate	浮力破壞率
부력파수	buoyancy wavenumber	浮力波數
부력플럭스	buoyancy flux	浮力-
부력플럭스연직분포	buoyancy flux profile	浮力-鉛直分布
부력효과	buoyancy effect	浮力效果
부르동관	Bourdon tube	-管
부르동관 온도계	Bourdon tube thermometer	-管溫度計
부르동온도계	Bourdon thermometer	-溫度計
부(기압)마루	minor ridge	副(氣壓)-
부목표역	subtarget area	副目標域
부문경계	sector boundary	部門境界
부방사부에코	sidelobe echo	副放射部-
부분가림	partial obscuration	部分-
부분가뭄	partial drought	部分-
부분거울면반사	partial specular reflection	部分-面反射
부분관통정	partially penetrating well	部分貫通井
부분-기간 계열	partial-duration series	部分期間系列
부분녹음	partial melting	部分-
부분바이어스	fractional bias	部分-
부분반사	partial reflection	部分反射
부분분수	partial fraction	部分分數
부분불확정성	fractional uncertainty	部分不確定性
부분브라운운동	fractional Brownian motion	部分-運動
부분비부피	partial specific volume	部分比-
부분비성질	partial specific property	部分比性質
부분빔차폐	partial beam blocking	部分-遮蔽
부분상관	partial correlation	部分相關
부분수열	subsequence	部分數列

부분시정	sector visibility	部分視程
부분식	partial eclipse	部分蝕
부분안정도	partial stability	部分安定度
부분온위	partial potential temperature	部分溫位
부분용적비	fractional volume abundance	部分容積比
부분장	subfield	部分場
부분적분(법)	integration by parts	部分積分(法)
부분전하	partial charge	部分電荷
부분제거	decimation	部分除去
부분조석	partial tide	部分潮汐
부분진공	partial vacuum	部分眞空
부분집합	subset	部分集合
부분폐색(의)	partly occluded	部分閉塞
부분한발	partial drought	部分旱魃
부분흐림	partly cloudy	部分-
부빙	floe ice, ice floe, ice pack, pack, pack ice	浮氷
부빙덩이	floe	浮氷-
부빙조각	calf, calved ice	浮氷-
부서지는 파도	comber	-波濤
부속구름	accessory cloud	附屬-
부속특징	supplementary feature	附屬特徵
부순환	minor circulation	副循環
부쉬선	Busch lemniscate	-線
부시네스크가정	Boussinesq assumption	-假定
부시네스크근사	Boussinesq approximation	-近似
부시네스크방정식	Boussinesq equation	-方程式
부시네스크수	Boussinesq number	-數
부식	corrosion	腐蝕
부식제거법	caustic scrubbing	腐蝕除去法
부싱어-다이어 관계	Businger-Dyer relationship	-關係
부싱어-다이어 바람프로파일	Businger-Dyer wind profile	
부영양화	eutrophication	富營養化
부유기구	floating balloon	浮遊氣球
부유대수층	perched aquifer	浮遊帶水層
부유라이시미터	floating lysimeter	浮遊-
부유물질	particulate matter	浮遊物質
부유미립자	particulate	浮遊微粒子
부유미립자유기물	particulate organic matter	浮遊微粒子有機物
부유생물	plankton	浮遊生物
부유성유공충	planktonic foraminifera	浮遊性有孔蟲
부유입자물질	suspended particulate matter	浮遊粒子物質
부유입자상물질	atmospheric particulate matter	浮遊粒子狀物質
부유자동기상관측소	floating automatic weather station	浮遊自動氣象觀測所
부이 기상관측소	buoy weather station	-氣象觀測所

부이자료	data buoy	-資料
부저기압	secondary cyclone, secondary depression, secondary low	副低氣壓
부적절 설정문제	ill-posed problem	不適切設定問題
부정형(구름)	informis	不定形-
부조일수	number of sunless days	不照日數
부조화난류	patchy turbulence	不調和亂流
부지선정	siting	敷地選定
부착	aggregation, attachment, clumping, coagulation, coherence	附着
부착력	adhesion	附着力
부착온도계	attached thermometer	附着溫度計
부착장력	adhesion tension	附着張力
부착효율	adhesion efficiency	附着效率
부채꼴빔안테나	fan-beam antenna	
부채꼴연기퍼짐	fanning	
부채형 플룸	fanning plume	-型-
부챗살 빛	backstay of the sun, crepuscular ray	
부척	vernier	副尺
부척눈금	vernier scale	副尺-
부침	heave	浮沈
부타디엔	butadiene	
부탄	butane	
부트스트랩	bootstrap	
부파	minor wave	副波
부패지수	deterioration index	腐敗指數
부표	buoy, drogue	浮標
부표자기우량계	float recording precipitation gauge	浮標自記雨量計
부표형우량계	float type rain gauge	浮標型雨量計
부표형자기기압계	float barograph	浮標型自記氣壓計
부피	bulk	
부피산란함수	volume scattering function	-散亂函數
부피속도처리	volume velocity processing	-速度處理
부피주사	volume scan	-走査
부피중앙지름	volume median diameter	-中央-
부피탄성률	bulk modulus	-彈性率
부피팽창	volume expansion	-膨脹
부피혼합층모형	bulk mixed-layer model	-混合層模型
부피흡수계수	volume absorption coefficient	-吸收係數
부하	loading	負荷
부호	symbol	符號
부호기호	code symbol	符號記號
부호무리	code group	符號-
부호문자	code letter	符號文字

부호설명	code specification	符號說明
부호송신라디오존데	code-sending radiosonde, code-type radiosonde	符號送信-
부호수	code figure	符號數
부호존데	code sonde	符號-
부호표	code table	符號表
부호해독	decoding	符號解讀
부호해독기	decoder	符號解讀器
부호형라디오존데	code-type radiosonde	符號型-
부호형식	code form	符號型式
부호화	coding	符號化
북극	Arctic Pole, North Pole	北極
북극(의)	arctic	北極
북극계	arctic system	北極系
북극고기압	arctic anticyclone, arctic high	北極高氣壓
북극고산	arctic-alpine	北極高山
북극광	aurora borealis, aurora polaris, northern lights	北極光
북극구역	north polar region	北極區域
북극권	arctic Circle	北極圈
북극권무전감쇠현상	arctic blackout	北極圈無電減衰現象
북극기단	arctic air (mass)	北極氣團
북극기상관측소	arctic weather station	北極氣象觀測所
북극기후	arctic climate, high artic climate	北極氣候
북극기후시스템연구	arctic climate system study	北極氣候-研究
북극기후영향평가	arctic climate impact assessment	北極氣候影響評價
북극김안개	arctic smoke	北極-
북극대	arctic zone, north frigid zone	北極帶
북극대류권계면	arctic tropopause	北極對流圈界面
북극대륙기단	arctic continental air (mass)	北極大陸氣團
북극바다김안개	arctic sea smoke	北極-
북극바다얼음	arctic sea ice	北極-
북극바람	arctic wind	北極-
북극부빙	arctic pack	北極浮氷
북극분지	arctic basin	北極盆地
북극사막	arctic desert	北極砂漠
북극성	north star, polar star, polaris	北極星
북극성기록계	pole-star recorder	北極星記錄計
북극성층권소용돌이	arctic stratospheric vortex	北極成層圈-
북극소용돌이관측	arctic vortex observation	北極-觀測
북극수목선	arctic tree line	北極樹木線
북극시스템	arctic system	北極-
북극연무	arctic haze	北極煙霧
북극엷은안개	arctic mist	北極-
북극유모	northern nanny	北極乳母
북극저층수	arctic bottom water, arctic surface water	北極底層水

북극전선	arctic front	北極前線
북극제트(기류)	arctic jet stream	北極-(氣流)
북극제트류	arctic jet stream	北極-流
북극중층수	arctic intermediate water	北極中層水
북극진동	arctic oscillation	北極振動
북극캐나다	Arctic Canada	北極-
북극하층성층권	arctic lower stratosphere	北極下層成層圈
북극한대	North Frigid Zone, artic zone	北極寒帶
북극한대전선	arctic polar front	北極寒帶前線
북극해양성기단	arctic maritime air mass	北極海洋性氣團
북극해양성기류	arctic maritime airstream	北極海洋性氣流
북극해빙	arctic sea ice	北極海氷
북극허리케인	arctic hurricane	北極-
북극화이트아웃	arctic whiteout	北極-
북녘의	boreal	
북대서양	North Atlantic Ocean	北大西洋
북대서양고기압	North Atlantic high	北大西洋高氣壓
북대서양심층수	North Atlantic deep water	北大西洋深層水
북대서양진동	North Atlantic oscillation	北大西洋振動
북대서양진동기록	North Atlantic oscillation record	北大西洋振動記錄
북대서양편류	North Atlantic drift, North Atlantic current	北大西洋偏流
북대서양해류	North Atlantic current	北大西洋海流
북대서양해양열대기단	North Atlantic mT air mass	北大西洋海洋熱帶氣團
북대서양해양한대기단	North Atlantic mP air mass	北大西洋海洋寒帶氣團
북대서양환류	North Atlantic gyre	北大西洋還流
북동	northeast	北東
북동강풍	northeaster	北東强風
북동무역풍	northeast trade wind, northeast trades	北東貿易風
북동폭풍	northeast storm	北東暴風
북반구	northern hemisphere	北半球
북반구아열대무풍대	calms of Cancer	北半球亞熱帶無風帶
북반구환상모드	northern annular mode	北半球環狀-
북벽	north wall	北壁
북부(의)	boreal	北部
북북동	north northeast	北北東
북북서	north northwest	北北西
북브라질해류	North Brazil current	北-海流
북빙양	arctic ocean	北氷洋
북상	northbound	北上
북서	northwest	北西
북서계절풍	northwest monsoon	北西季節風
북서오스트레일리아구름대	northwest Australian cloud band	北西-
북서퀸즐랜드	northwest Queensland	北西-
북아메리카	North America	北-

북아메리카고기압	North American anticyclone, North American high	北-高氣壓
북아메리카몬순	North American monsoon	北-季節風
북아메리카빙상	North American ice sheet	北-氷床
북아메리카스텝	North American steppe	北-
북아메리카여름몬순	North American summer monsoon	北-
북열대	north tropic	北熱帶
북온대	north temperate zone	北溫帶
북위	northern latitude	北緯
북적도해류	north equatorial current	北赤道海流
북점	north point	北點
북쪽통과경로	northern circuit	北-通過經路
북태평양	North Pacific Ocean	北太平洋
북태평양고기압	North Pacific high	北太平洋高氣壓
북태평양기단	North Pacific air mass	北太平洋氣團
북태평양동부해분	eastern North Pacific basin	北太平洋東部海盆
북태평양해양성열대기단	North Pacific mT air mass	北太平洋海洋性熱帶氣團
북태평양중층수	North Pacific intermediate water	北太平洋中層水
북태평양진동	North Pacific oscillation	北太平洋振動
북태평양해류	North Pacific current	北太平洋海流
북푄	north föhn (foehn)	北-
북풍	norther	北風
북한한류	North Korea cold current	北韓寒流
북해	North Sea	北海
북회귀선	Tropic of Cancer	北回歸線
분	minute	分
분계면	surface of separation	分界面
분계선	demarcation line	分界線
분광(의)	spectral	分光
분광계	spectrometer	分光計
분광광도계	spectrophotometer	分光光度計
분광기	spectroscope	分光器
분광반사율	spectral reflectance	分光反射率
분광법	spectral method, spectroscopy	分光法
분광복사기록	spectrobologram	分光輻射記錄
분광분석(법)	spectrometry	分光分析(法)
분광사진기	spectrograph	分光寫眞機
분광습도계	spectral hygrometer, spectroscopic hygrometer	分光濕度計
분광영역	spectral range	分光領域
분광요소법	spectral element method	分光要素法
분광일광계	spectroheliograph	分光日光計
분광일사계	spectropyrheliometer	分光日射計
분광전천일사계	spectral pyranometer	分光全天日射計
분광직달일사계	spectral pyrheliometer	分光直達日射計

분광휘도	spectral radiance	分光輝度
분극도	polarizability	分極度
분기	bifurcation	分岐
분기스트리머	junction streamer	分岐-
분기점	bifurcation	分岐點
분류	classification	分類
분류	diffluence, separation of flow	分流
분류	fractionation	分溜
분류구역	diffluence zone	分流區域
분류학	systematics	分類學
분류형(기압)골	diffluent trough	分流型(氣壓)-
분류형(기압)마루	diffluent ridge	分流型(氣壓)-
분류형온도골	diffluent thermal trough	分流型溫度-
분류형온도마루	diffluent thermal ridge	分流型溫度-
분류형온도제트	diffluent thermal jet	分流型溫度-
분류흐름	diffluent flow	分流-
분리	abstraction, decoupling, dissociation	分離
분리고기압	cut-off high	分離高氣壓
분리과정	breaking-off process, cutting-off process	分離過程
분리맴돌이	separation eddy	分離-
분리법	split method, splitting method	分離法
분리소용돌이	cut-off vortex	分離-
분리저기압	cut-off low	分離低氣壓
분리중심	cut-off center	分離中心
분리층운	detached stratus	分離層雲
분무역	spray region	噴霧域
분무탑	spray tower	噴霧塔
분배	distribution, dissemination	分配
분배법칙	distributive law	分配法則
분사물	effluent	噴射物
분산	dispersion, variance	分散
분산계	disperse system	分散系
분산계수	dispersion coefficient	分散係數
분산곡선	dispersion curve	分散曲線
분산관개	dispersed irrigation	分散灌漑
분산관계	dispersion relation, dispersion relationship	分散關係
분산도	dispersivity	分散度
분산매개변수화방안	dispersion parameterization scheme	分散媒介變數化方案
분산매질	dispersing medium, dispersion medium	分散媒質
분산모형	dispersion model	分散模型
분산변수	dispersion variable	分散變數
분산분석	analysis of variance, variance analysis	分散分析
분산비	variance ratio	分散比
분산선도	dispersion diagram	分散線圖

분산성매질	dispersive medium	分散性媒質
분산위상	dispersed phase	分散位相
분산-침적모형	dispersion-deposition model	分散沈積模型
분산파	dispersive wave	分散波
분산플럭스	dispersive flux	分散-
분석	diagnosis, diagnose	分析
분석·연구·훈련용 지구변화시스템	Global Change System for Analysis, Research and Training(GCSART)	分析·研究·訓練用地球變化-
분석도	analysed chart (or map)	分析圖
분석방법	analysis method	分析方法
분석센터	analysis center	分析-
분석증분	analysis increment	分析增分
분쇄	comminution	粉碎
분수	diversion of water	分水
분수	fraction	分數
분수계	water divide, water parting, watershed	分水界
분수공	outlet	噴水孔
분수령	divide, dividing crest, watershed	分水嶺
분수마루	dividing ridge	分水-
분수멱법칙	fractional power law	分數冪法則
분수방정식	fractional equation	分數方程式
분수부등식	factional inequality	分數不等式
분수선	divide line	分水線
분수식	fractional expression	分數式
분수식 빨대	bubbler	分數式-
분수함수	fractional function	分數函數
분압	partial pressure	分壓
분압법칙	law of partial pressure	分壓法則
분열	disintegration, fission	分裂
분열핵	fragmentation nucleus	分裂核
분열화	fragmentation	分裂化
분위수	quantile	分位數
분자	molecule, numerator	分子
분자 간 힘	intermolecular force	分子間-
분자결정	molecular crystal	分子結晶
분자구조	molecular structure	分子構造
분자규모온도	molecular-scale temperature	分子規模溫度
분자량	molecular mass, molecular weight	分子量
분자력	molecular force	分子力
분자류	molecular flow	分子流
분자모형	molecular model	分子模型
분자배열	molecular arrangement	分子配列
분자산란	molecular scattering	分子散亂
분자식	molecular formula	分子式

분자운동	molecular motion	分子運動
분자운동론	kinetic theory of molecule	分子運動論
분자이론	molecular theory	分子理論
분자전도	molecular conduction	分子傳導
분자점성	molecular viscosity	分子粘性
분자점성계수	coefficient of molecular viscosity, molecular viscosity coefficient	分子粘性係數
분자확산	molecular diffusion	分子擴散
분자확산계수	molecular diffusion coefficient	分子擴散係數
분점	equinox	分點
분지무	basin fog	盆地霧
분지안개	basin fog	盆地-
분출빙하	outlet glacier	噴出氷河
분포	distribution	分布
분포계수	distribution coefficient	分布係數
분포곡선	distribution curve	分布曲線
분포그래프	distribution graph	分布-
분포다각형	distribution polygon	分布多角形
분포표적(물)	distributed target	分布標的(物)
분포함수	distribution function	分布函數
분할	discretization	分割
분할각	contingency angle	分割角
분할스텝	fractional step	分割-
분할영상	sectorized image	分割映像
분할유선	dividing streamline	分割流線
분할저기압	breakaway low	分割低氣壓
분할표	contingency table	分割表
분할하이브리드주사	sectorized hybrid scan	分割-走査
분해	decomposition, disintegration, dissolution, resolution	分解
분해능	resolution	分解能
분해부피	resolution volume	分解-
분해세포	resolution cell	分解細胞
분해열	heat of decomposition	分解熱
분해자	decomposer	分解者
분홍눈	pink snow	粉紅-
분화구	crater	噴火口
분화운	volcanic eruption cloud	噴火雲
불규칙결정	irregular crystal	不規則結晶
불규칙고주파형공중전기	irregular high-frequency type atmospherics	不規則高周波型空中電氣
불규칙에코	irregular echo	不規則-
불규칙전선	irregular front	不規則前線
불균질매질	inhomogeneous medium	不均質媒質
불균질매체	inhomogeneous medium	不均質媒體

불꽃	spark	
불꽃방전	spark discharge	-放電
불꽃점화기관	spark ignition engine	-點火機關
불꽃지연	spark retardation	-遲延
불꽃집전기	flame collector	-集電器
불덩이유성	bolide	-流星
불명	obscuration	不明
불명하늘가림	obscured sky cover	不明-
불모지	barrens	不毛地
불변	invariable	不變
불변(의)	constant	不變
불빛현상	igneous meteor	-現象
불소침착증	fluorosis	弗素沈着症
불순도	impurity	不純度
불스아이 스콜	bull's eye squall	
불안정	unstable	不安定
불안정(도)	instability	不安定(度)
불안정(의)	labile	不安定
불안정공기	unstable air	不安定空氣
불안정공기층	unstable air layer	不安定空氣層
불안정기단	unstable air, unstable air mass	不安定氣團
불안정대기	unstable atmosphere	不安定大氣
불안정도	lability	不安定度
불안정도도표	instability chart	不安定度圖表
불안정선	instability line	不安定線
불안정소나기	instability shower	不安定-
불안정에너지	instability energy, labile energy	不安定-
불안정인자	instabilizing factor	不安定因子
불안정지수	index of instability	不安定指數
불안정진동	unstable oscillation	不安定振動
불안정파	unstable wave	不安定波
불안정패턴	instability pattern	不安定-
불안정평형	unstable equilibrium	不安定平衡
불안정한	astatic	不安定-
불안정흐름	unstable flow	不安定-
불연속(성)	discontinuity	不連續(性)
불연속(인)	discontinuous	不連續
불연속구역	zone of discontinuity	不連續區域
불연속난류	discontinuous turbulence	不連續亂流
불연속대류권계면	discontinuous tropopause	不連續對流圈界面
불연속면	discontinuity surface	不連續面
불연속면기울기	slope of surface discontinuity	不連續面-
불연속선	line of discontinuity	不連續線
불연속-세로좌표방법	discrete-ordinates method	不連續-座標方法

불연속스펙트럼	discrete spectrum	不連續-
불연속차수	order of discontinuity	不連續次數
불연속층	layer of discontinuity	不連續層
불연속푸리에변환	discrete Fourier transform	不連續-變換
불완전연소	incomplete combustion	不完全燃燒
불완전예보	imperfect forecast	不完全豫報
불용습성핵	insoluble wettable nucleus	不溶濕性核
불쾌지대	discomfort zone	不快地帶
불쾌지수	discomfort index	不快指數
불투명(의)	opaque	不透明
불투명고적운	altocumulus opacus	不透明高積雲
불투명고층운	altostratus opacus	不透明高層雲
불투명구름	opacus	不透明-
불투명도	opacity, opaqueness	不透明度
불투명얼음	opaque ice	不透明-
불투명층	opaque layer	不透明層
불투명층운	stratus opacus	不透明層雲
불투명층적운	stratocumulus opacus	不透明層積雲
불투명파도고층운	altostratus opacus undulatus	不透明波濤高層雲
불투수층	confining bed, watertight stratum	不透水層
불투열성	athermancy	不透熱性
불투열성(의)	athermous	不透熱性
불포화	unsaturation	不飽和
불포화결합	unsaturated bond	不飽和結合
불포화공기	nonsaturated air	不飽和空氣
불포화상태	unsaturated condition	不飽和狀態
불포화수리전도도	unsaturated hydraulic conductivity	不飽和水理傳導度
불포화습윤공기	nonsaturated moist air	不飽和濕潤空氣
불포화용액	unsaturated solution	不飽和溶液
불포화투수	unsaturated percolation	不飽和透水
불화물	fluoride	弗化物
불화바륨박막습도계	barium fluoride film hygrometer	弗化-薄膜濕度計
불확실영역퍼센트	percent of area of uncertainty	不確實領域-
불활성기체	noble gas, inert gas	不活性氣體
불활성화	deactivation	不活性化
불휘발성	non-degradation	不揮發性
붉은 눈	blood snow, red snow	
붉은 비	blood rain, red rain	
붉은 섬광	red flash	-閃光
붕괴	breakdown	崩壞
붕괴상수	decay constant	崩壞常數
붕괴장	breakdown field	崩壞場
붕괴침식	land slide erosion, land sliding erosion	崩壞浸蝕
붕괴퍼텐셜	breakdown potential	崩壞-

브라운운동	Brownian motion	-運動
브라운확산	Brownian diffusion	-擴散
브라운회전	Brownian rotation	-回轉
브라이어 점수	Brier score(BS)	-點數
브라질해류	Brazil current	-海流
브래그산란	Bragg scattering	-散亂
브래그조건	Bragg condition	-條件
브러시방전	brush discharge	-放電
브런트-더글러스 바람	Brunt-Douglas wind	
브런트-더글러스 변압풍	Brunt-Douglas isallobaric wind	-變壓風
브런트-바이살라 진동수	Brunt-Väisälä frequency	-振動數
브레드모드	bred mode	
브레드벡터	bred vector	
브로켄괴물	Brocken spectre	-怪物
브로켄사발	Brocken bowl	
브로켄스펙트럼	Brocken spectrum	
브롬	bromine	
브롬일산화물	bromine monoxide	-一酸化物
브롬화합물	bromine compound	-化合物
브뢰럽간빙기	Brørup interstadial	-間氷期
브루마	bruma	
브루부	brubu	
브루스터점	Brewster's point	-點
브루어-도브슨 순환	Brewer-Dobson circulation	-循環
브룩헤븐 분산다이어그램	Brookhaven dispersion diagram	-分散-
브룸	brume	
브뤼크너주기	Bruckner cycle	-週期
브리딩벡터	breeding vector(BV)	
브리슬콘소나무	bristlecone pine	
브리티시빙상	British ice sheet	-氷床
브리핑	briefing	
브후트	bhoot	
블라시우스 연직분포	Blasius profile	-鉛直分布
블라우트	blout, blouter, clowther, blowthir	
블라통공식	Blaton's formula	-公式
블래니-크리들 모형	Blaney-Criddle model	-模型
블랙카다르 혼합길이	Blackdar mixing length	-混合-
블런크	blunk	
블렌딩높이	blending height	
블로바이가스	blowby gas	
블록다이어그램	block diagram	
블록잼순평균법	bloxam	-旬平均法
블루노이즈	blue noise	
블루틸트	blue tilt	

블리니	blini	
블리커습도선도	Bleeker humidity diagram	-濕度線圖
비	rain	
비	ratio	比
비(의)	hyetal	
비(의)	specific	比
비가림재배	cultivation under rain shelter	-栽培
비가시복사	invisible radiation	非可視輻射
비가역과정	irreversible process	非可逆過程
비가역기관	irreversible engine	非可逆機關
비가역반응	irreversible reaction	非加逆反應
비가역적	irreversible	非可逆的
비가역현상	irreversible phenomenon	非可逆現象
비가우스강제	non-Gaussian forcing	非-強制
비가우스난류	non-Gaussian turbulence	非-亂流
비간섭레이더	noncoherent radar	非干涉-
비간섭산란기법	incoherent scatter technique	非干涉散亂技法
비간섭산란레이더	incoherent scatter radar	非干涉散亂-
비간섭신호	incoherent signal	非干涉信號
비간섭에코	noncoherent echo	非干涉-
비간섭적분	incoherent integration	非干涉積分
비간섭표적	noncoherent target	非干涉標的
비겔로증발공식	Bigelow's evaporation formula	-蒸發公式
비결정론적	non-deterministic	非決定論的
비교기상학	comparative meteorology	比較氣象學
비교레이벌	comparative rabal	比較-
비교성	comparability	比較性
비구름	nimbus, rain cloud	
비구형입자	nonspherical particle	非球形粒子
비국지일차종결	non-local first order closure	非局地一次終結
비국지정적안정도	non-local static stability	非局地靜的安定度
비국지종결	non-local closure	非局地終結
비국지종결모형	non-local closure model	非局地終結模型
비국지종결방식	non-local closure scheme	非局地終結方式
비국지플럭스	non-local flux	非局地-
비국지혼합	non-local mixing	非局地混合
비국지효과	non-local effect	非局地效果
비규정면	offlevel	非規定面
비균일빔채움	nonuniform beam filling	非均一-
비균일시정	nonuniform visibility	非均一視程
비균일하늘상태	nonuniform sky condition	非均一-狀態
비균질(의)	heterogeneous	非均質
비균질경계층	non-homogeneous boundary layer	非均質境界層
비균질계	heterogeneous system	非均質系

비균질구름핵화물질	inhomogeneous cloud nucleating material	非均質-核化物質
비균질권	heterosphere	非均質圈
비균질성	heterogeneity	非均質性
비균질유체	heterogeneous fluid	非均質流體
비균질체	heterogeneous substance	非均質體
비균질핵화	heterogeneous nucleation	非均質核化
비그늘	rain shadow	
비기록우량계	non-recording rain gauge	非記錄雨量計
비기상산란체	nonmeteorological scatter	非氣象散亂體
비기체상수	specific gas constant	比氣體常數
비나무	rain tree	
비뉴턴유체	non-Newtonian fluid	非-流體
비늘구름하늘	mackerel sky	
비늘권적운	cirrocumulus mackerel	-卷積雲
비늘살	louver	
비늘살백엽상	louvered thermometer screen	-百葉箱
비단계	rain stage	-段階
비단얼음	satin ice	緋緞-
비단열(의)	diabatic	非斷熱
비단열가열	diabatic heating	非斷熱加熱
비단열과정	diabatic process, non-adiabatic process	非斷熱過程
비단열냉각	diabatic cooling, non-adiabatic cooling	非斷熱冷却
비단열바람연직분포	diabatic wind profile	非斷熱-鉛直分布
비단열변화	non-adiabatic change	非斷熱變化
비단열비가역과정	non-adiabatic irreversible process	非斷熱非可逆過程
비단열영향함수	diabatic influence function	非斷熱影響函數
비단열온도변화	diabatic temperature change	非斷熱溫度變化
비단열지표층	diabatic surface layer	非斷熱地表層
비단열효과	diabatic effect	非斷熱效果
비달성지역	nonattainment area	非達成地域
비대류성플럭스	non-convective flux	非對流性-
비대칭(성)	asymmetry	非對稱(性)
비대칭(의)	anti-symmetrical	非對稱
비대칭도	skewness	非對稱度
비대칭순환패턴	asymmetric circulation pattern	非對稱循環-
비대칭인자	asymmetric factor	非對稱因子
비대칭중심	asymmetrical center	非對稱中心
비대칭텐서	anti-symmetric tensor	非對稱-
비동기결합	asynchronous coupling	非同期結合
비동기통신	asynchronous communication	非同期通信
비등	boil, boiling, ebullition	沸騰
비등방성	anisotropy	非等方性
비등방성(의)	anisotropic	非等方性
비등방성난류	non-isotropic turbulence	非等方性亂流

ㅂ

비등방성화	anisotropization	非等方性化
비등방수상체	anisotropic hydrometeor	非等方水狀體
비등압(의)	anisobaric	非等壓
비등점상승	elevation of boiling point	沸騰點上昇
비등점상승법	ebullioscopy	沸騰點上昇法
비등천	boiling spring	沸騰泉
비디오주파수	video frequency, vision frequency	-周波數
비디오증폭	video gain	-增幅
비디오통합처리기	video integrator and processor	-統合處理器
비디오통합처리수준	video integrator processor level, VIP level	-統合處理水準
비디콘	vidicon	
비디콘카메라시스템	vidicon camera system	
비례	proportion	比例
비례배분	proportional distribution	比例配分
비례부분	proportional part	比例部分
비례상수	proportional constant	比例常數
비례식	proportional expression	比例式
비례오차	fractional error	比例誤差
비망록	agenda	備忘錄
비메탄유기가스	nonmethane organic gas	非-有機-
비메탄탄화수소	nonmethane hydrocarbon	非-炭化水素
비문턱오염물질	nonthreshold pollutant	非門-汚染物質
비발산	non-divergence	非發散
비발산층	non-divergence layer	非發散層
비방출방정식	no-emission equation	非放出方程式
비버꼬리안테나	beavertail antenna	
비보라	rain storm	
비보존성	non-conservative property	非保存性
비복사강도	specific radiation intensity	比輻射強度
비복사강제	nonradiative forcing	非輻射強制
비부력성입자	nonbuoyant particle	非浮力性粒子
비부피편차	anomaly of specific volume, thermosteric anomaly	比-偏差
비분산적외선 가스분석계	nondispersive infrared analyzer, nondispersive infrared gas analyzer	非分散赤外線-分析計
비분산적외선 광도측정법	nondispersive infrared photometry	非分散赤外線光度測定法
비분산적외선 분광법	nondispersive infrared spectrometry	非分散赤外線分光法
비산란대기	non-scattering atmosphere	非散亂大氣
비산란방정식	no-scattering equation	非散亂方程式
비산재	fly ash	飛散-
비산출량	specific yield	比産出量
비상관거리	decorrelation distance	非相關距離
비상대피소	emergency depot	非常待避所
비상활주로	emergency landing strip	非常滑走路

비선택산란	nonselective scattering	非選擇散亂
비선형	nonlinear	非線形
비선형계산불안정	nonlinear computational instability	非線形計算不安定
비선형난류	nonlinear turbulence	非線形亂流
비선형마찰	nonlinear friction	非線形摩擦
비선형문제	nonlinear problem	非線形問題
비선형반응	nonlinear response	非線形反應
비선형불안정	nonlinear instability	非線形不安定
비선형성	nonlinearity	非線形性
비선형안정성	nonlinear stability	非線形安定性
비선형 정상모드 초기화	nonlinear normal mode initialization	非線形正常-初期化
비선형파	nonlinear wave	非線形波
비세척제거	rain washout	-洗滌除去
비쇄설성	nonclastic	非碎屑性
비숍고리	Bishop's ring	
비숍광환	Bishop's corona	-光環
비숍파	Bishop wave	-波
비스듬한 소용돌이도	crosswise vorticity	-度
비습	specific humidity	比濕
비습선	specific humidity line	比濕線
비실시간	non-real time	非實時間
비압축성	incompressibility, non-compression	非壓縮性
비압축성유체	incompressible fluid	非壓縮性流體
비야크네스 모형	Bjerknes model	-模型
비야크네스 선형화절차	Bjerknes linearization procedure	-線形化節次
비야크네스 순환정리	Bjerknes's circulation theorem	-循環定理
비야크네스 저기압모형	Bjerknes cyclone model	-低氣壓模型
비양(기구)	launching	飛揚
비얼음	glaze ice, glaze, glazed frost	
비얼음보라	glaze storm	
비에너지	specific energy	比-
비엔토로테리오	viento roterio	
비여과모형	unfiltered model	非濾過模型
비연동모형	offline model	非聯動模型
비열	specific heat	比熱
비열비	ratio of the specific heat	比熱比
비열용량	specific heat capacity	比熱容量
비엽면적	specific leaf area	比葉面積
비영순힘	nonzero net force	非零-
비예측	unpredictable	非豫測
비외연엔트로피	non-extensive entropy	非外延-
비용편익분석	cost benefit analysis	費用便益分析
비용함수	cost function	費用函數
비유도충전메커니즘	noninductive charging mechanism	非誘導充電-

비유사식생	no-analog vegetation	非類似植生
비유사집단	no-analog community	非類似集團
비유출량	discharge per unit drainage area, specific discharge	比流出量
비율	ratio	比率
비의도적 구름조절	inadvertent cloud modification	非意圖的-調節
비의도적 기상조절	inadvertent weather modification	非意圖的氣象調節
비의도적 기후개조	inadvertent climate modification	非意圖的氣候改造
비의도적 조절	inadvertent modification	非意圖的調節
비인법칙	Wien' law	-法則
비인변위법칙	Wien' displacement law	-變位法則
비인분포법칙	Wien' distribution law	-分布法則
비임계치오염물질	nonthreshold pollutant	非臨界値汚染物質
비장	bijang, specific tension	比張
비저항	resistivity	比抵抗
비전도율	specific conductivity	比傳導率
비전선성뇌우	non-frontal thunderstorm	非前線性雷雨
비전선성저기압	non-frontal low	非前線性低氣壓
비전선성조석분석	non-frontal tidal analysis	非前線性潮汐分析
비전선스콜선	non-frontal squall line	非前線-線
비전선저기압	non-frontal depression	非前線低氣壓
비점성유체	inviscid fluid, nonviscous fluid	非粘性流體
비정규체제	irregular regime	非正規體制
비정규항로	off-airway	非定規航路
비정시	offtime	非定時
비정시보고	offtime report	非定時報告
비정역학	non-hydrostatic	非靜力學
비정역학모형	non-hydrostatic model	非靜力學模型
비정역학압력	non-hydrostatic pressure	非靜力學壓力
비정적과정	non-static process	非靜的過程
비정지궤도위성	non-geostationary orbit satellite	非停止軌道衛星
비정질(의)	amorphous	非晶質
비정체전선면	non-stationary frontal surface	非停滯前線面
비정합수신기	incoherent receiver	非整合受信機
비정합적분	noncoherent integration	非整合積分
비종관관측	asynoptic observation	非綜觀觀測
비종관기상관측	asynoptic meteorological observation	非綜觀氣象觀測
비종관(기상)자료	asynoptic data	非綜觀(氣象)資料
비주기운동	aperiodic motion	非週期運動
비주기적	aperiodic	非週期的
비주기적 온도변화	nonperiodic temperature change	非周期的溫度變化
비주기진동	aperiodic oscillation	非週期振動
비주사계	nonscanner	非走査計
비중	specific gravity	比重

비중계	densitometer	比重計
비지균	ageostrophic	非地均
비지균모형	ageostrophic model	非地均模型
비지균방법	ageostrophic method	非地均方法
비지균성분	ageostrophic component	非地均成分
비지균운동	ageostrophic motion	非地均運動
비지균이류	ageostrophic advection	非地均移流
비지균이류항	ageostrophic advection term	非地均移流項
비지균풍	ageostrophic wind	非地均風
비지균풍성분	ageostrophic wind component	非地均風性分
비지균효과	ageostrophic effect	非地均效果
비직교적	nonorthogonal	非直交的
비차등위상	specific differential phase	比差等位相
비체적	specific volume	比體積
비체적이상	steric anomaly, specific volume anomaly	比體積異常
비초대형세포 토네이도	nonsupercell tornado	非超大型細胞-
비최적법	non-optimum method	非最適法
비추수량	specific yield	比秋收量
비침투대류	nonpenetrative convection	非浸透對流
비콘	beacon	
비클리우량계	Beekley gauge	-雨量計
비키니환초	Bikini atoll	-環礁
비탁계	nephelometer	比濁計
비탄성(의)	anelastic	非彈性
비탄성가정	anelastic assumption	非彈性假定
비탄성근사	anelastic approximation	非彈性近似
비탄성충돌	inelastic collision	非彈性衝突
비투수층	aquifuge	非透水層
비트	bit	
비트율	bit rate	-率
비틀림	distortion, torsion	
비틀림습도계	torsion hygrometer	-濕度計
비틀림파	distortional wave	-波
비파괴성	indestructibility	非破壞性
비평형통계역학	non-equilibrium statistical mechanics	非平衡統計力學
비행가능기구탐측	dirigible balloon ascent	飛行可能氣球探測
비행계획	flight plan	飛行計劃
비행고도	flight altitude, flight level	飛行高度
비행관제	flight control	飛行管制
비행권운	cirrus aviaticus	飛行卷雲
비행기관측	airplane observation	飛行機觀測
비행기기상기록계	airplane meteorograph	飛行機氣象記錄計
비행기상감시	flight meteorological watch	飛行氣象監視
비행기상정보시스템	flight weather information system	飛行氣象情報-

비행로기상자료철	flight folder	飛行路氣象資料綴
비행로기상정보구역	flight information region	飛行路氣象情報區域
비행로기상정보센터	flight information center	飛行路氣象情報-
비행로예보	flight forecast	飛行路豫報
비행로일기예보철	flight dossier, flight forecast-folder	飛行路日氣豫報綴
비행로정보	flight information	飛行路情報
비행로정보데이터베이스	flight information database	飛行路情報-
비행문서	flight document	飛行文書
비행불가기상조건	below minimum flight weather condition	飛行不可氣象條件
비행브리핑	flight briefing	飛行-
비행서류	flight documentation	飛行書類
비행선	aerostat	飛行船
비행시정	flight visibility, flying visibility	飛行視程
비행실험실	flying laboratory	飛行實驗室
비행예보철	flight document	飛行豫報綴
비행예보철지원시스템	Flight Document Service(FDS)	飛行豫報綴支援-
비행운	contrail	飛行雲
비행운분석곡선	contrail analysis curve	飛行雲分析曲線
비행운소산	dissipation trail, distrail	飛行雲消散
비행운형성곡선	contrail formation curve	飛行雲形成曲線
비행운형성그래프	contrail formation graph	飛行雲形成-
비행일기브리핑	flight-weather briefing, pilot briefing	飛行日氣-
비행장경보	aerodrome warning	飛行場警報
비행장고도	field-elevation pressure	飛行場高度
비행장공식표고	official elevation of the aerodrome	飛行場公式標高
비행장관제기상정보시스템	aerodrome control weather information system	飛行場管制氣象情報-
비행장관제탑	aerodrome control tower	飛行場管制塔
비행장기상관측시스템	aerodrome meteorological observation system	飛行場氣象觀測-
비행장기상최저값	aerodrome meteorological minimum	飛行場氣象最低-
비행장미규모폭풍	air base microburst	飛行場微規模暴風
비행장상당고도	equivalent altitude of aerodrome	飛行場相當高度
비행장색깔전문형식	airfield color code	飛行場-電文型式
비행장안전업무	aerodrome security service	飛行場安全業務
비행장예보	aerodrome forecast, terminal aerodrome forecast(TAF)	飛行場豫報
비행장주의보	aerodrome advisory	飛行場注意報
비행장특보	aerodrome special report	飛行場特報
비행장표고	aerodrome elevation	飛行場標高
비행 전 평가	preflight evaluation	飛行前評價
비행 중 평가	inflight evaluation	飛行中評價
비행 중 항공기 기상정보방송	VOLMET broadcast	飛行中航空機氣象情報放送
비행허가	clearance	飛行許可
비혼합액체	nonwetting liquid	非混合液體
비활동구름	passive cloud	非活動-

비활동전선	inactive front, passive front	非活動前線
비활성기체	inert gas	非活性氣體
비활성원소	inert element	非活性原素
비활성지역	inactive region	非活性地域
비활성화	inactivation	非活性化
비회전운동	irrotational motion	非回轉運動
비휘발성	nonvolatile	非揮發性
비흑체	nonblackbody	非黑體
비흡수형산란	conservative scattering	非吸收形散亂
비흡습성핵	non-hygroscopic nucleus	非吸濕性核
빈	bin	
빈대떡얼음	pancake ice	
빈도	frequency	頻度
빈도곡선	frequency curve	頻度曲線
빈도다각형	frequency polygon	頻度多角形
빈도분석	frequency analysis	頻度分析
빈도분포	frequency distribution	頻度分布
빈도스펙트럼	frequency spectrum	頻度-
빈도표시풍속계	frequency-meter anemometer	頻度表示風速計
빈스현상	Vince's phenomenon	-現象
빌엄스크린	Bilham screen	
빔	beam	
빔각	beam angle	-角
빔갇힘(레이더파의)	ducting	
빔발산	beam divergence	-發散
빔분산	beam spreading	-分散
빔블록킹	beam blocking	
빔수	beam number	-數
빔진동	beam swinging	-振動
빔(광)쪼개기	beamsplitter	-(光)-
빔채움	beam filling	
빔충만	beam filling	-充滿
빔충만 기상학적 표적	beam filling meteorological target	-充滿氣象學的標的
빔파	beam wave	-波
빔패턴	beam pattern	
빔퍼짐	beam broadening, beam spreading	
빔폭	beam width	-幅
빔폭 왜곡	beam width distortion	-幅歪曲
빗가시거리	oblique visual range	-可視距離
빗모양측운기	comb nephoscope	-模樣測雲器
빗물	rain water	
빗물량	rain water content	-量
빗물받이통	receiving bucket	-桶
빗방울	raindrop	

빗방울스펙트로그래프	raindrop spectrograph	
빗방울침식	raindrop erosion	-浸蝕
빗방울크기분포	raindrop size distribution	-分布
빗변	hypotenuse	-邊
빗시정	oblique visibility	-視程
빗원기둥	oblique circular cylinder	-圓-
빗원뿔	oblique (circular) cone	-圓-
빗입사	oblique incidence	-入射
빗줄기	rainshaft	
빙구	hummock, ice blister, ice mound, icing mound	氷丘
빙구온도	ice-bulb temperature	氷球溫度
빙극	ice pole	氷極
빙면습윤공기상대습도	relative humidity of moist air with respect to ice	氷面濕潤空氣相對濕度
빙모	ice cap	氷帽
빙붕	ice shelf	氷棚
빙산	iceberg	氷山
빙산가장자리 소용돌이	ice-edge vortex	氷山-
빙상	ice sheet, ice field	氷床
빙설경도계	ram penetrometer	氷雪硬度計
빙설권	cryosphere	氷雪圈
빙설기후	ice cap climate, ice climate	氷雪氣候
빙설대	cryochore	氷雪帶
빙설지역	ice desert	氷雪地域
빙설학	cryology	氷雪學
빙수량	ice water content	氷水量
빙원	ice field	氷原
빙원얼음	field ice	氷原-
빙점	ice point	氷點
빙정	ice crystal, poudrin	氷晶
빙정과정	ice crystal process	氷晶過程
빙정구름	ice crystal cloud	氷晶-
빙정설	ice crystal theory	氷晶說
빙정안개	ice crystal fog	氷晶-
빙정연무	ice crystal haze	氷晶煙霧
빙정핵	ice-forming nucleus, ice nucleus	氷晶核
빙정효과	ice crystal effect	氷晶效果
빙주	ice column	氷柱
빙착	adfreezing	氷着
빙침	spicule	氷針
빙퇴구	drumlin	氷堆丘
빙퇴석	moraine	氷堆石
빙하	basal ice, glacial, glacier	氷河
빙하고기압	glacial anticyclone, glacial high	氷河高氣壓
빙하고기압이론	glacial anticyclone theory	氷河高氣壓理論

빙하구름	glaciated cloud	氷河-
빙하기	glacial epoch, glacial period, glacial phase, glacier period	氷河期
빙하기저	basal ice	氷河基底
빙하기후	glacial climate	氷河氣候
빙하기후학	glacioclimatology	氷河氣候學
빙하바람	firn wind, glacier wind, glacial wind, glacier breeze, glacier wind	氷河-
빙하분리	calving	氷河分離
빙하성유수퇴적물	glaciofluvial sediment	氷河性流水堆積物
빙하성유수	glacial outwash	氷河性流水
빙하시대	ice age	氷河時代
빙하얼음	glacier ice	氷河-
빙하운반쇄설물	ice-rafted debris	氷河運搬碎屑物
빙하유역	glacial basin	氷河流域
빙하틈새	crack	氷河-
빙하이론	glacial theory	氷河理論
빙하작용	glaciation	氷河作用
빙하작용기간	glaciation period	氷河作用期間
빙하전면지형호수퇴적물	proglacial lacustrine sediment	氷河前面地形湖水堆積物
빙하전선	glacier front	氷河前線
빙하전진	glacial advance	氷河前進
빙하주변	periglacial	氷河周邊
빙하주변기후	periglacial climate	氷河周邊氣候
빙하증거	glacial evidence	氷河證據
빙하지진	glacial earthquake	氷河地震
빙하지질학	glacial geology	氷河地質學
빙하지형학	glacial geomorphology	氷河地形學
빙하질량균형관측	glacial mass balance observation	氷河質量均衡觀測
빙하침식	glacial erosion	氷河侵蝕
빙하침전층	glacial varves	氷河沈澱層
빙하코어	ice core	氷河-
빙하퇴적물	glacial sediment	氷河堆積物
빙하평원	glacier plain	氷河平原
빙하평형	glacial isostasy	氷河平衡
빙하폭포	glacier fall	氷河瀑布
빙하표류	glacial drift	氷河漂流
빙하표석	glacial erratic	氷河漂石
빙하학	glaciology	氷河學
빙하한계	glacial limit	氷河限界
빙하한계고도	glaciation limit	氷河限界高度
빙하해안선	glacial coastline	氷河海岸線
빙하해양퇴적물	glaciomarine sediment	氷河海洋堆積物
빙하핵	ice core	氷河核

ㅂ

빙하화	glaciation	氷河化
빙하후퇴	glacial retreat	氷河後退
빙호퇴적물	varved sediment	氷縞堆積物
빛간섭	light interference	-干涉
빛강도	light intensity	-强度
빛기둥	light pillar, sun pillar	
빛노출	light exposure	-露出
빛분산	dispersion of light	-分散
빛산란선도	light scattering diagram	-散亂線圖
빛산란표	light scattering table	-散亂表
빛에너지	light energy	
빛역진성	reversibility of light	-逆進性
빛입자모형	particle model of light	-粒子模型
빛입자설	corpuscular theory of light	-粒子說
빛현상	luminous meteor	-現象
빛회절	light diffraction	-廻折
빨대	syphon	
빼지는 수	minuend	-數
뺄셈	subtraction	
뻐꾸기폭풍	gowk storm	-暴風
뾰족구름패턴	cusp cloud pattern	
뾰족기압골	sharp trough	-氣壓-
뾰족기압마루	sharp ridge	-氣壓-
뿔모양복사체	horn radiator	-模樣輻射體
뿔모양안테나	horn antenna	-模樣-
뿔풍향판	horn card	-風向板

ㅅ

한글	영문	한자
사각기둥	quadrangular prism	四角-
사각뿔	quadrangular pyramid, square pyramid	四角-
사각수	quadrangular number	四角數
사각지시기	elevation position indicator	射角指示器
사각평면	square plane	四角平面
사각형	quadrangle	四角形
사각형그래프	rectangle graph	四角形-
사각형도파관	rectangular waveguide	四角形導波管
사건	episode	事件
사교좌표	oblique coordinate	斜交座標
사교카테시언좌표	oblique Cartesian coordinate	斜交-座標
사구	dune	砂丘
사나운 서풍	brave west wind	-西風
사다리꼴	trapezoid	
사다리모양구름	echelon cloud	-模樣-
사라짐	fade out	
사례기반추론	case-based reasoning	事例基盤推論
사례날씨	case weather	事例-
사례연구	case study	事例研究
사리	spring tide	
사막기후	desert climate	砂漠氣候
사막대	desert belt	砂漠帶
사막돌풍	desert wind squall	砂漠突風
사막바람	desert wind	砂漠-
사막스텝	desert steppe	砂漠-
사막신기루	desert mirage	砂漠蜃氣樓
사막알베도	desert albedo	砂漠-
사막지역	xerochore	砂漠地域
사막포도	deflation hollow	砂漠鋪道

사막화	desertification	砂漠化
사망률	mortality	死亡率
사면체	tetrahedron	四面體
사면체기구	tetroon	四面體氣球
사면포복	slope creep	斜面葡匐
사물기생	saprophytism	死物寄生
사바나	savanna(h)	
사바나기후	savanna climate	-氣候
사바르편광기	Savart polariscope	-偏光器
사변형	quadrilateral	四邊形
사분(면)	quadrant	四分(面)
사분(의)	quadrant	四分
사분스펙트럼	quadrature spectrum	四分-
사분시정	quadrant visibility	四分視程
사분위수	quartile	四分位數
사분전위계	quadrant electrometer	四分電位計
사분편차	quartile deviation	四分偏差
사상	mapping, morphism	寫像
사수	dead water	死水
사스트루기	sastrugi, zastrugi	
사십성폭풍	Forty Saints' storm	四十聖暴風
사암	sandstone	砂岩
사염화탄소	carbon tetrachloride	四鹽化炭素
사이아노미터	cyanometer	
사이클론	tropical cyclone	
사이클론규모	cyclone scale	-規模
사이클론분리기	cyclone separator	-分離機
사이클론스크러버	cyclone scrubber	
사이클론집진기	cyclone collector	-集塵器
사이클론집진장치	cyclone dust collector	-集塵裝置
사이펀기압계	siphon barometer	-氣壓計
사이펀수은기압계	siphon mercury barometer	-水銀氣壓計
사이펀식우량계	siphon-type rain gauge	-式雨量計
사이펀우량계	siphon rain gauge	-雨量計
사이펀우량기록계	siphon rainfall recorder	-雨量記錄計
사이펀자기기압계	siphon barograph	-自記氣壓計
사이펀자기우량계	siphon recording rain gauge	-自記雨量計
사인검류계	sine galvanometer	-檢流計
사인곡선	sine curve	-曲線
사인파	sine wave	-波
사인파온도패턴	sine wave thermal pattern	-波溫度-
사전분석	preanalysis	事前分析
사전이유	a priori reason	事前理由
사전확률	a priori probability	事前確率

사정풍	range wind	射程風
사중기록계	quadruple recorder	四重記錄計
사진경위의	photo theodolite	寫眞經緯儀
사진계측기	photogrammeter	寫眞計測器
사진기압계	photographic barograph	寫眞氣壓計
사진유성	photographic meteor	寫眞流星
사진일조계	photographic sunshine recorder	寫眞日照計
사진투영	photographic projection	寫眞投影
사진현미경	photomicroscope	寫眞顯微鏡
사차방정식	quartic equation	四次方程式
사체군집	death assemblage	死體群集
사출	ejection, emanation	射出
사출함	ejection chamber	射出函
사카가미불안정(도)	Sakagami's instability	-不安定(度)
사태	avalanche	沙汰
사태돌풍	avalanche blast	沙汰突風
사태바람	avalanche wind	沙汰-
사토	sand soil	沙土
사피어-심프슨 규모	Saffir-Simpson scale	-規模
사피어-심프슨 위험잠재규모	Saffir-Simpson damage-potential scale	-危險潛在規模
사피어-심프슨 허리케인 강도규모	Saffir-Simpson hurricane intensity scale	-强度規模
사하라먼지	Saharan dust	
사행경로	meandering course	蛇行經路
사후보고	de-briefing	事後報告
사후분석	post-analysis	事後分析
삭	new moon	朔
삭망	syzygy	朔望
삭망월	synodic month	朔望月
삭박작용	denudation	削剝作用
산	acid	酸
산간	mountainous area	山間
산곡풍	mountain and valley breeze, mountain-valley wind	山谷風
산골바람	mountain and valley wind	山-
산기슭빙하	piedmont glacier	山-氷河
산덩어리(스머트)	acid smut	酸-
산도	acidity	酸度
산도연직분포	acidity profile	酸度鉛直分布
산들바람	gentle breeze, zephyr	
산란	scatter, scattering	散亂
산란각	angle of scattering, scatter angle, scattering angle	散亂角
산란계	scatterometer	散亂計
산란계수	scattering coefficient	散亂係數

산란공률	scattered power	散亂工率
산란광	scattered light	散亂光
산란광선	scattered ray	散亂光線
산란단면	scattering cross-section	散亂斷面
산란면적계수	scattering area coefficient	散亂面積係數
산란면적비	scattering area ratio	散亂面積比
산란복사	scattered radiation	散亂輻射
산란복사계	diffusometer	散亂輻射計
산란시정계	scatter-type visibility meter	散亂視程計
산란율	scattering power	散亂率
산란일사계	diffusometer	散亂日射計
산란전달	scatter communication	散亂傳達
산란전파	scatter propagation	散亂傳播
산란지수	scattering index	散亂指數
산란체	scatterer	散亂體
산란폭	scattering amplitude	散亂幅
산란함수	scattering function	散亂函數
산란행렬	scattering matrix	散亂行列
산란효율	scattering efficiency	散亂效率
산란효율인자	scattering efficiency factor	散亂效率因子
산림개간	forest clearing	山林開墾
산림개조	forest conversion	山林改造
산림농업	agroforestry	山林農業
산림벌채	deforestation	山林伐採
산바람	berg wind, mountain breeze, mountain wind	山-
산발소나기	scattered shower	散發-
산발유성	sporadic meteor	散發流星
산불	forest fire, wildfire	山-
산불구름	forest-fire cloud	山-
산불기상학	forest-fire meteorology	山-氣象學
산사이바람	mountain-gap wind	山-
산사태	landslide	山沙汰
산사태바람	landslip wind	山沙汰-
산성	acidic	酸性
산성강수	acid precipitation	酸性降水
산성검댕	acid soot	酸性-
산성구름	acid cloud	酸性-
산성눈	acid snow	酸性-
산성박무	acid haze	酸性薄霧
산성변형	acid deformation	酸性變形
산성비	acid rain	酸性-
산성비분포	acid rain distribution	酸性-分布
산성비효과	acid rain effect	酸性-效果
산성서리	acid frost	酸性-

산성안개	acid fog	酸性-
산성오염	acid pollution	酸性汚染
산성우박	acid hail	酸性雨雹
산성이슬	acid dew	酸性-
산성이온	acid ion	酸性-
산성침적	acid deposition, acidic deposition	酸性沈積
산성화	acidification	酸性化
산소	oxygen	酸素
산소결핍(증)	anoxemia	酸素缺乏(症)
산소결핍증	anoxia	酸素缺乏症
산소공급	oxygenate	酸素供給
산소녹색선	green oxygen line	酸素綠色線
산소동위원소	oxygen isotope	酸素同位元素
산소동위원소방법	oxygen isotope method	酸素同位元素方法
산소동위원소비	oxygen isotope ratio	酸素同位元素比
산소동위원소 심해기록	oxygen isotope deep-sea record	酸素同位元素深海記錄
산소띠	oxygen band	酸素-
산소분자	oxygen molecule	酸素分子
산소-오존 시스템	oxygen-ozone system	酸素-
산소원자	atomic oxygen	酸素原子
산소흡수	oxygen absorption	酸素吸收
산수소탄소	oxyhydrocarbon	酸水素炭素
산술평균	arithmetic mean	算術平均
산술평균기온	arithmetic mean temperature	算術平均氣溫
산술평균법	arithmatic mean method	算術平均法
산악공기	mountain air	山岳空氣
산악관측	mountain observation	山岳觀測
산악관측소	high-altitude station, mountain station	山岳觀測所
산악기상	mountain weather	山岳氣象
산악기상관측소	mountain weather station	山岳氣象觀測所
산악기상학	mountain meteorology	山岳氣象學
산악기압계	mountain barometer	山岳氣壓計
산악기후	mountain climate	山岳氣候
산악난류	mountain turbulence	山岳亂流
산악뇌우	orographic thunderstorm	山岳雷雨
산악빙하	mountain glacier	山岳氷河
산악성	orographic	山岳性
산악융기	mountain uplift	山岳隆起
산악장벽	mountain barrier	山岳障壁
산악저기압	orographic cyclone	山岳低氣壓
산악지역작전기상	mountain area operation weather	山岳地域作戰氣象
산악지형	mountainous terrain	山岳地形
산악토크	mountain torque	山岳-
산악파	mountain wave	山岳波

산악파구름	mountain wave cloud	山岳波-
산악파난류	mountain wave turbulence	山岳波亂流
산악표준시(미국)	mountain standard time	山岳標準時
산악풍하파	mountain lee wave	山岳風下波
산악형강수	orographic precipitation	山岳形降水
산안개	hill fog, mountain fog	山-
산업구름	cloud from industry	産業-
산업기상학	industrial meteorology	産業氣象學
산업기후학	industrial climatology	産業氣候學
산월기류	airflow over mountain	山越氣流
산중성화능력	acid-neutralizing capacity	酸中性化能力
산지	mountainous area	山地
산지지형학	orography	山地地形學
산출	retrieval	算出
산출점	yield point	産出點
산타아나	Santa Ana	
산타아나바람	Santa Ana wind	
산타로사폭풍	Santa Rosa storm	-暴風
산탄소음	shot noise	散彈騷音
산포	scatter	散布
산포도	scatter diagram, scatter graph, scattergram	散布圖
산호	coral	珊瑚
산호띠	coral band	珊瑚-
산호백화작용	coral bleaching	珊瑚白化作用
산호초	coral reef	珊瑚礁
산화	oxidation	酸化
산화력	oxidizing power	酸化力
산화물	oxide	酸化物
산화반응	oxidation reaction	酸化反應
산화불꽃	oxidizing flame	酸化-
산화알루미늄제감습소자	aluminum oxide humidity element	酸化-感濕素子
산화제	oxidant	酸化劑
산화촉매	oxidation catalyst	酸化觸媒
산화탄화수소	oxygenated hydrocarbon	酸化炭化水素
산화환원반응	redox reaction	酸化還元反應
산화환원효소	oxidation reductase	酸化還元酵素
산화황	sulfur oxide	酸化黃
산화황농도계	sulfur oxide densitometer	酸化黃濃度計
산화효소	oxidase	酸化酵素
살생물제	biocide	殺生物劑
살수관개	springkler irrigation	撒水灌漑
살얼음	cat ice	
살인폭풍	killer storm	殺人暴風
살짝 결빙	light freeze	-結氷

살충제	insecticide	殺蟲劑
삼각깃	pennant	三角-
삼각비	trigonometric ratio	三角比
삼각절단	triangular truncation	三角切斷
삼각주	delta, delta region	三角洲
삼각측량	triangulation	三角測量
삼각함수	trigonometric function	三角函數
삼림기상학	forest meteorology	森林氣象學
삼림기후	forest climate	森林氣候
삼림미기상학	forest micrometeorology	森林微氣象學
삼림복사수지	forest radiation budget	森林輻射收支
삼림선	isohyle	森林線
삼림쇠퇴	forest decline	森林衰退
삼림한계	forest limit	森林限界
삼림환경	forest environment	森林環境
삼중근	triple root	三重根
삼중기록계	triple register	三重記錄計
삼중상태	triple state	三重狀態
삼중점	triple point	三重點
삼중점온도	triple point temperature	三重點溫度
삼차방정식	cubic equation	三次方程式
삼차부등식	cubic inequality	三次不等式
삼투수	osmotic water	滲透水
삼투압	osmosis	滲透壓
삼투압(력)	osmotic pressure	滲透壓(力)
삼투압계	osmometer	滲透壓計
삼투퍼텐셜	osmotic potential	慘透-
삼한사온	cycle of three cold and four warm days	三寒四溫
삿갓구름	cloud crest, crest cloud	
(위)상	phase	(位)相
상강	frost descent	霜降
상고대	rime	
상고대안개	rime fog	
상고대얼음	rime ice	
상고대착빙	rime icing	-着氷
상고대화	riming	-化
상공강수	precipitation aloft	上空降水
상관계수	coefficient of correlation, correlation coefficient	相關係數
상관곱	correlation product	相關-
상관기상학	correlative meteorology	相關氣象學
상관도	correlation diagram, correlogram	相關圖
상관법	correlation method	相關法
상관분석	correlation analysis	相關分析
상관비	correlation ratio	相關比

상관삼각형	correlation triangle	相關三角形
상관성	coherence	相關性
상관예보	correlation forecasting	相關豫報
상관인자	correlation factor	相關因子
상관텐서	correlation tensor	相關-
상관표	correlation table	相關表
상관함수	correlation function	相關函數
상(평형)선도	phase diagram	相(平衡)線圖
상단천정호	upper circumazenithal arc	上端天頂弧
상당경로	equivalent path	相當徑路
상당고도	equivalent height	相當高度
상당깊이	equivalent depth	相當-
상당남북바람	equivalent longitudinal wind	相當南北-
상당뒷바람	equivalent tail wind	相當-
상당레이더반사도	equivalent radar reflectivity	相當-反射度
상당레이더반사도인자	equivalent radar reflectivity factor	相當-反射度因子
상당맞바람	equivalent head wind	相當-
상당반사율	equivalent reflectance	相當反射率
상당반사율인자	equivalent reflectivity factor	相當反射率因子
상당순압고도	equivalent barotropic height, equivalent barotropic level	相當順壓高度
상당순압대기	equivalent barotropic atmosphere	相當順壓大氣
상당순압모형	equivalent barotropic model	相當順壓模型
상당온도	equivalent temperature	相當溫度
상당온위	equivalent potential temperature	相當溫位
상당온위도	equivalent potential temperature diagram	相當溫位圖
상당파장	equivalent wavelength	相當波長
상당폭	equivalent width	相當幅
상당흑체온도	equivalent blackbody temperature	相當黑體溫度
상대	relative	相對
상대각운동량	relative angular momentum	相對角運動量
상대경도류	relative gradient current	相對傾度流
상대고도	relative height	相對高度
상대고도분포	relative topography	相對高度分布
상대공기밀도	relative air density	相對空氣密度
상대굴절률	relative refractive index	相對屈折率
상대균질성	relative homogeneity	相對均質性
상대나선도	relative helicity	相對螺旋度
상대농도	relative concentration	相對濃度
상대등고선	relative contour, relative isohypse	相對等高線
상대론	theory of relativity	相對論
상대론적 질량	relativistic mass	相對論的質量
상대밀도	relative density	相對密度
상대바람	relative wind	相對-

상대발광효율	relative luminous efficiency	相對發光效率
상대발산	relative divergence	相對發散
상대빈도	relative frequency	相對頻度
상대산란강도	relative scatter intensity	相對散亂强度
상대산란함수	relative scattering function	相對散亂函數
상대생장	allometric growth	相對生長
상대성	relativity	相對性
상대성원리	relativity principle	相對性原理
상대성장률	relative growth rate	相對成長率
상대소용돌이도	relative vorticity	相對-度
상대소용돌이도이류	relative vorticity advection	相對-度移流
상대속도	relative velocity	相對速度
상대습도	relative humidity	相對濕度
상대습도분석	relative humidity analysis	相對濕度分析
상대시간	relative time	相對時間
상대시정	relative visibility	相對視程
상대안정도	relative stability	相對安定度
상대연령측정	relative dating	相對年齡測定
상대오차	relative error	相對誤差
상대우량계수	relative pluviometric coefficient	相對雨量係數
상대운동	relative motion	相對運動
상대운동량	relative momentum	相對運動量
상대유선	relative streamline	相對流線
상대유전율	relative dielectric constant	相對誘電率
상대이온층혼탁계	relative ionospheric opacitymeter	相對-層混濁計
상대일조	relative sunshine	相對日照
상대전하	relative charge	相對電荷
상대좌표계	relative coordinate system	相對座標系
상대증발	relative evaporation	相對蒸發
상대지구자전율	relative earth rotation rate	相對地球自轉率
상대지형학	relative topography	相對地形學
상대질량	relative mass	相對質量
상대측고(법)	relative hypsography	相對測高(法)
상대태양흑점수	relative sunspot number	相對太陽黑點數
상대피도	relative cover	相對皮度
상대헬리시티	relative helicity	相對-
상대환산	relative reduction	相對換算
상도	frost way	霜道
상륙군 목표기상	landing force objective weather	上陸軍目標氣象
상반방정식	reciprocal equation	相反方程式
상법칙	phase rule	相法則
상변화	change of phase, phase change	相變化
상변환	phase transformation	相變換
상보성원리	complementarity principle	相補性原理

상부개방형챔버	open-top chamber	上部開放型-
상부개방형챔버실험	open-top chamber experiment	上部開放型-實驗
상부대기경계층	outer boundary layer, upper atmospheric boundary layer	上部大氣境界層
상부대류권	upper troposphere	上部對流圈
상부대류권 접힘의 다니엘센 모형	Danielsen's model of upper tropospheric folding	上部對流圈-模型
상부마찰구역	upper frictional region	上部摩擦區域
상부성층권	upper stratosphere	上部成層圈
상부혼합층	upper mixing layer	上部混合層
상사(성)	similarity	相似(性)
상사기관	analogous organ	相似器管
상사매개변수	similarity parameter	相似媒介變數
상사이론	theory of similarity	相似理論
상사법칙	similarity law	相似法則
상세기상정보	detail weather information	詳細氣象情報
상쇄간섭	destructive interference	相殺干涉
상수	constant	常數
상수함수	constant function	常數函數
상수항	constant term	常數項
상승	ascent	上昇
상승(열수)설	ascension (hydrothermal) theory	上昇(熱水)說
상승곡선	ascent curve	上昇曲線
상승공기	ascending air	上昇空氣
상승궤도	ascending pass	上昇軌道
상승기류	ascending air current	上昇氣流
상승도	ascendent	上昇度
상승력	lifting force	上昇力
상승(기)류	updraft, updraught	上昇(氣)流
상승류강도	updraft strength	上昇流強度
상승(기)류바탕	updraft base	上昇(氣)流-
상승(기)류속도	updraft velocity	上昇(氣)流速度
상승률지시계	rate-of-climb indicator	上昇率指示計
상승부	rising limb	上昇部
상승시간	rise time	上昇時間
상승안개	lifting fog	上昇-
상승연직속도	upward vertical velocity	上昇鉛直速度
상승운동	ascending motion	上昇運動
상승응결고도	ascending condensation level, lifting condensation level	上昇凝結高度
상승작용	potentiation, synergism	相乘作用
상승중심	center of rise	上昇中心
상승지수	lifted index	上昇指數
상승한계	ceiling	上昇限界

상승형	lofting	上昇形
상아침	ivory point	象牙針
상아침고도	elevation of ivory point	象牙針高度
상업날씨	commercial weather	商業-
상업용위성	commercial satellite	商業用衛星
상영구동토층	pereletok, intergelisol	上永久凍土層
상용대수	common logarithm	常用對數
상용박명	civil twilight	常用薄明
상용법칙	common law	常用法則
상용소산계수	decimal coefficient of extinction	常用消散係數
상용수	common water	常用水
상용시	civil time	常用時
상용일	civil day	常用日
상자모형	box model	箱子模型
상자연	box kite	箱子鳶
상청액	supernatant liquid	上淸液
상층	upper level	上層
상층강수역	upper precipitation part	上層降水域
상층계	upper-level system	上層系
상층고기압	high aloft, upper air anticyclone, upper high, upper-level anticyclone, upper-level high	上層高氣壓
상층(기압)골	upper trough, upper-level trough	上層(氣壓)-
상층공기	superior air	上層空氣
상층구름	high cloud, upper cloud	上層-
상층기단	superior air mass	上層氣團
상층기압골	trough aloft, upper air trough	上層氣壓-
상층기압마루	upper air ridge	上層氣壓-
상층기온	upper air temperature	上層氣溫
상층(대기)기후학	upper air climatology	上層(大氣)氣候學
상층대기	upper air	上層大氣
상층대기순환	upper air circulation	上層大氣循環
상층대류	elevated convection	上層對流
상층(기)류	upper air current, upper-level flow	上層(氣)流
상층마루	upper ridge	上層-
상층(기압)마루	ridge aloft, upper ridge, upper-level ridge	上層(氣壓)-
상층물체온도	elevated body temperature	上層物體溫度
상층바람	upper wind, upper-level wind, wind aloft	上層-
상층바람관측	winds-aloft observation	上層-觀測
상층바람기록판	winds-aloft plotting board	上層-記錄板
상층바람기온예보	winds- and temperatures- aloft forecast	上層-氣溫豫報
상층바람보고	winds-aloft report	上層-報告
상층(대기)분석	upper air analysis	上層(大氣)分析
상층비행정보구역	upper flight information region	上層飛行情報區域
상층신기루	superior mirage, upper mirage	上層蜃氣樓

상층시스템	upper-level system	上層-
상층안정층	elevated stable layer	上層安定層
상층역전	upper layer inversion	上層逆轉
상층열원	elevated heat source	上層熱源
상층열흡원	elevated heat sink	上層熱吸源
상층오염원	elevated source	上層汚染源
상층온난전선	upper warm front	上層溫暖前線
상층온난혀	trowal, trowell	上層溫暖-
상층요란	upper air disturbance, upper-level disturbance	上層擾亂
상층운	upper cloud	上層雲
상층일기도	upper air chart, upper air weather chart, upper-level chart, upper-level map, upper-level weather chart	上層日氣圖
상층저기압	low aloft, upper air cyclone, upper cyclone, upper low, upper-level cyclone, upper-level low	上層低氣壓
상층전선	upper air front, upper front, upper-level front	上層前線
상층전선면	upper frontal surface	上層前線面
상층 푄	high foehn	上層-
상층풍	upper air wind	上層風
상층풍관측	upper wind observation	上層風觀測
상층한랭전선	upper cold front	上層寒冷前線
상태	state	狀態
상태곡선	state curve	狀態曲線
상태매개변수	state parameter	狀態媒介變數
상태방정식	equation of state	狀態方程式
상태변수	state variable, variable of state	狀態變數
상태변화	change of state	狀態變化
상태함수	state function	狀態函數
상한	least upper bound	上限
상향대기복사	upward atmospheric radiation	上向大氣輻射
상향복사	upward radiation	上向輻射
상향식 접근법	bottom-up approach	上向式接近法
상향운동	upward motion	上向運動
상향지구복사	upward terrestrial radiation	上向地球輻射
상향총복사	upward total radiation	上向總輻射
상현	first quarter (of the moon)	上弦
상혈	frost hole, frost hollow	霜穴
상호간섭함수	mutual coherence function	相互干涉函數
상호오염	cross contamination	相互汚染
상호유도	mutual induction	相互誘導
상호인덕턴스	mutual inductance	相互-
상호작용	interaction	相互作用
상호확산계수	coefficient of mutual diffusion	相互擴散係數
새공습	bird strike	-空襲

새끼거위돌풍	gosling blast	-突風
새끼거위폭풍	gosling storm	-暴風
새내리 바람	bird burst	
새벽공전음	dawn chorus	-空電音
새얼음	young ice	
새틸층운	feathery stratus	-層雲
색각	color sense	色覺
색깔비	colored rain	色-
색눈금	color bar	色-
색섬광	chromatic scintillation	色閃光
색소	colorant	色素
색소체	plastid	色素體
색소층	pigment layer	色素層
색수차	chromatic aberration	色收差
색온도	color temperature	色溫度
색인식	color perception	色認識
색인표	lookup table	索引表
색전하	color charge	色電荷
색지각	color perception	色知覺
색채기준	color criteria	色彩基準
색채합성	color composite	色彩合成
색채형성과정	color formation process	色彩形成課程
샘	source	
생(물)권	biosphere	生(物)圈
생기후	bioclimate	生氣候
생기후법칙	bioclimatic law	生氣候法則
생(물)기후학	biocimatology, bioclimatics	生(物)氣候學
생기후학 자료	bioclimatological data	生氣候學資料
생리기후학	phenology, physiological climatology	生理氣候學
생리생태학모형	ecophysiological model	生理生態學模型
생리적 가뭄	physiological drought	生理的-
생리적 건조	physiological dryness	生理的乾燥
생리적 설해	physiological snow damage	生理的雪害
생물검정	bio-assay	生物檢定
생물계절도	phenogram	生物季節圖
생물계절학	biophenology, phenology	生物季節學
생물계절학적 관측	phenological observation	生物季節學的觀測
생물군	biota	生物群
생물군계	biome	生物群系
생물군계모형	biome model	生物群系模型
생물권저장소	biosphere reserve	生物圈貯藏所
생물기간	biological period	生物期間
생물기상학	biometeorology	生物氣象學
생물기원미량기체	biogenic trace gas	生物起源微量氣體

생물기원오염물질	biogenic pollutant	生物起源汚染物質
생물기원퇴적물	biogenic sediment	生物起源堆積物
생물기원퇴적암	biogenic sedimentary rock	生物起源堆積巖
생물기후	biological climate	生物氣候
생물기후대	bioclimatic zone	生物氣候帶
생물기후도	bioclimatograph	生物氣候圖
생물농축	biomagnification	生物濃縮
생물누적	bioaccumulation	生物累積
생물다양성	biodiversity	生物多樣性
생물다양성열점	biodiversity hot spot	生物多樣性熱點
생물대	life zone	生物帶
생물-대기상호작용	biosphere-atmosphere interaction	生物大氣相互作用
생물대리요소	biotic proxy	生物代理要素
생물발광	bioluminescence	生物發光
생물발생설	biogenesis	生物發生說
생물산란체	biological scatterer	生物散亂體
생물생산	biological production	生物生産
생물생산량	biological yield	生物生産量
생물성빙정핵	biogenic ice neucleus	生物性氷晶核
생물안개	biofog	生物-
생물자원	biomass	生物資源
생물자원연료	biomass fuel	生物資源燃料
생물체량(바이오매스) 소각	biomass burning	生物體量燒却
생물학	biology	生物學
생물학적 산소요구량	biological oxygen demand(BOD)	生物學的酸素要求量
생물학적 스크러버	bioscrubber	生物學的-
생물학적 에어로졸	biological aerosol	生物學的-
생물학적 영점	biological zero point	生物學的零點
생물학적 오염물질	biological contaminant	生物學的汚染物質
생물학적 종	biological species	生物學的種
생물학적 진화	biological evolution	生物學的進化
생물학적 최저온도	biological minimum temperature	生物學的最低溫度
생물학적 필터	biofilter	生物學的-
생물학전쟁 기상	biological warfare weather	生物學戰爭氣象
생산구조	productive structure	生産構造
생산력	productivity	生産力
생성반응	formation reaction	生成反應
생육기	vegetative period	生育期
생육기간	duration of growing period	生育期間
생육도일	growing degree-day	生育度日
생장곡선	growth curve	生長曲線
생장기	period of growth	生長期
생장기간	period of growth	生長期間
생장운동	growth movement	生長運動

생장촉진기	accelerator	生長促進器
생장호흡	growth respiration	生長呼吸
생존온도	vital temperature	生存溫度
생지화학	biogeochemical, biogeochemistry	生地化學
생지화학순환	biogeochemical cycle	生地化學循環
생지화학플럭스	biogeochemical flux	生地化學-
생체공학	bionics	生體工學
생체량	biomass	生體量
생체배기	bioefflurnt	生體排氣
생체온도계	rafraichometer	生體溫度計
생층서학	biostratigraphic method	生層序學
생태계	ecosystem	生態系
생태계 관리	ecosystem management	生態系管理
생태계 다양성	ecosystem diversity	生態系多樣性
생태계 변화	ecosystem change	生態系變化
생태계영향	effect on ecosystem	生態系影響
생태계 평형	ecosystem equilibrium	生態系平衡
생태기후	ecoclimate	生態氣候
생태기후학	ecoclimatology, ecological climatology	生態氣候學
생태세	ecotax	生態稅
생태역	biochore	生態域
생태적 지위	niche	生態的地位
생태학	ecology, oecology	生態學
생태형	ecotype	生態型
생태효율	eco-efficiency	生態效率
생활계절학	human phenology	生活季節學
생활기후	domestic climate	生活氣候
생활기후학	domestic climatology	生活氣候學
샤를-게이-뤼삭 법칙	Charles-Gay-Lussac law	-法則
샤를법칙	Charles's law	-法則
샤말	shamal	
샤파렐	chaparral	
서리	frost, hoar	
서리갈라짐	frost splitting	
서리결정	hoar crystal	-結晶
서리경보	frost-freeze warning	-警報
서리계절	frost season	-季節
서리관측기	pagoscope	-觀測器
서리기후	frost climate	-氣候
서리깃	frost feather, ice feather	
서리꽃	frost flower	
서리나이테	frost ring	
서리내성	frost tolerance	-耐性
서리눈	frost snow	

서리도	degree of frost	-度
서리막이숲	frost prevent forest	
서리막이팬	frost fan	
서리박무	frost mist	-薄霧
서리방지	frost prevention	-防止
서리보호	frost protection	-保護
서리습도계	frost hygrometer	-濕度計
서리식물	frost plant	-植物
서리안개	frost fog, ice fog	
서리언덕	frost mound	
서리 없는 계절	frost-free season	-季節
서리연기(언 안개)	frost smoke	-煙氣
서리연무	frost haze	-煙霧
서리일수	number of frost day	-日數
서리저항	frost resistance	-抵抗
서리조각	frost flake, ice fog	
서리주머니	frost pocket	
서리주의보	frost advisory	-注意報
서리착빙	frost icing	-着氷
서리피해	frost damage, frost hazard	-被害
서림열	heat of condensation	-熱
서릿날	frost day	
서릿발	frost column, frost pillar, ice pillar, needle ice	
서릿점	frost point	-點
서릿점기법	frost-point technique	-點技法
서릿점습도계	frost-point hygrometer	-點濕度計
서릿점온도	frost-point temperature	-點溫度
서미스터	thermistor	
서미스터 열복사계	thermistor bolometer	-熱輻射計
서미스터 온도계	thermistor thermometer	-溫度計
서미스터 풍속계	thermistor anemometer	-風速計
서북태평양계절풍	Western North Pacific monsoon	西北太平洋季節風
서북태평양몬순	Western North Pacific monsoon	西北太平洋-
서술기상학	descriptive meteorology	敍述氣象學
서술기후학	descriptive climatology	敍述氣候學
서스효과	Suess effect	-效果
서식대	bryochore	棲息帶
서식지	habitat	棲息地
서아프리카 요란대	West African disturbance line	西-擾亂帶
서안기후	west coast climate	西岸氣候
서안아열대	west coast subtropical	西岸亞熱帶
서안아열대사막	west coast subtropical desert	西岸亞熱帶沙漠
서열변동율	intersequential variability	序列變動率
서지	surge	

서지번개	surge lightning	
서지선	surge line	-線
서지흐름	surge current	
서큘레이터	circulator	
서풍해류	west wind drift	西風海流
석면	asbestos	石綿
석면침착증	asbestosis	石綿沈着症
석양층적운	stratocumulus vesperalis	夕陽層積雲
석탄기(고생대)	Carboniferous	石炭期
석회석세정	limestone scrubbing	石灰石洗淨
선	ray	線
선형가속도	linear acceleration	線形加速度
선구식생	pioneer vegetation	先驅植生
선구파	fore-runner	先驅波
선너비	line width	線-
선단방송	Fleet Broadcast	船團放送
선대선계산	line-by-line calculation	線對線計算
선대칭	line symmetry	線對稱
선도	diagram	線圖
선도	leader	先導
선도낙뢰	leader storke	先導落雷
선도방전	leader streamer	先導放電
선두스트리머	pilot streamer	先頭-
선모양구름	line cloud	線模樣-
선모양발생원	line source	線模樣發生源
선밀도	linear density	線密度
선박결빙	ship icing	船舶結氷
선박관측	ship observation	船舶觀測
선박기압계	sea-barometer, ship-barometer	船舶氣壓計
선박보고	ship report	船舶報告
선박용 자동고층관측 프로그램	automated shipboard aerological program	船舶用自動高層觀測-
선박용 기압계	marine barometer	船舶用氣壓計
선박종관전문형식	ship synoptic code	船舶綜觀電文形式
선반구름	shelf cloud	
선반얼음	shelf ice	
선발	screen	選拔
선발관측선지점	selected ship (station)	選拔觀測船地點
선석	aragonite	霰石
선속도	linear velocity	線速度
선스콜	line squall	線-
선스펙트럼	line spectrum	線-
선신세기후	Pliocene climate	鮮新世氣候
선에코	line echo	線-

선에코파동형태	line echo wave pattern	線-波動形態
선적적 검토	pre-shipment review	船積的檢討
선진고분해능복사계	Advanced Very High Resolution Radiometer (AVHRR)	先進高分解能輻射計
선진비디콘카메라시스템	advanced vidicon camera system	先進-
선진지구관측시스템	advanced earth observing system	先進地球觀測-
선진지구관측위성	advanced earth observing satellite	先進地球觀測衛星
선진초단파탐측장치	advanced microwave sounding unit	先進超短波探測裝置
선진 TIROS-N	advanced TIROS-N	先進-
선체착빙	icing on a hull	船體着氷
선택규칙	selection rule	選擇規則
선택불안정(도)	selective instability	選擇不安定(度)
선택산란	selective scattering	選擇散亂
선택안정(도)	selective stability	選擇安定(度)
선택절단복사계	selective chopper radiometer	選擇切斷輻射計
선택흡수	selective absorption	選擇吸收
선특관측	selected special observation	選特觀測
선특기상보고	selected special meteorological report	選特氣象報告
선팽창	linear expansion	線膨脹
선퍼짐	line broadening	線-
선편파감소비	linear depolarization ratio	線編波減小比
선풍	whirl wind, whirls	旋風
선행(의)	antecedent	先行
선행강수지수	antecedent precipitation index	先行降水指數
선행시간	lead time	先行時間
선행토양수분	antecedent soil moisture	先行土壤水分
선험적 확률	a priori probability	先驗的確率
선형	linear	線形
선형가속기	linear accelerator	線形加速器
선형계산불안정	linear computational instability	線形計算不安定
선형대류	line convection	線形對流
선형미분방정식	linearized differential equation	線形微分方程式
선형방정식	linear equation	線形方程式
선형불안정도	linear instability	線形不安定度
선형상관	linear correlation	線形相關
선형성	linearity	線形性
선형수송	cyclostrophic transport	旋衡輸送
선형씨뿌리기	line-type seeding	線型-
선형안정성	linear stability	線形安定性
선형에코	linear echo	線形-
선형엔트로피 척도	linear entropy scale	線形-尺度
선형연산자	linear operator	線形演算子
선형온도장	linear field of temperature	線形溫度場
선형운동	cyclostrophic motion	旋衡運動

선형운동량	linear momentum	線形運動量
선형운동장	linear field of motion	線形運動場
선형이론	linear theory	線形理論
선형전이	linear transition	線形轉移
선형편파	linear polarization	線形偏波
선형평형	cyclostrophic balance	旋衡平衡
선형풍	cyclostrophic wind	旋衡風
선형풍발산	cyclostrophic divergence	旋衡風發散
선형풍수렴	cyclostrophic convergence	旋衡風收斂
선형함수	linear function	線形函數
선형함수	cyclostrophic function	旋衡函數
선형화	linearization	線形化
선형흐름	cyclostrophic flow	旋衡-
선회	spin	旋回
선회감소	spin down	旋回減少
선회감소시간	spin down time	旋回減少時間
선회안정	spin stabilization	旋回安定
선회증가	spin up	旋回增加
선회측운기	spindle nephoscope	旋回測雲器
선흡수	line absorption	線吸收
설계	snowdrift glacier	雪溪
설계기준	design criteria	設計基準
설계유량	design discharge	設計流量
설계폭풍	design storm	設計暴風
설계홍수	design flood	設計洪水
설계효과	layover effect	設計效果
설고막대	snow stick	雪高-
설량계	nivometer, snow gauge	雪量計
설림기후	snow forest climate	雪林氣候
설맹(증)	snow blindness	雪盲(症)
설면흑화	blackening of snow surface	雪面黑化
설명분산	explained variance	說明分散
설붕도	avalanche path	雪崩道
설상고기압	wedge high	舌狀高氣壓
설상당우량	snow water equivalent	雪相當雨量
설선	snow line	雪線
설식	nivation	雪蝕
설심	snow core	雪心
설진	snowquake, snow tremor	雪震
설편	snow flake	雪片
설해	snow damage	雪害
섬광	flash, flashing, glare, scintillation	閃光
섬광계	scintillometer	閃光計
섬광률	flash rate	閃光率

섬모	cilia	纖毛
섬유권운	cirrus fibratus, cirrus filosus	纖維卷雲
섬유권층운	cirrostratus fibratus, cirrostratus filosus	纖維卷層雲
섬유얼음	fibrous ice	纖維-
섭동	perturbation	攝動
섭동기법	perturbation technique	攝動技法
섭동량	perturbation quantity	攝動量
섭동론	perturbation theory	攝動論
섭동모형	perturbation model	攝動模型
섭동밀도	perturbation density	攝動密度
섭동방정식	perturbation equation	攝動方程式
섭동법	perturbation method	攝動法
섭동압력	perturbation pressure	攝動壓力
섭동온도	perturbation temperature	攝動溫度
섭동운동	perturbation motion	攝動運動
섭동조건식	conditional equation of perturbation	攝動條件式
섭동함수	perturbation function	攝動函數
섭씨	Celsius	攝氏
섭씨눈금	Celsius scale	攝氏-
섭씨온도	Celsius temperature	攝氏溫度
섭씨온도계	Celsius thermometer	攝氏溫度計
섭씨온도눈금	Celsius temperature scale, Celsius thermometric scale	攝氏溫度-
섭씨-절대온도환산	Celsius-Kelvin conversion	攝氏絕對溫度換算
섭씨-화씨온도환산	Celsius-Fahrenheit conversion	攝氏華氏溫度換算
성과기준	performance standard	成果基準
성긴 격자	coarse mesh	-格子
성긴 그물격자	coarse mesh grid	-格子
성루크여름	St. Luke's summer	聖-
성마틴여름	St. Martin's summer	聖-
성분	component	成分
성분벡터	component vector	成分-
성분속도	component velocity	成分速度
성숙	maturity, ripening	成熟
성숙기	mature stage	成熟期
성숙단계	mature stage	成熟段階
성에	window frost, window hoar, windowing	
성엘모불	Saint Elmo's fire, St. Elmo's fire	聖-
성우제거	rain-out	成雨除去
성운	nebula	星雲
성운가설	nebula hypothesis	星雲假說
성운이론	nebular theory	星雲理論
성장계절	growing season	成長季節
성장곡선	curve of growth	成長曲線

성촉절 이브 바람	Candlemas Eve wind	聖燭節-
성층	layering, stratification	成層
성층곡선	stratification curve	成層曲線
성층권	stratosphere	成層圈
성층권결합	stratospheric coupling	成層圈結合
성층권계면	stratopause	成層圈界面
성층권계절풍	stratospheric monsoon	成層圈季節風
성층권극소용돌이	stratospheric polar vortex	成層圈極-
성층권기구	stratostat	成層圈氣球
성층권낙진	stratospheric fallout	成層圈落塵
성층권-대류권 교환	stratosphere-troposphere exchange	成層圈-對流圈交換
성층권돌연승온	stratospheric sudden warming	成層圈突然昇溫
성층권보상	stratospheric compensation	成層圈補償
성층권복사	stratosphere radiation	成層圈輻射
성층권순환지수	stratospheric circulation index	成層圈循環指數
성층권승온	stratospheric warming	成層圈昇溫
성층권오존	stratospheric ozone	成層圈-
성층권접합	stratospheric coupling	成層圈接合
성층권제트류	stratospheric jet stream	成層圈-流
성층권지향	stratospheric steering	成層圈指向
성층권혼탁도	stratospheric turbidity	成層圈混濁度
성층유체	stratified fluid	成層流體
성층화	stratification	成層化
성층화산	stratovolcano	成層火山
세	epoch	世
세계공역예보시스템	World Area Forecast System	世界空域豫報-
세계기상감시	World Weather Watch(WWW)	世界氣象監視
세계기상감시계획	World Weather Watch Program	世界氣象監視計劃
세계기상감시프로그램	World Weather Watch Program	世界氣象監視-
세계기상관측망	réseau mondial	世界氣象觀測網
세계기상기구	World Meteorological Organization(WMO)	世界氣象機構
세계기상대회	World Meteorological Congress(WMC)	世界氣象大會
세계기상의 날	world meteorological day	世界氣象-
세계기후	world climate	世界氣候
세계기후프로그램	world climate programme	世界氣候-
세계보건기구	World Health Organization	世界保健機構
세계복사분석기준	World Radiometric Reference(WRR)	世界輻射分析基準
세계복사센터	World Radiation Center(WRC)	世界輻射-
세계시	universal time	世界時
세계일	universal day	世界日
세계지역예보센터	World Area Forecast Centre	世界地域豫報-
세계통신시스템	global telecommunication system	世界通信-
세계풍력협의회	global wind energy council	世界風力協議會
세계협정시	Universal Time Coordinated(UTC)	世界協定時

세로방향바람	longitudinal wind	
세로성분돌풍도	longitudinal gustiness	-成分突風度
세로좌표	ordinate	-座標
세로진동	longitudinal vibration	-振動
세로파	longitudinal wave	-波
세르	cers	
세설	light snow	細雪
세이시	seiche	
세정기	scrubber	洗淨器
세정집진장치	dust scrubber	洗淨集塵裝置
세정탈진장치	scrubber dedustor	洗淨脫塵裝置
세차운동	precession	歲差運動
세척기	scrubber	洗滌器
세척제거	washout	洗滌除去
세타그램	thetagram	
세타(θ)좌표계	theta (coordinate) system	-(座標)系
세포	cell	細胞
세포가설	cellular hypothesis	細胞假說
세포구조	cellular structure	細胞構造
세포설	cell theory	細胞說
세포윤곽	cell outline	細胞輪廓
세포자동모형	cellular automation model	細胞自動模型
세포형구름패턴	cellular cloud pattern	細胞型-
세포형대류	cellular convection	細胞型對流
세포형소용돌이	cellular vortex	細胞型-
세포형순환	cellular circulation	細胞型循環
세포형운동	cellular movement	細胞型運動
세포형태	cellular pattern	細胞形態
센 동서순환	high zonal circulation	-東西循環
센물	aggressive water	
센 바람	moderate gale, near gale	
센서	sensor	
센서계기	sensor instrument	-計器
센서관측각	sensor observation angle	-觀測角
센서기기	sensor instrument	-器機
센서민감도	sensor sensitivity	-敏感度
센서시야각	sensor viewing angle	-視野角
센서지연	sensor lag	-遲延
센서한계	sensor limitation	-限界
센서한계오차	sensor limitation error	-限界誤差
센서해상도	sensor resolution	-解像度
센 소나기	heavy shower	
센 스콜	heavy squall	
센 역풍	muzzler	-逆風

센티바	centibar	
센티포이즈	centipoise	
셀리니의 무리	Cellini's halo, heiligenschein	
셀신풍향계	selsyn wind vane	-風向計
셰지공식	Chézy formula	-公式
셰지방정식	Chézy equation	-方程式
소강상태	breathing space condition	小康狀態
소거비	cancellation ratio	消去比
소거율	cancellation ratio	消去率
소광	Quenching	消光
소광대후방산란비	backscatter to extinction ratio	消光對後方散亂比
소규모구조	small-scale structure	小規模構造
소규모치올림	small-scale lifting	小規模-
소금	salt	
소금물	brine, salt water	
소금물찌꺼기	brine slush	
소금바람	salt breeze, salty wind	
소금바람내성	salty wind tolerance	-耐性
소금바람피해	salty wind damage	-被害
소금씨뿌리기	salt seeding	
소금연무	salt haze	-煙霧
소금입자	salt particle	-粒子
소금피해	salt damage, salt injury	-被害
소나기	rain shower, shower, showery rain	
소나기공식	shower formula	-公式
소나기구름	shower cloud	
소나기선	line of showers	-線
소나탐지기상	sonar detection weather, sonar range weather	-探知氣象
소낙눈	flurry, showery snow, snow shower	
소낙성강수	showery precipitation	-性降水
소노라	sonora	
소노라날씨	sonora weather	
소다	sound detection and ranging(SODAR)	
소련F-2로켓	Soviet F-2 rocket	蘇聯-
소리	sound	
소리강도	sound intensity	-强度
소리구름	acoustic cloud	
소리그늘	sound shadow	
소리배열	acoustic array	-配列
소리변동	acoustical scintillation	-變動
소리선	sound ray	-線
소리에너지	sonic energy	
소리전파	sound propagation	-傳播
소리주파수	sound frequency	-周波數

소리흡수	sound absorption	-吸收
소만	grain full	小滿
소말리아제트	Somali jet	
소말리아해류	Somali current	-海流
소멸	dissipation	消滅
소멸기	dissipating stage	消滅期
소멸범위	dissipating range	消滅範圍
소멸성전선	frontolytical front	消滅性前線
소밀파	wave of condensation and rarefaction	疏密波
소빙산	bergy bit	小氷山
소빙하기	Little Ice Age(LIA)	小氷河期
소산	burn-off, dissipation, extinction	消散
소산계수	coefficient of extinction, dissipation coefficient, extinction coefficient	消散係數
소산길이규모	dissipation length scale	-規模
소산단계	dissipating stage	消散段階
소산단면	extinction cross-section	消散斷面
소산두께	extinction thickness	消散-
소산비행운	dissipation contrail	消散飛行雲
소산상수	dissipation constant	消散常數
소산영역	dissipation range	消散領域
소산율	dissipation rate	消散率
소산효율	extinction efficiency	消散效率
소산효율인자	extinction efficiency factor	消散效率因子
소서	minor heat	小暑
소설	minor snow	小雪
소속함수	membership function	所屬函數
소수	decimal	小數
소수	prime number	素數
소수성	hydrophobicity	疏水性
소스	source	
소용돌이	convolution, vortex	
소용돌이고리	vortex ring	
소용돌이관	vortex tube	-管
소용돌이구름길	vortex cloud street	
소용돌이꼬리	vortex trail, vortex street	
소용돌이늘임	vortex stretching	
소용돌이도	vorticity	-度
소용돌이도방정식	vorticity equation	-度方程式
소용돌이도보존	conservation of vorticity	-度保存
소용돌이도속	vorticity flux	-度束
소용돌이도수송가설	vorticity transport hypothesis	-度輸送假說
소용돌이도수송이론	vorticity transport theory	-度輸送理論
소용돌이도이류	vorticity advection	-度移流

소용돌이도이류법	vorticity advection method	-度移流法
소용돌이도장	field of vorticity	-度場
소용돌이도전이	vorticity transfer	-度轉移
소용돌이도최대	vorticity maximum	-度最大
소용돌이도플럭스	vorticity flux	-度-
소용돌이도효과	vorticity effect	-度效果
소용돌이비	vortex rain	
소용돌이선	vortex line	-線
소용돌이온도계	vortex thermometer	-溫度計
소용돌이운동	vortex motion	-運動
소용돌이정리	vortex theorem	-定理
소용돌이줄	vortex train	
소용돌이판(줄)	vortex sheet	-板
소용돌이필라멘트	vortex filament	
소용돌이후류난류	vortex wake turbulence	-後流亂流
소이작용제작전기상	incendiaries agent operation weather	燒夷作用劑作戰氣象
소자	cell	素子
소한	minor cold	小寒
소해작전기상	Minesweeping Operation weather	掃海作戰氣象
소행성	asteroid, minor planet, planetoid	小行星
소행성멸종가설	asteroid extinction hypothesis	小行星滅種假設
소행성충돌	asteroid impact	小行星衝突
소형증발계	small evaporimeter	小型蒸發計
속	packet	束
속도계	speedometer	速度計
속도반지름소용돌이	velocity-radius vortex, V-R vortex	速度-
속도발산	velocity divergence	速度發散
속도방위표시	velocity azimuth display	速度方位表示
속도분포	velocity distribution	速度分布
속도선택기	velocity selector	速度選擇器
속도손실법칙	velocity defect law	速度損失法則
속도압	velocity pressure	速度壓
속도연직분포	velocity profile	速度鉛直分布
속도장	velocity field	速度場
속도접힘	velocity aliasing, velocity folding	速度-
속도조절기	velocity governor	速度調節器
속도지시간섭적산계	velocity indication coherent integrator	速度指示干涉積算計
속도퍼텐셜	velocity potential	速度-
속도 펼침	velocity unfolding	速度-
속력	speed	速力
속밀도	power density	束密度
속서리	depth hoar	
손스웨이트 기후구분	Thornthwaite's classification of climate	-氣候區分
손스웨이트 수분지수	Thornthwaite moisture index	-水分指數

손스웨이트 열지수	Thornthwaite heat index	-熱指數
솔레노이드	solenoid	
솔레노이드장	field of solenoid, solenoidal field	-場
솔레노이드지수	solenoidal index	-指數
솜덩이구름	cotton ball cloud	
솜털	cilia	
송수신기	transponder, transmitter-responder	送受信器
송수신전환스위치	duplexer	送受信轉換-
송수신폭	transponder ranging	送受信幅
송신기	transmitter	送信器
송신존데	transosonde	送信-
송신출력	transmitted power	送信出力
송이(구름)	floccus	
송이고적운	altocumulus floccus	-高積雲
송이권운	cirrus floccus	-卷雲
송이권적운	cirrocumulus floccus	-卷積雲
송이층적운	stratocumulus floccus	-層積雲
쇄빙기	icebreaker	碎氷機
쇄빙선	icebreaker	碎氷船
쇄설류	debris flow	碎屑流
쇄파	breaker, surf, wave breaking	碎波
쇄파맥놀이	surf beat	碎波-
쇄파수심	breaker depth	碎波水深
쇠바람	iron wind	
쇠약성전선	frontolytical front	衰弱性前線
쇠퇴	fading	衰退
쇼월터안정지수	Showalter's stability index	-安定指數
쇼월터지수	Showalter index	-指數
수감계	sensing system	受感系
수감부	sensing element, sensor	受感部
수계	river system	水系
수관	crown	樹冠
수광대	euphotic zone	受光帶
수권	hydrosphere	水圈
수극	water gap	水隙
수농도	concentration of particle, number concentration	數濃度
수단	water mass	水團
수단형	Sudan type	-型
수동레이더	passive radar	受動-
수동마이크로파바람	passive microwave wind	受動-波-
수동마이크로파탐측	passive microwave sensing	受動-波探測
수동상승활주면	passive anafrontal surface	受動上昇滑走面
수동센서	passive sensor	受動-
수동스칼라	passive scalar	受動-

수동시료채취	passive sampling	受動試料採取
수동시스템	passive system	受動-
수동식 추적	manual tracking	手動式追跡
수동영구동토	passive permafrost, fossil permafrost	受動永久凍土
수동원격탐사	passive remote sensing	受動遠隔探査
수동태양열집열기	passive solar collector	受動太陽熱集熱器
수동하강활주면	passive katafrontal surface	受動下降滑走面
수동형가스채집기	passive gas sampler	手動型-採集器
수동형기기	passive instrument	受動型機器
수두	hydraulic head	水頭
수량	quantity	數量
수레바퀴형 위성	cartwheel satellite	-型衛星
수력	water power	水力
수력경도	hydraulic gradient	水力傾度
수력반경	hydraulic radius	水力半徑
수력반지름	hydraulic radius	水力-
수력유사	hydraulic analogy	水力類似
수력학	hydraulics, hydromechanics	水力學
수렴	convergence	收斂
수렴구역	convergence region	收斂區域
수렴기류	convergent air current	收斂氣流
수렴대	belt of convergence, convergence zone	收斂帶
수렴렌즈	convergence lens	收斂-
수렴모형	convergence model	收斂模型
수렴바람	convergent wind	收斂-
수렴선	convergence line, convergent line	收斂線
수렴성비	convergent rain	收斂性-
수렴수치방안	convergent numerical scheme	收斂數值方案
수렴장	convergence field, field of convergence	收斂場
수렴점	convergent point	收斂點
수렴점근선	asymptote of convergence	收斂漸近線
수렴주변부	convergent margin	收斂周邊部
수렴효과	convergent effect	收斂效果
수로강풍	channeled high wind	水路强風
수로구간	reach	水路區間
수로상강수(량)	channel precipitation	水路上降水(量)
수로유입	channel inflow	水路流入
수로이론	canal theory	水路理論
수로중앙선	thalweg	水路中央線
수로측량	hydrographic survey	水路測量
수로학	hydrography	水路學
수로화	canalization	水路化
수리기상학	mathematical meteorology	數理氣象學
수리기후	mathematical climate	數理氣候

수리기후대	mathematical climate zone	數理氣候帶
수리물리학	mathematical physics	數理物理學
수리전도도	hydraulic conductivity	水理傳導度
수리통계학	mathematical statistics	數理統計學
수리학	hydraulics	水理學
수마트라	sumatra	
수막현상	hydroplaning	水膜現象
수면습윤공기상대습도	relative humidity of moist air with respect to water	水面濕潤空氣相對濕度
수면온도	surface temperature of water	水面溫度
수면착수(우주선 등의)	splashdown	水面着水
수면포화	water saturation	水面飽和
수목기후	tree climate	樹木氣候
수목변형	tree deformation	樹木變形
수목한계선	timber line, tree line	樹木限界線
수문	sluice, flood gate	水門
수문계산	hydrologic accounting	水文計算
수문곡선	hydrograph	水文曲線
수문곡선분리	hydrograph separation	水文曲線分離
수문관측망	hydrometric network, hydrological network	水文觀測網
수문관측소	hydrographic station	水文觀測所
수문기상학	hydrometeorology	水文氣象學
수문기상학자	hydrometeorologist	水文氣象學者
수문기후단위	hydroclimatic unit	水文氣候單位
수문년	hydrologic year	水文年
수문단위	hydro unit	水文單位
수문순환	hydrological cycle	水文循環
수문자원	hydrological resource	水文資源
수문통계학	hydrological statistics	水文統計學
수문학	hydrology	水文學
수문학방정식	hydrologic equation	水文學方程式
수문학적 가뭄	hydrological drought	水文學的-
수문학적 모형	hydrologic model	水文學的模型
수문학적 예측	hydrologic forecasting	水文學的豫測
수문학적 홍수추적	hydrologic flood routing	水文學的洪水追跡
수밀도	number density	數密度
수반구	water hemisphere	水半球
수반모형	adjoint model	隨伴模型
수반민감도	adjoint sensitivity	隨伴敏感度
수반방정식	adjoint equation	隨伴方程式
수반복사전달방정식	adjoint radiative transfer equation	隨伴輻射傳達方程式
수반자료동화	adjoint assimilation	隨伴資料同化
수분	moisture	水分
수분경계값	moisture boundary	水分境界-

수분당량	moisture equivalent	水分當量
수분복원방안	moisture retrieval scheme	水分復原方案
수분상수	water constant	水分常數
수분수렴	moisture convergence	水分收斂
수분역전	moisture inversion	水分逆轉
수분연속방정식	moisture continuity equation	水分連續方程式
수분연직분포	moisture profile	水分鉛直分布
수분·온도 지수	moisture-temperature index	水分溫度指數
수분운동학	hygrokinematics	水分運動學
수분인자	moisture factor	水分因子
수분전도도	conductance for moisture	水分傳導度
수분조절	moisture adjustment	水分調節
수분지수	moisture index	水分指數
수분지시기	moisture indicator	水分指示器
수분특성곡선	moisture characteristic curve	水分特性曲線
수분평형	moisture equilibrium	水分平衡
수분함량	moisture content	水分含量
수산기방출	hydroxyl emission	水酸基放出
수산화암모늄	ammonium hydroxide	水酸化-
수상	hydrological phenomena	水象
수상	water phase	水相
수상당량비	snow water equivalent	水相當量比
수성기체	water gas	水性氣體
수소결합	hydrogen bond	水素結合
수소기구	hydrogen(-filled) balloon	水素氣球
수소선	hydrogen line	水素線
수소온도계	hydrogen thermometer	水素溫度計
수소온도 눈금	hydrogen scale of temperature	水素溫度-
수소이온농도	hydrogen-ion concentration	水素-濃度
수소이온농도서열	pH sequence	水素-濃度序列
수소이온농도척도	pH scale	水素-濃度尺度
수송기구시스템	carrier-balloon system	輸送氣球-
수송풍	transport wind	輸送風
수수구	orifice	受水口
수시관측	provisional observation	隨時觀測
수식	water erosion	水蝕
수신강도	receiving intensity	受信强度
수신공률	received power	受信工率
수신기	receiver	受信機
수신기증폭	receiver gain	受信機增幅
수압기	hydraulic press	水壓機
수에즈운하	Suez Canal	-運河
수역	hydrological basin	水域
수열	sequence	數列

수열형	incoming radiation type	受熱型
수온	water temperature	水溫
수온계	water thermometer	水溫計
수온약층	thermocline	水溫躍層
수용상자	accommodation box	收容箱子
수용성	water-solubility	水溶性
수용액	aqueous solution	水溶液
수용체	receptor	受容體
수위	stage, water level	水位
수위계	staff gauge	水位計
수위기록계	water-stage recorder	水位記錄計
수위상관	stage relation	水位相關
수위유량곡선	rating curve	水位流量曲線
수위유량관계	stage-discharge relation	水位流量關係
수위저하곡면	cone of impression	水位低下曲面
수위측정기	staff gauge	水位測定器
수은	mercury, quicksilver	水銀
수은기둥	mercury column	水銀-
수은기압계	mercurial barometer, mercury barometer	水銀氣壓計
수은기압기록계	mercurial barograph	水銀氣壓記錄計
수은압력계	mercury manometer	水銀壓力計
수은온도계	mercurial thermometer, mercury thermometer	水銀溫度計
수은자기기압계	mercurial barograph	水銀自記氣壓計
수은조	cistern	水銀槽
수은조조절식기압계	adjustable cistern barometer	水銀槽調節式氣壓計
수인성전염병	water-borne disease	水因性傳染病
수자원	water resource	水資源
수적	drop of water, shuidi	水滴
수적정렬방향	drop orientation	水滴整列方向
수정고도	correction line	修正高度
수정공역예보	Amended Area Forecast	修正空域豫報
수정굴절률	modified index of refraction, modified refractive index, refractive modulus	修正屈折率
수정비행장예보	amended aerodrome forecast	修正飛行場豫報
수정항공로예보	Amended Route Forecast	修正航空路豫報
수준	level	水準
수준기	level	水準器
수준점	bench mark	水準點
수준점관측소	bench mark station	水準點觀測所
수중파괴작전기상	UDT operation weather	水中破壞作戰氣象
수증기	aqueous vapour, vapo(u)r, water vapo(u)r	水蒸氣
수증기감률	hydrolapse	水蒸氣減率
수증기띠	water vapo(u)r band	水蒸氣-
수증기몰비	mole fraction of water vapo(u)r	水蒸氣-比

수증기밀도	water vapo(u)r density	水蒸氣密度
수증기보존	conservation of moisture	水蒸氣保存
수증기복사	water vapo(u)r radiation	水蒸氣輻射
수증기분광광도계	water vapo(u)r spectroscope	水蒸氣分光光度計
수증기압	water vapo(u)r pressure	水蒸氣壓
수증기영상	water vapo(u)r image	水蒸氣映像
수증기학	atmology	水蒸氣學
수증기함량	water vapo(u)r content	水蒸氣含量
수증기흡수	water vapo(u)r absorption	水蒸氣吸收
수지	budget	收支
수지년	budget year	收支年
수지연산자	budget operator	收支演算子
수직	normal	垂直
수직(의)	perpendicular	垂直
수직성분	normal component	垂直成分
수직엽형군락	vertical leaved canopy	垂直葉形群落
수직입사	normal incidence, perpendicular incidence	垂直入射
수직충격파	normal shock wave	垂直衝擊波
수질오염	water pollution	水質汚染
수집	acquisition	收集
수집기	receiver	收集機
수차	aberration	收差
수축	contraction, compactness, shrinking	收縮
수축장	contraction field	收縮場
수축축	axis of contraction, contraction axis	收縮軸
수치기호법	numerical notation	數值記號法
수치모델링	numerical modeling	數值-
수치모의	numerical simulation	數值模擬
수치모형	numerical model	數值模型
수치모형분석	numerical model analysis	數值模型分析
수치분산	numerical dispersion	數值分散
수치분산모형	numerical dispersion model	數值分散模型
수치분석	numerical analysis	數值分析
수치불안정	numerical instability	數值不安定
수치순차법	numerical step by step method	數值順次法
수치시뮬레이션	numerical simulation	數值-
수치실험	numerical experiment	數值實驗
수치안정	numerical stability	數值安定
수치역학불안정도	numerical dynamical instability	數值力學不安定度
수치예보	numerical forecast, numerical forecasting	數值豫報
수치예보모형	model of numerical weather prediction, numerical forecasting model	數值豫報模型
수치일기예보	numerical weather forecast	數值日氣豫報
수치일기예측모형	numerical weather prediction model	數值日氣豫測模型

수치일기예측산출물	numerical weather prediction output	數值日氣豫測産出物
수치일기예측	numerical weather prediction	數値日氣豫測
수치적분	numerical integration	數値積分
수크호베이	sukhovei	
수판기후대	Supan's climatic zone	-氣候帶
수평	horizon, horizontal, water balance	水平
수평각	horizontal angle	水平角
수평감지기	horizon sensor	水平感知器
수평거리	horizon distance	水平距離
수평격자엇갈림	staggering of horizontal grid	水平格子-
수평관성운동	horizontal inertia motion	水平慣性運動
수평기압경도력	horizontal pressure gradient force	水平氣壓傾度力
수평면	horizontal plane, level	水平面
수평면경각	dip of the horizon	水平面傾角
수평무지개	horizontal rainbow	水平-
수평바람벡터	horizontal wind vector	水平-
수평발산	horizontal divergence	水平發散
수평변동	horizontal variation	水平變動
수평분도원	horizontal circle	水平分度圓
수평분포	horizontal distribution	水平分布
수평불안정(도)	horizontal instability	水平不安定(度)
수평성분	horizontal component	水平成分
수평센서	horizon sensor	水平-
수평속도	horizontal velocity	水平速度
수평수렴	horizontal convergence	水平收斂
수평시단면	horizontal time section	水平時斷面
수평시정	horizontal visibility	水平視程
수평안정도	horizontal stability	水平安定度
수평압력	horizontal pressure force	水平壓力
수평압력경도	horizontal pressure gradient	水平壓力傾度
수평엽형군락	horizontal leaved canopy	水平葉型群落
수평온도경도	horizontal temperature gradient	水平溫度傾度
수평운동	horizontal motion	水平運動
수평운동방정식	horizontal motion equation	水平運動方程式
수평윈드시어	horizontal wind shear	水平-
수평이류	horizontal advection	水平移流
수평일사강도	horizontal insolation intensity	水平日射强度
수평접호	horizontal tangent arc	水平接弧
수평지진계	horizontal seismograph	水平地震計
수평탐측기술	horizontal sounding technique	水平探測技術
수평파	horizontal wave	水平波
수평편광	horizontal polarization	水平偏光
수평해상도	horizontal resolution	水平解像度
수평혼합	horizontal mixing	水平混合

수평흐름	horizontal flow	水平-
수풀바람	forest wind	
수풀한계선	forest line	-限界線
수학적 귀납법	mathematical induction	數學的歸納法
수학적 기댓값	mathematical expectation	數學的期待-
수학적 예보	mathematical forecasting	數學的豫報
수학적 확률	mathematical probability	數學的確率
수화물	hydrate	水化物
수화법	manual method	手話法
수확량구성요소	yield component	收穫量構成要素
수확량예측	crop forecast, yield prediction	收穫量豫測
수확성분	yield component	收穫成分
숙련도	skill score	熟練度
순	net	純
순간	moment	瞬間
순간가속도	instantaneous acceleration	瞬間加速度
순간관찰영역	instantaneous field of view	瞬間觀察領域
순간변화율	instantaneous rate of change	瞬間變化率
순간속도	instantaneous velocity	瞬間速度
순간속력	instantaneous speed	瞬間速力
순간시야	instantaneous field of view	瞬間視野
순간쌍극자모멘트	instantaneous dipole moment	瞬間雙極子-
순간적	instantaneous	瞬間的
순간폐색	instant occlusion	瞬間閉塞
순간풍속	instantaneous wind speed	瞬間風速
순강우	net rainfall	純降雨
순광합성	net photosynthesis	純光合成
순균형	net balance	純均衡
순록	caribou	馴鹿
순방출적외선	net outgoing IR	純放出赤外線
순복사	net radiation	純輻射
순복사강제	net radiative forcing	純輻射强制
순복사계	net radiometer	純輻射計
순복사냉각	net radiative cooling	純輻射冷却
순복사속	net radiation flux	純輻射-
순복사속밀도	net radiative flux density	純輻射-密度
순복사플럭스	net radiation flux	純輻射-
순복사플럭스밀도	net radiative flux density	純輻射-密度
순상향이동	net upward movement	純上向移動
순서	sequence	順序
순서도	flow chart	順序圖
순서분석	sequential analysis	順序分析
순서완화	sequential relaxation	順序緩和
순수공기	pure air	純粹空氣

순스톰비	net storm rain	純-
순시어	downshear	順-
순압(의)	barotropic	順壓
순압경계층	barotropic boundary layer	順壓境界層
순압계수	coefficient of barotropy	順壓係數
순압대기	barotropic atmosphere	順壓大氣
순압매질	barotropic medium	順壓媒質
순압모형	barotropic model	順壓模型
순압방정식	equation of barotropy	順壓方程式
순압불안정(도)	barotropic instability	順壓不安定(度)
순압불안정(성)	barotropic instability	順壓不安定(性)
순압상태	barotropic state	順壓狀態
순압성	barotropy	順壓性
순압소용돌이도 방정식	barotropic vorticity equation	順壓-度方程式
순압스티어링	barotropic steering	順壓-
순압안정도	barotropic stability	順壓安定度
순압압력함수	barotropic pressure function	順壓壓力函數
순압예보	barotropic forecast	順壓豫報
순압요란	barotropic disturbance	順壓搖亂
순압유체	barotropic fluid	順壓流體
순압장	barotropic field	順壓場
순압조건	barotropic condition	順壓條件
순압토크벡터	barotropic torque vector	順壓-
순압파	barotropic wave	順壓波
순압행성경계층	barotropic planetary boundary layer	順壓行星境界層
순에너지저장	net energy storage	純-貯藏
순위상관	rank correlation	順位相關
순응	acclimation	順應
순응결	net condensation	純凝結
순일차생산	net primary production	純一次生産
순전(풍향)	veering	順轉(風向)
순전바람	veering wind	順轉-
순전천복사계	net pyrradiometer	純全天輻射計
순전천일사계	net pyranometer	純全天日射計
순전천장파복사계	net pyrgeometer	純全天長波輻射計
순전파동복사	net all-wave radiation	純全波動輻射
순증발	net evaporation	純蒸發
순지구복사	net terrestrial radiation	純地球輻射
순태양복사	net solar radiation	純太陽輻射
순풍조	lee tide	順風潮
순항	cruising	巡航
순항거리	cruising range	巡航距離
순항고도	cruising level	巡航高度
순항성능	cruising efficiency	巡航性能

순항속도	cruising speed	巡航速度
순행궤도	prograde orbit	順行軌道
순허수	pure imaginary number	純虛數
순화	acclimation	馴化
순환	circulation	循環
순환계	circulatory system	循環系
순환규모	scale of circulation	循環規模
순환모양	circulation pattern	循環模樣
순환모형	circulation model	循環模型
순환세포	circulation cell	循環細胞
순환유형	pattern of circulation	循環類型
순환이상	circulation anomaly	循環異常
순환적분	circulation integral	循環積分
순환정리	circulation theorem	循環定理
순환지수	circulation index	循環指數
순환패턴	pattern of circulation	循環-
순환플럭스	circulation flux	循環-
순환형	circulation type	循環型
순환형풍동	closed circuit wind tunnel	循環型風洞
순힘	net force	純-
숨은 에너지	latent energy	
숨은 열	latent heat	-熱
숫자	figure	數字
숫자구름분포도	digitized cloud map	數字-分布圖
숫자레이더실험	digitized radar experiment	數字-實驗
숫자부호	figure code	數字符號
숫자코드	figure code	數字-
숫자화 장치	digitiser, digitizer	數字化裝置
숲캐노피	forest canopy	
쉼표구름꼬리	comma tail	-標-
쉼표머리	comma head	-標-
쉼표모양구름	comma cloud, comma-shaped cloud	-標模樣-
쉼표모양구름시스템	comma cloud system	-標模樣-
슈만-룽에 띠	Schumann-Runge band	
슈만-룽에 연속체	Schumann-Runge continuum	-連續體
슈와르츠차일드방정식	Schwarzchild's equation	-方程式
슈타르크효과	Stark effect	-效果
슈퍼관측	superobservation	-觀測
슈퍼앙상블	superensemble	
슈프룽자기기압계	Sprung barograph	-自記氣壓計
스넬법칙	Snell's law	-法則
스노우서버	snow server	
스노카메라	snow camera	
스메이즈	smaze, smoke and haze	

스모그	smog, smoke and fog	
스모그선	smog horizon	-線
스모크스	smokes	
스미스복귀방식	Smith's retrieval scheme	-復歸方式
스미스-울프 방식	Smith and Woolf scheme	-方式
스웨이	sway	
스웨트지수	SWEAT index	-指數
스카블러	skavler	
스카프구름	scarf cloud	
스칸디나비아 빙상	Scandinavian ice sheet	-氷床
스칼라	scalar	
스칼라곱	scalar product	
스칼라근사	scalar approximation	-近似
스칼라양	scalar quantity	-量
스칼라퍼텐셜	scalar potential	
스칼라함수	scalar function	-函數
스캐트로미터	scatterometer	
스코러수	scorer number	-數
스코틀랜드 박무	Scotch mist	-薄霧
스코프	scope	
스코프지시기	scope indicator	-指示器
스콜	squall	
스콜구름	squall cloud	
스콜면	squall surface	-面
스콜선	squall line	-線
스콜선뇌우	squall line thunderstorm	-線雷雨
스콜파	squall wave	-波
스큐티로그피 선도	skew T-log p diagram	-線圖
스타류트	starute, star and parachute	
스타트스코프	startscope	
스탠턴수	Stanton number	-數
스테라디안	steradian	
스테레오	stereo	
스테판법칙	Stefan's law	-法則
스테판-볼츠만 법칙	Stefan-Boltzmann law	-法則
스테판-볼츠만 상수	Stefan-Boltzmann constant	-常數
스텝	steppe	
스텝기후	steppe climate	-氣候
스토크	stoke	
스토크스근사	Stokes's approximation	-近似
스토크스법칙	Stokes's law	-法則
스토크스벡터	Stokes vector	
스토크스유선함수	Stokes's stream function	-流線函數
스토크스정리	Stokes's theorem	-定理

스토크스저항법칙	Stokes's law of resistance	-抵抗法則
스토크스파	Stokesian wave	-波
스톰법칙	law of storms	-法則
스튜브선도	Stüve diagram	-線圖
스트로할수	Strouhal number	-數
스트론튬동위원소	strontium isotope	-同位元素
스트리머	streamer	
스트리크	streak	
스티븐슨백엽상	Stevenson screen, Stevenson's shelter	-百葉箱
스틸브	stilb	
스파크	spark	
스펀지구역	spongy zone	-區域
스펀지얼음	spongy ice	
스펀지우박	spongy hail	-雨雹
스펙트럼	spectra, spectrum	
스펙트럼(의)	spectral	
스펙트럼강도	spectrum intensity	-强度
스펙트럼검출률	spectral detectivity	-檢出率
스펙트럼계열	spectrum series	-系列
스펙트럼광전도율	spectral photoconductivity	-光傳導率
스펙트럼농도	spectral concentration	-濃度
스펙트럼모형	spectral model	-模型
스펙트럼밀도	spectral density	-密度
스펙트럼법	spectral method	-法
스펙트럼분석	spectrum analysis	-分析
스펙트럼분해능	spectral resolution	-分解能
스펙트럼선	spectral line	-線
스펙트럼수치분석	spectral numerical analysis	-數值分析
스펙트럼수치예측	spectral numerical prediction	-數值豫測
스펙트럼채널	spectral channel	
스펙트럼폭	spectral width, spectrum width	-幅
스펙트럼함수	spectral function, spectrum function	-函數
스펠레오뎀	speleothem	
스푸트니크	Sputnik	
스푸트니크기상위성	Meteor sputnik	-氣象衛星
슬라이딩법	sliding method	-法
슬라이딩척도	sliding scale	-尺度
슬라이스법	slice method	-法
슬러리	slurry	
습구	wet bulb	濕球
습구빙점	wet-bulb zero	濕球氷點
습구빙점고도	wet-bulb zero height	濕球氷點高度
습구영점	wet-bulb zero	濕球零點
습구영점고도	wet-bulb zero height	濕球零點高度

습구온도	wet-bulb temperature	濕球溫度
습구온도계	wet-bulb globe thermometer, wet-bulb thermometer	濕球溫度計
습구온위	potential wet-bulb temperature, wet-bulb potential temperature	濕球溫位
습구효과	wet-bulb effect	濕球效果
습기	moisture	濕氣
습도	air humidity, humidity	濕度
습도감부	humidity sensor	濕度感部
습도계	humidometer, hygrometer	濕度計
습도계방정식	psychrometer equation	濕度計方程式
습도계산기	psychrometric calculator	濕度計算器
습도계산도	psychrometric chart	濕度計算圖
습도계산자	humidity slide-rule	濕度計算-
습도계수	coefficient of humidity, humidity coefficient	濕度係數
습도공식	hygrometric formula	濕度公式
습도기록계	hygrograph	濕度記錄計
습도기록곡선	hygrogram	濕度記錄曲線
습도방정식	hygrometric equation	濕度方程式
습도센서	humidity sensor	濕度-
습도소자	humidity strip, hygristor	濕度素子
습도요소	humidity element	濕度要素
습도지시기	atmidoscope	濕度指示器
습도측정법	hygrometry	濕度測定法
습도측정술	psychrometry	濕度測定術
습도학	hygrology	濕度學
습도환산표	psychrometric table	濕度換算表
습생식물	hygrophyte	濕生植物
습설	water snow	濕雪
습설	wet tongue	濕舌
습성침착	wet deposition	濕性沈着
습식 스크러버	wet scrubber	濕式-
습식전기집진장치	wet electrostatic dust precipitator	濕式電氣集塵裝置
습식 포집	wet scavenging, wet deposition	濕式捕集
습윤감률	humidity lapse rate	濕潤減率
습윤계절	wet season	濕潤季節
습윤공기	moist air, soft air, wet air	濕潤空氣
습윤공기밀도	density of moist air	濕潤空氣密度
습윤공기 빙면 포화수증기 몰비	mole fraction of saturation water vapour of moist air with respect to ice	濕潤空氣氷面飽和水蒸氣 -比
습윤공기 수면 포화수증기 몰비	mole fraction of saturation water vapour of moist air with respect to water	濕潤空氣水面飽和水蒸氣 -比
습윤구역	humidity province	濕潤區域
습윤기간	wet period	濕潤期間

습윤기후	humid climate, moist climate, wet climate	濕潤氣候
습윤눈송이	moist snow-flake	濕潤-
습윤단열(의)	wet adiabatic	濕潤斷熱
습윤단열감률	moist adiabatic lapse rate, wet-adiabatic lapse rate	濕潤斷熱減率
습윤단열과정	moist adiabatic process, wet-adiabatic process	濕潤斷熱過程
습윤단열냉각	moist adiabatic cooling, wet-adiabatic cooling	濕潤斷熱冷却
습윤단열률	wet-adiabatic rate	濕潤斷熱率
습윤단열변화	wet-adiabatic change	濕潤斷熱變化
습윤단열선	moist adiabat, wet adiabat	濕潤斷熱線
습윤단열온도차	wet-adiabatic temperature difference	濕潤斷熱溫度差
습윤대	humid zone	濕潤帶
습윤대류	moist convection	濕潤對流
습윤대륙기후	humid continental climate	濕潤大陸氣候
습윤미열기후	humid microthermal climate	濕潤微熱氣候
습윤불안정	wet unstable	濕潤不安定
습윤불안정(도)	wet instability	濕潤不安定(度)
습윤불안정도	moist-lability	濕潤不安定度
습윤불안정 에너지	moist-labile energy	濕潤不安定-
습윤상태	hygrometric state	濕潤狀態
습윤생장	wet growth	濕潤生長
습윤수	wet number	濕潤數
습윤아습기후	moist subhumid climate	濕潤亞濕氣候
습윤안개	wet fog	濕潤-
습윤열대	wet tropics	濕潤熱帶
습윤온대기후	humid temperate climate	濕潤溫帶氣候
습윤일	wet day	濕潤日
습윤전선	wetting front	濕潤前線
습윤정적에너지	moist static energy	濕潤靜的-
습윤중립	moist indifferent	濕潤中立
습윤중열기후	humid mesothermal climate	濕潤中熱氣候
습윤지속기간	wet spell	濕潤持續期間
습윤지수	humidity index	濕潤指數
습윤층	humid layer	濕潤層
습윤카타온도계	wet katathermometer	濕潤-溫度計
습윤표	hygrometric table	濕潤表
습윤하강돌풍	wet downburst	濕潤下降突風
습윤해	humid injury	濕潤害
습윤혀	moist tongue	濕潤-
습윤홍수방지	wet flood proofing	濕潤洪水防止
습지	swamp, wetland	濕地
습한 공기	damp air	濕-空氣
습한 눈	wet snow	濕-
습한 연무	damp haze	濕-煙霧

습해	wet injury	濕害
승강해수면	eustatic sea level	昇降海水面
승교점	ascending node	昇交點
승수	catch	承水
승온	temperature rising	昇溫
승적률	product-moment	乘積率
승화	sublimation	昇華
승화곡선	sublimation curve	昇華曲線
승화단열선	sublimation adiabat	昇華斷熱線
승화열	heat of sublimation	昇華熱
승화핵	sublimation nucleus	昇華核
시각	angle of view, visual angle	視角
시각분석요소	visual analysis element	視角分析要素
시간각	hour angle	時間角
시간간격라디오존데	time-interval radiosonde	時間間隔-
시간강수량	hourly precipitation	時間降水量
시간거리표	hourly distance scale	時間距離表
시간경과	time lapse	時間經過
시간규모	time scale	時間規模
시간기록기	chronograph	時間記錄器
시간단면	time section	時間斷面
시간단면도	time cross-section, time section	時間斷面圖
시간등온선	chronoisotherm	時間等溫線
시간면적깊이곡선	time-area-depth curve	時間面積-曲線
시간발전	time development	時間發展
시간변화	temporal change	時間變化
시간종속류	time-dependent flow	時間從屬流
시간중앙차분방식	centered-in-time scheme	時間中央差分方式
시간지연	time lag	時間遲延
시간척도	time scale	時間尺度
시간풍향등온선도	chronoanemoisothermal diagram	時間風向等溫線圖
시간해상도	temporal resolution	時間解像度
시감도	luminosity	視感度
시감도율	luminous efficiency	視感度率
시계	field of view, field of vision	視界
시계계기기상조건	visual and instrument weather condition	視界計器氣象條件
시계(비행)기상조건	visual meteorological condition	視界(飛行)氣象條件
시계 내 강수	precipitation within sight	視界內降水
시계명확선	clear line of sight	視界明確線
시계방향	clockwise, deasil	時計方向
시계방향바람	clockwise wind	時計方向-
시계별	clock-star	時計-
시계비행공항최저값	VFR terminal minimum	視界飛行空港最低-
시계비행규칙	visual flight rule	視界飛行規則

시계비행기상	VFR weather	視界飛行氣象
시계비행기상정보시스템	VFR flight weather information system	視界飛行氣象情報-
시계비행기상조건	visible flight weather condition	視界飛行氣象條件
시계비행날씨	contact weather	視界飛行-
시계식라디오존데	chronometric radiosonde	時計式-
시계열	time series	時系列
시계열분석	time series analysis	時系列分析
시계열상관	serial correlation	時系列相關
시계원뿔	cone of vision	視界圓-
시계장치풍속계	clockwork anemometer	時計裝置風速計
시계접촉높이	visual contact height	視界接觸-
시계형온도계	chronothermometer	時計形溫度計
시골바람	country breeze	
시공표본채취	space-time sampling	時空標本採取
시그마좌표	sigma coordinate	-座標
시그마좌표계	sigma coordinate system	-座標系
시그마-티	sigma-t	
시듦	wilting	
시듦계수	wilting coefficient	-係數
시듦점	wilting point	-點
시로코	scirocco, sirocco	
시면심곡선	time-area-depth curve	時面深曲線
시몬 지중온도계	Simon's earth thermometer	-地中溫度計
시뭄	simoon	
시뮬레이션	simulation	
시민소송	citizen suit	市民訴訟
시베리아고기압	Siberian anticyclone, Siberian high	-高氣壓
시베리아기단	Siberian air mass	-氣團
시베리아대륙기단	Siberian continental air mass	-大陸氣團
시베리아동토융해	Siberian permafrost melt	-凍土融解
시새트	Seasat	
시생대	Archean	始生代
시생대환경	Archean environment	始生代環境
시선범위	line-of-sight range	視線範圍
시선속도	radial velocity	視線速度
시선속도-방위각 표출	velocity azimuth display	視線速度方位角表出
시선전파	line-of-sight propagation	視線傳播
시스템	system	
시시변동성	interhourly variability	時時變動性
시야	field of view	視野
시야각	viewing angle	視野角
시야각오류	viewing angle error	視野角誤謬
시어	shear	
시어강도	shear strength	-强度

시어벡터	shear vector	
시어변형	shearing deformation	-變形
시어불안정도	shearing instability	-不安定度
시어소용돌이도	shear vorticity	-度
시어응력	shearing stress, shear stress	-應力
시어중력파	shear-gravity wave, shearing gravity wave	-重力波
시어층	shear layer	-層
시어파	shearing wave, shear wave	-波
시어패턴	shear pattern	
시어효과	shear effect	-效果
시에르조	cierzo	
시작일	key day, control day	始作日
시정	visibility	視程
시정감지기	visibility sensor	視程感知器
시정계	visibility meter	視程計
시정관측	observation of visibility	視程觀測
시정목표물	visibility marker, visibility object	視程目標物
시정분광	visible spectrum	視程分光
시정원뿔	cone of visibility	視程圓-
시정장애	obstruction to vision	視程障碍
시정지수	visibility index	視程指數
시정한계	limit of visibility	視程限界
시준의	alidade, collimator	視準儀
시차	parallax	時差
시추공기후학	borehole climatology	試錐孔氣候學
시카고학교(학파)	Chicago School	-學校(學派)
시클로헥산	cyclohexane	
시험비행기상	test flight weather	試驗飛行氣象
식	eclipse	蝕
식년	eclipse year	蝕年
식림	afforestation	植林
식물계절학	plant phenology	植物季節學
식물군	flora	植物群
식물군락	plant community	植物群落
식물기후	phytoclimate, plant climate	植物氣候
식물기후학	phytoclimatology, plant climatology	植物氣候學
식물대	botanical zone	植物帶
식물상해	plant injury	植物傷害
식물습성	plant behavior	植物習性
식물온도	plant temperature	植物溫度
식물이용환경측정법	phytometric method of environmental measurement	植物利用環境測定法
식물재해	plant damage	植物災害
식물지리학	phytogeography	植物地理學

식민지건설	colonization	植民地建設
식민지화	colonization	植民地化
식별	identification	識別
식생	vegetation	植生
식생감시	vegetation monitoring	植生監視
식생기간	vegetation period	植生期間
식생모니터링	vegetation monitoring	植生-
식생 바이오매스 응용	vegetation biomass application	植生-應用
식생반사율	vegetation reflectance	植生反射率
식생생장	vegetative growth	植生生長
식생수관	vegetation canopy	植生樹冠
식수철	planting season	植樹-
식스온도계	Six's thermometer	-溫度計
신기루	mirage	蜃氣樓
신드리아스기	younger Dryas	新-期
신뢰계수	confidence coefficient	信賴係數
신뢰구간	confidence interval, fiducial interval	信賴區間
신뢰대역	confidence band	信賴帶域
신뢰도	confidence degree, degree of confidence, reliability	信賴度
신뢰도	confidence figure	信賴圖
신뢰밴드	confidence band	信賴-
신뢰수준	confidence level	信賴水準
신뢰한계	confidence limit, fiducial limit	信賴限界
신문예보자료	newspaper forecast source	新聞豫報資料
신빙하	Neoglacial	新氷河
신빙하작용	neoglaciation	新氷河作用
신생대기후	Cainozoic climate, Cenozoic climate, Neogene climate	新生代氣候
신선공기	fresh air	新鮮空氣
신선기단	fresh air mass	新鮮氣團
신설사태	fresh snow avalanche	新雪沙汰
신월	new moon	新月
신장	dilatation, stretching	伸長
신장계	dilatometer	伸張計
신장장	dilatation field	伸張場
신장축	dilatation axis	伸張軸
신적설	fresh snow cover, new snow, snowfall	新積雪
신적설깊이	depth of snow fall	新積雪-
신진대사	metabolism	新陳代謝
신체화학	body chemistry	身體化學
신촉광(조도의 단위)	new candle	新燭光
신평형층	new equilibrium layer	新平衡層
신호	signal	信號

신호간섭	signal interference	信號干涉
신호강도	signal strength	信號强度
신호검출도	signal detectability	信號檢出度
신호발생기	signal generator	信號發生機
신호생성기	signal generator	信號生成機
신호속도	signal velocity	信號速度
신호잡음비	signal to noise ratio	信號雜音比
신호적분기술	signal integration technique	信號積分技術
신호조절	signal conditioning	信號調節
신호처리	signal processing	信號處理
신호처리기	signal processor	信號處理機
신호품질지수	signal quality index	信號品質指數
실권운	cirro-film (or thread)	-卷雲
실내공기	indoor air	室內空氣
실내공기분포	indoor air distribution	室內空氣分布
실내기류	indoor air flow	室內氣流
실내기후	cryptoclimate, house climate, indoor climate, kryptoclimate, room climate	室內氣候
실내기후학	cryptoclimatology, kryptoclimatology	室內氣候學
실내대기오염	indoor air pollution	室內大氣汚染
실내미기후	house microclimate	室內微氣候
실내온도	indoor temperature	室內溫度
실리콘검출기	silicon detector	-檢出器
실링	ceiling	
실링경보기	ceiling alarm	-警報器
실링영	ceiling zero	-零
실바람	light air	
실속마하수	stalling mach number	失速-數
실속속도	stalling velocity	失速速度
실시간	real time	實時間
실시간감시	real time monitoring	實時間監視
실시간송신시스템	real time transmission system	實時間送信-
실시간수액관측장비	real time liquid water observation equipment	實時間水液觀測裝備
실시간예측	real time prediction	實時間豫測
실시간전송	immediate transmission	實時間電送
실용단위	practical unit	實用單位
실용비행고도	service ceiling	實用飛行高度
실잠재불안정도	real latent instability	實潛在不安定度
실제고도	actual elevation	實際高度
실제관측시간	actual time of observation	實際觀測時間
실제기온	real air temperature	實際氣溫
실제압력	actual pressure	實際壓力
실제저기압	real of a depression	實際低氣壓
실제증발량	actual evaporation	實際蒸發量

실(유효)증발산(량)	actual evapotranspiration	實(有效)蒸發散(量)
실질도함수	substantial derivative	實質導函數
실질적 변화	individual change	實質的變化
실체파	body wave	實體波
실측	survey	實測
실험	experiment	實驗
실험기상학	experimental meteorology	實驗氣象學
실험실	laboratory	實驗室
실험위성	experimental satellite	實驗衛星
실험유역	experimental basin	實驗流域
실황예보	nowcast, nowcasting	實況豫報
실효반영함수	effective penumbra function	實效半影函數
실효방사온도	effective radiant temperature	實效放射溫度
실효습도	effective humidity	實效濕度
심사도법	gnomonic projection	心射圖法
심수관개	deep flood irrigation	深水灌漑
심수층	hypolimnion	深水層
심수파	deep-water wave	深水波
심야태양	midnight sun	深夜太陽
심층대류	deep convection	深層對流
심층대류지수	deep convective index	深層對流指數
심층순환	Great Ocean Conveyor Belt	深層循環
심프슨공식	Simpson's formula	-公式
심피조미터	sympiesometer	
심해굴착계획	deep-sea drilling project	深海掘鑿計畵
심해수온기록계	bathythermograph	深海水溫記錄計
심해수온기록계격자	bathythermograph grid	深海水溫記錄計格子
심해수온기록계슬라이드	bathythermograph slide	深海水溫記錄計-
심해수온기록계프린트	bathythermograph print	深海水溫記錄計-
심해저평원	abyssal plain	深海底平原
심해파	deep-sea wave	深海波
심혈관질환	cardiovascular disease	心血管疾患
심화(저기압의)	deepening	深化
심화단계	deepening stage	深化段階
심화저기압	deepening cyclone	深化低氣壓
십년	decade	十年
십년주기변동	decadal oscillation	十年週期變動
십분위수	decile	十分位數
십일	dekad	十日
십자무리	cross	十字-
십자햇무리	sun cross, cross	十字-
싱가폴지수	Singapore index	-指數
싱크	sink	
싸락눈	graupel, snow pellet, tapioca snow	

싹쓸바람	hurricane	
쌀쌀한	chilly	
쌀알눈	snow grain	
쌍경위의관측	double theodolite observation	雙經緯儀觀測
쌍경위의기술	double theodolite technique	雙經緯儀技術
쌍곡선	hyperbola	雙曲線
쌍곡점	hyperbolic point	雙曲點
쌍극비디오	bipolar video	雙極
쌍극순환형	bipolar circulation pattern	雙極循環型
쌍극시소	bipolar see-saw	雙極-
쌍극안테나	dipole antenna	雙極-
쌍극유형	bipolar pattern	雙極類型
쌍극자	dipole	雙極子
쌍극자모멘트	dipole moment	雙極子-
쌍금속온도계	bimetallic thermometer	雙金屬溫度計
쌍금속판자기온도계	bimetal thermograph	雙金屬板自記溫度計
쌍봉성	bimodality	雙峰性
쌍봉우리 분포	bimodal distribution	雙峰-分布
쌍봉우리 스펙트럼	bimodal spectrun	雙峰-
쌍봉우리 작은물방울분포	bimodal droplet distribution	雙峰-分布
쌍봉우리 진폭분포	bimodal amplitude distribution	雙峰-振幅分布
쌍상태	bistatic	雙狀態
쌍스펙트럼기술	bispectral technique	雙-技術
쌍안테나 레이더	bistatic radar	雙-
쌍우기	birainy	雙雨期
쌍우기기후	birainy climate	雙雨期氣候
쌍저기압	binary cyclone	雙低氣壓
쌍전자재결합	dielectronic recombination	雙電子再結合
썬포인팅	sun pointing	
썬포토미터	sun-photometer	
썰매다리비행기효과	ski plane effect	-飛行機效果
썰물	ebb, ebb current, falling tide, low tide	
썰물 조류	ebb tide	-潮流
쏠림	streamlining	
쏠림효과	streamlining effect	-效果
쐐기	wedge	
쐐기고기압	wedge high	-高氣壓
쐐기등압선	wedge isobar	-等壓線
쐐기선	wedge line	-線
쐐기토네이도	wedge tornado	
쓰나미	tsunami	
쓰나미경보	tsunami warning	-警報
쓰나미주의보	tsunami watch	-注意報
쓰나미특보	special report for tsunami	-特報

쓰시마해류	Tsushima current	-海流
쓸기	sweep	
쓸기길이	sweep length	
씨눈(물방울 또는 빙정의)	embryo	
씨뿌림	seeding	
씨뿌림가능성	seedability	-可能性
씨뿌림방법	seeding method	-方法
씨뿌림사업자	seeding agent	-事業者
씨뿌림수순	seeding subroutine	-手順
씨뿌림시기	seed time	-時期
씨뿌림율	seeding rate	-率
씨에이브이티(CAVT)표	CAVT table	-表
씻어내림	downwash	

ㅇ

한글	영문	한자
아간빙기	glacial interstadial, interstadial	亞間氷期
아건조	subarid	亞乾燥
아격자규모과정	subgrid-scale process	亞格子規模過程
아경도공기흐름	agradient airflow	亞傾度空氣-
아경도풍	subgradient wind	亞傾度風
아굴라스류	Agulhas stream	-流
아굴라스해류	Agulhas Current	-海流
아굴절	subrefraction	亞屈折
아궁화산분화	Agung eruption	-火山噴火
아(북)극기후	subarctic climate	亞(北)極氣候
아(북)극대	subarctic zone	亞(北)極帶
아(북)극해류	subarcric current	亞(北)極海流
아기후구	climatic subdivision	亞氣候區
아나디리해류	Anadyr Current	-海流
아나사지인	Anasazi people	-人
아날로그	analog(ue)	
아날로그기후모형	analog climate model	-氣候模型
아날로그데이터 전송	analogue data transmission	-電送
아날로그-디지털 변환	analog-to-digital conversion	-變換
아날로그디지털변환기	analog digital converter	-變換器
아날로그모형	analog model	-模型
아날로그방법	analog method	-方法
아날로그전산기	analog computer	-電算機
아날로그 프리필터링	analog prefiltering	
아네로이드	aneroid	
아네로이드공합	aneroid capsule	-空盒
아네로이드기압계	aneroid barometer	-氣壓計
아네로이드기압계 뒤짐현상	creeping of aneroid barometer	-氣壓計-現象
아네로이드기압계 지연	lag of aneroid barometer	-遲延

아네로이드기압기록	aneroidogram	-氣壓記錄
아네로이드소자	aneroid element	-素子
아네로이드자기기압계	aneroid barograph, aneroidograph	-自記氣壓計
아네모시네모그래프	anemo-cinemograph	
아라고거리	Arago distance	-距離
아라고나이트	aragonite	
아라고점	Arago point	-點
아라고중립점	Arago's neutral point	-中立點
아라비아바다	Arabian sea	
아라비아사막	Arabian desert	-砂漠
아라비아 숫자	cipher, cypher	-數字
아라비안동안류	East Arabian current	-東岸流
아라카와자코비안	Arakawa Jacobian	
아랄해	Aral Sea	-海
아래끝	lower limit	
아래무리해	undersun	
아래신기루	inferior mirage	-蜃氣樓
아레니우스방정식	Arrhenius expression	-方程式
아레니우스식	Arrhenius equation	-式
아령저기압	dumb-bell depression	啞鈴低氣壓
아로요	arroyo	
아르고프로그램	Argo Program	
아르고플로트	Argo float	
아르곤	argon	
아르키메데스부력	Archimedean buoyant force	-浮力
아르키메데스원리	Archimedes's principle	-原理
아마존강	Amazon River	-江
아마존강유역	Amazonia area	-江流域
아마존분지	Amazon basin	-盆地
아마존삼림	Amazon forest	-森林
아마추어기상관측소	amateur weather station	-氣象觀測所
아마추어예보	amateur forecast	-豫報
아메리플럭스	Ameri Flux	
아몽통법칙	Amontons's law	-法則
아보가드로가설	Avogadro's hypothesis	-假說
아보가드로법칙	Avogadro's law	-法則
아보가드로상수	Avogadro constant	-常數
아보가드로수	Avogadro's number	-數
아브레	avre	
아브롤로스스콜	Abrolhos squall, abroholos	
아산화질소	nitrous oxide	亞酸化窒素
아성층권	substratosphere	亞成層圈
아세토니트릴	acetonitrile	
아세톤	aceton	

아세트산	acetic acid	
아세트알데히드	acetaldehyde	
아세틸렌	acetylene	
아스만건습계	Assmann psychrometer	-乾濕計
아스만통풍건습계	Assmann ventilated psychrometer, Assmann's aspiration psychrometer	-通風乾濕計
아스완댐	Aswan Dam	
아스케시안학회	Askesian Society	-學會
아습윤(의)	subhumid	亞濕潤
아습윤기후	subhumid climate	亞濕潤氣候
아시아	Asia	亞細亞
아시아갈색연무	Asian Brown Cloud	亞細亞褐色煙霧
아시아몬순	Asian monsoon	亞細亞-
아시아한파	Asian cold wave	亞細亞寒波
아시아홍수	Asia flood	亞細亞洪水
아얄라스	ayalas	
아열대	subtropical belt, subtropics	亞熱帶
아열대(의)	semitropical, subtropical	亞熱帶
아열대고기압	horse latitude high, oceanic high, subtropical anticyclone, subtropical high	亞熱帶高氣壓
아열대고압대	subtropical high-pressure belt, subtropical high-pressure zone	亞熱帶高壓帶
아열대기단	subtropical air mass	亞熱帶氣團
아열대기압마루	subtropical ridge	亞熱帶氣壓-
아열대기후	subtropical climate	亞熱帶氣候
아열대무역풍역전	subtropical trade-wind inversion	亞熱帶貿易風逆轉
아열대무풍대	subtropical calm	亞熱帶無風帶
아열대세포	subtropical cell	亞熱帶細胞
아열대스톰	subtropical storm	亞熱帶-
아열대저기압	semitropical cyclone, subtropical cyclone	亞熱帶低氣壓
아열대저압부	subtropical depression	亞熱帶低壓部
아열대전선	subtropical front	亞熱帶前線
아열대제트류	subtropical jet stream	亞熱帶-流
아열대지역	subtropical zone	亞熱帶地域
아열대편동풍(대)	subtropical easterlies	亞熱帶偏東風(帶)
아열대편동풍지수	subtropical easterlies index	亞熱帶偏東風指數
아열대편서풍(대)	subtropical westerlies	亞熱帶偏西風(帶)
아열대폭풍우	subtropical storm	亞熱帶暴風雨
아염소산위축병	chlorotic dwarf disease	亞鹽素酸萎縮病
아온난기	interstadial	亞溫暖期
아웃라이어	outlier	
아음속(의)	subsonic	亞音速
아이스세인트	Ice Saints	
아이스하우스기후	Icehouse climate	-氣候

아이슬랜드 저기압	Icelandic low	-低氣壓
아이코날	eikonal	
아인슈타인 상대성이론	Einstein's theory of relativity	-相對性理論
아인슈타인표기법	Einstein notation	-表記法
아인슈타인합계약정	Einstein summation convention	-合計約定
아임계불안정	subcritical instability	亞臨界不安定
아적도대	subequatorial belt	亞赤道帶
아전선구름	subfrontal cloud	亞前線-
아조레스고기압	Azores anticyclone, Azores high	-高氣壓
아조레스해류	Azores current	-海流
아종관	subsynoptic	亞綜觀
아종관규모	subsynoptic scale	亞綜觀規模
아지균풍	subgeostrophic wind	亞地均風
아지랑이	atmospheric boil, atmospheric shimmer, heat shimmer, laurence, optical haze, shimmer, terrestrial scintillation	
아질산	nitrous acid	亞窒酸
아치구름	arch cloud, arcus	
아치적란운	cumulonimbus arcus	-積亂雲
아치적운	cumulus arcus	-積雲
아치형스콜	arched squall	-型-
아침고요	morning calm	
아침놀	morning glow	
아침위성	morning satellite	-衛星
아침조석	morning tide	-潮汐
아타카마사막	Atacama Desert	-砂漠
아토스 치내리바람	Athos fall wind	
아펠리오테스	Apheliotes	
아포스틸브	apostilb	
아표준굴절	substandard refraction	亞標準屈折
아표준전파	substandard propagation	亞標準傳播
아프가네츠	Afghanets	
아프리카가뭄	Africa drought	
아프리카기상개발응용센터	Africa Center of Meteorological Applications for Development	-氣象開發應用-
아프리카몬순다분야분석	African Monsoon Multidisciplinary Analysis	-多分野分析
아프리카제트	African jet	
아프리카파	African wave	-波
아프리카한발	Africa drought	-旱魃
아프토니아간빙기	Aftonian interglacial	-間氷期
아한대	subfrigid zone, subpolar belt, subpolar zone	亞寒帶
아한대(의)	subpolar	亞寒帶
아한대고기압	subpolar anticyclone, subpolar high	亞寒帶高氣壓
아한대공기덩이	subpolar cap	亞寒帶空氣-

아한대기단	subpolar air mass	亞寒帶氣團
아한대기후	subpolar climate	亞寒帶氣候
아한대빙하	subpolar glacier	亞寒帶氷河
아한대저기압	subpolar low	亞寒帶低氣壓
아한대저압대	subpolar low pressure belt	亞寒帶低壓帶
아한대편서풍(대)	subpolar westerlies	亞寒帶偏西風(帶)
아홉등 풍향풍속지시기	nine-light indicator	-燈風向風速指示器
아황산가스	sulfur dioxide, sulphur dioxide	亞黃酸-
악기류	rough air	惡氣流
악기상	severe weather, weather hazard	惡氣象
악성착빙	severe icing	惡性着氷
악취	malodor, odor	惡臭
악취문턱값	odor threshold	惡臭-
악취역치	odor threshold	惡臭閾値
악폭풍	barber	惡暴風
안개	fog	
안개강수	fog precipitation	-降水
안개걷힘	fog clearing	
안개결정	fog crystal	-結晶
안개계급	fog scale	-階級
안개계수	coefficient of haze	-係數
안개계절	dense fog season	-季節
안개관측시스템	fog observation system	-觀測-
안개구역	fog region	-區域
안개구조	fog structure	-構造
안개둑	fog bank	
안개막이숲	fog prevention forest	
안개모양(구름)	nebulosus	-模樣-
안개모양구름	neb	-模樣-
안개모형	fog model	-模型
안개무지개	fogbow	
안개미물리학	fog microphysics	-微物理學
안개바다	sea of fog	
안개바람	fog wind, nebewind	
안개방울	fog drop	
안개비	fog rain	
안개빈도수	fog frequency	-頻度數
안개소나기	fog shower	
안개소산	fog dispersion, fog dissipation	-消散
안개수평선	fog horizon	-水平線
안개숲	fog forest	
안개스트리머	fog streamer	
안개신호	fog signal	-信號
안개알갱이	fog droplet	

안개인공소산	fog dispersal	-人工消散
안개인공소산작업	fog dispersal operation	-人工消散作業
안개일	fog day	-日
안개적하	fog drip	-滴下
안개점	fog point	-點
안개조각	fog patch	
안데스광	Andes glow	-光
안데스번개	Andes lightning	
안데스빛	Andes light	
안데스산맥	Andes mountains	-山脈
안장부	col	鞍裝部
안장저압부	saddle	鞍裝低壓部
안장점	saddle point, col	鞍裝點
안장형갠구역	saddle-back	鞍裝形-區域
안전재배계절	safety cultivation season	安全栽培季節
안전재배기간	safety period of cultivation	安全栽培期間
안정공기	stable air	安定空氣
안정국부발진기	stable local oscillator	安定局部發振器
안정기단	stable air mass	安定氣團
안정기층	stable air layer	安定氣層
안정도	degree of stability, stability	安定度
안정도도	stability chart	安定度圖
안정도도표	stability diagram	安定度圖表
안정도매개변수	stability parameter	安定度媒介變數
안정도조건	stability condition	安定度條件
안정도지수	stability index	安定度指數
안정동위원소분석	stable isotope analysis	安定同位元素分析
안정등급	stability class	安定等級
안정력	stabilizing force	安定力
안정비	stability ratio	安定比
안정상태	stable state	安定狀態
안정성층	stable stratification	安定成層
안정운동	stable motion	安定運動
안정인자	stabilizing factor	安定因子
안정진동	stable oscillation	安定振動
안정층	stable layer	安定層
안정파	stable wave	安定波
안정평형	stable equilibrium	安定平衡
안정형	stable type	安定型
안정화	stabilization	安定化
안정화시스템	stabilization system	安定化-
안정흐름	stable flow	安定-
안테나	antenna	
안테나반사경	antenna reflector	-反射鏡

안테나반사체	antenna reflector	-反射體
안테나온도	antenna temperature	-溫度
안테나이득	antenna gain	-利得
안테나패턴	antenna pattern	
안테나피드	antenna feed	
안테나한계	antenna limit	-限界
안티로그	antilog	
안틸해류	Antilles current	-海流
알갱이눈	granular snow	
알고리즘	algorithm	
알고리즘언어	algorithmic language	-言語
알데히드	aldehyde	
알라드법칙	Allard's law	-法則
알래스카스트림	Alaskan Stream	
알래스카해류	Alaska current	-海流
알레뢰드기	Allerød period	-期
알레르기	allergy	
알렉산더의 검은 띠	Alexander's dark band	
알류샨저기압	Aleutian low	-低氣壓
알류샨해류	Aleutian current	-海流
알리솝기후구분	Alisov's classification of climate	-氣候區分
알메리아-오랑 전선	Almeria-Oran front	-前線
알메리아-오랑 효과	Almeria-Oran effect	-效果
알베도	albedo	
알베도값	albedo value	
알베도계	albedometer	-計
알베도기록계	albedograph	-記錄計
알베도-온도 되먹임	albedo-temperature feedback	-溫度-
알제리해류	Algerian current	-海流
알칼리도	alkalinity	-度
알케논	alkenone	
알케인	alkane	
알켄	alkane	
알코올	alcohol	
알코올바이오매스연료	alcohol biomass fuel	-燃料
알코올온도계	alcohol-in-glass thermometer,	-溫度計
	alcohol thermometer, spirit thermometer	
알킨	alkyne	
알킬납	lead alkyl	
알킬퍼록시라디칼	alkylperoxy radical	
알파붕괴	alpha decay	-崩壞
알파선	alpha-ray, α-ray	-線
알파입자	alpha-particle	-粒子
알프벤파	Alfvén wave	-波

암모늄	ammonium	
암모늄황산염	ammonium sulphate	-黃酸鹽
암모니아	ammonia	
암모니아생성	ammonification	-生成
암모니아이온	ammonium ion	
암석권	lithosphere	岩石圈
암석빙하	rock glacier	岩石氷河
암설	detritus	岩屑
암시적 방안	implicit scheme	暗示的方案
암시평활	implicit smoothing	暗示平滑
암윈드	almwind	
암페어	ampere	
암호	cipher, cypher	暗號
암호흡	dark respiration	暗呼吸
암화	darkening	暗化
압력	compression, pressure	壓力
압력경도	pressure gradient	壓力傾度
압력경향특성	characteristic of the pressure tendency	壓力傾向特性
압력계	manometer, pressure gauge	壓力計
압력계공합	pressure capsule	壓力計空盒
압력계수	pressure coefficient	壓力係數
압력고도	pressure altitude, pressure height	壓力高度
압력고도계	pressure altimeter	壓力高度計
압력관풍속계	pressure tube anemometer	壓力管風速計
압력관풍속기록계	anemo-biagraph, pressure tube anemograph	壓力管風速記錄計
압력넓어짐	pressure broadening	壓力-
압력녹음	pressure melting	壓力-
압력대수층	confined aquifer	壓力帶水層
압력변화	pressure change	壓力變化
압력분포	pressure distribution	壓力分布
압력얼음	pressure ice	壓力-
압력장	field of pressure	壓力場
압력중심	center of pressure, pressure center	壓力中心
압력지하수	confined groundwater	壓力地下水
압력편차	pressure anomaly	壓力偏差
압력확산	baro-diffusion	壓力擴散
압력확산계수	baro-diffusion coefficient	壓力擴散係數
압력확산비	baro-diffusion ratio	壓力擴散比
압밀	consolidation	壓密
압성	piezotropy	壓性
압성계수	coefficient of piezotropy, piezo coefficient	壓性係數
압성방정식	equation of piezotropy, piezometric equation	壓性方程式
압제적인	oppressive	壓制的
압축	compaction, compression	壓縮

압축계수	coefficient of compressibility, compressibility coefficient	壓縮係數
압축눈	penitent snow (columnar shape of compacted snow)	壓縮-
압축변형	compressional creep	壓縮變形
압축비	compression ratio	壓縮比
압축성	compressibility	壓縮性
압축성(의)	compressible	壓縮性
압축성유체	compressible fluid	壓縮性流體
압축얼음	penitent ice (columnar shape of compacted ice)	壓縮-
압축온난화	compressional warming	壓縮溫暖化
압축점화엔진	compression ignition engine	壓縮點火-
압축파	compressibility wave, compression wave	壓縮波
앙각	elevation angle, vertical angle	仰角
앙각계	clinometer	仰角計
앙각오차	elevation angle error	仰角誤差
앙각위치지시기	elevation position indicator	仰角位置指示器
앙골라-벵겔라 전선	Angola-Benguela front	-前線
앙골라해류	Angola current	-海流
앙상블	ensemble	
앙상블수문예측	ensemble hydrologic forecasting	-水文豫測
앙상블예보	ensemble forecast, ensemble forecasting, ensemble prediction	-豫報
앙상블예측계	ensemble prediction system	-豫測系
앙상블예측모형	ensemble prediction model	-豫測模型
앙상블예측시스템	ensemble prediction system	-豫測-
앙상블자료동화	ensemble data assimilation	-資料同化
앙상블칼만필터	ensemble Kalman filter	
앙상블평균	ensemble mean	-平均
앞바다	offing	
앞시정	forward visibility	-視程
애뉼러스(환대)실험	annulus experiment	-(環帶)實驗
(잉글랜드)애시처치실험	Ashchurch experiment	-實驗
애팔래치아산맥	Appalachian mountains	-山脈
애플턴층	Appleton layer	-層
애플턴-하트리 방정식	Appleton-Hartree equation	-方程式
액상	liquid state	液相
액수경로	liquid water path	液水經路
액체	liquid	液體
액체거품추적자	liquid-bubble tracer	液體-追跡子
액체공기	liquid air	液體空氣
액체기둥기압계	liquid-column barometer	液體-氣壓計
액체비중계	hydrometer	液體比重計
액체상	liquid phase	液體相

액체수	liquid water	液體水
액체수경로	liquid water path	液體水徑路
액체수온위	liquid water potential temperature	液體水溫位
액체수함량	liquid water content	液體水含量
액체수혼합비	liquid water mixing ratio	液體水混合比
액체온도계	liquid thermometer	液體溫度計
액체질소	liquid nitrogen	液體窒素
액화	liquefaction	液化
앤드히스	andhis	
앨버타저기압	Alberta cyclone, Alberta low	-低氣壓
앨버타클리퍼	Alberta clipper	
앰블선도	Amble diagram	-線圖
앵글리아빙하	Anglian glacial	-氷河
야간	night	夜間
야간경계층	nocturnal boundary layer	夜間境界層
야간경계층깊이	nocturnal boundary layer depth	夜間境界層-
야간경사류	nocturnal drainage flow	夜間傾斜流
야간냉각	nocturnal cooling	夜間冷却
야간뇌우	nocturnal thunderstorm	夜間雷雨
야간능선꼭대기 제트	nocturnal ridge-top jet	夜間-
야간대기경계층	nocturnal atmospheric boundary layer	夜間大氣境界層
야간도시습윤고도	nocturnal urban moisture level	夜間都市濕潤高度
야간복사	nocturnal radiation	夜間輻射
야간시계	night visual range	夜間視界
야간시정	night visibility	夜間視程
야간안정층	nocturnal stable layer	夜間安定層
야간역전	nocturnal inversion	夜間逆轉
야간제트	night jet, nocturnal jet	夜間-
야간최저온도	nocturnal minimum temperature	夜間最低溫度
야곱사다리	Jacob's ladder	
야광	nightglow	夜光
야광스펙트럼	night air-glow spectrum, nightglow spectrum	夜光-
야광운	luminous night cloud, noctilucent cloud	夜光雲
야기안테나	Yagi antenna	
야나이-마루야마 파	Yanai-Maruyama wave	-波
야외실험	field experiment	野外實驗
야장	log book, field book, field note	野帳
야전숙영기상	battle field billet weather	野戰宿營氣象
야외작전기상	field operation weather	野外作戰氣象
야코비안	Jacobian	
약결합수	loosely bound water	弱結合水
약뇌우	slight thunderstorm	弱雷雨
약도	outline map	略圖
약반향역	weak echo region	弱反響域

약식선박부호	abbreviated ship code	略式船舶符號
약에코역	weak echo region	弱-域
약한 난류	cobblestone turbulence	弱-亂流
약한 동서순환	low zonal circulation	弱-東西循環
약한 바람	slight air, slight breeze	弱-
약한 보라	borino	弱-
약한 비	light rain	弱-
약한 상호작용	weak interaction	弱-相互作用
약한 선근사	weak-line approximation	弱-線近似
약한 스콜	light squall	弱-
얇은 막	film	-膜
얇은 층	film	-層
양	quantity	量
양(의)	positive	陽
양(의) 구름-지상 뇌전	positive cloud to ground lightning	陽-地上雷電
양극	positive pole	陽極
양(의) 기울어짐기압골	positive-tilt trough	陽-氣壓-
양되먹임	positive feedback	陽-
양되먹임기작	positive feedback mechanism	陽-機作
양되먹임시스템	positive feedback system	陽-
양방향	positive direction	陽方向
양방향감쇠	two-way attenuation	兩方向減衰
양방향반사도분포함수	bi-directional reflectance distribution function	兩方向反射度分布函數
양방향반사율	bi-directional reflectance	兩方向反射率
양방향반사인자	bi-directional reflectance factor	兩方向反射因子
양방향반사함수	bi-directional reflection function	兩方向反射函數
양방향상호작용	two-way interaction	兩方向相互作用
양방향풍향계	bi-directional vane, bi-directional wind vane	兩方向風向計
양부력	positive buoyancy	陽浮力
양부호	positive sign	陽符號
양분충격장치	dichotomous impactor	兩分衝擊裝置
양서류	amphibia	兩棲類
양선형보간법	bilinear interpolation method	兩線形補間法
양성자	proton	陽性子
양성자권	protonosphere	陽性子圈
양(의) 기울어짐기압골	positive-tilt trough	陽-氣壓-
양식 내부구조	internal structure of regime	樣式內部構造
양영역	positive area	陽領域
양유리수	positive rational number	陽有理數
양이온	cation, positive ion	陽-
양이온치환용량	cation exchange capacity	陽-置換容量
양자론	quantum theory	量子論
양장습도계	sheepgut hydrometer	羊腸濕度計
양적예보	quantitative forecast	量的豫報

양전기비	positive rain	陽電氣-
양전하	positive charge	陽電荷
양정수	positive integer	陽整數
양지바른	sunny	陽地-
양지식물	sun plant	陽地植物
양쯔강고기압	Yangtze-River high	-江高氣壓
양쯔강기단	Yangtze-River air mass	-江氣團
양초모양 얼음	candle ice	-模樣-
양축	positive axis	陽軸
양털구름	fleecy cloud	
양털구름하늘	fleecy sky	
양털무늬(태양)	flocculi	
양파껍질벗기기	onion peeling	
얕은 근사	shallow assumption	-近似
얕은 습윤대	shallow moist zone	-濕潤帶
얕은 안개	shallow fog	
얕은 저기압	shallow low	-低氣壓
어는 강수	freezing precipitation	-降水
어는 계절	freezing season	-季節
어는 비	freezing rain	
어는 비주의보	freezing rain advisory	-注意報
어는 안개	freezing fog	
어는 온도	freezing temperature	-溫度
어는 이슬비	freezing drizzle	
어는 이슬비주의보	freezing drizzle advisory	-注意報
어는점	freezing point	-點
어는점 내림	freezing point depression	-點-
어는 점선	freezing point line	-點線
어두운 번개	dark lightning	
어두운 복사	dark radiation	-輻射
어두운 복사선	dark ray	-輻射線
어두운 빛	dark light	
어둠띠	dark band, dark segment	
어둠선	dark line	-線
어둠적응	dark adaptation	-適應
어란석	oolite, oolith	魚卵石
어린 양의 풍설	lamb blast	-羊-風雪
어림(수의)	rounding	-(數-)
어림수	round number	-數
어미구름	mother-cloud	
어미소용돌이	maternal vortex	
어업기상	meteorology for fisheries	漁業氣象
억수	rattler	
억제	damping	抑制

언 강수	frozen precipitation	-降水
언 눈표피	frozen snow crust	-表皮
언덕얼음	hummocked ice	
언면	frost table	-面
언 바다(빙해) 시험	frozen sea test	-(氷海)試驗
언 안개	frozen fog	
언비	frozen rain, sleet	
언 언덕	frost blister	
언 이슬	frozen dew, white dew	
언층	frost zone	-層
얼룩(레이더 에코의)	blob	
얼룩(구름)	maculosus	
얼룩층운	stratus maculosus	-層雲
얼림열	heat of freezing	-熱
얼음	freezing, ice	
얼음각기둥	ice prism	-角-
얼음갈라짐	ice run	
얼음결정	ice crystal	-結晶
얼음결정구름	ice crystal cloud	-結晶-
얼음결정안개	ice crystal fog	-結晶-
얼음결정연무	ice crystal haze	-結晶煙霧
얼음결정이론	ice crystal theory	-結晶理論
얼음결정핵	ice crystal nucleus	-結晶核
얼음결정효과	ice crystal effect	-結晶效果
얼음골짜기	ice gorge	
얼음과포화	supersaturation with respect to ice	-過飽和
얼음교란	cryoturbation	-攪亂
얼음구름	ice cloud	
얼음구멍	cryoconite hole, puddle	
얼음굴	ice cave	
얼음극	ice pole	-極
얼음기둥	ice column	
얼음기슭	ice foot	
얼음깃털	ice feather	
얼음껍질	ice rind	
얼음꽃	ice flower	
얼음덮개	ice cover	
얼음띠	ice band, ice belt, ice strip	
얼음리본	ice ribbon	
얼음마찰이론	ice friction theory	-摩擦理論
얼음만	ice bay, ice bight	-灣
얼음면도랑	rasvodye	-面-
얼음면포화	ice saturation	-面飽和
얼음모자	ice cap	-帽子

얼음바늘	ice needle	
얼음바람	icy wind	
얼음벌판	ice field	
얼음보라	ice storm	
얼음비계	ice fat	
얼음빛	ice blink	
얼음빛하늘	ice sky	
얼음생성조건	ice-formation condition	-生成條件
얼음생성핵	ice-forming nucleus	-生成核
얼음선반	ice shelf	
얼음섬	ice island	
얼음속기록기후변화	climate change recorded in ice	-記錄氣候變化
얼음순수포화(수)증기압	saturation vapo(u)r pressure in the pure phase with respect to ice	-純粹飽和(水)蒸氣壓
얼음습윤공기포화(수)증기압	saturation vapo(u)r pressure of moist air with respect to ice	-濕潤空氣飽和(水)蒸氣壓
얼음싸라기	grain of ice, ice pellet	
얼음쌓임	ice deposit, ice jam	
얼음안개	ice fog, pogonip	
얼음알갱이	droxtal(= drop + crystal), ice particle	
얼음언덕	hummock, ice blister, ice mound, icing mound	
얼음열량계	ice calorimeter	-熱量計
얼음입자	ice particle	-粒子
얼음입자 부착	aggregate of ice particle	-粒子附着
얼음전선	ice front	-前線
얼음점	ice point	-點
얼음정찰	ice reconnaissance	-偵察
얼음제거	ice clearing	-除去
얼음조각	ice splinter	
얼음조각뭉치	brash	
얼음줄무늬	ice fringe	
얼음증식	ice multiplication	-增殖
얼음지도	ice atlas	-地圖
얼음침	spicule	-針
얼음침식	cryoplanation	-浸蝕
얼음팩	ice pack, pack ice	
얼음평상	ice sheet, ice field	-平床
얼음포화혼합비	saturation mixing ratio with respect to ice	-飽和混合比
얼음폭포	ice cascade, ice fall	-瀑布
얼음표층	ice crust	-表層
얼음피복	ice cover	-被覆
얼음한계	ice limit	-限界
얼음혀	ice tongue	
얼음홍수	flooded ice	-洪水

얼음흐름	ice stream	
얽힌 구름	intortus	
엄동	severe winter	嚴冬
엄폐	occultation	掩蔽
엄폐침적	occult deposition	掩蔽沈積
엄한	killing freeze	嚴寒
업무예보	operational prediction	業務豫報
업무지시	operation directive	業務指示
엇갈림격자	staggered grid	-格子
엇갈림방식	stagger scheme	-方式
엉킨 권운	cirrus intortus	-卷雲
에그넬법칙	Egnell's law	-法則
에너지경로	energy pathway	-經路
에너지공급	energy supply	-供給
에너지교환	energy exchange	-交換
에너지균형	energy balance	-均衡
에너지균형모형	energy balance model	-均衡模型
에너지균형방법	energy balance method	-均衡方法
에너지기술	energy technology	-技術
에너지다단분산	cascade of energy	-多段分散
에너지론	energetics	-論
에너지밀도	energy density	-密度
에너지방정식	energy equation	-方程式
에너지방출	energy release	-放出
에너지변환	energy transformation	-變換
에너지보존	conservation of energy, energy conservation	-保存
에너지보존법칙	law of energy conservation	-保存法則
에너지보존원리	principle of conservation of energy	-保存原理
에너지분광기	energy dispersive spectrometer	-分光器
에너지분배	partition of energy	-分配
에너지분산	energy dispersion	-分散
에너지상태	energy state	-狀態
에너지선도	energy diagram	-線圖
에너지소산	dissipation of energy, energy dissipation	-消散
에너지소산율	dissipation rate of energy	-消散率
에너지소산자	energy dissipator	-消散子
에너지속	energy flux	-束
에너지수송	energy transport	-輸送
에너지수송모형	energy model of transport	-輸送模型
에너지수요	energy demand	-需要
에너지수지	energy balance, energy budget	-收支
에너지스펙트럼	energy spectrum	
에너지원	energy source	-源
에너지자원	energy resource	-資源

에너지저장	energy storage	-貯藏
에너지전달	energy transfer	-傳達
에너지전달률	energy transfer rate	-傳達率
에너지전달함수	energy transfer function	-傳達函數
에너지전환	energy conversion	-轉換
에너지준위	energy level	-準位
에너지준위도	energy level diagram	-準位圖
에너지캐스케이드	energy cascade	
에너지퇴화	degradation of energy	-退化
에너지플럭스	energy flux	
에너지학방법	energetics method	-學方法
에너지함유범위	energy-containing range	-含有範圍
에너지형태	energy form	-形態
에너지효율성	energy efficiency	-效率性
에너지흐름	energy flow	
에디	eddy	
에디캐스케이드	eddy cascade	
에르고드 가설	Ergodic hypothesis	-假說
에르고드 과정	Ergodic process	-過程
에르그	erg	
에르텔-로스비 불변량	Ertel-Rossby invariant	-不變量
에르텔 보존정리	Ertel's conservation theorem	-保存定理
에르텔 소용돌이도정리	Ertel vorticity theorem	-度定理
에르텔 소용돌이불변량	Ertel's vortex invariant	-不變量
에르텔 소용돌이정리	Ertel's vortex theorem	-定理
에르텔위치소용돌이도	Ertel potential vorticity	-位置-度
에마그램	Emagram	
에미안간빙기	Emian (Sangamonian) interglacial	-間氷期
에반스통신연구소	Evans Signal Laboratory	-通信研究所
에버셰드 효과	Evershed effect	-效果
에버트이온계수기	Ebert ion-counter	-計數器
에버트주사분광기	Ebert scanning spectrometer	-走査分光器
에버트-파스티 단색화장치	Ebert-Fastie monochromator	-單色化裝置
에보그램	evogram	
에스테그램	estegram	
에스토크 방정식	Estoque equation	-方程式
에스토크 이류선	Estoque advection line	-移流線
에스페란자 기지	Esperanza (Argentine station)	-基地
에스피-쾨펜 이론	Espy-Köppen theory	-理論
에어로그램	aerogram	
에어로알레르겐	aeroallergen	
에어로졸	aerosol, carbonaceous aerosol	
에어로졸간접효과	aerosol indirect effect	-間接效果
에어로졸광학깊이	aerosol optical depth	-光學-

에어로졸광학두께	aerosol optical depth	-光學-
에어로졸반직접효과	aerosol semi-direct effect	-半直接效果
에어로졸복사강제력	aerosol radiative forcing	-輻射強制力
에어로졸복사강제력효율	aerosol radiative forcing efficiency	-輻射強制力效率
에어로졸비대칭인자	aerosol asymmetry factor	-非對稱因子
에어로졸위상함수	aerosol phase function	-位相函數
에어로졸일차산란알베도	aerosol single-scattering albedo	-一次散亂-
에어로졸전기	aerosol electricity	-電氣
에어로졸지수	aerosol index	-指數
에어로졸직접복사효과	aerosol direct radiative effect	-直接輻射效果
에어로졸직접효과	aerosol direct effect	-直接效果
에어로졸질량소멸효율	aerosol mass extinction efficiency	-質量消滅效率
에어로졸크기분포	aerosol size distribution	-分布
에어로졸하중	aerosol loading	-荷重
에어리원판	Airy disk	-圓板
에어리함수	Airy function	-函數
에어포켓	air pocket	
에오세(신생대)기후최적	Eocene climatic optimum	-氣候最適
에오세말한랭화	end-Eocene cooling	-末寒冷化
에올	EOLE	
에이 디스플레이	A-display	
에이커풋	acre-foot	
에이킨계수기	Aitken counter	-計數器
에이킨먼지계수기	Aitken dust counter	-計數器
에이킨입자	Aitken particle	-粒子
에이킨핵	Aitken nucleus	-核
에이킨핵계수기	Aitken nucleus counter	-核計數器
에인절에코	angel echo	
에일리어싱	aliasing	
에카르트계수	Eckart coefficient	-係數
에코	blip, echo	
에코라디오존데	echo radiosonde	
에코상자	echo box	-箱子
에코선	line of echo	-線
에코세기	echo power	
에코신호	echo signal	-信號
에코전력기울기	echo power gradient	-電力-
에코전력비	echo power ratio	-電力比
에코정상	echo top	-頂上
에코정상고도	echo top height	-頂上高度
에코존데	echosonde	
에코주파수	echo frequency	-周波數
에코지역	echo area	-地域
에코진폭	echo amplitude	-振幅

에코펄스	echo pulse	
에크네피아스	ecnephias	
에크만경계층	Ekman boundary layer	-境界層
에크만깊이	Ekman depth	
에크만나선	Ekman spiral	-螺線
에크만모형	Ekman model	-模型
에크만방정식	Ekman equation	-方程式
에크만부피수송	Ekman volume transport	-輪送
에크만수	Ekman number	-數
에크만수송	Ekman transport	-輪送
에크만전향	Ekman turning	-轉向
에크만층	Ekman layer	-層
에크만층깊이	Ekman layer depth	-層-
에크만층불안정도	Ekman layer instability	-層不安定度
에크만펌핑	Ekman pumping	
에크만펌핑속도	Ekman pumping velocity	-速度
에크만평형	Ekman balance	-平衡
에크만힘	Ekman force	
에타놀	ethanol	
에타모형	Eta model	-模型
에탄	ethane	
에트나화산	Etna	-火山
에틸렌	ethylene	
에펠탑온도파문제	Eiffel Tower temperature wave problem	-塔溫度波問題
에플리 직달일사계	Eppley pyrheliometer	-直達日射計
엑스너함수	Exner function	-函數
엑스선분리	x-ray burst	-線分離
엑스선폭발	x-ray burst	-線爆發
엑스선회절	x-ray diffraction	-廻折
엔스트로피	enstrophy	
엔스트로피플럭스	enstrophy flux	
엔에이치시방법	NHC method	-方法
엔진착빙	engine icing	-着氷
엔탈피	enthalpy	
엔탈피플럭스	enthalpy flux	
엔트로피	entropy	
엔트로피상수	entropy constant	-常數
엔트로피생산	entropy production	-生産
엔트로피온위	entropy potential temperature	-溫位
엔트로피죽음	entropy death	
엔트로피증가원리	principle of increase of entropy	-增加原理
엔트로피펌프	entropy pump	
엘니뇨	El Niño	
엘니뇨남방진동	El Niño Southern Oscillation	-南方振動

엘니뇨-남방진동 진단토의	ENSO diagnostic discussion	-南方振動診斷討議
엘니뇨현상	El Niño event	-現象
엘류세라섬	Eleuthera	
엘모의 불	Elmo's fire, St. Elmo's fire	
엘사서복사도	Elsasser's radiation chart	-輻射圖
엘사서차트	Elsasser chart	
엘치촌화산	El Chichon volcano	-火山
엘치촌화산분화	El Chichon eruption	-火山噴火
엠간빙기	Eemian interglacial	-間氷期
여각	complementary angle	餘角
여과	filtration	濾過
여과기	filter	濾過器
여과복사계	filter radiometer	濾過輻射計
여광기	light filter	濾光器
여름	summer	
여름계절풍	summer monsoon	-季節風
여름극잔존얼음	dynamic ice	-極殘存-
여름날	summer day	
여름반구	summer hemisphere	-半球
여름반년	summer half year	-半年
여름작물	summer crop	-作物
여름잠	aestivation, estivation	
여름평균우량등치선	isothermobrose	-平均雨量等値線
여압	pressurization	與壓
여위도	co-latitude	餘緯度
여파	filtering	濾波
여파근사	filtering approximation	濾波近似
여파기	filter	濾波器
여파모형	filtered model	濾波模型
여파방정식	filter equation, filtered equation	濾波方程式
여파순압모형	filtered barotropic model	濾波順壓模型
여파조건	filter condition	濾波條件
여파지	filter paper	濾波紙
여파함수	filter function	濾波函數
역	converse, counter	逆
역경도속	upgradient flux	逆傾度束
역경도플럭스	upgradient flux	逆傾度-
역년	calendar year	曆年
역라그랑지	inverse Lagrangian	逆-
역류	back flow, reversing current	逆流
역류세정	backwash	逆流洗淨
역류온도계함	reverse flow thermometer housing	逆流溫度計函
역반사	retroreflection	逆反射
역반사체	retroreflector	逆反射體

역방향몬순기압골	reverse-oriented monsoon trough	逆方向-氣壓-
역방향추론	backward chaining	逆方向推論
역변동	inverse variation	逆變動
역변층	reversing layer	逆變層
역복사	back radiation, counterradiation	逆輻射
역사기후	historical climate	歷史氣候
역사적 순서	historical sequence	歷史的順序
역산모형	inverse model	逆算模型
역산문제	inverse problem	逆算問題
역상구름-지면번개	negative cloud-to-ground lightning, negative ground flash	逆狀-地面-
역상지면섬광	negative ground flash	逆狀地面閃光
역서	almanac	曆書
역세정	backwash	逆洗淨
역세포	reverse cell	逆細胞
역수	backwater	逆水
역수곡선	backwater curve	逆水曲線
역시어	upshear	逆-
역영상	inverted image	逆映像
역적	impulse	力積
역전	inversion	逆轉
역전구름	inversion cloud	逆轉-
역전돌출	inversion nose	逆轉突出
역전밑면	inversion base	逆轉-面
역전아래층	subinversion layer	逆轉-層
역전안개	inversion fog	逆轉-
역전연무	inversion haze	逆轉煙霧
역전층	inversion layer	逆轉層
역제곱법칙	inverse square law, inverse square rule	逆-法則
역조	reversed tide, rip tide, rip current	逆潮
역풍파	cross sea	逆風波
역학	dynamics, mechanics	力學
역학가열	dynamic heating	力學加熱
역학개념	dynamical concept	力學槪念
역학경계조건	dynamic boundary condition	力學境界條件
역학경계지면조건	dynamic boundary surface condition	力學境界地面條件
역학고기압	dynamic anticyclone	力學高氣壓
역학고도	dynamic height	力學高度
역학고도편차	anomaly of dynamic height	力學高度偏差
역학골	dynamic trough	力學-
역학구조	dynamical structure	力學構造
역학깊이	dynamic depth	力學-
역학난류	mechanical turbulence	力學亂流
역학냉각	dynamic cooling	力學冷却

역학단위	dynamic unit	力學單位
역학대류	dynamical convection	力學對流
역학반응	dynamic response	力學反應
역학불안정(도)	dynamic instability, mechanical instability	力學不安定(度)
역학상사	dynamic similarity	力學相似
역학상승	dynamic lift, dynamic lifting	力學上昇
역학성분	dynamic component	力學成分
역학승온	dynamic warming	力學昇溫
역학시스템	dynamical system	力學-
역학안정(도)	dynamic stability, mechanical stability	力學安定(度)
역학얼음	dynamic ice	力學-
역학에너지	mechanical energy	力學-
역학연직운동	dynamic vertical motion	力學鉛直運動
역학예보	dynamical forecast	力學豫報
역학오차	dynamic error	力學誤差
역학온도변화	dynamic temperature change	力學溫度變化
역학응결고도	mechanical condensation level	力學凝結高度
역학적 불안정한	mechanically unstable	力學的不安定-
역학적 에너지보존법칙	mechanical energy conservation law	力學的-保存法則
역학적운	dynamic cumulus	力學積雲
역학적 흡입	dynamic entrainment	力學的吸入
역학점성	dynamic viscosity	力學粘性
역학점성계수	coefficient of dynamic viscosity, dynamic coefficient of viscosity	力學粘性係數
역학조건	dynamic condition	力學條件
역학중립	dynamic indifference	力學中立
역학초기화	dynamic initialization	力學初期化
역학특성	dynamic characteristics	力學特性
역학항	dynamic term	力學項
역함수	inverse function	逆函數
역행	retrogression	逆行
역행궤도	retrograde orbit	逆行軌道
역행열방법	inverse matrix method	逆行列方法
역행저기압	retrograde depression	逆行低氣壓
역행파	retrograde wave	逆行波
역효과	adverse effect	逆效果
역V형	inverted V-pattern	逆-型
연감	almanac	年監
연개지평	smog horizon	煙-地平
연결도	connectivity	連結度
연결방전	connecting discharge	連結放電
연계열	annual series	年系列
연관측	kite observation	鳶觀測
연교차	annual range	年較差

연구소	laboratory	硏究所
연기	smoke	煙氣
연기가라앉음	fumigation	煙氣-
연기가라앉음사건	fumigation event	煙氣-事件
연기관리	smoke management	煙氣管理
연기구	kite balloon	鳶氣球
연기구름	fumulus, smoke cloud	煙氣-
연기덩이	puff	煙氣-
연기보	smoke pall	煙氣褓
연기분산	smoke dispersal	煙氣分散
연기뿜기	smudging	煙氣-
연기상승	plume rise	煙氣上昇
연기상승높이	plume rise height	煙氣上昇-
연기상승모형	plume rise model	煙氣上昇模型
연기선법	smoke wire method	煙氣線法
연기지평	smoke horizon	煙氣地平
연기피해	smoke damage	煙氣被害
연대학	chronology	年代學
연도기체	flue gas	煙道氣體
연돌	smokestack	煙突
연동모형	online model	聯動模型
연료공기비	fuel air ratio	燃料空氣比
연료수분	fuel moisture	燃料水分
연료연소원	fuel combustion source	燃料燃燒源
연료전지	fuel cell	燃料電池
연륜기후학	dendroclimatology	年輪氣候學
연륜연대기록	dendrochronological record	年輪年代記錄
연륜학	dendrochronology	年輪學
연립방정식	simultaneous equation	聯立方程式
연막	smoke screen	煙幕
연막작전기상	smoke screen operation weather	煙幕作戰氣象
연무	haze, smoke fog, smog	煙霧
연무경계선	haze line	煙霧境界線
연무고도	haze level	煙霧高度
연무방울	haze droplet	煙霧-
연무선	haze horizon	煙霧線
연무층	haze layer	煙霧層
연방과학산업연구원 (오스트레일리아)	commonwealth scientific and industrial research organization	聯邦科學産業硏究院
연방사선	soft radiation	軟放射線
연변	edge, side	緣邊
연변강화	edge enhancement	緣邊强化
연변동	annual variation	年變動
연변파	edge wave	緣邊波

연변화파	annual wave	年變化波
연보	annual report	年報
연산	operation	演算
연산자	operator	演算子
연산자법	operational calculus	演算子法
연상고대	soft rime	軟-
연설선	annual snow line	年雪線
연소	combustion	燃燒
연소기체유출	flue gas spillage	燃燒氣體流出
연소먼지	combustion dust	燃燒-
연소반응	combustion reaction	燃燒反應
연소배기가스	combustion exhaust gas	燃燒排氣-
연소배출 대기화학	atmospheric chemistry of combustion emission	燃燒排出大氣化學
연소지수	burning index	燃燒指數
연소핵	combustion nucleus	燃燒核
연속(성)강수	continuous precipitation	連續(性)降水
연속곡선	continuous curve	連續曲線
연속기록	continuous record	連續記錄
연속도	continuity chart	連續圖
연속방정식	continuity equation, equation of continuity	連續方程式
연속방정식법	method of continuity equation	連續方程式法
연속배출감시	continuous emission monitor	連續放出監視
연속배출점오염원	continuously emitting point source	連續排出汚染源
연속변수	continuous variable	連續變數
연속분포	continuous distribution	連續分布
연속(성)비	continuous rain	連續(性)-
연속선도(방전)	continuous leader	連續先導(放電)
연속성	continuity	連續性
연속성천둥번개	continuous thunder and lightning	連續性-
연속스펙트럼	continuous spectrum	連續-
연속전류	continuing current	連續電流
연속체	continuum	連續體
연속체가설	continuum hypothesis	連續體假說
연속체이론	continuum theory	連續體理論
연속체흡수	continuum absorption	連續體吸收
연속파	continuous wave	連續波
연속파레이더	continuous wave radar	連續波-
연속플랑크톤기록계	continuous plankton recorder	連續-記錄計
연속함수	continuous function	連續函數
연속확률분포	continuous probability distribution	連續確率分布
연속흡수	continuous absorption	連續吸收
연쇄규칙	chain rule	連鎖規則
연쇄반응	chain reaction	連鎖反應
연쇄반응길이	chain length	連鎖反應-

연쇄번개	chain lightning	連鎖-
연습자료집	training data set	演習資料集
연안갇힘파	coastally trapped wave	沿岸-波
연안대	coastal zone	沿岸帶
연안대색주사기	coastal zone color scanner	沿岸帶色走査機
연안류	littoral current, longshore current, shore current	沿岸流
연안물	coastal water	沿岸-
연안물예보	coastal water forecast	沿岸-豫報
연안바다	coastal sea	沿岸-
연안바람	shore wind	沿岸-
연안안개	coastal fog	沿岸-
연안얼음	shore ice	沿岸-
연안예보	nearshore forecast	沿岸豫報
연안용승	coastal upwelling	沿岸湧昇
연안작전	inshore operation	沿岸作戰
연안지역	coastal region	沿岸地域
연안풍	inshore wind	沿岸風
연안항로	coastal route	沿岸航路
연안항로예보	coastal route forecast	沿岸航路豫報
연안/호안 홍수	coastal/lakeshore flooding	沿岸/湖岸洪水
연안/호안 홍수감시보	coastal/lakeshore flood watch	沿岸/湖岸洪水監視報
연안/호안 홍수경보	coastal/lakeshore flood warning	沿岸/湖岸洪水警報
연안/호안 홍수주의보	coastal/lakeshore flood advisory	沿岸/湖岸洪水注意報
연안홍수	coastal flooding	沿岸洪水
연안홍수감시보	coastal flood watch	沿岸洪水監視報
연안홍수경보	coastal flood warning	沿岸洪水警報
연약권	asthenosphere	軟弱圈
연온도교차	annual temperature range	年溫度交差
연온도변화파	annual temperature wave	年溫度變化波
연우박	soft hail	軟雨雹
연유출	annual runoff	年流出
연장경계조건방법	extended boundary condition method	延長境界條件方法
연장예보	extended forecast	延長豫報
연저장	annual storage	年貯藏
연주기	annual cycle	年週期
연주기바람	anniversary wind	年週期-
연중반순	annual-basis pentad	年中半旬
연직강우레이더	vertical rain radar	鉛直降雨-
연직공기운동	vertical air motion	鉛直空氣運動
연직광선	vertical ray	鉛直光線
연직규모	vertical scale	鉛直規模
연직기압경도	vertical pressure gradient	鉛直氣壓傾度
연직기압속도	vertical p-velocity	鉛直氣壓速度
연직늘림	vertical stretching	鉛直-

연직단면	vertical profile	鉛直斷面
연직단면(도)	vertical cross section, vertical section	鉛直斷面(圖)
연직단면관측기	profiler	鉛直斷面觀測器
연직대류	vertical convection	鉛直對流
연직돌풍(성)	vertical gustiness	鉛直突風(性)
연직류	vertical airflow, vertical flow	鉛直流
연직류기록계	vertical current recorder	鉛直流記錄計
연직바람단면	VAD wind profile, vertical wind profile	鉛直-斷面
연직바람속도	vertical wind velocity	鉛直-速度
연직바람프로파일	VAD wind profile, vertical wind profile	鉛直-
연직발달구름	cloud with vertical development	鉛直發達-
연직복사선	vertical ray	鉛直輻射線
연직분포	vertical distribution	鉛直分布
연직불안정(도)	vertical instability	鉛直不安定(度)
연직빔레이더	vertical beam radar	鉛直-
연직빛살레이더	vertical beam radar	鉛直-
연직성분	vertical component	鉛直成分
연직소용돌이도	vertical vorticity	鉛直-度
연직속도	vertical velocity	鉛直速度
연직속도(분포)도	vertical velocity chart	鉛直速度(分布)圖
연직속도분포	vertical velocity ditribution	鉛直速度分布
연직속도예상도	vertical velocity prognostic chart	鉛直速度豫想圖
연직속력지시기	vertical speed indicator	鉛直速力指示器
연직순환	vertical circulation	鉛直循環
연직시어	vertical shear	鉛直-
연직시어관측기	profiler	鉛直-觀測器
연직시정	vertical visibility	鉛直視程
연직안정(도)	vertical stability	鉛直安定(度)
연직엽형수관	vertical leaved canopy	鉛直葉形樹冠
연직온도경도	vertical temperature gradient	鉛直溫度傾度
연직온도단면	vertical temperature profile	鉛直溫度斷面
연직온도분포	vertical temperature profile	鉛直溫度分布
연직온도분포복사계	vertical temperature profile radiometer	鉛直溫度分布輻射計
연직운동	vertical motion	鉛直運動
연직운동방정식	vertical motion equation	鉛直運動方程式
연직원	vertical circle	鉛直圓
연직바람시어	vertical wind shear	鉛直-
연직이류	vertical advection	鉛直移流
연직입사	normal incidence	鉛直入射
연직적분액체수량	vertical integrated liquid water content, vertically integrated liquid	鉛直積分液體水量
연직적산수액	vertically integrated liquid	鉛直積算水液
연직적외온도복사계	infrared temperature profile radiometer	鉛直赤外溫度輻射計
연직제트	vertical jet	鉛直-

연직좌표	vertical coordinate	鉛直座標
연직지향레이더	vertical pointing radar	鉛直指向-
연직차이도	vertical differential chart	鉛直差異圖
연직측심	vertical sounding	鉛直測深
연직측풍장비	wind profiler	鉛直測風裝備
연직층	vertical layer	鉛直層
연직탐측	vertical sounding	鉛直探測
연직편광	vertical polarization	鉛直偏光
연직풍속계	vertical anemometer	鉛直風速計
연직풍속분포	vertical wind profile	鉛直風速分布
연직풍향계	vertical anemoscope	鉛直風向計
연직프로파일	vertical profile	鉛直-
연직해상도	vertical resolution	鉛直解像度
연직혼합	vertical mixing	鉛直混合
연차근사법	method of successive approximation	連次近似法
연초과치계열	annual exceedance series	年超過值系列
연최대치계열	annual maximum series	年最大値系列
연최저치계열	annual minimum series	年最低値系列
연층	annual layer	年層
연탐측	kite-ascent, kite sounding	鳶探測
연편차	annual anomaly	年偏差
연평균	annual mean	年平均
연평균강수량	average annual precipitation	年平均降水量
연평균온도	mean annual temperature	年平均溫度
연평균온도교차	annual mean temperature range	年平均溫度交差
연필칼모양얼음	penknife ice, candle ice	鉛筆-模樣-
연한 얼음	grease ice	軟-
연합관측	combining observation	聯合觀測
연해	offshore	沿海
연해사주	offshore bar	沿海砂洲
연홍수계열	annual flood series	年洪水系列
연홍수위	annual flood	年洪水位
열	heat	熱
열(의)	thermal	熱
열거칠기	thermal roughness	熱-
열경도	thermal gradient	熱傾度
열경사	thermal slope	熱傾斜
열고기압	thermal high	熱高氣壓
열곡	rift valley	裂谷
열관성	thermal inertia	熱慣性
열교환	heat exchange	熱交換
열교환기	heat exchanger	熱交換器
열국지풍	thermally driven local wind	熱局地風
열권	thermosphere	熱圈

열권계면	thermopause	熱圈界面
열기관	heat engine	熱機關
열기구	hot-air balloon	熱氣球
열기류	thermal current	熱氣流
열기포	thermal	熱氣泡
열기포침식	erosion of thermal	熱氣泡侵蝕
열기후	thermal climate	熱氣候
열난류	thermal turbulence	熱亂流
열뇌우	heat thunderstorm, hot thunderstorm	熱雷雨
열단위	heat unit, thermal unit	熱單位
열당량	thermal equivalent	熱當量
열대	torrid zone, tropical belt, tropical zone, tropical, tropics	熱帶
열대강우측정임무	tropical rainfall measuring mission	熱帶降雨測定任務
열대강우측정위성	tropical rainfall measuring mission	熱帶降雨測定衛星
열대건조기후	tropical arid climate, tropical dry climate	熱帶乾燥氣候
열대계절풍기후	tropical monsoon climate	熱帶季節風氣候
열대고기압	tropical anticyclone	熱帶高氣壓
열대(대류)권계면	tropical tropopause	熱帶(對流)圈界面
열대기단	tropical air (mass)	熱帶氣團
열대기단안개	tropical air fog	熱帶氣團-
열대기상	tropical weather	熱帶氣象
열대기상전망	tropical weather outlook	熱帶氣象展望
열대기상학	tropical meteorology	熱帶氣象學
열대기후	tropical climate	熱帶氣候
열대기후학	tropical climatology	熱帶氣候學
열대다우기후	tropical rain climate, tropical rainy climate	熱帶多雨氣候
열대대륙기단	tropical continental air mass	熱帶大陸氣團
열대류	heat convection, thermal convection, tropical air current	熱對流
열대무풍대	tropical calm zone	熱帶無風帶
열대몬순기후	tropical monsoon climate	熱帶-氣候
열대불연속선	tropical discontinuity line	熱帶不連續線
열대비	thermal contrast	熱對比
열대사막기후	tropical desert climate	熱帶沙漠氣候
열대사바나기후	tropical savanna climate	熱帶-氣候
열대상부대류권(기압)골	tropical upper-tropospheric trough(TUTT)	熱帶上部對流圈(氣壓)-
열대성저기압	oraucan	熱帶性低氣壓
열대성저기압 이름	name of TC	熱帶性低氣壓-
열대수	tropical number, T-number	熱帶數
열대수렴대	intertropical convergence zone	熱帶收斂帶
열대습윤기후	tropical wet climate	熱帶濕潤氣候
열대아메리카 우기	invierno	熱帶-雨期
열대야	tropical night	熱帶夜

열대요란	tropical disturbance	熱帶擾亂
열대우림	tropical rainforest	熱帶雨林
열대우림기후	tropical rainforest climate	熱帶雨林氣候
열대일	tropical day	熱帶日
열대일기	tropical weather	熱帶日氣
열대일기전망	tropical weather outlook	熱帶日氣展望
열대저기압	tropical cyclone	熱帶低氣壓
열대저기압눈	eye of tropical cyclone	熱帶低氣壓-
열대저기압번호	number of TC	熱帶低氣壓番號
열대저기압위치 추정	tropical cyclone position estimate	熱帶低氣壓位置推定
열대저압부	tropical depression	熱帶低壓部
열대전선	intertropical front, tropical front	熱帶前線
열대주의보	tropical advisory	熱帶注意報
열대파	tropical wave	熱帶波
열대편동풍	tropical easterlies	熱帶偏東風
열대편동풍대	tropical easterly zone	熱帶偏東風帶
열대편동풍제트	tropical easterly jet	熱帶偏東風-
열대폭풍(우)	tropical storm	熱帶暴風(雨)
열대폭풍(우)경보	tropical storm warning	熱帶暴風(雨)警報
열대폭풍(우)감시보	tropical storm watch	熱帶暴風(雨)監視報
열대플룸	tropical plume	熱帶-
열대합류대	intertropical confluence zone	熱帶合流帶
열대해양기단	tropical maritime air mass	熱帶海洋氣團
열대회오리바람	breather	熱帶-
열량	calorific value	熱量
열량계	calorimeter	熱量計
열량보존법칙	law of heat conservation	熱量保存法則
열량측정법	calorimetry	熱量測定法
열로스비수	thermal Rossby number	熱-數
열류	heat flow	熱流
열류변환기	heat flow transducer	熱流變換機
열류판	heat flow plate	熱流板
열린각	opening angle	-角
열린계	open system	-系
열린곡선	open curve	-曲線
열린구간	open interval	-區間
열린눈금기압계	open scale barograph	-氣壓計
열린세포	open cell	-細胞
열린세포적운	open cell cumulus	-細胞積雲
열린집합	open set	-集合
열린회로	open circuit	-回路
열막풍속계	hot-film anemometer	熱膜風速計
열바람	hot wind	熱-
열방출스펙트럼	thermal emission spectrum	熱放出-

열번개	heat lightning	熱-
열변형	thermal distortion	熱變形
열복사	heat radiation, thermal radiation	熱輻射
열복사강도	caloradiance	熱輻射强度
열복사계	bolometer	熱輻射計
열분배	heat partition	熱分配
열분해	thermolysis	熱分解
열불안정	thermal instability	熱不安定
열사병	heat exhaustion, heat stroke	熱射病
열산악효과	thermal mountain effect	熱山岳效果
열산화	thermal oxidation	熱酸化
열상수	thermal constant	熱常數
열생성	heat generation	熱生成
열선	heat ray	熱線
열선풍속계	hot-wire anemometer	熱線風速計
열섬	heat island, thermal island	熱-
열섬효과	heat island effect	熱-效果
열성층	thermal stratification	熱成層
열소용돌이도	thermal vorticity	熱-度
열소용돌이도이류	thermal vorticity advection	熱-度移流
열속	heat flux	熱束
열수분출공	hydrothermal vent	熱水噴出孔
열수송	heat transport	熱輸送
열수지	heat budget	熱收支
열순환	heat cycle, thermal circulation	熱循環
열스트레스지수	heat stress index	熱-指數
열씨온도계	Réaumur thermometer	列氏溫度計
열씨온도눈금	Réaumur thermometric scale	列氏(溫度)-
열에너지	heat energy, thermal energy	熱-
열역학	thermodynamics	熱力學
열역학과정	thermodynamic process	熱力學過程
열역학법칙	law of thermodynamics	熱力學法則
열역학변수	thermodynamic variable	熱力學變數
열역학빙구온도	thermodynamic ice-bulb temperature	熱力學氷球溫度
열역학상태함수	thermodynamic function of state	熱力學狀態函數
열역학서리점온도	thermodynamic frost-point temperature	熱力學-點溫度
열역학선도	thermodynamic chart, thermodynamic diagram	熱力學線圖
열역학성층	thermodynamic stratification	熱力學成層
열역학솔레노이드	thermodynamic solenoid	熱力學-
열역학습구온도	thermodynamic wet-bulb temperature	熱力學濕球溫度
열역학에너지방정식	thermodynamic energy equation	熱力學-方程式
열역학온도	thermodynamic temperature	熱力學溫度
열역학온도눈금	thermodynamic temperature scale	熱力學溫度-
열역학이슬점온도	thermodynamic dew-point temperature	熱力學-點溫度

열역학제0법칙	zeroth law of thermodynamics	熱力學第零法則
열역학제1법칙	first law of thermodynamics, first principle of thermodynamics	熱力學第一法則
열역학제2법칙	second law of thermodynamics, second principle of thermodynamics	熱力學第二法則
열역학제3법칙	third law of thermodynamics	熱力學第三法則
열역학탐측기	thermodynamic profiler	熱力學探測器
열역학퍼텐셜	thermodynamical potential, thermodynamic potential	熱力學-
열역학프로파일러	thermodynamic profiler	熱力學-
열역학함수	thermodynamical function	熱力學函數
열역학효율	thermodynamic efficiency	熱力學效率
열염분순환	thermohaline circulation	熱鹽分循環
열영역	thermal regime	熱領域
열오염	thermal pollution	熱汚染
열용량	heat capacity, thermal capacity	熱容量
열용량지도제작임무위성	Heat Capacity Mapping Mission Satellite	熱容量地圖製作任務衛星
열운동	heat motion, thermal motion	熱運動
열원	heat source	熱源
열의 일(해)당량	mechanical equivalent of heat	熱-(該)當量
열이류	heat advection, thermal advection	熱移流
열이온관	thermionic tube	熱-管
열저기압	heat low, thermal low	熱低氣壓
열저기압발생	thermocyclogenesis	熱低氣壓發生
열저압부	thermal depression	熱低壓部
열저장체	heat reservoir	熱貯藏體
열적도	heat equator, thermal equator	熱赤道
열적산계	thermo-integrator	熱積算計
열적외선 영역	thermal infrared domain	熱赤外線領域
열적운	heat cumulus, thermic cumulus	熱積雲
열전기더미	thermopile	熱電氣-
열전달	heat transfer	熱傳達
열전달계수	heat transfer coefficient	熱傳達係數
열전대	thermo-junction	熱電對
열전대온도계	thermo-junction thermometer	熱電對溫度計
열전도	heat conduction	熱傳導
열전도계수	coefficient of heat conduction, coefficient of thermal conduction	熱傳導係數
열전도도	thermal conductivity	熱傳導度
열전도성	thermal conductivity	熱傳導性
열전도율	heat conductivity	熱傳導率
열전소자	thermo-element	熱電素子
열전(기)쌍	thermocouple	熱電(氣)雙
열전온도계	thermoelectric thermometer	熱電溫度計

열전일사계	thermoelectric actinometer	熱電日射計
열전자기일사계	thermoelectric actinograph	熱電自記日射計
열전직달일사계	thermoelectric pyrheliometer	熱電直達日射計
열전퇴	themopile	熱電堆
열점	hot spot	熱點
열제트	thermal jet	熱-
열조석	thermal tide	熱潮汐
열주의보	heat advisory	熱注意報
열중립	thermal neutrality, thermoneutrality	熱中立
열지수	heat index	熱指數
열지향	thermal steering	熱指向
열진공	thermal vacuum	熱眞空
열집진기	thermal precipitator	熱集塵器
열축적	heat accumulation	熱蓄積
열탑	hot tower	熱塔
열파	heat wave, hot wave	熱波
열판강우계측기	hotplate precipitation gauge	熱板降雨計測器
열팽창	thermal expansion	熱膨脹
열팽창계수	coefficient of thermal expansion	熱膨脹係數
열퍼텐셜	thermal potential	熱-
열편	lobe	裂片
열편구조	lobe structure	裂片構造
열평형	heat balance, thermal equilibrium	熱平衡
열폭풍	heat storm	熱暴風
열플럭스	heat flux	熱-
열함수	heat function	熱函數
열함유량	heat content	熱含有量
열확산율	thermal diffusivity	熱擴散率
열효율	thermal efficiency	熱效率
열효율비	thermal efficiency ratio	熱效率比
열효율지수	thermal efficiency index	熱效率指數
열흡수원	heat sink	熱吸收源
엷은막 물	film water	-膜-
엷은 서리	light frost	
엷은 안개	light fog, mist	
엷은 연무	slight haze	-煙霧
염	salt	鹽
염기도	alkalinity	鹽基度
염도	salinity	鹽度
염분	salinity	鹽分
염분계	salinometer	鹽分計
염분농도	salt concentration	鹽分濃度
염분약층	halocline	鹽分躍層
염분입자	salt particle	鹽分粒子

염분추출	salt extraction	鹽分抽出
염분축적	salinization	鹽分蓄積
염분측정기	salinity bridge	鹽分測定器
염소	chlorine	鹽素
염소도	chlorosity	鹽素度
염소량	chlorinity	鹽素量
염소산화물	chroline oxide	鹽素酸化物
염소원자	chlorine atom	鹽素原子
염소질산염	chlorine nitrate	鹽素窒酸鹽
염소화합물	chlorine compound	鹽素化合物
염수	esturine water	鹽水
염수오염	saltwater contamination	鹽水汚染
염수침해	salt-water invasion	鹽水沈害
염수해	salty water damage	鹽水害
염해	salt damage, salt injury	鹽害
염화리튬노점습도계	lithium chloride dew-point hygrometer	鹽化-露點濕度計
염화리튬노점온도계	lithium chloride dew-point thermometer	鹽化-露點溫度計
염화리튬이슬점습도계	lithium chloride dew-point hygrometer	鹽化-點濕度計
염화리튬이슬점온도계	lithium chloride dew-point thermometer	鹽化-點溫度計
염화불화탄소	chlorofluorocarbon	鹽化弗化炭素
염화암모늄	ammonium chloride	鹽化-
엽령	leaf age	葉齡
엽록소	chlorophyll	葉綠素
엽록체	chloroplast	葉綠體
엽면적밀도	leaf area density	葉面積密度
엽면적지수	leaf area index	葉面積指數
엽밀도	foliage density	葉密度
엽속도	foliage velocity	葉速度
엽중비	leaf weight ratio	葉重比
영구가뭄	permanent drought	永久-
영구고기압	permanent anticyclone, permanent high	永久高氣壓
영구기록부	permanent register	永久記錄簿
영구기체	constant gas, permanent gas	永久氣體
영구동결기후	perpetual frost climate	永久凍結氣候
영구동결면	pergélisol table, permafrost table	永久凍結面
영구동토	perennially frozen ground, pergélisol, permafrost, permanently frozen ground	永久凍土
영구동토기후	climate of eternal frost	永久凍土氣候
영구동토융해	permafrost melt	永久凍土融解
영구동토층	permafrost, pergéllisol	永久凍土層
영구동토풀림	depergelation	永久凍土-
영구동토화	pergelation	永久凍土化
영구바람	permanent wind	永久-
영구빙설기후	eternal frost climate	永久氷雪氣候

영구설선	névé line	永久雪線
영구순환	permanent circulation	永久循環
영구군락	permanent community	永久群落
영구 언땅	permafrost, pergéllisol	永久-
영구 언면	permafrost table	永久-面
영구 언섬	permafrost island	永久-
영구에코	permanent echo	永久-
영구위조	permanent wilting	永久萎凋
영구위조율	permanent wilting percentage	永久萎凋率
영구위조점	permanent wilting point	永久萎凋點
영구임계값이동	permanent threshold shift	永久臨界-移動
영구저기압	permanent low, permanent depression	永久低氣壓
영구청력변위	permanent threshold shift	永久聽力變位
영구파	permanent wave	永久波
영구흐름	perennial stream	永久-
영국-브라질 아마존 기후관측연구	Anglo-Brazilian climate observation study (ABRACOS)	英國-氣候觀測研究
영국열단위	British thermal unit	英國熱單位
영년기후관측	principal climatological observation	永年氣候觀測
영도층(토양)	zero curtain	(土壤)零度層
영(도)등온선	zero isotherm	零(度)等溫線
영면변위	zero-plane displacement	零面變位
영상	image, imagery	映像
영상강조	image enhancement	映像强調
영상강조기법	image enhancement technique	映像强調技法
영상개선	image enhancement	映像改善
영상경사	image inclination	映像傾斜
영상계구름	cloud from imager	映像計-
영상막	screen	映像幕
영상묘화	video mapping	映像描畵
영상변환/변조 전달기능	modulation transfer function	映像變換/變造傳達機能
영상보정	image correction	映像補正
영상분석 카메라시스템	image dissector camera system	映像分析-
영상신호	video signal	映像信號
영상애니메이션	animation of imagery	映像-
영상위치보정	Image Navigation and Registration	映像位置補正
영상위치보정시스템	Image Motion Compensation(IMC)	映像位置補正-
영상자료 취득 및 (전처리)관리시스템	Image Data Aquisition and Control System	映像資料取得-(前處理)管理-
영상적설계	image snow cover meter	映像積雪計
영상전달이론	image transfer theory	映像傳達理論
영상전처리시스템	Image Preprocessing Subsystem	映像前處理-
영상증폭	video gain	映像增幅
영상차분	image differencing	映像差分

영상촬영시스템	image filming system	映像撮影-
영상확대	image zoom	映像擴大
영시선속도선	zero-isodop	零視線速度線
영시정	zero visibility	零視程
영실링	zero ceiling	零-
영양생장	vegetative growth	營養生長
영양생장기	vegetative stage	營養生長期
영역	regime, domain	領域
영역방송	terrestrial broadcast	領域放送
영역평균	area average, area averaging	領域平均
영역평균문제	area averaging problem	領域平均問題
영절대소용돌이도	zero absolute vorticity	零絶對-度
영절대와도	zero absolute vorticity	零絶對渦度
영점기준	zero datum	零點基準
영차불연속	discontinuity of zero order, zero-order discontinuity	零次不連續
영층	null layer, zero layer	零層
영하온도	negative temperature	零下溫度
영향권원뿔	cone of influence	影響圈圓-
영향역	area of coverage	影響域
영향지역	area of influence	影響地域
영향함수	influence function	影響函數
영회사진법	shadowgraph	影繪寫眞法
옆굴절	lateral refraction	-屈折
옆돌풍도	lateral gustiness	-突風度
옆바람	beam wind, crosswind, side wind	
옆바람돌풍성	crosswind gustiness	-突風性
옆바람성분	crosswind component	-成分
옆신기루	lateral mirage	-蜃氣樓
옆접호	lateral tangent arcs	-接弧
옆혼합	lateral mixing	-混合
옆확산	lateral diffusion	-擴散
예년	normal year	例年
예리한 전선	sharp front	-前線
예보	forecast, precast	豫報
예보가이던스	forecast guidance	豫報-
예보가치	forecast value	豫報價値
예보검토	forecast verification	豫報檢討
예보관-사용자 협력	forecaster-user collaboration	豫報官-使用者協力
예보관서	forecast office	豫報官署
예보구역	forecast area, forecast district, forecast zone	豫報區域
예보기간	forecast period	豫報期間
예보기관	forecast organization	豫報機關
예보기술	forecast skill	豫報技術

예보대상구역	forecast target area	豫報對象區域
예보대상기간	forecast target period	豫報對象期間
예보도	forecast chart	豫報圖
예보모드	forecasting mode	豫報-
예보발표문	forecast bulletin	豫報發表文
예보법	forecasting method	豫報法
예보부서	forecast division	豫報部署
예보사후분석	forecast review	豫報事後分析
예보센터	forecasting center	豫報-
예보수정	forecast amendment	豫報修正
예보숙련도	skill score of forecast	豫報熟練度
예보시각	forecast time	豫報時刻
예보시스템연구소	Forecast Systems Laboratory	豫報-研究所
예보시점	forecast time	豫報時點
예보업무	forecast service	豫報業務
예보역검정	forecast reversal test	豫報逆檢定
예보오차	forecast error	豫報誤差
예보용어	forecast terminology	豫報用語
예보원	forecast circle	豫報員
예보유형	forecast type	豫報類型
예보유효시간	forecast valid time	豫報有效時間
예보인자	forecast parameter	豫報因子
예보자	forecaster	豫報者
예보자료	forecast data	豫報資料
예보정확도	forecast accuracy	豫報正確度
예보체계	forecast system	豫報體系
예보평가	forecast evaluation, forecast validation	豫報評價
예보품질	forecast quality	豫報品質
예비단계	preliminary phase	豫備段階
예비비행장일기예보	alternate forecast	豫備飛行場日氣豫報
예비항공작전기상	Preliminary Air Operations weather	豫備航空作戰氣象
예상	prog, prognosis	豫想
예상도	prog chart, prognostic chart	豫想圖
(아무도 믿지 않는)예상도	agnostic chart	豫想圖
예상방정식	prognostic equation	豫想方程式
예상분석	prognostic analysis	豫想分析
예상상층도	prognostic upper level chart	豫想上層圖
예상일기도	prognostic weather chart	豫想日氣圖
예측	foreshadow, prediction	豫測
예측가능성	predictability	豫測可能性
예측도	prediction diagram	豫測圖
예측량	predictand	豫測量
예측성	predictability	豫測性
예측인자	predictor	豫測因子

오	oe	
오가사와라 고기압	Ogasawara high	-高氣壓
오거소나기	Auger shower	
오닐실험	O'Neill experiment	-實驗
오더레이드간빙기	Odderade interstadial	-間氷期
오딘초보 간빙기	Odintsovo interstadial	-間氷期
오라	ora, aura	
오로라스펙트럼	polar band	
오로라영상	aurora image	-映像
오로치 바람(일본)	Oroshi wind	
오르도비스기(고생대)	Ordovician period	-期
오르도비스기 멸종	Ordovician extinction	-期滅種
오르도비스기-실루리아기 사건	Ordovician-Silurian event	-事件
오르수르	orsure	
오르-좀메르펠트 방정식	Orr-Sommerfeld equation	-方程式
오르트구름	Oort cloud	
오른나사법칙	law of clockwise screw	-法則
오른손회전	right-handed rotation	-回轉
오른스타인-우렌벡 과정	Ornstein-Uhlenbeck process	-課程
오른쪽(의)	dextral	
오리온성운	Orion Nebula	-星雲
오만	Oman	
오메가모양태양	omega sun	-模樣太陽
오메가방정식	omega equation	-方程式
오메가신호	omega signal	-信號
오메가울프법	Omega-Wolff method	-法
오메가존데	omega zonde	
오목면	concave	-面
오버샘플링	over sampling	
오버슈팅뇌우	overshooting thunderstorm	-雷雨
오버슛	overshoot	
오버슛(돌출)꼭대기	overshooting top	-(突出)-
오버행	overhang	
오버행에코	overhang echo	
오베르나제	auvergnasse	
오베르빈트	oberwind	
오보율	rate of forecast error	誤報率
오부코프길이	Obukhov length	
오브리언다항식	O'Brien polynomial	-多項式
오브리언3차다항식	O'Brien cubic polynomial	-三次多項式
오비탈 변이	orbital variation	-變異
오산화이질소	dinitrogen pentoxide	五酸化二窒素
오산화질소	nitrogen pentoxide	五酸化窒素
오세아니아	Oceania	

오센근사	Oseen's approximation	-近似
오센방정식	Oessn's equation	-方程式
오소스바람	Osos wind	
오스타쉬	Austausch	
오스타쉬 계수	Austausch coefficient	-係數
오스트레일리아 기상청	Australian Bureau of Meteorology	-氣象廳
오스트레일리아 몬순	Australian monsoon	
오스트레일리아 미개척지	Australian outback	-未開拓地
오스트레일리아 여름몬순	Australian summer monsoon	
오스트루	austru	
오스트리아	ostria, auster	
오실로그래프	oscillograph	
오실로스코프	oscilloscope	
오아시스	oasis	
오아시스상황	oasis situation	-狀況
오아시스효과	oasis effect	-效果
오야시오	Oyashio	
오야시오해류	Oyashio current	-海流
오얼라이트	oolite, oolith	
오염	contamination, pollution	汚染
오염공기	polluted air	汚染空氣
오염규제원리	philosophy of pollution control	汚染規制原理
오염물	contaminant	汚染物
오염물질	pollutant	汚染物質
오염지역	pollution area	汚染地域
오우아리	ouari	
오웬 먼지기록계	Owens dust recorder	-記錄計
오일	oil	
오일기질	oil substrate	-基質
오일러가속도	Eulerian acceleration	-加速度
오일러계	Eulerian system	-系
오일러규모	Eulerian scale	-規模
오일러기준계	Eulerian reference system	-基準系
오일러길이규모	Eulerian length scale	-規模
오일러난류통계	Eulerian turbulence statistics	-亂流統計
오일러도함수	Eulerian derivative	-導函數
오일러바람	Euler wind, Eulerian wind	
오일러바람자료	Eulerian wind data	-資料
오일러방법	Eulerian method	-方法
오일러방정식	Euler equation, Eulerian equation	-方程式
오일러변화	Eulerian change	-變化
오일러상관	Eulerian correlation	-相關
오일러상관함수	Eulerian correlation function	-相關函數
오일러속도	Eulerian velocity	-速度

오일러수	Euler number	-數
오일러스펙트럼	Eulerian spectrum	
오일러시간규모	Eulerian time scale	-時間規模
오일러시간규모변환	Eulerian time scale transformation	-時間規模變換
오일러자동상관계수	Eulerian autocorrelation coefficient	-自動相關係數
오일러적분규모	Eulerian integral scale	-積分規模
오일러적분 길이규모	Eulerian integral length scale	-積分-規模
오일러전개	Euler development	-展開
오일러접근	Eulerian approach	-接近
오일러좌표	Eulerian coordinates	-座標
오일러통계학	Eulerian statistics	-統計學
오일러함수	Euler's function	-函數
오일러확산	Eulerian diffusion	-擴散
오일러흐름	Eulerian flow	
오일상용기상	oil appropriate standard weather	-常用氣象
오일표준기상	oil appropriate standard weather	-標準氣象
오존	ozone	
오존감소	ozone depletion	-減少
오존감시	ozone monitoring	-監視
오존경보	ozone alert	-警報
오존경향	ozone trend	-傾向
오존고갈	ozone depletion	-枯渴
오존고갈물질	ozone-depleting substance	-枯渴物質
오존고갈퍼텐셜	ozone-depleting potential	-枯渴-
오존광해리율	ozone photodissociation rate	-光解離率
오존구름	ozone cloud	
오존구멍	ozone hole	
오존권	ozonosphere	-圈
오존권계면	ozonopause	-圈界面
오존농도	ozone concentration	-濃度
오존등치선도표	ozone isopleth plot	-等値線圖表
오존모양함수	ozone shape function	-模樣函數
오존반응	ozone reaction	-反應
오존분광광도계	ozone spectrophotometer	-分光光度計
오존분리	ozone cutoff	-分離
오존분자	ozone molecule	-分子
오존-산소 주기	ozone-oxygen cycle	-酸素週期
오존손실	ozone loss	-損失
오존전달계수	ozone transfer coefficient	-傳達係數
오존전량	column ozone	-全量
오존존데	ozone-sonde	
오존주의보	ozone advisory	-注意報
오존차폐	ozone shield	-遮蔽
오존층	ozone layer	-層

오존층유효온도	ertor	-層有效溫度
오존활동일	ozone action day	-活動日
오존흡수	ozone absorption	-吸收
오즈미도프 규모	Ozmidov scale	-規模
오차	error	誤差
오차검출부호	error-detecting code	誤差檢出符號
오차법칙	law of error	誤差法則
오차보정	error correction	誤差補正
오차분석	error analysis	誤差分析
오차분포	error distribution	誤差分布
오차정규곡선	normal curve of error	誤差正規曲線
오차한계	limit of error	誤差限界
오차함수	error function	誤差函數
오크나무	oak tree	
오클랜드동안류	East Auckland current	-東岸流
오토존데	autosonde	
오팔껍질	opal shell	
오호츠크해	Okhotsk sea, sea of Okhotsk	-海
오호츠크해 고기압	Okhotsk high	-海高氣壓
오호츠크해 기단	Okhotsk sea air mass	-海氣團
오후혼합두께	afternoon mixing depth	午後混合-
오후효과	afternoon effect	午後效果
옥수수열지수	corn heat unit	-熱指數
옥타브밴드	octave band	
옥탄	octane	
옥탄가	octane number, octane rating	-價
온난고기압	warm anticyclone, warm high	溫暖高氣壓
온난(기압)골	warm trough	溫暖(氣壓)-
온난기단	warm air mass	溫暖氣團
온난대	thermal belt	溫暖帶
온난류	warm current	溫暖流
온난비	warm rain	溫暖-
온난안개	warm fog	溫暖-
온난역	warm sector	溫暖域
온난웅덩이	warm pool	溫暖-
온난이류	warm advection	溫暖移流
온난저기압	warm cyclone, warm low	溫暖低氣壓
온난전(선)면	warm frontal surface	溫暖前(線)面
온난전선	warm front	溫暖前線
온난전선강수	warm front precipitation	溫暖前線降水
온난전선뇌우	warm front thunderstorm	溫暖前線雷雨
온난전선면	anaphalanx, warm front surface	溫暖前線面
온난전선비	warm front rain	溫暖前線-
온난전선파(동)	warm-front wave	溫暖前線波(動)

온난전선폐색	warm front occlusion	溫暖前線閉塞
온난전선형폐색	warm-front-type occlusion	溫暖前線型閉塞
온난지대	thermal zone	溫暖地帶
온난컨베이어벨트	warm conveyor belt	溫暖-
온난파(동)	warm wave	溫暖波(動)
온난폐색	warm occlusion	溫暖閉塞
온난폐색전선	warm occluded front	溫暖閉塞前線
온난하강풍	warm downslope wind	溫暖下降風
온난핵고기압	warm-core anticyclone, warm-core high	溫暖核高氣壓
온난핵저기압	warm-core cyclone, warm-core low	溫暖核低氣壓
온난허	warm tongue	溫暖-
온난형 폐색	warm-type occlusion	溫暖型閉塞
온난형 폐색전선	warm-type occluded front	溫暖型閉塞前線
온난화	warming	溫暖化
온대	extratropical belt (or zone), extratropics, temperate belt, temperate zone	溫帶
온대(의)	extratropical	溫帶
온대계절풍기후	temperate monsoon climate	溫帶季節風氣候
온대기후	temperate climate	溫帶氣候
온대낙엽수림	temperate deciduous forest	溫帶落葉樹林
온대다우기후	temperate rainy climate	溫帶多雨氣候
온대대류권계면	extratropical tropopause	溫帶對流圈界面
온대동우기후	temperate climate with winter rain	溫帶冬雨氣候
온대북반구기후	extratropical northern hemisphere climate	溫帶北半球氣候
온대빙하	temperate glacier	溫帶氷河
(온대성)저기압	cyclone	(溫帶性)低氣壓
온대성펌프	extratropical pump	溫帶性-
온대습윤기후	temperate humid climate	溫帶濕潤氣候
온대우림	temperate rainforest	溫帶雨林
온대저기압	extratropical cyclone, extratropical low	溫帶低氣壓
온대저기압모형	extratropical cyclone model	溫帶低氣壓模型
온대전이	extratropical transition	溫帶轉移
온대편서풍	temperate westerlies	溫帶偏西風
온대편서풍지수	temperate westerlies index	溫帶偏西風指數
온대폭풍	extratropical storm	溫帶暴風
온대하우기후	temperate climate with summer rain	溫帶夏雨氣候
온대허리케인	extratropical hurricane	溫帶-
온도	temperature	溫度
온도감률	lapse rate of temperature	溫度減率
온도경도	temperature gradient	溫度傾度
온도계	thermometer	溫度計
온도계구부	thermometer bulb	溫度計球部
온도계수	temperature coefficient	溫度係數
온도골	thermal trough	溫度-

온도교차	range of temperature, temperature range	溫度較差
온도극값	temperature extreme	溫度極-
온도기압곡선	temperature pressure curve	溫度氣壓曲線
온도눈금	temperature scale, thermometric scale	溫度-
온도대	temperature belt, temperature zone	溫度帶
온도마루	thermal ridge	溫度-
온도변동	temperature variation	溫度變動
온도변화과정	march of temperature	溫度變化過程
온도보정	temperature correction, temperature reduction	溫度補正
온도분석	temperature analysis	溫度分析
온도분포	temperature distribution	溫度分布
온도상수	thermometric constant	溫度常數
온도수감부	temperature sensing element	溫度受感部
온도-염도곡선	temperature-salinity curve	溫度-鹽度曲線
온도-염도선도	temperature-salinity diagram	溫度-鹽度線圖
온도영역	temperature province	溫度領域
온도 절대연교차	absolute annual range of temperature	溫度絶對年較差
온도이류변화	advective change of temperature	溫度移流變化
온도이상	temperature anomaly	溫度異常
온도일교차	daily range of temperature	溫度日較差
온도장	temperature field	溫度場
온도전도도	temperature conductivity	溫度傳導度
온도전도율	thermometric conductivity	溫度傳導率
온도제어	temperature control	溫度制御
온도조건	temperature condition	溫度條件
온도조절	temperature control	溫度調節
온도조절기	thermostat	溫度調節器
온도존데	thermosonde	溫度-
온도차	temperature difference	溫度差
온도측정	temperature measurement	溫度測定
온도측정기	thermoscope	溫度測定器
온도측정법	thermometry	溫度測定法
온도층	temperature layer	溫度層
온도파	temperature wave	溫度波
온도편차	temperature departure	溫度偏差
온도풍	thermal wind	溫度風
온도풍방정식	thermal wind equation	溫度風方程式
온도효과	temperature effect	溫度效果
온도효율	temperature efficiency	溫度效率
온도효율비	temperature efficiency ratio	溫度效率比
온도효율지수	temperature efficiency index	溫度效率指數
온몸순환	systemic circulation	-循環
온사거-카시미르 상반관계	Onsagar-Casimir reciprocity relation	-相反關係
온수로	water warming channel	溫水路

온수못	water warming pond	溫水-
온수지	water warming pond	溫水池
온습계	thermohygrometer	溫濕計
온습도	hythergraph	溫濕圖
온습도계	hygrothermoscope	溫濕度計
온습도기록계	hygrothermograph	溫濕度記錄計
온습도기록곡선	hygrothermogram	溫濕度記錄曲線
온습도적외복사계	temperature humidity infrared radiometer	溫濕度赤外輻射計
온습선도	temperature moisture diagram	溫濕線圖
온습지수	temperature humidity index, temperature moisture index	溫濕指數
온실기체	greenhouse gas	溫室氣體
온실기체방출목록	greenhouse gas inventory	溫室氣體放出目錄
온실기후	glass-house (or greenhouse) climate	溫室氣候
온실날씨	greenhouse weather	溫室-
온실효과	greenhouse effect	溫室效果
온실효과강화	enhanced greenhouse effect	溫室效果强化
온/오프 기기	on and off instrument	-器機
온위	potential temperature	溫位
온위좌표계	potential temperature coordinate system	溫位座標系
온화한	mild	溫和
온흐림	overcast	
온흐림날	overcast day	
온흐림원	overcast circle	-圓
온흐림하늘	overcast sky	
올더 드리아스	Older Dryas	
올랜드사이클	Olland cycle	
올랜드원리	Olland principle	-原理
올레핀류	Olefins	-類
올리고세(신생대 제3기)	Oligocene	
올려본 각	angel of elevation	-角
옴	ohm	
옴법칙	Ohm's law	-法則
옹스트룀	Ångström	
옹스트룀방정식	Ångström equation	-方程式
옹스트룀보상	Ångström compensation	-補償
옹스트룀보상직달일사계	Ångström compensation pyrheliometer	-補償直達日射計
옹스트룀지구복사계	Ångström pyrgeometer	-地球輻射計
옹스트룀지수	Ångström exponent	-指數
와	vortex	渦
와도	vorticity	渦度
와도방정식	vorticity equation	渦度方程式
와도속	vorticity flux	渦度束
와도수송가설	vorticity transport hypothesis	渦度輸送假說

와도수송이론	vorticity transport theory	渦度輸送理論
와도이류	vorticity advection	渦度移流
와도전이	vorticity transfer	渦度轉移
와도최대	vorticity maximum	渦度最大
와도플럭스	vorticity flux	渦度-
와도효과	vorticity effect	渦度效果
와상관법	eddy correlation method	渦相關法
와위보존	conservation of potential vorticity	渦位保存
와위보존칙	law of potential vorticity conservation	渦位保存則
와일드바람막이(우량계의)	wild fence	
와일드증발계	Wild's evaporimeter	-蒸發計
와정리	vortex theorem	渦定理
완두콩수프안개	pea-soup fog	
완만경사	low slope	緩慢傾斜
완전결빙	freeze-up	完全結氷
완전기체	perfect gas, ideal gas	完全氣體
완전기체법칙	perfect gas law	完全氣體法則
완전동결	complete freeze-up	完全凍結
완전반사거울	perfect mirror	完全反射-
완전복사	perfect radiation	完全輻射
완전복사체	perfect radiator	完全輻射體
완전연소	complete(perfect) combustion	完全燃燒
완전예단	perfect prognostic	完全豫斷
완전예단방법	perfect prognostic method	完全豫斷方法
완전예보	perfect forecast	完全豫報
완전예상	perfect prognostic	完全豫想
완전유체	perfect fluid	完全流體
완전탄성충돌	complete(perfect) elastic collision	完全彈性衝突
완전해	exact solution	完全解
완전혼합	perfect mixing	完全混合
완전확산반사체	perfectly diffuse reflector	完全擴散反射體
완전확산복사체	perfectly diffuse radiator	完全擴散輻射體
완전흑체	perfect black body	完全黑體
완충용량	buffering capacity	緩衝容量
완충인자	buffer factor	緩衝因子
완충지대	buffer zone	緩衝地帶
완화법	relaxation method	緩和法
완화시간	relaxation time	緩和時間
왕복조류	reversing tidal current	往復潮流
왜곡수	distorted water	歪曲水
왜곡후류기류	distorted wake	歪曲後流氣流
왜도	skewness	歪度
왜도계수	coefficient of skewness	歪度係數
외각	exterior angle	外角

외계중력	extra-terrestrial gravitational force	外界重力
외기권	exosphere	外氣圈
외기단파복사	extra-terrestrial short-wave radiation	外氣短波輻射
외기복사	extra-terrestrial radiation	外氣輻射
외기온도	outside air temperature	外氣溫度
외기태양스펙트럼	extra-terrestrial solar spectrum	外氣太陽-
외력	external force	外力
외르스테드	oersted	
외부검정목표	external calibration target	外部檢定目標
외부경계층	external boundary layer	外部境界層
외부물순환	external water circulation	外部-循環
외부일	external work	外部-
외부전리권	outer ionosphere	外部電離圈
외부중력파	external gravity wave	外部重力波
외부파	external wave	外部波
외삽	extrapolation	外揷
외삽법	extrapolation method	外揷法
외압	external pressure	外壓
외적	cross product, outer product	外積
외전	abduction	外轉
외접	circumscription	外接
외접다각형	circumscribed polygon	外接多角形
외접무리	circumscribed halo	外接-
외접원	circumscribed circle	外接圓
외청역	outer audible zone, outer zone of audibility	外聽域
외해수	offshore water	外海水
외해예보	high seas forecast	外海豫報
외해초계기상	offshore patrol weather	外海哨戒氣象
외핵	outer core	外核
외행성	superior planet	外行星
요각류	copepod	境脚類
요동	air bump, fluctuation, perturbation	搖動
요동속도	fluctuation velocity	搖動速度
요드화은 에어로졸 발생장치	Alecto unit	-銀-發生裝置
요란	disturbance	擾亂
요란관리	disturbance management	擾亂管理
요란선	disturbance line	擾亂線
요소	element	要素
요소자료	element data	要素資料
요오드화은	silver iodide	-化銀
요오드화은 씨뿌리기	silver iodide seeding	-化銀-
요지저류량	depression storage	凹地貯溜量
욕탕플러그모양소용돌이	bath plug vortex	浴湯-模樣-
용량	capacity	容量

용량-반응	dose-response	用量反應
용량보정	capacity correction	容量補正
용량성소자	capacitive element	容量性素子
용량차원	capacity dimension	容量次元
용량형우량계	capacitance rain gauge	容量型雨量計
용매	solvent	溶媒
용수량	duty of water	用水量
용승(류)	upwelling	湧昇(流)
용액	solution	溶液
용액특성	colligative property	溶液特性
용오름	dragon, spout, tornado, twister, waterspout	龍-
용오름대	tornado belt	龍-帶
용적	bulk	容積
용적강우율	volumetric rain rate	容積降雨率
용적속도처리	volume velocity processing	容積速度處理
용적수분함량	volumetric water content	容積水分含量
용제	solvent	溶劑
용존산소량(DO)	dissolved oxygen amount	溶存酸素量
용질	solute	溶質
용해	dissolution	溶解
용해도	solubility	溶解度
용해도곱	solubility product	溶解度-
용해열	heat of dissolution, heat of solution	溶解熱
용해평형	equilibrium of dissolution	溶解平衡
우각호	oxbow lake	牛角湖
우계	rainy season	雨季
우기	rainy period, rainy season, wet period	雨期
우라늄연대측정법	uranium-series dating	-年代測定法
우량	rainfall	雨量
우량계	hyetometer, pluviometer, pluvioscope, rain gauge	雨量計
우량계바람막이	rain-gauge shield	雨量計-
우량계수	hyetal coefficient, pluviometric coefficient	雨量係數
우량계용바람막이	Alter shield	雨量計用-
우량관측소	rainfall station	雨量觀測所
우량구역	hyetal region	雨量區域
우량되	rain measuring glass	雨量-
우량-면적 곡선	depth-area curve	雨量面積曲線
우량-면적 공식	depth-area formula	雨量面積公式
우량-면적-지속기간 분석	depth-area-duration(DAD) analysis	雨量面積持續期間分析
우량비	pluviometric quotient	雨量比
우량승	rain measuring glass	雨量升
우량인자	rain factor	雨量因子
우량적도	hyetal equator	雨量赤道
우량-지속 곡선	depth-duration curve	雨量持續曲線

우량-지속-빈도 곡선	depth-duration-frequency curve	雨量持續頻度曲線
우량지수	pluvial index	雨量指數
우량측정법	pluviometry	雨量測定法
우림	rainforest, rain tree	雨林
우림기후	rainforest climate	雨林氣候
우박	hail	雨雹
우박경로	hail path	雨雹經路
우박구역	hailswath	雨雹區域
우박단계	hail stage	雨雹段階
우박대포	hail cannon	雨雹大砲
우박덩이	hailstone	雨雹-
우박돌기	hail lobe	雨雹突起
우박못	hail spike	雨雹-
우박방재	hail prevention	雨雹防災
우박보라	hail storm	雨雹-
우박생성역	hail generation zone	雨雹生成域
우박선	hailstreak	雨雹線
우박성장	hail growth	雨雹成長
우박성장역	hail growth zone	雨雹成長域
우박소나기	hail shower	雨雹-
우박소산	hail mitigation	雨雹消散
우박스톰	hail storm	雨雹-
우박신호	hail signal	雨雹信號
우박억제	hail suppression	雨雹抑制
우박억제로켓	anti-hail rocket	雨雹抑制-
우박억제포	anti-hail gun	雨雹抑制砲
우박연구실험	hail research experiment	雨雹研究實驗
우박오염	hail contamination	雨雹汚染
우박지대	hail fallout zone	雨雹地帶
우박지수	hail index	雨雹指數
우박측정판	hailpad	雨雹測定板
우박피해	hail damage	雨雹被害
우박핵	hail embryo	雨雹核
우발가뭄	accidental drought	偶發-
우수	rain water	雨水
우수량	rain water content	雨水量
우수시정	excellent visibility	優秀視程
우(세)시정	prevailing visibility	優(勢)視程
우식	rain erosion	雨蝕
우역	rain field	雨域
우연오차	accidental error	偶然誤差
우온비	pluviothermic ratio	雨溫比
우윳빛얼음	milky ice	牛乳-
우윳빛하늘	milky weather	牛乳-

우적	raindrop	雨滴
우적계	disdrometer	雨滴計
우적침식	raindrop erosion	雨滴浸蝕
우주	cosmos	宇宙
우주(공간)	space	宇宙(空間)
우주공간	aerospace	宇宙空間
우주근거하위계	space-based subsystem	宇宙根據下位系
우주기상	space weather	宇宙氣象
우주기상학	cosmic meteorology, cosmical meteorology, space meteorology	宇宙氣象學
우주기원방사성핵종	cosmogenic radionuclide	宇宙起源放射性核種
우주도킹	spacelink	宇宙-
우주라이다	space lidar	宇宙-
우주먼지	cosmic dust	宇宙-
우주복사	cosmic radiation	宇宙輻射
우주비행체	spaceflight	宇宙飛行體
우주선	cosmic ray	宇宙線
우주선소나기	cosmic ray shower	宇宙線-
우주선이온화	cosmic ray ionization	宇宙線-化
우주성방사성동위원소	cosmogenic radioisotope	宇宙性放射性同位元素
우주연구	space research	宇宙研究
우주왕복선	space shuttle	宇宙往復船
우주의학	space medicine	宇宙醫學
우주전력	space power	宇宙電力
우주전쟁	space warfare	宇宙戰爭
우주지원	space support	宇宙支援
우주지원작전기상	space support operations weather	宇宙支援作戰氣象
우주추적	space tracking	宇宙追跡
우주탐측기	space probe	宇宙探測器
우주통제작전기상	space control operations weather	宇宙統制作戰氣象
우주항행학	astronautics	宇宙航行學
우주환경감시기	space environmental monitor	宇宙環境監視器
우주환경센터	space environment center	宇宙環境-
우측원형편파	right circular polarization	右側圓形偏波
우측(오른쪽)이동뇌우	right-moving thunderstorm	右側移動雷雨
우표분포도	stamp map	郵票分布圖
운고	ceiling height, cloud ceiling, cloud height	雲高
운고계	ceilometer, cloud height meter	雲高計
운고관측등	ceiling light	雲高觀測燈
운고기구	ceiling balloon	雲高氣球
운고분류	ceiling classification	雲高分類
운고지시기	ceiling height indicator	雲高指示器
운고측정법	cloud height measurement method	雲高測定法
운고탐지기	ceiling detector	雲高探知器

운고투광기	ceiling projector	雲高投光器
운동	motion	運動
운동규모	scale of motion	運動規模
운동기관	locomotive organ	運動器官
운동량	momentum	運動量
운동량능률	moment of momentum	運動量能率
운동량방정식	momentum equation	運動量方程式
운동량보존법칙	law of momentum conservation, momentum conservation law	運動量保存法則
운동량보존	conservation of momentum	運動量保存
운동량전달	momentum transfer	運動量傳達
운동량플럭스	momentum flux	運動量-
운동마찰계수	coefficient of kinetic friction	運動摩擦係數
운동마찰력	force of kinetic friction	運動摩擦力
운동방정식	equation of motion	運動方程式
운동법칙	law of motion	運動法則
운동에너지	kinetic energy	運動-
운동에너지보존	conservation of kinetic energy	運動-保存
운동온도	kinetic temperature	運動溫度
운동전선	kinematic front	運動前線
운동점성계수	coefficient of kinematic viscosity	運動粘性係數
운동제1법칙	first law of motion	運動第一法則
운동학	kinematics	運動學
운동학방법	kinematic method	運動學方法
운동학적 경계조건	kinematic boundary condition	運動學的境界條件
운동학적 분석	kinematic analysis	運動學的分析
운동학적 조건	kinematic condition	運動學的條件
운동학적 플럭스	kinematic flux	運動學的-
운량	amount of clouds, cloud amount, cloud cover	雲量
운량감소	clearance, clearing	雲量減少
운량계	nephometer	雲量計
운량조건	operational cloud amount standard	雲量條件
운립자계수기	cloud particle counter	雲粒子計數器
운립자영상기	cloud particle imager	雲粒子映像機
운립자존데	cloud particle sonde	雲粒子-
운립자채집기	cloud particle sampler	雲粒子採集器
운립자카메라	cloud particle camera	雲粒子-
운반	conveyance	運搬
운석	aerolite, meteorite	隕石
운석우	meteorite shower	隕石雨
운영기상정보	operational meteorological information	運營氣象情報
운영일정	operation schedule	運營日程
운저고도	height of cloud base, level of cloud base	雲底高度
운적	cloud drop	雲滴

운적토	transported soil	運積土
운정	cloud top, top of cloud	雲頂
운정고도	cloud top height, level of cloud top	雲頂高度
운정기압	cloud top pressure	雲頂氣壓
운정온도	cloud top temperature	雲頂溫度
운정유입불안정도	cloud-top entrainment instability	雲頂流入不安定度
운정지시기	cloud-top indicator	雲頂指示器
운중	inside of cloud	雲中
운해	cloud sea	雲海
운형	cloud form	雲形
울로아고리	Ulloa's circle, Ulloa's ring	
울부짖는 60도	shrieking sixties	
움츠림	shrinking	
움케르효과	Umkehr effect	-效果
웅대구름	congestus	雄大-
웅대적운	cumulus congestus, cumulus congestus cloud	雄大積雲
워커순환	Walker circulation	-循環
워크먼-레이놀즈 효과	Workman-Reynolds effect	-效果
워터 스크러버	water scrubber	
원거리우주탐사체	deep-space probe	遠距離宇宙探査體
원거리장	far field	遠距離場
원거리통신작전기상	telecommunication operations weather	遠距離通信作戰氣象
원격관측소	observatory-remote	遠隔觀測所
원격광도측정법	telephotometry	遠隔光度測定法
원격명령	telecommand	遠隔命令
원격명령거리측정	telemetry command and ranging	遠隔命令距離測定
원격모드	remote mode	遠隔-
원격상관	teleconnection	遠隔相關
원격상관지수	teleconnection index	遠隔相關指數
원격제어	telecontrol	遠隔制御
원격제어장치	remote control equipment	遠隔制御裝置
원격탐사	remote sensing	遠隔探査
원격탐사센서	remote sensor	遠隔探査-
원격탐사신호	remote sensing signal	遠隔探査信號
원격탐사위성	remote sensing satellite	遠隔探査衛星
원격탐사응용	remote sensing application	遠隔探査應用
원격탐사전문조직	remote sensing professional organization	遠隔探査專門組織
원격탐측	remote sounding	遠隔探測
원격통신	telecommunication	遠隔通信
원궤점	apocenter	遠軌點
원기둥극좌표	cylindrical polar coordinates	圓-極座標
원기둥복사계	cylindrical radiometer	圓-輻射計
원기둥좌표	circular cylindrical coordinates, cylindrical coordinates	圓-座標

원기둥함수	cylindrical function	圓-函數
원대칭	circular symmetry	圓對稱
원둘레	circumference	圓-
원둘레속도	circumferential velocity	圓-速度
원리	principle	原理
원부	original record	原簿
원뿔	cone	圓-
원뿔각	cone angle	圓-角
원뿔곡선	conic section	圓-曲線
원뿔복사계	cone radiometer	圓-輻射計
원뿔빔	conical beam	圓-
원뿔주사	conical scanning	圓-走査
원뿔형수위강하	cone of depression	圓-型水位降下
원뿔형형성	coning	圓-型形成
원생대기후	Proterozoic climate	原生代氣候
원소	element	元素
원소탄소	elemental carbon	元素炭素
원소탄소미세분진	particulate elemental carbon	元素炭素微細粉塵
원소탄소에어로졸	elemental carbon aerosol	元素炭素-
원시대기	primitive air	原始大氣
원시방정식	primitive equation	原始方程式
원시방정식계	primitive equation system	原始方程式系
원시방정식모형	primitive equation model	原始方程式模型
원시생명체	primitive organism	原始生命體
원시인류	primitive man, protothropic man	原始人類
원시자료	crude data	原始資料
원식생	original vegetation	原植生
원심가속도	centrifugal acceleration	遠心加速度
원심력	centrifugal force	遠心力
원심분리기	centrifuge	遠心分離機
원심불안정	centrifugal instability	遠心不安定
원심퍼텐셜	centrifugal potential	遠心-
원심함수당량	centrifuge moisture equivalent	遠心含水當量
원심효과	centrifugal effect	遠心效果
원운동	circular motion	圓運動
원인-결과 관계	cause-effect relationship	原因結果關係
원일점	aphelion	遠日點
원자	atom	原子
원자결정	atomic crystal	原子結晶
원자구조	atomic structure	原子構造
원자기호	atomic symbol	原子記號
원자량	atomic weight	原子量
원자력	nuclear power	原子力
원자료	raw data	原資料

원자모형	atomic model	原子模型
원자번호	atomic number	原子番號
원자질량	atomic mass	原子質量
원적외복사	far-infrared radiation	遠赤外輻射
원적외선	far-infrared	遠赤外線
원점	origin	原點
원주각	angle of circumference, inscribed angle	圓周角
원지점	apogee	遠地點
원지점바람	apogean wind	遠地點-
원지점조석	apogean tide	遠地點潮汐
원지점조차	apogean range	遠地點潮差
원진동	circular oscillation	圓振動
원진동수	circular frequency	圓振動數
원천	source	源泉
원천함수	source function	源泉函數
원추곡선회전면	toroid	圓錐曲線回轉面
원추형 주사	conical scan	圓錐形走査
원추형 플룸	coning plume	圓錐形-
원칙	principle	原則
원통도법	cylindrical projection	圓筒圖法
원통좌표계	cylindrical coordinate system	圓筒座標系
원통형설량계	cylindrical snow gauge	圓筒形雪量計
원통형우량계	cylindrical rain gauge	圓筒形雨量計
원판경도계	disk hardness-gauge	圓板硬度計
원편광	circular polarization	圓偏光
원편광회복률	circular depolarization ratio	圓偏光回復率
원편파	circular polarized wave	圓偏波
원형도파관	circular waveguide	圓形導波管
원형변수	circular variable	圓形變數
원형소용돌이	circular vortex	圓形-
원형소용돌이 불안정도	instability of circular vortex	圓形-不安定度
원형수평운동	circular horizontal motion	圓形水平運動
원형씨뿌리기	circle-type seeding	圓形-
원형의 두터운 구름	central dense overcast	圓形-
원형편파비	circular polarization ratio	圓形偏波比
월간예보	monthly forecast, one-month forecast	月間豫報
월강수량	monthly precipitation	月降水量
월류	overflow	越流
월리스-코스키파	Wallace-Kousky wave	-波
월별기록	monthly record	月別記錄
월별 일평균최고온도	mean daily maximum temperature for a month	月別日平均最高溫度
월별 일평균최저온도	mean daily minimum temperature for a month	月別日平均最低溫度
월별평년일기도	normal monthly mean weather chart	月別平年日氣圖
월보	monthly bulletin	月報

월식	lunar eclipse	月食
월절대최고온도	absolute monthly maximum temperature	月絶對最高溫度
월절대최저온도	absolute monthly minimum temperature	月絶對最低溫度
월조간격	lunitidal interval	月潮間隔
월총강수량	monthly total precipitation	月總降水量
월최고기온	monthly maximum temperature	月最高氣溫
월최저기온	monthly minimum temperature	月最低氣溫
월평균	monthly mean	月平均
월평균등치선	isomenal	月平均等值線
월평균온도	mean monthly temperature	月平均溫度
월평균최고온도	mean monthly maximum temperature	月平均最高溫度
월평균최저온도	mean monthly minimum temperature	月平均最低溫度
웨렌스키올드선도	Werenskiold diagram	-線圖
웨버법칙	Weber's law	-法則
웨버수	Weber number	-數
웨버-페히너 법칙	Weber-Fechner law	-法則
위갠드시정계	Wigand visibility meter	-視程計
위급농도	emergency concentration	危急濃度
위단열	pseudo-adiabatic	僞斷熱
위단열감률	pseudo-adiabatic lapse rate	僞斷熱減率
위단열과정	pseudo-adiabatic process	僞斷熱過程
위단열대류	pseudo-adiabatic convection	僞斷熱對流
위단열변화	pseudo-adiabatic change	僞斷熱變化
위단열선	pseudo-adiabat	僞斷熱線
위단열선도	pseudo-adiabatic chart, pseudo-adiabatic diagram	僞斷熱線圖
위단열팽창	pseudo-adiabatic expansion	僞斷熱膨脹
위도	latitude	緯度
위도-경도 격자	latitude-longitude grid	緯度-經度格子
위도편각	argument of latitude	緯度偏角
위도변동	latitudinal fluctuation	緯度變動
위도효과	latitude effect	緯度效果
위불안정	pseudo-instability	僞不安定
위상각	phase angle	位相角
위상감지기	phase detector	位相感知器
위상검출기	phase detector	位相檢出器
위상검파	phase detection	位相檢波
위상경로	phase path	位相經路
위상공간	phase space	位相空間
위상당온도	pseudo-equivalent temperature	僞相當溫度
위상당온위	pseudo-equivalent potential temperature	僞相當溫位
위상도	phase diagram	位相圖
위상배열	phase array, phased array	位相配列
위상배열레이더	phase array radar	位相配列-

위상배열안테나	phased array antenna	位相配列-
위상벡터선도	phasor diagram	位相-線度
위상변이	phase shift	位相變移
위상변조	phase modulation	位相變調
위상변화	phase change	位相變化
위상보정	phase calibration	位相補正
위상분산방식	phase diversity method	位相分散方式
위상불안정	phase instability	位相不安定
위상상수	phase constant	位相常數
위상속도	phase velocity, phase speed	位相速度
위상속력	phase speed, phase velocity	位相速力
위상스펙트럼	phase spectrum	位相-
위상시변조	phase time modulation	位相時變調
위상신호	burst, U burst	位相信號
위상유체	phase fluid	位相流體
위상 이그러짐	phase distortion	位相-
위상일치	in-phase(I-component)	位相一致
위상잡음	phase noise	位相雜音
위상전선(면)	phase front	位相前線(面)
위상점	phase point	位相點
위상정합 도플러 라이다	coherent Doppler lidar	位相整合-
위상정합형	coherent type	位相整合型
위상지연	phase delay, phase lag	位相遲延
위상평균	phase averaging	位相平均
위상함수	phase function	位相函數
위상행렬	phase matrix	位相行列
위상흐름	phase flow	位相-
위생기상학	hygiene meteorology, hygieno meteorology	衛生氣象學
위생제거지대	sanitary clearance zone	衛生除去地帶
위성	satellite	衛星
위성관측	satellite observation	衛星觀測
위성구름사진	satellite cloud picture	衛星-寫眞
위성궤도	satellite orbit	衛星軌道
위성기상관측	satellite meteorological observation	衛星氣象觀測
위성기상학	satellite meteorology	衛星氣象學
위성대역	satellite band	衛星帶域
위성도표화	satellite mapping	衛星圖表化
위성-레이더 조합영상	combined satellite and radar imagery	衛星-組合映像
위성메인컴퓨터	spacecraft computer unit	衛星-
위성바람	satellite wind	衛星-
위성발사 및 초기 궤도 　진입 운영단계	Launch and Early Operation Phase	衛星發射-初期 軌道進入 運營段階
위성발사 연직 대기분포	launch profile	衛星發射鉛直大氣分布
위성밴드	satellite band	衛星-

위성버스	satellite bus	衛星-
위성사진	satellite picture	衛星寫眞
위성사진항법	satellite picture navigation	衛星寫眞航法
위성수명	lifetime of satellite	衛星壽命
위성수문프로그램	satellite hydrology program	衛星水文-
위성영상	satellite image, satellite imagery	衛星映像
위성영상분석시스템	satellite image analysis system	衛星映像分析-
위성위치잡기(배치)	positioning of satellite	衛星位置-(配置)
위성이미지	satellite image	衛星-
위성자료분배시스템	satellite data distribution system	衛星資料分配-
위성자료중계	satellite relay of data	衛星資料中繼
위성자세	attitude of satellite	衛星姿勢
위성적외분광계	satellite infrared spectrometer	衛星赤外分光計
위성직하점	subsatellite point	衛星直下點
위성직하점궤도	subsatellite point track	衛星直下點軌道
위성천정각	satellite zenith angle	衛星天頂角
위성추적안테나	satellite tracking antenna	衛星追跡-
위성출현	satellite appearance	衛星出現
위성탐측	satellite sounding, sounding by satellite	衛星探測
위성탑재체 접속장치	modular payload interface unit	衛星搭載體接續裝置
위성투사	satellite projection	衛星投射
위스콘신빙하작용	Wisconsinian (Weichselian, Wurm) glaciation	-氷河作用
위습구온도	pseudo wet-bulb temperature	僞濕球溫度
위습구온위	pseudo wet-bulb potential temperature, wet-bulb pseudo-potential temperature	僞濕球溫位
위신호	aliasing	僞信號
위신호보정	dealiasing	僞信號補整
위신호오차	aliasing error	僞信號誤差
위온위	pseudo-potential temperature	僞溫位
위잠재불안정	pseudo-latent instability	僞潛在不安定
위전선	pseudo front	僞前線
위축	stooping	萎縮
위치	position, potential	位置
위치감각	sense of position	位置感覺
위치굴절률	potential index of refraction, potential refractive index	位置屈折率
위치밀도	potential density	位置密度
위치벡터	position vector	位置-
위치불안정도	kinetic instability	位置不安定度
위치상당온도	potential equivalent temperature	位置相當溫度
위치소용돌이도	potential vorticity	位置-度
위치소용돌이도보존	conservation of potential vorticity	位置-度保存
위치습구온도	potential wet-bulb temperature	位置濕球溫度
위치에너지	potential energy	位置-

위치온도	potential temperature	位置溫度
위치위상당온도	potential pseudo-equivalent temperature	位置僞相當溫度
위치위습구온도	potential pseudo-wet-bulb temperature	位置僞濕球溫度
위치정보시스템	Global Positioning System(GPS)	位置情報-
위치지시기	position indicator	位置指示器
위치체적	potential volume	位置體積
위치통보	position report	位置通報
위 캘리브레이션	miscalibration	僞-
위폐색	pseudo-occlusion	僞閉塞
위한랭전선	pseudo cold-front	僞寒冷前線
위해대기오염물질 국가배출기준	National Emissions Standards for Hazardous Air Pollutants	危害大氣汚染物質 國家排出基準
위험간섭	dangerous interference	危險干涉
위험국지폭풍우	severe local storm	危險局地暴風雨
위험국지폭풍우단위	severe local storm unit	危險局地暴風雨單位
위험기상	severe weather, weather hazard	危險氣象
위험기상가능성보고서	severe weather potential statement	危險氣象可能性報告書
위험기상관리상담	hazardous weather management consultation	危險氣象管理相談
위험기상보고서	severe weather statement	危險氣象報告書
위험기상분석	severe weather analysis	危險氣象分析
위험기상예보	severe weather forecast	危險氣象豫報
위험기상요소	severe weather element	危險氣象要素
위험기상위협지수	severe weather threat index	危險氣象威脅指數
위험기상전망	hazardous weather outlook	危險氣象展望
위험기상정보	severe weather information, SIGMET information	危險氣象情報
위험기상징후	severe weather signature	危險氣象徵候
위험기상확률	severe weather probability	危險氣象確率
위험날씨	hazardous weather	危險-
위험뇌우	severe thunderstorm	危險雷雨
위험뇌우경보	severe thunderstorm warning	危險雷雨警報
위험뇌우전망	severe thunderstorm outlook	危險雷雨展望
위험뇌우감시보	severe thunderstorm watch	危險雷雨監視報
위험도	risk	危險度
위험도평가	hazard assessment, risk assessment	危險度評價
위험바다감시보	hazardous seas watch	危險-監視報
위험바다경보	hazardous seas warning	危險-警報
위험반원	dangerous half, dangerous semicircle	危險半圓
위험비행기상주의보서비스	hazardous in-flight weather advisory service	危險飛行氣象注意報-
위험선	danger line	危險線
위험열대폭풍우	severe tropical storm	危險熱帶暴風雨
위험일기	weather hazard	危險日氣
위험폭풍우	severe rain storm	危險暴風雨
위협점수	threat score	威脅點數
윈드시어	wind shear	

윈드시어경계	wind shear alert	-警戒
윈드시어경보	wind shear alert, wind shear warning	-警報
윈드시어연직분포	wind shear profile	-鉛直分布
윌리-윌리	willy-willy	
윌슨구름상자	Wilson cloud-chamber	-箱子
윌슨주기	Wilson cycle	-周期
윗줄	overbar	
유계성(해의)	boundedness	有界性
유공충	coccolith	有孔蟲
유관	stream tube	流管
유기과산화물	organic peroxide	-有機過酸化物
유기구름씨뿌리기물질	organic cloud seeding material	有機-物質
유기물	organic matter	有機物
유기물소나기	shower of organic matter	有機物-
유기물질	organic matter	有機物質
유기산	organic acid	有機酸
유기염소류	organochlorine	有機鹽素類
유기염소화합물	organochlorine compound	有機鹽素化合物
유기인제	organophosphate	有機燐劑
유기질산염	organic nitrate	有機窒酸鹽
유기질산염화합물	organic nitrate compound	有機窒酸鹽化合物
유기탄소 부주기	organic carbon subcycle	有機炭素副週期
유기탄소	organic carbon	有機炭素
유기지화학적 프록시	organic geochemical proxy	有機地化學的-
유기화합물	organic compound	有機化合物
유도기전력	induced electromotive force	誘導起電力
유도단위	derived unit	誘導單位
유도돌풍속도	derived gust velocity	誘導突風速度
유도뢰	inductive lightning	誘導雷
유도방출	stimulated emission	誘導放出
유도법	induction method	誘導法
유도시험	induction experiment	誘導試驗
유도전류	induced current	誘導電流
유도전하	induced charge	誘導電荷
유도체	derivative	誘導體
유라시아고기압	Eurasian high	-高氣壓
유라시아 습윤대륙기후	Eurasian humid continental climate	-濕潤大陸氣候
유량계	flow meter, stream gauge	流量計
유량지속곡선	flow duration curve	流量持續曲線
유량측정	discharge measurement	流量測定
유럽 극궤도위성	MetOp	-極軌道衛星
유럽기상용통신망	Meteorological Operational Telecommunications Network Europe	-氣象用通信網
유럽 기상위성	Meteosat	-氣象衛星

유럽중기예보센터	European Centre for Medium-range Weather Forecast(ECMWF)	-中期豫報-
유령에코	phantom echo	幽靈-
유로	channel	流路
유로흐름	channel flow	流路-
유리관 수은온도계	mercury-in-glass thermometer	-管水銀溫度計
유리관액체 자기온도계	liquid-glass thermograph	-管液體自記溫度計
유리저기압	breakaway depression	遊離低氣壓
유막(수면의)	oil slick	油膜
유발기구	triggering mechanism	誘發機構
유방(구름)	mamma	乳房-
유방고적운	altocumulus mamma	乳房高積雲
유방고층운	altostratus mamma	乳房高層雲
유방구름	mammato cloud, mammatus, pocky cloud	乳房-
유방권운	cirrus mamma	乳房卷雲
유방권적운	cirrocumulus mamma	乳房卷積雲
유방적란운	cumulonimbus mammatus	乳房積亂雲
유방적운	mammato cumulus	乳房積雲
유방층적운	stratocumulus mammatus	乳房層積雲
유빙	drift ice	遊氷
유사	quasi-	類似
유사	shifting sand	流砂
유사(물)	analog(ue)	類似(物)
유사(성)	similarity	類似(性)
유사모형	analogue model, analogus model	類似模型
유사법	analogue method	類似法
유사성	resemblance, similitude	類似性
유사예보	analog forecasting	類似豫報
유사일기도	analogous map, similarity weather chart	類似日氣圖
유선	streamline	流線
유선곡률	streamline curvature	流線曲率
유선도	streamline chart	流線圖
유선류	streamline flow	流線流
유선분석	streamline analysis	流線分析
유선장	streamline field	流線場
유선저울우량계	wire-weight gauge	有線-雨量計
유선진폭	streamline amplitude	流線振幅
유선패턴	streamline pattern	流線-
유선함수	stream function	流線函數
유선형	streamlined shape	流線形
유선형	streamline pattern	流線型
유성	meteor, shooting star	流星
유성먼지	meteoric dust	流星-
유성소나기	meteoric shower	流星-

유성우	meteor shower	流星雨
유성체	meteoroid	流星體
유성흔적	meteor trail	流星痕迹
유속계	current meter	流速計
유속단면	velocity cross section	流速斷面
유속연직분포	current profile	流速鉛直分布
유속측정막대기	current pole	流速測定-
유수	cascading water	流水
유수지	retarding basin	留水池
유숙기	milk ripe stage	乳熟期
유역	basin, catchment, catchment area, drainage area, watershed	流域
유역경계	catchment boundary	流域境界
유역계산	basin accounting	流域計算
유역면적	catchment basin, drainage area, drainage basin, watershed	流域面積
유역반응	basin response	流域反應
유역배출구	basin outlet	流域排出口
유역분수계	drainage divide	流域分水界
유역지체	basin lag	流域遲滯
유역총유출양	catchment yield	流域總流出量
유역평균고도	mean elevation of basin	流域平均高度
유역평균폭	mean width of basin	流域平均幅
유역함양	basin recharge	流域涵養
유연성	plasticity	柔軟性
유의고도	significant level	有意高度
유의도	significance	有意度
유의성	significance	有意性
유의성검정	test of significance, significance test	有意性檢定
유의수준	significance level	有意水準
유의파	significant wave	有意波
유의파고	significant wave height	有意波高
유인관측소	attended station	有人觀測所
유인기구	manned balloon	有人氣球
유인비	trainment ratio	誘引比
유입	entrainment	流入
유입(량)	inflow	流入(量)
유입각	angle of indraft	流入角
유입계면층	entrainment interfacial layer	流入界面層
유입계수	entrainment coefficient	流入係數
유입매개변수	entrainment parameter	流入媒介變數
유입비	entrainment ratio	流入比
유입상수	entrainment constant	流入常數
유입역	entrainment zone	流入域

유입-저수-유출곡선	inflow-storage-discharge curve	流入貯水流出曲線
유입축	axis of inflow	流入軸
유잉-돈이론	Ewing-Donn theory	-理論
유적	trajectory	流跡
유적곡률	trajectory curvature	流跡曲率
유적분석	trajectory analysis	流跡分析
유적선	path line	流跡線
유적선방정식	equation of trajectory	流跡線方程式
유전	oil field	油田
유전물질	deelectric material	誘電物質
유전상수	deelectric constant	誘電常數
유전율	permittivity, dielectric constant, dielectric factor	誘電率
유전체	dielectric, dielectric material	誘電體
유조선	oil tanker	油槽船
유지호흡	maintenance respiration	維持呼吸
유질혈암	oil shale	油質頁巖
유체	fluid	流體
유체광도계	hydrophotometer	流體光度計
유체덩이	fluid parcel	流體-
유체동압력	hydrodynamic pressure	流體動壓力
유체동역학	hydrokinetics	流體動力學
유체면	fluid surface	流體面
유체부피	fluid volume	流體-
유체선속	fluid flux	流體線束
유체역학	fluid dynamics, fluid mechanics, hydrodynamics, hydromechanics	流體力學
유체역학 불안정	hydrodynamic instability	流體力學不安定
유체역학 안정	hydrodynamic stability	流體力學安定
유체역학적 거친면	hydrodynamically rough surface	流體力學的-面
유체요소	fluid element	流體要素
유체자기방출	hydromagnetic emission	流體磁氣放出
유체자기역학	hydromagnetics	流體磁氣力學
유체저항	fluid resistance	流體抵抗
유체적분	fluid integral	流體積分
유체정압	hydrostatic pressure	流體靜壓
유체정역학	hydrostatics	流體靜力學
유체지구	fluid earth	流體地球
유체포유물	fluid inclusion	流體包有物
유체흐름	fluid flow	流體-
유체흐름형	fluid flow type	流體-型
유출	detrainment, discharge, effluent, efflux, outflow, runoff	流出
유출(량)	discharge, outflow, runoff	流出(量)
유출경계	outflow boundary	流出境界

유출계수	coefficient of runoff, discharge coefficient, flood coefficient	流出係數
유출고	depth of run off	流出高
유출곡선	discharge curve	流出曲線
유출나선	outflow spiral	流出螺線
유출량곡선	discharge mass curve	流出量曲線
유출량구분선	discharge section line	流出量區分線
유출물	effluent	流出物
유출바람	outflow wind	流出-
유출속도	efflux velocity	流出速度
유출수문곡선	outflow hydrograph	流出水文曲線
유출수분산	overwater dispersion	流出水分散
유출순환	runoff cycle	流出循環
유출역	discharge area	流出域
유출인자	runoff factor	流出因子
유출제트	outflow jet	流出-
유출천	effluent stream	流出川
유출축	axis of outflow	流出軸
유출층	outflow layer	流出層
유출함양삼출	effluent seepage	流出涵養滲出
유클리드	Euclid	
유클리드공간	Euclidian space	-空間
유클리드표준	Euclidian norm	-標準
유한수열	finite sequence	有限數列
유한요소법	finite element method	有限要素法
유한집합	finite set	有限集合
유한차분	finite difference, finite differencing	有限差分
유한차분근사	finite-difference approximation	有限差分近似
유한차분법	method of finite difference	有限差分法
유한체	finite field	有限體
유해대기오염물질	air toxic, hazardous air pollutant	有害大氣汚染物質
유해분진	hazardous dust	有害粉塵
유해항력	parasite drag	有害抗力
유해효과	adverse effect	有害效果
유형	pattern	類型
유형인식	pattern recognition	類型認識
유효가강수량	effective precipitable water	有效可降水量
유효강수량	effective precipitation	有效降水量
유효관측속도	Nyquist velocity	有效觀測速度
유효습도	effective humidity	有效濕度
유형상관	pattern correlation	類型相關
유효강우(량)	effective rainfall amount	有效降雨(量)
유효강우도표	effective rainfall hyetograph	有效降雨圖表
유효공극률	effective porosity	有效孔隙率

유효광학두께	effective optical thickness	有效光學-
유효구간	effective range	有效區間
유효구경	effective aperture	有效口徑
유효굴뚝높이	effective stack height	有效-
유효기간	period of validity	有效期間
유효높이	effective height	有效-
유효눈녹음	effective snow melt	有效-
유효대기	effective atmosphere	有效大氣
유효돌풍속도	effective gust velocity	有效突風速度
유효등방복사능	effective isotropic radiation power	有效等方輻射能
유효레이더반사도	effective radar reflectivity	有效-反射度
유효레이더반사도인자	effective radar reflectivity factor	有效-反射度因子
유효면적	effective area	有效面積
유효물함량	effective liquid water content	有效-含量
유효바람	effective wind	有效-
유효반경	effective radius	有效半徑
유효방사온도	effective radiating temperature	有效放射溫度
유효방출높이	effective release height	有效放出-
유효복사	effective radiation	有效輻射
유효산란단면적	effective cross section	有效散亂斷面積
유효생육계절	effective growing season	有效生育季節
유효수분	available moisture	有效水分
유효수분함량	available water capacity	有效水分含量
유효숫자	significant digit	有效數字
유효시	valid time	有效時
유효시간	valid time	有效時間
유효시정	effective visibility	有效視程
유효야간복사	effective nocturnal radiation	有效夜間輻射
유효오염원높이	effective source height	有效汚染源-
유효온도	effective temperature	有效溫度
유효우량	effective rainfall amount	有效雨量
유효운량	effective cloud amount	有效雲量
유효장파복사	effective longwave radiation	有效長波輻射
유효적산온도	effective accumulated temperature	有效積算溫度
유효전천일사계	effective pyranometer	有效全天日射計
유효중력	effective gravity	有效重力
유효증발산	effective evapotranspiration	有效蒸發散
유효지구반경	effective earth radius, effective radius of the earth	有效地球半徑
유효지구반지름	effective earth radius, effective radius of the earth	有效地球半-
유효지구복사	effective terrestrial radiation	有效地球輻射
유효지형도	effective topography	有效地形圖
유효충돌	effective collision	有效衝突
유효풍속	effective wind speed	有效風速
유효풍향	effective downwind direction	有效風向

유효하중	payload	有效荷重
유효항력계수	effective drag coefficient	有效抗力係數
유효핵전하	effective nuclear charge	有效核電荷
육각기둥	hexagonal column	六角-
육각판	hexagonal platelet	六角板
육방계	hexagonal system	六方系
육분의 지표막대	index arm of sextant	六分儀指標-
육상관측소	land station	陸上觀測所
육상관측항공기	observation landplane	陸上觀測航空機
육상광역예보구역	domestic regional forecast	陸上廣域豫報區域
육상국지예보구역	domestic local forecast	陸上局地豫報區域
육상기상경보구역	region for land forecast	陸上氣象警報區域
육상기압계	land barometer	陸上氣壓計
육상단기예보	domestic short range forecast	陸上短期豫報
육상예보	land forecast	陸上豫報
육상일일예보	domestic daily forecast	陸上日日豫報
육상정찰기	patrol landplane	陸上偵察機
육상특보	land weather warning	陸上特報
육상특보구역	region for land weather warning	陸上特報區域
육성층	continental sediment	陸成層
육안관측	macroscopic observation	肉眼觀測
육종법	breeding method	育種法
육지관측소	land station	陸地觀測所
육지바람전선	land breeze front	陸地-前線
육지안개	land fog	陸地-
육지얼음	land ice	陸地-
육지용오름	landspout	陸地龍-
육풍	land breeze	陸風
육풍순환	land breeze circulation	陸風循環
육풍전선	land breeze front	陸風前線
윤달	leap month	閏-
윤일	intercalary day	閏日
윤회성퇴적작용(윤회층)	cyclic sedimentation	輪廻性堆積作用(輪廻層)
율	modulus, power	率
융빙수퇴적물	glaciofluvial sediment	融氷水堆積物
융빙유수퇴적물	outwash	融氷流水堆積物
융삭대	zone of ablation	融削帶
융삭영역	zone of ablation	融削領域
융설	melting of snow	融雪
융설촉진	acceleration of melting snow	融雪促進
융설홍수	snowmelt flooding	融雪洪水
융에 에어로졸층	Junge aerosol layer	-層
융점강하	lowering melting point	融點降下
융합	merging	融合

융해	fusion	融解
융해땅	talik	融解-
융해열	heat of fusion	融解熱
융해잠열	latent heat of fusion	融解潛熱
융해층	melting layer	融解層
으르렁40도	roaring forties	-度
은량분석	silver analysis	銀量分析
은반일사계	silver-disk pyrheliometer	銀盤日射計
은상고대	silver thaw	銀-
은서리	silver frost	銀-
은폭풍	silver storm	銀暴風
은하수	milky way	銀河水
음	sound	音
음경사골	negative-tilt trough	陰傾斜-
음극	negative pole	陰極
음극선관	cathode-ray tube	陰極線管
음극선무선방위계	cathode-ray radiogoniometer	陰極線無線方位計
음극선방향결정	cathode-ray direction finding	陰極線方向決定
음극선오실로그래프	cathode-ray oscillograph	陰極線-
음극선오실로스코프	cathode-ray oscilloscope	陰極線-
음대전비	negative rain	陰帶電-
음되먹임	negative feedback	陰-
음되먹임기작	negative feedback mechanism	陰-機作
음되먹임시스템	negative feedback system	陰-
음등온소용돌이도이류	negative isothermal vorticity advection	陰等溫-度移流
음력	lunar calendar	陰曆
음부력	negative buoyancy	陰浮力
음부호	negative sign	陰符號
음상관관계	negative correlation	陰相關關係
음성주파수	voice frequency	音聲周波數
음소용돌이도	negative vorticity	陰-度
음소용돌이도이류	negative vorticity advection	陰-度移流
음속	sonic speed, sound speed, sound velocity, speed of sound	音速
음압수준	sound pressure level	音壓水準
음압수준계	sound pressure level meter	音壓水準計
음영역	negative area	陰領域
음이온	anion, negative ion	陰-
음전기	negative electricity, resinuous electricity	陰電氣
음전자	negative electron, negatron	陰電子
음전하	negative charge	陰電荷
음점성	negative viscosity	陰粘性
음축	negative axis	陰軸
음파	acoustic wave, sonic wave, sound wave	音波

음파가온도	sound virtual temperature	音波假溫度
음파강도	acoustic intensity	音波强度
음파공명	acoustic resonance	音波共鳴
음파굴절	acoustic refraction	音波屈折
음파높낮이	acoustic topography	音波-
음파레이더	acoustic detection and ranging, acoustic radar	音波-
음파렌즈	acoustic lens	音波-
음파반사율	acoustic reflectivity	音波反射率
음파반사탐사	acoustic reflecting profiling	音波反射探查
음파반향	acoustic reverberation	音波反響
음파발생기	acoustic frequency generator	音波發生機
음파산란	acoustic scattering	音波散亂
음파속도계	acoustic velocimeter	音波速度計
음파송수신기	acoustic transponder	音波送受信機
음파신호	acoustic signature	音波信號
음파압력	acoustic pressure	音波壓力
음파에코측심기	acoustic echo sounding	音波-測深器
음파영상	acoustic imaging	音波映像
음파온도계	acoustic thermometer, sonic thermometer	音波溫度計
음파우량계	acoustic rain gauge	音波雨量計
음파측심장치	echo sounding apparatus	音波測深裝置
음파탐사센서	air acoustic ranging sensor	音波探查-
음파탐측	acoustic sounding	音波探測
음파탐측기	acoustic sounder	音波探測機
음파풍속계	sonic anemometer	音波風速計
음파풍속온도계	sonic anemometer thermometer	音波風速溫度計
음파해류계	acoustic ocean current meter	音波海流計
음향반사운	acoustic cloud	音響反射雲
음향변환기	acoustic transducer	音響變換機
음향분산	acoustic dispersion	音響分散
음향섬광	acoustic scintillation	音響閃光
음향임피던스	acoustic impedance	音響-
음향중력파	acoustic gravity wave	音響重力波
음향출력	sound power	音響出力
음향학	acoustics	音響學
음향후방산란	acoustic backscattering	音響後方散亂
응결	condensation, congelation	凝結
응결고도	condensation level	凝結高度
응결과정	condensation process	凝結過程
응결단계	condensation stage	凝結段階
응결단열선	condensation adiabat	凝結斷熱線
응결성	condensability	凝結性
응결성장과정	condensation growth process	凝結成長過程
응결수	sweat	凝結水

응결압력	condensation pressure	凝結壓力
응결열	heat of condensation, heat of congelation	凝結熱
응결온도	condensation temperature	凝結溫度
응결자국	condensation trail	凝結-
응결층	condensation layer	凝結層
응결파	condensation wave	凝結波
응결한계	condensation limit	凝結限界
응결핵	condensation nucleus	凝結核
응결핵계수기	condensation nucleus counter	凝結核計數器
응결효율	condensation efficiency	凝結效率
응고	solidification	凝固
응고열	heat of solidification	凝固熱
응고점	solidifying point	凝固點
응답기	responder	應答器
응력	stress	應力
(응력)변형	strain	(應力)變形
응력변형텐서	strain tensor	應力變形-
응력일정층	constant stress layer	應力一定層
응력텐서	stress tensor	應力-
응용기상	applied weather	應用氣象
응용기상학	applied meteorology	應用氣象學
응용기술위성	application technology satellite	應用技術衛星
응용기후학	applied climatology	應用氣候學
응용기후학자	applied climatologist	應用氣候學者
응용수문학	applied hydrology	應用水文學
응용종관기후학연구	applied synoptic climatological study	應用綜觀氣候學研究
응집	cohesion	凝集
응집구조	coherent structure	凝集構造
응회석	tuff	凝灰石
의료기상학	medical meteorology	醫療氣象學
의료기후학	medical climatology	醫療氣候學
의복기후	clothing climate	衣服氣候
의사결정	decision making	意思決定
의학지리	geomedicine	醫學地理
이가원소	dyad	二價元素
이가원소(의)	dyadic	二價元素
이극관	diode	二極管
이극관레이저	diode laser	二極管-
이년바람진동	biennial wind oscillation	二年-振動
이년진동	biennial oscillation	二年振動
이동	migration	移動
이동관측소	moving observation station	移動觀測所
이동기상관측소	mobile weather station	移動氣象觀測所
이동도	mobility	移動度

이동목표물지시기	moving target indicator	移動目標物指示器
이동발생원	mobile source	移動發生源
이동보정	removal correction	移動補正
이동선박	mobile ship	移動船舶
이동선박관측소	mobile ship station	移動船舶觀測所
이동성	migratory, mobility	移動性
이동성고기압	migratory anticyclone, migratory high, travelling anticyclone	移動性高氣壓
이동성안장부	migratory col	移動性鞍裝部
이동성온대저기압	migratory extratropical anticyclone	移動性溫帶低氣壓
이동성저기압	migratory cyclone	移動性低氣壓
이동식기상관측시스템	mobile weather observation system	移動式氣象觀測-
이동자기압계	movable-scale barometer	移動-氣壓計
이동취주거리	moving fetch	移動吹走距離
이동평균	consecutive mean, moving average, running mean	移動平均
이동형레이더	mobile radar	移動形-
이디문제	Eady problem	-問題
이랑관개	furrow irrigation	-灌漑
이력(현상)	hysteresis	履歷(現想)
이론기상학	theoretical meteorology	理論氣象學
이류	advection	移流
이류가설	advective hypothesis	移流假說
이류경계층	advective boundary layer	移流境界層
이류규모	advection scale	移流規模
이류모형	advection model, advective model	移流模型
이류방정식	advection equation	移流方程式
이류변화	advective change	移流變化
이류서리	advection frost	移流-
이류성뇌우	advective thunderstorm	移流性雷雨
이류솔레노이드	advection solenoid	移流-
이류시간규모	advection time scale	移流時間規模
이류안개	advection fog	移流-
이류압력경향	advective pressure tendency	移流壓力傾向
이류역	advection region, advective region	移流域
이류역전	advectional inversion	移流逆轉
이류역전층	advection inversion layer	移流逆轉層
이류중력 흐름	advective-gravity flow	移流重力-
이류층	advection layer, advective region, advective layer	移流層
이류플럭스	advective flux	移流-
이류항	advection term, advective term	移流項
이류-확산 방정식	advection-diffusion equation	移流擴散方程式
이류활승구름	advection upslope cloud	移流活昇-
이류효과	advective effect	移流效果
이륙기상	take-off weather	離陸氣象

이륙예보	take-off forecast	離陸豫報
이륙활주거리	take-off distance	離陸滑走距離
이르밍거해류	Irminger current	-海流
이른 서리	early frost	
이면각	dihedral angle	二面角
이면표적	dihedral target	二面標的
이산화염소	chlorine dioxide	二酸化鹽素
이산화질소	nitrogen dioxide	二酸化窒素
이산화질소 감소	nitrogen dioxide reduction	二酸化窒素減少
이산화질소 경향	nitrogen dioxide trend	二酸化窒素傾向
이산화탄소	carbon dioxide	二酸化炭素
이산화탄소대기농도	carbon dioxide atmospheric concentration	二酸化炭素大氣濃度
이산화탄소띠	carbon dioxide band	二酸化炭素-
이산화탄소배출	carbon dioxide emission	二酸化炭素排出
이산화탄소보상점	carbon dioxide compensation point, CO_2 compensation point	二酸化炭素補償點
이산화탄소분자	carbon dioxide molecule	二酸化炭素分子
이산화탄소시비	carbon dioxide fertilization	二酸化炭素施肥
이산화탄소시비효과	carbon dioxide fertilizing effect, CO_2 fertilization effect	二酸化炭素施肥效果
이산화탄소정보분석센터	carbon dioxide information analysis center	二酸化炭素情報分析-
이산화탄소환산	carbon dioxide equivalence	二酸化炭素換算
이산확률분포	discrete probability (random) distribution	離散確率分布
이상	abnormal, abnormality, anomaly	異常
이상(의)	anomalous	異常
이상감률	abnormal lapse rate	異常減率
이상강수	excessive precipitation	異常降水
이상강우	excess rain, excessive rain	異常降雨
이상건조	abnormal dryness	異常乾燥
이상경도풍	anomalous gradient wind	異常傾度風
이상군락	ideal foliage	理想群落
이상굴절	abnormal refraction, anomalous refraction	異常屈折
이상기상	abnormal weather	異常氣象
이상기체	ideal gas	理想氣體
이상기체	perfect gas, ideal gas	理想氣體
이상기체방정식	ideal gas equation	理想氣體方程式
이상기체법칙	ideal gas law, perfect gas law	理想氣體法則
이상기후	abnormal climate	異常氣候
이상기후	ideal climate	理想氣候
이상도	abnormality	異常度
이상보정	anomaly correction	異常補正
이상분산	anomalous dispersion	異常分散
이상상관관계	anomaly correlation	異常相關關係
이상(광)선	extraordinary ray	異常(光)線

이상수평	ideal horizon	理想水平
이상순환	anomalous circulation	異常循環
이상액체	ideal liquid	理想液體
이상(구름)열	anomalous (cloud) line	異常-列
이상유체	ideal fluid	理想流體
이상음파전파	anomalous sound propagation	異常音波傳播
이상전자기파전파	anomalous radio propagation	異常電磁氣波傳播
이상전파	anomalous propagation	異常傳播
이상청역	zone of abnormal audibility	異常聽域
이상파	extraordinary wave	異常派
이상풍	overnormal wind	異常風
이상한파	abnormal cold wave	異常寒波
이상회파	angel	異常回波
이소프렌	isoprene	
이슬	dew	
이슬량계	dew gauge, drosometer	-量計
이슬막이	dew cap	
이슬무지개	dewbow	
이슬비	drizzle	
이슬비방울	drizzle drop	
이슬비안개	drizzling fog	
이슬점	dew point	-點
이슬점감률	dew point lapse rate	-點減率
이슬점계	dew cell, dew point instrument	-點計
이슬점공식	dew point formula	-點公式
이슬점기록계	dew point recorder	-點記錄計
이슬점라디오존데	dew point radiosonde	-點-
이슬점불연속선	dew point line	-點不連續線
이슬점습도계	dew point hygrometer, hygrometer of dew point	-點濕度計
이슬점온도	dew point temperature	-點溫度
이슬점온위	potential dew-point temperature	-點溫位
이슬점위치온도	potential dew-point temperature	-點位置溫度
이슬점전선	dew point front	-點前線
이슬점차	dew point depression	-點差
이슬점측정계	dew point apparatus	-點測定計
이슬트랩	dew trap	
이식	transplanting	移植
이식기	transplanting period	移植期
이심각	eccentric angle	離心角
이심궤도 지구물리관측위성	eccentric-orbiting geophysical observatory	離心軌道地球物理觀測衛星
이심률	eccentricity	離心率
이심이상	eccentric anomaly	離心異常
이안류	offshore current, rip current	離岸流

이안류~연안류 분산모형	offshore and coastal dispersion model	離岸流沿岸流分散模型
이안바람	offshore breeze	離岸-
이안저기압골	offshore trough	離岸低氣壓-
이안풍	offshore breeze	離岸風
이앙	transplanting	移秧
이앙기	transplanting period	移秧期
이오노그램	ionogram	
이온	ion	
이온가속기	ion accelerator	-加速器
이온결정	ionic crystal	-結晶
이온결합	ionic bond	-結合
이온계수기	ion counter	-計數器
이온교환과정	ion exchange process	-交換過程
이온교환수지	ion exchange resin	-交換樹脂
이온구름	ion cloud	
이온권	ionosphere	-圈
이온권계면	ionopause	-圈界面
이온기둥	ion column	
이온농도	ion concentration, ion density	-濃度
이온모형	ionic model	-模型
이온밀도	ion density	-密度
이온반응식	ionic equation	-反應式
이온쌍	ion pair	-雙
이온이동도	ion mobility, ionic mobility	-移動度
이온전도	ionic conduction	-傳導
이온전류	ionic current	-電流
이온존데	ionosonde	
이온층	ionization layer, ionospheric layer	-層
이온층골	ionospheric trough	-層-
이온층기록계	ionospheric recorder	-層記錄計
이온층기울기	ionospheric tilt	-層-
이온층난류	ionospheric turbulence	-層亂流
이온층바람	ionospheric wind	-層-
이온층보라	ionospheric storm	-層-
이온층스톰	ionospheric storm	-層-
이온층조석	ionospheric tide	-層潮汐
이온층진행파	travelling ionospheric disturbance	-層進行波
이온크로마토그래프	ion chromatograph	
이온탐측기	topside sounder	-探測器
이온평균수명	ion mean life	-平均壽命
이온포착설	ion-capture theory	-捕捉說
이온화	ionization	-化
이온화상자	ionization chamber	-化箱子
이온화평형	ionization equilibrium	-化平衡

이온화-해리 교환작용	ionization-dissociation interaction	-化解離交換作用
이월	carry-over	移越
이월효과	carry-over effect	移越效果
이유기유황염	dimethyl sulfide	二有機硫黃鹽
이전토양수분	antecedent soil moisture	以前土壤水分
이종검색	hybrid retrieval	異種檢索
이종변광성	hybrid variable	異種變光星
이중관온도계	sheathed thermometer	二重管溫度計
이중권계면	double tropopause	二重圈界面
이중기록계	double register	二重記錄計
이중눈패턴	double eye pattern	二重-
이중도트곱	double-dot product	二重-
이중도플러	dual Doppler	二重-
이중도플러처리	dual Doppler processing	二重-處理
이중성	duality	二重性
이중수신계	dual receiver system	二重受信系
이중수신시스템	dual receiver system	二重受信-
이중스칼라곱	double scalar product	二重-
이중안테나 레이더	bistatic radar	二重-
이중적분	double integral	二重積分
이중적산분석	double mass analysis	二重積算分析
이중진동수	dual frequency	二重振動數
이중채널차방법	dual channel difference method	二重-差方法
이중파	double wave	二重波
이중파장레이더	dual wavelength radar	二重波長-
이중편파레이더	dual polarization radar	二重偏波-
이중확산	double diffusion	二重擴散
이진변수	binary variable	二進變數
이진보편양식표기	binary universal format representation	二進普遍樣式表記
이진부호분류	binary encoding classification	二進符號分類
이진분류검증	binary class validation	二進分類檢證
이진수-십진수 변환기	binary-to-decimal converter	二進數十進數變換器
이진연산	binary operation	二進演算
이차곡면	quadratic surface	二次曲面
이차곡면근사	quadratic surface approximation	二次曲面近似
이차곡선	quadratic curve	二次曲線
이차방정식	quadratic equation	二次方程式
이차부등식	quadratic inequality	二次不等式
이차식	quadratic expression	二次式
이차함수	quadratic function	二次函數
이차항	quadratic term	二次項
이착륙기상	take-off and landing weather	離着陸氣象
이탄	peat	泥炭
이탈고도	level of escape	離脫高度

이탈극류	ab-polar current	離脫極流
이탈속도	escape velocity	離脫速度
이탈속력	escape speed	離脫速力
이탈역	region of escape	離脫域
이탈온실효과	runaway greenhouse effect	離脫溫室效果
이탈파	departing wave	離脫波
이항계수	binomial coefficient	二項係數
이항방정식	binomial equation	二項方程式
이항분포	binomial distribution	二項分布
이항식	binomial expression	二項式
이항정리	binomial theorem	二項定理
이항확률변수	binomial random variable	二項確率變數
이행연귀	couplet	二行連句
이환율	morbidity	罹患率
이황화탄소	carbon disulfide	二黃化炭素
인간-기계 혼합	man-machine mix	人間機械混合
인간기후	human climate	人間氣候
인간생기상학	human biometeorology	人間生氣象學
인간생(물)기후학	human bioclimatology	人間生(物)氣候學
인간진화 기후가설	climatic hypothesis of human evolution	人間進化氣候假說
인공강설	artificial snowfall	人工降雪
인공강수	artificial precipitation	人工降水
인공강우	artificial rainfall, artificially induced rainfall, precipitation stimulation, rain making	人工降雨
인공구름	artificial cloud	人工-
인공기후	artificial climate	人工氣候
인공기후실	climatic chamber	人工氣候室
인공발광운	artificial chemiluminescence cloud	人工發光雲
인공방사능	artificial radioactivity	人工放射能
인공보조	artificial assistance	人工補助
인공빙정핵	artificial ice nucleus	人工氷晶核
인공수평선	artificial horizon	人工水平線
인공신경망	artificial neural network	人工神經網
인공열	artificial heat	人工熱
인공위성	artificial satellite	人工衛星
인공재충전	artificial recharge	人工再充電
인공조절	artificial control	人工調節
인공증설	artificial snow enhancement	人工增設
인공증우	artificial increasing of rain, artificial rain increasing, artificial rain enhancement	人工增雨
인공지능	artificial intelligence	人工知能
인공지능계	artificial intelligence system	人工知能系
인공지능시스템	artificial intelligence system	人工知能-
인공핵	artificial nucleus	人工核

인공핵생성	artificial nucleation	人工核生成
인과관계	causality	因果關係
인광	phosphorescence	燐光
인도계절풍	Indian monsoon	印度季節風
인도 동안류	East Indian current	印度東岸流
인도양쌍극자모드	Indian Ocean dipole mode	印度洋双極子-
인도여름몬순	India summer monsoon	印度-
인도-파키스탄 저기압	Indo-Pakistan low	印度-低氣壓
인디언서머	Indian summer	
인력	attraction	引力
인력퍼텐셜	attractive potential	引力-
인류기후학	anthropoclimatology	人類氣候學
인류세	anthropocene	人類世
인수	factor	因數
인수분해	factorization	因數分解
인수정리	factor theorem	因數定理
인순환	phosphorus cycle	燐循環
인위적 강제	anthropogenic forcing	人爲的强制
인위적 과정	anthropogenic process	人爲的過程
인위적 기후변화	anthropogenic climatic change	人爲的氣候變化
인위적 기후효과	anthropogenic climatic effect	人爲的氣候效果
인위적 방출	anthropogenic effluent, anthropogenic release	人爲的放出
인위적 배출	anthropogenic emission	人爲的排出
인위적 에어로졸	anthropogenic aerosol	人爲的-
인위적 열	anthropogenic heat	人爲的熱
인위적 열입력	anthropogenic heat input	人爲的熱入力
인위적 이산화탄소 증가	anthropogenic CO_2 increase	人爲的二酸化炭素增加
인위적 지구온난화	anthropogenic global warming	人爲的地球溫暖化
인자	factor	因子
인접점시험	adjacent spot test	隣接點試驗
인조구름	cloudier	人造-
인지과제분석	cognitive task analysis	認知課題分析
인체보온지수	clo	人體保溫指數
인편모충	coccolith	鱗鞭毛蟲
일가열	diurnal heating	日加熱
일강수량	daily precipitation	日降水量
일관측	daily observation	日觀測
일광	daylight, sunbeam, sunlight	日光
일광계	heliograph	日光計
일광절약시간	daylight saving time	日光節約時間
일광지대	sunlight zone	日光地帶
일광효과	photoperiodism	日光效果
일교차	daily range, daily temperature range, diurnal range, diurnal temperature range	日較差

일극값	daily extreme	日極-
일기경계	weather divide	日氣境界
일기계	weather system	日氣系
일기기호	symbol of weather, weather sign, weather symbol	日氣記號
일기도	chart, weather chart, weather map	日氣圖
일기도규모	weather map scale	日氣圖規模
일기도기입	chart (or map) plotting, map plotting, map spotting	日氣圖記入
일기도분석	weather chart analysis	日氣圖分析
일기도축척	weather map scale	日氣圖縮尺
일기도형	weather map type	日氣圖型
일기모사전송	weather fax	日氣模寫電送
일기분석	weather analysis	日氣分析
일기분석기호	symbol for weather analysis	日氣分析記號
일기분석자료	weather analysis data	日氣分析資料
일기속담	meteorological proverb, weather lore, weather maxim	日氣俗談
일기순환	weather cycle	日氣循環
일기시스템	weather system	日氣-
일기악화보고	deterioration report	日氣惡化報告
일기예보	weather forecast, weather forecasting	日氣豫報
일기예보관	weather forecaster	日氣豫報官
일기예보자	weather forecaster	日氣豫報者
일기예측	weather prediction	日氣豫測
일기요소	weather element	日氣要素
일기이상	weather anomaly	日氣異常
일기조절	weather control	日氣調節
일기진단	weather diagnosis	日氣診斷
일기팩스	weather fax	日氣-
일기평가과정	weather evaluation process	日氣評價過程
일기표시도	weather depiction chart	日氣標示圖
일기회복보고	improvement report	日氣回復報告
일기효과	weather effect	日氣效果
일냉각	diurnal cooling	日冷却
일년생 작물	annual crop	一年生作物
일대일 기후상담	one-on-one climate consultation	一對一氣候相談
일대일 예보상담	one-on-one forecaster consultation	一對一豫報相談
일도	day degree, degree day	日度
일라이트	illite	
일련번호	serial number	一連番號
일몰	sunset	日沒
일반경압소용돌이정리	general baroclinic vortex theorem	一般傾壓-定理
일반기상학	general meteorology	一般氣象學

일반기후관측소	ordinary climatological station	一般氣候觀測所
일반농업기상관측소	ordinary agricultural meteorological station, ordinary agrometeorological station	一般農業氣象觀測所
일반류	general current	一般流
일반방향	general direction	一般方向
일반복사선측정소	ordinary radiation station	一般輻射線測定所
일반세포	ordinary cell	一般-
일반예보	general forecast	一般豫報
일반화규모불변성	generalized scale invariance	一般化規模不變性
일반화선형파동방정식	generalized linear wave equation	一般化線形波動方程式
일반화연직속도	generalized vertical velocity	一般化鉛直速度
일반화연직좌표	generalized vertical coordinate	一般化鉛直座標
일반화정역학방정식	generalized hydrostatic equation	一般化靜力學方程式
일반화좌표	generalized coordinates	一般化座標
일반화투과함수	generalized transmission function	一般化透過函數
일반화흡수계수	generalized absorption coefficient	一般化吸收係數
일변동	daily variation, diurnal variation	日變動
일변화	diurnal change	日變化
일변화경향	diurnal trend	日變化傾向
일변화바람	diurnal wind	日變化-
일변화율	daily rate	日變化率
일변화파	diurnal wave	日變化波
일별종관연속	daily synoptic series	日別綜觀連續
일보상점	daily compensation point	日補償點
일본해류	Japan current	日本海流
일부등성	diurnal inequality	日不等性
일사	bright sunshine, insolation, solar radiation	日射
일사계	actinometer, solarimeter	日射計
일사기록	actinogram	日射記錄
일사량	quantity of solar radiation	日射量
일사병	sunstroke	日射病
일사보정	correction of (solar) insolation	日射補正
일사시간	bright sunshine duration, insolation duration	日射時間
일사존데	solar radiation sonde	日射-
일사측정학	actinometry	日射測定學
일사풍	heliotropic wind	日射風
일산화염소	chlorine monoxide	一酸化鹽素
일산화염소라디칼	chlorine monoxide radical	一酸化鹽素-
일산화염소이합체	chlorine monoxide dimer	一酸化鹽素二合體
일산화질소	nitric oxide	一酸化窒素
일산화탄소	carbon monoxide	一酸化炭素
일생주기	life cycle	一生週期
일시기후반응	transient climate response	一時氣候反應
일시문턱변위	temporary threshold shift	一時門-變位

일시변동	transient variation	一時變動
일시변화	temporal change	一時變化
일시시력상실	temporary blindness	一時視力喪失
일시에디	transient eddy	一時-
일시역치변위	temporary threshold shift	一時閾値變位
일시적 고요	lull	一時的-
일시파	transient wave	一時波
일시해상도	temporal resolution	一時解像度
일식	solar eclipse	日蝕
일식바람	eclipse wind	日食-
일식저기압	eclipse cyclone	日食低氣壓
일액	guttation	溢液
일연직이동	diurnal vertical migration	日鉛直移動
일예보	daily forecast	日豫報
일온도교차	daily temperature range	日溫度較差
일일조	diurnal tide	一日潮
일잉여생산량	daily surplus production	日剩餘生産量
일저류량	daily storage	日貯流量
일정고도	constant altitude	一定高度
일정고도평면위치	constant altitude plan position	一定高度平面位置
일정고도 PPI	constant altitude PPI	一定高度-
일정바람	constant wind	一定-
일정성(바람의)	constancy	一定性
일정절대소용돌이도	constant absolute vorticity	一定絶對-度
일정절대소용돌이도궤적	constant absolute vorticity trajectory	一定絶對-度軌跡
일정절대소용돌이도궤적법	constant absolute vorticity trajectory method	一定絶對-度軌跡法
일정절대소용돌이도궤적표	CAVT table	一定絶對-度軌跡表
일정플럭스층	constant flux layer	一定-層
일정한	constant	一定-
일조	sunshine	日照
일조계	sunshine recorder	日照計
일조극광	sunlit aurora	日照極光
일조기록(지)	sunshine record	日照記錄(紙)
일조시간	duration of sunshine, insolation duration, sunshine duration, sunshine hour, sunshine time	日照時間
일조율	rate of sunshine	日照率
일조적분계	sunshine integrator	日照積分計
일주기	diurnal cycle	日周期
일주기성	daily periodism, diurnal periodicity	日週期性
일지표면에너지균형	diurnal surface energy balance	日地表面-均衡
일진광풍	wind flurry	一陣狂風
일진폭	diurnal amplitude	日振幅
일차변환	linear transformation	一次變換
일차부등식	linear inequality	一次不等式

일차불연속	discontinuity of first order,	一次不連續
	first-order discontinuity	
일차식	linear expression	一次式
일차종속	linear dependence	一次從屬
일차항	linear term	一次項
일최고온도	daily maximum temperature	日最高溫度
일최저기온	minimum daily temperature	日最低氣溫
일최저온도	daily minimum temperature	日最低溫度
일출	sunrise	日出
일치수치방식	consistent numerical scheme	一致數値方式
일치평균	consensus average	一致平均
일치평균검사	consensus average check	一致平均檢查
일치평균화	consensus averaging	一致平均化
일평균(값)	daily mean	日平均
일평균습도	daily mean humidity	日平均濕度
일평균온도	daily mean temperature, mean daily temperature	日平均溫度
일평균운량	daily mean amount of cloud	日平均雲量
일평균최고온도	mean daily maximum temperature	日平均最高溫度
일평균최저온도	mean daily minimum temperature	日平均最低溫度
일평균풍속	daily mean wind speed	日平均風速
일함수	work function	-函數
읽기(눈금)	reading	
읽어내기	read out	
임계각	critical angle	臨界角
임계감률	critical lapse rate	臨界減率
임계값	critical value	臨界-
임계경도	critical gradient	臨界傾度
임계고도	critical level	臨界高度
임계고도불안정	critical level instability	臨界高度不安定
임계깊이	critical depth	臨界-
임계깊이조절	critical depth control	臨界-調節
임계높이	critical height	臨界-
임계레이놀즈수	critical Reynolds number	臨界-數
임계레일리수	critical Rayleigh number	臨界-數
임계리처드슨수	critical Richardson number	臨界-數
임계물방울반지름	critical drop radius	臨界-
임계반경	critical radius	臨界半徑
임계방전	critical discharge	臨界放電
임계배풍거리	critical downwind distance	臨界背風距離
임계부피	critical volume	臨界-
임계상수	critical constant	臨界常數
임계상태	critical state	臨界狀態
임계샘플링	critical sampling	臨界-
임계생장기간	critical period of growth	臨界生長期間

임계성공지수	critical success index	臨界成功指數
임계속도	critical velocity	臨界速度
임계시간	critical time	臨界時間
임계압	critical pressure	臨界壓
임계역	critical area	臨界域
임계영역	critical region	臨界領域
임계온도	cardinal temperature, critical temperature	臨界溫度
임계점	critical point	臨界點
임계조건	critical condition	臨界條件
임계조위기준면	critical tidal level	臨界潮位基準面
임계주파수	critical frequency	臨界周波數
임계질량	critical mass	臨界質量
임계탈출고도	critical level of escape	臨界脫出高度
임계파장	critical wave length	臨界波長
임계폭발점	critical bursting point	臨界爆發點
임계플럭스리처드슨수	critical flux Richardson number	臨界-數
임계함수량	critical liquid water content	臨界含水量
임계현상	critical phenomenon	臨界現象
임계회전수	critical rotation number	臨界回轉數
임계흐름	critical flow	臨界-
임관	leaf canopy	林冠
임관구조	canopy architecture	林冠構造
임의매체	random medium	任意媒體
임의추출(법)	random sampling	任意抽出(法)
임의표본	random sample	任意標本
임학	forestry	林學
입구역	entrance region	入口域
입도	particle size	粒度
입도분포	particle size distribution	粒度分布
입동	start of winter	立冬
입력	input	入力
입방결정계	cubic system	立方結晶系
입사	incidence	入射
입사각	angle of incidence	入射角
입사광	incident light	入射光
입사광선	incident ray	入射光線
입사면	plane of incidence	入射面
입사복사	incident radiation, incoming radiation	入射輻射
입사복사형	incoming radiation type	入射輻射型
입사점	incident point	入射點
입사태양복사	incoming solar radiation	入射太陽輻射
입사파	incident wave	入射波
입사펄스	incident pulse	入射-
입자	particle, particulate	粒子

입자-높이확률분포	particle-height probability distribution	粒子-確率分布
입자대입자 상호작용	particle-particle interaction	粒子對粒子相互作用
입자도함수	particle derivative	粒子導函數
입자모형	particle model	粒子模型
입자물질	particle matter	粒子物質
입자방출률	particle emission rate	粒子放出率
입자분산	particle dispersion	粒子分散
입자불변성조건	condition of particle invariance	粒子不變性條件
입자상물질	particulate matter	粒子狀物質
입자속도	particle velocity	粒子速度
입자전하	particle charge	粒子電荷
입자채집시스템	particle collection system	粒子採集-
입자추적모형	particle trajectory model	粒子追跡模型
입자침적속도	particle deposition velocity	粒子沈積速度
입자크기	particle size	粒子-
입자크기매개변수	particle size parameter	粒子-媒介變數
입자크기모수	particle size parameter	粒子-母數
입자하중	particulate loading	粒子荷重
입자환경모니터	Particle Environment Monitor	粒子環境-
입체	solid, stereo	立體
입체각	solid angle	立體角
입체도	stereogram	立體圖
입체사진측량법	stereophotogrammetry	立體寫眞測量法
입체전기상	multi-dimensional warfare weather	立體戰氣象
입체투영	stereographic projection	立體投影
입추	start of autumn	立秋
입춘	start of spring	立春
입하	start of summer	立夏
잎각도분류	leaf angle distribution	-角度分類
잎광학특성	leaf optical property	-光學特性
잎구름 패턴	leaf cloud pattern	
잎구조 특성	leaf structural property	-構造特性
잎맥간황백화	interveinal chlorosis	-脈間黃白化
잎성분	leaf component	-成分
잎수분량	leaf moisture content	-水分量
잎응력	leaf stress	-應力
잎파장별 특성	leaf spectral signature	-波長別特性

ㅈ

한글	영문	한자
자	rule	
자개구름	nacreous cloud	
자광	purple glow	紫光
자극	dip pole, magnetic pole	磁極
자금기상서비스	funding meteorological service	資金氣象-
자기강도	magnetic intensity	磁氣强度
자기건습계	psychrograph	自記乾濕計
자기경위의	recording theodolite	自記經緯儀
자기고도계	altigraph	自記高度計
자기고요일	magnetically quiet day	磁氣-日
자기고층계	aerograph	自記高層計
자기공분산	auto-convariance	自己共分散
자기공분산함수	autocovariance function	自己共分散函數
자기권	magnetosphere	磁氣圈
자기권계면	magnetopause	磁氣圈界面
자기극	magnetic pole	磁氣極
자기극세기	strength of magnetic pole	磁氣極-
자기금융예보서비스	self-finance forecasting service	自己金融豫報-
자기기록	autographic record	自己記錄
자기기록계원통	drum (of self-recording instrument)	自己記錄計圓筒
자기기압계	recording barometer, self-recording barometer, self-registering barometer	自記氣壓計
자기기압기록	barograph trace	自記氣壓記錄
자기기압온도계	barothermograph	自記氣壓溫度計
자기기압온도습도계	barothermohygrograph	自記氣壓溫度濕度計
자기냉각계	recording frigorimeter	自記冷却計
자기모멘트	magnetic moment	磁氣-
자기모발습도계	hair hygrograph	自記毛髮濕度計
자기미변계	variograph	自記微變計

자기미압계	statoscope	自記微壓計
자기방향제어코일	magnetic orientation control coil	磁氣方向制御-
자기변동	magnetic variation	磁氣變動
자기북극	magnetic north pole	磁氣北極
자기산란복사계	diffusograph	自記散亂輻射計
자기상관	autocorrelation	自己相關
자기상관계수	autocorrelation coefficient	自己相關係數
자기상관함수	autocorrelation function	自己相關函數
자기설량계	recording snow gauge	自記雪量計
자기성	magnetism	磁氣性
자기습도계	self-recording hygrometer	自記濕度計
자기알베도계	recording albedometer	自記-計
자기에너지	magnetic energy	磁氣-
자기온도계	self-registering thermometer, thermograph	自記溫度計
자기온도계보정표	thermograph correction card	自記溫度計補正表
자기온도기록	thermogram	自記溫度記錄
자기온습계	thermohygrograph	自記溫濕計
자기온습도기록	thermohygrogram	自記溫濕度記錄
자기요란	magnetic disturbance	磁氣擾亂
자기요소	magnetic element	磁氣要素
자기우량계	automatic rain recording gauge, pluviograph, recording rain gauge, self-recording rain gauge	自己雨量計
자기위도	magnetic latitude	磁氣緯度
자기유도	magnetic induction	磁氣誘導
자기유체역학	magnetohydrodynamics	磁氣流體力學
자기유체역학파	magnetohydrodynamic wave	磁氣流體力學波
자기유체파	hydromagnetic wave	磁氣流體波
자기이슬량계	drosograph	自記-計
자기이온이론	magneto-ionic theory	磁氣-理論
자기이중굴절	magnetic double refraction	磁氣二重屈折
자기일사계	actinograph, solarigraph	自記日射計
자기자오선	magnetic meridian	磁氣子午線
자기장	magnetic field	磁氣場
자기장강도	magnetic field intensity	磁氣場强度
자기장세기	magnetic field strength	磁氣場勢氣
자기적도	magnetic equator	磁氣赤道
(지)자기적도	aclinic line	(地)磁氣赤道
자기전위계	recording potentiometer	自記電位計
자기전천일사계	pyranograph	自記全天日射計
자기전천일사곡선	solarigram	自記全天日射曲線
자기증발계	evaporograph, recording evaporation pan, recording evaporimeter	自記蒸發計
자기지온계	soil thermograph	自記地溫計
자기지중온도계	recording earth thermometer	自記地中溫度計

자기직달일사계	pyrheliograph	自記直達日射計
자기층	magnetism layer	磁氣層
자기크로치트	magnetic crotchet	磁氣-
자기토양온도계	soil thermograph	自記土壤溫度計
자기특성수	magnetic character figure	磁氣特性數
자기편각	magnetic declination	磁氣偏角
자기편향	magnetic deflection	磁氣偏向
자기폭풍	magnetic storm, magstorm	磁氣暴風
자기폭풍일	magnetically disturbed day	磁氣暴風日
자기폭풍지수	Dst index	磁氣暴風指數
자기풍력계	self-registering anemometer	自記風力計
자기풍속계	anemograph, recording anemometer	自記風速計
자기풍향	magnetic wind direction	磁氣風向
자기풍향계	recording wind vane	自記風向計
자기학	magnetism	磁氣學
자기회귀과정	autoregressive process	自己回歸過程
자기회귀모형	autoregressive model	自己回歸模型
자기회귀법	autoregression method	自己回歸法
자기회귀수열	autoregressive series	自己回歸數列
자남	magnetic south	磁南
자남극	magnetic southern pole	磁南極
자동감도조절	automatic sensibility control	自動感度調節
자동검색	automated search	自動檢索
자동검조기	automatic tide gauge	自動檢潮器
자동경보장치	auto-alarm	自動警報裝置
자동관측소관리	automatic station keeping	自動觀測所管理
자동교정	autocorrection	自動校正
자동기구비양장치	automated balloon launcher	自動氣球飛揚裝置
자동기록계	automatic recorder	自動記錄計
자동기상관측소	automatic meteorological observing station, automatic weather station	自動氣象觀測所
자동기상관측시스템	automatic weather observation system	自動氣象觀測-
자동기상관측장비	Automatic Weather System	自動氣象觀測裝備
자동기후관측소	automatic climatological station	自動氣候觀測所
자동다중주파수이온권기록계	automatic multifrequency ionospheric recorder	自動多重周波數-圈記錄計
자동대류	autoconvection	自動對流
자동대류감률	autoconvective lapse rate	自動對流減率
자동대류경도	autoconvection gradient	自動對流傾度
자동대류불안정	autoconvective instability	自動對流不安定
자동무선기상계기	automatic radiometeorograph	自動無線氣象計器
자동무선우량계	automatic radio rain gauge	自動無線雨量計
자동변환	autoconversion	自動變換
자동순압	autobarotropic	自動順壓
자동순압(의)	autobarotropy	自動順壓

ㅈ

자동순압대기	autobarotropic atmosphere	自動順壓大氣
자동순압조건	condition of autobarotropy	自動順壓條件
자동식표준원격자기관측소	automatic standard magnetic observatory-remote	自動式標準遠隔磁氣觀測所
자동오차교정장치	automatic error request equipment	自動誤差校正裝置
자동원격측정장치	remote observing system automation	自動遠隔測定裝置
자동이득제어	automatic gain control	自動利得制御
자동이온화반응	autoionization	自動-化反應
자동자료교환시스템	automatic data exchange system	自動資料交換-
자동자료처리	automatic data processing	自動資料處理
자동자료편집시스템	automatic data editing and switching system	自動資料編輯-
자동적설계	automatic snow meter	自動積雪計
자동적설측정망	snotel, snow telemeter	自動積雪測定網
자동존데	autosonde	自動-
자동주파수제어	automatic frequency control	自動周波數制御
자동증발관측소	automatic evaporation station	自動蒸發觀測所
자동지상관측시스템	automated surface observing system	自動地上觀測-
자동진동수제어	automatic frequency control	自動振動數制御
자동차대수	automobile number	自動車臺數
자동차배기가스	automobile exhaust	自動車排氣-
자동차배출기준	motor vehicle emission standard	自動車排出基準
자동차 통기장치	positive crankcase ventilation	自動車 通氣裝置
자동추적	automatic tracking	自動追跡
자동표준자기관측소	automatic standard magnetic observatory	自動標準磁氣觀測所
자동해양기상관측부이	automatic meteorological oceanographic buoy	自動海洋氣象觀測-
자동화상운송	automatic picture transmission	自動畵像運送
자람벡터	breeding vector(BV)	
자력계	magnetometer	磁力計
자력기록	magnetogram	磁力記錄
자력기록계	magnetograph	磁力記錄計
자력선	magnetic line of force	磁力線
자력풍속계	magneto anemometer	磁力風速計
자료	data	資料
자료군	data series	資料群
자료동화	data assimilation	資料同化
자료동화계	data assimilation system	資料同化系
자료동화문제	data assimilation problem	資料同化問題
자료동화시스템	data assimilation system	資料同化-
자료명부	data directory	資料名簿
자료배열	format	資料配列
자료보정	data correction	資料補正
자료분석센터	data analysis center	資料分析-
(위성)자료수신/송신부시스템	data acquisition and transmission subsystem	(衛星)資料受信送信副-
자료수집	data acquisition, data collection	資料蒐集
자료수집대	data collection platform	資料蒐集臺

자료수집시설	data acquisition facility	資料蒐集施設
자료수집장치	data acquisition system	資料蒐集裝置
자료수집체계	data acquisition system	資料蒐集體系
자료열대수	data tropical number	資料熱帶數
자료요구	data requirement	資料要求
자료윈도	data window	資料-
자료유형구분자	data block indentifier	資料類型區分子
자료은행	data bank	資料銀行
자료이용소	data utilization station	資料利用所
자료저장	data archive	資料貯藏
자료전송	data transmission	資料傳送
자료집록장치	data logger	資料集錄裝置
자료처리	data processing	資料處理
자료크기	data volume	資料-
자료표본	data sample	資料標本
자료함수	data function	資料函數
자발결빙	spontaneous freezing	自發結氷
자발대류	spontaneous convection	自發對流
자발승화	spontaneous sublimation	自發昇華
자발응결	spontaneous condensation	自發凝結
자발핵화	spontaneous nucleation	自發核化
자북	magnetic north	磁北
자북극	north magnetic pole	磁北極
자분우물	artesian well	自噴-
자분정	artesian well, flowing well	自噴井
자분지하수	artesian ground water	自噴地下水
자생대전	autogenous electrification	自生帶電
자세	attitude	姿勢
자연	nature	自然
자연계절	natural season	自然季節
자연계절현상	natural seasonal phenomenon	自然季節現象
자연대류	natural convection	自然對流
자연대수	natural logarithm	自然對數
자연도태	natural selection	自然淘汰
자연돌연변이	natural mutation	自然突然變異
자연방사능	natural radioactivity	自然放射能
자연방제	natural control	自然防除
자연식생	natural vegetation	自然植生
자연에어로졸	natural aerosol	自然-
자연오염원	natural source	自然汚染源
자연우박눈	natural hail embryo	自然雨雹-
자연원인	natural cause	自然原因
자연유하량	natural flow	自然流下量
자연재해	natural disaster	自然災害

ㅈ

자연적 발생(의)	biogenic	自然的發生-
자연제거과정	natural removal process	自然除去過程
자연종관계절	natural synoptic season	自然綜觀季節
자연종관기간	natural synoptic period	自然綜觀期間
자연종관영역	natural synoptic region	自然綜觀領域
자연좌표	natural coordinates	自然座標
자연좌표계	natural coordinate system	自然座標系
자연주기	natural period	自然週期
자연증설	natural snow enhancement	自然增雪
자연진동	natural oscillation	自然振動
자연통풍식 냉각탑	natural-draft cooling tower	自然通風式冷却塔
자연현상	natural phenomenon	自然現象
자연황순환	natural sulfur cycle	自然黃循環
자연흐름	natural flow	自然-
자오(의)	meridian	子午
자오권	meridian, meridian circle	子午圈
자오면	meridian plane	子午面
자오면교환	meridional exchange	子午面交換
자오면류	meridional flow	子午面流
자오면세포	meridional cell	子午面細胞
자오면순환	meridional circulation	子午面循環
자오면전선	meridional front	子午面前線
자오선	meridian	子午線
자오선바람	meridional wind	子午線-
자외(선)	ultraviolet	紫外(線)
자외광	ultraviolet light	紫外光
자외복사	ultraviolet radiation	紫外輻射
자외복사지수	ultraviolet radiation index	紫外輻射指數
자외선	ultraviolet ray	紫外線
자외선광도측정법	UV photometry	紫外線光度測定法
자외선습도계	ultraviolet ray hygrograph	紫外線濕度計
자외선차단지수	sun protection factor	紫外線遮斷指數
자외선측정계	UV biometer, UV dosimeter	紫外線測定計
자외선측정기	dosimeter	紫外線測定器
자외선-A	UV-A	紫外線-
자외선-B	UV-B	紫外線-
자외선-C	UV-C	紫外線-
자외스펙트럼	ultraviolet spectrum	紫外-
자외역	ultraviolet region	紫外域
자외일사계	ultraviolet pyrheliometer	紫外日射計
자외(선)지수	ultraviolet index	紫外(線)指數
자외파장	ultraviolet wavelength	紫外波長
자원소	daughter isotope	子元素
자유경로	free path	自由經路

자유고공기구	free aerostat	自由高空氣球
자유공간	free space	自由空間
자유기	free radical	自由基
자유기구	free balloon	自由氣球
자유기류마하수	free-stream Mach number	自由氣流-數
자유낙하	free falling	自由落下
자유낙하가속도	free fall acceleration	自由落下加速度
자유낙하운동	motion of free fall	自由落下運動
자유난류	free turbulence	自由亂流
자유대기	free air, free atmosphere	自由大氣
자유대기중규모현상	free atmosphere mesoscale phenomenon	自由大氣中規模現象
자유대류	free convection	自由對流
자유대류 규모화 온도	free convection scaling temperature	自由對流規模化溫度
자유대류고도	level of free convection	自由對流高度
자유대류층	free convection layer	自由對流層
자유도	degree of freedom	自由度
자유면조건	free surface condition	自由面條件
자유산소원자	free oxygen atom	自由酸素原子
자유상승	free lift	自由上昇
자유상승력	free lifting force	自由上昇力
자유스펙트럼역	free spectral range	自由-域
자유에너지	free energy	自由-
자유엔탈피	free enthalpy	自由-
자유유선	free streamline	自由流線
자유전자	free electron	自由電子
자유주기	free period	自由週期
자유진동	free oscillation	自由振動
자유파	free wave	自由波
자유팽창	free expansion	自由膨脹
자유 푄	free foehn	自由-
자유표면	free surface	自由表面
자유표면파	free surface wave	自由表面波
자이로주파수	gyro-frequency	-周波數
자이로참조시스템	gyro reference system	-參照-
자전	rotation	自轉
자전각속도	angular velocity of rotation	自轉角速度
자전주사구름카메라	spin scan cloud camera	自轉走查-
자정	midnight	子正
자체교정	self-calibration	自體矯正
자체기록계	self recorder	自體記錄計
자체유도	self-induction	自體誘導
자체일관성	self-consistency	自體一關性
자침	magnetic needle	磁針
작계	cropping season	作季

작기	cultivation period	作期
작동상태	operating condition	作動狀態
작동주기	duty cycle	作動週期
작물계수	crop coefficient	作物係數
작물기간	crop period	作物期間
작물기상	crop weather	作物氣象
작물기후	crop climate	作物氣候
작물기후한계	climatic limit of crop	作物氣候限界
작물력	crop calendar	作物曆
작물성장	crop growth	作物成長
작물종류	crop type	作物種類
작용	action	作用
작용변수	action variable	作用變數
작용선	line of action	作用線
작용온도	operative temperature	作用溫度
작용적분	action integral	作用積分
작용점	point of application	作用點
작용중심	action center, center of action	作用中心
작은 구름	small cloud	
작은 구름방울	cloud droplet, droplet	
작은 구름방울상	droplet phase	-相
작은 구름방울수집기	droplet collector	-收集器
작은 구름방울스펙트럼	droplet spectrum	
작은 구름방울크기분포	droplet-size distribution	-分布
작은 부빙	glacon	-浮氷
작은빗방울채집기	rain droplet collector	-採集器
작은 빙산	growler	-氷山
작은 순환	minor circulation	-循環
작은 우박	small hail	-雨雹
작은 이온	small ion	
작은 제트	jetlet	
작은 태풍눈	distinct small eye	-颱風-
작은 파	wavelet	-波
작전기상	operation weather	作戰氣象
작전운량조건	operational cloud amount standard	作戰雲量條件
작전운층시계	operational cloud layer visibility	作戰雲層視界
작전지원공수기상	operational support airlift weather	作戰支援空輸氣象
작황	crop situation	作況
작황예보	crop forecasting	作況豫報
작황예보시험	yield forecast test	作況豫報試驗
잔광	afterglow	殘光
잔류	residual	殘留
잔류광물	residual mineral	殘留鑛物
잔류분산	residual variance	殘留分散

잔류수분	residual moisture	殘留水分
잔류시간	lifetime	殘留時間
잔류층	residual layer	殘留層
잔류편차정책	offset policy	殘留偏差政策
잔물결	ripple, wavelet	
잔물결권운	cirro-ripple	-卷雲
잔물결바다	slight sea	
잔상	afterimage	殘像
잔서	lingering summer heat	殘暑
잔인가설	overkill hypothesis	殘忍假設
잔적토	residual soil	殘積土
잔존저기압	remnant low	殘存低氣壓
잔차	residual	殘差
잠시변동	transient variation	暫時變動
잠열	latent heat	潛熱
잠열수송	latent heat transfer	潛熱輸送
잠재	potential	潛在
잠재결정중심	potential center of crystallization	潛在結晶中心
잠재굴절률	potential index of refraction, potential refractive index	潛在屈折率
잠재불안정	latent instability	潛在不安定
잠재불안정도	potential instability	潛在不安定度
잠재생산력	potential productivity	潛在生産力
잠재소용돌이도	potential vorticity	潛在-度
잠재저기압발생	potential cyclogenesis	潛在低氣壓發生
잠재증발	latent evaporation	潛在蒸發
잡음	grass, noise	雜音
잡음가이드라인	noise guideline	雜音-
잡음능	noise power	雜音能
잡음대역폭	noise bandwidth	雜音帶域幅
잡음문턱값	noise threshold	雜音門-
잡음비	noise ratio	雜音比
잡음상당온도변위(영상의)	noise equivalent delta temperature	雜音相當溫度變位
잡음수준	noise level	雜音水準
잡음여과	noise filtering	雜音濾過
잡음온도	noise temperature	雜音溫度
잡음인자	noise factor	雜音因子
잡음지수	noise figure	雜音指數
잡음파워	noise power	雜音-
잡종	hybrid	雜種
장	field	場
장거리 월경대기오염 제네바 협약	Geneva Convention on Long-Range Transboundary Air Pollution	長距離越境大氣汚染-關-協約
장거리이동	long-range transport	長距離移動

장거리통신	telecommunication	長距離通信
장구 시간	deep time	長久時間
장기예보	long-range forecast	長期豫報
장기우량기록계	long-range pluviograph	長期雨量記錄計
장기육지동결	taryn	長期陸地凍結
장기자기우량계	long period rain recorder	長期自記雨量計
장동	nutation	章動
장력	tension	張力
장력계수	coefficient of tension	張力係數
장마	Baiu, Changma, Mei-yu, plum rain, Tsuyu	
장마강우량	rainfall amount of Changma	-降雨量
장마기간	period of Changma	-期間
장마날씨	rainy weather	
장마부활	revival of Changma	-復活
장마시작	beginning of Changma	-始作
장마저온	low temperature during Changma	-低溫
장마전선	Changma front, Mei-yu front	-前線
장마전선호우	heavy rain associated with Changma front	-前線豪雨
장마제트류	Changma jet stream	-流
장마종료	ending of Changma	-終了
장마지수	Changma index	-指數
장마철	rainy wet season	
장마휴식	break of Changma	-休息
장막극광	dramundan	帳幕極光
장맛비	mold rain	
장밝기	field brightness	場-
장방형배수형	rectangular drainage pattern	長方形排水形
장벽	barrier	障壁
장벽설	barrier theory	障壁說
장벽제트	barrier jet	障壁-
장벽층	barrier layer	障壁層
장변수	field variable	場變數
장세기	field intensity, field strength	場-
장소선정	siting	場所選定
장소특화날씨경보	site specific weather alert	場所特化-警報
장애물	obstacle	障碍物
장애흐름징후	obstacle flow signature	障碍-徵候
장우	long spell of rain	長雨
장일식물	long-day plant	長日植物
장주기	long period	長週期
장축선(타원의)	line of apsis	長軸線
장치	installation	裝置
장파	long wave	長波
장파곡	long wave trough	長波谷

장파공식	long wave formula	長波公式
장파기압골	long wave trough	長波氣壓-
장파방정식	long wave equation	長波方程式
장파복사	long wave radiation	長波輻射
장파 비등방성인자	long wave anisotropic factor	長波非等方性因子
장파역행	retrogression of long wave	長波逆行
장파채널	long wave channel	長波-
장표고	field elevation	場標高
장휘도	field luminance	場輝度
재	ash	灰
재결합	recombination	再結合
재결합계수	recombination coefficient	再結合係數
재결합에너지	recombination energy	再結合-
재난	disaster	災難
재난구조	disaster aid	災難救助
재래식레이더	conventional radar	在來式-
재생	regeneration, renewable, reproduction	再生
재생산성	reproducibility	再生産性
재생성장	reproductive growth	再生成長
재생에너지	renewable energy	再生-
재설정	reset	再設定
재소나기	shower of ash	
재진입	re-entry	再進入
재해	damage	災害
재해위험지역	area subject to disaster	災害危險地域
재향군인병	legionnaires' disease	在鄕軍人病
재현	playback, recurrence, reproduction	再現
재현공식	recurrence formula	再現公式
재현기간	recurrence interval, return period	再現期間
재현빈도	recurrence frequency	再現頻度
재현자기폭풍	recurrent magnetic storm	再現磁氣暴風
재활성화	reactivation	再活性化
재흡수	reabsorption	再吸收
저기압	low	低氣壓
저각도 상승폭격기상	low angle loft bombing weather	低角度上昇爆擊氣象
저고도침투기상	low-level terrain flight weather	低高度浸透氣象
저고도폭격기상	low altitude bombing weather	低高度爆擊氣象
저공기습기상	pop-up weather	低空奇襲氣象
저공비행폭격기상	skip bombing weather	低空飛行爆擊氣象
저공해자동차	low-emission vehicle	低公害自動車
저궤도위성	non-geostationary orbit satellite	低軌道衛星
저기압	barometric depression, barometric low, low atmospheric pressure	低氣壓
저기압	pressure depression	低氣壓

저기압가족	cyclone family, depression family, family of cyclones, family of depressions	低氣壓家族
저기압강우	cyclone rain	低氣壓降雨
저기압경로	cyclone path, cyclone track	低氣壓經路
저기압경보	cyclone warning	低氣壓警報
저기압구름계	cloud system of an depression	低氣壓-系
저기압규모	cyclone scale, cyclonic scale	低氣壓規模
저기압다발구역	cyclone-prone area	低氣壓多發區域
저기압대류설	convection theory of cyclone	低氣壓對流說
저기압매몰	filling of a depression	低氣壓埋沒
저기압모형	cyclone model	低氣壓模型
저기압발산이론	divergence theory of cyclone	低氣壓發散理論
저기압발생	cyclogenesis	低氣壓發生
저기압발생발산설	divergence theory of cyclogenesis	低氣壓發生發散說
저기압발생파동설	wave theory of cyclogenesis	低氣壓發生波動說
저기압생성 대류이론	convective theory of cyclogenesis	低氣壓生成對流理論
저기압성	contra solem, cyclonic, cyclonicity	低氣壓性
저기압성강수	cyclone precipitation, cyclonic precipitation	低氣壓性降水
저기압성곡률	cyclonic curvature	低氣壓性曲率
저기압성공기운동	cyclonic air motion	低氣壓性空氣運動
저기압성뇌우	cyclonic thunderstorm	低氣壓性雷雨
저기압성단계	cyclonic phase	低氣壓性段階
저기압성바람	cyclonic wind	低氣壓性-
저기압성방향	cyclonic sense	低氣壓性方向
저기압성보라	cyclonic bora	低氣壓性-
저기압성비	cyclonic rain	低氣壓性-
저기압성상승	cyclonic lifting	低氣壓性上昇
저기압성소용돌이도	cyclonic vorticity	低氣壓性-度
저기압성시어	cyclonic shear	低氣壓性-
저기압성온도퇴화	cyclonic thermal involution	低氣壓性溫度退化
저기압성운동	cyclonic motion	低氣壓性運動
저기압성중위도폭풍연구계획	cyclonic extra-tropical storms project	低氣壓性中緯度暴風研究計劃
저기압성지형류	cyclo-geostrophic current	低氣壓性地衡流
저기압성치올림	cyclonic lifting	低氣壓性-
저기압성토네이도	cyclonic tornado	低氣壓性-
저기압성파	cyclonic wave	低氣壓性波
저기압성폭풍우	cyclonic storm	低氣壓性暴風雨
저기압성회전	cyclonic rotation	低氣壓性回轉
저기압성흐름	cyclonic flow	低氣壓性-
저기압소멸	cyclolysis	低氣壓消滅
저기압순환	cyclonic circulation	低氣壓循環
저기압심화	deepening of a depression	低氣壓深化
저기압유출	outflow from the cyclone	低氣壓流出
저기압이동	cyclone movement	低氣壓移動

저기압이론	theory of cyclone	低氣壓理論
저기압장벽설	barrier theory of cyclone, drop theory	低氣壓障壁說
저기압재생	regeneration of a depression	低氣壓再生
저기압중심	center of cyclone	低氣壓中心
저기압축	axis of depression, axis of low	低氣壓軸
저기압축경사	inclination of the axis of a cyclone	低氣壓軸傾斜
저기압파	cyclone wave	低氣壓波
저기압형보라	bora scura	低氣壓型-
저기준신호	low-reference signal	低基準信號
저녁놀	afterglow, evening glow	
저녁뜸	evening calm	
저녁무풍	evening calm	-無風
저류	impound, storage	貯留
저류경로	storage routing	貯留經路
저류계수	coefficient of storage	貯留係數
저류방정식	storage equation	貯留方程式
저서서식지	benthic habitat	底棲棲息地
저서유공충	benthic foraminifera	底棲有孔蟲
저속 위성자료 전송	low-rate information transmission	低速衛星資料電送
저수(위)	low water	低水(位)
저수량	pondage	貯水量
저수심경보	low depth of water warning	低水深警報
저수위	low water level	低水位
저수위표지	low water mark	低水位標識
저수지	reservoir	貯水池
저스펙트럼분해능	poor spectral resolution	低-分解能
저시정	low visibility	低視程
저시정등급	low visibility level	低視程等級
저(기)압	barometric minimum, low pressure	低(氣)壓
저압대	depression belt, low pressure zone	低壓帶
저압부	depression, low pressure area, pressure depression	低壓部
저에너지입자	low energy particle	低-粒子
저온가지배열에타중간자분광기	cryogenic limb array etalon spectrometer	低溫-配列-中間子分光器
저온기	cryogenic period	低溫期
저온기후	microthermal climate	低溫氣候
저온대	microthermen	低溫帶
저온습도계	cryogenic hygrometer	低溫濕度計
저온습도측정(법)	low temperature hygrometry	低溫濕度測定(法)
저온식생형	microtherm, microthermal type	低溫植生型
저온요구성	chilling requirement	低溫要求性
저온플랑크톤	cryoplankton	低溫-
저용량 공기 샘플러	low volume air sampler	低容量空氣-
저울강수계	weighing gauge	-降水計

저울계수기	weighing gauge	-計數器
저울설량계	weighing snow-gauge	-雪量計
저울우량계	weighing rain-gauge	-雨量計
저울형 강수계	weighing-type precipitation gauge	-型降水計
저잡음증폭기	low-noise amplifier	低雜音增幅器
저장	storage	貯藏
저장우량계	storage raingauge	貯藏雨量計
저장자료	archived data	貯藏資料
저조	low tide, low water	低潮
저조위표지	low water mark	低潮位標識
저주파수	low frequency	低周波數
저주파진동자	local frequency oscillator, low frequency oscillator	低周波振動子
저지	block, blocking	沮止
저지계	blocking system	沮止系
저지고기압	blocking high	沮止高氣壓
저지구역	blocking area	沮止區域
저지류	blocked flow	沮止流
저지발진기	blocking oscillator	沮止發振器
저지상황	blocking situation	沮止狀況
저지시스템	blocking system	沮止-
저지작용	blocking action	沮止作用
저지저기압	blocking low	沮止低氣壓
저지파	blocking wave	沮止波
저지현상	blocking, blocking phenomenon, damming	沮止現象
저지효과	blocking effect	沮止效果
저진공	low vacuum	低眞空
저첨도	platykurtosis	底尖度
저층구름소용돌이	low-level cloud vortex	低層-
저층기상예보	ground air meteorological forecast	底層氣象豫報
저층난류	low-level turbulence	低層亂流
저층난류대	lower turbulent zone	低層亂流帶
저층마찰	bottom friction	底層摩擦
저층바람시어경보시스템	low-level wind shear alert system	低層-警報-
저층수	bottom water	底層水
저층악기상정보	airman's meteorological information	底層惡氣象情報
저층윈드시어	low-level wind shear	低層-
저층해류	bottom current	底層海流
저폭	base width	底幅
저항	resistance	抵抗
저항력	force of resistance	抵抗力
저항온도계	resistance thermometer	抵抗溫度計
저항전류	ohmic current	抵抗電流
저해물질	inhibitor	沮害物質

저흔	sole mark	底痕
적경	right ascension	赤經
적도	equator	赤道
적도가속도	equatorial acceleration	赤道加速度
적도건조대	equatorial dry zone	赤道乾燥帶
적도기압골	equatorial trough	赤道氣壓-
적도골	equatorial trough	赤道-
적도궤도위성	equatorial orbiting satellite	赤道軌道衛星
적도기단	equatorial air (mass)	赤道氣團
적도기후	equatorial climate	赤道氣候
적도대	equatorial zone	赤道帶
적도대륙기단	equatorial continental air (mass)	赤道大陸氣團
적도로스비파	equatorial Rossby wave	赤道-波
적도림	equatorial forest	赤道林
적도면	equatorial plane	赤道面
적도무풍	equatorial calm	赤道無風
적도무풍대	doldrum equatorial trough, doldrums, equatorial calm belt	赤道無風帶
적도반류	equatorial countercurrent	赤道反流
적도베타면	equatorial beta-plane	赤道-面
적도부풂	equatorial bulge	赤道-
적도성층권	equatorial stratosphere	赤道成層圈
적도소용돌이	equatorial vortex	赤道-
적도수렴대	equatorial belt of convergence, equatorial convergence zone	赤道收斂帶
적도완충대	equatorial buffer zone	赤道緩衝帶
적도우림	equatorial rainforest	赤道雨林
적도저기압	equatorial low	赤道低氣壓
적도저압대	equatorial low pressure belt, equatorial low pressure zone	赤道低壓帶
적도전선	equatorial front	赤道前線
적도전선대	equatorial frontal zone	赤道前線帶
적도조석	equatorial tide	赤道潮汐
적도지구반경	equatorial radius of earth	赤道地球半徑
적도파	equatorial wave	赤道波
적도판	equatorial plate	赤道板
적도편동풍	equatorial easterlies	赤道偏東風
적도편서풍	equatorial westerlies	赤道偏西風
적도해양기단	equatorial maritime air (mass)	赤道海洋氣團
적란운	cumulonimbus, nimbus	積亂雲
적란운 무리	Cb cluster, organized Cb cluster	積亂雲-
적분가능	integrable	積分可能
적분	integration	積分
적분상수	integral constant	積分常數

적분학	integral calculus	積分學
적분한계	limit of integration	積分限界
적빙	ice cover	積氷
적빙기간	duration of ice cover	積氷期間
적산	accumulation	積算
적산계	totalizer, totalizator	積算計
적산대	accumulation zone	積算帶
적산면적	accumulative area	積算面積
적산설량계	snowfall totalizer, totalizer snow gauge	積算雪量計
적산온도	accumulated temperature	積算溫度
적산우량계	accumulative rain gauge, rainfall totalizer, totalizer raingauge	積算雨量計
적산우량곡선	mass rainfall curve	積算雨量曲線
적색경보	red flag warning	赤色警報
적색깃발	red flag	赤色-
적색잡음	red noise	赤色雜音
적색층	red bed	赤色層
적설	accumulated snow	積雪
적설(량)	snow accumulation	積雪(量)
적설	snow depth, snow cover	積雪
적설계	snow cover meter, snow meter	積雪計
적설기록계	snow recorder	積雪記錄計
적설기어내림	snow creep	積雪-
적설깊이	depth of snow, depth of snow cover	積雪-
적설도	snowcover chart	積雪圖
적설물당량	water equivalent of snow cover	積雪-當量
적설-삭설 모형	snow accumulation and ablation model	積雪-朔雪模型
적설상당수량	water equivalent of snow cover	積雪相當水量
적설선	snowcover line	積雪線
적설윤곽	snow cap profile	積雪輪廓
적설융해	snow cover melting	積雪融解
적설일(수)	day of snow lying	積雪日(數)
적설침강력	snow settling force	積雪沈降力
적설판	snow board, snow measuring plate	積雪板
적외(선)	infrared	赤外(線)
적외간섭분광계	infrared interferometer spectrometer	赤外干涉分光計
적외광선	infrared ray	赤外光線
적외대기창	infrared atmospheric window	赤外大氣窓
적외복사	infrared radiation	赤外輻射
적외복사측정	infrared radiometry	赤外輻射測定
적외선	infrared ray	赤外線
적외선가스분석계	infrared gas analyzer	赤外線-分析計
적외선강조기법	enhanced IR technique	赤外線强調技法
적외선대역	infrared band	赤外線帶域

적외선망원경	infrared telescope	赤外線望遠鏡
적외선사진	infrared picture	赤外線寫眞
적외선영역부근	near-infrared region	赤外線領域附近
적외선영역부근 스펙트럼	near-infrared spectrum	赤外線領域附近-
적외선창	infrared window	赤外線窓
적외선파장	infrared wavelength	赤外線波長
적외스펙트럼	infrared spectrum	赤外-
적외역	infrared region	赤外域
적외열방사	infrared thermal radiation	赤外熱放射
적외영상	infrared image, infrared imagery, infrared imaging	赤外映像
적외위성영상	infrared satellite imagery	赤外衛星映像
적외지구센서	Infra-Red Earth Sensor	赤外地球-
적외화상기	infrared imager	赤外畵像器
적외흡수습도계	infrared absorption hygrometer	赤外吸收濕度計
적운	cumulus, cumulus cloud	積雲
적운단계	cumulus stage	積雲段階
적운대류	cumulus convection	積雲對流
적운마찰	cumulus friction	積雲摩擦
적운모양비구름	nimbus cumuliformis	積雲模樣-
적운모형	cumulus model	積雲模型
적운밑면	cumulus base	積雲-
적운사이구역	intercumulus region	積雲-區域
적운성	cumulogenitus	積雲性
적운장모형	cumulus field model	積雲場模型
적운형	cumuliform	積雲形
적운형구름	cumuliform cloud	積雲形-
적운형권층운	cirrostratus cumulosus	積雲形卷層雲
적위	declination	赤緯
적응	adaptation	適應
적응격자	adaptive grid	適應格子
적응도	adaptability	適應度
적응밝기	adaptation brightness	適應-
적응병	adaptive disease	適應病
적응복사	adaptive radiation	適應輻射
적응성	adaptability	適應性
적응수렴	adaptive convergence	適應收斂
적응수준	adaptation level	適應水準
적응장	adaptation field	適應場
적응전략	adaptation strategy	適應戰略
적응조도	adaptation illuminance	適應照度
적응조절	adaptive control	適應調節
적응형	adaptation type	適應型
적응휘도	adaptation luminance	適應輝度

ㅈ

적재(량)	loading	積載(量)
적정	titration	滴定
적정곡선	titration curve	滴定曲線
적정오차	titration error	滴定誤差
적조	algal bloom, red tide	赤潮
적출	avulsion	摘出
적합도	goodness of fit	適合度
적합도검정	goodness-of-fit test	適合度檢定
적합성조건	compatibility condition	適合性條件
적합지수	fitness figure	適合指數
전개섭동	expanding perturbation	展開攝動
전개(식)	expansion	展開(式)
전격	lightning stroke, stroke	電擊
전격사이변화	interstroke change	電擊-變化
전구분포	global distribution	全球分布
전구에너지수지	global energy budget	全球-收支
전구평균온도변동	global-mean temperature variation	全球平均溫度變動
전극효과	electrode effect	電極效果
전기	electricity	電氣
전기데스펀안테나(인공위성의)	electrical despun antenna	電氣-
전기도체	electric conductor	電氣導體
전기력	electric force	電氣力
전기력선	electric line of force	電氣力線
전기습도계	electric hygrometer, electrical hygrometer	電氣濕度計
전기쌍극자모멘트	electric dipole moment	電氣雙極子-
전기에너지	electrical energy	電氣-
전기온도계	electrical theromometer	電氣溫度計
전기요란	electric disturbance	電氣擾亂
전기운동퍼텐셜	electrokinetic potential	電氣運動-
전기장	electric field	電氣場
전기장기록	electrogram	電氣場記錄
전기장세기	electric field intensity, electric field strength, electric intensity	電氣場-
전기적 중성	electrical neutrality	電氣的中性
전기전도	electrical conduction	電氣傳導
전기전도도	electric conductivity	電氣傳導度
전기전도도계	conductivity meter	電氣傳導度計
전기전도율	electrical conductivity, electric conductivity	電氣傳導率
전기진동	electric oscillation	電氣振動
전기진자	electric pendulum	電氣振子
전기컵풍속계	electric cup anemometer	電氣-風速計
전기투석	electrodialysis	電氣透析
전기폭발장치	electro explosive device	電氣爆發裝置
전기풍속계	electrical anemometer	電氣風速計

전기현상	electrical phenomenon	電氣現象
전기화학	electrochemistry	電氣化學
전기화학당량	electrochemical equivalent	電氣化學當量
전기화학존데	electrochemical sonde	電氣化學-
전기회로	electric circuit	電氣回路
전단	shear	剪斷
전단효과	shear effect	剪斷效果
전달	conveyance	傳達
전도	conduction	傳導
전도도	conductance	傳導度
전도도-온도-깊이 프로파일	conductivity-temperature-depth profile	傳導度溫度-
전도되	tipping bucket	轉倒-
전도성	conductivity	傳導性
전도성전류	conductivity current	傳導性電流
전도식우량계	rain-gauge tipping-bucket type	轉倒式雨量計
전도온도계	reversing thermometer	轉倒溫度計
전도용량	conductive capacity	傳導容量
전도전류	conduction current	傳導電流
전도전자	conduction electron	傳導電子
전도평형	conductive equilibrium	傳導平衡
전도함수	total derivative	全導函數
전도형우량계	tipping bucket rain gauge	轉倒形雨量計
전동주사마이크로파복사계	electrically scanning microwave radiometer	電動走査-波輻射計
전력	electric power	電力
전력기상업무	meteorological service for electric power system	電力氣象業務
전력분리기	power divider	電力分離機
전력분배기	power splitter, divider	電力分配機
전력손실	power dissipation	電力損失
전력수송	electric power transmission	電力輸送
전류	current, electric current	電流
전류계	ammeter	電流計
전리권	ionosphere	電離圈
전리방사선	ionizing radiation	電離放射線
전리층	ionospheric layer	電離層
전망	outlook	展望
전망기	outlook period	展望期
전망브리핑	outlook briefing	展望-
전문기상예보기술사	professional engineer weather forecaster	專門氣象豫報技術士
전문기상학감정	professional meteorological advice	專門氣象學鑑定
전미분	total differential	全微分
전방경로확률	forward path probability	前方經路確率
전방공간차분근사	forward-in-space difference approximation	前方空間差分近似
전방모형	forward model	前方模型
전방산란	forward scattering	前方散亂

전방산란체	forward scatter	前方散亂體
전방산란CW 레이더	forward scatter continuous wave radar	前方散亂-
전방상류차분	forward-upstream difference	前方上流差分
전방시간차분	forward time difference	前方時間差分
전방차분	forward difference	前方差分
전방향복사계	omnidirectional radiometer	全方向輻射計
전방향성	omnidirectional	全方向性
전보	message	電報
전보교신	telegraph exchange	電報交信
전보교환시설	message switching facility	電報交換施設
전복형우량계	tilting bucket rain gauge	顚覆形雨量計
전산구속사항	computational constraint	電算拘束事項
전산역학	computational mechanics	電算力學
전산자원	computational resource	電算資源
전선	front, frontal line	前線
전선강도	frontal strength	前線强度
전선(성)강수	frontal precipitation	前線(性)降水
전선(성)강우	frontal rainfall	前線(性)降雨
전선경계	frontal boundary	前線境界
전선경사	frontal slope	前線傾斜
전선계	frontal system	前線系
전선골	frontal trough	前線-
전선구름	frontal cloud	前線-
전선구름대	frontal cloud band	前線-帶
전선구름생성	frontal cloud formation	前線-生成
전선구조	frontal structure	前線構造
전선기단	frontal mass	前線氣團
전선-기압골시스템모형	front-trough system model	前線-氣壓-模型
전선기울기	frontal slope, slope of a front	前線-
전선난류	frontal turbulence	前線亂流
전선(성)날씨	frontal weather	前線(性)-
전선뇌우	frontal thunderstorm	前線雷雨
전선대	frontal strip, frontal zone	前線帶
전선뒤안개	post-frontal fog	前線-
전선등고선	frontal contour	前線等高線
전선등고선도	frontal contour chart	前線等高線圖
전선론	frontology	前線論
전선면	frontal surface	前線面
전선면높낮이	frontal topography	前線面-
전선면형태	frontal topography	前線面形態
전선모형	frontal model	前線模型
전선물리	physics of front	前線物理
전선발달	frontogenesis	前線發達
전선발달역	frontogenetical area	前線發達域

전선발생	frontogenesis	前線發生
전선발생선	line of frontogenesis	前線發生線
전선발생역	frontogenetical area, frontogenetical sector	前線發生域
전선발생함수	frontogenetical function	前線發生函數
전선발생효과	frontogenetic effect	前線發生效果
전선변질	front modification	前線變質
전선분류	classification of front, frontal classification	前線分類
전선분석	frontal analysis	前線分析
전선소멸	frontal decay, frontolysis	前線消滅
전선소멸역	frontolytical area	前線消滅域
전선소멸지역	region of frontolysis	前線消滅地域
전선속력	frontal speed	前線速力
전선쇠약	frontolysis	前線衰弱
전선쇠약역	frontolytical area, frontolytical sector	前線衰弱域
전선순환	frontal circulation	前線循環
전선시어선	front shear line	前線-線
전선쐐기	frontal wedging	前線-
전선안개	frontal fog	前線-
전선앞뇌우	pre-frontal thunderstorm	前線-雷雨
전선앞스콜선	pre-frontal squall line	前線-線
전선앞안개	pre-frontal fog	前線-
전선역전	frontal inversion	前線逆轉
전선역전층	frontal inversion layer	前線逆轉層
전선이론	frontal theory	前線理論
전선저기압	frontal cyclone, frontal depression, frontal low	前線低氣壓
전선제트	frontal jet	前線-
전선차단	frontal trapping	前線遮斷
전선층	frontal layer	前線層
전선치올림	frontal lifting	前線-
전선통과	frontal passage	前線通過
전선통과안개	frontal passage fog, front passage fog	前線通過-
전선파	frontal wave	前線波
전선파동	frontal wave motion	前線波動
전선파모형	frontal wave model	前線波模型
전선파복	looping of the front	前線波腹
전선폐색	frontal occlusion	前線閉塞
전선프로파일	frontal profile	前線-
전선활동	frontal action	前線活動
전선활승	frontal upgliding	前線滑昇
전송	transmission	傳送
전시	display	展示
전신방정식	telegraphic equation	電信方程式
전신인쇄기	teleprinter	電信印刷機
전신타자기	teletype, teleprinter	電信打字機

전압	voltage	電壓
전압강하	potential(voltage) drop	電壓降下
전압계	voltmeter	電壓計
전압조정기	voltage regulator	電壓調整器
전엽	front lobe	前葉
전오존량	total ozone	全-量
전운량	total cloud cover, total sky cover	全雲量
전(력)원	power source	電(力)源
전원장치	power supply	電源裝置
전원풍	country breeze	田園風
전위	electric potential, potential	電位
전위경도	electric potential gradient, potential gradient	電位傾度
전위계	electrometer	電位計
전위존데	eletrosonde	電位-
전위차	electric potential difference	電位差
전위차계	potentiometer	電位差計
전이궤도	transfer orbit	轉移軌道
전이기단	transitional air-mass	轉移氣團
전이기후	transitional climate	轉移氣候
전이상태	transition state	轉移狀態
전이역	zone of transition	轉移域
전이온도	transition temperature	轉移溫度
전이원소	transition element	轉移元素
전이음속(의)	transonic	轉移音速
전이표면	transitional surface	轉移表面
전이함수	transfer function	轉移函數
전이흐름	transitional flow	轉移-
전자	electron	電子
전자각	electron shell	電子殼
전자강수	electron precipitation	電子降水
전자경위의	electronic theodolite	電子經緯儀
전자공여체	electron donor	電子供與體
전자공학	electronics	電子工學
전자구름	electron cloud	電子-
전자구름모형	electron cloud model	電子-模型
전자궤도	electron orbit	電子軌道
전자기가스기상	electromagnetic gas weather	電磁氣-氣象
전자기간섭	electromagnetic interference	電磁氣干涉
전자기단위	electromagnetic unit(emu)	電磁氣單位
전자기방사법칙	electromagnetic radiation law	電磁氣放射法則
전자기방사전파	electromagnetic radiation propagation	電磁氣放射傳播
전자기복사	electromagnetic radiation	電磁氣輻射
전자기복사법칙	electromagnetic radiation law	電磁氣輻射法則
전자기복사전파	electromagnetic radiation propagation	電磁氣輻射傳播

전자기스펙트럼	electromagnetic spectrum	電磁氣-
전자기에너지	electromagnetic energy	電磁氣-
전자기유도	electromagnetic induction	電磁氣誘導
전자기이론	electromagnetic theory	電磁氣理論
전자기장	electromagnetic field	電磁氣場
전자기적 상호작용	electromagnetic interaction	電磁氣的相互作用
전자기적 진동	electromagnetic oscillation	電磁氣的振動
전자기파반사	reflection of electomagnetic wave	電磁氣波反射
전자기학	electromagnetics, electromagnetism	電磁氣學
전자기흡수	electromagnetic absorption	電磁氣吸收
전자껍질	electron shell	電子-
전자볼트	electron volt	電子-
전자비전하	electron specific charge	電子比電荷
전자빔	electron beam	電子-
전자사태	electron avalanche	電子沙汰
전자석	electromagnet	電磁石
전자수치적분계산기	Electronic Numerical Integrator And Calculator (ENIAC)	電子數値積分計算機
전자온도계	electronic thermometer	電子溫度計
전자이동	electron transfer	電子移動
전자읽음장치	electronic readout equipment	電子-裝置
전자저울	electronic balance	電子-
전자전이	electronic transition	電子轉移
전자전쟁기상	electronic warfare weather	電子戰爭氣象
전자전하	electronic charge	電子電荷
전자지상지원장비	electrical ground support equipment	電子地上支援裝備
전자친화성	electron affinity	電子親和性
전자(기)파	electromagnetic wave	電磁(氣)波
전자(기)파스펙트럼	electromagnetic wave spectrum	電磁(氣)波-
전자포획	electron capture	電子捕獲
전장기후	battlefield climate	戰場氣候
전장요란	electric storm	電場擾亂
전쟁기상학	meteorology of war	戰爭氣象學
전접컵풍속계	cup contact anemometer	電接-風速計
전조	fore-runner, precursor, symbol	前兆
전조공전	precursor	前兆空電
전조하늘(저기압 접근의)	emissary sky	前兆-
전지구관측시스템	global observation system, global observing system	全地球觀測-
전지구기상정찰	global weather reconnaissance	全地球氣象偵察
전지구기후	global climate	全地球氣候
전지구기후변화	global climate change	全地球氣候變化
전지구기후온난화	global climate warming	全地球氣候溫暖化
전지구냉각화	global cooling	全地球冷却化

전지구모형	global model	全地球模型
전지구물순환	global water cycle	全地球-循環
전지구배출목록감시 모델링시스템	Global Inventory Monitoring and Modeling Systems	全地球排出目錄監視-
전지구범위	global coverage	全地球範圍
전지구복사평형	global radiative equilibrium	全地球輻射平衡
전지구수문순환	global hydrologic cycle	全地球水文循環
전지구순환	global circulation	全地球循環
전지구순환계	global circulation system	全地球循環系
전지구순환모형	global circulation model	全地球循環模型
전지구예보체계	global forecast system	全地球豫報體系
전지구위치측정 존데	Global Positioning System(GPS) sonde	全地球位置測定-
전지구자료동화예측시스템	global data assimilation and prediction system	全地球資料同化豫測-
전지구자료처리관측시스템	global data processing observation system	全地球資料處理觀測-
전지구자료처리시스템	global data processing system	全地球資料處理-
전지구지상온도	global near-surface temperature	全地球地上溫度
전지구평균에너지수지	global-mean energy balance	全地球平均-收支
전지구해양탐험자계획	Global Area Coverage Oceans Pathfinder Project	全地球海洋探險者計劃
전진운동	progressive motion	前進運動
전진파	progressive wave	前進波
전처리	preprocessing	前處理
전처리과정	preprocessor	前處理過程
전처리기	preprocessor	前處理機
전천광도계	all-sky photometer	全天光度計
전천단파복사	global shortwave radiation	全天短波輻射
전천복사	global radiation	全天輻射
전천복사계	pyrradiometer	全天輻射計
전천복사스펙트럼	global radiation spectrum	全天輻射-
전천복사평균강도지수	heliometric index	全天輻射平均强度指數
전천일사	global solar radiation	全天日射
전천일사계	pyranometer	全天日射計
전천일사기록	pyranogram	全天日射記錄
전천일사평균강도계	lucimeter	全天日射平均强度計
전천카메라	all-sky camera	全天-
전천후공항	all-weather airport	全天候空港
전천후착륙	all-weather landing	全天候着陸
전체지수	total index	全體指數
전체파수	total index	全體波數
전치증폭기	preamplifier	前置增幅器
전투기용전문형식	combat aircraft code	戰鬪機用電文型式
전파	dissemination, propagation	傳播
전파	radio wave	電波
전파간섭	radio interference	電波干涉
전파거리측정기	radio distance finder	電波距離測定器

전파경위의	radio theodolite	電波經緯儀
전파고도계	radio altimeter	電波高度計
전파관	radio duct	電波管
전파관측	radiosounding	電波觀測
전파구멍(난청)	radio hole	電波-(難聽)
전파굴절률	radio refractive index	電波屈折率
전파굴절률구조모수	radio refractive index structure parameter	電波屈折率構造母數
전파굴절보정	radio refraction correction	電波屈折補正
전파극광	radio aurora	電波極光
전파기상기록계	radio meteorograph	電波氣象記錄計
전파기상학	radioelectric meteorology, radio meteorology	電波氣象學
전파기후학	radio climatology	電波氣候學
전파두절	radio blackout	電波杜絶
전파바람	propagated wind	傳播-
전파방출	radio emission	電波放出
전파방향기록계	radio goniograph	電波方向記錄計
전파방향탐지	radio direction finding	電波方向探知
전파방향탐지기	radio direction finder, radio goniometer	電波方向探知器
전파방향탐지법	radio goniometry	電波方向探知法
전파빔	radio beam	電波-
전파상수	propagation constant	傳播常數
전파소멸	radio fadeout	電波消滅
전파속도	propagation velocity	傳播速度
전파수평선	radio horizon	電波水平線
전파신기루	radio mirage	電波蜃氣樓
전파실링	radio ceiling	電波-
전파실링계	radio ceiling	電波-計
전파에너지	radio energy	電波-
전파예보	propagation forecasting	傳播豫報
전파위치측정기	radiolocator	電波位置測定器
전파유성	radio meteor	電波流星
전파장방출복사체	full radiator	全波長放出輻射體
전파추적	radio tracking	電波追跡
전파측풍기구	radio pilot balloon	電波測風氣球
전파탐측	radio wave sounding	電波探測
전파팩스	radio fax	電波-
전파팩시밀리	radio facsimile	電波-
전하	charge, electrical charge, electric charge	電荷
전하결합소자(CCD)	charge coupled device	電荷結合素子
전하량	quantity of electric charge	電荷量
전하밀도	charge density	電荷密度
전하보존	conservation of charge	電荷保存
전하(량)보존법칙	electric charge conservation law	電荷(量)保存法則
전하분리	charge separation, separation of charge	電荷分離

전하수	charge number	電荷數
전해물	electrolyte	電解物
전해박판	electrolytic strip	電解薄板
전해질	electrolyte	電解質
전향	deflection, recurvature	轉向
전향가속도	deflecting acceleration	轉向加速度
전향력	deflecting force, deflection force	轉向力
전향인자	deflecting factor	轉向因子
전향점	point of recurvature, turning point	轉向點
전화구름	mutatus	轉化-
전환	conversion	轉換
전환고도	transition altitude	轉換高度
전환인자	conversion factor	轉換因子
절기	solar term	節氣
절단산각	truncated spur	切斷山脚
절단오차	truncation error	切斷誤差
절대가뭄	absolute drought	絕對-
절대가속도	absolute acceleration	絕對加速度
절대각운동량	absolute angular momentum	絕對角運動量
절대각운동량보존	conservation of absolute angular momentum	絕對角運動量保存
절대각운동량보존법칙	conservation law of absolute angular momentum	絕對角運動量保存法則
절대경도류	absolute gradient current	絕對傾度流
절대계기	absolute instrument	絕對計器
절대고도	absolute altitude	絕對高度
절대공동복사계	absolute cavity radiometer	絕對空洞輻射計
절대굴절률	absolute index of refraction, absolute refractive index	絕對屈折率
절대극값	absolute extreme	絕對極-
절대기압	absolute pressure	絕對氣壓
절대기준틀	absolute reference frame	絕對基準-
절대꽃가루빈도	absolute pollen frequency	絕對-頻度
절대높낮이	absolute topography	絕對-
절대눈금	absolute scale	絕對-
절대단위	absolute unit	絕對單位
절대등고선	absolute isohypse	絕對等高線
절대변동(률)	absolute variability	絕對變動(率)
절대복사규모	absolute radiation scale	絕對輻射規模
절대부등식	absolute inequality	絕對不等式
절대불안정	absolute instability	絕對不安定
절대빈도	absolute frequency	絕對頻度
절대선형운동량	absolute linear momentum	絕對線形運動量
절대소용돌이도	absolute vorticity	絕對-度
절대소용돌이도보존	conservation of absolute vorticity	絕對-度保存
절대소용돌이도최대	absolute vorticity maximum	絕對-度最大

절대속도	absolute velocity	絶對速度
절대수렴	absolute convergence	絶對收斂
절대순환	absolute circulation	絶對循環
절대습도	absolute humidity	絶對濕度
절대습도계	absolute hygrometer	絶對濕度計
절대시차	absolute parallax	絶對視差
절대실링	absolute ceiling	絶對-
절대안정도	absolute stability	絶對安定度
절대압력	absolute pressure	絶對壓力
절대연대측정	absolute dating	絶對年代測定
절대영도	absolute zero	絶對零度
절대오차	absolute error	絶對誤差
절대온도	absolute temperature, Kelvin temperature	絶對溫度
절대온도극값	absolute temperature extreme	絶對溫度極-
절대온도눈금	absolute temperature scale, Kelvin scale, Kelvin temperature scale	絶對溫度-
절대와도보존법칙	law of absolute vorticity conservation	絶對渦度保存法則
절대운동량	absolute momentum	絶對運動量
절대운동량보존	conservation of absolute momentum	絶對運動量保存
절대위치소용돌이도	absolute potential vorticity	絶對位置-度
절대좌표계	absolute coordinate system	絶對座標系
절대지형	absolute topography	絶對地形
절대직달일사계	absolute pyrheliometer	絶對直達日射計
절대진공	absolute vacuum	絶對眞空
절대표준기압계	absolute standard barometer	絶對標準氣壓計
절댓값	absolute value, modulus	絶對-
절수기술	water-saving technology	節水技術
절연	insulation, isolation	絶緣
절정공률	peak power	絶頂工率
절정돌풍	peak gust	絶頂突風
절정바람	peak wind	絶頂-
절정전력	peak power	絶頂電力
절정풍속	peak wind speed	絶頂風速
절편매개변수	intercept parameter	切片媒介變數
점검	check	點檢
점검관측	check observation	點檢觀測
점근관측	asymptotic observation	漸近觀測
점근극한	asymptotic limit	漸近極限
점근상사이론	asymptotic similarity theory	漸近相似理論
점근선	asymptote	漸近線
점근전개	asymptotic expansion	漸近展開
점근점	asymptotic point	漸近點
점근조화	asymptotic matching	漸近調和
점근혼합길이	asymptotic mixing length	漸近混合-

ㅈ

점대다중점통신	point-to-multipoint communication	點對多重點通信
점대점통신	point-to-point communication	點對點通信
점도	viscosity	粘度
점빈도해석	point frequency analysis	點頻度解析
점성	viscosity	粘性
점성계수	coefficient of viscosity, viscosity coefficient	粘性係數
점성력	viscous force	粘性力
점성류	viscous flow	粘性流
점성소산	viscous dissipation	粘性消散
점성유체	viscous fluid	粘性流體
점성응력	viscous stress	粘性應力
점성저항	viscosity resistance	粘性抵抗
점성체	viscous body	粘性體
점성항	viscosity term	粘性項
점성항력	viscous drag	粘性抗力
점액	mucin	粘液
점우량	point rainfall	點雨量
점원	point source	點源
점전하	point charge	點電荷
점진평균방법	method of progressive average	漸進平均方法
점착	agglutination, coherence	粘着
점표적	point target	點標的
점화	firing	點火
접근속도	approach velocity	接近速度
접근수로	approach channel	接近水路
접근시정	approach visibility	接近視程
접선	tangent line	接線
접선가속도	tangential acceleration	接線加速度
접선길이	length of tangent segment	接線-
접선력	tangential force	接線力
접선바람	tangential wind	接線-
접선선형모형	tangent linear model	接線線形模型
접선성분	tangential component	接線成分
접선속도	tangential velocity	接線速度
접선응력	tangential stress	接線應力
접속조절문서	interface control document	接續調節文書
접시안테나	parabolic antenna	
접안얼음	landfast ice	接岸-
접점컵풍속계	contact-cup anemometer	接點-風速計
접지	earth grounding, earthing	接地
접지기후	climate near the ground	接地氣候
접지역전	ground inversion, surface inversion	接地逆轉
접지(경계)층	surface boundary layer	接地(境界)層
접착수	adhesive water	接着水

접촉각	contact angle	接觸角
접촉결빙과정	contact freezing process	接觸結氷過程
접촉냉각	contact cooling	接觸冷却
접촉면	contact surface	接觸面
접촉비행	contact flight	接觸飛行
접촉식풍속계	contact anemometer	接觸式風速計
접촉저항	contact resistance	接觸抵抗
접촉핵	contact nucleus	接觸核
접평면	tangent (tangential) plane	接平面
접합	coupling	接合
접합기	synapsis stage	接合期
접호	tangent arcs	接弧
접호 22도 무리	tangent arcs 22° halo	接弧-度-
접호 46도 무리	tangent arcs 46° halo	接弧-度-
접힌대류권계면	folding tropopause	-對流圈界面
접힘	folding	
접힘횟수	Nyquist folding number	-回數
정(의)	positive	正
정고도기구	constant altitude balloon, constant-level balloon, Moby Dick balloon	定高度氣球
정고도면	constant-height surface, constant-level surface	定高度面
정고도면도	constant-height chart, constant-level chart	定高度面圖
정규	normal	正規
정규과정	normal process	正規過程
정규관측소	authorized station	正規觀測所
정규도	normal map	正規圖
정규모드	normal mode	正規-
정규모드초기화	normal mode initialization	正規-初期化
정규모드해	normal mode solution	正規-解
정규방정식	normal equation	正規方程式
정규분산	normal dispersion	正規分散
정규분포	normal distribution	正規分布
정규분포곡선	Gaussian curve, normal distribution curve	正規分布曲線
정규분포표	normal distribution table	正規分布表
정규세계일	Regular World Days	正規世界日
정규식생지수	normalized difference vegetation index	正規植生指數
정규오차법칙	normal law of error	定規誤差法則
정규원형분포	normal circular distribution	正規圓形分布
정규함수	normal function	正規函數
정규화	normalization	正規化
정규화레이더반사면적	normalized radar cross section	正規化-反射面積
정규화진동수	normalized frequency	正規化振動數
정굿값	norm	正規-
정기상학	statical meteorology	靜氣象學

정기후학	static climatology	靜氣候學
정기후학기후대	climatic zone by statical climatology	靜氣候學氣候帶
정다각형	regular polygon	正多角形
정다면체	regular polyhedron	正多面體
정단세포	apical cell	頂端細胞
정단층	normal fault	正斷層
정량강수예보	quantitative precipitation forecast	定量降水豫報
정량적강수추정	quantitative precipitation estimation	定量的降水推定
정량분석	quantitative analysis	定量分析
정량예보	quantitative forecast	定量豫報
정렬	alignment	整列
정류	rectification	整流
정류관	rectifying tube	整流管
정류기	commutator, rectifier	整流器
정류벽	baffle	整流壁
정류식라디오존데	commutator radiosonde	整流式-
정류자	commutator	整流子
정류작용	rectifying action	整流作用
정류회로	rectifier circuit	整流回路
정립파	standing wave	定立波
정밀도	precision	精密度
정밀분석	precision analysis	精密分析
정밀아네로이드기압계	precision aneroid barometer	情密-氣壓計
정반사	regular reflection, specular reflection	正反射
정반사체	regular reflector, specular reflector	正反射體
정보교환센터	clearing house	情報交換-
정사각형	regular quadrilateral	正四角形
정사면체	regular tetrahedron	正四面體
정사투영	orthogonal projection, orthographic projection	正射投影
정삼각형	equilateral triangle	正三角形
정상	normal	定常
정상감률	normal lapse	定常減率
정상값	normal value	定常-
정상경도	normal gradient	定常傾度
정상곡률	normal curvature	定常曲率
정상관	positive correlation	正相關
정상광선	ordinary ray	正常光線
정상극성	normal polarity	定常極性
정상기압계	normal barometer	定常氣壓計
정상기온	normal temperature	定常氣溫
정상기온감률	normal lapse rate	定常氣溫減率
정상년 강우량 비율법	normal ratio method	定常年降雨量比率法
정상대기	normal atmosphere	定常大氣
정상도	steadiness	定常度

정상류	stationary flow, steady flow	定常流
정상모드	normal mode	定常-
정상미만	below normal	正常未滿
정상상태	normal state, steady state	定常狀態
정상속도	steady velocity	定常速度
정상순환	normal circulation	定常循環
정상시어	normal shear	定常-
정상연강수량	normal annual precipitation	定常年降水量
정상운동	steady motion	定常運動
정상월평균기온	normal monthly temperature	定常月平均氣溫
정상이상	above normal	正常以上
정상일평균기온	normal daily temperature	定常日平均氣溫
정상장	steady field	定常場
정상적 작동조건	normal operating condition	定常的作動條件
정상전류	stationary electric current	定常電流
정상중력	normal gravity	定常重力
정상파	clapotis, ordinary wave, steady wave	定常波
정상파영역	steady wave regime	定常波領域
정상환기	normal aeration	定常換氣
정성관측	qualitative observation	定性觀測
정수	integer	整數
정수압면	potentiometric surface	靜水壓面
정수위	still-water level	靜水位
정수통	still well	靜水桶
정습도	constant humidity	定濕度
정습도선	constant humidity line	定濕度線
정시	on time	定時
정시관측	routine observation	定時觀測
정시등치	isotimic	定時等値
정시등치면	isotimic surface	定時等値面
정시등치선	isotimic line	定時等値線
정시방송	fixed time broadcast, regular broadcast	定時放送
정시비행장예보	routine terminal aerodrome forecast	定時飛行場豫報
정시항로예보	routine route forecast	定時航路豫報
정십이면체	regular dodecahedron	正十二面體
정압	constant pressure, static pressure	定壓
정압기구	constant pressure balloon	定壓氣球
정압기체온도계	constant pressure gas thermometer	定壓氣體溫度計
정압면	constant pressure surface	定壓面
정압면도	constant pressure chart, constant pressure map	定壓面圖
정압면일기도	constant pressure level chart	定壓面日氣圖
정압비열	specific heat at constant pressure	定壓比熱
정압패턴비행	constant pressure pattern flight	定壓-飛行
정역학	hydrostatic, statics	靜力學

정역학가정	hydrostatic assumption	靜力學假定
정역학근사	hydrostatic approximation	靜力學近似
정역학모형	hydrostatic model	靜力學模型
정역학방정식	hydrostatic equation	靜力學方程式
정역학불안정(도)	hydrostatic instability	靜力學不安定(度)
정역학안정(도)	hydrostatic stability	靜力學安定(度)
정역학에너지	hydrostatic energy	靜力學-
정역학조절	hydrostatic adjustment	靜力學調節
정역학평형	hydrostatic balance, hydrostatic equilibrium	靜力學平衡
정역학평형방정식	equation of hydrostatic equilibrium	靜力學平衡方程式
정영역	positive area	正領域
정오	noon, midday, solar noon	正午
정온기	alcyone days, halcyon days	靜穩期
정유소	oil refinery	精油所
정육면체	cube, regular hexahedron	正六面體
정율희석검량	constant-rate dilution gauging	定率稀釋檢量
정이십면체	regular icosahedron	正二十面體
정일경	heliostat	定日鏡
정적	static	靜的
정적과정	static process	靜的過程
정적구름씨뿌리기	static cloud seeding	靜的-
정적기구	constant volume balloon	定積氣球
정적기체온도계	constant volume gas thermometer	定積氣體溫度計
정적분	definite integral	定積分
정적불안정(도)	static instability	靜的不安定(度)
정적비열	specific heat at constant volume	定積比熱
정적수위	static level	靜的水位
정적안정(도)	static stability	靜的安定(度)
정적평형	static equilibrium	靜的平衡
정전	blackout	停電
정전기	static electricity	靜電氣
정전기 가려막기	electrostatic shielding	靜電氣-
정전기 가리기	electrostatic screening	靜電氣-
정전기 단위	electrostatic unit(esu)	靜電氣單位
정전기력	electrostatic force	靜電氣力
정전기 발생장치	electrostatic generator	靜電氣發生裝置
정전기병합	electrostatic coalescence	靜電氣倂合
정전기 유도	electrostatic induction	靜電氣誘導
정전기 현상	electrostatic phenomenon	靜電氣現象
정전용량	electrostatic capacity	靜電容量
정전전위계	electrostatic electrometer	靜電電位計
정전집진기	electrostatic filter	靜電氣集塵機
정전집진장치	electrostatic precipitator	靜電集塵裝置
정전필터	electrostatic filter	靜電-

정전하	electrostatic charge	靜電荷
정점관측선	fixed ship station	定點觀測船
정점바람	spot wind	定點-
정준방정식	canonical equation	正準方程式
정준패턴	canonical pattern	正準-
정지	stationary	靜止
정지궤도	geostationary orbit, geosynchronous orbit, stationary orbit	靜止軌道
정지궤도위성	geostationary orbit satellite	靜止軌道衛星
정지궤도해색센서	geostationary ocean color imager	靜止軌道海色-
정지기상위성	geostationary meteorological satellite, synchronous meteorological satellite	靜止氣象衛星
정지기상위성계	geostationary meteorological satellite system	靜止氣象衛星系
정지대기	stationary atmosphere	靜止大氣
정지마찰	static friction	靜止摩擦
정지마찰계수	coefficient of static friction	靜止摩擦係數
정지마찰력	force of static friction	靜止摩擦力
정지실용환경위성	geostationary operational environmental satellite	靜止實用環境衛星
정지에너지	rest energy	靜止-
정지위성	geostationary satellite, geosynchronous satellite, stationary satellite	靜止衛星
정지전류	static current	靜止電流
정지전압	halting voltage	靜止電壓
정지지구복사수지위성	Geostationary Earth Radiation Budget Satellite	靜止地球輻射收支衛星
정진동	seiche	靜振動
정찰	reconnaissance	偵察
정체	stationary	停滯
정체가우스과정	stationary Gaussian process	停滯-過程
정체가우스시계열	stationary Gaussian time series	停滯-時系列
정체공기	dead air, stagnant air	停滯空氣
정체구름	standing cloud	停滯-
정체기	stationary phase	停滯期
정체기압골	anchor(ed) trough	停滯氣壓-
정체렌즈고적운	Ac standing lenticular	停滯-高積雲
정체로스비파	stationary Rossby wave	停滯-波
정체류	slack water	停滯流
정체빙하	stagnant glacier	停滯氷河
정체상태	stationary state	停滯狀態
정체수	dead water, slack water	停滯水
정체시계열	stationary time series	停滯時系列
정체압	stagnation pressure	停滯壓
정체에디	standing eddy	停滯-
정체운동	stationary motion	停滯運動
정체저기압	stationary cyclone	停滯低氣壓

정체전선	stationary front	停滯前線
정체지역	stagnation area	停滯地域
정체지하수	stagnant ground water	停滯地下水
정체파	stationary wave	停滯波
정체파길이	stationary wave length	停滯波-
정촉매	positive catalyst	正觸媒
정합신호	coherent signal	整合信號
정합적분	coherent integration	整合積分
정합필터	matched filter	整合-
정현검류계	sine galvanometer	正弦檢流計
정현곡선	sine curve, sinusoid	正弦曲線
정현파	sine wave	正弦波
정형도	orthomorphic map	正形圖
정화	scavenging	淨化
정확도	accuracy	正確度
정확도분산그래프	accuracy scatter graph	正確度分散-
정확도평가	accuracy assessment	正確度評價
젖빛광	opalescence	-光
젖빛혼탁도	opalescent turbidity	-混濁度
젖은 눈	cooking snow	
제1권계면	primary tropopause	第一圈界面
제1자색광	first purple light	第一紫色光
제2권계면	secondary tropopause	第二圈界面
제2대류권계면	second tropopause	第二對流圈界面
제2보랏빛노을	second purple light	第二-
제2열대권계면	secondary tropical tropopause	第二熱帶圈界面
제2종대류불안정도	convective instability of the second kind	第二種對流不安定度
제2종조건부불안정(성)	conditional instability of the second kind	第二種條件附不安定(性)
제2종중력습윤대류불안정	gravitational moist-convective instability of the second kind	第二種重力濕潤對流不安定
제3기 후기 기후지표	Late Tertiary climate indicator	第三紀後紀氣候指標
제4간빙기	Quaternary interglacial epoch	第四間氷期
제4기	Quaternary, Quaternary period	第四紀
제4기기후	Quaternary climate	第四紀氣候
제4기기후전이	Quaternary climate transition	第四紀氣候轉移
제4기말 거대홍수	Late Quaternary megaflood	第四紀末巨大洪水
제4기말 홀로세 식생모델링	Late Quaternary-Holocene vegetation modeling	第四紀末-植生-
제4기몬순	Quaternary monsoon	第四紀-
제4기식생분포	Quaternary vegetation distribution	第四紀植生分布
제4기이전	Pre-Quaternary	第四紀以前
제4기이전몬순	Pre-Quaternary monsoon	第四紀以前-
제4기이전밀란코비치 주기	Pre-Quaternary Milankovitch cycle	第四紀以前-週期
제4빙하시대	Pleistocene glacial epoch, Quaternary Ice Age	第四氷河時代
제곱	square	

제곱근	square root	-根
제곱근표	table of square root	-根表
제곱비	duplicate ratio	-比
제곱수	square number	-數
제노아 저기압	Genoa cyclone	-低氣壓
제노아형 저기압	Genoa-type depression	-型低氣壓
제동깊이	damping depth	制動-
제동복사효과	Bremsrahlung effect	制動輻射效果
제동압력판	bridled pressure plate	制動壓力板
제동컵풍속계	bridled-cup anemometer	制動-風速計
제동풍속계	bridled anemometer	制動風速計
제방	embankment	堤防
제번스효과	Jevons effect	-效果
제비폭풍	swallow storm, peesweep storm	-暴風
제빙	deice	除氷
제빙기	deicer	除氷器
제빙실 한랭기후	icehouse climate	製氷室寒冷氣候
제빙장치	deicing device	除氷裝置
제습	dehumidification	除濕
제습건조	exsiccation	除濕乾燥
제습기	dehumidifier	除濕器
제습환기	dehumidifying ventilation	除濕換氣
제어	control	制御
제어격자	control grid	制御格子
제어단위모듈	control unit module	制御單位-
제어작용	control action	制御作用
제운기	nepheloscope	製雲器
제작준비 검토	manufacturing readiness review	製作準備檢討
제타퍼텐셜	zeta potential	
제트(류)	jet	-(流)
제트기류	jet stream	-氣流
제트류	jet flow, jet stream	-流
제트류구름	jet stream cloud	-流-
제트류권운	jet stream cirrus	-流卷雲
제트류비행	jet stream flight	-流飛行
제트류전선	jet stream front	-流前線
제트류중심부	jet stream core	-流中心部
제트류축	jet stream axis	-流軸
제트스트리크	jet streak	
제트스트리크 상호작용	jet streak interaction	-相互作用
제트스트리크입구	entrance to jet streak	-入口
제트축	jet axis	-軸
제트평행비행	along-jet flight	-平行飛行
제트핵	jet core	-核

ㅈ

제트횡단비행	across-jet flight	-橫斷飛行
제트효과바람	jet effect wind	-效果-
제퍼슨지수	Jefferson index	-指數
제피로스	Zephyros	
제한구역예보모형	limited-area forecast model	制限區域豫報模型
제한 남반구영역	limited Southern Hemisphere	制限南半球領域
제한상세격자망모형	limited fine-mesh model	制限詳細格子網模型
제한 약한 에코 영역	bounded weak echo region	制限弱-領域
제한층	confining bed	制限層
젠틸리 기후분류	Gentilli's climate classification	-氣候分類
조각	patch	
조각고층운	altostratus fractus	-高層雲
조각구름	fractus, scud	
조각난운	fracto-nimbus	-亂雲
조각난층운	nimbostratus pannus	-亂層雲
조각유빙	brash ice	-流氷
조각적운	cumulus fractus, fracto-cumulus	-積雲
조각층운	fracto-stratus, stratus fractus	-層雲
조건부대칭불안정	conditional symmetric instability	條件附對稱不安定
조건부등식	conditional inequality	條件不等式
조건부보존성	conditional conservatism	條件附保存性
조건부불안정(성)	conditional instability	條件附不安定(性)
조건부불안정공기	conditionally unstable air	條件附不安定空氣
조건부불안정대기	conditionally unstable atmosphere	條件附不安定大氣
조건부샘플링	conditional sampling	條件附-
조건부안정(성)	conditional stability	條件附安定(性)
조건부운고	ceiling	條件附雲高
조건부입자기법	conditioned particle technique	條件附粒子技法
조건부평형	conditional equilibrium	條件附平衡
조건부확률	conditional probability	條件附確率
조건부확률밀도함수	conditional probability density function	條件附確率密度函數
조건분포	conditional distribution	條件分布
조건평균	conditional mean	條件平均
조금	neap tide	潮-
조금차	neap range	潮-差
조금 흐린하늘	sky slightly clouded	
조기경보체계	early warning system	早期警報體系
조기낙하	early fallout	早期落下
조대입자	coarse particle	粗大粒子
조도	roughness	粗度
조도	illuminance, intensity of illumination	照度
조도계	illuminometer	照度計
조도기후	illumination climate	照度氣候
조령	age of tide	潮齡

조류	algae	藻類
조류곡선	current curve	潮流曲線
조류발전	tidal power	潮流發電
조류에코	bird echo	鳥類-
조류타원	current ellipse	潮流楕圓
조르단 일조계	Jordan sunshine recorder	-日照計
조림	afforestation	造林
조명	illumination	照明
조명도 코사인법칙	cosine law of illumination	照明度-法則
조명원	circle of illumination	照明圓
조사의정서	investigative protocol	調査議定書
조사적량	dosage	照射適量
조산운동	orogeny	造山運動
조상해	early frost damage	早霜害
조석	tide	潮汐
조석격류	tidal rip	潮汐激流
조석균열	tide crack	潮汐龜裂
조석노정	tidal excursion	潮汐路程
조석류	tidal current, tidal stream	潮汐流
조석바람	tidal breeze, tidal wind	潮汐-
조석성분	constituent of tide	潮汐成分
조석예상장치	tide-predicting machine	潮汐豫想裝置
조석예측	tide prediction	潮汐豫測
조석일	tidal day	潮汐日
조석진폭	tide amplitude	潮汐振幅
조석차	tidal range, tide range	潮汐差
조석파	tidal wave	潮汐波
조석표	current table, tide table	潮汐表
조석프리즘	tidal prism	潮汐-
조석해일	tidal bore	潮汐海溢
조성	composition	組成
조성탐측	composition sounding	組成探測
조위	tide level	潮位
조작원	operator	操作員
조절	adjustment, modulation	調節
조절계수	accommodation coefficient	調節係數
조절과정	adjustment process	調節過程
조절기	regulator	調節器
조절기후	conditioned climate	調節氣候
조절사례	control case	調節事例
조절요인	regulatory factor	調節要因
조절유전자	regulatory gene	調節遺傳子
조정장치	regulating device	調整裝置
조종사브리핑	pilot briefing	操縱士-

ㅈ

조종사일기보고	pilot weather report	操縱士日氣報告
조직	tissue	組織
조직화 큰맴돌이	organized large eddy	組織化-
조차	range of tide, tidal range, tide range	潮差
조초	algal reef	藻礁
조파	harmonic	調波
조풍	briny wind	潮風
조합	combination	組合
조해성	deliquescence	潮解性
조화급수	harmonic series	調和級數
조화다이얼	harmonic dial	調和-
조화량	harmonics	調和量
조화복사	harmonic radiation	調和輻射
조화분석	harmonic analysis	調和分析
조화분석기	harmonic analyser	調和分析器
조화운동	harmonic motion	調和運動
조화진동	harmonic oscillation	調和振動
조화진동자방정식	equation of harmonic oscillator	調和振動子方程式
조화함수	harmonic function	調和函數
조회기록위치확인시스템	interrogation recording and location system	照會記錄位置確認-
존데	sonde	
존슨-윌리엄스 수액계	Johnson-Williams liquid water meter	-水液計
좁은 각 카메라	narrow angle camera	-角-
좁은 띠 복사	narrow-band radiation	-輻射
좁은 띠 센서	narrow-band sensor	
좁은 빔무선방향 탐지기	narrow beam radiogoniometer	-無線方向探知器
좁은 전선구름대	narrow frontal cloud band	-前線-帶
좁은 지역 기록계	narrow-sector recorder	-地域記錄計
좁은 파장역 복사	narrow-band radiation	-波長域輻射
좁은 파장역 센서	narrow-band sensor	-波長域-
종결	closure	終結
종결가정	closure assumption	終結假定
종결기법	closure technique	終結技法
종결난류	closure turbulence	終結亂流
종결문제	closure problem	終結問題
종결방안	closure scheme	終結方案
종관(의)	synoptic	綜觀
종관개황	synoptic situation	綜觀槪況
종관고층기상학	synoptic aerology	綜觀高層氣象學
종관관측	synoptic observation	綜觀觀測
종관관측소	synoptic station	綜觀觀測所
종관규모	synoptic scale, weather-map scale	綜觀規模
종관규모강제	synoptic-scale forcing	綜觀規模强制
종관규모순환	synoptic-scale circulation	綜觀規模循環

종관기상관측시스템	synoptic weather observation system	綜觀氣象觀測-
종관기상학	synoptic meteorology	綜觀氣象學
종관기후학	synoptic climatology	綜觀氣候學
종관(일기)도	synoptic chart, synoptic map	綜觀(日氣)圖
종관모형	synoptic model	綜觀模型
종관(기상)보고	synoptic report	綜觀(氣象)報告
종관분석	synoptic analysis	綜觀分析
종관시간	synoptic hour	綜觀時間
종관예보	synoptic forecast, synoptic forecasting	綜觀豫報
종관일기개황	synoptic weather situation	綜觀日氣槪況
종관일기도	synoptic weather chart	綜觀日氣圖
종관일기부호	synoptic weather code	綜觀日氣符號
종관전문형식	synoptic code	綜觀電文型式
종관파랑도	synoptic wave chart	綜觀波浪圖
종관형태	synoptic type	綜觀形態
종단(낙하)속도	terminal (fall) velocity	終端(落下)速度
종말점	end point	終末點
종발산	longitudinal divergence	縱發散
종속	dependence	從屬
종속변수	dependent variable	從屬變數
종속사건	dependent event	從屬事件
종점	terminal point	終點
종합적해충관리	integrated pest management	綜合的害蟲管理
종횡비	aspect ratio	縱橫比
좌우대칭	bilateral symmetry	左右對稱
좌이동뇌우	left-moving thunderstorm	左移動雷雨
좌측 원형편파	left circular polarization	左側圓形偏波
좌표	coordinate	座標
좌표계	coordinate system	座標系
좌표공간	coordinate space	座標空間
좌표단순화	coordinate simplification	座標單純化
좌표변환	coordinate transformation	座標變換
좌표선	coordinate line	座標線
좌표원점	origin of coordinates	座標原點
좌표축	axis of coordinate, coordinate axis	座標軸
좌표평면	coordinate plane	座標平面
좌표회전	coordinate rotation	座標回轉
주간선회로	main trunk circuit	主幹線回路
주간예보	mid-term forecast, one-week forecast	週間豫報
주강수중심	main precipitation core	主降水中心
주거오염	dwelling pollution	住居汚染
주(기압)골	major trough	主(氣壓)-
주관예보	subjective forecast	主觀豫報
주기	cycle, period	週期

ㅈ

주기강제	cyclic forcing	週期强制
주기과정	cycle process	週期過程
주기도	periodogram	週期圖
주기도분석	periodogram analysis	週期圖分析
주기바람	periodic wind	週期-
주기변동	periodic variation	週期變動
주기변동성	periodic variability	週期變動性
주기성	periodicity	週期性
주기운동	periodic motion	週期運動
주기율	periodic law	週期律
주기적	periodical	週期的
주기적 경계조건	periodic boundary condition	週期的境界條件
주기적 성질	periodic property	週期的性質
주기주파수	cyclic frequency	週期周波數
주기진동	periodic oscillation	週期振動
주기함수	periodic function	週期函數
주기후관측소	principal climatological station	主氣候觀測所
주농업기상관측소	principal agricultural meteorological station	主農業氣象觀測所
주뇌격	main stroke	主雷擊
주레이더	primary radar	主-
주류	main current	主流
주름통	bellow	-桶
주리나현상	Jurine's phenomenon	-現象
주(기압)마루	major ridge	主(氣壓)-
주문제작예보	customized forecast	注文製作豫報
주방사부	main lobe	主放射部
주방어지역기상	main defense area weather	主防禦地域氣象
주방향	principal direction	主方向
주변	ambient	周邊
주변공기	ambient air	周邊空氣
주변공기질가이드라인	ambient air quality guideline	周邊空氣質-
주변공기질표준	ambient air standard	周邊空氣質標準
주변광	aureole	周邊光
주변광선	marginal ray	周邊光線
주변기온	ambient air temperature	周邊氣溫
주변물함량	ambient liquid water content	周邊-含量
주변복사휘도	limb radiance	周邊輻射輝度
주변복사휘도 역산복사계	limb radiance inversion radiometer	周邊輻射輝度逆算輻射計
주변압력	ambient pressure	周邊壓力
주변온도	ambient temperature	周邊溫度
주변장	environmental flow	周邊場
주변흐름	environmental flow	周邊-
주빔	main beam	主-
주변빙하지형학	periglacial geomorphology	周邊氷河地形學

주사	scan	走査
주사각	scan angle	走査角
주사계	scanner	走査計
주사관측	scanning observation	走査觀測
주사다주파마이크로파복사계	scanning multifrequency microwave radiometer	走査多周波-波輻射計
주사마이크로파분광계	scanning microwave spectrometer	走査-波分光計
주사방식	scanning	走査方式
주사복사계	scanning radiometer	走査輻射計
주사선	raster line, scan line	走査線
주사원	scan circle	走査圓
주사유형	scan pattern	走査類型
주사전자현미경	scanning electron microscope	走査電子顯微鏡
주사점	scan spot	走査點
주사폭	swath width	走査幅
주수대수층	perched aquifer	宙水帶水層
주수지하수	perched groundwater	宙水地下水
주수하천	perched stream	宙水河川
주순환	primary cycle	主循環
주양자수	principal quantum number	主量子數
주열편	major lobe	主裂片
주요기상대	main meteorological office	主要氣象臺
주위공기	ambient air	周圍空氣
주육상관측소	principal land station	主陸上觀測所
주의보	advisory, advisory forecast, subwarning, watch	注意報
주의보구역	watch area	注意報區域
주의보영역	advisory area	注意報領域
주입온도	injection temperature	注入溫度
주저기압	primary cyclone, primary depression, primary low	主低氣壓
주전선	principal front	主前線
주전투지역기상	main battle area weather	主戰鬪地域氣象
주종관관측	principal synoptic observation	主綜觀觀測
주통신망	main telecommunication network	主通信網
주파	major wave	主波
주파수	frequency	周波數
주파수대	frequency band, range	周波數帶
주파수반송파	frequency carrier	周波數搬送波
주파수변조	frequency modulation	周波數變調
주파수변조 레이더	frequency-modulated radar	周波數變調-
주파수변조 연속파 레이더	frequency-modulated continuous-wave radar	周波數變調連續波-
주파수편이	frequency shift	周波數偏移
주파수평활	frequency smoothing	周波數平滑
죽은 빙하	dead glacier	-氷河
준	quasi-	準

준경보	subwarning	準警報
준광역방송	subregional broadcast	準廣域放送
준균질	quasi-homogeneous	準均質
준대수층	aquitard	準帶水層
준등압과정	quasi-isobaric process	準等壓過程
준라그랑지좌표	quasi-Lagrangian coordinates	準-座標
준무발산	quasi-nondivergent	準無發散
준보존	quasi-conservative	準保存
준불연속	quasi-discontinuity	準不連續
준비발산	quasi-nondivergent	準非發散
준수평운동	quasi-horizontal motion	準水平運動
준안정	metastable	準安定
준안정도	metastability	準安定度
준역산법	quasi-inverse method	準逆算法
준영구저기압	quasi-permanent low	準永久低氣壓
준2년주기성	quasi-biennial periodicity	準二年週期性
준2년주기진동	quasi-biennial oscillation	準二年週期振動
준작용중심	quasi-center of action	準作用中心
준정상(의)	quasi-steady	準定常
준정역학근사	quasi-hydrostatic approximation	準靜力學近似
준정적과정	quasi-static process	準靜的過程
준정체(의)	quasi-stationary	準停滯
준정체섭동	quasi-stationary perturbation	準停滯攝動
준정체시계열	quasi-stationary time series	準停滯時系列
준정체전선	quasi-stationary front	準停滯前線
준주기(의)	quasi-periodic	準週期
준중심시	parafoveal vision	準中心視
준지균근사	pseudo-geostrophic approximation, quasi-geostrophic approximation	準地均近似
준지균근사모형	quasi-geostrophic approximation model	準地均近似模型
준지균방정식	quasi-geostrophic equation	準地均方程式
준지균조절	quasi-geostrophic adjustment	準地均調節
준지균평형	quasi-geostrophic balance, quasi-geostrophic equilibrium	準地均平衡
준지균풍계	quasi-geostrophic wind system	準地均風系
준지균풍소용돌이도방정식	vorticity equation of quasi-geostrophic wind	準地均風-度方程式
준지균풍와도방정식	vorticity equation of quasi-geostrophic wind	準地均風渦度方程式
준층류저항	quasi-laminar resistance	準層流抵抗
준평원	peneplain	準平原
줄	joule	
줄기마름병	die back	-病
줄기잎선도	stem-and-leaf diagram	-線圖
줄기흐름	plume	
줄무늬	streak, striation	

줄무늬번개	streak lightning	
줄무늬선	streak line	-線
줄법칙	Joule's law	-法則
줄상수	Joule's constant	-常數
중각카메라	medium angle camera	中角-
중간	medium	中間
중간(구름)	mediocris	中間-
중간권	mesosphere	中間圈
중간권계면	mesopause	中間圈界面
중간권구름	mesospheric cloud	中間圈-
중간권제트류	mesospheric jet	中間圈-流
중간권 최고온도점	mesopeak	中間圈最高溫度點
중간규모	intermediate scale	中間規模
중간규모	medium scale	中間規模
중간규모난류	medium-scale turbulence	中間規模亂流
중간규모순환	mesoscale circulation	中間規模循環
중간규모요란	medium-scale disturbance	中間規模擾亂
중간단계 해양-통계대기 접합모형	intermediate ocean-statistical atmosphere coupled model	中間段階海洋統計大氣 接合模型
중간식물	intermediate plant	中間植物
중간이온	intermediate ion	中間-
중간일기도시각	intermediate standard time	中間日氣圖時刻
중간입력	intermediate insertion	中間入力
중간자	meson	中間子
중간잡종	intermediate hybrid	中間雜種
중간 적외영역	mid-infrared region	中間赤外領域
중간적운	cumulus mediocris	中間積雲
중간종관관측	intermediate synoptic observation	中間綜觀觀測
중간주파수	intermediate frequency	中間周波數
중간주파수증폭기	intermediate frequency amplifier	中間周波數增幅器
중간층	intermediate layer	中間層
중간형식	intermediate form	中間形式
중간흐름	interflow	中間-
중거리낙진	intermediate fallout	中距離落塵
중고도폭격기상	medium-altitude bombing weather	中高度爆擊氣象
중국연안류	China coastal current	中國沿岸流
중규모	mesoscale	中規模
중규모고기압	mesohigh	中規模高氣壓
중규모구조	mesoscale structure	中規模構造
중규모난류	mesoscale turbulence	中規模亂流
중규모대류계	mesoscale convective system	中規模對流系
중규모대류복합체	mesoscale convective complex	中規模對流複合體
중규모모형	mesoscale model	中規模模型
중규모범주	mesoscale category	中規模範疇

중규모분석	mesoanalysis, mesoscale analysis	中規模分析
중규모분석도	mesochart	中規模分析圖
중규모순환	mesoscale circulation	中規模循環
중규모스톰	mesoscale storm	中規模-
중규모요란	mesoscale disturbance	中規模搖亂
중규모운동	mesoscale motion	中規模運動
중규모저기압	mesocyclone, mesolow	中規模低氣壓
중규모전선	mesoscale front	中規模前線
중규모제트류	mesojet	中規模-流
중규모폭풍	mesoscale storm	中規模暴風
중규모현상	mesoscale phenomenon	中規模現象
중기상학	mesometeorology	中氣象學
중기선신세 온난기	mid-Pliocene warming period	中期鮮新世溫暖期
중기수문예보	medium-term hydrological forecast	中期水文豫報
중기예보	medium-range forecast, mid-range forecast	中期豫報
중기후	mesoclimate	中氣候
중기후학	mesoclimatology	中氣候學
중대경보	severe warning	重大警報
중력	force of gravity, gravity, gravity force, gravitation, gravitational force	重力
중력가속도	acceleration of gravity, gravitational acceleration	重力加速度
중력경도안정화	gravity gradient stabilization	重力傾度安定化
중력기압계	weight barometer	重力氣壓計
중력단위계	gravitational unit system	重力單位系
중력대류	gravitational convection	重力對流
중력류	gravity current	重力流
중력바람	gravity wind	重力-
중력벡터	gravity vector	重力-
중력보정	gravity correction	重力補正
중력부하	g-load	重力負荷
중력분리	gravitational separation	重力分離
중력불안정	gravitational instability	重力不安定
중력상수	gravitational constant	重力常數
중력상호작용	gravitational interaction	重力相互作用
중력수	gravitational water, gravity water	重力水
중력수축	gravitational contraction	重力收縮
중력안정	gravitational stability	重力安定
중력안정도	gravitational stability	重力安定度
중력압	gravitational pressure	重力壓
중력위치에너지	gravitational potential energy	重力位置-
중력자기기압계	weight barograph	重力自記氣壓計
중력장	gravitational field	重力場
중력조석	gravitational tide	重力潮汐
중력중심	center of gravity	重力中心

중력질량	gravitational mass	重力質量
중력침강	gravitational settling	重力沈降
중력파	gravitational wave, gravity wave	重力波
중력퍼텐셜	gravitational potential	重力-
중력평형	gravitational equilibrium	重力平衡
중립	neutrality	中立
중립경계층	neutral boundary layer	中立境界層
중립공기	neutral air	中立空氣
중립기단	neutral air mass	中立氣團
중립대기	neutral atmosphere	中立大氣
중립대기경계층	neutral atmospheric boundary layer	中立大氣境界層
중립면	neutral plane, neutral surface	中立面
중립모드	neutral mode	中立-
중립선	neutral line	中立線
중립성층	neutral stratification	中立成層
중립안정(도)	neutral stability	中立安定(度)
중립안정도	indifferent stability	中立安定度
중립안정도바람	neutral stability wind	中立安定度-
중립열대저기압	neutercane	中立熱帶低氣壓
중립온도	neutral temperature	中立溫度
중립온도연직분포	neutral temperature profile	中立溫度鉛直分布
중립이류	neutral advection	中立移流
중립저기압	neutral cyclone	中立低氣壓
중립점	neutral point	中立點
중립조건	neutral condition	中立條件
중립진동	neutral oscillation	中立振動
중립파	neutral wave	中立波
중립평형	indifferent equilibrium, neutral equilibrium	中立平衡
중립폐색	neutral occlusion	中立閉塞
중립항력계수	neutral drag coefficient	中立抗力係數
중복	duplication	重複
중부표준시간(미국)	Central Standard Time	中部標準時間
중분해 적외복사계	medium resolution infrared radiometer	中分解赤外輻射計
중생대기후	Mesozoic climate	中生代氣候
중성	neutrality	中性
중성권	neutrosphere	中性圈
중성권계면	neutropause	中性圈界面
중성기단	indifferent air mass	中性氣團
중성입자	neutral particle	中性粒子
중성자	neutron	中性子
중성자산란법	neutron-scattering method	中性子散亂法
중성자탐측기	neutron probe	中性子探測器
중성자토양수분계	neutron moisture meter	中性子土壤水分計
중세 온난기	Medieval warm period	中世溫暖期

중수소	deutrium	重水素
중수소과잉	deutrium excess	重水素過剩
중수소함량	deutrium content	重水素含量
중심	centroid	重心
중심각	central angle	中心角
중심거리	distance between centers	中心距離
중심경향	central tendency	中心傾向
중심극한정리	central limit theorem	中心極限定理
중심도약	center jump	中心跳躍
중심력	central force	中心力
중심력장	central force field	中心力場
중심모멘트	central moment	中心-
중심무풍구역	central calm	中心無風區域
중심선	central line	中心線
중심압력	central pressure	中心壓力
중심원리	central dogma	中心原理
중심저기압	central depression	中心低氣壓
중심체	centrosome	中心體
중심핵	central core	中心核
중앙값	median, mid-value	中央-
중앙기상대	main meteorological office	中央氣象臺
중앙반경	median radius	中央半徑
중앙반지름	median radius	中央-
중앙부피반경	median volume radius	中央-半徑
중앙부피반지름	median volume radius	中央-
중앙부피지름	median volume diameter	中央-
중앙어림	central estimate	中央-
중앙예보소	central forecasting office	中央豫報所
중앙유한차분근사	central finite-difference approximation	中央有限差分近似
중앙잉글랜드 기후기록	Central England climate record	中央-氣候記錄
중앙잉글랜드 온도	Central England temperature	中央-溫度
중앙차분	centered difference	中央差分
중앙첨도	mesokurtosis	中央尖度
중앙첨도(의)	mesokurtic	中央尖度
중앙파수	central wavenumber	中央波數
중온기후	mesothermal climate	中溫氣候
중온대	mesothermen	中溫帶
중온식생(형)	mesotherm	中溫植生(型)
중온식생형	mesothermal type	中溫植生型
중위도	middle latitude	中緯度
중위도고압대	high pressure belt of middle latitudes	中緯度高壓帶
중위도기단	middle latitude air	中緯度氣團
중위도대류권계면	middle tropopause	中緯度大流圈界面
중위도저기압	middle latitude cyclone	中緯度低氣壓

중위도편서풍	middle latitude westerlies	中緯度偏西風
중위도편서풍대	middle latitude westerlies zone	中緯度偏西風帶
중이온	heavy ion	重-
중적분	multiple integral	重積分
중주파수	medium frequency	中周波數
중첩	overlap, superposition	重疊
중첩원리	principle of superposition	重疊原理
중첩평균	overlapping average (mean)	重疊平均
중층구름	medium-level cloud	中層-
중층대기	middle atmosphere	中層大氣
중층대기연구계획	middle atmosphere program	中層大氣研究計畵
중층운	medium cloud, middle cloud, middle-level cloud	中層雲
중합	polymerization	重合
중화	neutralization	中和
쥐라기기후	Jurassic climate	-紀氣候
즉시전송	immediate transmission	卽時電送
증가	increasing	增加
증가변수	increase variable	增加變數
증가상태	increasing state	增加狀態
증가함수	increasing function	增加函數
증강	reinforcement	增强
증강지수	build-up index	增强指數
증기	steam, vapo(u)r, water vapo(u)r	蒸氣
증기기관	steam engine	蒸氣機關
(수)증기농도	vapo(u)r concentration	(水)蒸氣濃度
(수)증기밀도	vapo(u)r density	(水)蒸氣密度
(수)증기선	vapo(u)r line	(水)蒸氣線
(수)증기압	vapo(u)r pressure	(水)蒸氣壓
(수)증기압 강하	vapo(u)r pressure lowering	(水)蒸氣壓降下
(수)증기압곡선	vapo(u)r pressure curve	(水)蒸氣壓曲線
(수)증기압편차	vapo(u)r pressure deficit	(水)蒸氣壓偏差
(수)증기장력	vapo(u)r tension	(水)蒸氣張力
증기터빈	steam turbine	蒸氣-
증대	swelling	增大
증량(빙하의)	alimentation	增量
증류	distillation	蒸溜
증발	evaporation	蒸發
증발가능률	evaporation opportunity	蒸發可能率
증발계	atmidometer, atmometer, evaporation gauge, evaporimeter	蒸發計
증발계계수	evaporimeter coefficient, pan coefficient	蒸發計係數
증발계조정계수	pan adjustment coefficient	蒸發計調整係數
증발계증발	pan evaporation	蒸發計蒸發

증발곡선	evaporation curve	蒸發曲線
증발기록	evaporogram	蒸發記錄
증발기록계	evaporograph	蒸發記錄計
증발냉각	evaporative cooling	蒸發冷却
증발능	evaporation capacity	蒸發能
증발량	amount of evaporation	蒸發量
증발력	evaporation power, evaporative power	蒸發力
증발률	evaporation rate, evaporativity	蒸發率
증발배출	evaporative emission	蒸發排出
증발산	evapotranspiration, flyoff	蒸發散
증발산계	evapotranspirometer	蒸發散計
증발산량계	lysimeter	蒸發散量計
증발서리	evaporation frost	蒸發-
증발안개	evaporation fog	蒸發-
증발연못	evaporation pond	蒸發-
증발열	evaporating heat, heat of evaporation, heat of vaporization	蒸發熱
증발열조절	enaporative heat regulation	蒸發熱調節
증발접시	evaporating dish, evaporation pan	蒸發-
증발조절기	evapotron	蒸發調節器
증발조정역	zone of evaporative regulation	蒸發調整域
증발측정법	atmidometry, atmometry	蒸發測定法
증발측정용 후크게이지	evaporation hook gauge	蒸發測定用-
증발탱크	evaporation tank	蒸發-
증발혼합안개	evaporation-mixing fog	蒸發混合-
증발흔적	evaporation trail	蒸發痕跡
증보후지타강도척도	enhanced Fujita intensity scale(EF-scale)	增補-强度尺度
증분	increment	增分
증분분석갱신	incremental analysis update	增分分析更新
증분율	rate of increment	增分率
증산계	phytometer, potometer	蒸散計
증산계수	transpiration coefficient	蒸散係數
증산비	transpiration ratio	蒸散比
증산율	water-use ratio, transpiration ratio	蒸散率
증산작용	transpiration	蒸散作用
증설	increasing of snow	增雪
증수부	rising limb	增水部
증식형 원자로	breeder reactor	增殖型原子爐
증우	enhancement of rainfall	增雨
증폭	amplification, gain	增幅
증폭인자	gain factor	增幅因子
증폭행렬	amplification matrix, gain matrix	增幅行列
지각기후	perceptual climate	知覺氣候
지각변동	diastrophism	地殼變動

지각평형	isostasy	地殼平衡
지-공전류	earth-air current	地-空電流
지구각속도	angular velocity of the earth	地球角速度
지구감시프로그램	earthwatch program	地球監視-
지구계	earth system	地球系
지구곡률보정	earth curvature correction	地球曲率補正
지구공전궤도이심률	eccentricity of the earth's orbit	地球公轉軌道離心率
지구공학	geoengineering	地球工學
지구과학	earth sciences	地球科學
지구관측계	earth observing system	地球觀測係
지구관측계프로그램	earth observing system program	地球觀測係-
지구관측시스템	earth observing system	地球觀測-
지구관측시스템프로그램	earth observing system program	地球觀測-
지구광	earthshine, earthlight	地球光
지구굴절	terrestrial refraction	地球屈折
지구권	geosphere	地球圈
지구궤도	earth orbit	地球軌道
지구규모	earth scale	地球規模
지구규모확산	global-scale diffusion, global-scale dispersion	地球規模擴散
지구규모확산모형	global-scale diffusion model	地球規模擴散模型
지구그림자	earth shadow, shadow of the earth	地球-
지구기상관측시스템	global meteorological observing system	地球氣象觀測-
지구기상실험	Global Weather Experiment(GWE)	地球氣象實驗
지구기상위성시스템	global meteorological satellite system	地球氣象衛星-
지구기후관측시스템	Global Climate Observing System(GCOS)	地球氣候觀測-
지구기후학	earth climatology	地球氣候學
지구대기	earth atmosphere	地球大氣
지구대기감시	Global Atmosphere Watch(GAW)	地球大氣監視
지구대기복사수지	earth atmosphere radiation budget	地球大氣輻射收支
지구대기시스템	earth atmosphere system	地球大氣-
지구대기연구계획	Global Atmospheric Research Program(GARP)	地球大氣研究計劃
지구대기연구프로젝트	Global Atmospheric Research Project	地球大氣研究-
지구대기오염	global air pollution	地球大氣汚染
지구동기궤도	geosynchronous orbit	地球同期軌道
지구동시궤도	earth synchronous orbit	地球同時軌道
지구되쬐임	earthlight	地球-
지구물리관측궤도위성	orbiting geophysical observatory	地球物理觀測軌道衛星
지구물리년	Geophysical Year	地球物理年
지구물리유체역학	geophysical fluid dynamics	地球物理流體力學
지구물리유체역학연구소	Geophysical Fluid Dynamics Laboratory	地球物理流体力學研究所
지구물리일	geophysical day	地球物理日
지구물리학	geophysics	地球物理學
지구물리학자	geophysicist	地球物理學者
지구물수지	global water budget	地球-收支

지구물수지방정식	global water balance equation	地球-收支方程式
지구반사광	earthlight	地球反射光
지구반사율	global albedo	地球反射率
지구반사일사	reflected global radiation	地球反射日射
지구변화시나리오	global change scenario	地球變化-
지구복사	earth radiation, terrestrial (surface) radiation	地球輻射
지구복사계	pyrgeometer	地球輻射計
지구복사수지	earth radiation budget, global radiation budget	地球輻射收支
지구복사스펙트럼	spectrum of global radiation	地球輻射-
지구복사온도계	terrestrial radiation thermometer	地球輻射溫度計
지구복사평형	terrestrial radiation balance	地球輻射平衡
지구수문센터	Global Hydrology Center	地球水文-
지구수평탐측기술	global horizontal sounding technique	地球水平探測技術
지구시스템	earth system	地球-
지구시스템과학	earth system science	地球-科學
지구시스템연구센터	Earth System Research Center	地球-研究-
지구식생역학모형	dynamic global vegetation model	地球植生力學模型
지구안정화	earth stabilization	地球安定化
지구알베도	albedo of the earth, global albedo	地球-
지구에너지물순환실험	Global Energy and Water Cycle Experiment	地球-循環實驗
지구역사기후학네트워크	Global Historical Climatology Network	地球歷史氣候學-
지구역학고도	geodynamic height	地球力學高度
지구열수지	earth heat budget	地球熱收支
지구오존관측시스템	Global Ozone Observing System(GOOS)	地球-觀測-
지구온난화	global warming	地球溫暖化
지구온난화기간	global warming period	地球溫暖化期間
지구온난화논쟁	global warming controversy	地球溫暖化論爭
지구온난화잠재력	global warming potential	地球溫暖化潛在力
지구위치시스템	Global Positioning System(GPS)	地球位置-
지구인력	gravitational attraction	地球引力
지구자기장	terrestrial magnetic field	地球磁氣場
지구자원정보데이터베이스	Global Resource Information Database	地球資源情報-
지구자원탐사위성	Earth Resource Technology Satellite	地球資源探查衛星
지구자전	earth rotation	地球自轉
지구자전속력	earth rotational speed	地球自轉速力
지구자전율	rotation rate of earth	地球自轉率
지구저궤도	low earth orbit	地球低軌道
지구전기회로	global electric circuit	地球電氣回路
지구정거장 마이크로파 영상기	geostation microwave imager	地球停車場-波映像器
지구정거장 마이크로파 탐측기	geostation microwave sounder	地球停車場-波探測器
지구정상회담	earth summit	地球頂上會談
지구 코로나	geocorona	地球-

지구타원체	earth ellipsoid	地球楕圓體
지구-태양 관계	earth-sun relationship	地球-太陽關係
지구-해양-대기계	earth-ocean-atmosphere system	地球海洋大氣系
지구-해양-대기시스템	earth-ocean-atmosphere system	地球海洋大氣-
지구화학 기후모형	geochemical climate model	地球化學氣候模型
지구화학 대용물	geochemical proxy	地球化學代用物
지구화학 대용자료	geochemical proxy data	地球化學代用資料
지구화학 모형	geochemical model	地球化學模型
지구화학 미량기체	geochemical tracer	地球化學微量氣體
지구화학자	geochemist	地球化學者
지구환경	global environment	地球環境
지구환경감시시스템	Global Environmental Monitoring System	地球環境監視-
지구환경감시지수	global environment monitoring index	地球環境監視指數
지구환경시설	global environment facility	地球環境施設
지권	geosphere	地圈
지균(의)	geostrophic	地均
지균값	geostrophic value	地均-
지균강제력	geostrophic forcing	地均强制力
지균거리	geostrophic distance	地均距離
지균계수	geostrophic coefficient	地均係數
지균고도	geostrophic height	地均高度
지균궤적	geostrophic trajectory	地均軌跡
지균균형	geostrophic balance	地均均衡
지균근사	geostrophic approximation	地均近似
지균난류	geostrophic turbulence	地均亂流
지균류	geostrophic current	地均流
지균리처드슨수	geostrophic Richardson number	地均-數
지균발산	geostrophic divergence	地均發散
지균방정식	geostrophic equation	地均方程式
지균변형	geostrophic deformation	地均變形
지균성	geostrophy	地均性
지균소용돌이도	geostrophic vorticity	地均-度
지균속도	geostrophic velocity	地均速度
지균수송	geostrophic transport	地均輸送
지균시어	geostrophic shear	地均-
지균운동	geostrophic motion	地均運動
지균유선함수	geostrophic streamfunction	地均流線函數
지균이류	geostrophic advection	地均移流
지균이류규모	geostrophic advection scale	地均移流規模
지균전선발생	geostrophic frontogenesis	地均前線發生
지균조절	geostrophic adjustment	地均調節
지균좌표	geostrophic coordinates	地均座標
지균편차	geostrophic departure, geostrophic deviation	地均偏差
지균평형	geostrophic equilibrium	地均平衡

ㅈ

지균풍	geostrophic wind	地均風
지균풍고도	geostrophic wind level	地均風高度
지균풍관계	geostrophic wind relation	地均風關係
지균풍 디바이더	geostrophic divider	地均風-
지균풍벡터	geostrophic wind vector	地均風-
지균풍속	geostrophic wind velocity	地均風速
지균풍시어	geostrophic wind shear	地均風-
지균풍척도	geostrophic wind scale, geostrophical wind scale	地均風尺度
지균풍편차	geostrophic wind departure	地均風偏差
지균플럭스	geostrophic flux	地均-
지균항력	geostrophic drag	地均抗力
지균항력계수	geostrophic drag coefficient	地均抗力係數
지균흐름	geostrophic flow	地均-
지도	geographic map	地圖
지도규모	mapping scale	地圖規模
지도인자	map factor	地圖因子
지도책	atlas	地圖册
지도축척	map scale	地圖縮尺
지리격자	geographic grid	地理格子
지리경도	geographic(al) longitude	地理經度
지리도	geographic map	地理圖
지리범위	geographic scope	地理範圍
지리분류	geographical classification	地理分類
지리위도	geographic(al) latitude	地理緯度
지리위치	geographic location, geographic position	地理位置
지리자오선	geographic meridian	地理子午線
지리정보계	geographic information system	地理情報系
지리정보시스템	geographic information system	地理情報-
지리좌표	geographic(al) coordinates	地理座標
지리지평선(면)	geographic horizon	地理地平線(面)
지리평형	geographic balance	地理平衡
지리학	geography	地理學
지리효과	geographic effect	地理效果
지면	surface	地面
지면감부	ground sensor	地面感部
지면거칠기	surface roughness	地面-
지면기원대류	surface-based convection	地面起源對流
지면낙진도	ground fallout plot	地面落塵圖
지면동결	ground frost	地面凍結
지면류	overland flow	地面流
지면반사에코	ground clutter	地面反射-
지면복사	terrestrial (surface) radiation	地面輻射
지면복사수지	surface radiation budget	地面輻射收支
지면상태	state of ground	地面狀態

지면센서	ground sensor	地面-
지면약화경보	warning for ground-loosening	地面弱化警報
지면얼음	stone ice, ground ice	地面-
지면연화경보	warning for ground-loosening	地面軟化警報
지면온도	surface temperature	地面溫度
지면온도계	surface thermometer	地面溫度計
지면융해지수	surface thawing index	地面融解指數
지면일조량 감소	global dimming	地面日照量減少
지면저류(수)	surface storage	地面貯留(水)
지면정체수	surface retention	地面停滯水
지면체류수	surface detention	地面滯留水
지면현상	ground phenomenon	地面現象
지면흐름	overland flow	地面-
지문법	fingerprint method	指紋法
지방(의)	local	地方
지방기상대	District Meteorological Office	地方氣象臺
지방기상청	regional meteorological office	地方氣象廳
지방날씨(일기)통보	local weather report	地方-(日氣)通報
지방상용시	local civil time	地方常用時
지방시	local time	地方時
지방족화합물	aliphatic compound	脂肪族化合物
지방진시	local true time	地方眞時
지방태양시	local solar time	地方太陽時
지방평균시	local mean time	地方平均時
지방표준시	local standard time	地方標準時
지방항성시	local sidereal time	地方恒星時
지배방정식	governing equation	支配方程式
지상	surface	地上
지상강설	ground snowfall	地上降雪
지상결빙지수	air freezing-index	地上結氷指數
지상고도이상	above ground level	地上高度以上
지상관제접근	ground-controlled approach	地上管制接近
지상관측	surface air observation, surface observation	地上觀測
지상관측소	ground station	地上觀測所
지상기상관측	surface weather observation	地上氣象觀測
지상기압	surface pressure	地上氣壓
지상기압골	surface trough	地上氣壓-
지상기온	surface air temperature, surface temperature	地上氣溫
지상기온분포	surface temperature distribution	地上氣溫分布
지상난류	surface turbulence	地上亂流
지상(일기)도	surface chart	地上(日氣)圖
지상동결지수	surface freezing index	地上凍結指數
지상마찰	surface friction	地上摩擦
지상밀도	surface density	地上密度

지상바람	surface wind	地上-
지상발생기	ground generator	地上發生器
지상부근난류과정	near-surface turbulence process	地上附近亂流過程
지상부근대류권	near-surface troposphere	地上附近對流圈
지상분석도	surface analysis chart	地上分析圖
지상불연속면	surface of discontinuity	地上不連續面
지상사태	ground avalanche	地上沙汰
지상서리	air frost, surface hoar	地上-
지상수신 및 통제소	ground acquisition and command station	地上受信-統制所
지상습도	surface humidity	地上濕度
지상시정	ground visibility, surface visibility	地上視程
지상실험	ground experiment	地上實驗
지상에너지	surface energy	地上-
지상연소기	ground burner	地上燃燒器
지상예보도	surface forecast chart	地上豫報圖
지상예상도	surface prognostic chart	地上豫想圖
지상예상일기도	prebaratic chart	地上豫想日氣圖
지상오존	surface ozone	地上-
지상온도계산	surface temperature calculation	地上溫度計算
지상요란	surface disturbance	地上擾亂
지상일기도	surface map, surface weather chart, surface weather map	地上日氣圖
지상자유에너지	surface free energy	地上自由-
지상전선	surface front	地上前線
지상점검	ground check	地上點檢
지상점검상자	ground check chamber	地上點檢箱子
지상종관관측소	surface synoptic station	地上綜觀觀測所
지상종관기상실황전문	SYNOP message	地上綜觀氣象實況電文
지상종관기상실황전문형식	SYNOP code	地上綜觀氣象實況電文型式
지상착빙	ground icing	地上着氷
지상참값	ground truth	地上-
지상파	ground wave, surface wave	地上波
지상풍속	surface wind speed	地上風速
지상해빙지수	air thawing-index	地上解氷指數
지상활주	ground run	地上滑走
지상흐름	surface flow	地上-
지세	relief	地勢
지속가능과학	sustainability science	持續可能科學
지속경향	persistence tendency	持續傾向
지속기간곡선	duration curve	持續期間曲線
지속법	persistence method	連續法
지속성	persistence	持續性
지속성비	persistence ratio	持續性比
지속성예보	persistence forecast	持續性豫報

지속성전선비	prolonged frontal rain	持續性前線-
지속성진동	persistent oscillation	持續性振動
지속속도	sustained speed	持續速度
지속파레이더	CW radar	持續波-
지속풍	sustained wind	持續風
지속풍속	sustained wind speed	持續風速
지수	exponent, index, index number	指數
지수대기	exponential atmosphere	指數大氣
지수방정식	exponential equation	指數方程式
지수법칙	exponential law, law of exponent	指數法則
지수부등식	exponential inequality	指數不等式
지수분포	exponential distribution	指數分布
지수순환	index cycle	指數循環
지수커넬근사	exponential kernel approximation	指數-近似
지수함수	exponential function	指數函數
지시고도	indicated altitude	指示高度
지시공기속도	indicated airspeed	指示空氣速度
지시기	indicator	指示器
지시기온	indicated air temperature	指示氣溫
지시등	indicator lamp	指示燈
지시선도	indicator diagram	指示線圖
지심경도	geocentric longitude	地心經度
지심수평	rational horizon	地心水平
지심위도	geocentric latitude	地心緯度
지심좌표계	geocentric reference system	地心座標計
지심지평	geocentric horizon	地心地平
지심천정	geocentric zenith	地心天頂
지역기본종관네트워크	regional basic synoptic network	地域基本綜觀-
지역기상(통신)센터	regional meteorological center	地域氣象(通信)-
지역기상전문형식	regional code	地域氣象電文形式
지역기상통신네트워크	Regional Meteorological Telecommunication Network	地域氣象通信-
지역모형	regional model	地域模型
지역방송	regional broadcast	地域放送
지역소음표준	community noise standard	地域騷音標準
지역스펙트럼모형	regional spectral model	地域-模型
지역예보	area forecast, regional forecast	地域豫報
지역예보센터	area forecast center, Regional Area Forecast Centre	地域豫報-
지역예보전문형식	area forecast code	地域豫報電文型式
지역운영기상정보회보 교환	Regional OPMET Bulletin Exchange	地域運營氣象情報會報交換
지역지정기상센터	Regional Specialized Meteorological Centre	地域指定氣象-
지역코드	regional code	地域-
지역통신허브	Regional Telecommunication Hub	地域通信-

지역표준기압계	regional standard barometer	地域標準氣壓計
지역홍수빈도분석	regional flood frequency analysis	地域洪水頻度分析
지연	delay, lag, retardation	遲延
지연계수	lag coefficient	遲延係數
지연낙진	delayed fallout	遲延落塵
지연발진기이론	delayed oscillator theory	遲延發振器理論
지연보고	delayed report	遲延報告
지연상관	lag correlation	遲延相關
지연선	delay line	遲延線
지연시간	lag time	遲延時間
지연위치	lag position	遲延位置
지연자동이득제어	delayed automatic gain control	遲延自動利得制御
지연형냉해	cool summer damage by delayed growth	遲延型冷害
지연효과	retarding effect	遲延效果
지열가열	geothermal heating	地熱加熱
지열경도	geothermal gradient	地熱傾度
지열류	geothermal heat flow	地熱流
지열에너지	geothermal energy	地熱-
지열자원	geothermal resource	地熱資源
지열플럭스	geothermal flux	地熱-
지오다이나믹미터	geodynamic meter	
지오이드	geoid	
지오이드면	geoidal surface	-面
지오퍼텐셜	geopotential	
지오퍼텐셜고도	geopotential altitude, geopotential height	-高度
지오퍼텐셜고도경도	geopotential height gradient	-高度傾度
지오퍼텐셜고도면	geopotential height surface	-高度面
지오퍼텐셜높낮이	geopotential topography	
지오퍼텐셜두께	geopotential thickness	
지오퍼텐셜면	geopotential surface	-面
지오퍼텐셜미터	geopotential meter	
지오퍼텐셜장	geopotential field	-場
지오퍼텐셜좌표	geopotential coordinates	-座標
지오퍼텐셜차 편차	anomaly of geopotential difference	-差偏差
지온	ground temperature, soil temperature	地溫
지온경도	geothermic step	地溫傾度
지온계	ground thermometer, soil thermometer	地溫計
지온불변층	invariable stratum of soil temperature	地溫不變層
지온불역층	constant earth temperature layer	地溫不易層
지온상승깊이	geothermic depth	地溫上昇-
지위	geopotential	地位
지위고도	geopotential altitude, geopotential height	地位高度
지위높낮이	geopotential topography	地位-
지위두께	geopotential thickness	地位-

지위면	geopotential surface	地位面
지위장	geopotential field	地位場
지위좌표	geopotential coordinates	地位座標
지의계측(법)	lichenometry	地衣計測(法)
지자기	earth magnetism, geomagnetism, terrestrial magnetism	地磁氣
지자기경각계	dip circle	地磁氣傾角計
지자기극	geomagnetic pole	地磁氣極
지자기도	geomagnetic map	地磁氣圖
지자기위도	geomagnetic latitude	地磁氣緯度
지자기자오선	geomagnetic meridian	地磁氣子午線
지자기장	geomagnetic field	地磁氣場
지자기적도	geomagnetic equator	地磁氣赤道
지자기좌표	geomagnetic coordinates	地磁氣座標
지자기지수	geomagnetic index	地磁氣指數
지자기폭풍	geomagnetic storm	地磁氣暴風
지적선	ground track	地跡線
지점	solstice	至點
지점간전송	point-to-point transmission	地點間傳送
지점번호	index number of station	地點番號
지점(색인)번호	station (index) number	地點(索引)番號
지점색인	station index	地點索引
지점시간권	solstitial colure	至點時間圈
지점우량	point rainfall	地點雨量
지점원	station circle	地點圓
지점조석	solstitial tide	至點潮汐
지정기압면	mandatory surface	指定氣壓面
지정층	mandatory layer	指定層
지중온도	earth temperature	地中溫度
지중온도계	earth thermometer, geothermometer	地中溫度計
지중온도기록계	earth thermograph	地中溫度記錄計
지중해계절풍	Etesian winds	地中海季節風
지중해기후	Etesian climate, Mediterranean climate	地中海氣候
지중해전선	Mediterranean front	地中海前線
지중해전선대	Mediterranean frontal zone	地中海前線帶
지지대	mast	支持臺
지진	seism	地震
지진계	seismograph, seismoscope	地震計
지진공백	seismic gap	地震空白
지진관측	seismological observation	地震觀測
지진기록	seismic record	地震記錄
지진단층촬영	seismic tomography	地震斷層撮影
지진조사	seismic surveying	地震調査
지진파	seismic wave	地震波

지진학	seismology	地震學
지진해일	tsunami	地震海溢
지진해일경보	tsunami warning	地震海溢警報
지진해일특보	special report for tsunami	地震海溢特報
지진해파	seismic sea wave	地震海波
지질기	geologic period	地質期
지질대	geologic era, geological era	地質代
지질세	geologic epoch	地質世
지질시대	geologic age, geological age	地質時代
지질시대기후	geological climate	地質時代氣候
지질연대규모	geological time scale	地質年代規模
지질연대	geologic time	地質年代
지질학	geology	地質學
지질학-지구화학 대용물	geological-geochemical proxy	地質學-地球化學代用物
지축	axis of earth, earth axis	地軸
지평	horizon	地平
지표거칠기	land surface roughness	地表-
지표로스비수	surface Rossby number	地表-數
지표면	earth surface, ground level, land surface	地表面
지표면방출률	emittance of the earth's surface	地表面放出率
지표면유출	overland runoff	地表面流出
지표면조도	land surface roughness	地表面粗度
지표바람	surface wind	地表-
지표바람응력	surface wind stress	地表-應力
지표반사	surface reflection	地表反射
지표보정	index correction	指標補正
지표생물	indicator organism	指標生物
지표수	surface water	地表水
지표수흐름	surface streamflow	地表水-
지표식물	indicator plant	指票植物
지표오차	index error	指標誤差
지표온도	land surface temperature	地表溫度
지표유출	flow-off	地表流出
지표저항	surface resistance	地表抵抗
지표지형	surface topography	地表地形
지표층	ground layer, surface layer	地表層
지표(경계)층	surface boundary layer	地表(境界)層
지표피복	land cover	地表被服
지표피복매핑	land cover mapping	地表被服-
지표피복분류체계	Land Cover Classification System	地表被服分類體系
지하대피실	cyclone cellar	地下待避室
지하류	subsurface flow, underrun	地下流
지하배수	subsurface drainage	地下排水
지하빙	underground ice	地下氷

지하수	ground water, phreatic water, subterranean water, underground water	地下水
지하수감소곡선	ground water depletion curve	地下水減少曲線
지하수감퇴	ground water recession	地下水減退
지하수면	ground water table, phreatic surface, water table	地下水面
지하수위	ground water level, undergorund water level, water table	地下水位
지하수이동	stream piracy	地下水移動
지하수층	phreatic zone, zone of saturation	地下水層
지하얼음	underground ice	地下-
지하유출	subsurface runoff, underground runoff	地下流出
지향	orientation, steering	指向
지향고도	steering level	指向高度
지향류	steering current	指向流
지향면	steering surface	指向面
지향법	steering method	指向法
지향법칙	steering law	指向法則
지향선	steering line	指向線
지향성	directivity	指向性
지향흐름	steering flow	指向-
지형	relief	地形
지형(학)	topography	地形(學)
지형강우(량)	orographic rainfall	地形降雨(量)
지형기후학	topoclimatology	地形氣候學
지형뇌우	orographic thunderstorm	地形雷雨
지형등압선	orographic isobar	地形等壓線
지형미기후	contour microclimate	地形微氣候
지형분석	terrain analysis	地形分析
지형설선	orographic snow-line	地形雪線
지형성	orographic	地形性
지형성강수	orographic precipitation	地形性降水
지형성강우증가	orographic rainfall enhancement	地形性降雨增加
지형성고기압	orographic high	地形性高氣壓
지형성구름	orographic cloud	地形性-
지형성바람	orographic wind	地形性-
지형성바람흐름	orographic wind flow	地形性-
지형성비	orographic rain	地形性-
지형성상승	orographic lifting, orographic uplift	地形性上昇
지형성소용돌이	orographic vortex	地形性-
지형성스톰	orographic storm	地形性-
지형성저기압	orographic cyclone, orographic depression	地形性低氣壓
지형성저기압발생	orographic cyclogenesis	地形性低氣壓發生
지형성파	orographic wave	地形性波

ㅈ

지형성폐색	orographic occlusion	地形性閉塞
지형성폭풍	orographic storm	地形性暴風
지형저기압발생	orographic cyclogenesis	地形低氣壓發生
지형전선발생	topographic frontogenesis	地形前線發生
지형조건	orographic condition	地形條件
지형학	geomorphology, physiography	地形學
지형효과	orographic effect	地形效果
지형흐름	orographic flow	地形-
직각	right angle	直角
직경	caliber	直徑
직교(의)	orthogonal	直交
직교곡선좌표	orthogonal curvilinear coordinates	直交曲線座標
직교다항식	orthogonal polynomial	直交多項式
직교선	orthogonal line	直交線
직교성	orthogonality	直交性
직교안테나	orthogonal antenna	直交-
직교좌표	rectangular coordinates	直交座標
직교좌표(계)	orthogonal (Cartesian) coordinates (system)	直角座標(系)
직교좌표계	rectangular coordinate system	直交座標系
직교편파	orthogonal polarization	直交偏波
직교함수	orthogonal function	直交函數
직교확장	orthogonal expansion	直交擴張
직달복사	direct radiation	直達輻射
직달빔복사	direct beam radiation	直達-輻射
직달일사	direct solar radiation	直達日射
직달일사계	pyrheliometer	直達日射計
직달일사기록	pyrheliogram	直達日射記錄
직달일사량	quantity of direct solar radiation	直達日射量
직달일사스케일	pyrheliometric scale	直達日射-
직달태양복사	direct solar radiation	直達太陽輻射
직독측기	direct reading instrument	直讀測器
직렬모형	in-tandem model	直列模型
직류	direct current	直流
직류회로	DC circuit	直流回路
직사	perpendicular incidence	直射
직선	straight line	直線
직선운동	straight line motion	直線運動
직시측운기	direct vision nephoscope	直視測雲器
직시프리즘	direct vision prism	直視-
직접가열법	direct heating method	直接加熱法
직접곱	direct product	直接-
직접방식	direct mode	直接方式
직접상호작용근사	direct interaction approximation	直接相互作用近似
직접세포	direct cell	直接細胞

직접수치모의	direct numerical simulation	直接數值模擬
직접순환	direct circulation	直接循環
직접유출	direct runoff	直接流出
직접유출수문곡선	direct runoff hydrograph	直接流出水文曲線
직접전송계	direct transmission system	直接傳送系
직접전송시스템	direct transmission system	直接傳送-
직접포착	direct capture	直接捕捉
직접흐름	direct flow	直接-
직판풍속계	normal-plate anemometer	直板風速計
직하관측	nadir viewing	直下觀測
직항로	direct route	直航路
직행조	direct tide	直行潮
진공	vacuum	眞空
진공계	vacuum ga(u)ge	眞空計
진공관전위계	vacuum-tube electrometer	眞空管電位計
진공방전	vacuum discharge	眞空放電
진공보정	vacuum correction	眞空補正
진공자외선	vacuum ultraviolet ray	眞空紫外線
진균독	mycotoxin	眞菌毒
진눈깨비	mixed rain and snow, rain and snow mixed, sleet	
진눈깨비경보	sleet warning	-警報
진단	diagnosis, diagnose	診斷
진단결합	diagnostic coupling	診斷結合
진단모형	diagnostic model	診斷模型
진단방정식	diagnostic equation	診斷方程式
진단변수	diagnostic variable	診斷變數
진동	oscillation	振動
진동(기압계수은면의)	pumping (of barometer)	面振動(氣壓計水銀)
진동급수	oscillation series	振動級數
진동수	frequency, number of vibration	振動數
진동수계	frequency meter	振動數計
진동수반송	frequency carrier	振動數搬送
진동수반응	frequency response	振動數反應
진동수방정식	frequency equation	振動數方程式
진동수영역	frequency domain	振動數領域
진동수조건	frequency condition	振動數條件
진동수편이	frequency shift	振動數偏移
진동수함수	frequency function	振動數函數
진동운동	vibrational motion	振動運動
진동자	oscillator	振動子
진동자메커니즘	oscillator mechanism	振動子-
진동전류	oscillating electric current	振動電流
진동전이	vibrational transition	振動轉移

진동주기	period of oscillation	振動週期
진동체	oscillating body	振動體
진동파	oscillatory wave, wave of oscillation	振動波
진동판	diaphragm	振動板
진동판압력계	diaphragm manometer	振動板壓力計
진동판풍속계	swinging plate anemometer	振動板風速計
진동회로	oscillating circuit	振動回路
진로	track	進路
진방위	true azimuth	眞方位
진앙	epicenter	震央
진원	focus	震源
진입등 교신고도	approach-light contact height	進入燈交信高度
진자	pendulum	振子
진자일	pendulum day	振子日
진잠재불안정	real latent instability	眞潛在不安定
진전방정식	evolution equation	進展方程式
진주모운	mother of pearl cloud	眞珠母雲
진주목걸이번개	pearl-necklace lightning	
진태양일	true solar day	眞太陽日
진폭	amplitude	振幅
진폭동요	amplitude vacillation	振幅動搖
진폭변조	amplitude modulation	振幅變調
진폭변조지시기	amplitude modulated indicator	振幅變調指示器
진폭분포	amplitude distribution	振幅分布
진폭스펙트럼	amplitude spectrum	振幅-
진풍	flaw	陣風
진행방정식	progression equation	進行方程式
진행운동	progressive motion	進行運動
진행파	progressive wave, travelling wave	進行波
진행파통로	travelling wave tube	進行波通路
진화	evolution	進化
진화반응	evolutionary response	進化反應
진화설	evolution theory	進化說
진흙사태	mud avalanche	-沙汰
진흙증발계	clay atmometer	-蒸發計
질량	mass	質量
질량결손	mass defect	質量缺損
질량농도	mass concentration	質量濃度
질량발산	mass divergence	質量發散
질량보존	conservation of mass	質量保存
질량보존방정식	equation of mass conservation	質量保存方程式
질량보존법칙	law of mass conservation	質量保存法則
질량분석기	mass spectrometer	質量分析器
질량비	mass ratio	質量比

질량사출계수	mass emission coefficient	質量射出係數
질량소산계수	mass extinction coefficient	質量消散係數
질량수	mass number	質量數
질량수렴	mass convergence	質量收斂
질량수송	mass transport	質量輸送
질량중심속도	barycentric velocity	質量中心速度
질량평균기온	mass mean temperature	質量平均氣溫
질량흡수계수	mass absorption coefficient	質量吸收係數
질산	nitric acid	窒酸
질산과산화아세틸	peroxyacethyl nitrate	窒酸過酸化-
질산삼수화물	nitric acid trihydrate	窒酸三水化物
질산암모늄	ammonium nitrate	窒酸-
질산염	nitrate	窒酸鹽
질산염라디칼	nitrate radical	窒酸鹽-
질산염이온	nitrate ion	窒酸鹽-
질소	nitrogen	窒素
질소고정	nitrogen fixation	窒素固定
질소고정식물	nitrogen-fixing plant	窒素固定植物
질소동위원소	nitrogen isotope	窒素同位元素
질소동화	nitrogen assimilation	窒素同化
질소분자	nitrogen molecule	窒素分子
질소산화물	nitrogen oxides, oxides of nitrogen, NOx	窒素酸化物
질소산화물배출	nitrogen oxides emission	窒素酸化物排出
질소순환	nitrogen cycle	窒素循環
질소축적량	nitrogenous deposition	窒素蓄積量
질소침적	nitrogen deposition	窒素沈積
질소포화	nitrogen saturation	窒素飽和
질소화합물	nitrogen compound	窒素化合物
질점	material point, particle	質點
질퍽눈	slush, snow slush	
질화	nitrification	窒化
짐기구	Jimsphere	-氣球
집게번개	pinched lightning	
집결시간	concentration time	集結時間
집단	collective	集團
집수면적	accumulative area	集水面積
집수역	catchment area, drainage basin, watershed	集水域
집적구역	zone of accumulation	集積區域
집적차분화	compact differencing	集積差分化
집중호우	gully washer, localized torrential rainfall	集中豪雨
집중화장치	multiplexer	集中化裝置
집진기	collector	集塵機
집진장치	dust collector	集塵裝置
집행위원회	executive committee	執行委員會

징조	precursor	徵兆
짙은 권운	cirrus densus, cirrus nothus, cirrus spissatus	-卷雲
짙은 상층운	dense upper cloud	-上層雲
짙은 안개	dense fog, heavy fog	
짙은 흐림	heavy overcast	
짙음(구름)	densus	
짝수	even number	-數
짝수점	even point	-數點
짧은 마루파	short-crested wave	-波
짧은 펄스	short pulse	
쪽집게기상	pin point weather	-氣象
쬐임	irradiation	

ㅈ

ㅊ

한글	영문	한자
차가움	cold	
차광디스크	shading disk	遮光-
차광띠	shadow band	遮光-
차광환	shade ring	遮光環
차기레이더	next generation radar	-次期-
차녹공식	Charnock's formula	-公式
차녹관계	Charnock's relation	-關係
차녹상수	Charnock's constant	-常數
차녹접근법	Charnock approach	-接近法
차니-드라진 범주	Charney-Drazin criterion	-範疇
차단	interception	遮斷
차단우량계	interceptometer	遮斷雨量計
차단주파수	cut-off frequency	遮斷周波數
차단지역기상	Interdiction Area weather	遮斷地域氣象
차단효과	block off, cut-off effect	遮斷效果
차등가열	differential heating	差等加熱
차등낙하속력	differential fall speed	差等落下速力
차등반사도	differential reflectivity	差等反射度
차등온도이류	differential thermal advection	差等溫度移流
차등위상	differential phase	差等位相
차등위상차	differential phase	差等位相差
차등이류	differential advection	差等移流
차량형기상관측시스템	automotive weather observation system	車輛型氣象觀測-
차례	sequence	順序
차례어림	step-by-step (or successive) approximation	
차르듀이	charduy	
차분방정식	difference equation, finite-difference equation	差分方程式
차분법	finite-difference method, method of finite difference	差分法

차분비	finite-difference ratio	差分比
차분아날로그	difference analogue	差分-
차분흡수법	differential absorption	差分吸收法
차분흡수법라이다	differential absorption lidar	差分吸收法-
차상샘플러	dichotomous sampler	叉狀-
차세대도플러레이더망(미국)	next generation doppler weather radar network	次世代-網
차원방정식	dimensional equation	次元方程式
차원분석	dimensional analysis	次元分析
차원행렬	dimensional matrix	次元行列
차이	contrast	差異
차이도	differential chart	差異圖
차폐	beam blocking, obscuration, shield	遮蔽
차폐계수	sheltering coefficient	遮蔽係數
차폐전선	masked front	遮蔽前線
차폐현상	obscuring phenomenon	遮蔽現像
착륙기상	landing weather	着陸氣象
착륙속도	landing speed	着陸速度
착륙예보	landing forecast	着陸豫報
착륙지역기상	Landing Area weather	着陸地域氣象
착륙활주거리	landing roll	着陸滑走距離
착모역전층	capping inversion	着帽逆轉層
착빙	ice accretion, icing	着氷
착빙강도	icing intensity	着氷強度
착빙계	icing meter	着氷計
착빙고도	icing level	着氷高度
착빙률계	icing-rate meter	着氷率計
착빙성 폭풍우경보	ice storm warning	着氷性暴風雨警報
착빙억제	anti-icing	着氷抑制
착빙지수	icing index	着氷指數
착빙지시기	ice accretion indicator	着氷指示器
착빙환경	icing environment	着氷環境
착설	snow accretion	着雪
찬 공기	cold air	-空氣
찬 공기 가두기	cold air damming	-空氣-
찬 공기강하	cold air drop, cold pool	-空氣降下
찬 공기깔때기	cold air funnel	
찬 공기댐	cold air dam	-空氣-
찬 공기돔	cold air dome	-空氣-
찬 공기사태	cold air avalanche	-空氣沙汰
찬 공기싱크	cold air sink	-空氣-
찬 공기유입	cold air injection	-空氣流入
찬 공기이류	cold air advection	-空氣移流
찬 공기터져나감	cold air outbreak	-空氣-
찬 공기풀	cold air pool, cold pool	-空氣-

찬 공기호수	cold air lake	-空氣湖水
찬 공기흐름	cold air flow	-空氣-
찬 물산호	cold water coral	-珊瑚
찬 북동풍	bise	-北東風
찬비	cold rain	
찬 이슬	cold dew	
찬정분지	artesian basin	鑽井盆地
참겨울날	ice day	
참고문서	reference document	參考文書
참고자료	reference data	參考資料
참고틀	reference frame	參考-
참여유도	nudging	參與誘導
참여유도계수	nudging coefficient	參與誘導係數
참지평선	real horizon	實地平線
찹퓌스	Chappuis	
찹퓌스띠	Chappicus band	
창끝에코	spearhead echo	槍-
창분리법	split window method	窓分離法
창살하계	trellis drainage	窓-河系
창서리	window frost	窓-
창얼음	window ice, windowing	窓-
창(영)역	window region	窓(領)域
창온도계	window thermometer	窓溫度計
창함수	window function	窓函數
찾아보기 번호	index number	-番號
채널	channel	
채널링	channelling	
채널흐름	channel flow	
채설기	snow sampler	採雪器
채운	iridescent cloud	彩雲
채움	filling up	
채집	collection	捕捉
채집기	collector	採集器
채집법	collecting method	採集法
채집효율	collection efficiency	捕捉效率
채층	chromosphere	彩層
채택역	acceptance region	採擇域
채택용량	acceptance capacity	採擇容量
채프먼메커니즘	Chapman mechanism	
채프먼순환	Chapman cycle	-循環
채프먼역	Chapman region	-域
채프먼오존화학	Chapman ozone chemistry	-化學
채프먼층	Chapman layer	-層
채프먼프로파일	Chapman profile	

ㅊ

책상보구름	tablecloth	册床褓-
책상빙산	table iceberg	册床氷山
챈들러 흔들림	Chandler wobble	
처리장치	processor	處理裝置
처서	limit of heat	處暑
척도	measure, scale	尺度
천공	canopy	天空
천공비	sky view factor	天空比
천구	celestial dome, celestial sphere	天球
천구극	celestial pole	天球極
천구적도	celestial equator	天球赤道
천구지평선	celestial horizon	天球地平線
천년단위 기후변동성	millennial climate variability	千年單位氣候變動性
천둥	thunder	
천둥번개	thunder and lightning	
천둥번개공포	astraphobia	-恐怖
천둥번개공포증	astrapophobia	-恐怖症
천둥번개기록계	ceraunograph	-記錄計
천둥번개측정계	ceraunometer	-測定計
천둥소나기	thunder shower	
천문강제	astronomical forcing	天文强制
천문굴절	astronomical refraction	天文屈折
천문단위	astronomical unit	天文單位
천문대	observatory	天文臺
천문박명	astronomical twilight	天文薄明
천문섬광	astronomical scintillation	天文閃光
천문시상	astronomical seeing	天文視像
천문위도	astronomical latitude	天文緯度
천문이론	astronomical theory	天文理論
천문일	astronomical day	天文日
천문조석	astronomical tide	天文潮汐
천문주기성	astronomical periodicities	天文週期性
천문지평선	astronomical horizon	天文地平線
천문학	astronomy	天文學
천수	meteoric water	天水
천수파	shallow water wave	淺水波
천수효과	shallow water effect	淺水效果
천식	asthma	喘息
천연가스	natural gas	天然-
천연방사능	natural radioactivity	天然放射能
천유방정식	shallow fluid equation	淺流方程式
천이	succession	遷移
천이수정법	successive correction method	遷移修正法
천저	nadir	天底

천저각	nadir angle	天底角
천저온도탐측	nadir sounding of temperature	天底溫度探測
천정	zenith	天頂
천정각	zenith angle	天頂角
천정거리	zenith distance	天頂距離
천정호	circumzenithal arc	天頂弧
천체	celestial body, heavenly body	天體
천체기상학	astrometerorology	天體氣象學
천체력	ephemeris	天體歷
천해	shallow sea, shallow water	淺海
천해산호	shallow water coral	淺海珊瑚
천해파	shallow water wave	淺海波
천해효과	shallow water effect	淺海效果
철새비행통로	flyways	-飛行通路
첨단방전	point discharge	尖端放電
첨단방전류	point discharge current	尖端放電流
첨도	kurtosis	尖度
첨두유출량	peak rate of runoff	尖頭流出量
첩보위성	spy satellite	諜報衛星
첫관측일	first observation date	-觀測日
첫눈	first snow	
첫단계	preliminary phase	-段階
첫돌풍	first gust	-突風
첫서리	first frost, first hoar frost	
첫서리날	first hoar frost day	
첫속도	initial velocity	-速度
첫시간단계	first time step	-時間段階
첫얼음	first freezing	
첫읽기	initial reading	
청결대기	clear and pure atmosphere	淸潔大氣
청도계	audibility meter	聽度計
청동기시대	Bronze Age	靑銅器時代
청록불꽃	blue-green flame	靑綠-
청록색전파	blue-green propagation	靑綠色傳播
청록색펄스	blue-green pulse	靑綠色-
청명	ceiling and visibility unlimited, pure brightness	淸明
청색편이	blue shift	靑色偏移
청신경성외상	acoustic trauma	聽神經性外傷
청우계	barometer, weather glass, weather gauge	晴雨計
청정개발체제	clean development mechanism	淸淨開發體制
청정대기법개정안	Clear Skies Act	淸淨大氣法改正案
청정도(기체의)	clarity(of gas)	淸淨度
청천	clear air, sky clear	晴天
청천되돌이에코	clear-air return echo	晴天-

차

청천공기에코	clear-air return echo	晴天空氣-
청천난류	clear air turbulence	晴天亂流
청천대기레이더	clear air radar	晴天大氣-
청천대기에코	clear air echo	晴天大氣-
청천복사	clear sky radiation	晴天輻射
청천비율	ratio of clear line of sight	晴天比率
체감기후	sensible climate	體感氣候
체감온도	sensible temperature	體感溫度
체계	regime	體系
체계적 접근	systematic approach	體系的接近
체노모그램	Tse nomogram	
체류	retention	滯留
체류시간	residence time	滯留時間
체르구이	chergui	
체방법	Tse method	-方法
체온계	rafraichometer	體溫計
체온저하	hypothermia	體溫低下
체자로 합	Cesàro sum	-合
체적밀도	volumetric density	體積密度
체적팽창	volume expansion	體積膨脹
체적팽창계수	coefficient of cubical expansion	體積膨脹係數
첼퍼드간빙기	Chelford interstadial	-間氷期
쳐오름파	swash	-波
초경도	supergradient	超傾度
초경도풍	supergradient wind	超傾度風
초경압(의)	hyperbaroclinic	超傾壓
초고분해복사계	very high resolution radiometer	超高分解輻射計
초고속도	hyper-velocity	超高速度
초고압기구	anchor balloon	超高壓氣球
초고주파	very high frequency	超高周波
초고주파수	ultra high frequency	超高周波數
초고층대기	ultra-upper atmosphere	超高層大氣
초과간격	exceedance interval	超過間隔
초과강우량	rainfall excess	超過降雨量
초관구조	canopy architecture	草冠構造
초굴절	superrefraction	超屈折
초기	initial	初期
초기값설정	initialization	初期値設定
초기강수과정	early rainfall process	初期降水過程
초기경로복구	first path retrieval	初期徑路復舊
초기경로탐측	first path sounding	初期徑路探測
초기궤도시험	In-Orbit Test	初期軌道試驗
초기단계	initial step, first time step	初期段階
초기단계 드보락분석법	early stage Dvoral analysis	初期段階-分析法

초기상승	initial lift	初期上乘
초기상태	initial state	初期狀態
초기속도	initial velocity	初期速度
초기손실	initial abstraction, initial loss	初期損失
초기시듦점	primary wilting point	初期-點
초기에오세기후최적	early Eocene climatic optimum	初期-氣候最適
초기역류	initial detention	初期逆留
초기위상	initial phase	初期位相
초기위조점	primary wilting point	初期萎凋點
초기응결단계	initial condensation phase	初期凝結段階
초기저기압	nascent cyclone	初期低氣壓
초기조건	initial condition	初期條件
초기추정	first guess	初期推定
초기추정장	first guess field	初期推定場
초기침투율	initial infiltration rate	初期浸透率
초기화	initialization	初期化
초깃값문제	initial value problem	初期值問題
초난류	superturbulent flow	超亂流
초단기예보	very-short-range forecast	超短期豫報
초단열	superadiabat	超斷熱
초단열감률	superadiabatic lapse rate	超斷熱減率
초단열상태	superadiabatic state	超斷熱狀態
초대형세포분리	splitting of supercells	超大型細胞分離
초록눈	green snow	草綠-
초록달	green moon	草綠-
초록태양	green sun	草綠太陽
초록하늘	green sky	草綠-
초미립자	ultrafine particle	超微粒子
초미세먼지	ultrafine particle	超微細-
초분광감부	hyperspectral sensor	超分光感部
초분광분류	hyperspectral classification	超分光分類
초분광센서	hyperspectral sensor	超分光-
초산	acetic acid	硝酸
초상온도	grass temperature	草上溫度
초상최저	grass minimum	草上最低
초상최저온도	grass minimum temperature	草上最低溫度
초상최저온도계	grass minimum theromometer	草上最低溫度計
초생수	juvenile water, magmatic water	初生水
초설	first snowfall	初雪
초설일	first snowfall day	初雪日
초성층권	superstratosphere	超成層圈
초시류	beetles	鞘翅類
초압기구	superpressure balloon	超壓氣球
초원	grassland, steppe	草原

초원기후	grassland climate, steppe climate	草原氣候
초원실험	grassland experiment	草原實驗
초음속(의)	supersonic	超音速
초음파(의)	ultrasonic	超音波
초음파 도플러 유속측정기	acoustic Doppler current profile	超音波-流速測定器
초음파적설계	ultrasonic snow cover meter	超音波積雪計
초음파풍속계	supersonic anemometer, ultrasonic anemometer	超音波風速計
초장기예보	extra long-range forecast	超長期豫報
초장파	ultra long wave	超長波
초저음	infrasound	超低音
초저음(의)	infrasonic	超低音
초저주파	very low frequency	超低周波
초저주파방출	VLF emission, very low frequency emission	超低周波放出
초점	focus	焦點
초점거리	focal distance	焦點距離
초점깊이	focal depth	焦點-
초정상굴절	supernormal refraction	超定常屈折
초지	grassy soil, poëchore	草地
초지균풍	supergeostrophic wind	超地均風
초콜릿 큰바람	chocolate gale	
초타버르사트	chota bursat	
초표준굴절	superstandard refraction	超標準屈折
초표준전파	superstandard propagation	超標準傳播
초헤테로다인	superheterodyne	超-
초형	plant type	草型
촉광	candle, candle power	燭光
촉매	catalyst, catalytic agent	觸媒
촉매변환기	catalytic converter	觸媒變換機
촉매소각로	catalytic incinerator	觸媒燒却爐
촉매소각장치	catalytic incineration	觸媒燒却裝置
촉매순환	catalytic cycle	觸媒循環
촉매컨버터	catalytic converter	觸媒-
촙	chop	
촙퍼	chopper	
총강설지역	amount of snow cover	總降雪地域
총광합성	gross photosynthesis	總光合成
총대류지수	total totals index	總對流指數
총분산	total variance	總分散
총생산량	gross production	總生産量
총수액량	total liquid water path	總水液量
총체공기역학방법	bulk aerodynamic method	總體空氣力學方法
총체공기역학항력공식화	bulk aerodynamic drag formulation	總體空氣力學抗力公式化
총체기공저항	bulk stomatal resistance	總體氣孔抵抗
총체난류규모	bulk turbulence scale	總體亂流規模

총체-리처드슨 수	bulk-Richardson number	總體-數
총체법	bulk method	總體法
총체변수	bulk variable	總體變數
총체열플럭스	bulk heat flux	總體熱-
총체적 안정경계성장	bulk stable boundary growth	總體的安定境界成長
총체전달	bulk transfer	總體傳達
총체전달계수	bulk transfer coefficient	總體傳達係數
총체전달법칙	bulk transfer law	總體傳達法則
총체점성	bulk viscosity	總體粘性
총체접근방법	bulk approach	總體接近方法
총체토양밀도	bulk density of soil	總體土壤密度
총체평균	bulk average	總體平均
총합조화예보	ensemble forecasting	總合調和豫報
총휘발성유기화합물 계산이론	TVOC theory	總揮發性有機化合物計算理論
촬영경위(의)	kinetheodolite	撮影經緯
촬영학	photology	撮影學
최고	maximum	最大
최고수위	high water	最高水位
최고수위계	crest gauge	最高水位計
최고온도	highest temperature, maximum temperature	最高溫度
최고온도계	maximum thermometer	最高溫度計
최고조위	high water	最高潮位
최고최저온도계	maximum and minimum thermometer	最高最低溫度計
최고태양활동	solar maximum	最高太陽活動
최다강수역	zone of maximum precipitation	最多降水域
최다풍향	most frequent wind direction	最多風向
최단시간항로	least-time track, least-time track, minimum time track, minimal flight path	最短時間航路
최단시정	geodesic visibility, minimum visibility, shortest visibility	最短視程
최대	maximum	最大
최대가능강수(량)	maximum possible precipitation	最大可能降水(量)
최대가능강수량	maximum probable precipitation(MPP)	最大可能降水量
최대가능일광기간	maximum possible sunshine duration	最大可能日光期間
최대가능홍수량	maximum probable flood(MDF)	最大可能洪水量
최대강수량	maximum precipitation	最大降水量
최대공률	peak power	最大工率
최대돌풍	peak gust	最大突風
최대돌풍감쇠	maximum gust lapse	最大突風減衰
최대돌풍감쇠간격	maximum gust lapse interval	最大突風減衰間隔
최대돌풍감쇠시간	maximum gust lapse time	最大突風減衰時間
최대마찰력	maximum frictional force	最大摩擦力
최대방전	peak discharge	最大放電
최대복사온도계	maximum radiation thermometer	最大輻射溫度計

최대빙하기	glacial maximum	最大氷河期
최대빙하기 말	Last Glacial maximum	最大氷河期末
최대빙하장소	glacial maximum	最大氷河場所
최대상승기류	maximum updraft	最大上昇氣流
최대설탐지속도	maximum unambiguous velocity	最大設探知速度
최대속도	maximum speed	最大速度
최대수증기장력	maximum water vapour tension	最大水蒸氣張力
최대순간바람	peak instantaneous wind	最大瞬間-
최대순간풍속	maximum instantaneous wind speed,	最大瞬間風速
	peak instantaneous wind speed	
최대용수량	maximum water holding capacity	最大用水量
최대유량법	peak flow method	最大流量法
최대유효거리	maximum unambiguous range	最大有效距離
최대유효속도	maximum unambiguous velocity	最大有效速度
최대이륙하중	maximum take-off load	最大離陸荷重
최대적외복사	maximum infrared radiance	最大赤外輻射
최대전력	peak power	最大電力
최대전류	peak current	最大電流
최대주파수	peak frequency	最大周波數
최대증기압	maximum vapour pressure	最大蒸氣壓
최대진동수	peak frequency	最大振動數
최대탐지거리	maximum detectable range	最大探知距離
최대파고	maximum wave height	最大波高
최대풍	peak wind	最大風
최대풍고도	maximum wind level	最大風高度
최대풍높낮이	maximum wind topography	最大風-
최대풍도	maximum wind chart	最大風圖
최대풍속	maximum wind speed, peak wind speed	最大風速
최대풍속반경	radius of maximum wind	最大風速半經
최대풍속반지름	radius of maximum wind	最大風速半-
최대풍향풍속	maximum wind	最大風向風速
최대허용농도	maximum acceptable concentration	最大許容濃度
최대혼합깊이	maximum mixing depth	最大混合-
최대DAD자료	maximum depth-area-duration data	最大-資料
최댓값	maximum value	最大-
최댓값검증	maximum value test	最大-檢證
최량선형 비편향추정	Best Linear Unbiased Estimation	最良線形非偏向推定
최빈값	mode	最頻-
최빈값반경	mode radius	最頻-半徑
최빈값반지름	mode radius	最頻-
최빈크기	most frequent size	最頻-
최소	minimum	最小
최소가시영역휘도	minimum visible radiance	最小可視領域輝度
최소공배수	least common multiple(LCM)	最小公倍數

최소배지	minimal media	最少培地
최소분산법	minimum variance method	最小分散法
최소이온화속도	minimum ionizing speed	最小-化速度
최소자승법	method of least squares	最小自乘法
최소정보추정법	minimum information method	最小情報推定法
최소제곱	least squares	最小-
최소제곱근법	least square method	最小-根法
최소제곱내삽법	least squares interpolation	最小-內揷法
최소제곱법	method of least squares	最小-法
최소취송거리	minimum fetch	最小吹送距離
최소탐지거리	minimum detectable range	最小探知距離
최소탐지신호	min detectable signal, minimum detectable signal	最小探知信號
최소태양활동	solar minimum	最小太陽活動
최소편향	minimum deviation	最小偏向
최소편향각	angle of minimum deviation	最小偏向角
최소항로	aerologation	最小航路
최솟값	minimum value	最小-
최심신적설	maximum fresh snow depth	最深新積雪
최심적설	maximum snow depth	最深積雪
최우도법	maximum likelihood method	最尤度法
최우도접근법	maximum likelihood approach	最尤度接近法
최우수시정	exceptional visibility	最優秀視程
최우앙상블필터	maximum likelihood ensemble filter	最尤-
최저	minimum	最低
최저감광한계	limen	最低感光限界
최저계기비행규칙	low instrument flight rule	最低計器飛行規則
최저고도	minimum altitude	最低高度
최저구역고도	highest minimum sector	最低區域高度
최저기상조건	weather minimum	最低氣象條件
최저습도	minimum humidity	最低濕度
최저시계비행규칙	marginal visual flight rule	最低視界飛行規則
최저안전고도	minimum safe altitude	最低安全高度
최저온도	lowest temperature, minimum air temperature, minimum temperature	最低溫度
최저온도계	minimum thermometer	最低溫度計
최저조건미만	below minimums	最低條件未滿
최적가중치	optimal weight	最適加重値
최적경로	best track	最適經路
최적관측	adaptive observation	最適觀測
최적관측망	adaptive observational network	最適觀測網
최적기후	optimum climate	最適氣候
최적내삽	optimum interpolation	最適內揷
최적내삽 해수면온도	optimum interpolation sea surface temperature	最適內揷海水面溫度
최적비행	optimum flight	最適飛行

최적서식지역	bioclim	最適棲息地域
최적섭동	optimal perturbation	最適攝動
최적수	best number	最適數
최적습도	optimum humidity	最適濕度
최적양수량	optimal yield	最適揚水量
최적온도	optimum temperature	最適溫度
최적채수량	optimal yield	最適採水量
최적해	optimum solution	最適解
최적화	optimization	最適化
최종수량일정의 법칙	law of constant final yield	最終收量一定法則
최종승온	final warming	最終昇溫
최한기등수온선	isocryme	最寒期等水溫線
최한월	coldest month	最寒月
최한월 월평균등온선	isoryme	最寒月月平均等溫線
최후빙하기 종료	Last Glacial termination	最後氷河期終了
추라다	churada	
추라오후	chou lao hu	
추력마력	thrust horse power	推力馬力
추력풍속계	thrust-anemometer	推力風速計
추분	autumnal equinox, fall equinox	秋分
추분기	autumn equinoctial period	秋分期
추분조석	autumn equinox tide	秋分潮汐
추세제거	detrending	趨勢除去
추운 계절	cold sector	-季節
추월한랭전선	overrunning cold front	追越寒冷前線
추위	cold	
추위막이	cold protection	
추이대	ecotone	推移帶
추적	tracking	追跡
추적자	tracer	追跡子
추적자법	tracer method	追跡子法
추적장치	tracking system	追跡裝置
추정	estimate	推定
추정강도	estimated intensity	推定强度
추정오차	error of estimate	推定誤差
추정장	guess field	推定場
추정표준오차	standard error of estimate	推定標準誤差
추진력	driving force	推進力
추축고기압	Pivoting high	樞軸高氣壓
추출	abstraction	抽出
추측항법	dead reckoning	推測航法
축	axis	軸
축경사	axial tilt	軸傾斜
축대칭	axisymmetric	軸對稱

축대칭(의)	axial symmetry	軸對稱
축대칭난류	axisymmetric turbulence	軸對稱亂流
축대칭류	axisymmetric flow	軸對稱流
축대칭언덕	axisymmetric hill	軸對稱丘
축대칭흐름	axially symmetric flow, axisymmetric flow	軸對稱-
축받이대	pedestal	
축산기상학	livestock meteorology	畜産氣象學
축세차	axial precession	軸歲差
축소	compactness	縮小
축소 없는 시나리오	business as usual scenario	
축차어림	step-by-step (or successive) approximation	逐次-
축차회귀법	stagewise regression procedure	逐次回歸法
축척	scale	縮尺
축척인자	scale factor	縮尺因子
춘분	vernal equinox	春分
춘분(점)	spring equinox, vernal equinox	春分(點)
춘추분	equinox	春秋分
춘추분경선	equinoctial colure	春秋分經線
춘추분스톰	line storm, equinoctial gale	春秋分-
춘추분우기	equinoctial rains	春秋分雨期
춘추분조석	equinoctial tide	春秋分潮汐
춘추분 큰바람	equinoctial gale	春秋分-
춘추분폭풍	line gale, line storm, equinoctial gale, equinoctial gale	春秋分暴風
춘추분폭풍우	equinoctial storm	春秋分暴風雨
춘추분폭풍우효과	equinoctial storm effect	春秋分暴風雨效果
출구역	delta region, exit region	出口域
출력	output	出力
출력밀도	power density	出力密度
출력밀도 스펙트럼	power density spectrum	出力密度-
출수예측	forecast for the heading time	出穗豫測
출아	emergence	出芽
출엽	leaf-emergence	出葉
충	opposition	衝
충격	bump	衝激
충격	impetus, impulse, impingement	衝擊
충격계수	duty factor	衝擊係數
충격도	bumpiness	衝激度
충격력	impulsive force	衝擊力
충격씨뿌리기	shock seeding	衝擊-
충격파	shock wave	衝擊波
충돌	collision, conflict, impact, impingement	衝突
충돌들뜸	collisional excitation	衝突-
충돌병합과정	collision coalescence process	衝突併合過程

충돌붕괴	collisional deactivation	衝突崩壞
충돌빈도	collision frequency	衝突頻度
충돌식우박측정기	hail impactor	衝突式雨雹測定器
충돌이론	collision theory	衝突理論
충돌적분	collision integral	衝突積分
충돌채집기	impactor	衝突採集器
충돌확장	collision broadening	衝突擴張
충돌효율	collision efficiency	衝突效率
충분조건	sufficient condition	充分條件
충적(의)	alluvial	沖積
충적대수층	alluvial aquifer	沖積帶水層
취송거리	wind drift distance	吹送距離
취송기간	wind duration	吹送期間
취송류	wind drift, wind-driven current	吹送流
취송해양순환	wind-driven oceanic circulation	吹送海洋循環
취수구	outlet	取水口
취식와지	deflation hollow	吹蝕窪地
취주거리	fetch, fetch length	吹走距離
취주거리효과	fetch effect	吹走距離效果
(바람의)취주량	capacity of the wind	吹走量
측각계	goniometer	測角計
측각방향탐지	goniometry direction finding	測角方向探知
측고계	hypsometer	測高計
측고공식	barometric formula, hypsometric formula	測高公式
측고방정식	hypsometric equation	測高方程式
측고법	altimetry, hypsography, hypsometry	測高法
측고평균온도	barometric mean temperature	測高平均溫度
측광(학)	photometry	測光(學)
측광법	photometry	測光法
측기내구년수	instrument outage	測器耐久年數
측기보정	compensation of instrument	測器補正
측기상수	instrument constant	測器常數
측도가속도	metric acceleration	測度加速度
측면경계조건	lateral boundary condition	側面境界條件
측방관측	side-looking observation	側方觀測
측심법	bathymetry	測深法
측우기	Chukwookee, Korean raingauge, Korean rain gauge	測雨器
측우대	Chukwoodae, a pedestal of Korean raniguage	測雨臺
측우학	ombrology	測雨學
측운기	nepheloscope, nephoscope	測雲器
측정	measure, measurement	測定
측정값	measured value, observed value	測定-
측정실링	measured ceiling	測定-

측정오차	error of measurement	測定誤差
측정요소	measuring element	測定要素
측정척	measuring stick	測定尺
측지곡선	geodetic curve	測地曲線
측지선	geodesic line	測地線
측지학	geodesy, geodetics	測地學
측풍	crosswind	測風
측풍경위(의)	balloon theodolite	測風經緯
측풍기구	pibal, pilot balloon	測風氣球
측풍기구경위(의)	pilot-balloon theodolite	測風氣球經緯
측풍기구계산자	pilot-balloon slide rule	測風氣球計算尺
측풍기구관측	pilot-balloon observation	測風氣球觀測
측풍기구관측소	pilot-balloon station	測風氣球觀測所
측풍기구기입판	pilot-balloon plotting board	測風氣球記入板
측풍기구띄움	pilot-balloon ascent	測風氣球-
측풍기구자기경위(의)	pilot-balloon self-recording theodolite	測風氣球自記經緯
측풍용경위(의)	aerological theodolite	測風用經緯
측풍유탄	grenade	側風榴彈
측풍트랜짓	balloon transit	測風-
측풍학	anemology	測風學
층	layer, level	層
층간	between layers	層間
층고적운	altocumulus stratiformis	層高積雲
층과 층 사이	between layers	
층권적운	cirrocumulus stratiformis	層卷積雲
층깊이	layer depth	層-
층류	laminar flow, laminated current	層流
층류경계층	laminar boundary layer	層流境界層
층류하층	laminar sub-layer	層流下層
층모양강수	stratiform precipitation	層模樣-降水
층모양구름	stratiform cloud, stratiformis	層模樣-
층모양비구름	pallio-nimbus, pallium	層模樣-
층모양에코	stratiform echo	層模樣-
층모양층적운	stratocumulus stratiformis	層模樣層積雲
층밀리기 변형	shear strain	層-變形
층밀림	shear	
층밀림변형	shearing deformation	-變形
층밀림선	shear line	-線
층밀림응력	shearing stress	-應力
층밀림중력파	shear-gravity wave	-重力波
층밀림층	shear layer	-層
층밀림파	shearing wave	-波
층상구조	lamella structure	層狀構造
층서학	stratigraphy	層序學

층운	stratus	層雲
층적운	stratocumulus	層積雲
층차분석	differential analysis	層差分析
층후	thickness	層厚
층후류	thickness advection	層厚流
층후변화	thickness change	層厚變化
층후변화이류	thickness change advection	層厚變化移流
층후상대소용돌이도	thickness relative vorticity	層厚相對-度
층후선	thickness line	層厚線
층후선도	thickness chart	層厚線圖
층후척도	thickness scale	層厚尺度
층후편차	thickness anomaly	層厚偏差
층흐름	laminary flow	層-
치누크	Chinook	
치누크바람	Chinook wind	
치누크벽구름	Chinook wall cloud	-壁-
치누크아치	Chinook arch	
치무홍수	Chimu flood	-洪水
치밀화작용	compaction	緻密化作用
치올림	lift	
치올림안개	lifting fog	
치올림온도	lifting temperature	-溫度
치올림응결고도	lifting condensation level	-凝結高度
치올림지수	lift index, lifted index	-指數
치올림탐지	lifting detection	-探知
치우침	bias	
칙쇼루브충돌	Chicxulub impact	-衝突
친기	atmophile	親氣
친수성	hydrophilicity	親水性
칠레해류	Chile current	-海流
칠리	chili	
침강	deposition, downwelling, sink, sinking, subsidence	沈降
침강류	downwelling, sinking stream	沈降流
침강면	surface of subsidence	沈降面
침강속도	settling velocity	沈降速度
침강(온도)역전	subsidence (temperature) inversion	沈降(溫度)逆轉
침루	percolation	浸漏
침상빙	acicular ice	針狀氷
침수	submergence	沈水
침수우발지역	flood-prone area	浸水偶發地域
침식	erosion	浸蝕
침식기선	base level of erosion	浸蝕基線
침식마루	erosion ridge	浸蝕-

침식지	encroachment	浸蝕池
침식취약성	erosional susceptibility	侵蝕脆弱性
침엽(의)	coniferous	針葉
침윤전선	wetting front	浸潤前線
침입	incursion, invasion, intrusion	侵入
침입공기	invading air	侵入空氣
침적	deposition	沈積
침전물	sediment	沈澱物
침전작용	sedimentation	沈澱作用
침착	deposition	沈着
침착량측정기	deposit gauge	沈着量測定器
침착속도	deposition velocity	沈着速度
침착핵	deposition nucleus	沈着核
침출	leaching	浸出
침출구역	zone of leaching	浸出區域
침출천	effluent stream	浸出川
침출침윤	effluent seepage	浸出浸潤
침투	infiltration, intrusion, penetration	浸透
침투거리	penetration range, night visual range	浸透距離
침투경로	percolation path	浸透經路
침투계	infiltrometer	浸透計
침투계수	infiltration coefficient	浸透係數
침투깊이	penetration depth	浸透深度
침투꼭대기	penetrating top	浸透頂上
침투능	infiltration capacity	浸透能
침투능곡선	infiltration capacity curve	浸透能曲線
침투대	percolation zone	浸透帶
침투대류	penetrative convection	浸透對流
칩로그	chip log	
칭량식우량계	rain-gauge weighing of balance type	秤量式雨量計

ㅊ

ㅋ

한글	영문	한자
카나리해류	Canaries current	-海流
카드뮴	cadmium	
카라데오도리원리	principle of Caratheodory	-原理
카라부란	karaburan	
카라비네	carabine	
카라세넷	caracenet	
카르노기관	Carnot engine	-機關
카르노순환	Carnot cycle	-循環
카르노순환과정	Carnot's cycle process	-循環過程
카르노정리	Carnot's theorem	-定理
카르노효율	Carnot efficiency	-效率
카르만상수	Kármán constant	-常數
카르만소용돌이행렬	Kármán vortex street	-行列
카르만온도	Kármán temperature	-溫度
카르보닐화합물	carbonyl compound	-化合物
카르복실화	carboxylation	-化
카리부	caribou	
카리브해류	Caribbean current	-海流
카만차카	camanchaca, garúa	
카바버드	kavaburd, cavaburd	
카발리에	cavaliers	
카본블랙	carbon black	
카비	kaavie	
카세그레인거울	Cassegrainian mirror	
카세그레인망원경	Cassegrainian telescope	-望遠鏡
카세그레인반사망원경	Cassegrainian reflector	-反射望遠鏡
카셀라의 사이폰식 우량기록계	Casella's siphon rainfall recorder	-式雨量記錄計
카스피해	Caspian Sea	-海

카심보	cacimbo	
카우-퀘이커	cow-quaker	
카이제곱검정	chi-square test	-檢定
카이제곱분포	chi-square distribution	-分布
카주의 소나기	cajú rains	
카타리나 맴돌이	Catalina eddy	
카타온도계	catathermometer, katathermometer	-溫度計
카테시안좌표	Cartesian coordinates	-座標
카테시안텐서	Cartesian tensor	
칸델라	candela	(燭光單位)
칸토르집합	Cantor set	-集合
칼라오페인터	Callao painter	
칼럼농도	column concentration	-濃度
칼럼최대치	column max	-最大値
칼레마	kaléma	
칼로리양방정식	caloric equation	-量方程式
칼만-부시 필터	Kálmán-Bucy filter	
칼만필터	Kalman filter	
칼만필터링	Kalman filtering	
칼바이사히	kal Baisakhi	
캄신	khamsin, khamseen, chamsin	
캐나다굳기측정계	Canadian hardness gauge	-測定計
캐나다기후모형	Canadian Climate Model	-氣候模型
캐노피	canopy	
캐노피구조	canopy structure	-構造
캐노피높이	canopy height	
캐노피매개변수	canopy parameters	-媒介變數
캐노피밀도	canopy density	-密度
캐노피온도	canopy temperature	-溫度
캐노피저항	canopy resistance	-抵抗
캐노피항력	canopy drag	-抗力
캐바버드	cavaburd, kavaburd	
캐스케이드과정	cascade process	-過程
캐올리나이트	kaolinite	
캔터베리북서풍	Canterbury northwester	-北西風
캘런더효과	callendar effect	-效果
캘리나	calina	
캘리포니아방법	California method	-方法
캘리포니아북풍	California norther	-北風
캘리포니아안개	California fog	
캘리포니아해류	California current	-海流
캘리포니아확률도시법	California plotting position	-確率圖視法
캠벨-스토크스 일조계	Campbell-Stokes sunshine recorder	-日照計
캣포우	cat's paw	

커네인도시위치	Cunnane plotting position	-都市位置
커닝엄미끄러짐보정	Cunningham slip correction	-補正
커닝엄보정계수	Cunningham correction factor	-補正係數
커스터드바람	custard wind	
커튼	curtain	
커튼극광	curtain aurora	-極光
커티스-고드슨 근사	Curtis-Godson approximation	-近似
컨베이어벨트	conveyor belt	
컨베이어벨트모형	conveyor belt model	-模型
컬	curl	
컬러바	color bar	
컴퓨터 모의	computer simulation	-模擬
컵결정	cup crystal	-結晶
컵결정서리	cup crystal frost	-結晶-
컵발전풍속계	cup generator anemometer	-發電風速計
컵풍속계	cup anemometer	-風速計
컵풍속계과속	cup anemometer overspeeding	風杯風速計過速
케널리-헤비사이드 층	Kennelly-Heaviside layer	-層
케른호	Kern's arc	-弧
케이프스크럽	cape scrub	
케임	Kames	
케플러방정식	Kepler's equation	-方程式
케플러법칙	Kepler's laws	-法則
켈빈	kelvin	
켈빈눈금	Kelvin scale	
켈빈-섭씨 변환	Kelvin-Celsius conversion	-變換
켈빈순환정리	Kelvin's circulation theorem	-循環定理
켈빈식	Kelvin's equation	-式
켈빈온도	Kelvin temperature	-溫度
켈빈온도눈금	Kelvin temperature scale	-溫度-
켈빈파	Kelvin wave	-波
켈빈-헬름홀츠 물결구름	Kelvin-Helmholtz billows	
켈빈-헬름홀츠 불안정	Kelvin-Helmholtz instability	-不安定
켈빈-헬름홀츠 파	Kelvin-Helmholtz wave	-波
켈셔	kelsher	
켤레멱	conjugate power	-冪
켤레멱법칙	conjugate power law	-冪法則
켤레복소수	conjugate of complex number	-複素數
켤레상	conjugate image	-像
켤레좌표	conjugate coordinate	-座標
코나스톰	kona storm	
코나저기압	kona cyclone	-低氣壓
코니컬 스캔	conical scan	
코로나	corona	

코로나구멍	coronal holes	
코로나광휘	streamer	-光輝
코로나그래프	coronagraph	
코로나방법	corona method	-方法
코로나방전	corona discharge, corposant	-放電
코로나전류	corona current	-電流
코로나질량방출	coronal mass injection	-質量放出
코르도나조	Cordonazo	
코르디예라	Cordillera	
코르디예라빙상	Cordillera ice sheet	-氷床
코리올리가속도	Coriolis acceleration	-加速度
코리올리매개변수	Coriolis parameter	-媒介變數
코리올리인자	Coriolis factor	-因子
코리올리편향	Coriolis deflection	-偏向
코리올리효과	Coriolis effect	-效果
코리올리힘	Coriolis force	
코만차카	comanchaca	
코사인반응	cosine response	-反應
코사인테이퍼링	cosine tapering	
코슈미더법칙	Koschmieder's law	-法則
코슈미더시정공식	Koschmieder's visibility formula	-視程公式
코스모스위성시리즈	cosmos satellite series	-衛星-
코시-리만 방정식	Cauchy-Riemann equation	-方程式
코시분포	Cauchy distribution	-分布
코시수	Cauchy number	-數
코어분석	core analysis	-分析
코어시료	core sample	-試料
코펜하겐해수	Copenhagen water	-海水
코히어런트 수신기	coherent receiver	-受信機
코히어런트 신호	coherent signal	-信號
콕아이밥	cockeyed bob	
콘크리트최저온도	concrete minimum temperature	-最低溫度
콘테사 디 벤토	contessa di vento	
콘트라스테	contraste	
콜로라도저기압	Colorado cyclone, Colorado low	-低氣壓
콜로라도증발계	Colorado sunken pan	-蒸發計
콜로라도파동저기압	Colorado wave cyclone	-波動低氣壓
콜로모그로브 스펙트럼	Kolomogrov spectrum	
콜로이드	colloid	
콜로이드계	colloidal system	-系
콜로이드기상학	colloid meteorology	-氣象學
콜로이드부유	colloidal suspension	-浮游
콜로이드분산	colloidal dispersion	-分散
콜로이드불안정(도)	colloidal instability	-不安定(度)

콜로이드안정(도)	colloidal stability	-安定(度)
콜로이드용액	colloidal solution	-溶液
콜로이드적안정	colloidal stable, colloidally stable	-的安定
콜로이드준안정	colloidal metastable	-準安定
콜모고로프 유사가설	Kolmogoroff's similarity hypothesis	-類似假說
콜모고로프 캐스케이드	Kolmogoroff cascade	
콤마머리	comma head	
콤바부호	COMBAR code	-符號
콤프턴전자	Compton electron	-電子
콤프턴효과	Compton effect	-效果
콩태풍	midget	豆颱風
쾌적곡선	comfort curve	快適曲線
쾌적기준	comfort standard	快適基準
쾌적대	comfort zone	快適帶
쾌적도표	comfort chart	快適圖表
쾌적온도	comfortable temperature	快適溫度
쾌적지수	comfort index	快適指數
쾨펜 구분	Köppen classification	
쾨펜 기후분류	Köppen's classification of climates,	-氣候分類
	Köppen's climatic classification	
쾨펜-수판선	Köppen-Supan line	-線
쿠랑수	Courant number	-數
쿠랑-프리데리흐스-레위 안정도기준	Courant-Friedrichs-Lewy stability criterion	-安定度基準
쿠랑-프리데리흐스-레위 조건	Courant-Friedrichs-Lewy condition	-條件
쿠로시오	Kuroshio	
쿠로시오반류	Kuroshio countercurrent	-反流
쿠로시오속류	Kuroshio extension	-屬流
쿠로시오해류계	Kuroshio system	-海流系
쿠에테흐름	Couette flow	
쿡(채표) 방법	Cook method	-(採表)方法
쿨롱	coulomb	
퀴리	curie	
퀴리온도	curie temperature	-溫度
퀴리원리	Curie's principle	-原理
퀵버드위성	QuickBird satellite	-衛星
큐벡터	Q-vector	
큐벡터강제력	Q-vector forcing	-强制力
큐약호	Q-code	-略號
큐코드	Q-code	
큐티클증산	cuticular transpiration	-蒸散
큐형기압계	Kew (pattern) barometer	-型氣壓計
크기	magnitude	
크기매개변수	size parameter	-媒介變數

크기분포	size distribution	-分布
크기순서	order of magnitude	-順序
크기차수	order of magnitude	-次數
크노이드파	cnoidal wave	
크놀렌베르크 프로브	Knollenberg probe	
크누센 표	Knudsen's tables	-表
크라이오제니아기	cryogenian period	-期
크라친	crachin	
크라카토아 바람	Krakatoa wind	
크랭크-니콜슨 방법	Crank-Nicholson method	-方法
크랭크-니콜슨 방안	Crank-Nicholson scheme	-方案
크러스트	crust	
크레바스	crevasse	
크레바스벽	crevasse wall	-壁
크레바스서리	crevasse hoar	
크레바스흔적	crevasse traces	-痕迹
크레이터	crater	
크로노그래프	chronograph	
크로노미터	chronometer	
크로메리아 간빙기	Cromerian interglacial	-間氷期
크롬웰해류	Cromwell current	-海流
크루츠버그기후대	Creutzburg's climatic zone	-氣候代
크리게바이라디칼	Criegee biradicals	
크리게중간단계	Criegee intermediate	-中間段階
크리깅	kriging	
크리스토펠기호	Christoffel symbol	-記號
크리프	creep	
크리핑흐름	creeping flow	
크세논	xenon	
큰 노을	blind roller	
큰눈	heavy snow	
큰 바람	fresh gale, gale	
큰 바람감시	gale watch	-監視
큰 바람경보	gale warning	-警報
큰 바람신호	gale signal	-信號
큰 바람표지	gale cone	-標識
큰 붉은 점	Great Red Spot	-點
큰비	heavy rain	
큰센바람	strong gale	
큰 우박	heavy hail	
큰 이온	large ion	
큰 입자	large particle	-粒子
큰잎모형	big leaf model	-模形
큰 재앙	catastrophe	大災殃

ㅋ

큰 칼로리	large calorie	
큰 태풍눈	distinct large eye	-颱風-
큰 핵	large nucleus	-核
큰 햇무리	large halo	
클라우드스트리트	cloud street	
클라우시우스부등식	Clausius inequality	-不等式
클라우시우스상태방정식	Clausius equation of state	-狀態方程式
클라우시우스원리	Clausius theorem	-原理
클라우시우스-클라페이론 방정식	Clausius-Clapeyron equation	-方程式
클라우시우스-키르히호프 관계	Clausius-Kirchhoff relation	-關係
클라이모그래프	climograph	
클라이모그램	climogram	
클라이트론	Klystron	
클라크궤도	Clarke orbit	-軌道
클라페이론방정식	Clapeyron equation	-方程式
클라페이론선도	Clapeyron's diagram	-線圖
클라포티	clapotis	
클락의 호두까기	Clark's nutcrackers	
클러스터분석	cluster analysis	-分析
클러스터이온	cluster ion	
클러스터일기도	cluster map	-日氣圖
클러스터확산	cluster diffusion	-擴散
클러터	clutter	
클러터제거	clutter rejection	-除去
클러터지도	clutter map	-地圖
클러터필터	clutter filter	
클레이든효과	Clayden effect	-效果
클로	clo	
클로로포름	chloroform	
클론커리저기압	Cloncurry low	-低氣壓
클리도노그래프	klydonograph	
키르히호프 법칙	Kirchhoff's law	-法則
키르히호프 복사법칙	Kirchhoff's radiation law	-輻射法則
키작은 고기압	short high	-高氣壓
키치-바이람 가뭄지수	Keetch-Byrum Drought Index	-指數
키큰 고기압	tall high	参天高氣壓
킬로 파스칼	kilopascal	

E

한글	영문	한자
타당성조사	feasibility study	妥當性調査
타래법	tuft method	-法
타우값	tau-value	
타운센드받침대	Townsend support	-臺
타원	ellipse	橢圓
타원기하학	ellipse geometry	橢圓幾何學
타원면	ellipsoid	惰圓面
타원상승	elliptic ascent	橢圓上昇
타원형편광	elliptical polarization	橢圓形偏光
타이가기후	taiga climate	北方針葉林氣候
타이노	taino	
타이로스	television and infrared observing satellite(TIROS)	
타이로스연직탐측계	TIROS operational vertical sounder	-鉛直探測計
탁도계	nephelometer	濁度計
탁월편동풍	prevailing easterlies	卓越偏東風
탁월편서풍	prevailing westerlies	卓越偏西風
탁월풍	dominant wind, prevailing wind	卓越風
탁월풍계	prevailing wind system	卓越風系
탁월풍향	prevailing wind direction	卓越風向
탄도밀도	ballistic density	彈道密度
탄도바람	ballistics wind, equivalent constant wind	彈道-
탄도온도	ballistics temperature	彈道溫度
탄도풍	ballistic wind	彈導風
탄도학	ballistics	彈道學
탄력	resilience	彈力
탄산	carbonic acid	炭酸
탄산염바위	carbonate rock	炭酸鹽-
탄산염보상심도	carbonate compensation depth	炭山鹽補償深度
탄산염보존깊이	carbonate conservation depth	炭酸鹽保存-

탄성	elasticity	彈性
탄성(의)	elastic	彈性
탄성기압계	elastic barometer	彈性氣壓計
탄성력	elastic force	彈性力
탄성률	elastic modulus	彈性率
탄성반발	elastic rebound	彈性反撥
탄성변형	elastic deformation	彈性變形
탄성체	elastic body	彈性體
탄성충돌	elastic collision	彈性衝突
탄성파	elastic wave	彈性波
탄성한계	elastic limit	彈性限界
탄소가루 씨뿌리기	carbon black seeding	炭素-
탄소강도	carbon intensity	炭素強度
탄소격리	carbon sequestration	炭素隔離
탄소결합메커니즘	carbon bond mechanism	炭素結合-
탄소교환	carbon exchange	炭素交換
탄소균형	carbon balance	炭素均衡
탄소동위원소	carbon isotope	炭素同位元素
탄소동화	assimiliation of carbon	炭素同化
탄소동화작용	carbon assimilation	炭素同化作用
탄소막습도계소자	carbon-film hygrometer element	炭素膜濕度計素子
탄소무역	carbon trading	炭素貿易
탄소발생원	carbon source	炭素發生源
탄소발자국	carbon footprint	炭素-
탄소방출	carbon emission	炭素放出
탄소세	carbon tax	炭素稅
탄소순환	carbon cycle	炭素循環
탄소에어로졸	carbonaceous aerosol	炭素-
탄소연대측정법	carbon-dating	炭素年代測定法
탄소저장	carbon storage	炭素貯藏
탄소풀	carbon pool	炭素-
탄소화합물	carbon compound	炭素化合物
탄소흡(수)원	carbon sink	炭素吸(收)源
탄화불소	fluorocarbon	炭火弗素
탄화수소	hydrocarbons	炭化水素
탈빙하	deglaciation	脫氷河
탈질	denoxification	脫窒
탈착	desorption	脫着
탈출원뿔	cone of escape	脫出圓-
탈활성	Quenching	脫活性
탈황	desulfurization	脫黃
탐사구름	clouds from sounders	探査-
탐색	searchlighting	探索
탐색구조	search and rescue	探索救助

탐색구조작전	search and rescue operation	探索救助作戰
탐색레이더	search radar	探索-
탐조	searchlighting	探照
탐조등	search light	探照燈
탐지	detection	探知
탐지기	detector	探知機
탐지기법	detection technique	探知技法
탐측	sounding	探測
탐측기구	balloonsonde, sounding balloon	探測氣球
탐측기상기록계	sounding meteorograph	探測氣象記錄計
탐측도	sounding chart	探測圖
탐측로켓	sounding rocket	探測-
탐측장치	probe	探測裝置
탑고적운	altocumulus castellanus	塔高積雲
탑구름	turreted cloud	塔-
탑권운	cirrus castellanus	塔卷雲
탑권적운	cirrocumulus castellanus	塔卷積雲
탑모양(구름)	castellanus	塔模樣-
(관제)탑시정	tower visibility	(管制)塔視程
탑재량	carrying capacity	搭載量
탑재체접속판	payload Interface Panel	搭載體接續板
탑적운	towering cumulus	塔積雲
탑층적운	stratocumulus castellanus	塔層積雲
태양	sun	太陽
태양거리	solar distance	太陽距離
태양경사	solar declination	太陽傾斜
태양계	solar system	太陽界
태양고도	altitude of the sun, elevation angle of the sun, solar altitude, sun's altitude	太陽高度
태양고도각	solar elevation angle	太陽高度角
태양공기질량	solar air mass	太陽空氣質量
태양관측궤도위성	orbiting solar observatory	太陽觀測軌道衛星
태양광	sunbeam	太陽光
태양광도계	sun-photometer	太陽光度計
태양광스펙트럼	solar spectrum	太陽光-
태양기압변화	solar barometric variation	太陽氣壓變化
태양기후	solar climate	太陽氣候
태양-기후연관성	sun-climate connection	太陽氣候聯關性
태양년	solar year	太陽年
태양대기	solar atmosphere	太陽大氣
태양대기조석	solar atmospheric tide	太陽大氣潮汐
태양동기궤도	sun synchronous orbit	太陽同期軌道
태양동기상관	sun synchronous correlation	太陽同期相關
태양동기위성	sun synchronous satellite	太陽同期衛星

E

태양동기준극궤도	quasi-polar sun-synchronous orbit	太陽同期準極軌道
태양동력	solar power	太陽動力
태양미립자	solar corpuscle	太陽微粒子
태양미립자설	solar corpuscular theory	太陽微粒子說
태양반짝임	sun glint	太陽-
태양변동	solar variation	太陽變動
태양변동성	solar variability	太陽變動性
태양복사	solar radiation	太陽輻射
태양복사감쇠	attenuation of solar radiation	太陽輻射減衰
태양복사관측	solar radiation observation	太陽輻射觀測
태양복사발산도	solar radiant exitance	太陽輻射發散度
태양복사수지	balance of solar radiation	太陽輻射收支
태양복사스펙트럼	solar radiation spectrum	太陽輻射-
태양복사온도계	solar radiation thermometer	太陽輻射溫度計
태양복사차폐	solar radiation shield	太陽輻射遮蔽
태양복사플럭스	flux of solar radiation	太陽輻射-
태양복사환경	solar radiation environment	太陽輻射環境
태양복사흡수	absorption of solar radiation	太陽輻射吸收
태양분광	solar spectrum	太陽分光
태양분광복사	spectral solar radiation	太陽分光輻射
태양상수	solar constant	太陽常數
태양섬광에코	sun strobe echo	太陽閃光-
태양수감부	sun sensor	太陽受感部
태양스펙트럼조도	solar spectral irradiance	太陽-照度
태양시	solar time	太陽時
태양양자감시기	solar proton monitor, sun proton monitor	太陽陽子監視器
태양양자소동	solar proton event	太陽陽子騷動
태양엄폐	solar occultation	太陽掩蔽
태양에너지	solar energy	太陽-
태양열에너지	solar thermal energy	太陽熱-
태양온도	solar temperature	太陽溫度
태양일	solar day	太陽日
태양입자환경	solar particle environment	太陽粒子環境
태양잎새	sun leaf	太陽-
태양자기방출역	emerging flux region	太陽磁氣放出域
태양자외복사	solar ultraviolet radiation	太陽紫外輻射
태양자외선복사감시기	monitor of ultraviolet solar radiation	太陽紫外線輻射監視器
태양전지	solar battery, solar cell	太陽電池
태양전지판	solar paddle	太陽電池板
태양조석	solar tide	太陽潮汐
태양좌표계	solar coordinates	太陽座標系
태양주기	solar cycle	太陽週期
태양지구관계	solar terrestrial relationship	太陽地球關係
태양-지구기하학	sun-earth geometry	太陽地球幾何學

E

태양지구물리학	solar terrestrial physics	太陽地球物理學
태양진행방향	deasil	太陽進行方向
태양천정각	solar zenith angle	太陽天頂角
태양풍	solar wind	太陽風
태양플레어	solar flare	太陽-
태양플레어요란	solar flare disturbance	太陽-擾亂
태양홍염	solar prominence	太陽紅焰
태양활동	solar activity	太陽活動
태양흑점	sunspot	太陽黑點
태양흑점상대수	sunspot relative number	太陽黑點相對數
태양흑점수	sunspot number	太陽黑點數
태양흑점주기	sunspot cycle, sunspot periodicity	太陽黑點週期
태음년	lunar year	太陰年
태음월	lunation	太陰月
태음일	lunar day	太陰日
태음조	lunar tide	太陰潮
태평양	Pacific Ocean	太平洋
태평양고기압	Pacific anticyclone, Pacific high	太平洋高氣壓
태평양기단	Pacific air mass	太平洋氣團
태평양 및 인도양 일반수	Pacific and Indian-ocean common water	太平洋印度洋一般水
태평양-북아메리카 진동	Pacific-North American oscillation	太平洋-北美振動
태평양심층수	Pacific deep water	太平洋深層水
태평양십년진동	Pacific decadal oscillation	太平洋十年振動
태평양아북극해류	Pacific subarctic current	太平洋亞北極海流
태평양판	Pacific Plate	太平洋板
태평양표준시(미국)	Pacific Standard Time	太平洋標準時
태평양해령	Pacific ridge	太平洋海嶺
태평양해양한대기단	Pacific mP air masses	太平洋海洋寒帶氣團
태풍	typhoon, t'aifung	颱風
태풍강도	typhoon intensity	颱風強度
태풍강도예보	typhoon intensity forecast	颱風強度豫報
태풍강도지수	typhoon intensity number	颱風強度指數
태풍강풍지역반경	radius of strong wind area of typhoon	颱風強風地域半徑
태풍경로	typhoon track	颱風經路
태풍경로예보	typhoon track forecast	颱風經路豫報
태풍경보	typhoon warning	颱風警報
태풍계절	typhoon season	颱風季節
태풍구름벽	typhoon bar	颱風-壁
태풍눈	eye of typhoon, hurricane eye, typhoon eye	颱風-
태풍모형	typhoon model	颱風模型
태풍번호	serial number of typhoon	颱風番號
태풍보거싱	bogusing, typhoon bogussing	颱風-
태풍스콜	typhoon squall	颱風-
태풍예보	typhoon forecast	颱風豫報

E

태풍예보시스템	typhoon forecast system	颱風豫報-
태풍위원회	typhoon committee	颱風委員會
태풍위치확률반경	probability radius of typhoon location	颱風位置確率半徑
태풍의 눈	banding eye, ragged eye	颱風-
태풍의 상륙	landfall	颱風上陸
태풍이름	name of typhoon	颱風名
태풍정보	typhoon information	颱風情報
태풍주의보	hurricane watch, typhoon advisory	颱風主意報
태풍통과	pass	颱風通過
털보구름	capillatus	
테일러급수	Taylor series	-級數
테일러기둥	Taylor column	
테일러선도	Taylor diagram	-線圖
테일러소용돌이	Taylor's vortex	
테일러수	Taylor number	-數
테일러정리	Taylor's theorem	-定理
테일러-프라우드맨 정리	Taylor-Proudman theorem	-定理
테일러효과	Taylor effect	-效果
테프라연대학	tephrochronology	-年代學
테피그램	tephigram	
텐서	tensor	張量
텐서 2차곡면	tensor quadric	張量二次曲面
텐시오미터	tensiometer	
텔레비전기상방송	television weather broadcasting	-氣象放送
텔레비전적외관측위성	television and infrared observing satellite(TIROS)	-赤外觀測衛星
텔레타이프	teletype, teleprinter	
텔레프린터	teleprinter	
텔렉스	telex, telegraph exchange	
템포랄레스	temporales	
토네이도	twister, tornado	
토네이도가족	family of tornadoes	-家族
토네이도강도분류	tornado intensity category	-强度分類
토네이도뇌우	tornadic thunderstorm	-雷雨
토네이도대	tornado belt	-帶
토네이도대피소	tornado cave	-待避所
토네이도돌발	tornado outbreak	-突發
토네이도 소용돌이 신호	tornado vortex signature	-信號
토네이도에코	tornado echo	
토네이도저기압	tornado cycone	-低氣壓
토로이드	toroid	
토론	thoron	
토류	earthflow	土流
토르	torr	
토리첼리	Torricelli	

토리첼리관	Torricelli's tube	-管
토리첼리진공	Torricellian vacuum	-眞空
토막고층운	altostratus pannus	-高層雲
토막구름	pannus, scud	
토막적운	cumulus pannus	-積雲
토막적란운	cumulonimbus pannus	-積亂雲
토사류	mud flow	土砂流
토석류	earthflow	土石流
토양	soil	土壤
토양가뭄	soil drought	土壤-
토양경도	soil hardness	土壤硬度
토양공극	soil pore	土壤孔隙
토양공기	soil air	土壤空氣
토양공기온도	soil air temperature	土壤空氣溫度
토양구조	soil structure	土壤構造
토양기어내림	soil creep	土壤-
토양기원질병	soil borne diseases	土壤起源疾病
토양기체	soil gas	土壤氣體
토양기후	soil climate	土壤氣候
토양기후학	soil climatology	土壤氣候學
토양내풍식성	wind resistance of soil	土壤耐風蝕性
토양물수지	soil water balance	土壤-收支
토양밀도	density of soil	土壤密度
토양반사특성	soil reflectance properties	土壤反射特性
토양온도계	soil thermometer	土壤溫度計
토양상대수분	relative moisture of the soil	土壤相對水分
토양수분	soil moisture, soil water	土壤水分
토양수분계	soil moisture meter	土壤水分計
토양수분부족(량)	soil moisture deficit	土壤水分不足(量)
토양수분응력	soil moisture tension	土壤水分應力
토양수분응력계	soil moisture tensiometer	土壤水分應力計
토양수분퍼텐셜	soil water potential, water potential of soil	土樓-水分-
토양수분함유량	soil moisture content	土壤水分含有量
토양수평선	soil horizon	土壤水平線
토양실제밀도	actual density of the soil	土壤實際密度
토양염류장해	soil salt injury	土壤鹽類障害
토양온도	soil atmosphere, soil temperature	土壤溫度
토양전도도	soil conductivity	土壤傳導度
토양절대수분함량	absolute moisture of the soil	土壤絶對水分含量
토양증발	soil evaporation	土樓蒸發
토양증발계	soil evaporimeter	土壤蒸發計
토양질감	soil texture	土壤質感
토양침식	soil erosion	土壤浸蝕
토양통기	soil aeration	土壤通氣

E

토양특성	soil property	土壤特性
토양표면온도	temperature of the soil surface	土壤表面溫度
토양학	pedology	土壤學
토양호흡	soil respiration	土壤呼吸
토양흐름	soil flow	土壤-
토우사공식	Toussaint's formula	-公式
토제	earth mound	土堤
토지이용유형	land use type	土地利用類型
토지표면고도	height of land surface	土地表面高度
토지피복	land cover	土地被覆
토지피복분류체계	Land Cover Classification System	土地被覆分類體系
토지피복지도제작	land cover mapping	土地被覆地圖製作
토지효과	edaphic effect	土地效果
토크	torque	
토탄	peat	土炭
통계	statistics	統計
통계가중	statistical weight	統計加重
통계강조	statistical enhancement	統計强調
통계기후학	statistical climatology	統計氣候學
통계독립성	statistical independence	統計獨立性
통계동역학예측	statistical dynamic prediction	統計動力學豫測
통계동화법	statistical assimilation method	通計同化法
통계매개변수	statistical parameter	統計媒介變數
통계법	statistical method	統計法
통계열역학	statistical thermodynamics	統計熱力學
통계예보	statistical forecast	統計豫報
통계예보법	statistical methods of forecasting	統計豫報法
통계유의성검정	statistical significance test	統計有意性檢定
통계의존성	statistical dependence	統計依存性
통계파라미터	statistical parameter	統計-
통계표	statistical table	統計表
통계학	statistics	統計學
통계확률	statistical probability	統計確率
통곗값	statistic	統計-
통기계수	coefficient of air permeability	通氣係數
통기구역	zone of aeration	通氣區域
통기대	vadose zone	通氣帶
통기성	air permeability	通氣性
(자동차의)통기장치	crackcase ventilation	通氣裝置
통로	channel	通路
통로가설	gateway hypothesis	通路假說
통보	message	通報
통보대상기관	organizations to deliver massage	通報對象機關
통보시각	delivering time of message	通報時刻

통보점	reporting point	通報點
통신센터	communication center	通信-
통신소	communication station	通信所
통신위성	communication satellite	通信衛星
통신중추	telecommunication hub	通信中樞
통신해양기상위성	communication oceanic and meteorological satellite	通信海洋氣象衛星
통제자료수신소	command and data acquisition station	統制資料受信所
통풍	aeration, ventilation	通風
통풍건습계	aspiration psychrometer, ventilated psychrometer	通風乾濕計
통풍구	blowholes	通風口
통풍기	aspirator	通風器
통풍기상기록계	aspiration meteorograph	通風氣象記錄計
통풍습도계	aspirated hygrometer	通風濕度計
통풍온도계	aspirated thermometer, aspiration thermometer, ventilated thermometer	通風溫度計
통풍자기온도계	aspiration thermograph	通風自記溫度計
통풍지수	ventilation index	通風指數
통풍콘덴서	aspiration condenser	通風-
통합비행장기상시스템	integrated terminal weather system	統合飛行場氣象-
통합안정도척도	integrated indicators of stability	統合安定度尺度
통합억제	integral constraint	統合抑制
통합적 시료채취	integrated sampling	統合的試料採取
통합해저시추프로그램	Integrated Ocean Drilling Program	統合海底試錐-
퇴빙	deglaciation	退氷
퇴빙 2단계	deglacial two-step	退氷二段階
퇴적과정	banking process, pile-up process	堆積過程
퇴적구름	heap cloud	堆積-
퇴적물저장한계	sediment storage capacity	堆積物貯藏限界
퇴적상	sedimentary facies	堆積相
퇴적암	sedimentary rock	堆積岩
퇴적작용	sedimentation	堆積作用
퇴적침전물	sedimentary deposit	堆積沈澱物
퇴화	degeneration	退化
투과	transmission	透過
투과거리	transmission range	透過距離
투과계수	coefficient of transmission, transmission coefficient	透過係數
투과곡선	transmission curve	透過曲線
투과도	transmittance	透過度
투과도계	penetrometer	透過度計
투과성	hydraulic conductivity, transmissibility, permeability	透過性
투과성계수	permeability coefficient	透過性係數
투과성막	permeable membrane	透過性膜

투과성 튜브	permeation tube	透過性-
투과손실	transmission loss	透過損失
투과율	hydraulic conductivity, permeability, transmissivity, transmittancy	透過率
투과율계	permeameter, transmissometer, transmittance meter	透過率計
투과측정술	transmissiometry	透過測定術
투과함수	transmission function	透過函數
투구구름	helm cloud	鬪狗-
투명계수	coefficient of transparency	透明係數
투명도	transparency	透明度
투명도판	Secchi disk, Secchi disc	透明度板
투명체	transparent body	透明體
투명층	clear layer	透明層
투석	dialysis	透析
투석막	dialysis membrane	透析膜
투수대	pervious zone	透水帶
투수율	permeability, hydraulic conductivity	透水率
투수율계수	permeability coefficient	透過率係數
투수층	aquifer	透水層
투열성	diathermancy	透熱性
투영	projection	投影
투하연소시스템	droppable pyrotechnic flare system	投下燃燒-
툰드라	tundra	
툰드라기후	tundra climate	凍土氣候
툰드라사막	tundra desert	凍土沙漠
트라이아스기	Triassic, Triassic period	-期
트라이아스기말사건	end-Triassic event	-期末事件
트라이아스기소산	Triassic extinction	-期消散
트랙교차방향	cross-track direction	-交叉方向
트랙교차해상도	cross-track resolution	-交叉解像度
트로포그램	tropogram	
특별관측	special observation	特別觀測
특별기상보고	special meteorological report	特別氣象報告
특별기상통보	special weather report	特別氣象通報
특별기후변화기금	special climate change fund	特別氣候變化基金
특별눈사태경보	special avalanche warning	特別-沙汰警報
특별세계간격	special world interval	特別世界間隔
특별열대요란발표문	special tropical disturbance statement	特別熱帶擾亂發表文
특별일기통보	special weather report	特別日氣通報
특별해상경보	special marine warning	特別海上警報
특별화재기상	special fire weather	特別火災氣象
특보구역	pertinent area of special report	特報區域
특보문	message of special report	特報文

특보사후분석	post-analysis of special report	特報事後分析
특성	characteristics	特徵
특성가온도	characteristic virtual temperature	特性假溫度
특성값	characteristic value	特性-
특성값문제	characteristic value problem	特性-問題
특성곡선	characteristic curve, characteristics	特性曲線
특성규모	characteristic scale	特性規模
특성길이	characteristic length	特性-
특성방정식	characteristic equation	特性方程式
특성법	method of characteristics	特性法
특성비습	characteristic specific humidity	特性比濕
특성선	characteristic line	特性線
특성속도	characteristic velocity	特性速度
특성수	characteristic number	特性數
특성온도	characteristic temperature	特性溫度
특성점	characteristic point	特性點
특성지상에너지학	characteristic surface energetics	特性地上-學
특성진동수	characteristic frequency	特性振動數
특성함수	characteristic function	特性函數
특성형	characteristic form	特性型
특성환산주파수	characteristic reduced frequency	特性換算周波數
특수관측소	special station	特殊觀測所
특수목적기후관측소	climatological station for specific purposes	特殊目的氣候觀測所
특수목적농업기상관측소	agricultural meteorological station for specific purpose, agrometeorological station for specific purposes	特殊目的農業氣象觀測所
특수예보	specific forecast	特殊豫報
특수한	specific	特殊-
특이값분해	singular value decomposition	特異値分解
특이구름	significant cloud	特異-
특이기상	significant weather	特異氣象
특이기상도	significant weather chart	特異氣象圖
특이기상예보	significant weather forecast	特異氣象豫報
특이기상예상도	significant weather prognostic chart	特異氣象豫想圖
특이기상전망	significant weather outlook	特異氣象展望
특이대응점	singular corresponding point	特異對應點
특이벡터	singular vector	特異-
특이상태	calendaricity	特異狀態
특이선	singular line	特異線
특이성	singularity	特異性
특이이류	singular advection	特異移流
특이점	significant point, singular point	特異點
특정관리해역	sea area specially cared	特定管理海域
특정에너지	certain energy	特定-

E

특징	characteristics	特徵
틀	frame	
틈새	fissure	
틈새(구름)	perlucidus	
틈새고적운	altocumulus perlucidus	-高積雲
틈새모형	gap model	-模型
틈새채우기	gap filling	
틈새층적운	stratocumulus perlucidus	-層積雲
틈새형	interstitial	-型
틈흐림	breaks in overcast	
튕고도	broken level	-高度
튕흐림	broken, broken cloud	
튕흐림하늘	broken sky	
티벳고기압	Tibetan high (pressure)	-高氣壓
티센(다각형)법	Thiessen (polygon) method	-(多角形)法
틴달로미터	Tyndallometer	

E

ㅍ

한글	영문	한자
파	wave, wavelet	波
파감쇠	decay of wave	波減衰
파고	wave height	波高
파고계	wave gauge, wave height meter	波高計
파고자	state of sea scale	波高-
파고측정기둥	wave pole, wave staff	波高測定棒
파괴율	destruction rate	破壞率
파기록계	wave recorder	波記錄計
파꼭대기	wave crest	波-
파나마	Panama	
파나마분지	Panama Basin	-盆地
파나마운하	Panama Canal	-運河
파나스오에타라	panas oetara	
파노라마비틀림	panoramic distortion	
파다발	wave packet	波-
파도고적운	altocumulus undulatus	波濤高積雲
파도고층운	altostratus undulatus	波濤高層雲
파도구름	undulatus, wave cloud, wind cloud	波濤-
파도권적운	cirrocumulus undulatus	波濤卷積雲
파도권층운	cirrostratus undulatus	波濤卷層雲
파도적운	cumulus undulatus	波濤積雲
파도층운	stratus undulatus	波濤層雲
파도층적운	stratocumulus undulatus	波濤層積雲
파동	fluctuation, undulation, wave, wave motion	波動
파동가파르기	wave steepness	波動-
파동경사	wave steepness	波動傾斜
파동골	wave trough	波谷
파동광학	wave optics	波動光學
파동나이	wave age	波動年齡

파동대	belt of fluctuation	波動帶
파동모형	wave model	波動模型
파동방정식	equation of wave motion, wave equation	波動方程式
파동양식	wave regime	波動樣式
파동역학	wave mechanics	波動力學
파동열	wave train	波動列
파동요란	wave disturbance	波動搖亂
파동이론	wave theory	波動理論
파동저기압	wave cyclone, wave depression	波動低氣壓
파동전선	wave front	波動前線
파동전선법	wave front method	波動前線法
파동전파	wave propagation	波動傳播
파동주파수	wave frequency	波動周波數
파동체계	wave system	波動体系
파동함수	wave function	波動函數
파동해	wave solution	波動解
파라카스 하층제트	Paracas low-level Jet	-下層-
파라핀	paraffins	
파란트로푸스	Paranthropus	
파랑에코	sea clutter	波浪-
파랑예보	wave forecasting	波浪豫報
파력	wave power	波力
파력계	wave dynamometer	波力計
파머가뭄심도지수	Palmer Drought Severity Index(PDSI)	-深度指數
파머수문가뭄지수	Palmer Hydrological Drought Index(PHDI)	-水文-指數
파민첩성	wave celerity	波敏捷性
파바닥	wave base	波-
파봉	wave crest	波峰
파브리-페로 간섭계	Fabry-Pérot interferometer	-干涉計
파블로프스키 근사	Pavlovsky's approximation	-近似
파생구름	genitus	派生-
파속	wave packet	波束
파속도	wave celerity, wave velocity	波速度
파속력	wave speed	波速力
파쇄구름	pannus	破碎-
파쇄구역	zone of fracture	破碎區域
파수	wave number	波數
파스칼	Pascal	
파스칼법칙	Pascal's law	-法則
파스퀼-기퍼드 범주	Pasquill-Gifford category	-範疇
파스퀼-기퍼드 안정도	Pasquill-Gifford stability	-安定度
파스퀼-기퍼드 안정도등급	Pasquill-Gifford stability class	-安定度等級
파스퀼범주	Pasquill category	-範疇
파스퀼베타	Pasquill's beta	

파스퀼 안정도	Pasquill's stability	-安定度
파스퀼 안정도등급	Pasquill stability class	-安定度等級
파스퀼 안정도범주	Pasquill stability category	-安定度範疇
파스타그램	pastagram	
파스펙트럼	wave spectrum	波分光
파슬법	parcel method	-法
파시발정리	Parseval's theorem	-定理
파에사	paesa	
파에사노	paesano	
파열	blowout, burst, buster, splintering	破裂
파열고도	bursting height	破裂高度
파열전선	split front	破裂前線
파열점	bursting point	破裂點
파열현상	burst phenomena	破裂現象
파엽	wavelet	波葉
파엽분석	wavelet analysis	小波分析
파원	wave source	波源
파의 소산	dissipation of wave	波消散
파이발	pibal	
파이어니어단열대	pioneer fracture zone	-斷裂帶
파이정리	pi-theorem	-定理
파이토미터	phytometer	
파장	wavelength	波長
파종법	seeding method	播種法
파주기	wave cycle, wave period	波周期
파진폭	wave amplitude	波振幅
파타고니아	Patagonia	
파타 모르가나(신기루)	fata morgana	(蜃氣樓)
파파가요	papagayo	
파편	debris	破片
파편구름	shred cloud	破片-
파푸아뉴기니	Papua New Guinea	
파프미분형식	Pfaff's differential form, Pfaffian differential form	-微分型式
파향	wave direction	波向
파형	waveform	波形
판게아(초대륙)	Pangaea	
판결정	platelet, plate crystal	板結晶
판구름	sheet cloud	板-
판구조론	plate tectonics	板構造論
판구조이론	plate tectonics theory	板構造理論
판모양결정	plate crystal	板模樣結晶
판모양결정서리	plate crystal frost	板模樣結晶-
판번개	sheet lightning	板-
판별기준	criterion	判別基準

Ⅱ

판별분석	discriminant analysis	判別分析
판서리	sheet frost	板-
판얼음	sheet ice	板-
판정지침서	criteria document	判定指針書
판흐름	sheet flow	板-
(신생대 제3기)팔레오세	Paleocene	
팔로서	palouser	
팔루크다이어그램	Paluch diagram	
팜파스	pampas	
팜페로	pampero	
팝콘적란운군	popcorn cumulonimbi	-積亂雲群
패돔	fathom	
패러데이기호법	Faraday notation	-記號法
패러데이회전	Faraday rotation	-回轉
패러드	farad	(靜電容量單位)
패리호	Parry arcs	-弧
패터슨법칙	Patterson's rule	-法則
패턴	pattern	
패턴상관	pattern correlation	-相關
패턴인식	pattern recognition	-認識
팩스	fax	
팩스도	fax chart, fax map	-圖
팩시밀리	facsimile	
팩시밀리기록계	facsimile recorder	-記錄計
팩시밀리도	facsimile chart	-圖
팩시밀리복사장비	facsimile copier equipment	-複寫裝備
팩시밀리장비	facsimile equipment	-裝備
팩시밀리전송	facsimile transmission	-電送
팬풍속계	fan anemometer	-風速計
팬핸들 훅(폭풍)	Panhandle Hook	
팽창	dilatation, expansion	膨脹
팽창계수	coefficient of expansion, expansion coefficient	膨脹係數
팽창냉각	expansion cooling, expansional cooling	膨脹冷却
팽창률	expansibility	膨脹率
팽창일	expansion work	膨脹-
팽창축	axis of dilatation	膨脹軸
팽창파	expansion wave	膨脹波
퍼가	poorga, purga	
퍼밀	per mille	
퍼스일조계	Pers sunshine recorder	-日照計
퍼옥시라디칼	peroxy radicals	
퍼옥시알킬라디칼	peroxyalkyl radical	
퍼옥실라디칼	peroxyl radical	
퍼지논리	fuzzy logic	-論理

퍼지해석	defuzzification	-解釋
퍼지화	fuzzification	-化
퍼진 F층	spread F	-層
퍼텐셜에너지	potential energy	
퍼펙트스톰	perfect storm	
펀치기	punching machine	
펀치카드	punched card	
펄스	pulse	
펄스간 시간간격	interpulse period	-時間間隔
펄스간코딩	pulse-to-pulse coding	-間-
펄스길이	pulse length	
펄스길이변조	pulse length modulation	-變調
펄스너비	pulse width	
펄스너비변조	pulse width modulation	-變調
펄스도플러레이더	pulse Doppler radar	
펄스레이더	pulsed radar	
펄스반복빈도	pulse recurrence frequency	-反復頻度
펄스반복시간	pulse repetition time	-反復時間
펄스반복주파수	pulse repetition frequency	-反復周波數
펄스변조	pulse modulation	-變調
펄스변조기	pulse modulator	-變造器
펄스부피	pulse volume	
펄스부호변조	pulse coded modulation	-符號變調
펄스분해체적	pulse resolution volume	-分解體積
펄스빛	pulsed light	
펄스수변조	pulse number modulation	-數變調
펄스시간변조	pulse time modulation	-時間變調
펄스시간변조라디오존데	pulse-time-modulated radiosonde	-時間變調-
펄스쌍처리	pulse-pair processing	-雙處理
펄스압축	pulse compression	-壓縮
펄스위상변조	pulse phase modulation	-位相變調
펄스위치변조	pulse position modulation	-位置變調
펄스일사계	pulse actinometer	-日射計
펄스적분기	pulse integrator	-積分器
펄스주파수	pulse frequency	-周波數
펄스주파수변조	pulse frequency modulation	-周波數變調
펄스지속변조	pulse duration modulation	-持續變調
펄스지속시간	pulse duration	-持續時間
펄스진폭변조	pulse amplitude modulation	-振幅變調
펄스진폭부호변조	pulse-amplitude-code modulation	-振幅符號變調
펄스집적기	pulse integrator	-集積器
페렐법칙	Ferrel's law	-法則
페렐세포	Ferrel cell	-細胞
페렐순환	Ferrel circulation	-循環

Ⅱ

페루비안이슬	Peruvian dew	
페루비안페인트	Peruvian paint	
페루해류	Peru current	-海流
페르마원리	Fermat's principle	-原理
페르시아만	Persian Gulf	-灣
페르시아만수	Persian Gulf Water	-灣水
(고생대)페름기	Permian Period	-期
페름기말멸종	end-Permian extinction	-期末滅種
페름기-트라이아스기멸종	Permian-Triassic extinction	-期-期滅種
페리	perry	
페이딩	fading	
페인터	painter	
페클렛수	Péclet number	-數
페터슨파속방정식노모그램	Petterssen wave speed equation nomogram	-波速方程式-
페터슨파속방정식법	Petterssen wave speed equation method	-波速方程式法
페테르센규칙	Petterssen's rule	-規則
페테르센발달방정식	Petterssen's development equation	-發達方程式
페티 생 베르나르	petit St. Bernard	
펜만공식	Penman formula	-公式
펜만-몬티스 방정식	Penman-Monteith equation	-方程式
(고생대)펜실베이니아기	Pennsylvanian Period	-期
펑센	Fengshen (typhoon)	
편각	angle of declination, declination, deviation	偏角
편광	polarization	偏光
편광계	polarimeter	偏光計
편광기	polariscope	偏光器
편광도	degree of polarization	偏光度
편광등기울기선	polarization isocline	偏光等傾斜線
편광방사계	polar-radiometer	偏光放射計
편광변조요소	polarization modulation factor	偏光變造要素
편광자	polarizer	偏光子
편구형우적	oblates raindrop	偏球形雨滴
편극	polarization	偏極
편극소거비	depolarization ratio	偏極消去比
편극효과	polarization effect	偏極效果
편도함수	partial derivative	偏導函數
편동무역풍	easterly trade wind	偏東貿易風
편동편차	easterly dip	偏東偏差
편동풍	easterlies, easterly wind	偏東風
편동풍골	easterly trough	偏東風-
편동풍대	easterly belt	偏東風帶
편동풍제트	easterly jet	偏東風-
편동풍파	easterly wave	偏東風波
편류	drift current	偏流

편류계	drift meter	偏流計
편류계수	drift coefficient	偏流係數
편류보정	drift correction	偏流補正
편류조정각	drift correction angle	偏流調整角
편리	schistosity	片離
편미분(법)	partial differentiation	偏微分(法)
편미분계수	partial differential coefficient	偏微分係數
편빙	lolly ice	片氷
편상관	partial correlation	偏相關
편서풍	westerlies	偏西風
편서풍(기압)골	westerly trough	偏西風(氣壓)-
편서풍다우대	westerly rain belt	偏西風多雨帶
(편)서풍돌진	westerly wind burst	(偏)西風突進
편서풍대	westerly belt, westerly zone	偏西風帶
편서풍파	westerly wave	偏西風波
편심PPI스코프	off-center PPI scope	偏心-
편의	bias	偏倚
편의분산	biased variance	偏倚分散
편장(의)	prolate	扁長
편전각	angle of deflection	偏轉角
편중도	bias	偏重度
편차	anomaly, departure, deviation	偏差
편차도	deviation chart	偏差圖
편파다양성	polarization diversity	偏波多樣性
편파레이더	polarization radar	偏波-
편파복사	polarized radiation	偏波輻射
편파전환성	polarization agility	偏波轉換性
편파특징	polarization signature	偏波特徵
편평구름	humilis	扁平-
편평도	flattening	扁平度
편평타원체	oblate spheroid	偏平楕圓體
편향각	angle of deviation	偏向角
편향력	deviating force	偏向力
편향변조지시기	deflection-modulated indicator	偏向變調指示器
편향풍속계	deflection anemometer	偏向風速計
평가	assessment	評價
평균	average, mean, norm	平均
평균가속도	average acceleration	平均加速度
평균값	average, average value, mean value	平均-
평균값정리	mean value theorem	平均値定理
평균거리	mean distance	平均距離
평균결합길이	mean bonding distance	平均結合-
평균고도장	mean height field	平均高度場
평균기온	mean temperature	平均氣溫

Ⅱ

평균기온감률	mean temperature lapse rate	平均氣溫減率
평균남북순환	mean meridional circulation	平均南北循環
평균단면(도)	mean cross section	平均斷面(度)
평균대류권계면	mean tropopause	平均對流圈界面
평균도	mean chart, mean map	平均圖
평균도플러속도	mean Doppler velocity	平均-速度
평균바람속도	mean wind velocity	平均-速度
평균반응속도	mean reaction rate	平均反應速度
평균반경	mean radius	平均半徑
평균반지름	mean radius	平均半-
평균변동률	average variability	平均變動率
평균변화율	average(mean) rate of change	平均變化率
평균복사온도	mean radiant temperature	平均輻射溫度
평균부피반경	mean-volume radius	平均-半徑
평균부피반지름	mean-volume radius	平均-
평균분자량	mean molecular weight	平均分子量
평균분점	mean equinox	平均分點
평균삭망월	mean synodic lunar month	平均朔望月
평균상층기류	mean upper-air flow	平均上層氣流
평균소산율	average dissipation rate	平均消散率
평균속도	mean velocity	平均速度
평균속력	average speed	平均速力
평균수명	mean life	平均壽命
평균순균형	average net balance	平均純均衡
평균시	mean time	平均時
평균시선속도	mean radial velocity	平均視線速度
평균시차	mean parallax	平均視差
평균연직온도	mean vertical temperature	平均鉛直溫度
평균오차	average error, mean error	平均誤差
평균온도	mean temperature, temperature average	平均溫度
평균온도연교차	mean annual range of temperature	平均溫度年較差
평균우량	average rainfall	平均雨量
평균우량깊이	rainfall depth	平均雨量-
평균우량깊이-유역면적관계	rainfall depth-area relationship	平均雨量-流域面積關係
평균운동상수	mean motion constant	平均運動常數
평균운동에너지	average kinetic energy, mean kinetic energy	平均運動-
평균유효지름	mean effective diameter	平均有效-
평균이상	mean anomaly	平均異常
평균일별변동	mean day-to-day variation	平均日別變動
평균일별변동률	mean interdiurnal variability	平均日別變動率
평균일별변동성	mean interdiurnal variability	平均日別變動性
평균자오면순환	mean meridional circulation	平均子午面循環
평균자유행로	mean free path	平均自由行路
평균장	mean field	平均場

평균적도일	mean equatorial day	平均赤道日
평균전력	average power	平均電力
평균절대오차	mean absolute error	平均絶對誤差
평균정오	mean noon	平均正午
평균제곱오차	mean-square error	平均-誤差
평균조위(면)	half-tide level	平均潮位(面)
평균지각수심	mean sphere depth	平均地殼水深
평균차	mean difference	平均差
평균최대풍속	maximum sustained wind	平均最大風速
평균커널	averaging kernel	平均-
평균태양	mean sun	平均太陽
평균태양년	mean solar year	平均太陽年
평균태양시	mean solar hour	平均太陽時
평균태양시간	mean solar time	平均太陽時間
평균태양일	mean solar day	平均太陽日
평균태음일	mean lunar day	平均太陰日
평균틈새속도	average interstitial velocity	平均-速度
평균편차	average departure, average deviation, mean anomaly, mean deviation	平均偏差
평균표피온도	mean skin temperature	平均表皮溫度
평균풍속	average wind speed	平均風速
평균항력계수	average drag coefficient	平均抗力係數
평균항성일	mean sidereal day	平均恒星日
평균해면	mean sea level	平均海面
평균해면(기압)경정	reduction to mean sea level	平均海面(氣壓)更正
평균해면기압	mean sea level pressure	平均海面氣壓
평균해면온도경정	reduction of temperature to mean sea level	平均海面溫度更正
평균해면패턴	mean sea level pattern	平均海面-
평균해발고도이상	above mean sea level	平均海拔高度以上
평균해수면기압	average sea level pressure	平均海水面氣壓
(영국)평균해수면상설서비스	Permanent Service for Mean Sea Level	平均海水面常設-
평균핵	averaging kernel	平均核
평균화	averaging	平均化
평균화시간	averaging time	平均化時間
평균환경바람	mean environmental wind	平均環境-
평년	normal year, ordinary year	平年
평년값	climatic normal, normal, normal value	平年-
평년강우	normal rainfall	平年降雨
평년공간평균도	normal space mean chart	平年空間平均圖
평년기후	normal climate	平年氣候
평년도	normal chart, normal map	平年圖
평년작	normal crop	平年作
평년편차도	anomaly chart	平年偏差圖
평면곡선	plane curve	平面曲線

평면구조	planar structure	平面構造
평면기하(학)	plane geometry	平面幾何(學)
평면나무가지모양결정	plane-dendritic crystal	平面-模樣結晶
평면대기파	plane atmospheric wave	平面大氣波
평면도형	plane figure	平面圖形
평면벡터	plane vector	平面-
평면시어지시기	plane shear indicator	平面-指示器
평면위치지시기	plane-position indicator	平面位置指示器
평면파	plane wave	平面波
평면편광기	plane polarization indicator	平面偏光器
평범해	trivial solution	平凡解
평수구역	smooth sea water region	平水區域
평원	plain	平原
평탄(원뿔)복사계	flat-plate (cone) radiometer	平坦(-)輻射計
평탄저기압	flat low	平坦低氣壓
평탄형반사기	flat-plate reflector	平坦型反射器
평탄형최고	flat maximum	平坦形最高
평탄형태양집광기	flat-plate solar collector	平坦型太陽集光器
평판	plate	平板
평판결정	tabular crystal	平板結晶
평판경계	plate boundaries	平版境界
평판빙산	barrier berg, barrier iceberg, tabular iceberg	平板氷山
평판풍속계	plate anemometer	平板風速計
평행	parallelism	平行
평행사변형	parallelogram	平行四邊形
평행선	paralle lines	平行線
(평행)이동	translation	(平行)移動
평형	equilibrium, neutralization	平衡
평형고도	equilibrium level	平衡高度
평형과포화	equilibrium supersaturation	平衡過飽和
평형근사모형	balanced approximation model	平衡近似-
평형기후	equilibrium climate	平衡氣候
평형농도	equilibrium concentration	平衡濃度
평형대류	equilibrate convection	平衡對流
평형력	equilibrant	平衡力
평형류	equilibrium flow	平衡流
평형바람	equilibrium wind	平衡-
평형상전환	equilibrium phase transformation	平衡狀轉換
평형상태	equilibrium state, state of equilibrium	平衡狀態
평형선	equilibrium lines	平衡線
평형속도	equilibrium velocity	平衡速度
평형수위강하	equilibrium drawdown	平衡水位降下
평형시각	equilibrium time	平衡時刻
평형온도	equilibrium temperature	平衡溫度

Ⅱ

평형온도면	equilibrium temperature level	平衡溫度面
평형이동	equilibrium shift	平衡移動
평형점	equilibrium point	平衡點
평형조석	equilibrium tide	平衡潮汐
평형증기	equilibrium vapour	平衡蒸氣
평형증기압	equilibrium vapour pressure	平衡蒸氣壓
평형증발량	equilibrium evaporation	平衡蒸發量
평형지표유출량	equilibrium surface discharge	平衡地表流出量
평형타원체	equilibrium spheriod	平衡楕圓體
평형포물체	equilibrium paraboloid	平衡抛物體
평형풍	equilibrium wind	平衡風
평형해	equilibrium solution	平衡解
평형혼합층	equilibrium mixed layer	平衡混合層
평형흐름	equilibrium flow	平衡-
평활	smoothing	平滑
평활과정	smoothing process	平滑過程
평활태양흑점수	smoothed sunspot number	平滑太陽黑點數
폐색	occlusion	閉塞
폐색과정	occlusion process	閉塞過程
폐색소멸	occlusion decay	閉塞消滅
폐색저기압	occluded cyclone, occluded depression	閉塞低氣壓
폐색전선	occluded front, occlusion front	閉塞前線
폐색전선계	occluded frontal system	閉塞前線系
폐색점	point of occlusion	閉塞點
폐색중규모저기압	occluded mesocyclone	閉塞中規模低氣壓
포근한	mild	
포기적운	cumuliformis floccus	-積雲
포도당	glucose	葡萄糖
포라칸	foracan	
포락선	envelope	包絡線
포락선지형	envelope orography	包絡線地形
포렐척도	Forel scale	-尺度
포르틴기압계	Fortin barometer	-氣壓計
포르틴 수은기압계	Fortin mercury barometer	-水銀氣壓計
포마(POMAR) 전보식	POMAR code	-電報式
포물면	paraboloid	抛物面
포물면반사기	parabolic reflector	抛物面反射器
포물면안테나	parabolic antenna	抛物面-
포물선	parabola	抛物線
포물선궤도	parabolic orbit	抛物線軌道
포물선운동	motion of projectile	抛物線運動
포사체	projectile	抛射體
포이즈	poise	
포인트	point	

포일채취기	foil impactor	-採取機
포장수분부족량	field moisture deficiency	圃場水分不足量
포장최대용수량	field maximum moisture capacity	圃場最大容水量
포차	saturation deficit (or deficiency), vapo(u)r pressure deficit	飽差
포착	capture	捕捉
포착과정	capture process	捕捉過程
포착단면(적)	capture cross-section	捕捉斷面(積)
포착률	capture rate	捕捉率
포커-플랑크 방정식	Fokker-Planck equation	-方程式
포클랜드해류	Falkland Current	-海流
포토멀티플라이어관	photomultiplier tube	-管
포함오차	commission error	包含誤差
포화	saturation	飽和
포화가온도	saturated virtual temperature	飽和假溫度
포화경계	boundary of saturation	飽和境界
포화곡선	saturation curve	飽和曲線
포화공기	saturated air	飽和空氣
포화구역	zone of saturation	飽和區域
포화단열	saturation adiabatic	飽和斷熱
포화단열감률	saturated-adiabatic lapse (rate), saturation adiabatic lapse (rate)	飽和斷熱減率
포화단열과정	saturation adiabatic process	飽和斷熱過程
포화단열냉각	saturation adiabatic cooling (rate)	飽和斷熱冷却
포화단열변화	saturation adiabatic change	飽和斷熱變化
포화단열선	saturated adiabat, saturation adiabat, saturation adiabatics	飽和斷熱線
포화대	phreatic zone, zone of saturation	飽和帶
포화도	degree of saturation	飽和度
포화비습	saturation specific humidity	飽和比濕
포화비율	fraction of saturation	飽和比率
포화수리전도(도)	saturated hydraulic conductivity	飽和水理傳導(度)
포화수증기	saturated vapo(u)r	飽和水蒸氣
포화수증기압	saturation vapo(u)r pressure, saturation water vapo(u)r pressure	飽和水蒸氣壓
포화신호	saturation signal	飽和信號
포화용수량	saturated water capacity	飽和容水量
포화용액	saturated solution	飽和溶液
포화점	point of saturation, saturation point	飽和點
포화토양	saturated soil	飽和土壤
포화토양용수량	field capacity	飽和土壤容水量
포화혼합비	saturation mixing ratio	飽和混合比
포화혼합비선	saturation mixing ratio line	飽和混合比線
포획사육	captive breeding	捕獲飼育

폭발구름	cloud from explosions, explosion cloud	爆發-
폭발성분화	Peléean eruption	暴發性噴火
폭발승온	explosive warming, impulsive warming	爆發昇溫
폭발저기압	explosive cyclone	爆發低氣壓
폭발저기압발생	explosive cyclogenesis	爆發低氣壓發生
폭발적 발달	explosive deepening	爆發的 發達
폭발파	blast wave, explosion wave	爆發波
폭발풍	blast	爆發風
폭발하한계	Lower Explosion Limit	爆發下限界
폭발효과	blast effect	爆發效果
폭설	heavy snowfall	暴雪
폭염	excessive heat	暴炎
폭염경보	excessive heat warning, heat wave warning	暴炎警報
폭염전망	excessive heat outlook	暴炎展望
폭염주의보	excessive heat watch, heat wave advisory	暴炎注意報
폭우	cloudburst, downpour, gush, hard rain, heavy rainfall, pouring rain, rain gush, rain gust	暴雨
폭우 선탐색	breadth-first search	幅優先探索
폭탄저기압	bomb cyclone	爆彈低氣壓
폭포	waterfall	瀑布
폭포구름	cloud from waterfall	瀑布-
폭포운	clouds from water falls	瀑布雲
폭포효과	waterfall effect	瀑布效果
폭풍	storm, wind storm	暴風
폭풍(우)경로	storm path, storm track	暴風(雨)經路
폭풍(우)경보	storm warning	暴風(雨)警報
폭풍경보서비스	storm alert service	暴風警報-
폭풍경보표지	visual storm warning	暴風警報標識
폭풍(우)기간	storm period	暴風(雨)期間
폭풍눈	eye of storm	暴風-
폭풍(우)대피소	storm cellar	暴風(雨)待避所
폭풍(우)돌연시작	storm sudden commencement	暴風(雨)突然始作
폭풍(우)모형	storm model	暴風(雨)模型
폭풍(우)발생대	storm feeder band	暴風(雨)發生帶
폭풍법칙	law of storms	-法則
폭풍(우)세포	storm cell	暴風(雨)細胞
폭풍신호	cone	暴風信號
폭풍(우)역	storm area	暴風(雨)域
폭풍우	rain storm, storm	暴風雨
폭풍(우)이동	storm transposition	暴風(雨)移動
폭풍(우)파	storm wave	暴風(雨)波
폭풍(우)표지	storm cone	暴風(雨)標識
폭풍(우)해일	storm surge	暴風(雨)海溢
폭풍우완화	rainstorm easing	暴風雨緩和

Ⅱ

폭풍일수	number of days with storm	暴風日數
폭풍전환경	prestorm environment	暴風前環境
폭풍조석	storm tide	暴風潮汐
폭풍조석경보	storm tide warning	暴風潮汐警報
폭풍표시신호	visual storm signal	暴風表示信號
폭풍해일경보	storm surge warning	暴風海溢警報
폭풍해일주의보	storm surge advisory	暴風海溢注意報
폰카르만법칙	von Kármán's law	-法則
폰카르만상수	von Kármán's constant	-常數
폴리니아	polyn'ya	
폴리에틸렌기구	polyethylene balloon	-氣球
폴리염화비페닐	polychlorinated biphenyls	
푄	foehn, föhn, favogn	
푄골	foehn trough	
푄공기	foehn air	-空氣
푄구름	foehn cloud	
푄구름틈새	foehn break, foehn gap	
푄기간	foehn period	-期間
푄단계	foehn phase	-段階
푄둑	foehn bank	
푄멈춤	sturmpause, foehn pause	
푄바람	foehn wind	
푄벽	foehn wall	-壁
푄섬	foehn island	
푄쉼	foehn pause	
푄저기압	foehn cyclone	-低氣壓
푄정지	foehn pause	-靜止
푄코	foehn nose	
푄파	foehn wave	-波
푄폭풍우	foehn storm	-暴風雨
푄효과	foehn effect	-效果
표고	elevation	標高
표류	drift	漂流
표류기	drift epoch	漂流期
표류병	bottle post, drift bottle	漂流甁
표류부이	drifting buoy	漂流-
표류부표	drifting buoy	漂流浮漂
표류빙산	floeberg	漂流氷山
표류전류	drift current	漂流電流
표면	surface	表面
표면마찰	skin friction	表面摩擦
표면마찰계수	skin friction coefficient	表面摩擦係數
표면반사율	surface albedo	表面反射率
표면분광반사율	surface spectral reflectance	表面分光反射率

표면설붕	surface avalanche	表面雪崩
표면열저항	surface thermal resistance	表面熱抵抗
표면온도	skin temperature	表面溫度
표면온도검정	surface temperature test	表面溫度檢定
표면온도계	armoured thermometer, surface thermometer	表面溫度計
표면유출	surface runoff	表面流出
표면응결	surface condensation	表面凝結
표면장력	surface tension	表面張力
표면적	surface area	表面積
표면전하	surface charge	表面電荷
표면조성화상복사계	surface composition mapping radiometer	表面造成畵像輻射計
표면(경계)층	surface boundary layer	表面(境界)層
표면파	surface wave	表面波
표면후방산란변수	surface backscatter variable	表面後方散亂變數
표본	sample	標本
표본공간	sample space	標本空間
표본부피	sample volume	標本(體積)
표본분산	sample variance	標本分散
표본분포	sampling distribution	標本分布
표본비율	sample rate	標本比率
표본수집기	sampler	標本收集機
표본수집소	sampling station	標本收集所
표본수집주기	sampling frequency	標本收集週期
표본시간	sample time	標本時間
표본오차	sampling error	標本誤差
표본조사	sample survey	標本調査
표본채취	sampling	標本採取
표본채취기	sampler	標本採取機
표본채취소	sampling station	標本採取所
표본채취주기	sampling frequency	標本採取週期
표본평균	sample mean	標本平均
표본표준편차	sample standard deviation	標本標準偏差
표적부피	target volume	標的體積
표적신호	target signal	標的信號
표준	criterion	標準
표준가시거리	standard visual range	標準可視距離
표준계기	standard instrument	標準計器
표준계획홍수량	standard project flood	標準計劃洪水量
표준관측시각	standard time of observation	標準觀測時刻
표준굴절	standard refraction	標準屈折
표준기상관측소	standard weather observatory	標準氣象觀測所
표준기압	standard pressure	標準氣壓
표준기압계	standard barometer	標準氣壓計
표준기압고도	standard pressure altitude	標準氣壓高度

ㅍ

표준기압면	standard pressure level	標準氣壓面
표준깊이	standard depth	標準-
표준단위	standard unit	標準單位
표준대기	standard atmosphere	標準大氣
표준대기고도	standard atmosphere altitude	標準大氣高度
표준대기기온감률	standard atmosphere lapse rate, standard atmospheric lapse rate	標準大氣氣溫減率
표준대기압	standard atmosphere pressure	標準大氣壓
표준대기표	table of standard atmosphere	標準大氣表
표준등압면	standard isobaric surface	標準等壓面
표준면기압경정	reduction of pressure to a standard level	標準面氣壓更正
표준모집단	normal population	標準母集團
표준밀도	standard density	標準密度
표준밀도고도	standard density altitude	標準密度高度
표준방법	standard method	標準方法
표준방정식	normal equation	標準方程式
표준브리핑	standard briefing	標準-
표준상태	normal state, standard state	標準狀態
표준시	standard time	標準時
표준시정	standard visibility	標準視程
표준오차	standard error	標準誤差
표준온도	standard temperature	標準溫度
표준온도계	standard thermometer	標準溫度計
표준온도압력	normal temperature and pressure	標準溫度壓力
표준우량계	standard raingauge	標準雨量計
표준유리구슬	standard glass beads	標準琉璃-
표준자오선	standard meridian	標準子午線
표준작업온도	standard operative temperature	標準作業溫度
표준전파	standard propagation	標準傳播
표준정규분포	standard normal distribution	標準正規分布
표준중력	standard gravity	標準重力
표준탄도구역	standard artillery zone	標準彈道區域
표준탄도대기	standard artillery atmosphere	標準彈道大氣
표준편차	standard deviation	標準偏差
표준표적	standard target	標準標的
표준해면온도	standard sea level temperature	標準海面溫度
표준해수	normal water, standard sea water	標準海水
표준형	standard(normal) form	標準形
표준화	standardization	標準化
표차	tabular difference	表差
표출	display	表出
표출분해능	display resolution	表出分解能
표피마찰계수	coefficient of skin friction	表皮摩擦係數
표해수대	epipelagic zone	表海水帶

Ⅱ

푸대집	baghouse	
푸른 달	blue moon	
푸른 띠	blue band	
푸른 섬광	blue flash	-閃光
푸른 얼음	blue ice	
푸른 얼음 지역	blue ice area	-地域
푸른 제트	blue jets	
푸른 태양	blue sun	-太陽
푸른 하늘	blue sky	
푸른 하늘 척도	blue sky scale	-尺度
푸름도측정법	cyanometry	-測定法
푸리에계수	Fourier coefficient	-係數
푸리에급수	Fourier series	-級數
푸리에법칙	Fourier's law	-法則
푸리에변환	Fourier transform	-變換
푸리에분석	Fourier analysis	-分析
푸리에수	Fourier number	-數
푸리에스펙트럼	Fourier spectrum	
푸리에열전도법칙	Fourier's law of heat conduction	-熱傳導法則
푸리에적분	Fourier integral	-積分
푸리에코사인급수	Fourier cosine series	-級數
푸시부름 스캐너	push-broom scanner	
푸아송방정식	Poisson's equation	-方程式
푸아송분포	Poisson distribution	-分布
푸아송상수	Poisson constant	-常數
푸아죄유흐름	Poiseuille flow	
푸아죄-하겐 법칙	Poiseuille-Hagen law	-法則
풀구라이트	fulgurite	
품질	quality	品質
품질계수	quality indicator	品質係數
품질관리	quality control	品質管理
품질보증	quality assurance	品質保證
품질플래그	quality flags	品質-
풋램버트	footlambert	
풋촉광	foot-candle	-燭光
풍건토	air dried soil	風乾土
풍계	system of winds, wind divide, wind system	風系
풍동	rotor cloud, wind tunnel	風洞
풍동풍속	wind tunnel velocity	風洞風速
풍랑	wind sea, wind wave	風浪
풍랑경보	high wind wave warning	風浪警報
풍랑주의보	high wind wave advisory, wind wave advisory	風浪注意報
풍력	wind force, wind power	風力
풍력감쇠	abatement of wind	風力減衰

ㅍ

풍력계급	scale of wind-force, wind force scale, wind scale	風力階級
풍력발전기	wind generator	風力發電機
풍력발전소	wind power station	風力發電所
풍력자원(지)도	wind resource map	風力資源(地)圖
풍매	anemophily	風媒
풍배도	wind rose	風配圖
풍상	upwind	風上
풍상(의)	windward	風上
풍상경사면	pstream slope	風上傾斜面
풍상쪽	upstream	風上-
풍상효과	upwind effect	風上效果
풍성	eolian	風成
풍성(질)	aeolian	風成(質)
풍성과정	eolian process	風成過程
풍성순환	wind driven circulation	風成循環
풍성우량지수	driving rain index	風成雨量指數
풍성퇴적물	eolian sediment	風成堆積物
풍속	wind speed, wind velocity	風速
풍속계	anemometer	風速計
풍속계고도	anemometer level	風速計高度
풍속계수기	wind speed counter, wind velocity counter	風速計數器
풍속계유효높이	effective height of anemometer	風速計有效高度
풍속방정식	wind velocity equation	風速方程式
풍속연직분포	wind velocity profile	風速鉛直分布
풍속지시기	wind velocity indicator	風速指示器
풍속측정법	anemometry	風速測定法
풍수해	wind and flood damage	風水害
풍수해보험	insurance damage from storm and flood	風水害保險
풍식돌	ventifact	風蝕-
풍압	velocity pressure, wind pressure	風壓
풍압계	pressure anemometer	風壓計
풍압판풍속계	pressure plate anemometer	風壓板風速計
풍역	wind zone	風域
풍작용	wind action	風作用
풍적설	driven snow	風積雪
풍정	run of wind, wind run	風程
풍정풍속계	runoff wind anemometer	風程風速計
풍차	aerogenerator, wind turbine	風車
풍차풍속계	windmill anemometer	風車風速計
풍차형 풍향풍속계	combined recording wind vane and fan-anemograph	風車型風向風速計
풍파	wind wave	風波
풍하(쪽)	down wind	風下
풍하기압골	lee trough	風下氣壓-

풍하맴돌이	lee eddies	風下-
풍하열도	leeward islands	風下列島
풍하저기압	lee depression	風下低氣壓
풍하저기압발달	lee cyclogenesis	風下低氣壓發達
풍하중	wind load	風荷重
풍하측	downstream, down wind, lee side, leeward side, leeward	風下側
풍하측발달	downstream development	風下側發達
풍하파	lee wave	風下波
풍하파시스템	lee wave system	風下波-
풍하파지역	lee wave region	風下波地域
풍하편류	leeway	風下偏流
풍해	wind damage	風害
풍향	wind direction	風向
풍향각	wind angle	風向角
풍향계	anemoscope, vane, weather vane, wind vane	風向計
풍향급변	wind shift	風向急變
풍향급변선	wind shift line	風向急變線
풍향기록계	wind direction recorder	風向記錄計
풍향반전	wind backing	風向反轉
풍향살(기호)	wind-direction shaft	風向-(記號)
풍향순전	wind veering	風向順轉
풍향자루	cone	風向-
풍향지시기	wind direction indicator	風向指示器
풍향풍속계	anemovane, vane anemometer, wind vane and anemometer	風向風速計
(프로펠러식)풍향풍속계	aerovane	風向風速計
풍화	weathering	風化
프라운호퍼 띠	Fraunhofer band	
프라운호퍼 선	Fraunhofer's line	-線
프란틀수	Prandtl number	-數
프랙탈변동성	fractal variability	-變動性
프랙탈불안정도	fractal instability	-不安定度
프랙탈차원	fractal dimension	-次元
프레네-세레 공식	Frenet-Serret formula	-公式
프레넬대	Fresnel zone	-帶
프레넬산란모형	Fresnel scattering model	-散亂模型
프레넬산란체	Fresnel scatter	-散亂體
프레드홀름 적분방정식	Fredholm integral equation	-積分方程式
프레셰 도함수	Fréchet derivative	-導函數
프레온	freon	
프로미넌스	prominence	
프로파일	profile	
프로파일항력	profile drag	形狀抗力

프로펠러풍속계	propeller anemometer	-風速計
프로펠러효율	propeller efficiency	-效率
프로펠러후류	slip stream	-後流
프록시	proxy	
프루드수	Froude number	-數
프리아젬	friagem, vriajem	
프리즘	prism	
프리치-차펠 적운 매개변수화방식	Fritsch-Chappell cumulus parameterization scheme	-積雲媒介變數化方式
플라우바람	plow/plough wind	
플라이스토세	Pleistocene	
플라이스토세 기후	Pleistocene climate	-氣候
플라이스토세 시대	Pleistocene epoch	-時代
플라이스토세말 멸종	end-Pleistocene extinction	-末滅種
플라이스토신 빙하시대	Pleistocene glacial epoch, Quaternary Ice Age, Pleistocene Ice Age	-氷河時代
플라이스토신 빙하작용	Pleistocene glaciation	-氷河作用
플라이오세 기후	Pliocene climate	-氣候
플라즈마	plasma	
플라즈마권	plasmasphere	-圈
플라즈마권계면	plasmapause	-圈界面
플란드리아 간빙기	Flandrian interglacial	-間氷期
플랑드르 폭풍	Flanders storm	-暴風
플랑크법칙	Planck's law	-法則
플랑크복사법칙	Planck's radiation law	-輻射法則
플랑크상수	Planck's constant	-常數
플랑크의 복사체	Planckian radiator	-輻射體
플랑크톤	plankton	
플랑크함수	Planck's function	-函數
플러싱시간	flushing time	-時間
플럭스	flux	束
플럭스결정	flux determination	-決定
플럭스경도관계	flux-gradient relationship	-傾度關係
플럭스경도상사	flux-gradient similarity	-傾度相似
플럭스-리처드슨수	flux-Richardson number	-數
플럭스밀도	flux density	-密度
플럭스밀도문턱값	flux density threshold	-密度門-
플럭스발산	flux divergence	-發散
플럭스분산관계	flux-variance relation	-分散關係
플럭스분산상사	flux-variance similarity	-分散相似
플럭스수렴	flux convergence	-收斂
플럭스온도	flux temperature	-溫度
플럭스조정	flux adjustment	-調整
플럭스측정	flux measurement	-測定

플럭스판	fluxplate	-板
플레어	flare	
플레어 에코	flare echo	
플레어 에코 레이더 징후	flare echo radar signature	-徵候
플레어산화	flare oxidation	-酸化
플로리다해류	Florida Current	-海流
플룸	plume	
플리스트	flist	
피네만측운기	Fineman's nephoscope	-測雲器
피뢰침	lightning arrester, lightning conductor, lightning rod	避雷針
피뢰판	lightning protector plate	避雷板
피복역	area of converage	被覆域
피사시	peesash	
피셔-포터 우량계	Fischer-Porter rain gauge	-雨量計
피스윕폭풍	peesweep storm	-暴風
피압대수층	artesian aquifer, confined aquifer	被壓帶水層
피압지하수	confined groundwater	被壓地下水
피압지하수면	piezometric surface	被壓地下水面
피어슨분포	Pearson distribution	-分布
피외르토프방정식	Fjørtoft equation	-方程式
피외르토프조건	Fjørtoft condition	-條件
피외르토프지향법	Fjørtoft steering method	-指向法
피이티엠	Paleocene-Eocene Thermal Maximum	-(-最高溫期)
피즈	pieze	
피체증발계	Pich evaporimeter	-蒸發計
피츠로이 기압계	FitzRoy barometer	-氣壓計
피치각	pitch angle	-角
피치축	pitch axis	-軸
피토관	Pitot tube	-管
피토관풍속계	Pitot tube anemometer	-管風速計
피토정압관	Pitot-static tube	-靜壓管
피토트론	phytotron	
피피선도	phipigram, piphigram	-線圖
피흡착질	adsorbate	被吸着質
픽방정식	Fickian equation, Fick's equation	-方程式
픽법칙	Fick's law	-法則
픽셀	pixel	
픽확산	Fickian diffusion	-擴散
픽확산방정식	Fickian diffusion equation	-擴散方程式
픽확산법칙	Fick's diffusion law	-擴散法則
핀다이센-베르게론 핵형성과정	Findeisen-Bergeron nucleation process	-核形成過程
핀들레터 제트	Findlater jet	

필드밀	field mill	
필라멘트	filament	
필라멘트전류	filamental flow, filament current	-電流
필라멘트채널	filament channel	
필름	film	
필름크러스트	film crust	
필리핀판	Philippine plate	-板
필리핀해	Philippine sea	-海
필리핀해류	Philippines current	-海流
필립보정	Philip correction	-補正
필립스바람	Philips wind	
필머	pilmer	
필요환기량	necessary ventilation amount	必要換氣量
필터	filter	
필터링	filtering	
필터쐐기형분광계	filter wedge spectrometer	-形分光計
필터포착	filter-capture	-捕捉
(남극대륙)필히너빙붕	Filchner ice shelf	-氷棚
핍	pip	
핑고	pingo	

Ⅱ

ㅎ

한글	영문	한자
하강	descent	下降
하강(의)	descendent	下降
하강궤도	descending pass	下降軌道
하강급류	downrush	下降急流
하강기류	descending air current, descending flow, down current, downdraught, downdraft	下降氣流
하강돌풍	downburst	下降突風
하강류	descending current	下降流
하강류속도	downdraft velocity	下降流速度
하강부	falling limb	下降部
하강점동	creep	下降漸動
하강중심	center of falls	下降中心
하겐-푸아죄유 흐름	Hagen-Poiseuille flow	
하구	estuary	河口
하구역	esturine zone	河口域
하늘	heaven, sky	天
하늘강	atmospheric river	
하늘겉보기모양	apparent form of the sky	-模樣
하늘눈물	serein	
하늘모양	shape of the sky	-模樣
하늘밝기온도	sky brightness temperature	-溫度
하늘복사	sky radiation	-輻射
하늘복사계	sky radiometer	-輻射計
하늘빛	skylight	
하늘상태	sky condition, state of sky	-狀態
하늘지도	sky map	-地圖
하늘차폐	opaque sky cover	-遮蔽
하늘푸름	blue of the sky	
하늘휘도	sky radiance	-輝度

하단천정호	lower circumzenithal arc	下端天頂弧
하단호	lower arc	下端弧
하도저류	channel storage	河道貯溜
하라카나	haracana	
하로우카나	haroucana	
하루살이 내	ephemeral stream	
하루아	harrua	
하루의	diurnal	
하르	haar	
하마탄	harmattan	
하면	aestivation, estivation	夏眠
하부구름	lower cloud	下部-
하부마찰층	lower friction layer	下部磨擦層
하부브	haboob, habbub, haboub, hubbob, hubbub	
하부성층권	lower stratosphere	下部成層圈
하부신기루	lower mirage	下部蜃氣樓
하부축대칭영역	lower symmetric regime	下部軸對稱領域
하부측면접호	infralateral tangent arcs	下部側面接弧
하빙풍	off-ice wind	河氷風
하안	bank	河岸
하안저류	bank storage	河岸貯流
하얀 서리	hoarfrost	
하우드	haud	
하우르비츠-헬름홀츠 파	Haurwitz-Helmholtz wave	-波
하이드로플루오로카본	hydrofluorocarbons	
하이드록실라디칼	hydroxyl radical	
하이브리드검색	hybrid retrievals	-檢索
하이브리드변광성	hybrid variables	-變光星
하이서그래프	hythergraph	
하이퍼론	hyperon	
하인리히사건	Heinrich event	-事件
하인스지수	Haines index	-指數
하일리겐샤인	Heiligenschein, Cellini's halo	
하전	charge	荷電
하전입자	charged particle	荷電粒子
하중평균속도	weighted mean velocity	荷重平均速度
하지	summer solstice	夏至
하천수위	river stage	河川水位
하천수위계	river gauge	河川水位計
하천안개	river fog	河川-
하천예보	river forecast	河川豫報
하천(면)온도	river (surface) temperature	河川(面)溫度
하천유수	streamflow	河川流水
하천유수추적	streamflow routing	河川流水追跡

(하천)유역	river basin	(河川)流域
하천차수	stream order	河川次數
하천학	potamology	河川學
하층대기	lower atmosphere	下層大氣
하층류	undercurrent	下層流
하층예상도	low-level prog chart	下層豫想圖
하층운	low cloud, low-level cloud	下層雲
하층제트	low-level jet	下層-
하층제트류	low-level jet stream	下層-流
하틀리띠	Hartley band	
하틀리-허긴스 띠	Hartley-Huggins band	
하한	greatest lower bound	下限
하향단파복사	downward shortwave radiation	下向短波輻射
하향대기복사	atmospheric counter radiation	下向大氣輻射
하향력	downward force	下向力
하향방사	downward radiation	下向放射
하향(전)복사	downward (total) radiation	下向(全)輻射
하향운동	downward motion	下向運動
하향장파복사	downward longwave radiation	下向長波輻射
하향지구복사	downward terrestrial radiation	下向地球輻射
한가을	depth of autumn	
한겨울	depth of winter	
한계	limit	限界
한계각	limiting angle	限界角
한계유선	limiting streamline	限界流線
한극	cold pole	寒極
한기범람	polar outbreak, cold-air outbreak	寒氣氾濫
한기쇄도	cold surge	寒氣殺到
한기침입	polar invasion	寒氣侵入
한낮	midday	
한냉전선면	kataphalanx	寒冷前線面
한대	frigid zone, polar band, polar belt, polar zone	寒帶
한대고기압	polar anticyclone, polar high	寒帶高氣壓
한대(기압)골	polar trough	寒帶氣壓-
한대권계면	polar tropopause	寒帶圈界面
한대기	boreal period	寒帶期
한대기단	polar air (mass)	寒帶氣團
한대기류	polar current	寒帶氣流
한대기상학	polar meteorology	寒帶氣象學
한대기후	boreal climate, cold polar climate, polar climate	寒帶氣候
한대대륙기단	polar continental air (mass)	寒帶大陸氣團
한대수림	boreal forest	寒帶樹林
한대야간제트	polar night jet	寒帶夜間-
한대야간제트류	polar night jet stream	寒帶夜間-流

ㅎ

한대연무	polar air haze	寒帶煙霧
한대전선	polar-front	寒帶前線
한대전선대	polar-front zone	寒帶前線帶
한대전선모형	polar-front model	寒帶前線模型
한대전선이론	polar-front theory	寒帶前線理論
한대전선제트	polar-front jet	寒帶前線-
한대전선제트류	polar-front jet stream	寒帶前線-流
한대제트	polar jet	寒帶-
한대제트류	polar jet stream	寒帶-流
한대지역	boreal region, boreal zone	寒帶地域
한대침엽수림지대	boreal woodland	寒帶針葉樹林地帶
한대태평양기단	polar Pacific air mass	寒帶太平洋氣團
한대편서풍	polar westerlies	寒帶偏西風
한대해양기단	polar marine air (mass), polar maritime air mass	寒帶海洋氣團
한랭	chill, coldness	寒冷
한랭고기압	cold anticyclone, cold core anticyclone, cold core high, cold high	寒冷高氣壓
한랭골	cold trough	寒冷-
한랭공기깔때기	cold air funnel	寒冷空氣-
한랭구름	cold cloud	寒冷-
한랭기	cold period	寒冷期
한랭기간	cold spell	寒冷期間
한랭기단	cold air-mass	寒冷氣團
한랭기후	cold climate	寒冷氣候
한랭뇌우	cold thunderstorm	寒冷雷雨
한랭대	cold belt, cold zone	寒冷帶
한랭돔	cold dome	寒冷-
한랭번개	cold lightning	寒冷-
한랭벽	cold wall	寒冷壁
한랭사막	cold desert	寒冷砂漠
한랭소용돌이	cold core vortex, cold vortex	寒冷-
한랭쐐기	cold wedge	寒冷-
한랭역	pool of cold air	寒冷域
한랭원	cold source	寒冷源
한랭이류	cold advection	寒冷移流
한랭인자	chill factor	寒冷因子
한랭저기압	cold core cyclone, cold core low, cold cyclone, cold low	寒冷低氣壓
한랭전선	cloud front, cold front	寒冷前線
한랭전선뇌우	cold front thunderstorm	寒冷前線雷雨
한랭전선면	cataphalanx, cold frontal surface	寒冷前線面
한랭전선비	cold front rain	寒冷前線-
한랭전선앞스콜선	pre-cold-frontal squall line	寒冷前線-線
한랭전선파	cold front wave	寒冷前線波

ㅎ

한랭전선폐색	cold front occlusion	寒冷前線閉塞
한랭전선형 강수	cold front precipitation	寒冷前線型降水
한랭전선형폐색	cold-front-type occlusion	寒冷前線型閉塞
한랭지	cold district	寒冷地
한랭지수	coldness index	寒冷指數
한랭철	cold season	寒冷-
한랭컨베이어벨트	cold conveyor belt	寒冷-
한랭편향	cold bias	寒冷偏向
한랭폐색	cold occlusion	寒冷閉塞
한랭폐색저기압	cold occlusion depression	寒冷閉塞低氣壓
한랭하강풍	cold downslope wind	寒冷下降風
한랭한	hibernal, winter	寒冷-
한랭한대기후	cold polar climate	寒冷寒帶氣候
한랭핵	cold core	寒冷核
한랭핵고리	cold core ring	寒冷核-
한랭핵뇌우	cold core thunderstorms	寒冷核雷雨
한랭핵시스템	cold-cored system	寒冷核-
한랭혀	cold tongue	寒冷-
한랭형 폐색	cold-type occlusion	寒冷型閉塞
한랭형 폐색전선	cold-type occluded front	寒冷形閉塞前線
한랭호우	cold soak	寒冷豪雨
한로	cold dew	寒露
한류	cold current	寒流
한발	drought	旱魃
한발계속일수	continuous drought days	旱魃繼續日數
한밤	midnight	
한방향 상호작용	one-way interaction	一方向相互作用
한방향 유입과정	one-way entrainment process	一方向流入課程
한방향 질량전달모형	one-way mass transfer model	一方向質量傳達模型
한 번씩 소나기	occasional showers	
한봄	depth of spring	
한 소광두께	one extinction thickness	一消光-
한여름	depth of summer, mid-summer	
한쪽차분	one-sided difference	-差分
한파	cold wave	寒波
한파경보	cold wave warning	寒波警報
한파주의보	cold wave advisory	寒波注意報
한풍해	cold wind injury, damage	寒風害
한해	cold injury, drought disaster, drought injury	寒旱
한후기	cold half year	寒候氣
할니비아트르	halny wiatr	
함대중앙기상부	Fleet Weather Central	艦隊中央氣象部
함량	concentration, content, loading	含量
함박눈	heavy snow with large flakes	

ㅎ

함수량	water content	含水量
함수행렬식	functional determinant	函數行列式
함수화물	hydrates	含水化物
함양	recharge	涵養
함유량분석	content analysis	含有量分析
합	conjunction, resultant	合
합동공정작전훈련기상	Joint Airborne Training Weather	合同空挺作戰訓練氣象
합동조건	congruence condition	合同條件
합동태풍경보센터	Joint Typhoon Warning Center	合同颱風警報-
합류	confluence, confluent flow	合流
합류구역	confluence zone	合流區域
합류변형	confluent deformation	合流變形
합류설	confluent theory	合流說
합류제트	confluent jet	合流-
합류형온도골	confluent thermal trough	合流型溫度-
합류형온도마루	confluent thermal ridge	合流型溫度-
합리수평	rational horizon	合理水平
합리식	rational formula, rational runoff formula	合理式
합병얼음	consolidated ice	合併-
합산원칙	summation principle	合算原則
합성	resultant	合成
합성개구레이더	synthetic aperture radar	合成開口-
합성단위수문도	synthetic unit hydrograph	合成單位水文圖
합성도	composite chart, composite map	合成圖
합성량	resultant	合成量
합성력	resultant force	合成力
합성바람	resultant wind	合成-
합성반사율	composite reflectivity	合成反射率
합성배리오그램	composite variogram	合成-
합성벼락	composite stroke	合成-
합성분석	composite analysis	合成分析
합성빙하	composite glacier	合成氷河
합성속도	resultant velocity	合成速度
합성수문곡선	composite hygrograph	合成水文曲線
합성수문도	synthetic hydrograph	合成水文圖
합성수샘플	composite water sample	合成水-
합성연직단면(도)	composite vertical cross-section	合成鉛直斷面(圖)
합성예보도	composite forecast chart	合成豫報圖
합성예상도	composite prognostic chart	合成豫想圖
합성파	composite wave	合成波
합성편파알고리즘	synthetic polarimetric algorithm	合成偏波-
합성풍속	resultant wind velocity	合成風速
합성풍향	resultant wind direction	合成風向
항공	aviation	航空

항공거대강수입자스펙트로미터	high volume precipitation spectrometer probe	航空巨大降水粒子-
항공고시보	notice to airmen	航空告示報
항공고정통신망	aeronautical fixed telecommunication network	航空固定通信網
항공관제	air traffic control	航空管制
항공관제레이더	air traffic control radar	航空管制-
항공관측	apob, aviation observation	航空觀測
항공교통관제	air traffic control	航空交通管制
항공교통업무	aeronautical traffic service	航空交通業務
항공구름씨뿌리기	aerial cloud seeding	航空-
항공기	aircraft	航空機
항공기관측	aircraft observation	航空機觀測
항공기궤적	aircraft trail, aircraft trajectory	航空機軌跡
항공기기상정찰	aircraft weather reconnaissance	航空機氣象偵察
항공기난류	aircraft turbulence	航空機亂流
항공기대전	aircraft electrification	航空機帶電
항공기보고	air report, aircraft report	航空機報告
항공기상	aviation weather	航空氣象
항공기상관측	aeronautical meteorological observation, aviation weather observation	航空氣象觀測
항공기상관측소	aeronautical meteorological station, aircraft meteorological station	航空氣象觀測所
항공기상실황전문	avationroutine meteorological report	航空氣象實況電文
항공기상업무	aeronautical meteorological service, aviation meteorological service	航空氣象業務
항공기상예보	aviation weather forecast	航空氣象豫報
항공기상요소	aviation meteorological element	航空氣象要素
항공기상전문	aviation meteorological message	航空氣象電文
항공기상전문형식	AERO code	航空氣象電文型式
항공기상정보	aviation weather information	航空氣象情報
항공기상청	Aviation Meteorological Agency	航空氣象廳
항공기상특보	special report for aviation weather	航空氣象特報
항공기상특보구역	special report regions for aviation weather	航空氣象特報區域
항공기상학	aeronautical meteorology, aviation meteorology	航空氣象學
항공기실링	aircraft ceiling	航空機-
항공기안테나	aeroplane antenna	航空機-
항공기온도측정법	aircraft thermometry	航空機溫度測定法
항공기운항정보교신시스템	aircraft communications addressing and reporting system	航空機運航情報交信-
항공기-위성자료중계	aircraft-to-satellite data relay	航空機衛星資料中繼
항공기전력공급시스템	aircraft power distribution system	航空機電力供給-
항공기정찰	aircraft reconnaissance	航空機偵察
항공기착빙	aircraft ice accretion, aircraft icing, airframe icing	航空機着氷
항공기측정	aircraft measurement	航空機測定

ㅎ

항공기탐측	aircraft sounding	航空機探測
항공기탑재(의)	airborne	航空機搭載
항공기탑재감지시스템	airborne sensing system	航空機搭載感知-
항공기탑재관측	airborne observation	航空機搭載觀測
항공기탑재기상레이더	airborne weather radar	航空機搭載氣象-
항공기탑재남극오존실험	airborne antarctic ozone experiment	航空機搭載南極-實驗
항공기탑재남반구오존실험	airborne southern hemisphere ozone experiment	航空機搭載南半球-實驗
항공기탑재북극성층권탐사	airborne arctic stratospheric expedition	航空機搭載北極成層圈探査
항공기탑재센서	aircraft-borne sensor	航空機搭載-
항공기탑재측방감시레이더	side-looking airbone radar	航空機搭載側方監視-
항공기탑재탐사레이더	airborne search radar	航空機搭載探査-
항공기탑재형	aircraft impactor	航空機搭載型
항공기통제	aircraft control	航空機統制
항공기투하식 수심수온계	airborne expendable bathythermograph	航空機投下式水深水溫計
항공기플랫폼	aircraft platform	航空機-
항공기후학	aeronautical climatology, aviation climatology	航空氣候學
항공난류	aviation turbulence	航空亂流
항공레이더	airborne radar	航空-
항공로	airway, route	航空路
항공로관측	airways observation	航空路觀測
항공로단면	route cross-section	航空路斷面
항공로백엽상	airways shelter	航空路百葉箱
항공로부호	airways code	航空路符號
항공로성분	route component	航空路成分
항공로예보	airway forecasting, route forecast, rofor	航空路豫報
항공로일기	airway weather	航空路日氣
항공로코드	airways code	航空路-
항공보안무선시설	aeronautical safe radio aids	航空保安無線施設
항공보안시설	aeronautical safe aids	航空保安施設
항공복사온도계	airborne radiation thermometer	航空輻射溫度計
항공사진	aerial photos	航空寫眞
항공사진술	aerophotography	航空寫眞術
항공실험	aerial experiment	航空實驗
항공예보	aviation forecast	航空豫報
항공예보관서	organization responsible for aviation forecast	航空豫報官署
항공예보구	aviation forecast zone	航空豫報區
항공예보발표시각	issuing time of aviation forecast	航空豫報發表時刻
항공운항정보시스템	aircraft data collection system	航空運航情報-
항공의학	aviation medicine	航空醫學
항공이동업무	aeronautical mobile service	航空移動業務
항공정보간행물	aeronautical information publication	航空情報刊行物
항공정보업무	aeronautical information service	航空情報業務
항공정보업무기관	aeronautical information service unit	航空情報業務機關
항공통신소	aeronautical telecommunication station	航空通信所

ㅎ

항등식	identity	恒等式
항력	drag, drag force, resistance	抗力
항력가속(도)	drag acceleration	抗力加速
항력계수	drag coefficient	抗力係數
항력위기	drag crisis	抗力危機
항력판	drag plate	抗力板
항로	airpath, air route	航路
항로기상단면도	flight cross section	航路氣象斷面圖
항로기상정보이용	weather routing	航路氣象情報利用
항로도	pilot chart	航路圖
항로보정	air line correction	航路補正
항로주의보서비스	enroute advisory service	航路注意報-
항로탐측	air line sounding	航路探測
항만기상연락관	port meteorological liaison officer	港灣氣象連絡官
항만얼음	bay ice	港灣-
항법	navigation method	航法
항복점	yield point	降伏點
항성	star	恒星
항성년	sidereal year	恒星年
항성시	sidereal hour	恒星時
항성일	sidereal day	恒星日
항속거리	cruising range	航續距離
항속성능	cruising efficiency	航續性能
항아리 곰팡이	chytrid fungus	
항적	trail	航跡
항적난류	wake turbulence	航跡亂流
항적바람	track wind	航跡-
항해	navigation	航海
항해박명	nautical twilight	航海薄明
항해시스템	nautical system	航海-
항해여명	nautical dawn	航海黎明
항해황혼	nautical dusk	航海黃昏
항행	navigation	航行
해	sun dog	
해	resolution	解
해구	ocean deep, ocean trench	海溝
해군특수전기상	Naval Special Warfare weather	海軍特殊戰氣象
해군해양전술정보단	naval sea tactical information wing	海軍海洋戰術情報團
해기둥	sun pillar	
해도	chart, sea chart, sea map	海圖
해도기준	chart datum	海圖基準
해동바람	aberwind	解凍-
해동풍	alpach, aberwind, aperwind	解凍風
해들리 단일대순환	one-cell Hadley circulation	-單一大循環

해들리세포	Hadley cell	-細胞
해들리순환	Hadley circulation	-循環
해들리영역	Hadley regime	-領域
해들리원리	Hadley's principle	-原理
해령	ocean ridges	海嶺
해류	current, ocean current	海流
해류단면	current cross section	海流斷面
해류도	current chart	海流圖
해류장미	current rose	海流薔薇
해륙풍	land and sea breeze	海陸風
해리	nautical mile	海里
해리	decoupling	解離
해리곡선	dissociation curve	解離曲線
해리상수	dissociation constant	解離常數
해리에너지	dissociation energy	解離-
해리율	dissociation rate	解離率
해리파장	dissociation wavelength	解離波長
해먹	bummock	
해면	sea surface	海面
해면경정	reduction to sea level	海面更正
해면기압	atmospheric pressure of sea level	海面氣壓
해면상태	sea state	海面狀態
해면에코	sea clutter	海面-
해면온도	sea surface temperature	海面溫度
해면온도분포도	sea surface temperature chart	海面溫度分布圖
해면온도편차도	sea surface temperature anomaly chart	海面溫度偏差圖
해명	oceanic noise	海鳴
해무	sea fog	海霧
해묵은 눈	firn, firn snow	
해묵은 눈밭	firn field	
해묵은 눈얼음	firn ice	
해묵은 눈한계	firn limit	-限界
해묵은 언눈	ice firn	
해발고도	elevation, sea level elevation	海拔高度
해발고도이상	above sea level	海拔高度以上
해빙	breakup, dabacle, deglaciation	解氷
해빙	sea ice	海氷
해빙(기)	thaw	解氷(期)
해빙감소	sea ice thinning	海氷減少
해빙계절	thawing season	解氷季節
해빙수	thaw water	海氷水
해빙주기	breakup period	解氷周期
해빙지수	thawing index	解氷指數
해빙철	breakup season	解氷-

해상	resolution	解像
해상고층기상실황전문	TEMPSHIP message	海上高層氣象實況電文
해상관측소	sea station	海上觀測所
해상관측항공기	observation seaplane	海象觀測航空機
해상광역예보	regional forecast over sea area	海上廣域豫報
해상광역예보구역	regions for sea area forecast	海上廣域豫報區域
해상국지예보	local area forecast over sea area	海上局地豫報
해상국지예보구역	local areas for sea area forecast	海上局地豫報區域
해상단기예보	marine short range forecast	海上短期豫報
해상도	resolution	解像度
해상안개	marine fog	海上-
해상예보구역	regions for marine forecast	海上豫報區域
해상정찰기	patrol seaplane	海上偵察機
해상특보	marine special report	海上特報
해상특보구역	regions for marine special report	海上特報區域
해상풍	ocean surface wind	海上風
해석위상	interpretation phase	解釋位象
해수	sea water	海水
해수담수화	desalination of seawater	海水淡水化
해수면	ocean surface, sea level	海水面
해수면경정	sea level correction	海水面更正
해수면고도계	sea level indicator	海水面高度計
해수면기압	sea level pressure	海水面氣壓
해수면기압도	sea level pressure chart	海水面氣壓圖
해수면기압패턴	sea level pressure pattern	海水面氣壓-
해수면도	sea level chart	海水面圖
해수면변화	sea level change	海水面變化
(전지구적)해수면변화	eustacy	海水面變化
해수면상승	sea level rise	海水面上昇
해수면지표	sea level indicator	海水面指標
해수온도	ocean temperature, sea temperature, sea water temperature	海水溫度
해수온도계	marine thermometer, sea water thermometer	海水溫度計
해수욕장예보	forecast for bathing beach	海水浴場豫報
해시계	sundial	-時計
(고대)해시계	gnomon	-時計
해안	bank	海岸
해안경계층	coastal boundary layer	海岸境界層
해안기후	coastal climate, littoral climate	海岸氣候
해안밤안개	coastal night fog	海岸-
해안사막	coastal desert	海岸砂漠
해안사막기후	coastal desert climate	海岸沙漠氣候
해안선	coastline	海岸線
해안선길이	coastline length	海岸線-

ㅎ

해안선변화	coastline change	海岸線變化
해안소형선박주의보	small craft advisory	海岸小形船舶注意報
해안에서 바다로 부는 바람	offshore wind	
해안영향	coastal impact	海岸影響
해안전선	coastal front	海岸前線
해안제트	coastal jet	海岸-
해안집단	coastal community	海岸集團
해안침식	coastal erosion	海岸浸蝕
해안침투기상	coastal infiltration weather	海岸浸透氣象
해안평야	coastal plains	海岸平野
해안폭풍	coastal storm	海岸暴風
해안확산	coastal diffusion	海岸擴散
해안환경	coastal environments	海岸環境
해안효과	coastal effect	海岸效果
해양	ocean	海洋
해양고기압	oceanic anticyclone, oceanic high, subtropical high	海洋高氣壓
해양공기	maritime air	海洋空氣
해양관측	oceanographical observation	海洋觀測
해양(기상)관측	marine observation	海洋(氣象)觀測
해양관측선	ocean station vessel	海洋觀測船
해양관측소	marine observatory, ocean station	海洋觀測所
해양관측위성	marine observation satellite	海洋觀測衛星
해양구름	maritime cloud	海洋-
해양기단	marine air (mass), maritime air mass, oceanic air mass, ocean air(mass)	海洋氣團
해양기단안개	maritime air fog	海洋氣團-
해양기상관측	marine weather observation	海洋氣象觀測
해양기상관측선	ocean weather ship, ocean weather station vessel	海洋氣象觀測船
해양기상관측소	oceanic weather station, ocean weather station	海洋氣象觀測所
해양기상구(역)	marine meteorological district	海洋氣象區(域)
해양기상연구	oceanic meteorological research	海洋氣象研究
해양기상학	marine meteorology, maritime meteorology, naval meteorology, oceanic meteorology	海洋氣象學
해양기후	marine climate, maritime climate	海洋氣候
해양기후학	marine climatology	海洋氣候學
해양높이변동	ocean height variation	海洋-變動
해양대기	ocean air(mass)	海洋大氣
해양-대기결합모형	coupled ocean-atmosphere model	海洋大氣結合模型
해양-대기시스템	ocean-atmosphere system	海洋大氣-
해양-대기열플럭스	ocean-atmosphere heat flux	海洋大氣熱-
해양대순환	ocean general circulation	海洋大循環
해양대순환모형	oceanic general circulation models	海洋大循環模型
해양도	oceanity, oceanicity	海洋度

해양도각	marine angle	海洋道角
해양물리학	oceanophysics	海洋物理學
해양반구	oceanic hemisphere, water hemisphere	海洋半球
해양분지	ocean basin	海洋盆地
해양비생물기원 퇴적물	marine non-biogenic sediments	海洋非生物起源堆積物
해양산성화	ocean acidification	海洋酸性化
해양상공성층권	oceanic stratosphere	海洋上空成層圈
해양생물기원퇴적물	marine biogenic sediments	海洋生物起源堆積物
해양섬	ocean islands	海洋島
해양성기후	ocean climate, oceanic climate	海洋性氣候
해양성열대기단	maritime tropical air mass	海洋性熱帶氣團
해양성온난	oceanic moderate	海洋性溫暖
해양성툰드라	maritime tundra	海洋性-
해양성한대기단	maritime polar air mass	海洋性寒帶氣團
해양순환	ocean circulation	海洋循環
해양순환시스템	oceanic circulation system	海洋循環-
해양시추프로그램	Ocean Drilling Program	海洋試錐-
해양시추프로젝트	Ocean Drilling Project	海洋試錐計劃
해양안개	ocean fog	海洋-
해양알베도	ocean albedo	海洋-
해양에어로졸	marine aerosol	海洋-
해양열관성	ocean heat inertia	海洋熱慣性
해양열수송	oceanic heat transport	海洋熱輪送
해양열수송가설	ocean heat transport hypothesis	海洋熱輪送假說
해양영향	ocean influence	海洋影響
해양(일기)예보	marine forecast	海洋(日氣)豫報
해양예상도	marine prognostic chart	海洋豫想圖
해양의 고생산력	ocean paleoproductivity	海洋古生産力
해양의 고온도	ocean paleotemperature	海洋古溫度
해양잡음	oceanic noise	海洋雜音
해양저류	undertows	海洋底流
해양저장소	ocean reservoir	海洋貯藏所
해양전선	oceanic front	海洋前線
해양점토광물	marine clay minerals	海洋粘土鑛物
해양조	oceanic tide	海洋潮
해양지각	ocean crust, oceanic crust	海洋地殼
해양지역	oceanic area	海洋地域
해양지형실험	Ocean Topography Experiment	海洋地形實驗
해양지형학실험위성	TOPEX Poseidon satellite	海洋地形學實驗衛星
해양컨베이어벨트	ocean conveyor belt	海洋-
해양탄소 지구화학적 성질	marine carbon geochemistry	海洋炭素地球化學的性質
해양탄소 펌프가설	ocean carbon pump hypothesis	海洋炭素-假說
해양통로	ocean gateway	海洋通路
해양퇴적물	ocean sediment	海洋堆積物

ㅎ

해양표면풍	ocean surface wind	海洋表面風
해양표층	ocean surface	海洋表層
해양표층혼합층	oceanic surface mixed layer	海洋表層混合層
해양풍	ocean wind	海洋風
해양학	oceanography	海洋學
해양항공교통통제	oceanic air traffic control	海洋航空交通統制
해양형	ocean type	海洋型
해양혼합	ocean mixing	海洋混合
해양혼합층	oceanic mixed layer	海洋混合層
해양혼합층모형	ocean mixed layer model	海洋混合層模型
해양화학	ocean chemistry	海洋化學
해연	ocean deep	海淵
해염에어로졸	sea salt aerosol	海鹽-
해염입자	sea salt particle	海鹽粒子
해염핵	sea salt nucleus	海鹽核
해왕성	Neptune	海王星
해저	ocean floor, seafloor	海底
해저지각시추프로그램	Ocean Drilling Program	海底地殼試錐-
해저지형도	bathymetric chart	海底地形圖
해저퇴적	seafloor sediment	海底堆積
해저확장	ocean floor spreading, seafloor spreading	海底擴張
해제시각	time of removal	解除時刻
해코로나	solar corona	
해터라스곶저기압	Cape Hatteras low	-低氣壓
해파	ocean wave, sea wave	海波
해파모형	sea wave model	海波模型
해풍	onshore breeze, sea breeze	海風
해풍수렴대	sea breeze convergence zone	海風收斂帶
해풍순환	sea breeze circulation	海風循環
해풍전선	sea breeze front	海風前線
핵	kernel, nucleus	核
핵가을	nuclear autumn	核-
핵겨울	nuclear winter	核-
핵결합에너지	nuclear binding energy	核結合-
핵계수기	Kern counter, dust counter, nuclei counter, nucleus counter	核計數器
핵공	nuclear pore	核孔
핵교환효과	nuclear exchange effect	核交換效果
핵구조	nuclear structure	核構造
핵기공필터	nuclepore filter	核氣孔-
핵동중체	nuclear isobars	核同重體
핵력	nuclear force	核力
핵막	nuclear membrane	核膜
핵모드	nuclei mode	核-

핵무기	nuclear weapon	核武器
핵물리학	nuclear physics	核物理學
핵발전소	nuclear power	核發電所
핵변환	nuclear transmutation	核變換
핵분열	nuclear fission	核分裂
핵분열생성물	fission product	核分裂生成物
핵생성	nucleation	核生成
핵생성문턱	nucleation threshold	核生成門-
핵생성제	nucleant, nucleating agent	核生成劑
핵심감지	core sensor	核心感知
핵에너지	nuclear energy	核-
핵융합	fusion, nuclear fusion	核融合
핵폭탄실험	atomic bomb tests	核爆彈實驗
핵화염	nuclear fireball	核火焰
핼시온기간	halcyon day, alcyone days	-期間
햇무리	solar halo	
햇빛기간	sunny interval	-其間
행동모형	behavioral model	行動模型
행렬	matrix	行列
행렬식	determinant	行列式
행로	path	行路
행성	planet	行星
행성경계층	planetary boundary layer	行星境界層
행성계	planetary system	行星系
행성규모	planetary scale	行星規模
행성규모바람	planetary scale wind	行星規模-
행성규모시스템	planetary scale system	行星規模-
행성대기	planetary atmosphere	行星大氣
행성바람	planetary wind	行星-
행성반사율	planetary albedo	行星反射率
행성소용돌이도효과	planetary vorticity effect	行星-度效果
행성순환	planetary circulation	行星循環
행성알베도	planetary albedo	行星-
행성온도	planetary temperature	行星溫度
행성운동	planetary motion	行星運動
행성원자모형	planetary atomic model	行星原子模型
행성지구연구	mission to planet earth	行星地球研究
행성파	planetary wave	行星波
행성파공식	planetary wave formula	行星波公式
행진문제	marching problem	行進問題
향사도수정	angularity correction	向斜度修正
향안풍	onshore breeze	向岸風
향열모형	thermotropic model	向熱模型
허긴스띠	Huggins bands	

ㅎ

허드슨만저기압	Hudson bay low	-低氣壓
허리케인	furacana, furacan(e), furicane, furicano, jimmycane, hurricane, ox's eye	
허리케인경보	hurricane warning	-警報
허리케인경보체계	hurricane warning system	-警報體系
허리케인구름	hurricane cloud	
허리케인급 바람	hurricane force wind	
허리케인급 바람경보	hurricane force wind warning	-警報
허리케인기구	hurricane balloon	-氣球
허리케인눈	hurricane eye	
허리케인둑	hurricane bar	
허리케인띠	hurricane band	
허리케인미진	hurricane microseisms	-微震
허리케인비컨	hurricane beacon	
(미국)허리케인센터	National Hurricane Center	
허리케인씨뿌리기	hurricane seeding	
허리케인앞스콜선	pre-hurricane squall line	-線
허리케인주의보	hurricane watch	-注意報
허리케인진로예상법	hurricane analog technique	-進路類似法
허리케인추적	hurricane tracking	-追跡
허수	imaginary number	虛數
허수단위	imaginary unit	虛數單位
허수부분	imaginary part	虛數部分
허용농도한계	admissible concentration limit	許容濃度限界
허용된 전이	allowed transition	許容轉移
허용오차	allowable error, tolerance	許容誤差
허용함수	admittance function	許容函數
헐로프슨선도	Herlofson diagram	-線圖
헐리-벌리	hurly-burly	
헛돌음	backlash	
헤르츠	hertz	
헤르츠베르크띠	Herzberg band	
헤르츠선도	Hertz's diagram	-線圖
헤벨리우스무리	halo of Hevelius	
헤비사이드층	Heaviside layer	-層
헤빌리우스무리	Hevelian halo	
헤빌리우스무리해	Hevelius's parhelia	
헤이스터	haster	
헤이즈미터	hazemeter	
헤이즌법	Hazen method	-法
헤테로다인	heterodyne	
헥토	hecto-	
헥토파스칼	hectopascal(hPa)	
헬름홀츠방정식	Helmholtz equation	-方程式

헬름홀츠불안정도	Helmholtz instability	-不安定度
헬름홀츠상호성원리	Helmholtz reciprocity principle	-相互性原理
헬름홀츠자유에너지	Helmholtz free energy	-自由-
헬름홀츠중력파	Helmholtz gravitational wave	-重力波
헬름홀츠파	Helmholtz wave	-波
헬름홀츠함수	Helmholtz function	-函數
헬리시티	helicity	
헬리오스태트	heliostat	定日鏡
헬리오층	heliosphere	-層
현미경사진술	microphotography	顯微鏡寫眞術
현상	phenomenon	現象
현상계수	phenomenological coefficient	現象係數
현상일수	number of days with phenomena	現象日數
현상학이론	phenomenological theory	現象學理論
현생누대	Phanerozoic Eon	現生累代
현생이언	Phanerozoic Eon	現生-
현업선주사시스템	operational linescan system	現業線走査-
현업수치모형	operational numerical model	現業數值模型
현업예보	operational prediction	現業豫報
현업용	operational use	現業用
현업위성	operational satellites	現業衛星
현업일기한계	operational weather limits	現業日氣限界
현열	sensible heat	顯熱
현장채취	in situ	現場採取
현재날씨	present weather	現在天氣
현재날씨분석	present weather analysis	實況分析
현재이동속도	present movement speed	現在移動速度
현재일기	current weather	現在日氣
현존곡물량	standing crop	現存穀物量
현천예보	nowcasting	現天豫報
현탁	suspension	懸濁
현탁단계	suspended phase	懸濁段階
현탁상	suspended phase	懸濁相
현탁액	slurry	懸濁液
현탁질	suspensoid	懸濁質
혈류	blood flow	血流
혐광식물	heliophobe	嫌光植物
혐기성(의)	anaerobic	嫌氣性
혐기성박테리아	anaerobic bacteria	嫌氣性-
혐기성조건	anaerobic condition	嫌氣性條件
혐기성처리장치	anaerobic digestors	嫌氣性處理裝置
혐기호흡	anaerobic respiration	嫌氣呼吸
협곡	canyon	峽谷
협곡바람	canyon wind, gorge wind, ravine wind	峽谷-

ㅎ

협곡침식	gully erosion	峽谷浸蝕
협력관측자	cooperative observer	協力觀測者
협만	fiord, fjord	峽灣
협심증	angina pectoris	狹心症
형광	fluorescent light	螢光
형광복사	fluorescence radiation	螢光輻射
형상인자	shape factor	形象因子
형성호흡	constructive respiration	形成呼吸
형식	format	形式
형체마찰	form friction	形體摩擦
형체항력	form drag	形體抗力
형태변화	change of type	形態變化
형판	template	型板
혜성	comet	彗星
호광식물	heliophil(e)	好光植物
호도	radian	弧度
호도그래프	hodograph	
호로그래피법	holography method	-法
호모다인	homodyne	
호소(생물)기상학	limnological meteorology	湖沼(生物)氣象學
호소퇴적물	lacustrine sediment	湖沼堆積物
호소학	limnology	湖沼學
호수면	lake level	湖水面
호수면변화	lake level variation	湖水面變化
호수면온도	lake temperature, lake surface temperature	湖水面溫度
호수바람	lake breeze	湖水-
호수증발	lake evaporation	湖水蒸發
호수효과	lake effect	湖水效果
호수효과 눈스톰	lake effect snowstorm	湖水效果-
호숙기	dough ripe stage	糊熟期
호아지중해수	Austalasian mediterranean water	濠亞地中海水
호우	heavy rain, pouring rain	豪雨
호우경보	heavy rain warning	豪雨警報
호우주의보	heavy rain advisory	豪雨注意報
호이헨스원리	Huygens's principle	-原理
호이헨스 작은 파	Huygens's wavelets	-波
호흡	respiration	呼吸
호흡가능입자	respirable particle	呼吸可能粒子
호흡기관	respiratory organ	呼吸器官
호흡기질	respiratory substrate	呼吸基質
호흡색소	respiratory pigment	呼吸色素
호흡지수	respiration quotient	呼吸指數
혹한	killing freeze, severe cold	酷寒
혹한해	severe cold damage	酷寒害

ㅎ

혼돈	chaos	混沌
혼돈거동	chaotic behavior	混沌擧動
혼돈이론	chaos theory	混沌理論
혼성	hybrid	混成
혼탁계수	coefficient of turbidity, turbidity coefficient	混濁係數
혼탁도	turbidity	混濁度
혼탁도계	turbidimeter	混濁度計
혼탁매질	turbid medium	混濁媒質
혼탁인자	turbidity factor	混濁因子
혼합	mixing	混合
혼합경로	path of mixing	混合經路
혼합구름	mixed cloud	混合-
혼합기단	mixing air mass	混合氣團
혼합길이	mixing length	混合-
혼합깊이	mixing depth	混合-
혼합냉각	mixing cooling	混合冷却
혼합높이	mixing height	混合-
혼합로스비중력파	mixed Rossby-gravity wave	混合-重力波
혼합무	mixed cloud	混合霧
혼합물	mixture	混合物
혼합비	mixing ratio	混合比
혼합안개	mixing fog	混合-
혼합응결	mixing condensation	混合凝結
혼합응결고도	mixing condensation level	混合凝結高度
혼합조	mixed tide	混合潮
혼합지형	blended orography	混合地形
혼합착빙	mixed icing	混合着氷
혼합층	mixed layer, mixing layer	混合層
혼합층깊이	mixed layer depth	混合層-
혼합층높이	mixed layer height	混合層-
혼합하지 않는 액체	nonwetting liquid	混合-液體
혼합핵	mixed nucleus	混合核
혼합형 여름냉해	mixed type cool summer damage	混合型-障害
혼합형스톰	hybrid storm	混合形-
홀러리드장치	hollerith system	-裝置
홀로그래피	holography	
홀로그램	hologram	
홀로세기후	holocene climate	沖積世氣候
홀로세수목한계선변동	holocene treeline fluctuations	-樹木限界線變動
홀로세지도작성프로젝트	Cooperative Holocene Mapping Project	-地圖作成-
홀수	odd number	-數
홀수산소	odd oxygen	-酸素
홀수산소시스템	odd oxygen system	-酸素-
홀수수소	odd hydrogen	-水素

ㅎ

홀수염소	odd chlorine	-鹽素
홀수점	odd point	-數點
홀수질소	odd nitrogen	-窒素
홀수질소산화물	NOy, odd nitrogen	-窒素酸化物
홀효과	Hall effect	-效果
홉스이론	Hobb's theory	-理論
홉필드띠	Hopfield bands	
홍수	flood, inundation	洪水
홍수가능성전망	flood potential outlook	洪水可能性展望
홍수가설	diluvial hypothesis	洪水假說
홍수경보	flood warning	洪水警報
홍수공시	flood statement	洪水公示
홍수로	flood way	洪水路
홍수방류	flood discharge	洪水放流
홍수방류수위	flood discharge level	洪水放流水位
홍수방어벽	flood wall	洪水防禦壁
홍수방지법	flood proofing	洪水防止法
홍수부분시계열	floods-above-base series	洪水部分時系列
홍수분류(계급)	flood category	洪水分類(階級)
홍수빈도	flood frequency	洪水頻度
홍수빈도곡선	flood frequency curve	洪水頻度曲線
홍수수문곡선	flood hydrograph	洪水水門曲線
홍수예방	flood prevention	洪水豫防
홍수예보	flood forecast	洪水豫報
홍수위	flood stage	洪水位
홍수재발간격	flood recurrence interval	洪水再發間隔
홍수재해	flood damage	洪水災害
홍수정점	flood crest	洪水頂點
홍수조절	flood control	洪水調節
홍수조절문	flood gate	洪水調節門
홍수조절용량	flood control storage	洪水調節容量
홍수조절저수지	flood control reservoir	洪水調節貯水池
홍수주의보	flood watch	洪水注意報
홍수지도	flood atlas	洪水地圖
홍수추산	flood routing	洪水推算
홍수파	flood wave	洪水波
홍수프로파일	flood profile	洪水-
홍수피해저감대책	flood loss reduction measure	洪水被害低減對策
홍염	prominence	紅焰
홍엽	red coloring of leaves	紅葉
홍엽일	date of red coloring of leaves	紅葉日
홍토	laterite	紅土
화력발전	thermal power generation	火力發電
화분분석	pollen analysis	花粉分析

ㅎ

화분예보	pollen forecast	花紛豫報
화분학	palynology	花粉學
화산	volcano	火山-
화산가스	volcanic gas	火山-
화산구름	cloud due to volcanic eruption	火山-
화산뇌우	volcanic thunderstorm	火山雷雨
화산대량 황화합물침적	volcanogenic massive sulfide deposit	火山大量黃化合物沈積
화산먼지	volcanic dust	火山-
화산모래	volcanic sand	火山-
화산바람	volcanic wind	火山-
화산번개	volcanic lightning	火山-
화산분석	volcanic cinder	火山噴石
화산분출	volcanic eruption	火山噴出
화산분출구름	cloud from volcanic eruptions, volcanic eruption cloud	火山噴出-
화산섬고리	volcanic island chain	火山-
화산에어로졸	volcanic aerosol	火山-
화산용암지대	channeled scablands	火山鎔岩地帶
화산작용	volcanism	火山作用
화산재	volcanic ash	火山灰
화산재구름	volcanic ash cloud	火山灰-
화산재연대학	tephrochronology	火山災年代學
(화산)재 회오리	ash devils	(火山)灰-
화산천둥	volcanic thunder	火山天動
화산탄	bomb, volcanic bomb	火山彈
화산폭발	volcanic eruption	火山爆發
화산폭발구름	volcanic eruption cloud	火山爆發-
화산폭풍	volcanic storm	火山暴風
화산활동	volcanic activity	火山活動
화살선도방전	dart leader	-先導放電
화살표머리 도표	arrowhead chart	-圖表
화생방전기상	NBC Warfare weather	化生放戰氣象
화석기록	fossil record	化石記錄
화석꽃가루	fossil pollen	化石-
화석난류	fossil turbulence	化石亂流
화석얼음	fossil ice	化石-
화석연료	fossil fuel	化石燃料
화석연료에너지	fossil fuel energy	化石燃料-
화석연료연소	fossil fuel burning, fossil fuel combustion	化石燃料燃燒
화석영구 언땅	fossil permafrost	化石永久-
화성	Mars	火星
화성암	igneous rock	火成巖
화소	pixel	畵素
화소크기	pixel size	畵素-

ㅎ

화숫값	pixel value	畵素-
화씨	Fahrenheit	華氏
화씨눈금	Fahrenheit scale	華氏-
화씨온도	Fahrenheit temperature	華氏溫度
화씨온도계	Fahrenheit thermometer	華氏溫度計
화씨온도눈금	Fahrenheit temperature scale	華氏溫度-
화이트아웃	whiteout	
화이팅안정(도)지수	whiting stability index	-安定(度)指數
화재구름	cloud from fire	火災-
화재기상학	fire meteorology	火災氣象學
화재날씨	fire weather	火災-
화재위험계	fire-danger meter	火災危險計
화재위험지수	fire danger index(FDI)	火災危險指數
화재재해	fire hazard	火災災害
화재폭풍	fire storm	火災暴風
화학결합	chemical bond	化學結合
화학권	chemosphere	化學圈
화학권계면	chemopause	化學圈界面
화학기작	chemical mechanism	化學機作
화학량계수	stoichiometric coefficient	化學量係數
화학량비	stoichiometric ratio	化學量比
화학량수	stoichiometric number	化學量數
화학미량원소	chemical tracers	化學微量元素
화학반응	chemical reaction	化學反應
화학반응률	chemical reaction rate	化學反應率
화학반응식	chemical equation	化學反應式
화학발광	chemiluminescence	化學發光
화학변수	chemical variable	化學變數
화학성분	chemical composition	化學成分
화학수명	chemical lifetime	化學壽命
화학수송모형	chemical transport model	化學輸送模型
화학수용기	chemoreceptor	化學受容器
화학습도계	chemical hygrometer	化學濕度計
화학식량	formula weight	化學式量
화학에너지	chemical energy	化學-
화학연료추진체계	chemical propulsion system	化學燃料推進體系
화학운동학	chemical kinetics	化學運動學
화학적 잔류자성	chemical remanent magnetism	化學的殘留磁性
화학적침식	chemical erosion	化學的浸蝕
화학적풍화(작용)	chemical weathering	化學的風化(作用)
화학전달물질	chemical transmitter	化學傳達物質
화학전지	chemical cell(battery)	化學電池
화학종가족	family of chemical species	化學種家族
화학친화력	chemical affinity	化學親和力

ㅎ

화학침전물	chemical sediment	化學沈澱物
화학침전암	chemical sedimentary rock	化學沈澱巖
화학탐사	chemical prospecting	化學探查
화학퍼텐셜	chemical potential	化學-
화학평형	chemical equilibrium	化學平衡
화합물	compound	化合物
화합열	heat of combination	化合熱
확대경	magnifying glass	擴大鏡
확대비	ratio of expand	擴大比
확대인자	magnification factor	擴大因子
확률	probability	確率
확률(적)	stochastic	確率(的)
확률공간	probability space	確率空間
확률과정	stochastic process	確率過程
확률기댓값	chance expectation	確率期待-
확률론	probability theory	確率論
확률밀도함수	probability density function	確率密度函數
확률보행	random walk	確率步行
확률분포	probability distribution	確率分布
확률분포함수	probability distribution function	確率分布函數
확률역학예보	stochastic dynamic prediction	確率力學豫報
확률예보	probability forecast	確率豫報
확률오차	probable error, random error	確率誤差
확률우량	probable rainfall amount	確率雨量
확률적분	probability integral	確率積分
확률질량함수	probability mass function	確率質量函數
확률함수	probability function	確率函數
확률화	randomization	確率化
확산	diffusion	擴散
확산경계면	diffuse boundary	擴散境界面
확산계산	diffusion computation	擴散計算
확산계수	coefficient of diffusion, diffusion coefficient	擴散係數
확산공식	diffusion formula	擴散公式
확산과정	diffusion process	擴散過程
확산광	diffuse light, diffused light	擴散光
확산근사	diffuse approximation	擴散近似
확산단파복사	diffuse short-wave radiation	擴散短波輻射
확산력	diffusive force	擴散力
확산모형	diffusion model	擴散模型
확산문제	diffusion problem	擴散問題
확산방정식	diffusion equation	擴散方程式
확산방정식한계	diffusion equation limit	擴散方程式限界
확산법칙	law of diffusion	擴散法則
확산복사	diffuse radiation	擴散輻射

ㅎ

확산분리	diffusive separation	擴散分離
확산선도	diffusion diagram	擴散線圖
확산속도	diffusion velocity	擴散速度
확산습도계	diffusion hygrometer	擴散濕度計
확산실험	diffusion experiment	擴散實驗
확산유사물	diffusion analogue	擴散類似物
확산율	diffusivity	擴散率
확산율방정식	diffusivity equation	擴散率方程式
확산이론	diffusion theory	擴散理論
확산조명	diffuse illumination	擴散照明
확산태양복사	diffuse solar radiation	擴散太陽輻射
확산평형	diffusive equilibrium	擴散平衡
확산플럭스	diffusion flux	擴散-
확산하늘복사	diffuse sky radiation	擴散-輻射
확산하늘빛	diffuse skylight	擴散-
확신율	certainty factor	確信率
확장 칼만필터링	extended Kalman filtering	擴張-
확정예보	determinate forecast	確定豫報
환경	environment, environments	環境
환경감률	environmental lapse rate, sounding lapse rate	環境減率
환경과학	environmental sciences	環境科學
환경관측위성	environmental satellite	環境觀測衛星
환경기상학	environmental meteorology	環境氣象學
환경론자	environmentalist	環境論者
환경문제	environmental issue, environmental problems	環境問題
환경바람	environmental wind	環境-
환경범죄	environmental crime	環境犯罪
환경변화	environmental change	環境變化
환경보전그룹	environmental integrity group	環境保全-
환경보호	environmental protection	環境保護
환경붕괴	environmental degradation	環境崩壞
환경소음	environmental noise	環境騷音
환경영향	environmental impact	環境影響
환경영향평가	environmental impact assessment	環境影響評價
환경예보서비스	environmental forecasting services	環境豫報-
환경오염	environmental pollution	環境汚染
환경온도	environmental temperature	環境溫度
환경온도연직탐측	environmental temperature sounding	環境溫度鉛直探測
환경요인	environmental factor	環境要因
환경운동	environmental action	環境運動
환경위협	environmental threat	環境威脅
환경자료	environmental data	環境資料
환경장	environmental flow	環境場
환경저항	environmental resistance	環境抵抗

ㅎ

환경조절금지조약	Environmental Modification Convention	環境調節禁止條約
환경질표준	environmental quality standard	環境質標準
(미국)환경청	Environmental Protection Agency	環境廳
환경탐사위성	environmental survey satellite	環境探査衛星
환경효과	environmental effect	環境效果
환기	aeration, ventilation	換氣
환기지수	ventilation index	換氣指數
환등	delinescope	幻燈
환산계수	reduction factor	換算係數
환산기압	reduced pressure	換算氣壓
환산온도	reduced temperature	換算溫度
환산중력	reduced gravity	換算重力
환상(고리모양)모드	annular mode	環狀-
환수평호	circumhorizontal arc	環水平弧
환원	retrieval	還元
환원력	reducing power	還元力
환원모듈	retrieval module	還元-
환원문제	retrieval problem	還元問題
환원방법	retrieval method	還元方法
환원촉매	reduction catalysts	還元觸媒
환월	paraselene	幻月
환일	parhelia, sun dog	幻日
환초	coral atoll	環礁
환태양복사	circumsolar radiation	環太陽輻射
활강	downslide	滑降
활강면	downslide surface	滑降面
활강바람	downslope wind, katabatic wind	滑降-
활강운동	downglide motion, downslope motion	滑降運動
활강전선	catafront, katafront, katabatic front, katafront	滑降前線
활강풍	downslope wind, slope wind	滑降風
활강흐름	downslope airflow, downslope flow	滑降-
활공비행	gliding flight	滑空飛行
활공평균	gliding mean	滑空平均
활동빙하	living glacier	活動氷河
활동성구름	active cloud	活動性-
(채층의)활동성망상조직	active network	活動性網狀組織
활동성분지지역	active basin area	活動性盆地地域
활동성빙하	active glacier	活動性氷河
활동성영구 언땅	active permafrost	活動性永久-
활동성전선	active front	活動性前線
활동영역	active region	活動領域
활동적산온도	active accumulated temperature	活動積算溫度
활동중심	active center	活動中心
활동층	active layer	活動層

활동표면	active surface	活動表面
활모양 강수띠	bow echo	-模樣降水-
활모양구름	arc cloud	
활모양구름선	arc cloud line	-線
활모양방전	arc discharge	-放電
활모양선	arc lines	弧線
활성도	activity	活性度
활성도계수	activity coefficient	活性度係數
활성복합체	activated complex	活性複合體
활성복합체 이론	activated complex theory	活性複合體理論
활성부위	active site	活性部位
활성산소	oxygen free radical, reactive oxygen	活性酸素
활성질소	active nitrogen	活性窒素
활성착물	activated complex	活性錯物
활성탄	activated charcoal	活性炭
활성탄소	activated carbon	活性炭素
활성탄소과정	activated carbon process	活性炭素過程
활성탄화수소류	reactive hydrocarbons	活性炭化水素類
활성화	activation	活性化
활성화에너지	activation energy	活性化-
활승(의)	anabatic	滑昇
활승구름	anabatic cloud	滑昇-
활승류	anaflow, upslope flow	滑昇流
활승면	upslide surface	滑昇面
활승바람	anabatic wind, upslope wind	滑昇-
활승안개	upslope fog	滑昇-
활승운	upglide cloud	滑昇雲
활승전선	anabaric front, anafront	滑昇前線
활승중심	anabaric center	滑昇中心
활승효과	upslope effect	滑昇效果
활승흐름	anabatic flow	滑昇-
활에코	bow echo	
활주로	runway	滑走路
활주로가시거리	runway visual range	滑走路可視距離
활주로결빙	ice on runway	滑走路結氷
활주로관측	runway observation	滑走路觀測
활주로시정	runway visibility	滑走路視程
활주로온도	runway temperature	滑走路溫度
활주로표고	runway elevation	滑走路標高
황도	ecliptic, zodiac	黃道
황도경사	obliquity of the ecliptic	黃道傾斜
황도광	zodiacal light	黃道光
황도대	zodiacal band	黃道帶
황도대일조	zodiacal counterglow	黃道對日照

ㅎ

황도면	plane of the elliptic	黃道面
황도원뿔	zodiacal cone	黃道圓-
황도피라미드	zodiacal pyramid	黃道-
황동위원소	sulfur isotope	黃同位元素
황백화(현상)	chlorosis	黃白化(現象)
황사	yellow sand	黃砂
황사경보	yellow sand warning	黃砂警報
황사바람	yellow wind	黃砂-
황사주의보	yellow sand advisory	黃砂注意報
황사현상	Asian dust phenomenon	黃砂現象
황산	sulfuric acid, sulphuric acid	黃酸
황산미스트	sulfuric mist, sulphuric mist	黃酸-
황산비	sulphur rain	黃酸-
황산실안개	sulfuric mist, sulphuric mist	黃酸-
황산암모늄	ammonium sulfate	黃酸-
황산암모늄박무	ammonium sulfate haze	黃酸-薄霧
황산염	sulfate	黃酸鹽
황숙기	yellow ripe stage	黃熟期
황야폭풍	moor-gallop	荒野暴風
황엽	yellow coloring of leaves	黃葉
황엽일	date of yellow coloring of leaves	黃葉日
황토	loess	黃土
황토침전물	loess deposits	黃土沈澱物
황폐화	overrunning	荒廢化
황화병	chlorosis	黃化病
황화카르보닐	carbonyl sulfide	黃化-
회귀	regression	回歸
회귀방정식	regression equation	回歸方程式
회귀변수	regressor	回歸變數
회귀분석	regression analysis	回歸分析
회귀선	tropics	回歸線
회귀식	regression equation	回歸式
회귀예측	regression estimation	回歸豫測
회귀함수	regression function	回歸函數
회로	circuit	回路
회보	bulletin	會報
회복	retrieval	回復
회복능	resilience	回復能
회분식모의	batch simulation	回分式模擬
회색광	ash-grey light	灰色光
회색권층운	dim cloud	灰色券層雲
회색대기	gray atmosphere	灰色大氣
회색도	shades of gray	灰色度
회색스모그	gray smog	灰色-

ㅎ

회색체	graybody, greybody	灰色體
회색체방법	graybody method	灰色體方法
회색체복사	greybody radiation	灰色體輻射
회색체흡수	graybody absorption	灰色體吸收
회색흡수체	grey absorber	灰色吸收體
회선	convolution	回旋
회수율	recovery factor	回收率
회오리	twister	
회오리바람	whirl wind, whirls	
회유형	closed circuit wind tunnel	回流型
회이관순환정리	Höiland's circulation theorem	-循環定理
회전	revolution, rotation, spin	回轉
회전각	angle of rotation	回轉角
회전능률	moment of rotation	回轉能率
회전다중원통	rotating multicylinder	回轉多重圓筒
회전대	rotating band	回轉帶
회전레이놀즈수	rotating Reynolds number	回轉-數
회전류	rotary current	回轉流
회전모멘트	rotation moment	回轉-
회전불안정(도)	rotational instability	回轉不安定(度)
회전빔운고계	rotating-beam ceilometer	回轉-雲高計
회전속도계	tachometer	回轉速度計
회전수조	rotating annulus, rotating dishpan	回轉水槽
회전수조실험	rotating annulus experiment, rotating dishpan experiment	回轉水槽實驗
회전에너지	rotational energy	回轉-
회전운고계	rotating ceilometer	回轉雲高計
회전운동	rotational motion	回轉運動
회전원판실험	dishpan experiment	回轉圓板實驗
회전유체	revolving fluid, rotating fluid	回轉流體
회전유체역학	rotating fluid dynamics	回轉流體力學
회전이동	rotatory translation	回轉移動
회전자기빈도	gyromagenetic frequency	回轉磁氣頻度
회전장	rotational field	回轉場
회전전이	rotational transition	回轉轉移
회전조류	rotary tidal current	回轉潮流
회전중심	center of rotation	回轉中心
회전진동스펙트럼	rotation-vibration spectrum	回轉振動-
회전집진기	mechanical collector	回轉集塵器
회전체	solid of revolution	回轉體
회전축	rotational axis	回轉軸
회전타원체	ellipsoid of revolution, spheroid	回轉楕圓體
회전폭풍우	revolving storm	回轉暴風雨
회전풍	rotating wind	回轉風

회전풍속계	rotation anemometer	回轉風速計
회전프루드수	rotation Froude number	回轉-數
회전하강돌풍	rotor downburst	回轉下降突風
회전행렬	rotation matrix	回轉行列
회절	diffraction	回折
회절격자	diffraction lattice	回折格子
회절광선	diffracted ray	回折光線
회절무늬	diffraction fringe	回折-
회절분광기	grating spectroscope	回折分光器
회절스펙트럼	diffraction spectrum	回折-
회절역	diffraction zone	回折域
회절패턴	diffraction pattern	回折-
회절현상	diffraction phenomenon	回折現象
회합주기	synodic period	會合週期
획득조건	acquisition condition	獲得條件
횡발산	transverse divergence	橫發散
횡압력	transverse force	橫壓力
횡파	transversal wave	橫波
후굴온난전선	back-bent warm front	後屈溫暖前線
후굴폐색	back-bent occlusion, bent-back occlusion, recurved occlusion	後屈閉塞
후기고생대 기후	Late Paleozoic climate	後期古生代氣候
후기 제4기 기후	Late Quaternary climate	後期第四紀氣候
후두두비	sprinkle	
후류	wake	後流
후류난류	wake turbulence	後流亂流
후류저기압	wake depression	後流低氣壓
후류포착	wake capture	後流捕捉
후방방정식	backward equation	後方方程式
후방복사	backward radiation	後方輻射
후방산란	back-scattering, backward scattering	後方散亂
후방산란계수	back-scattering coefficient, backward scattering coefficient	後方散亂係數
후방산란단면	back-scattering cross-section	後方散亂斷面
후방산란신호	backscattered signal	後方散亂信號
후방산란위상변이	back-scattering phase shift	後方散亂位相變移
후방산란인자	back-scattering parameter	後方散亂因子
후방산란자	backward scatter	後方散亂子
후방산란자외선복사계	back-scattering ultraviolet radiometer	後方散亂紫外線輻射計
후방산란자외선분광계	back-scattering ultraviolet spectrometer	後方散亂紫外線分光計
후방산란차등위상	back-scattering differential phase	後方散亂差等位相
후방산란체	backscatter	後方散亂體
후방산란편극 소거기법	backscattered depolarization technique	後方散亂偏極消去技法
후방전파	backpropagation	後方傳播

ㅎ

후방차분	backward difference	後方差分
후방차분법	backward differencing	後方差分法
후방층밀린모루	back-sheared anvil	後方層-
후빙기	postglacial age	後氷期
(안테나)후엽	backlobe	後葉
후지와라효과	Fujiwhara effect	-效果
후지타강도등급	Fujita intensity scale	-強度等級
후지타규모	F scale	-規模
후지타등급(토네이도)	Fujita scale	-等級
후지타토네이도강도등급	Fujita tornado intensity scale	-強度等級
후지타토네이도등급	Fujita tornado scale	-等級
후지타-피어슨 등급	Fujita-Pearson scale	-等級
후진(기압)골	retrogressive trough	後進(氣壓)谷
후진(기압)마루	retrogressive ridge	後進(氣壓)稜
후처리과정	post-processing	後處理過程
훅(부는)바람	puff	
훅법칙	Hooke's Law	-法則
훈증사건	fumigation event	熏蒸事件
훔볼트해류	Humboldt current	-海流
휘도	brightness values, luminance	輝度
휘도대조	contrast of luminance, luminance contrast	輝度對照
휘돌이건습계	sling psychrometer, whirling psychrometer	-乾濕計
휘돌이온도계	sling thermometer, whirling thermometer	-溫度計
휘발	volatilization	揮發
휘발성	volatility	揮發性
휘발성 유기화합물	volatile organic compound	揮發性有機化合物
휘젓기	stirring	
휘파람바람	whistling wind	
휘파람유성	whistling meteor	-流星
휴대용복사계	portable radiometers	携帶用輻射計
휴대용컵풍속계	portable cup anemometer	携帶式風向風速計
휴대용풍속계	hand anemometer	携帶用風速計
휴면기	diapause	休眠期
휴면홍염	Quiescent Prominence	休眠紅焰
휴지시간	idle time	休止時間
휴지시간	dwell time	休止時間
휴화산	dormant volcano	休火山
흐르는 모래	shifting sand	
흐름	current, flow	
흐름가시화	flow visualization	-可視化
흐름겹침	flow folding	
흐름곡률	flow curvature	-曲率
흐름깊이	flow depth	
흐름박리현상	bluff body	-剝離現象

흐름박리현상효과	bluff body effect	-剝離現象效果
흐름벡터	flow vector	
흐름분리	flow separation	-分離
흐름분지	flow splitting	-分枝
흐름안정도	stability of flow	-安定度
흐름영역	flow regime	-領域
흐름의 가시화법	law of flow visualization	-可視化法
흐름적분	flow integral	-積分
흐름패턴	flow pattern	
흐름함수	current function, flow function	-函數
흐름회오리	stream devil	
흐린 날	cloudy day	
흐린 날 대류	cloudy convection	-對流
흐린 날씨	cloudy weather	
흐린 하늘	cloudy sky	
흐림	cloudy	
흑구온도계	black-bulb thermometer	黑球溫度計
흑사병	Black Death	黑死病
흑색종	melanoma	黑色腫
흑체	black body	黑體
흑체복사	black body radiation	黑體輻射
흑체복사곡선	black body radiation curve	黑體輻射曲線
흑체복사이론	black body radiation theory	黑體輻射理論
흑체스펙트럼	black body spectrum	黑體-
흑체온도	black body temperature	黑體溫度
흑해홍수가설	Black Sea flood hypothesis	黑海洪水假說
흔들바람	fresh breeze	
흔들이풍속계	pendulum anemometer	-風速計
흔적	trace	痕迹
(강수)흔적기록계	trace recorder	(降水)痕跡記錄計
흘러오는쪽	upstream	
흘러오는쪽 경사면	upstream slope	-傾斜面
흙비	mud rain	
흡광도	absorbance, optical density	吸光度
흡수	absorption, assimilation, imbibition	吸收
흡수계	potometer	吸收計
흡수계수	absorption coefficient, coefficient of absorption	吸收係數
흡수광학깊이	absorption optical depth	吸收光學-
흡수광학두께	absorption optical thickness	吸收光學-
흡수기	absorber	吸收器
흡수능	absorptive power	吸收能
흡수단면(적)	absorption cross-section	吸收斷面(積)
흡수대	absorption band	吸收帶
흡수도	absorptance	吸收度

흡수띠	absorption band	吸收-
흡수매질	absorbing medium	吸收媒質
흡수면	absorbing surface	吸收面
흡수물	absorbate	吸收物
흡수방식	absorption schemes	吸收方式
흡수분광법	absorption spectroscopy	吸收分光法
흡수선	absorption line	吸收線
흡수선량	absorbed dose	吸收線量
흡수성	hygroscopic	吸收性
흡수손실	absorption loss	吸收損失
흡수수	absorption number	吸收數
흡수스펙트럼	absorption spectrum	吸收-
흡수습도계	absorption hygrometer	吸收濕度計
흡수율	absorptance, absorptivity	吸收率
흡수인자	absorption factor	吸收因子
흡수제	absorber, sorbents	吸收劑
흡수지수	absorptive index, index of absorption	吸收指數
흡수체	absorber	吸收體
흡수층	absorbing layer	吸收層
흡수타워	absorption tower	吸收塔
흡수태양복사	absorbed solar radiation	吸收太陽輻射
흡수특징분석	absorption features analysis	吸收特徵分析
흡수효율	absorption efficiency	吸收效率
흡습도	hygroscopicity	吸濕度
흡습성	hygroscopic	吸濕性
흡습성입자	hygroscopic particle	吸濕性粒子
흡습성핵	hygroscopic nucleus, hygroscopic wettable nucleus	吸濕性核
흡습성핵물질	hygroscopic nucleus	吸濕性核物質
흡습성핵물질 구름씨뿌리기	hygroscopic nucleus cloud seeding	吸濕性核物質-
흡습수	hygroscopic water	吸濕水
흡습수분	hygroscopic moisture	吸濕水分
흡습운동	hygroscopic movement	吸濕運動
흡열반응	endothermic reaction	吸熱反應
흡원	sink	吸原
흡인	attraction	吸引
흡인자	attractor	吸引子
흡입	entrainment, imbibition	吸入
흡입계수	entrainment coefficient	吸入係數
흡입관	suction tube	吸入管
흡입률	entrainment rate	吸入率
흡입성입자상 물질	Inhalable Particulate Matter	吸入性粒子狀物質
흡입소용돌이	suction vortices	吸入-
흡입압	suction pressure	吸入壓

ㅎ

흡입착빙	induction icing	吸入着氷
흡입풍속계	suction anemometer	吸入風速計
흡착	sorption	吸着
흡착(작용)	adsorption	吸着(作用)
흡착등온선	adsorption isotherm	吸着等溫線
흡착수	hygroscopic water, pellicular water	吸着水
흡착제	absorbent, adsorbent, chemisorbent, sorbents	吸着劑
희미한 전선	diffuse front	-前線
희석	dilution	稀釋
희소자료지역	sparse data area	稀少資料地域
희소파	rarefaction wave	稀疏波
희유기체	rare gas	稀有氣體
흰 무지개	mistbow, white rainbow	
흰 서리	white frost	
흰 스콜	white squall	
흰 얼음	white ice	
흰 얼음층	white band	-層
흰 파도	white rainbow	
히그	hig, ig	
히스테리시스	hysteresis	
히스테리시스 특성	hysteresis characteristic	-特性
히스토그램	histogram	
힘	force	力
힘균형	force balance	-均衡
힘의 분해	resolution of force	-分解
힘의 성분	component of force	-成分
힘의 작용선	line of action of the force	-作用線
힘의 중력단위	gravitational unit of force	-重力單位
힘의 크기	magnitude of force	
힘의 합성	composition of force	-合成
힘평형	equilibrium of force	-平衡
힙시서멀	Hypsithermal	

ㅎ

기타

한글	영문	한자
0도 자오선시간	Zulu(Z) Time	零度子午線時間
0의 기호	cipher, cypher	零-記號
1.5차수종결	one-and-a-half order closure	-次數終結
10μ 미세입자	particulate matter 10(PM10)	-微細粒子
10바	decibar	
100만	mega-	-萬
1월 해빙	January thaw	一月解氷
1차계량단순화	first metric simplification	一次計量單純化
1차대기	primordial atmosphere	一次大氣
1차대기미량가스	primary trace atmospheric gases	一次大氣微量-
1차모멘트	first moment	一次-
1차무지개	primary bow, primary rainbow	一次-
1차산란	primary scattering	一次散亂
1차생산	primary production	一次生産
1차순환	primary circulation	一次循環
1차오염물질	primary pollutant	一次汚染物質
1차원구름탐측기	one-dimensional cloud probe	一次元-探測機
1차원모형	one-dimensional model	一次元模型
1차원보딘모형	one-dimensional Bodin's model	一次元-模型
1차원스펙트럼밀도텐서	one-dimensional spectral density tensor	一次元-密度-
1차원습도스펙트럼	one-dimensional humidity spectrum	一次元濕度-
1차원온도스펙트럼	one-dimensional temperature spectrum	一次元溫度-
1차전선	primary front	一次前線
1차종결	first-order closure	一次終結
1차종결모형	first-order closure model	一次終結模型
1차종결방식	first-order closure scheme	一次終結方式
1차천이	primary succession	一次遷移
1차표준	primary standard	一次標準
1차표준물질	primary standard substance	一次標準物質

1차 GARP 전지구실험	First GARP Global Experiment	一次-全地球實驗
22도 무리	halo of 22°	
22도할로무리해	parhelia associated with the 22° halo	
2급기후관측소	second-order climatological station	二級氣候觀測所
2진법범용형식표현	binary universal format representation	二進法汎用型式表現
2차경로추출	second path retrieval	二次經路抽出
2차경로탐측	second path sounding	二次經路探測
2차대기	secondary atmosphere	二次大氣
2차대기오염물질	secondary air pollutant	二次大氣汚染物質
2차레이더	secondary radar	二次-
2차모멘트	second moment	二次-
2차무지개	secondary rainbow	二次-
2차미량대기가스/입자	secondary trace atmospheric gases or particle	二次微量大氣-/粒子
2차방출	secondary emission	二次放出
2차산란	secondary scattering	二次散亂
2차생산	secondary production	二次生産
2차순환	secondary circulation	二次循環
2차에어로졸	secondary aerosol	二次-
2차에코	secondary wave	二次-
2차(레이더)에코	second trip (radar) echo	二次-
2차오염물질	secondary pollutant	二次汚染物質
2차온난전선	secondary warm front	二次溫暖前線
2차원 난류	two-dimensional turbulence	二次元亂流
2차원 입자영상기	two-dimensional stereo imager	二次元粒子映像機
2차유기에어로졸	secondary organic aerosol	二次有機-
2차전선	secondary front	二次前線
2차제곱근	root mean square	二次-根
2차제곱근오차	root-mean-square error	二次-根誤差
2차천이	secondary succession	二次遷移
2차표준기	secondary standard	二次標準器
2차표준대기질	secondary ambient air quality standard	二次標準大氣質
2차한랭전선	secondary cold-front	二次寒冷前線
2차회절	secondary diffraction	二次回折
2차흐름	secondary flow	二次-
3배풍속계	three-cup anemometer	三盃風速計
3중도플러	triple Doppler	三重-
3중도플러 레이더	triple Doppler radar	三重-
46도 무리	halo of 46°	
46도할로무리해	parhelia associated with the 46° halo	
4방위바람	cardinal winds	四方位-
4방위점	cardinal point	四方位點
4분위범위	inter-quartile range	四分位範圍
4월 소나기	April shower	

기타

4차원변수자료동화구성틀	four-dimensional variational data assimilation framework	四次元變數資料同化構成-
4차원분석	four-dimensional analysis	四次元分析
4차원자료동화	four-dimensional data assimilation	四次元資料同化
5-10 시스템	five-and-ten system	
5년간	lustrum	-年間
5렌즈공중카메라	five-lens aerial camera	五-空中-
5일파	five-day wave	五日波
5채널주사복사계	five-channel scanning radiometer	五-走査輻射計
8방위	octant	八方位
8분위	octant, okta	八分位
8분의 1	octa, okta	
90도위상차성분	quadrature component (Q-component)	-位相差成分
A급증발계	class-A pan	-級蒸發計
ABC 두레박	ABC bucket	(表面水溫測定用)
ARDC 모형대기	ARDC Model Atmosphere	-模型大氣
AVHHR 영상	AVHHR imagery	-映像
B 곡선	B-curve	-曲線
B단위	B-unit	-單位
B 디스플레이	B-display	
B스코프	B scope	
BLAG 가설	BLAG hypothesis	-假說
BM 방안	Betts Miller scheme	-方案
BPI증발계	BPI pan	-蒸發計
C3 경로	C3 pathway	-經路
C4 경로	C4 pathway	-經路
C날씨	C weather	
C밴드	C-band	
C수	C figure	-數
C지수	C index	-指數
cgs계	centimeter-gram-second system	-系
CLIMAT 메시지	CLIMAT message	
CLIMAT 방송	CLIMAT broadcast	-放送
CRSTER 모형	CRSTER model	-模型
D값	D-value	
D-분석	D-analysis	-分析
D영역	D-region	-領域
D층	D-layer	-層
DAD값	DAD value	
DAD분석	DAD analysis	-分析
DAD해석	DAD analysis	-解析
dBZ	dBZ	
DDA값	DDA value	
DDA 분석	depth-duration-area(DDA) analysis	-分析

기타

E₂층	E₂-layer	-層
e-감쇠시간	e-folding time	-減衰時間
E영역	E-region	-領域
E층	E-layer	-層
ELF방출	ELF emission	-放出
F₁ 층	F₁-layer	-層
F₂ 층	F₂-layer	-層
F검정	F-test	-檢定
F-비	F-ratio	-比
F영역	F-region	-領域
F층	F-layer	-層
FGGE 실험	First GARP Global Experiment	-實驗
f-p-s 단위계	foot-pound-second system	-單位系
FSW의 실행프로그램	interpreted program	-實行-
G-구역	G-region	-區域
GARP 대서양열대실험	GARP Atlantic Tropical Experiment	-大西洋熱帶實驗
GATE 실험	GARP Atlantic Tropical Experiment	-實驗
GCA 최저기상조건	GCA minimum	-最低氣象條件
GHOST 기구계획	GHOST Balloon Project	-氣球計劃-
GOES 위성	GOES Satellite	-衛星
HCMM 위성	HCMM Satellite	-衛星
HHLL법	HHLL method	-法
HLLH법	HLLH method	-法
ICAO표준대기	ICAO standard atmosphere	-標準大氣
IFR목적지 최저기상조건	IFR terminal minimum	-目的地最低氣象條件
IFR비행	IFR flight	-飛行
Insat위성	Insat	-衛星
J 스코프	J scope	
J-W 계	J-W meter	-計
K-바람	K-wind	
K-밴드	K-band	
K 지수	K index	-指數
K-T 경계층	Cretaceous-Tertiary(K-T) boundary	-境界層
L-밴드	L-band	
LRIT 및 HRIT 자료생성 부시스템	L/HRIT Generation Subsystem	-資料生成 副-
M-곡선	M-curve	-曲線
M-단위	M-unit	-單位
M 미터	M meter	
M영역	M-region	-領域
MI 접속장치	MI Interface Unit	-接續裝置
MONT 전보식	MONT code	-電報式
mts단위계	meter-ton-second system of unit	-單位系
MU레이더	MU radar	

기타

N-곡선	N-curve	-曲線
N날씨	N weather	
n년사건	n-year event, T-year event	-年事件
N-단위	N-unit	-單位
NACA표준대기	NACA standard atmosphere	-標準大氣
NASA 스캐터로미터	NASA scatterometer	
NAVTEX 예보	NAVTEX Forecast	-豫報
NEBUL 전보식	NEBUL code	-電報式
NEXRAD 기초자료	NEXRAD base data	-基礎資料
Nimbus N 마이크로파분광계	Nimbus N Microwave Spectrometer	-極超短波分光計
NOAA급 위성	NOAA-class satellites	-級衛星
NOAA날씨자료전달서비스	NOAA Weather Wire Service	-資料傳達-
NOAA 위성시리즈	NOAA satellite series	-衛星-
NRM 풍력등급	NRM wind scale	-風力等級
N-S 계수	N-S coefficient	-係數
NWS 협력관측자망	NWS Cooperative Observer Network	-協力觀測者網
P파	Primary wave, P-wave	-波
PAR 스코프	PAR scope	
P-E몫	P-E quotient(precipitation-evaporation quotient)	
P-E지수	P-E index(precipitation-effectiveness index, precipitation-evaporation index)	-指數
pH도	pH	酸度
PPI스코프	PPI scope	
PSI	plan shear indicator	
PT(패턴열대)수	pattern tropical number	-(-熱帶)數
Q띠	Q-band	-帶
Q인자	Q-factor	-因子
QuiKSCAT위성	QuiKSCAT satellite	-衛星
R-미터	R-meter	
R스코프	R scope	
RECCO 전문형식	RECCO code	-電文形式
RHI스코프	RHI scope (range-height indicator scope)	
S값	S-value	
S-곡선법	S-curve method	-曲線法
S밴드	S-band	
S밴드레이더	S-band radar	-波-
S-수문곡선	S-hydrograph summation curve	-水文曲線
S파	S wave, secondary wave	-波
S파레이더	S-band radar	-波-
SFAZI 전문형식	SFAZI code	-電文形式
SFAZU 전문형식	SFAZU code	-電文形式
SI-단위	SI units	-單位
SIRS-위성적외분광계	SIRS-satellite infrared spectrometer	-衛星赤外分光計
SYNOP 코드	SYNOP code	

기타

't' 검증	Student's 't' test	't' 檢證
t-검정	t-test	-檢定
T꼴바람개비	wind tee	
T-수	T-number, Tropical number	
T-존데	T-sonde	
TAD곡선	time-area-depth curve	-曲線
T-E비	T-E ratio	-比
T-E지수	T-E index(Temperature-Efficiency Index)	-指數
TEMPAL전문형식	TEMPAL code	-電文形式
U지수	U figure, U index	-指數
U파열	U burst	-破裂
U형태	U figure	-形態
V-띠	V-band	
V모양등압선	V-shaped isobar	-模樣等壓線
V모양저압부	V-shaped depression	-模樣低壓部
V점	V point, radiant point	-點
VAD바람단면	VAD wind profile	-斷面
VETH 선도	VETH chart	-線圖
VIP수준	VIP level	-水準
VOLMET방송	VOLMET broadcast	-放送
X띠	X-band	
X선	x-ray	-線
Z시스템	Z-system	
Z-R관계	Z-R relationship(radar reflectivity(Z)-rain rates(R) relationship)	-關係
Zulu타임	Zulu(Z) Time	

기타

부록

■ 상용약어일람표

약어	원어	한글	한자
AABW	Antarctic Bottom Water	남극심층수	南極深層水
AAMS	Asian Aeronautical Meteorology Service	아시아항공기상서비스	-航空氣象-
AOE	Airborne Antarctic Ozone Experiment	대기남극오존실험	大氣南極-實驗
AAOS	Automatic Agriculture Observing System	자동농업기상관측시스템	自動農業氣象觀測-
AASE	Airborne Arctic Stratosphere Expedition	대기북극성층권 탐사	大氣北極成層圈探査
AATSR	Advanced Along Track Scanning Radiometer	궤도를 따라 주사하는 진보된 복사계	軌道走査進步輻射計
ABC	Atmospheric Brown Clouds	대기갈색구름	大氣-
ABI	Advanced Baseline Imager	신진화 기준영상기	先進化基準映像機
ABL	Atmospheric Boundary Layer	대기경계층	大氣境界層
ACARS	Aircraft Communications Addressing and Reporting System	항공기운항정보 교신시스템	航空機運港情報交信-
ACC	Antarctic Circumpolar Current	남극순환해류	南極循環海流
ACCAD	Advisory Committee on Climate Applications and Data, WCDMP, WCASP	기후응용·자료에 관한 자문위원회	氣候應用資料諮問委員會

부록-I

ATOVS	Advanced TIROS Operational Vertical Sounder	첨단타이로스	尖端
ATSR	Along Track Scanning Radiometer	궤도추적주사복사계	軌道追跡走査輻射計
AUV	Autonomous Underwater Vehicle	자율무인잠수정	自律無人潛水艇
AVAWOS	AViation Automatic Weather Observing System	항공자동기상관측시스템	航空自動氣象觀測-
AVHRR	Advanced Very High Resolution Radiometer	초해상도복사계	超高解像度輻射計
AWACS	Airborne Warning And Control System	항공경보관제시스템	航空警報管制-
AWC	Atlantic Water Core	북대서양수문중심	北大西洋水分場中心
AWPG	Aviation Weather Products Generator	항공기상자료생산	航空氣象資料生産
AWS	Automatic Weather Station	자동기상관측소	自動氣象觀測所
BAPMoN	Background Air Pollution Monitoring Network(WMO)	배경 대기오염 모니터링 감시망	背景大氣汚染-監視網
BAS	British Antarctic Survey(UK)	영국 남극연구소	英國南極研究所
BipAG	Bipolar Action Group	양극연구협력과학위원회	兩極研究協力科學委員會
BL	Boundary Layer	경계층	境界層
BoM	Australian Bureau of Meteorology	호주 기상청	-氣象廳
BS	Brier Score	브라이어 지수	-指數
BSRN	Baseline Surface Radiation Network, Switzerland	국제기준면복사네트워크	國際基準面輻射-
BSS	Brier Skill Score	브라이어숙련도	-熟練度
BTD	Brightness Temperature Difference	휘도온도차별	輝度溫度差法
BV	Breeding Vector	브리딩 벡터	
BWER	Bounded Weak Echo Region	약한 에코 영역	弱-領域
CAA	Civil Aviation Authority	민간항공국	民間航空局
CAC	Climate Analysis Center, NOAA	기후해석센터	氣候解釋-
CAEA	Climate and Atmospheric Environment Activities(trust fund)	기후 및 대기환경활동	氣候·大氣環境活動
CAEM	Center on Antarctic Environmental Monitoring	남극환경감시센터	南極環境監視-
CAeM	Commission for Aeronautical Meteorology(WMO)	항공기상위원회	航空氣象委員會
CAgM	Commission for Agricultural Meteorology(WMO)	농업기상전문위원회	農業氣象專門委員會
CAM	Community Atmosphere Model, NCAR	대기모형공동체	大氣模型共同體
CAPE	Convective Available Potential Energy	대류가용잠재에너지	對流可用潛在-
CAPPI	Constant Altitude Plan Position Indicator	등고도 평면위치지시기	等高度平面位置指示器

부-록

약어	영문	국문	한자
CAPS/OU	Center for Analysis and Prediction of Storms, University of Oklahoma	악기상 예보분석센터, 오클라호마대학교	惡氣象豫報分析-
CARS	Climate Applications and Referral System, WCASP	기후응용검색서비스	氣候應用檢索
CAS	Commission for Atmospheric Sciences(WMO)	WMO 대기과학위원회	-大氣科學委員會
CASO	Climate of Antarctica and the Southern Ocean(IPY)	남극과 남반구해양기후	南極南半球海洋氣候
CAT	Clear-Air Turbulence	청천난류	晴天亂流
CAVT	Constant Absolute Vorticity Trajectory	일정절대 소용돌이도궤적	一定絕對-度遺跡
CBHWR	Capacity Building in Hydrology and Water Resources Management (WMO Office)	수문 및 수자원관리	水文水資源管理
CBS	Commission for Basic Systems(WMO)	WMO 기본체계위원회	-基本體系委員會
CCA	Climate Coordination Activities	기후조정활동	氣候調整活動
CCAIS	Climate Change Adaptation Intelligence System	기후변화적응정보시스템	氣候變化適應情報-
CCCPR	Center for Climate/Environment Change Prediction Research	기후·환경변화예측연구센터	氣候環境變化豫測硏究-
CCCSN	Canadian Climate Change Scenarios Network	캐나다 기후변화시나리오망	-氣候變化網
CCD	Charge Coupled Device	전하결합소자	電荷結合素子
CCDP	Climate Change Detection Project, WCDMP	기후변화탐지계획	氣候變化探知計劃
CCIC	Climate Change Information Center	기후변화정보센터	氣候變化情報-
CCl	Commission for Climatology	기후위원회	氣候委員會
CCL	Convective Condensation Level	대류응결고도	對流凝結高度
CCM	Chemistry Climate Model(WCRP)	화학기후모형	化學氣候模型
CCMVal	Chemistry Climate Model Validation Activity(SPARC)	화학기후모형검증	化學氣候模型檢證
CCN	Cloud Condensation Nuclei	구름응결핵	-凝結核
CCS	Carbon Capture and Storage	탄소 포집·저장기술	炭素捕獲貯藏技術
CCSR	Center for Climate System Research, University of Tokyo, Japan	동경대학 기후시스템연구센터	東京大學氣候-硏究-
CDBMS	Climate Database Management System	기후 데이터베이스 관리시스템	氣候-管理
CDO	Central Dense Overcast	(태풍)중심밀부근처	(颱風)中心部近處
CDW	Circumpolar Deep Water	심층수극순환	深層水極循環
CEOP	Coordinated Enhanced Observing Period(WWW)	종합(지구 물 순환) 강화관측기간 프로젝트	綜合(地球-循環)强化觀測期間-

CEOS	Committee on Earth Observation Systems	지구관측시스템위원회	地球觀測委員會
CERES	Clouds and the Earth's Radiant Energy System	구름 및 지구복사에너지 시스템	-地球輻射-
CFC	ChloroFluoroCarbon	염화불화탄소	鹽化弗化炭素
CFD	Computational Fluid Dynamics(Modeling)	전산유체역학(모형화), 계산유체역학	電算流體力學(模型化), 計算流體力學
CFMIP	Cloud Feedback Model Intercomparison Project	구름되먹임모형 상호비교프로그램	-模型相互比較-
CFS	Climate Forecast System	기후예보체계	氣候豫報體系
CGCM	Coupled General Circulation Model	대기-해양 대순환 접합모형	大氣-海洋大循環接合模型
CGMS	Coordination Group for Meteorological Satellites	기상위성 조정그룹	氣象衛星調整-
CHL	Cold Halocline Layer	저온염분약층	低溫鹽分躍層
CHy	Commission for Hydrology(CHy)	WMO 수문위원회	-水文委員會
CIAP	Climate Impact Assessment Program	기후영향평가계획	氣候影響評價計劃
CIB	International Council for Building Research Studies and Documentation	국제건물연구정보협의회	國際建物硏究情報協議會
CIC(NAS)	Climate Impact Committee(NAS)	기후영향위원회	氣候影響委員會
CICG	Geneva International Conference Centre	제네바 국제회의센터	-國際會議-
CIDA	Canadian International Development Agency	캐나다 국제개발단	-國際開發團
CIE	Commission Internationale de l'Eclairage	국제조명위원회	國際照明委員會
CIIFEN	International Center on Research 'El Niño'	엘니뇨국제연구센터	-國際硏究-
CILSS	Permanent Inter-State Committee on Drought Control in the Sahel	사하라지역 가뭄통제국가간위원회	-地域統制國家間委員會
CIMH	Caribbean Institute for Meteorology and Hydrology	카리브기상수문연구소	-氣象水文硏究所
CIMO	Commission for Instruments and Methods of Observation(WMO)	WMO 기상측기 및 관측법위원회	-氣象測機·測法委員會
CIMSS	Cooperative Institute for Meteorological Satellite Studies	기상위성연구협동기구	氣象衛星硏究協同機構
CIPM	International Committee for Weights and Measures	국제도량형위원회	國際度量衡委員會
CIRA/CSU	Cooperative Institute for Research in the Atmosphere, CSU	대기연구협동연구소	大氣硏究協同硏究所
CIRM	International Radio Maritime Committee(SAUO)	국제해상무선위원회	國際海上無線委員會
CISK	Conditional Instability of the Second Kind	제2종조건부불안정	第二種條件附不安定
CliC	Climate and Cryosphere(WCRP)	기후와 설빙권	氣候-雪氷圈
CLICOM	Climate-Computing system(WCDMP)	기후-전산화시스템	氣候電算化-
CLIMAG	Climate Prediction for Agriculture	농업기후예측	農業氣候豫測

부록

약어	영문	국문	漢字
CLIMAP	Climate Long-range Investigation, Mapping and Prediction	장기적 기후조사, 지도제작과 예측 (고기후복원프로젝트)	長期的氣候調査, 地圖製作·豫測 (古氣候復原-)
CliPAS	Climate Prediction and its Application to Society	기후예측 및 사회응용	氣候豫測-社會應用
CLIPS	Climate Information and Prediction Services(WCP)	기후정보 및 예측서비스	氣候情報-豫測-
CLIVAR	Climate Variability and Predictability	기후변동성 및 예측	氣候變動性-豫測
CLPA	Climate Prediction and Adaptation(WMO Branch)	기후변화적응	氣候變化適應
CLW	Climate and Water(WMO Department)	WMO 기후수문과	-氣候水文課
CMA	China Meteorological Administration	중국 기상청	中國氣象廳
CMAQ	Community Multiscale Air Quality model	군집다중규모대기질모형	群集多衆規模大氣質模型
CMC/EnvCanada	Meteorological Service of Canada/Environment Canada	캐나다 기상청	-氣象廳
CMDL	Climate Monitoring and Diagnostics Laboratory, NOAA	기후감시진단연구소	氣候監視診斷研究所
CMDPS	COMS Meteorological Data Processing System	통신해양기상위성 기상자료처리시스템	通信海洋氣象衛星氣象資料處理-
CMIP	Coupled Model Intercomparison Projects(WVRP)	결합기후모형상호비교사업	結合氣候模型相互比較事業
CMM	Commission for Marine Meteorology	해상기상위원회	海上氣象委員會
CMO	Caribbean Meteorological Organization	카리브 기상기구	-氣象機構
CNES	Centre national d'études spatiales(National Space Research Centre) (France)	프랑스 국립우주연구소	-國立宇宙研究所
CNRS	Centre national de la recherche scientifique(France)	프랑스 국립과학연구센터	-國立科學研究-
COAPS/FSU	Center for Ocean-Atmospheric Prediction Studies, Florida State Univ.	(플로리다대)해양대기예측연구센터	海洋大氣豫測研究-
COHMAP	Cooperative Holocene Mapping Project	홀로세 기후복원프로젝트	-氣候復原-
COLA	Center for Ocean-Land-Atmosphere Studies	해양-지표-대기연구소(미국)	海洋地表大氣研究所
COMET	Cooperative Programme for Operational Meteorology Education and Training	기상운영교육훈련공동프로그램	氣象運營敎育訓練協同-
COMS	Communication, Ocean and Meteorological Satellite	통신해양기상위성, 천리안위성	通信海洋氣象衛星, 千里眼衛星
COP	Conference of the Parties	당사국총회	當事國總會
COPS	Convective and Orographically induced Precipitation Study(AREP)	대류융기수강수연구	對流隆起水研究
CoSAMC	Commission for Special Applications of Meteorology and Climatology	기상기후특별응용위원회	氣象氣候特別應用委員會

COSMIC	Constellation of Satellites for Meteorology, Ionosphere and Climate	기상, 이온층, 기후관측 위성군	氣象、-層、氣候觀測衛星群
COSNA	Composite Observing System for the North Atlantic	북대서양복합관측체계구축	北大西洋觀測體系構築
COST	European Cooperation in the Field of Scientific and Technical Research	유럽 과학기술연구 협력사업	-科學技術研究協力事業
CPTEC	Centro de Previso de Tempo e Estudos Climticos	브라질 기후연구센터	-氣候研究-
CRED	Centre for Research on the Epidemiology of Disasters	재난역학연구센터	災難力學研究-
CRIDA	Central Research Institute for Dryland Agriculture(India)	인도 건조지역 농업중앙연구소	-乾燥地域農業中央研究所
CrIS	Cross-track Infrared Sounder	크로스트랙 적외선탐측기	-赤外線探測機
CRT	Cathode-Ray Tube	음극선관	陰極線管
CryOS	Cryosphere Observing System	지구빙권 관측시스템	地球水圈觀測-
CSAQ	Community Scale Air Quality	군집규모대기질	群集規模大氣質
CSD	Commission on Sustainable Development(UN)	지속개발위원회	持續開發委員會
CSIRO	Commonwealth Scientific and Industrial Research Organization	오스트레일리아 연방과학산업조사사청	-聯邦科學産業調査廳
CSM	Climate System Monitoring, WCDMP	기후계감시	氣候系監視
CTBTO	Comprehensive Nuclear Test Ban Treaty Organization	포괄적핵실험금지조약기구	包括的核實驗禁止條約機構
CTD	Conductivity(salinity), Temperature and Depth	수심수온염분기록계	水深水溫鹽分記錄計
CTI	Climate Technology Initiative(OECD)	기후변화시범사업	氣候變化示範事業
CW	Continuous-wave	연속파	連續波
CZCS	Coastal Zone Color Scanner	해안지역 파장별조사	海岸地域波長別走査
DAD analysis	Depth-Area-Duration analysis	강우량·면적·지속기간 분석	降雨量面積持續期間分析
DAD value	Depth-Area-Duration value	강우량·면적·지속기간 산정	降雨量面積持續期間算定
DARE	Data Rescue programme, WCDMP	데이터 레스큐 계획	-計劃
DART	Deep Ocean Assessment and Reporting Tsunamis	심해쓰나미 평가보고	深海-(津波)評價報告
DBCP	Data Buoy Cooperation Panel(WMO/IOC)	자료 부이 협력위원회	資料-協力委員會
dBZ	decibels of Z	레이저Z 반사도값	-反射度-
DC	Data-Collection platform	데이터수집플랫폼	-收集-
DCPC	Data Collection or Product Centre(WIS)	데이터수집 및 생산센터	-收集·生産-
DCR	Development Cooperation and Regional Activities (WMO Department)	WMO 개발협력지역활동	-開發協力地域活動

ECE	Economic Commission for Europe(UN)	유럽경제위원회	-經濟委員會
EC	Executive Council(WMO)	행정위원회	行政委員會
ECMWF	European Centre of Medium Range Weather Forecast	유럽 중기기상예보센터	-中期氣象豫報-
ECO	Economic Cooperation Organization	경제협력기구	經濟協力機構
ECOSOC	Economic and Social Council(UN)	유엔경제사회이사회	-經濟社會理事會
ECOWAS	Economic Community of West African States	서아프리카경제공동체	西-經濟共同體
ECV(s)	Essential Climate Variable(s)	필수기후변수	必須氣候變數
EEA	European Environment Agency	유럽환경청	-環境廳
EHF	Extremely High Frequency	극고주파	極高周波
EHP	Elevated Heat Pump Effect	고조된 열펌프효과	高潮熱-效果
EIA	Environmental Impact Assessment	환경영향평가	環境影響評價
ELF	Extremely Low Frequency	극저주파	極低周波
ELI/PAC	Environmental Law and Institutions Programme Activity Centre	환경법기구 프로그램 활동센터	環境法機構-活動-
ELLFB	Experimental long-lead Forecast Bulletin	경험적 장기예측게시	經驗的長期豫測啓示
EM	Electromagnetic Modeling	전자모형화	電磁模型化
EML	Environmental Measurements Laboratory, USA	환경측정실험실	環境測定實驗室
ENH	Extended Northern Hemisphere	북반구확장영역	北半球擴張領域
ENSO	El Niño/Southern Oscillation	엘니뇨/남방진동현상	-南方振動現象
EnvCanada	Environment Canada	캐나다환경부	-環境部
EO	Earth Observation	지구관측	地球觀測
EOF	Empirical Orthogonal Function	경험직교함수	經驗直交函數
EOS	Earth Observing System	지구관측시스템	地球觀測-
EPA	Environmental Protection Agency, U.S.A.	(미국)환경보호청	環境保護廳
ERA	Emergency Response Activities	비상대응활동	非常對應活動
ERB	Earth Radiation Budget	지구복사수지	地球輻射收支
ERBE	Earth Radiation Budget Experiment	지구복사수지실험	地球輻射收支實驗
ERBS	Earth Radiation Budget Satellite	지구복사수지위성	地球輻射收支衛星
ESA	Electric/Electronic Sub Assembly	전자기기 작동부분석	電磁機器作動分析
ESA	European Space Agency	유럽우주기구	-宇宙機構

부록

ESCAP	Economic and Social Commission for Asia and the Pacific(UN)	아시아태평양 경제사회이사회	-太平洋經濟社會理事會
ESM	Earth System Model	지구시스템모형	地球-模型
ESRL	Earth System Research Laboratory, NOAA	지구시스템연구실	地球-研究室
ESSP	Earth System Science Partnership	지구시스템과학협력	地球-科學協力
ETM	Enhanced Thematic Mapper	ETM 센서	
ETR	Education and Training(WMO Department)	교육훈련사무국	教育訓鍊事務局
ETRP	Education and Training Programme	교육·훈련프로그램	教育訓鍊
ETWS	Expert Team on Wind wave and Storm surge	폭풍해일전문가팀	暴風海溢專門家-
EUMETNET	Network of European Meteorological Services	유럽기상서비스망	-氣象-網
EUMETSAT	European Organisation for the Exploitation of Meteorological Satellites	유럽기상위성개발기구	-氣象衛星開發機構
FAA	Federal Aviation Administration	(미국)연방항공청	聯邦航空廳
FAO	Food and Agriculture Organization of the United Nations	유엔식량농업기구	-食糧農業機構
FASTEX	Fronts and Atlantic Storm Track Experiment	전선과 북대서양폭풍우추적실험	前線-北大西洋暴風雨追跡實驗
FCCC	UN Framework Convention on Climate Change	기후변화협약	氣候變化協約
FDP	Forecast Demonstration Project(WMO/WWRP)	예보시범프로젝트	豫報示範
FFMP	Flash Flood Monitoring & Prediction	돌발홍수감시 및 예측	突發豪雨監視-豫測
FGGE	First GARP Global Experiment	1차 GARP 전지구실험	一次-全地球實驗
FIR	Flight Information Region	비행정보구역	飛行情報區域
FNMOC	Fleet Numerical Meteorology and Oceanography Center	미함대 수치기상해양센터	美艦隊數値氣象海洋-
FRM	Federal Reference Method	연방참조방법	聯邦參照方法
FSS	Federal Snow Sampler	연방눈채집기	聯邦-採集器
FT-IR	Fourier Transform Infrared Spectroscopy	푸리에변환 적외분광법	-變換赤外分光法
FWCC	First World Climate Conference	제1차세계기후회의	第一次世界氣候會議
FW-CW Radar	Frequency-modulated continuous-wave radar	주파수변조 지속파레이다	周波數變造持續波-
FWS	Federal Weather Service	(미국)연방기상대	聯邦氣象臺
GAGE	Global Atmospheric Gases Experiment	지구대기실험	地球大氣實驗
GAMET	Ground Air METeorological forecast	저층공역예보	低層空域豫報
GAO	Global Atmospheric Office	전지구기후사무국	全地球氣候事務局

약어	영어	한국어	한자
GARP	Global Atmospheric Research Programme	지구대기 연구계획	地球大氣研究計劃
GASP	Global Analysis and Prediction, BoM	지구분석 및 예측	地球分析·豫測
GATE	GARP Atlantic Tropical Experiment	지구대기연구사업-열대대서양실험	地球大氣研究事業熱帶大西洋實驗
GAW	Global Atmosphere Watch(WMO/AREP)	지구대기감시	地球大氣監視
GAWSIS	GAW Station Information System	GAW 관측소정보시스템	-觀測所情報-
GCA	Ground-Controlled Approach	지상관제접근	地上管制接近
GCAs	Geographic Coordinating Areas	지리적좌표지역	地理的座標地域
GCIP GEWEX	GEWEX Continental Scale International Project, GEWEX/WCRP	대륙스케일 상호작용실험	大陸-相互作用實驗
GCM	General Circulation Model	대기대순환모형	大氣大循環模型
GCMs	Global Climate Models	전지구기후모형	全地球氣候模型
GCOS	Global Climate Observing System	전지구 기후 관측시스템	全地球氣候觀測-
GCSS GEWEX	GEWEX Cloud System Study, GEWEX/WCRP	구름시스템연구	-研究
GDAPS	Global Data Assimilation and Prediction System	전지구자료동화및예보시스템	全地球資料同化豫報-
GDAS	Global Data Assimilation System, NCMRWF	전지구자료동화시스템	全地球資料同化體系
GDP	Gross Domestic Product	국내총생산	國內總生産
GDPFS	Global Data-processing and Forecast System(WMO/WWW)	전지구자료처리및예보시스템	全地球資料處理·豫報-
GDPS	Global Data Processing System	전지구자료처리시스템	全地球資料處理-
GEF	Global Environment Facility(UNDP)	지구환경금융	地球環境金融
GEM	Global Environmental Multiscale, CMC/EnvCanada	지구환경다중규모(캐나다)	地球環境多重規模
GEMS	Global Environment Monitoring System(UN)	지구환경감시체계	地球環境監視體系
GEO	Group on Earth Observations	지구관측그룹	地球觀測-
GEOS-FV-CUBE	Geodetic Earth Orbiting Satellite-Finite Volume-Cube, NASA	측지지구궤도위성	測地地球軌道衛星
GEOSS	Global Earth Observing System of Systems	전지구관측시스템	全地球觀測-
GERB	Geostationary Earth Radiation Budget	정지(위성)지구복사수지	靜止(衛星)地球輻射收支
GEWEX	Global Energy and Water cycle Experiment(WCRP)	지구·에너지·물순환실험	地球·循環實驗
GFCS	Global Framework for Climate Services	전지구기후서비스체계	全地球氣候-體系
GFDL	Geophysical Fluid Dynamics Laboratory, NOAA	지구물리유주체역학연구소	地球物理流體力學研究所
GFS	Global Forecast System, NCEP	전구예측시스템	全球豫測-
GHCN	Global Historical Climate Network	전지구역사적기후망	全地球歷史的氣候網

GRID	Global Resource Information Database	지구자원정보데이터베이스	地球資源情報-
GRIMs	Global/Regional Integrated Model system, YSU	전지구(지역)통합모형시스템	全地球(地域)統合模型
GSFC	Goddard Space Flight Center, NASA	고다드 미항공우주국우주항공센터	-美航空宇宙局宇宙航空-
GSM	Global Spectral Model, JMA	일본기상청 전지구분광모형	日本氣象廳全地球分光模型
GSN	GCOS Surface Network	GCOS 지상기상관측망	-地上氣象觀測網
GSWP	Global Soil Wetness Project	전구토양수분프로젝티	全球土壤水分-
GTN-G	GCOS Terrestrial Network for Glaciers	GCOS 지구빙하관측망	-地球氷河觀測網
GTN-H	GCOS Terrestrial Network-Hydrology	GCOS 지구수문관측망	-地球水文觀測網
GTN-P	GCOS Terrestrial Network for Permafrost	GCOS 지구영구동토 관측망	-地球永久凍土觀測網
GTOS	Global Terrestrial Observing System(UNEP, UNESCO, ICSU)	세계육지관측시스템	世界陸地地面觀測-
GTS	Global Telecommunication System(WWW/WMO)	세계기상자료통신망	世界氣象資料通信網
GUAN	GCOS Upper Air Network	GCOS 상층대기관측망	-上層大氣觀測網
GURME	GAW Urban Research Meteorology and Environment	GAW 도시기상 및 환경연구 프로젝트	都市氣象-環境硏究-
GWP	General Warming Potential Index	잠재적 대기온난화지수	潛在的大氣溫暖化指數
GWRF	Global WRF	전지구일기연구예보	全地球日氣硏究豫報
HCFCs	hydrochlorofluorocarbons	수소화염화불화탄소	水素化鹽化弗化炭素
HDP	Human Dimensions of Global environmental change(ICSU)	지구환경변화의 인간(석)차원 연구계획	地球環境變化-人間的次元硏究計劃
HELCOM	Baltic Marine Environment Protection Commission	발틱해양환경보호위원회	-海洋環境保護委員會
HFWR	Hydrological Forecasting(WMO Division)	WMO 수문예보국	-水文豫報局
HHWS	Heat and Health Warning System	열파 및 건강경보시스템	熱波健康警報-
HiRID	High Resolution Imager Data	고해상도이미지자료	高解像度-資料
HIRS	High Resolution Infrared Sounder	고분해적외선탐측기	高分解赤外探測機
HMEI	Association of Hydro-Meteorological Equipment Industry	고단위환경모델링방법	高單位環境-方法
HOMS	Hydrological Operational Multipurpose System(WMO/HWRP)	수문운영다목적시스템	水文運營多目的-
HRLDAS	High Resolution Land Data Assimilations Systems	고해상지표자료동화체계	高解像地表資料同化體系
HTAP	Hemispheric Transport of Air Pollution	반구오염물수송	牛球汚染物輸送
HWRP	Hydrology and Water Resources Programme(WMO)	수문 및 수자원프로그램	水文-水資源-
HYCOS	Hydrological Cycle Observing System(HWRP)	수문순환관측체계	水文循環觀測體系
IABM	International Association of Broadcast Meteorology	국제방송기상협의체	國際放送氣象協議體

IABP	International Arctic Buoy Programme(DBCP)	국제북극부이 프로그램	國際北極-
IAC	International Analysis Code	국제분석전보식	國際分析電報式
IAEA	International Atomic Energy Agency(UN)	국제원자력에너지사무국	國際原子力事務局
IAF	International Astronautical Federation	국제우주비행사연합	國際宇宙飛行士聯合
IAHR	International Association of Hydraulic Engineering and Research	국제수력공학연구협의회	國際水力工學硏究協議會
IAHS	International Association of Hydrological Sciences(IUGG)	국제수문과학협의회	國際水文科學協議會
IAI	Inter-American Institute for Global Change Research	지구변화연구아메리카내부기구	地球變化硏究-內部機構
IAM	Integrated Assessment Model	통합적 평가모형	統合的評價模型
IAMAS	International Association of Meteorology and Atmospheric Sciences (IUGG)	기상 및 대기과학국제연맹	氣象大氣科學國際聯盟
IAOOS	Integrated Arctic Ocean Observing System(IPY)	통합북극해양관측체계	統合北極海洋觀測體系
IAPSO	International Association for the Physical Sciences of the Ocean (IUGG)	국제해양물리학연맹	國際海洋物理學聯盟
IASI	Infrared Atmospheric Sounding Interferometer	적외선대기탐측기	赤外線大氣探測機
IAS	Indicated Air Speed	지시공기속도	指示空氣速度
IATA	International Air Transport Association	국제항공운송협회	國際航空運送協會
IAU	International Astronomical Union	국제천문연맹	國際天文聯盟
IAUC	International Association for Urban Climate	도시기후국제연맹	都市氣候國際聯盟
IAVW	International Airways Volcanic Watch	국제항공로화산감시	國際航空路火山監視
IAWGD	Inter-Agency Working Group on Desertification	사막화에 관한 기관간 작업그룹	沙漠化-機關間作業-
ICALPE	International Centre for Alpine Environment	국제알프스환경센터	國際-環境-
ICAO	International Civil Aviation Organization(UN)	국제민간항공기구	國際民間航空機構
ICARP	International Conference on Arctic Research Planning	북극연구기획 국제회의	北極硏究企劃國際會議
ICAS	Interdepartmental Committee on Atmospheric Sciences (former Federal Council for Science and Technology)	부처간대기과학위원회	部處間大氣科學委員會
ICH-ITSU	International Coordination Group for the Tsunami Warning System in the Pacific(of IOC)	태평양지진해일 경보체제 국제조정그룹	太平洋地震海溢警報體制國際調整-
ICHM	International Commission on History of Meteorology	기상역사국제위원회	氣象歷史國際委員會
ICID	International Commission on Irrigation and Drainage	관개수배출수국제위원회	灌漑水排出水國際委員會

ICIMOD	International Centre for Integrated Mountain Development	종합산악개발국제센터	綜合山岳開發國際-
ICL	International Consortium on Landslides	산사태국제협회	山沙汰國際協會
ICPAC	IGAD Climate Prediction and Applications Centre	IGAD 기후예측응용센터	-氣候豫測應用-
ICSU	International Council of Scientific Union	국제과학위원회	國際科學委員會
ICT	Information & Communication Technology	정보통신기술	情報通信技術
IDB	Inter-American Development Bank	미주개발은행	美洲開發銀行
IDNDR	International Decade for Natural Disaster Reduction(UN)	국제자연재해감소 10개년계획	國際自然災害減少十個年計劃
IDRC	International Disaster Reduction Conference	국제재해저감회의	國際災害低減會議
IEE	Initial Environmental Evaluation	초기환경평가	初期環境評價
IEEP	International Environmental Education Programme	국제환경교육프로그램	國際環境敎育-
IEM	Integrated Environmental Management	통합환경관리	統合環境管理
IETC	International Environmental Technology Centre	국제환경기술센터	國際環境技術-
IF	Intermediate Frequency	중간주파수	中間周波數
IFAD	International Fund for Agricultural Development(UN)	(유엔)국제농업개발기금	國際農業開發基金
IFALPA	International Federation of Air Line Pilots Associations	항공기항로파일럿협회국제연합	航空機航路-協會國際聯合
IFAP	International Federation for Agricultural producers	농업생산을 위한 국제연합	農業生産-國際聯合
IFR	Instrument Flight Rules	계기비행규칙	計器飛行規則
IFS	Integrated Forecast System, ECMWF	ECMWF의 종합예보체계	-綜合豫報體系
IFU	Franhofer-Institut fuer Atmosphaerische Umweltforschung, Germany	독일 프라운호퍼 대기과학연구소	獨逸-大氣科學硏究所
IGAC	International Global Atmospheric Chemistry Programme (ICSU, IGBP)	전구대기화학국제공동연구계획	全球大氣化學國際共同硏究計劃
IGACO	International Global Atmospheric Chemistry Observations strategy (WMO/GAW)	국제전지구대기화학관측전략	國際全地球大氣化學觀測戰略
IGADD	Intergovernmental Authority on Drought and Development	가뭄과 개발에 관한 정부간 책임당국	-開發-關-政府間責任當局
IGAD	Intergovernmental Authority on Development	개발에 관한 정부간 책임당국	開發-政府-責任當局
IGBP	International Geosphere-Biosphere Programme, ICSU	국제지권생물권프로그램	國際地圈生物圈-
IGDDS	Integrated Global Data Dissemination Service	종합적전지구자료보급서비스	綜合的全地球資料補給-
IGM-WCP	Intergovernmental Meeting on the World Climate Programme	세계 기후 프로그램에 관한 정부 간 회의	世界氣候-關-政府間會議
IGOS	Integrated Global Observing Strategy	통합지구관측전략	統合地球觀測戰略

부록

MOPITT	Measurements of Pollution in the Troposphere	대류권오염측정	對流圈污染測定
MOR	Meteorological Optical Range	기상광학거리	氣象光學距離
MOS	Model Output Statistics	모형출력통계	模型出力統計
MPAS	Model for Prediction Across Scales, NCAR	규모예보모형	規模豫報模型
MPDI	Microwave Polarization Difference Index	마이크로파편광차지수	-波偏光差指數
MPF	Maximum Probable Flood	가능최대홍수량	可能最大洪水量
MPI	Max Planck Institute, Germany	막스플랑크연구소	-研究所
MPP	Maximum Probable Precipitation	가능최대강수량	可能最大降水量
MRF	Medium-Range Forecast Model, NCEP	중기예보모형	中期豫報模型
MRR	Micro Rain Radar	마이크로파강수 레이더	-波降雨-
MSA	Metropolitan Statistical Area	도심통계지역	都心統計地域
MSC	Meteorological Service of Canada	캐나다 기상청	-氣象廳
MSG	Meteosat Second Generation	2세대기상위성	二世代氣象衛星
MSL	Mean Sea Level	평균해면	平均海面
MSS	Multi-Spectral Scanner	다중분광주사기	多重分光走査器
MTI	Moving Target Indication	이동목표표시	移動目標表示
MTN	Main Telecommunication Network	주요장거리통신망	主要長距離通信網
MTSAT	Multifunctional Transport Satellite	다목적위성	多目的衛星
MU radar	Middle and Upper Atmosphere Radar	MU 레이더	
MWO	Meteorological Watch Office	기상감시사무국	氣象監視事務局
MWR	Microwave Radiometer	마이크로파 라디오미터	-波-
MWRP	Microwave Radiometric Profilers	마이크로파 수직복사계	-波垂直輻射計
N weather	iNstrument weather	계기일기	計器日氣
NAAQS	National Ambient Air Quality Standards	국립표준대기질	國立標準大氣質
NABOS	Nansen and Amundsen Basins Observational System	난센아문센기지관측체계	-基地觀測體系
NACA	National Advisory Committee for Aeronautics	미국항공자문위원회	美國航空諮問委員會
NADP	US National Atmospheric Deposition Program	미국 국립대기침적 프로그램	美國國立大氣沈積-
NADW	North Atlantic Deep Water	북대서양심층수	北大西洋深層水
NAM	Northern Annular Mode	북반구극진동	北半球極振動

록부

약어	영문	국문	漢字
NWS	National Weather Service	(미국)국립기상대	國立氣象臺
OBS	Observing and Information Systems(WMO Department)	WMO 관측정보체계	-觀測情報體系
OceanSITES	Ocean Sustained Interdisciplinary Time Series Environmental Observation System	해양학적적지시계연환경관측시스템	海洋學制的持續時系列環境觀測-觀測系
ODP	Ocean Drilling Program	해저지각시추프로그램	海底地殼試錐-
ODS	Ozone-Depleting Substance(s)	오존고갈물질	-枯渴物質
OECD	Organization for Economic Cooperation and Development	경제협력개발기구	經濟協力開發機構
OFCM	Office of the Federal Coordinator for Meteorology	연방기상협력청	聯邦氣象協力局
OHR	Operational Hydrology Report(WMO)	현업수문보고	現業水文報告
OIS	Operational Information Service(WWW)	현업정보서비스	現業情報-
OLR	Outgoing Longwave Radiation	외향장파복사	外向長波輻射
OMPS	Ozone Mapping and Profiler Suite	오존 매핑 및 프로파일러 장치	-裝置
ONI	Oceanic Nino Index	해양니노지수	海洋-指數
OOSDP	Ocean Observing System Development Panel	해양관측시스템개발패널	海洋觀測-開發-
OPEC	Organization of the Petroleum Exporting Countries	석유수출국기구	石油輸出國機構
OTA	Office of Technology Assessment	미국 기술평가국	美國技術評價局
OWS	Ocean Weather Station	해양기상관측선	海洋氣象觀測船
PACE	Permafrost and Climate in Europe(European Union)	유럽 영구인망과 기후	-永久-氣候
PAGES	Past Global change	과거 지구변화, 페이지스	過去地球變化
PAR	Precision Approach Radar	정밀진입레이더	精密進入-
PBL	planetary boundary layer	행성경계층	行星境界層
PCCWMR	Physics and Chemistry of Clouds and Weather Modification Research Programme	구름물리·화학 및 기상조절연구계획	物理化學-氣象調節硏究計劃
PCMDI	Program for Climate Model Diagnosis and Intercomparison	기후모형분석과 상호비교 프로그램	氣候模型分析-相互比較-
PDA	Personal Digital Assistant	개인휴대정보 단말기	個人携帶情報端末機
P-E index	Precipitation Effectiveness Index	강수효과지수	降水效果指數
P-E QUOTIENT	Precipitation-Evaporation Quotient	강수증발몫	降水蒸發-
P-E RATIO	Precipitation-Evaporation Ratio	강수증발비	降水蒸發比
PE	Permanent Echo	고정에코	固定-

PM	Particular Matter	입자	粒子
PMO	Port Meteorological Officers	항만기상관	港灣氣官
PNA	Pacific North America	북태평양지수	北太平洋指數
PNNL	Pacific Northwest National Laboratory	퍼시픽 노스웨스트 국립연구소	-國立硏究所
POES	Polar Operational Environmental Satellite(USA)	극궤도환경위성	極軌道環境衛星
POLDER	POLarization and Directionality of the Earth's Reflectances	지구반사도의 편광성과 직달성	地球反謝度-偏光性-直達性
PoP	Probability of Precipitation	강수확률	降水確率
PPI scope	Plan-Position Indicator scope	평면위치지시스코프	平面位置指示-
PR	Permanent Representative(WMO)	상주대표	常駐代表
PRF	Pulse Repetition Frequency	펄스반복주파수	-反復周波數
PSC	Polar Stratospheric Cloud	극성층권운	極成層圈雲
PST	Pacific Standard Time	태평양표준시	太平洋標準時
PWS	Public Weather Services	일반기상지원서비스	一般氣象支援-
PWSP	Public Weather Services Programme	공공기상서비스 프로그램	公共氣象-
QBO	Quasi-Biennial Oscillation	준격년주기진동	準隔年週期振動
QMS	Quality Management System	품질관리시스템	品質管理-
QPE	Quantitative Precipitation Estimation	정량적 강수량추정	定量的降水量推定
QPF	Quantitative precipitation forecasting	정량적 강수예측	定量的降水豫測
RAFC	Regional Area Forecast Centre(ICAO/WMO)	지역공역예보센터	地域空域豫報-
RAIDOM	RAdar Image DigitalizatiOn Method	레이더 영상티지털변환법	-映像-變換法
RAM	Regional Office for the Americas(WMO)	아메리카 지역사무소	-地域事務所
RANS	Raynolds Averaged Navier-Stokes	레이놀즈평균나비에스토크스	-平均-
RA	Regional Association(WMO)	지역연합	地域聯合
RAOB	RAwinsonde OBservation	라오브(레윈존데관측)	-(觀測)
RASS	Radio acoustic sounding system	고도별무선음향측정계	高度別無線音響測定計
RBDT	RiBbon Display Terminal	자료표출장치	資料表出裝置
RBSN	Regional Basic Synoptic Networks	지역기반종관망	地域基盤綜觀網
RCC	Regional Climate Centre	지역기후센터	地域氣候-
RCD	Regional Climate Downscaling	지역기후규모축소	地域氣候規模縮小

RCM	Radiative-Convective Model	복사대류모형	輻射對流模型
RCMs	Regional Climate Models	지역기후모형	地域氣候模型
RCP	Representative Concentration Pathway	대표농도경로	代表濃度徑路
RDAPS	Regional Data Assimilation Prediction System	지역자료동화예측시스템	地域資料同化像測-
RDP	Research and Development Project(WMO/AREP/WWRP)	연구개발계획	研究開發計劃
RES	Research(WMO Department)	연구	硏究
RF	Radio Frequency	무선주파수	無線周波數
RFC	Radio Frequency Coordination	무선주파수제어기	無線周波數制御器
RGB image	Red/Green/Blue images	RGB 영상	-映像
RHI scope	Range-Hight Indicator scope	거리고도지시스코프	距離高度指示-
RIAMOM	Regional Integrated Assessment Model Ocean Model	지역종합평가모형해양모형	地域綜合評價模型海洋模型
RIC	Regional Instrument Centre	지역측기센터	地域測器機
RMC	Regional Meteorological Centre(WWW)	지역기상중추	地域氣象中樞
RMSE	Root Mean Square Error	평방근오차	平方根誤差
RMTN	Regional Meteorological Telecommunication Network	지역기상장거리통신망	地域氣象長距離通信網
ROSE	Radar Ocean SEnsing	레이더해양관측	-海洋觀測
RP	Regional Programme(WMO)	지역프로그램	地域-
RSL	Roughness Sublayer	거친 중간층	-中間層
RSMC	Regional Specialized Meteorological Centre(WWW)	지역특별기상센터	地域特別氣象-
RTC	Radiative Transfer Code	복사전달코드	輻射傳達-
RTC	Regional Training Centre(WMO)	지역훈련센터	地域訓練-
RTH	Regional Telecommunication Hub(WWW)	지역장거리통신축	地域長距離通信軸
RVR	Runway Visual Range	활주로가시거리	滑走路可視距離
RWOS	Road Weather Observation System	도로기상관측시스템	道路氣象觀測-
SACEP	South Asia Co-operative Environment Programme	남아시아공동환경계획	南-共同環境計劃
SAC/WCIRP	Scientific Advisory Committee, WCIRP	과학자문위원회	科學諮問委員會
SADC	Southern African Development Community	남부아프리카개발공동체	南部-開發共同體
SADIS	SAtellite DIStribution System(United Kingdom)	위성분포체계	衛星分類體系
SAG	Scientific Advisory Group	과학자문위원회	科學諮問委員會

부록

SAM	Southern Hemisphere Annular Mode	남반구환상모드	南半球環狀-
SAR	Synthetic Aperture Radar	합성개구레이더	合性開口-
SBI	Subsidiary Body for Implementation	기후변화협약 이행보조기구	氣候變化協約履行補助機構
SBOS	Space Based Observing(WMO Division)	우주기반관측	宇宙基盤觀測
SBSTA	Subsidiary Body for Scientific and Technological Advice (UNFCCC and UNCCD)	과학기술자문부속기구	科學技術諮問附屬機構
SBUV	Solar Backscatter UltraViolet	태양후방산란자외선	太陽後方亂風紫外線
SCAN	System for Convection Analysis & Nowcasting	대류성뇌우실황분석시스템	對流性雷雨實況分析-
SCAR	Scientific Committee on Antarctic Research(ISCU)	남극과학연구위원회	南極科學研究委員會
SCIAMACHY	Scanning Imaging Absorption Spectrometer for Atmospheric Chartography	대기 중 온실가스 측정	大氣中溫室가스-測定
SCPP	Seasonal-to-interannual Climate Prediction Program	단기기후예측프로그램	短期氣候豫測-
SDSWs	Sand and Dust Storms Warning System(WMO/AREP)	황사경보체계	黃砂警報體系
SEASAT	SEAfaring SATellite	미국해양관측위성	美國海洋觀測衛星
SeaWiFS	Sea-viewing Wide Field-of-view Sensor	해양관측전문센서	海洋觀測專門-
SEB	surface energy balance	지표에너지균형	地表-均衡
SEM	Scanning Electron Microscope	주사전자현미경	走査電子顯微鏡
SHF	Super High Frequency	초고주파	超高周波
SICA	System for Central American Integration	중미통합체제	中美統合體制
SID	Sudden Ionospheric Disturbance	돌연전리층요란	突然電離層搖亂
SIRS	Satellite InfRared Spectrometer	위성적외분광계	衛星赤外分光計
SL-AV	Semi-Lagrangian based on Absolute Vorticity equation, HMC	준라그랑지안절대와도방정식	準-絶對渦度方程式
SMOS	Soil Moisture and Ocean Salinity	토양수분 및 해양염분 (해양염분 관측 업무를 수행하는 위성)	土壤水分-海洋鹽分
SMS	Short Message Service	단문전달서비스	單文傳達-
SNUGCM	Seoul National University General Circulation Model	서울대학교 대기대순환모형	-大學校大氣大循環模型
SODAR	SOnic Detection And Ranging	음파기상탐지기	音波氣象探知機
SOI	Southern Oscillation Index	남방진동지수	南方振動指數
SOLAS	International Convention for the Safety of Life at Sea	국제해상인명안전협약	國際海上人命安全協約

TCOP	Technical Cooperation Programme	기술협력 프로그램	技術協力-
TCP	Tropical Cyclone Programme(WMO Division and programme)	열대저기압 프로그램	熱帶低氣壓-
TC	Technical Committee(WMO)	기술위원회	技術委員會
TC	Ton of Carbon Equivalent	탄소환산톤	炭素換算-
TC	Tropical Cyclone	열대저기압	熱帶低氣壓
TC	Typhoon Committee	태풍위원회	颱風委員會
TD	Tropical Depression	열대저압부	熱帶低壓部
TDWR	Terminal Doppler Weather Radar	공항기상레이더	空港氣象-
T-E index	Temperature Efficiency index	온도효율지수	溫度效率指數
T-E ratio	Temperature-Efficiency ratio	온도효율비	溫度效率比
THORPEX	The Observing System Research and Predictability Experiment (WMO/AREP)	관측시스템연구 및 예측실험	觀測-研究-豫測實驗
TIROS	Television and Infrared Observation Satellite(USA)	텔레비전과 적외관측위성 (미국 기상위성)	-赤外觀測衛星 (美國氣象衛星)
TLFDP	Typhoon Landfall Forecast Demonstration Project	태풍육상예보실험계획	颱風陸上豫報實驗計劃
TMI	TRMM Microwave Imager	TRMM 마이크로웨이브 영상기	-映像機
TMR	Tropical Meteorology Research(WMO/AREP)	열대기상연구	熱帶氣象硏究
TMRP	Tropical Meteorology Research Programme	열대기상연구 프로그램	熱帶氣象硏究-
TN	Technical Note(WMO)	기술보고	技術報告
TNO	Netherlands Organization for Applied Scientific Research	네덜란드응용과학연구기구	-應用科學硏究機構
TOE	Ton of Energy	에너지 환산톤	-換算-
TOGA	Tropical Ocean and Global Atmosphere Programme(WCRP)	열대해양·대기 지구대기계획	熱帶海洋地球大氣計劃
TOGA-COARE TOGA	Coupled Ocean-Atmosphere Response Experiment, TOGA/WCRP	해양·대기 결합응답실험	海洋大氣結合應答實驗
TOMS	Total Ozone Mapping Spectrometer	지구전체오존량측정 분광계	地球全體-量測定分光計
TOVS	TIROS Operational Vertical Sounder	대기연직구조 탐측계 (NOAA시리즈위성에 탑재)	大氣鉛直構造探測械
T-PARC	THORPEX Pacific Asian Regional Campaign	태풍국제공동 특별관측실험	颱風國際共同特別觀測實驗
TREND	Working Group on Training, the Environment and New Developments in Aeronautical Meteorology	항공기상환경, 기술개발, 훈련실무반	航空氣象環境, 技術開發, 訓練實務班

URSI	Union Radio Scientifique Internationale - International Union of Radio Science - Also, ISRU - Belgium	국제전파과학연합	國際電波科學聯合
USGS	United States Geological Survey	미국 지질조사소	美國地質調查所
USOAP	Universal Safety Oversight Audit Programme(ICAO)	항공안전종합평가	航空安全綜合評價
UTC	Coordinated Universal Time	협정세계시	協定世界時
UV	Ultra Violet	자외선	紫外線
UV-B	Ultra-Violet B	B영역 자외선	-領域紫外線
UZ	Uncorrected Reflectivity	보정전 반사도(단위 : dBz)	補正前反射度
VAAC	Volcanic Ash Advisory Centre	영국 화산재예보센터	英國火山豫報-
VCP	Voluntary Cooperation Programme(WMO)	자발적 협력프로그램	自發的協力-
VETH chart	Ventilation Evaporation Temperature Humidity chart	VETH 선도	-線圖
VHF	Very High Frequency	초단파	超短波
VHRR	Very High Resolution Radiometer	첨단고해상복사계	尖端高解像輻射計
VIIRS	Visible/Infrared Imager/Radiometer Suite	가시광선/적외선 촬영기	可視光線/赤外線撮影機
VIRS	Visible and Infrared Scanner	가시적외주사	可視赤外走査
VLF	Very Low Frequency	초저주파	超低周波
VOCs	Volatile Organic Compounds	휘발성유기화합물	揮發性有機化合物
VTL	Virtual Training Library	가상교육자료실	假想敎育資料室
WAFC	World Area Forecast Centres(ICAO/WMO)	세계공역예보센터	世界空域豫報
WAFS	World Area Forecast System(ICAO/WMO)	세계공역예보시스템	世界空域豫報
WAMIS	World AgroMeteorological Information Service	세계농업기상정보서비스	世界農業氣象情報-
WCASP	World Climate Applications and Services Programme, WCP/WMO	세계기후응용 및 서비스 프로그램	世界氣候應用-
WCC	World Calibration Centre	세계표준센터	世界標準-
WCCOS	World Calibration Center for Ozone Sondes	세계오존측비검정센터	世界校正-
WCDMP	World Climate Data and Monitoring Programme	세계기후자료 및 감시프로그램	世界氣候資料監視-
WCIRP	World Climate Impact Assessment and Response Strategies Programme	세계기후영향평가 및 대응전략 프로그램	世界氣候影響評價對應戰略-
WCP	World Climate Programme(WMO)	세계기후프로그램	世界氣候-
WCRP	World Climate Research Programme(WMO/IOC/ICSU)	세계기후연구 프로그램	世界氣候研究-

부록

WOAP	WCRP Observations and Assimilation Panel	WCRP 관측 및 동화패널	-觀測-同化-
WOCE	World Ocean Circulation Experiment(WCRP)	세계해양대순환실험	世界海洋大循環實驗
WOUDC	World Data Centre for Ozone and UV	오존/자외선 세계자료센터	-紫外線世界資料-
WRC	World Radiation Centres	세계복사센터	世界輻射-
WRC	World Radiocommunication Conference	세계무선통신협의회	世界無線通信協議會
WRDC	World Radiation Data Centre	대기복사 세계자료센터	大氣輻射世界資料-
WRF	Weather Research and Forecasting model	일기연구예보모형	日氣硏究豫報模型
WRF/Chem model	Weather Research and Forecasting/Chemistry model	일기연구 및 예보/화학모형	日氣硏究豫報/化學模型
WRF-NMM	Weather Research and Forecasting-Nonhydrostatic Mesoscale Model	일기연구 및 예보-비정역학 중규모모형	日氣硏究豫報-非靜力學中規模模型
WSSD	World Summit on Sustainable Development(UN)	세계지속가능발전 정상회담	世界持續可能發展頂上會談
WTO	World Trade Organization	세계무역기구	世界貿易機構
WWF	World Wildlife Fund	세계야생보호기금	世界野生保護基金
WWIS	World Weather Information Service(WMO)	세계기상정보서비스	世界氣象情報
WWRP	World Weather Research Programme(WMO/AREP)	세계기상연구프로그램	世界氣象硏究-
WWW	World Weather Watch(WMO Programme)	세계기상감시	世界氣象監視
WWWDM	WWW Data Management	WWW 자료관리	-資料管理
YONUAGCM	YONsei University Atmospheric General Circulation Model	연세대학교 대기대순환모형	延世大學校大氣大循環模型
3DOI	3-Dimension Optimum Interpolation	3차원 최적내삽법	三次元最適內揷法
4DVAR	4-Dimensional variational method	4차원 변분법	四次元變分法

■ 보퍼트(Beaufort) 풍력계급표

풍력계급	명칭		10m 높이에서의 풍속	
	영문	한글	m/sec	knot
0	calm	고요	0.0~0.2	1 미만
1	light air	실바람	0.3~1.5	1~3
2	light breeze	남실바람	1.6~3.3	4~6
3	gentle breeze	산들바람	3.4~5.4	7~10
4	moderate breeze	건들바람	5.5~7.9	11~16
5	fresh breeze	흔들바람	8.0~10.7	17~21
6	strong breeze	된바람	10.8~13.8	22~27
7	near gale	센바람	13.9~17.1	28~33
8	gale	큰바람	17.2~20.7	34~40
9	strong gale	큰센바람	20.8~24.4	41~47
10	storm	노대바람	24.5~28.4	48~55
11	violent storm	왕바람	28.5~32.6	56~63
12	hurricane	싹쓸바람	32.7 이상	64 이상

■ 구름분류표(10종)

영문	한글	약기호
cirrus	권운	Ci
cirrocumulus	권적운	Cc
cirrostratus	권층운	Cs
altocumulus	고적운	Ac
altostratus	고층운	As
nimbostratus	난층운	Ns
stratocumulus	층적운	Sc
stratus	층운	St
cumulus	적운	Cu
cumulonimbus	적란운	Cb

■ 구름변종일람표

영문	한글
altocumulus cumulogenitus	적운성고적운
altocumulus cumulomutatus	적운계고적운
cirrocumulus altostratomutatus	고층운계권적운
cirrocumulus cirromutatus	권운계권적운
cirrocumulus cirrostratomutatus	권층운계권적운
cirrostratus altostratomutatus	고층운계권층운
cirrostratus cirrocumulogenitus	권적운성권층운
cirrostratus cirrocumulomutatus	권적운계권층운
cirrostratus cirromutatus	권운계권층운
cirrostratus cumulonimbogenitus	적란운성권층운
cirrus altocumulogenitus	고적운성권운
cirrus cirrocumulogenitus	권적운성권운
cirrus cirrostratomutatus	권층운계권운
cirrus cumulonimbogenitus	적란운성권운
cumulonimbus altocumulogenitus	고적운성적란운
cumulonimbus altostratogenitus	고층운성적란운
cumulonimbus cumulogenitus	적운성적란운
cumulonimbus cumulomutatus	적운계적란운
cumulonimbus nimbostratogenitus	난층운성적란운
cumulonimbus stratocumulogenitus	층적운성적란운
cumulus altocumulogenitus	고적운성적운
cumulus stratocumulogenitus	층적운성적운
cumulus stratocumulomutatus	층적운계적운
cumulus stratomutatus	층운계적운
nimbostratus altocumulomutatus	고적운계난층운
nimbostratus altostratomutatus	고층운계난층운
nimbostratus cumulogenitus	적운성난층운
nimbostratus cumulonimbogenitus	적난운성난층운
nimbostratus stratocumulomutatus	층적운계난층운
stratocumulus altostratogenitus	고층운성층적운
stratocumulus cumulogenitus	적운성층적운
stratocumulus nimbostratogenitus	난층운성층적운
stratocumulus nimbostratomutatus	난층운계층적운
stratocumulus stratomutatus	층운계층적운
stratus cumulogenitus	적운성층운
stratus cumulonimbogenitus	적난운성층운
stratus nimbostratogenitus	난층운성층운
stratus stratocumulomutatus	층적운계층운

■ 국지바람일람표

영문	한글
aberwind(= aperwind)	해동바람(알프스)
Afer(= Africino, Africo, Africuo, Africus ventus)	아퍼(이탈리아에서 부는 강한 남서풍)
afghanets	아프가네츠(아프가니스탄으로부터 불어오는 아무다리아강 상류의 강풍)
allerheiligenwind	성풍(오스트리아 티롤지방의 늦은 가을 바람)
almwind	암윈드(폴란드 크라코의 남쪽 타트라산맥을 가로질러 불어오는 푄바람)
aloegoe	알로에고에(북수마트라섬의 바람)
angin	앤진(말레이시아의 해륙풍)
angin-darat	앤진다라트(말레이시아의 육풍)
angin-laut	앤진라우트(말레이시아의 해풍)
aspre	아스프레(프랑스 중부 산간지대의 북동 푄)
Athos fall wind	아토스치내리바람(그리스 아토스산맥 부근)
aura(= ora)	오라(이탈리아 가르다호의 산골바람)
auster(= ostria)	오스터(불가리아 연안의 따뜻한 남풍)
autan(= altanus)	오탄(프랑스 남부의 남동쪽으로부터 불어오는 센 바람)
auvergnasse(= auvergnac)	오베르나제(프랑스 중부 산간지대의 찬 북서풍)
avalaison(= avalasse)	아발레이즈(프랑스 서부의 탁월한 서풍)
avre	아브레(겨울철의 따뜻한 바람과 여름철의 서늘한 바람이 교대로 부는 바람의 프랑스 방언)
ayalas	아얄라스(프랑스 중부 산간지대에서 남동쪽으로부터 불어오는 대륙성 열풍)
bali	발리(이탈리아 가르다호의 호수바람)
Bali wind	발리바람(자바 동부에서 부는 강한 동풍의 일종)
ban-gull	반걸(스코틀랜드의 여름철 해풍)
barat	바라트(메나도 부근 셀레베스 북쪽 해안의 강한 북서풍 또는 서풍)
barber	바버(미국과 캐나다 지역의 냉강풍)
barih(= shamal, shemaal, shimal, shumal)	→ shamal
barinés	바리네스(동부 베네수엘라의 서풍)
bayamo	바야모(쿠바섬 남해안의 돌풍)
belat	벨랏(아라비아 남해안의 강한 북서 먼지 바람)
bentu de soli	벤투드솔리(이탈리아 사르디니아 해안에서 부는 동풍)
berber	→ barber
berg wind	산바라(남아프리카 남해안 열풍의 일종)
bhut(= bhoot)	부트(인도 먼지보라의 일종)
birazon(= virazon)	비라존(안데스산맥의 태평양 연안 경사면으로 불어오는 바람)
bise(= bize)	비즈(프랑스와 스위스의 산간지역에서 부는 차고 건조한 북 ~북동풍)
bise brume	비즈브룸(남유럽의 산악지대를 넘어가는 차고 습한 북풍)
black buran(= karaburan)	블랙부란(중앙아시아에서 여름과 봄에 부는 강한 북동풍)
black northeaster	블랙북동폭풍(오스트레일리아 남동부에서 여름에 부는 북동

	큰 바람)
black southeaster	블랙남동폭풍(남아프리카 테이블만에서 부는 남동풍)
Blackthorn winds	블랙손바람(영국 템즈계곡에서 부는 3~4월의 차고 건조한 바람)
black wind(= reshabar)	검은바람(페르시아 남부 쿠르디스탄 산맥에서 불어내리는 강한 북동풍)
blaze	블레이즈(갑자기 부는 건조한 바람의 스코틀랜드말)
bliffart(= bluffart, bliffert)	블리파트(스코틀랜드의 폭풍설을 동반한 돌풍 또는 스콜)
blinter	블린터(돌풍의 스코틀랜드말)
blizzard	블리저드(북미대륙에서 겨울철에 강풍과 눈보라를 동반한 한랭한 북서풍)
bochorno	보초르노(스페인 에브로계곡의 무더운 바람)
bohorok	보호로크(북부 수마트라 평원에서 부는 푄)
bolon	볼론(북수마트라 토바호의 바람)
boorga(= burga)	부르가(시베리아 연안과 베링해 연안의 눈보라와 폭풍을 동반하는 북동풍)
bora	보라(아드리아해 북동부 달마티안 해안에서 겨울철에 부는 강하고 찬 북동풍)
boraccia	→ bora
bordelais	보드레이(프랑스 남서부의 서풍)
bornan	보르난(스위스 제네바호 중앙에 부는 바람)
boulbie	볼비(프랑스 남부 알리에계곡의 강한 북풍)
brenner	브레너(영국 해상에 갑자기 부는 돌풍)
breva	브레바(북부 이탈리아 코모호 부근의 낮에 부는 강한 계곡 바람)
brickfielder	브릭필더(오스트레일리아 남부의 북~북서풍)
brisa(= briza)	브리사(무역풍 계절에 남아메리카 연안에 부는 북동풍, 푸에르토리코의 동풍, 필리핀의 북동 계절풍)
brisa carabinera(= carabiné)	브리사카라비네라(프랑스나 스페인의 돌연성 강풍)
brisote	브리소테(쿠바의 강한 북동풍)
broeboe	브로보(인도네시아 스퍼문데 아키펠라고 북부의 돌풍)
brubu	브루부(인도 동부의 회오리 바람의 일종)
brughierous	브루지로스(프랑스 중부 산악지방의 남풍)
brüscha	브루샤(스위스 엥가딘계곡의 북동풍)
buran	부란(러시아 및 중앙아시아 지역의 북동 한랭 강풍)
burga(= boorga)	부르가(시베리아해 연안이나 베링해 연안의 취설을 동반한 북동풍)
buria	부리아(bora의 불가리아말)
burraxka silch	부라카실치(지중해 말타섬 부근의 우박보라)
burster(= southerly burster, buster, southerly buster)	버스터(오스트레일리아의 차고 매서운 남동풍)
camsin(= khamsin)	캄신(이집트와 홍해의 건조하고 먼지가 섞인 사막바람)
Candlemas Eve winds(= Candlemas-crack)	캔들마스이브바람(영국에서 2월 또는 3월에 부는 강풍)
Canterbury northwester	캔터베리북서풍(알프스산맥을 넘어서 뉴질랜드 남쪽섬 캔터베리 평원으로 부는 북서의 강한 푄바람)
Cape doctor	케이프닥터(남아프리카 남해안에 부는 강한 남동풍)
carabiné	카라비네(프랑스와 스페인의 돌발 강풍)
carcenet	카스넷(동부 피레네 부근에 부는 대단히 차고 격렬한 협곡

부록

	바람)
cat's nose	캣노즈(영국지방의 북서 냉풍)
caver(= kaver)	케이버(스코틀랜드 서쪽 헤브리데스 부근의 온화한 바람)
challiho	챌리호(인도지역에서 봄에 40일간 부는 강한 남풍)
chamsin	→ camsin
chanduy(= chandui)	챈두이(에콰도르 과야킬지방의 건조한 계절에 부는 찬 하강풍)
chergui	체르구이(모로코 북부 사막지역의 동~남동풍)
chibli(= ghibli, gebli, gibleh, gibli, kibli)	치블리(리비아 지중해 연안의 덥고 먼지가 많은 푄과 비슷한 사막바람)
chichili	치칠리(남부 알제리 칠리의 명칭)
chili	칠리(튜니지아의 온난하고 건조한 하강풍으로서 시로코와 유사)
chinook	치누크(로키산맥 동쪽에 부는 푄)
chocolatero	초콜라테로(멕시코만 지역에 부는 온화한 북풍)
chocolatta north	초콜라타노스(서인도제도의 북서 강풍)
chom	촘(시로코의 북아프리카 방언)
chubasco	추바스코(중앙아메리카 서해안의 뇌우를 동반하는 강한 돌풍)
cierzo	시에르조(스페인 에브로계곡의 강한 돌풍)
cissero	시세로(프랑스 로스강 삼각주의 덥고 비를 동반하는 남동 시로코바람)
colla	콜라(필리핀의 남~남서풍)
collada	콜라다(캘리포니아만에 부는 강한 북풍)
coromell	코로멜(캘리포니아만 가까이의 라파즈에서 일어나는 육풍으로서 11~5월 사이의 밤에 시작하여 9~10시경까지 지속됨)
cowshee(= kaus, quas, sharki)	카우시(페르시아만에서 악천우와 스콜을 동반하는 남동 강풍)
cow storm	카우스톰(캐나다의 앨즈미르섬에서 일어나는 강풍)
criador	크리아도르(북부 스페인에 비를 가져다주는 서풍)
crivetz(= crivăt, krivu)	크리베츠(루마니아의 보라와 비슷한 북~동풍)
custard winds	커스터드바람(영국 북동 연안의 찬 편동풍)
dadur	다두르(히말라야의 남쪽 언덕으로부터 불어내리는 바람)
dahatoe	다하토(수마트라 토바호 부근의 바람)
deaister	→ doister
depéq	데페크(남서 계절풍 기간에 수마트라나 동인도지방에서 부는 강한 바람)
dimmerfoehn	딤머푄(상층에 강한 남풍이 불고 있을 때 알프스지방에서 일어나는 푄)
doctor	독터(자메이카섬에서 저녁 때 부는 활강 바람, 또는 아프리카의 하마탄)
doister	도이스터(스코틀랜드의 바다로부터 부는 격렬한 폭풍)
Düseneffekt wind(= düsen wind)	(1) 듀센바람(다르다넬즈해협에서 에게해로 불어오는 강한 동 북동풍) (2) 급류풍
dyster	→ doister
eissero	아이세로(나일강 삼각주에 따뜻하고 많은 비를 동반하는 남 동풍 또는 시로코)
elephanta	엘레판타(인도에서 9~10월 사이의 남서 계절풍이 끝날 무렵 남서 해안에 불어 닥치는 심한 뇌우와 큰 비를 동반하는 강 한 남동풍)
elerwind	엘러윈드(오스트리아 티롤지방 선계곡의 바람)

elvegust(= sno)	엘베거스트(노르웨이 북부의 좁은 만에서 일어나는 찬 하강류)
embata	엠바타(카나리아제도의 풍하쪽에서 북동 무역풍의 역전에 의해 생기는 남서풍)
etesians	에테시안즈(지중해 동부 특히 에게해에서 여름철에 부는 북풍)
euraquilo(= euroaquilo, euroclydon)	유라퀼로(아라비아와 근동에서 부는 강한 북동 또는 북북동풍)
Euros	유로스(비를 동반하는 남동풍의 그리스말)
favogn	파본(푄의 스위스말)
flan	플란(스코틀랜드의 육지로부터의 돌풍 또는 스콜)
forano	포라노(이탈리아 나폴리 부근의 해풍)
fowan	포완(영국 맨섬에서 일어나는 건조한 열풍)
furiani	푸리아니(이탈리아 포(Po)강 유역에 발생하는 짧은 시간의 풍향 변화를 동반하는 강풍)
gaine	게인(이탈리아에서 산꼭대기에 구름을 형성하면서 산을 넘는 바람)
galerne(= galerna, galerno)	갈런(영국 해협, 프랑스, 스페인 앞바다에서 발달하는 저기압 후면에 일어나는 차고 습한 소나기를 동반하는 북서풍)
gallego	갈레고(스페인과 포르투갈 북부에서 일어나는 살갗을 오려내는 듯한 찬 북풍)
garbin(= garbi, garbino, garbis)	가빈(남서해풍〈남서 프랑스〉, 동쪽 해풍〈북동 스페인〉)
gargal	→ gregale
gebil(= kibil)	게빌(북아프리카 리비아의 지중해 연안에 발생하는 푄과 비슷한 모래먼지를 포함한 열풍)
gending	겐딩(자바섬 북부에서 부는 푄과 비슷한 국지 바람)
gharbi	가르비(모로코 해양의 신선한 서풍)
gharra	가라(리비아에서 북동쪽으로부터 닥쳐오는 뇌우를 동반하는 격렬한 스콜)
ghaziyah	→ bora
gherbine	→ garbin
ghibli(= gibleh, gibli)	기블리(리비아의 트리폴리타니아에서 일어나는 푄과 비슷한 덥고 먼지 섞인 사막바람)
giboulee	→ galerne
glaves	글레이브스(페로군도의 푄과 비슷한 바람)
golfada	골파다(지중해의 심한 바람)
Grand Vent	그랜드 벤트(프랑스 오파니의 강한 비를 동반한 서풍)
grécale	그레케일(프랑스 코르시카섬의 건조하고 찬 바람)
greco	그레코(지중해 서부의 북동풍)
gregale	그리게일(지중해 중부와 서부 및 유럽 연안에 부는 강한 북동풍)
gregori wind	그레고리바람(오스트리아에서 3~4월에 부는 강풍)
guba	구바(뉴기니섬 남쪽 해안의 강한 돌풍)
gully-squall	걸리스콜(중앙아메리카 안데스산맥 서쪽 협곡의 돌풍)
guxen	국센(스위스 알프스의 찬 바람)
guzzle	구즐(셰틀랜드섬에서 한바탕 부는 왕바람)
hale de mars	할드마(프랑스 몰반 중앙산괴의 3월경의 건조한 bise풍)
halny wiatr	할니비아트르(높새바람의 폴란드말)
harmattan(= harmatan, harmetan, hermitan)	하마탄(12월부터 2월에 걸쳐 사하라 지방에서 아프리카 서해안을 향하여 부는 건조하며 모래를 함유한 열풍)
haugull(= havgul, havgula)	하우길(스코틀랜드와 노르웨이에서 여름에 바다로부터 불어

오는 차갑고 습한 바람)

helm wind	헬름바람(영국 북부의 크로스펠산맥의 서쪽 사면에서 에덴 협곡으로 불어내리는 차가운 북동풍)
hupe	휴프(오스트레일리아 다히치섬의 밤바람)
Ibe wind	아이베바람(중국 서부 Dzungarian문을 불어 지나가는 강한 국지 바람)
inverna	인베르나(이탈리아 마조리 호수의 남동풍)
iron wind	쇠바람(중앙아메리카 북동풍)
Jaloque(= xaloque)	얄로커(시로코바람의 스페인말)
jauk(= jauch)	조크(오스트레일리아의 쿠라겐홀트 분지의 푄)
Jochwinde	요크바람(알프스의 타엘은산 고갯길의 산골바람)
Joran(= Juran)	요단(스위스의 쥐라산맥에서 쥬네브호수를 향해 북서방향으로 부는 바람)
junk wind	정크바람(정크의 항해에 알맞은 아시아의 남동 몬순)
junta	장타(안데스산맥을 불어나가는 큰 바람)
kachchan	카츠찬(실론섬 동쪽 바티카로아 부근에서 부는 남서의 열풍)
kahamsin	카함신(겨울철에 이집트 및 홍해에서 부는 동~남동의 열풍)
kaikias	카이키아스(북동풍, 북북동풍의 그리스말)
kamsin	→ khamsin
karaburan	카라부란(봄부터 여름에 걸친 중앙아시아의 거센 북동풍)
karajol(= qarajel)	카라졸(불가리아 해안에서 비를 동반하고 1~3일 계속되는 서풍)
karema wind	카레마바람(아프리카의 탄자니아 호수에서 부는 거센 동풍)
karif(= kharif)	카리프(아프리카 소말리아의 아덴만에서 부는 강한 남서풍)
karstbora	카르스트보라(유고슬라비아 연안의 보라풍)
kaus	카우스(페르시아만의 남동풍)
kaver	케이버(스코틀랜드 서부 해브리디스제도의 고요한 바람)
khamasseen	→ khamsin
khamsin	→ camsin
kibil(= gebil)	키빌(지중해의 리비아 해안에서 부는 남동~남의 열풍)
kloof wind	클루프바람(남아프리카 시몬즈만의 차가운 남서풍)
knif wind	니프바람(알래스카 파모지방의 남서풍)
knik wind	니크바람(알래스카 마타누스카협곡의 파루마 부근에서 부는 남동풍)
koembang	코엠뱅(자바의 추리본이나 테가루에서 부는 남동~남의 건조한 푄 같은 바람)
Kol Baisakgi	콜베이사키(남서 몬순이 시작할 무렵 벵갈만 북부연안에서 일어나는 북서의 돌풍)
kona	코나(하와이에서 폭풍우를 동반하는 남서~남남서풍)
kossava(= kosava, koschawa)	코사바(베오그라드 동방 도나우강의 아니안게트협곡에서 칼파트산맥을 가로질러 동~남동으로부터 불어내리는 차고 무서운 바람)
Krakatao wind	크라카토아바람(열대지방에 있어서 18~24km의 고도에서 부는 편동풍)
krivu	크리부(루마니아와 소련 남부의 북동으로부터 부는 보라풍)
labbé(= labé)	라베(프랑스 남동부 프로방스지방에서 주로 3월에 부는 강한 남서풍)

labech	라베(알프스산의 labbé)
laheimar	라헤이마(아라비아지방의 10, 11월의 강풍)
langkisau	랑키샤(수마트라섬 등 동인도제도에서 일사가 강한 맑은 날에 부는 푄바람)
lansan	란산(뉴헤브리디즈와 동인도제도의 강한 남동 무역풍)
Lausan	라우산(뉴헤브리디즈제도 부근의 강한 남동 무역풍)
laveche(= leveche)	라베체(사하라 사막으로 이동해 오는 저기압에 앞서 스페인에서 부는 남풍)
leste	레스테(북아프리카, 마데이라제도 및 카나리아제도에서, 사하라 사막으로부터 불어오는 동~남동풍의 스페인말)
levant	르방(레반테의 프랑스말)
levant blanc	→ levante
levante	레반테(프랑스 남부로부터 지브롤터해협에 이르러 스페인 연안으로 부는 동~북동풍⟨스페인말⟩)
levanter	레반터(레반테의 영국말)
levantera	레반테라(흐린 날씨를 동반한 아드리아해와 지브롤터해협의 지속적 동풍)
levanto	레반토(북아프리카 카나리아제도의 남동 열풍)
leveche	레베체(스페인의 지중해 연안에서 부는 아프리카 사막으로부터 불어오는 편동풍의 스페인말)
libeccio	리베시오(남서풍의 이탈리아말)
liberator	리베라토르(리브롤터해협의 서풍)
Lips	립스(그리스 수도 아테네에서의 남서 해풍)
ljuka	류카(유고슬라비아 북서쪽 카린디아의 푄바람)
llebetjado	레베트야도(북동 스페인 카탈로니아에서 피레네산맥까지 2~3시간 계속 부는 더운 흔들바람)
lodos	로다스(불가리아 흑해 연안의 남풍)
loehis	→ aloegoe
lombarde	롬바르데(프랑스와 이탈리아 국경의 롬바르디로부터 불어오는 지배적인 동풍)
long shore wind	긴 해안바람(① 인도의 마드라스에서 부는 습하고 불쾌한 남풍, ② 스리랑카에서 부는 야간의 북동풍)
loo(= lu, loo marna)	루(인도 서부로부터 불어오는 더운 바람)
luganot	루가놋(이탈리아 가르다호의 강한 남~남동의 바람)
maestro	마에스트로(아드리아해에 있어서 특히 여름에 불며, 서안에서 가장 많은 북서풍)
magnofango	마그노팡고(프랑스 남부의 프로방스에서 부는 미스트랄)
Maloja wind	말로야바람(스위스 앵가딘지구의 조용한 바람)
mamatele(= mamaliti, mamatili)	마마텔레(이탈리아 시칠리아섬의 북서풍)
marenco	마렌코(이탈리아 막지오레호지방의 동남동풍)
marin	마린(프랑스 지중해 연안에서 알프스산맥으로 향하는 따뜻하고 습한 남동풍)
marinada	마리나다(스페인 카탈루냐와 프랑스 호시옹 연안을 부는 해풍)
marin blanc	→ marin
Matanuska wind	마타누스카바람(알래스카 파머 근처에서 겨울철에 때때로 일어나는 강하고 돌발적인 북동풍)
matinal	마티날(프랑스 모루방 중앙산맥의 여름철 편동풍)

matinière	마티니레(알프스산맥의 서쪽 사면에 있어서의 겨울철 내리바람)
medina	메디나(스페인 카디스지방의 겨울철 육풍)
meltém(= meltémi)	멜템(① 불가리아 연안과 보스포러스해협 지방의 여름철에 부는 동~북동풍, ② 에테시안 바람의 터키말)
mergozzo	메르고조(이탈리아 막지오레호수 지방의 북서풍)
Michael-riggs(= rig)	미카엘리그(영국의 9월 하순의 큰바람)
midnight wind	심야바람(독일 바이에른고지나 뷰름에 알프스산으로부터 불어오는 남풍)
miejour	미주르(프랑스 프로방스 남쪽에 부는 습하고 더운 바다 바람)
minuano	미누아노(브라질 남부의 6~9월에 부는 한랭한 남서풍)
mistral(= mystral)	미스트럴(지중해 북쪽해안의 가장 건조하고 찬 북서 또는 북풍)
mitgjorn	미루오에렌(프랑스 리용으로부터 피레네산맥으로 불어오는 건조하고 약한 남풍)
Moazagotl wind	모아자고틀바람(모아자고틀 구름을 형성하는 강한 바람)
molan	몰란(아르브강으로부터 제네바호를 향해 부는 실바람)
morget(= morgeasson)	모겟(스위스 제네바호에서 밤에 부는 육풍)
Morning glory	아침글로리(오스트레일리아 카펀테리아만 먼 바다에서 이른 아침의 정적을 깨고 돌연 나타나는 바람)
muerto	무에르토(멕시코에서 여름철에 부는 강한 북풍)
narbonnais(= narbonés)	나르보네스(프랑스의 나르본으로부터 불어오는 바람)
nashi(= n'aschi)	나시(페르시아만에서 겨울에 일어나는 북동풍의 아라비아말)
nemere	네메레(헝가리의 차고 강한 치내리바람)
nevada	네바다(산악의 설원 또는 빙원으로부터 불어내리는 찬바람의 스페인말)
nirta	니르타(북수마트라 토바호의 바람)
Nopsae	높새(한국의 태백산맥 동쪽과 서쪽 기슭에서 늦가을~초봄에 나타나는 건조하고 따뜻한 바람의 한국말)
Norta	노르타(여름철 칠레연안의 강한 북풍)
nortada	노르타다(필리핀에서 부는 강한 북풍)
norte	노르테(스페인의 겨울철 북풍)
norther	노더(미국 남부의 강하고 찬 바람)
northern nanny	북극유모(잉글랜드의 우박을 동반하는 한랭한 북풍)
oberwind	오버윈드(오스트리아 살츠캄머구트의 바람)
oe	오(덴마크령 페로제도의 회오리 바람)
ora	오라(이탈리아 가르다호의 규칙적인 골바람)
orsure	오르수르(프랑스 리용만의 왕바람)
Osos wind	오소스바람(미국 캘리포니아 주의 로스오소스골짜기로부터 샌루이스골짜기까지 부는 강한 북서풍)
ostria(= auster)	오스트리아(불가리아 연해의 따뜻한 남풍)
ouari	오우아리(소말리아의 남풍)
paesa	파에사(이탈리아 가르다호의 격렬한 북북동풍)
paesano	파에사노(이탈리아 가르다호에서 밤에 북쪽 산으로부터 불어 내리는 실바람)
pampero	팜페로(남아메리카 아르헨티나와 우루과이 대초원의 한랭하고 험악한 남~남서풍)
pampero seco	팜페로세코(무강수형 팜페로바람)
pampero sucio	팜페로수시오(모래먼지형 팜페로바람)

panas oetara	파나스오에타라(인도네시아에서 2월에 부는 온난 건조한 강한 북풍)
papagayo(= popogaio)	파파가요(중앙아메리카 니카라구아와 과테말라의 태평양 연안에서 부는 강한 북동풍)
peesash(= peshash, pisachee, pisachi)	피사시(인도의 덥고 건조한 모래먼지 바람)
petit St. Bernard	페티 성 베르나르(프랑스 Haute Tarentaire의 산바람)
piner	파이너(영국의 북쪽 또는 북동쪽에서 불어오는 강풍)
pisachee	→ peesash
polacke(= polake)	폴래크(알바니아 보히미아의 차고 건조한 북동 활강풍)
ponente	포넨테(지중해의 서풍)
poniente	포니엔테(지브롤터해협의 서풍)
pontias	퐁티아스(프랑스 리옹의 산바람)
poorga	→ purga
popogaio	→ papagayo
poroiaz	포리아즈(흑해의 북동 왕바람)
Porlezzina	폴레지나(이탈리아 루가노호의 동풍)
prester	프리스터(지중해와 그리스에서 나타나는 번개를 동반한 회오리바람 또는 용오름)
puelche	푸엘체(남아메리카 안데스산맥을 횡단하는 동풍)
punos	푸노스(남아메리카 푸나지역의 한랭건조한 남 또는 남서풍)
pyrhener wind	피레너바람(오스트리아 알프스의 푄)
qarajel	→ karajol
qaus	→ kaus
quara	→ karajol
raffiche(= refoli)	라피체(지중해지방에서 산으로부터 불어오는 돌풍, 거센 돌풍성 보라)
rageas	→ ragut
raggiatura	라기아투라(이탈리아 고지대의 협곡으로부터 불어내리는 육지 스콜)
ragut(= rageas, ghaziyah)	라구트(지중해 이스켄데론만의 강한 치내리바람)
rebat	레바트(스위스 제네바호의 호수바람)
reshabar(= rushabar)	레샤바(터키 쿠르디스탄의 보라)
ribut	리붓(말레이반도에서 5~11월에 나타나는 지속시간의 짧은 스콜)
rig(= Michael-riggs)	리그(영국의 9월 하순의 강한 바람)
rosau	로사우(프랑스의 론강 골짜기에서 낮에 부는 서~남서풍)
rotenturm wind	로텐텀바람(트렌실베이니아 알프스산맥의 로텐텀고개를 불어나가는 따뜻한 남풍)
rushabar	→ reshabar
sahel	사헬(모로코의 강한 먼지 사막바람)
sansar(= sarsar, shamsir)	산사르(페르시아의 북서풍)
Santa Ana	산타아나(미국 캘리포니아 산타아나의 강 계곡에서 북동 또는 동쪽으로부터 불어오는 덥고 건조한 푄과 같은 사막바람)
saoet	사오트(북부 수마트라 토바호의 바람)
sàrca	사르카(이탈리아 가르다호의 격렬한 북풍)
savet	사베트(북부 수마트라 토바호의 바람)
scharnitzer	샤르니처(오스트리아 티롤에서 오랫동안 부는 찬 북풍)

scirocco(= sirocco)	시로코(남부 지중해 또는 북아프리카를 가로질러 동쪽으로 이동하는 저압부에 앞서 부는 따뜻한 남풍 또는 남동풍)
seca	세카(브라질의 건조한 바람)
sécaire	세케르(지중해 북안의 한랭건조한 북서풍이나 북풍)
sechard	세차드(스위스 제네바호의 푄 현상)
seistan	세이스탄(이란 동부에서 여름철에 부는 북서~북북서의 강한 바람, 120일간 지속)
selatan	셀라탄(인도네시아의 건조한 남쪽 계통의 강한 바람)
shahali	샤할리(알제리지방의 시로코와 같은 건조한 열풍)
shaluk	샬룩(아라비아 사막지방의 건조한 열풍)
shamal(= shemaal, shimal, shumal)	샤말(페르시아만과 티그리스 및 유프라테스계곡의 열기와 먼지를 포함한 건조한 북서풍)
sharki(= kaus)	샤키(페르시아만의 비가 섞인 남동 · 회오리바람)
siffanto	시판토(이탈리아 아드리아해의 남서풍)
si giring giring	시기링기링(북수마트라 토바호의 바람)
sigua	시구아(필리핀의 8급 이상의 계절풍)
simm(= simoom)	심(아프리카와 아라비아 사막지역의 건조 열풍)
sirocco	시로코(유럽 남부지방의 푄)
sirocco di levante	시로코 디 레반테(그리스 서남부지방의 푄)
Skiron	스키론(겨울철에 한랭하고 여름철에 고온 건조한 북서풍의 그리스말)
slatan(= selatan)	슬라탄(인도네시아의 건조하고 강한 남풍)
sno(= elvegust)	스노(노르웨이 피오르드의 북부에서 발생하는 한랭한 하강 스콜)
solaire(= soulédras, souléare)	솔레르(프랑스 중남부의 동풍)
solano	솔라노(스페인 남동해안 지방에서 여름철에 부는 더운 남동풍 또는 동풍)
solaure(= solore)	솔레우르(프랑스 남동부 드로메강 경로를 따라 야간에 부는 한랭한 산바람)
sondo	→ zonda
sopero(= sover)	소페로(이탈리아 가르다호 부근 계곡 아래로 불어내리는 야간 실바람)
soulaire	→ solaire
southerly burster(= burster)	→ burster
sover	→ sopero
Stikine wind	스티킨바람(캐나다 스티킨강 부근의 돌발성 동북동풍)
Struma fall wind	스트루마내리바람(불가리아 스트루마계곡에서 부는 차고 강한 내리바람)
suahili	수아힐리(페르시아만의 강한 남서풍)
suchovei(= sukhovei)	수하베이(남부 러시아지역의 건조하고 더운 먼지바람)
sudois	수도이스(스위스 제네바호의 남서풍)
suer	수에르(이탈리아 가르다호의 격렬한 북북서풍)
suestada	수에스타다(아르헨티나, 우루과이, 남부 브라질 연안의 강한 남동풍)
sukhovei	→ suchovei
sumatra	수마트라(남서 계절풍 기간 중의 말라카해협의 강풍)
sur	수르(브라질의 찬 바람)

suraçon	수라송(볼리비아의 찬 비바람)
suroet	수레(프랑스 서해안의 비를 동반하는 남서풍)
syzygy	시지지(여름철 북서 계절풍에 앞서 뉴기니와 오스트레일리아 사이의 바다에서 부는 서풍)
Table Mountain southeaster	테이블마운틴남동풍(남아프리카 테이블만의 남동풍)
Taku wind	타쿠바람(알래스카 타쿠강 하류의 강한 동북동풍)
tamboen	탐보엔(북수마트라 토바호의 바람)
tarantata	타란타타(지중해의 강한 북서풍)
tehuantepecer	테완테페세르(멕시코만으로부터 테완테펙지역에 불어오는 겨울철 북~북북동풍)
temporale	템포레일(미국 태평양 연안의 비를 동반하는 남서풍)
tenggara	텡가라(스퍼몬데 아키펠라고에서 동쪽 몬순이 계속되는 동안에 부는 강하고 건조한 동풍 또는 남동풍)
tereno(= terrenho)	테레노(인도의 건조하고 차가운 육풍)
terral	테랄(안데스 산맥의 강한 북동 육풍)
terral levante	테랄레반테(때때로 푄의 북서 스콜의 성질을 나타내는 스페인과 브라질의 육풍)
terre altos	테레알토스(리오데자네이로의 경사면을 불어내리는 차고 강한 북서 내리바람)
terrenho	→ tereno
thalwind	탈윈드(프랑스 알자스지방의 냉풍)
tivano	티바노(이탈리아 코모호 부근 계곡을 불어내리는 야간 실바람)
tofan(= tufon, tufan)	토판(인도네시아 산간지대의 봄철 폭풍)
tongara	통가라(인도네시아 마카사해협의 연무 낀 남동풍)
tormento	토르멘토(아르헨티나의 바람, pampero와 비슷한 바람)
toscà	토스카(이탈리아 가르다호의 남서풍)
touriello	토우리엘로(프랑스 피레네산맥으로부터 불어내리는 푄과 같은 남풍)
traersu	트라에르수(이탈리아 가르다호의 격렬한 동풍)
tramontana	트라몬타나(이탈리아 서해안과 북부 코르시카의 한랭한 북동풍 또는 북풍)
traverse	트라베르세(프랑스 중부의 서풍)
traversia	트라베르시아(바다로부터 불어오는 서풍에 대한 남아메리카 선원들의 용어)
traversier	트라베르시에르(지중해에서 항구로 직접 불어오는 위험한 바람)
tromba	트롬바(말타의 회오리바람)
unterwind	운터윈드(오스트리아 살츠캄메르구트호수 위의 낮은 실바람)
valais wind	발라이스바람(스위스 남서부 로누지역의 골바람)
varatrazo	바라트라조(마다가스카르 유러파섬의 육풍)
vardar(= vardarac)	바르다(그리스 바르다계곡의 차가운 내리바람)
vaudaire(= vauderon)	바우다이레(제네바호 남쪽으로부터의 강한 푄 바람)
vendaval	벤다발(지브롤터해협에서 때때로 발생하는 남서 폭풍)
vent da Mùt	벤트다무트(이탈리아 가르다호의 습윤한 바람)
vent des dames	벤트데다메(프랑스 론삼각주 동부 지중해 연안의 해풍)
vent du midi	벤트두미디(프랑스 중앙고지와 세베네스산맥 남부의 남풍)
vento di sotto	벤토디소토(이탈리아 가르다호의 실바람)
vesine	비안(프랑스 드롬의 골바람)

viento zonda	비엔토존다(아르헨티나의 뜨거운 북풍)
vinessa	비네사(이탈리아 가르다호의 바람)
virazon(= birazon)	비라존(남아메리카 안데스산맥의 태평양쪽 경사면을 불어올라가는 바람의 스페인말)
virazones	비라조네스(해륙 실바람의 스페인말)
visentina	비센티나(이탈리아 가르다호의 강한 동북동~동풍)
viuga	비우가(러시아 스텝지대에 부는 찬 북 또는 북동 폭풍)
vuthan	부단(남아메리카 남부의 강한 폭풍)
wam Braw	왐브로(기니아 북부의 건조 열풍)
warm braw	웜브로(뉴기니아 북부의 슈텐섬에서 부는 푄의 일종)
Wasatch winds	워새치바람(미국 서부의 워새치산맥에서 부는 강한 동풍)
Washoe zephyr	워쇼제피르(미국 네바다 주와 오리건 주의 해안 일대의 따뜻한 바람, 치누크 바람)
whirlies	휠리즈(남극 아델리랜드 방면의 눈을 불어올리는 회오리바람)
white buran	→ buran
williwaw(= willyway, williwau, willie-wa, willywaa)	윌리와(남아메리카 남부 대서양쪽의 마젤란해협에서 발생하는 강한 스콜)
wind of 120 days	→ seistan
wisperwind	위스퍼윈드(독일 위스퍼강 계곡의 찬 밤바람)
xaloch(= xaloque, xaroco)	시로코(사하라사막이나 아라비아반도 사막의 모래 먼지를 동반한 바람의 스페인말)
yalca	얄카(페루 북부 안데스산맥의 봉우리에서 일어나는 강한 바람)
yamase	야마세(일본 삼육지방의 국지 바람)
Yangkanjipung	양간지풍[한국 동해안 중북부지방(양양과 간성 사이)에서 부는 강한 바람의 한국말]
youg	유그(지중해 지역의 여름철 열풍)
zobaa	조바(이집트에서 고속으로 움직이는 기둥과 같은 강한 모래 회오리바람)
zonda	존다(아르헨티나 안데스산맥 동쪽 대초원의 열풍)